Late Quaternary Environments of the Soviet Union

Late Quaternary Environments of the Soviet Union

A. A. Velichko, Editor

H. E. Wright, Jr., and C. W. Barnosky, Editors of the English-Language Edition

Longman
London

Longman Group Limited
Longman House, Burnt Mill, Harlow
Essex CM20 2JE, England
Associated companies throughout the world

Published in the United States of America
by the University of Minnesota Press

This edition first published by Longman Group Limited 1984

British Library Cataloguing in Publication Data

Late quaternary environments of the Soviet Union.
1. Geology, Stratigraphic—Quaternary
2. Geology—Soviet Union
I. Velichko, A. A. II. Wright, H. E.
III. Barnosky, C. W.
551.7'9'0947 QE696

ISBN 0-582-30125-4
Printed in the United States of America

Contents

Contributors to This Volume

S. A. Arkhipov, *Institute of Geology and Geophysics, USSR Academy of Sciences, Siberian Department, Novosibirsk 90*

V. V. Baulin, *Scientific Research and Production Institute for Engineering Studies in Building, Okruzhnoy proyezd 18, Moscow 105058*

Ye. B. Belopukhova, *Scientific Research and Production Institute for Engineering Studies in Building, Moscow*

V. G. Bespalyy, *North-East Scientific Research Institute, Far-East Scientific Center, USSR Academy of Sciences, Portovaya 16, Magadan 685010*

A. B. Bogucki, *L'vov Polytechnical Institute, Prospekt Mira 12, L'vov*

O. K. Borisova, *Institute of Geography, USSR Academy of Sciences, Staromonetniy per. 29, Moscow 109017*

A. L. Chepalyga, *Institute of Geography, USSR Academy of Sciences, Moscow*

N. S. Danilova, *Scientific Research and Production Institute for Engineering Studies in Building, Moscow*

P. M. Dolukhanov, *Institute of Archaeology of the USSR Academy of Sciences, Leningrad Department, Dvortsovaya Naberezhnaya 18, Leningrad 41*

M. A. Faustova, *Institute of Geography, USSR Academy of Sciences, Moscow*

V. P. Grichuk, *Institute of Geography, USSR Academy of Sciences, Moscow*

Ye. Ye. Gurtovaya, *Institute of Geography, USSR Academy of Sciences, Moscow*

L. L. Isayeva, *Institute of Geography, USSR Academy of Sciences, Moscow*

T. N. Kaplina, *Scientific Research and Production Institute for Engineering Studies in Building, Moscow*

T. A. Khalcheva, *Institute of Geography, USSR Academy of Sciences, Moscow*

V. G. Khodakhov, *Institute of Geography, USSR Academy of Sciences, Moscow*

N. A. Khotinskiy, *Institute of Geography, USSR Academy of Sciences, Moscow*

S. V. Kiselev, *Moscow State University, Geographical Department, Leninskiye Gory, Moscow 117234*

V. A. Klimanov, *Institute of Geography, USSR Academy of Sciences, Moscow*

I. Ye. Kuz'mina, *Zoological Institute of the USSR Academy of Sciences, Universitetskaya Naberezhnaya I, Leningrad*

A. A. Lazarenko, *Geological Institute of the USSR Academy of Sciences, Pyzhevskiy per. 7, Moscow 109017*

I. M. Lebedeva, *Institute of Geography, USSR Academy of Sciences, Moscow*
A. V. Lozhkin, *North-East Scientific Research Institute, Far-East Scientific Center, USSR Academy of Sciences, Magadan*
V. M. Makeyev, *Arctic and Antarctic Scientific Research Institute, Fontanka 34, Leningrad 192104*
A. K. Markova, *Institute of Geography, USSR Academy of Sciences, Moscow*
G. G. Matishov, *Murmansk Marine Biological Institute, Kil'skiy Department of the USSR Academy of Sciences, Dal'niye Zelentsy, Murmansk Oblast'*
A. V. Minervin, *Moscow State University, Geological Department, Leninskiye Gory, Moscow 117234*
T. D. Morozova, *Institute of Geography, USSR Academy of Sciences, Moscow*
V. I. Nazarov, *Institute of Geochemistry and Geophysics of the Byelorussian Academy of Sciences, Leninskiy Prospekt 68, Minsk 220072*
V. P. Nechayev, *Institute of Geography, USSR Academy of Sciences, Moscow*
M. I. Neustadt, *Institute of Geography, USSR Academy of Sciences, Moscow*
N. D. Praslov, *Institute of Archaeology of the USSR Academy of Sciences, Leningrad Department*
S. S. Savina, *Institute of Geography, USSR Academy of Sciences*
L. R. Serebryanny, *Institute of Geography, USSR Academy of Sciences, Moscow*
S. V. Tomirdiaro, *North-East Scientific Research Institute, Far-East Scientific Center, USSR Academy of Sciences, Magadan*
A. I. Tsatskin, *Institute of Geography, USSR Academy of Sciences, Moscow*
V. P. Udartsev, *Institute of Geography, USSR Academy of Sciences, Moscow*
A. A. Velichko, *Institute of Geography, USSR Academy of Sciences, Moscow*
N. K. Vereshchagin, *Zoological Institute of the USSR Academy of Sciences, Leningrad*
I. A. Volkov, *Institute of Geology and Geophysics, USSR Academy of Sciences, Siberian Department, Novosibirsk*
E. M. Zelikson, *Institute of Geography, USSR Academy of Sciences, Moscow*
V. S. Zykina, *Institute of Geology and Geophysics, USSR Academy of Sciences, Siberian Department, Novosibirsk*

Preface

H. E. Wright, Jr., and C. W. Barnosky

The US-USSR Bilateral Agreement on Cooperation in the Field of Environmental Protection, Working Group VIII on Quaternary Paleoclimates, called for small conferences of American and Soviet scientists, of which four have been held. The Soviet delegations were headed by I. P. Gerasimov and A. A. Velichko, and the American by John Imbrie and Alan D. Hecht. In order to bring the discussions to a wider audience, it was decided to prepare monographs summarizing the recent research results on Late Quaternary environments in the two countries, for simultaneous publication in English and Russian. The two American volumes were published in English late in 1983: one on the late Pleistocene edited by S. C. Porter, the other on the Holocene edited by H. E. Wright, Jr. Russian editions of these two volumes are in preparation in Moscow.

The present volume on the Late Quaternary of the Soviet Union was prepared in Russian by Soviet specialists and edited by A. A. Velichko of the Institute of Geography of the USSR Academy of Sciences, with the assistance of Ye. Ye. Gurtovaya. It covers the time range from the last interglaciation through the various phases of the last glacation and up to the present time, dealing not only with the history of ice-sheet and mountain glaciation but also with the loess deposits and permafrost features of the periglacial areas, the complex history of the inland seas, the sequence of vegetation, the distribution of mammal and insect faunas, the development of human cultures, and the reconstruction of climatic changes made possible by quantitative estimates based on various lines of evidence.

The U.S. National Science Foundation and the U.S. National Oceanic and Atmospheric Administration provided funds for the translation of the Russian manuscript into English. The translations were prepared for publication by the present editors, and the edited versions were checked by the chapter authors with the assistance of I. I. Spasskaya of the Institute of Geography of the USSR Academy of Sciences. The introduction to this English edition by the U.S. editors provides a synopsis of the book. They express their appreciation to A. A. Velichko, Ye. Ye. Gurtovaya, I. I. Spasskaya, and the Soviet authors for their cooperation in the preparation of this edition and to Alan D. Hecht for assistance in completion of the project.

H. E. Wright, Jr.
C. W. Barnosky

Introduction to the English Edition

H. E. Wright, Jr., and C. W. Barnosky

The Soviet Union is a vast and varied land, with an area larger than that of the United States and Canada combined, and with an equal diversity in physiography and climate. It stretches from the temperate Russian Plain in the west across the Ural Mountains to the West Siberian Plain and the arid mountains of Central Asia, and thence to the mountainous Northeast and the oceanic Far East of Kamchatka. In the north is the Arctic Ocean, and in the south the subtropical shores of the Caspian and Black Seas and the semi-deserts of Kazakhstan and the Himalayan foothills. The vegetation is strictly zonal in its broad aspects in Siberia (east of the Urals). Tundra lies in the north, various facies of conifer forest (taiga) across the center, and steppe in the south. The temperate broad-leaved forest exists as a wedge coming in from Europe in the west between taiga and steppe. The Far East vegetation is distinctive because of its oceanic character.

The broad scope of the country permits the study of large-scale natural processes, not only as they are active today but as they have operated in the past under conditions of changing climate, and this monograph is concerned with the geologic record of environmental changes from the last interglacial period through the various fluctuations of the last glaciation and up to the present. Because these major climatic changes were global, the events of the Soviet land can be correlated with those of Western Europe, as indicated at least for the last 40,000 years by radiocarbon dating. Yet the findings in this large area are not simply an extension of those to the west or those of America, because the configuration of the landmass influenced the extent and style of glaciation, and the secondary effects of the ice sheet—such as the spread of proglacial lakes—created unique circumstances. The vast size of the continent engendered climatic conditions much more extreme than those in North America, even though the ice sheets were less extensive. Thus permafrost was much more widespread and reached greater depths, and loess deposits were thicker.

Soviet research on Quaternary environmental history has been expanded during the last few decades in association with the great mineral explorations of Siberia, so that something is now known about most of the country, and the synthesis represented by the present volume is timely. Although many more localities must be studied to confirm the postulated correlations from one part of the country to another and to fill in the gaps, a broad framework is now available for evaluation not only by Soviet scientists by also by others who deal with global environmental history or who wish to compare Late Quaternary events in one half of the northern Hemisphere with those in the other.

The introduction to the volume by A. A. Velichko provides an outline of the formal classification of the Late Quaternary for the glaciated areas of European USSR and for Siberia, with a table giving the names of the various glacial, interglacial, and interstadial intervals for different sections of the country. It also presents his hypothesis for asymmetry of glacial growth during the last glacial phase (Valdai = Wisconsin)—initial climatic cooling first affected the interior of the landmass in Siberia, initiating glaciation in that area. As cooling increased, glaciation extended westward into the European sector, whereas Siberia became too cold for the invasion of moist air masses from the west, and glaciers there diminished while permafrost expanded over virtually the entire country. The effects of the cold Siberian anticyclone during the glacial maximum were manifested as well by the dominance of loess deposition and by the widespread development of tundra and steppe at the expense of forest.

This introduction to the English edition provides a synopsis of the contents of the monograph, with some commentary on the scope and emphasis of the research involved. The reader is referred to the introductory chapter by A. A. Velichko and especially to his correlation table to obtain an overall impression of the monograph and to keep track of the stratigraphic terminology for different parts of the country.

Glaciation

The Scandinavian ice sheet reached into the northern part of the Russian Plain during the last glaciation, subdivided into the Early Valdai and Late Valdai separated by the Middle Valdai nonglacial interval, which includes several climatic fluctuations recorded by pollen evidence for alternating tundra and birch-conifer forests (Faustova, chap. 1). The ice sheet near the margin was differentiated into a number of ice lobes fronted by proglacial lakes, and the eastern portion was confluent with ice from Novaya Zemlya. As the ice sheet began its retreat 16,000 to 15,000 years ago, distinct fluctuations of the margins occurred on the southwestern sector in Western Europe, with push moraines and other indications of active ice, but farther east on the more continental Russian Plain the marginal deposits were those of stagnant ice. Retreat was more rapid after 15,000 years ago, and especially after 12,000 years ago. The late-glacial Salpausselkä moraines of Finland extend eastward into Karelia and the Kola Peninsula, where ice persisted until about 9500 years ago. The Novaya Zemlya glacier also broke up about this time. In the West Siberian Plain in the lower Ob' and Yenisey River valleys, the last glaciation is similarly subdivided into two glacial phases and an intervening complex interstadial (Arkhipov, chap. 2). The taiga of today was converted to tundra or tundra-steppe.

Farther east in north-central Siberia the Middle Zyryanka nonglacial interval, dated between 50,000 and 24,000 years ago, is represented by several phases in which trees invaded areas that are now tundra, implying warmer conditions than today (Isayeva, chap. 3). The Late Zyryanka Glaciation (Sartan, Late Valdai) that followed was centered on different mountain groups and plateaus in northeastern Siberia as well (Bespalyy, chap. 4). Periglacial loess and syngenetic ice-wedge features in alluvium indicate a cold and windy climate. Tundra or forest-tundra extended 1000 km from the ice margin.

The extent of ice sheets on the Arctic shelf of Eurasia is of special interest (Velichko et al., chap. 5). Submarine mapping through echo sounding, photography, and bathymetry reveals relief forms that resemble moraine features as well as glacial-erosional trenches similar to those on the land surface. The Barents Sea shelf probably was not occupied by an independent ice dome merging with the Scandinavian ice sheet. Rather, small ice caps may have occupied islands and plateaus on the shelf, which was reached by ice lobes from Scandinavia, Spitzbergen, Franz Josef Land, Novaya Zemlya, and other ice sheets, as indicated by concentric zones of moraines.

Some authors postulate that the Kara Sea shelf was occupied by an ice sheet 2500 km in diameter, joined to a Barents Sea ice sheet. It carried glacial debris southward across the mountains of northern Siberia and formed moraines that dammed the northward flow of the great Siberian rivers (Ob', Yenisey, and Irtysh), producing huge lakes in the West Siberian Lowland that overflowed southward through the Turgay Hollow to the Aral and Caspian Seas, and from there to the Black Sea and the Mediterranean Sea, thereby transecting the continent.

This major reconstruction, which involves the recognition of new ice sheets larger than the Scandinavian ice sheet itself, is controversial, however. Some of the critical land areas east of the Ob' River provide no firm evidence for glaciation, and some of the submarine features attributed to glaciation may in fact be non-glacial in origin. An alternative reconstruction, favored by Velichko (chap. 5), proposes separate small ice sheets on the Polar Ural Mountains and other highlands as well as on islands and peninsulas projecting onto the Arctic shelf (e.g., Taimyr, Yamal). During the glacial maximum several of these ice masses merged, and the ice center shifted far enough to the north so that glacial debris was transported south over the mountains to produce the observed directional features and frontal moraines. All the ice masses did not link up, however, and the lower Ob' River was not blocked.

Because of uncertainties about the extent of ice cover in this region, the maps presented show both a maximum variant, with almost complete coverage of the Kara Sea shelf and the adjacent parts of Siberia, and a minimum variant, with largely separate ice masses on uplands and a less extensive cover on the shelf.

Besides the northern ice sheets, extensive Late Pleistocene glaciers occurred in the dozen or so major mountain ranges that fringe the southern and eastern margins of the Soviet Union, ranging from the Carpathians in the southwest to the mountains of the extreme northeast (Serebryanny, chap. 6). The older geomorphic studies have recently been supplemented by quantitative analyses of the lithology, mineralogy, fabric, biostratigraphy, and geochronology of the deposits, especially in the Caucasus, the Pamir, and Tien-Shan.

Interpretations of the paleoclimatic sequence in the mountains and the nearby loess areas differ, however, for some authors think that the glacial episodes were cold and wet, with conifer forests descending on the mountains, whereas others believe they were cold and dry. The relations were complicated by continued mountain uplift, especially in the Pamir, resulting in progressively drier conditions. In several areas two phases of glaciation are recognized, interrupted by an interval with higher percentages of tree and shrub pollen than below or above, when steppe or desert-steppe prevailed. During the last glaciation the lower treeline descended about 1000 m below its present position in Tien-Shan.

Holocene glacial fluctuations in the mountains have been studied especially in the Caucasus, where progressive retreat is recorded even though the pollen evidence suggests warmer conditions during the Middle Holocene than later. The apparent contradiction may reflect different critical climatic effects at different elevations, e.g., snowfall may have decreased in the high firn basins, while precipitation increased in the forest belt. Or the seasonal distribution of precipitation may have changed. There is no evidence that glaciers disappeared from the High Caucasus during the Middle Holocene. Lichen dating suggests 10 episodic movements during the last 700-800 years of ice retreat—thus cycles of about 80 years.

In the mountains of Central Asia, major ice advances are indicated in the earliest Holocene and in the Middle Holocene, as well as numerous small advances that interrupted the general retreat. Little Ice Age advances were recorded in the 12th through 13th and the 18th through 19th centuries, as in the High Caucasus.

Calculation of former glacier regimes provides a basis for paleoclimatic reconstructions by Lebedeva and Khodakhov (chap. 7), who carefully estimated the critical parameters. The surface profile was assumed to be comparable to that of a modern glacier. Ablation was determined from the summer air temperature above the ice surface, calculated from the normal lapse rate with elevation combined with the cooling effect of the ice surface. Calculation of snow accumulation was more difficult, especially in the mountains, where the orographic effect is dominant. Cooling by the ice surface decreased the moisture content of the air, but the greater turbulence tends to offset this effect. The proportion of snow to rain increases with altitude. The contribution of wind drift and avalanching was also included in the calculation. For ice sheets, precipitation increased with altitude and thus distance from the margin, but ice sheets blocked the passage of cyclones and intensified the cooling of the air, so the net result is uncertain. The mass balance also included a factor for the freezing of meltwater and rainwater.

The mass-balance equations, after being checked on different mountain glaciers and the Greenland ice sheet, were applied to the reconstruction of the Scandinavian ice sheet. With a regional depression of mean annual temperature of 8°C for northwestern Europe during the last major period of glacier growth (Late Valdai), snow accumulated first in the Scandinavian mountains about 35,000 years ago. The terminal area on the Russian Plain was reached after about 17,000 years. Subsequently the air temperature was increased by 5°C or 6°C, and the glacier mass was heated as much as 20°C by the freezing of meltwater and the addition of rain. A thinning of 35% may have resulted from this warming, in combination with increased flow velocity and thus a final "degradation advance" of the margin. With further thinning the accumulation area was eliminated, and the degradation became catastrophic, completing the wastage in a total of 10,000 years.

The same set of equations was used for the reconstruction of representative Late Pleistocene mountain glaciers in the Caucasus, Kamchatka, the Pamir, and the Wasatch Mountains (USA). A value was first estimated from independent paleo-environmental evidence for the depression of mean annual air temperature, and the altitude of the equilibrium line was estimated from end-moraine patterns. Results show that the depression of summer temperature was 9°C for the Wasatch Range, 6°C for the Caucasus, and 3°C for the Pamir and Kamchatka—inversely proportional to the distance from the "cold center" represented by the ice sheet. The critical factor in the mass balance was ablation and thus summer temperature; for the Wasatch, the accompanying change in snowfall compensated for only 30% of the change in ablation.

The firn-line depression was greater in a temperate region (Wasatch) than in a continental region (Pamir), because the greater snowfall in the former caused the glacier to extend to lower elevations, where summer temperature and ablation are greater and more sensitive to change. The effects in a maritime region (Kamchatka) should be even greater, but here the glacier expanded into a large piedmont lobe, and the firn line was not depressed very much.

Another effect of the firn-line depression was the increase in run-off, resulting from the decrease in evaporation off the ice-covered surface, as well as from the lower air temperatures everywhere. Calculations for the Pamir and the Wasatch indicate an increase by four times. Lake Bonneville expanded below the Wasatch Range almost to the level of the glacier margin.

Permafrost

The dynamics involved in the formation of ice complexes can be viewed in recent floodplain alluvium in northern Siberia, where ground temperatures are lower than −7°C (Baulin and Danilova, chap. 8). The ice complexes of the last glacial epoch are larger, however, with ice-wedge polygons that are thicker (3-5 m), more closely spaced (5 to 10 m apart), and covering larger areas. Calculations from the dimensions in the complex indicate a mean annual air temperature of about 10°C below the present.

Pseudomorphs of ice-wedge polygons formed in contemporaneous alluvium during the Zyryanka Glaciation occur as far south as latitude 52°N, those of the Karginskiy Interglaciation latitude 56° to 57°N, and those of the Sartan Glaciation latitude 48° to 49°N.

Calculations based on heat exchange, air-temperature fluctuations, moisture content of the ground, and geothermal flux indicate that in eastern Siberia the permafrost is thawing from beneath at a rate of 1 to 3 cm per year and that farther west relict permafrost occurs at depths of 50 to 150 m and 200 to 400 m. The average annual ground temperature near the surface in the north was −20°C, compared with −10°C to −12°C today.

In the Russian Plain, permafrost thawed completely during the Mikulino Interglaciation, and forest covered the entire area (Velichko and Nechayev, chap. 9). During the Valdai Glaciation intervals of permafrost development are recorded by three horizons of polygonal ice pseudomorphs, fine-crack systems, and solifluction and involution layers in stratigraphic sections of loess and fossil soils. The cryogenic structures of the Early Valdai disrupt the interstadial soil complexes. The most intense interval of cryogenesis was the Late Valdai. This started in western Russia with a thin active layer, producing plastic deformation of the buried soil by seasonal freezing over a rigid base of permafrost. Then with the drier, more continental climate of the glacial maxima, the active layer thickened and solifluction dominated. Farther east the process was more intensive, and fine polygonal cracking occurred. With the final, more humid phase of Late Valdai loess accumulation, the formation of polygonal ice veins was intensified; they formed in moraines and glaciofluvial deposits on the Russian Plain as well as in loess, as they do today in areas of continuous permafrost where ground temperatures are less than −5°C to −7°C. These veins reached a depth of 5 m in the north, or 2.5 m in the south, and the polygons were 15 to 20 m wide. The volume of ice veins was only 8 to 12% of the mass—much less than for contemporaneous features in Siberia.

In the Holocene the southern boundary of permafrost shifted northward by 20° of latitude in the Russian Plain, 13° in western Siberia, and 6° in central Siberia, as annual temperatures rose by 12 or 13°C, and areas of thawed ground and thermokarst increased (Baulin et al., chap. 10). In the Early Holocene ice-wedge pseudomorphs still formed in areas 4° of latitude south of their present limit, but in the Middle Holocene partial melting of permafrost occurred, with refreezing in the Late Holocene. Ground temperatures were 1°C to 1.5°C higher than at present during the warm interval; because of the change of state in the thawing process, the air-temperature change may have been 2°C to 3°C. Permafrost thawed to a depth of 50 m or more, producing depressions in which peat accumulated, although locally permafrost persisted at shallow depth under the dark taiga. During the Late Holocene frost-cracks and ice wedges formed in the Holocene peat and recent alluvium. Calculations, modeling, and analysis of relict permafrost indicate that between 150 and 200 m of Late Pleistocene permafrost thawed during the Holocene in areas of permeable sedimentary rocks, and up to 300 m more in hard rocks. The relict permafrost in the Early Holocene had a latitudinal breadth of 700 km in the Russian Plain and 1200 km in western Siberia.

Loess

Although loesslike sediments locally originate as alluvium or colluvium, most loess is eolian, as determined by studies of the physical, mineralogical, and chemical properties of the loess and the loessial soils that are interbedded in stratigraphic complexes.

In the East European Plain most of the Mezin soil complex represents the Mikulino Interglaciation (Velichko et al., chap. 11). Regional variations are evident: forest soils and forest-steppe soils extended farther south than they do today, and soil types of Western Europe extended into Eastern Europe, indicating a less continental climate than today, with less severe winters and wetter summers.

The upper part of the Mezin soil complex was deformed by frost processes and includes some loess. It was modified to the chernozemlike Krutitsa soil, formed under steppe conditions in an Early Valdai interstadial interval, with a strongly continental climate. The soil was then cryogenically deformed, with frost polygons indicating permafrost conditions south to latitude 51°N.

Loess I (Middle Valdai) is more poorly sorted than the Late Valdai loess and is only 1.5 to 2 m thick over a very large area. Sand grains are variably rounded by wind abrasion.

The Bryansk soil (32,000 to 22,000 yr B.P.), which is generally gleyed, has no modern analogues, because of the deformation by contemporaneous frost processes. In the humus horizon freezing produced rounded aggregates of clay, which had annular microstructure formed by segregation ice.

Loess II and III, separated by a weak soil, are well-sorted and up to 8 m in total thickness, with only slightly weathered mineral grains. Sand grains are rounded and dulled, indicating strong and long-term wind abrasion. Pollen content and remains of fossil mammals indicate tundra-steppe conditions (as in central Yakutiya today). Ice-wedge pseudomorphs dated 18,000 to 16,000 years ago and extending to latitude 48°N suggest an increase in moisture during the final phase of the Late Valdai Glaciation, and winter temperatures 20°C lower than today. Late-glacial (Alleröd) warming resulted in degradation of many ice-wedge systems, which were later revived in the Younger Dryas.

In southwestern Siberia (Volkov and Zykina, chap. 12) eolian features include deflation surfaces with areas of sand dunes dating from the main Sartan (Late Valdai) Glaciation. In the loess areas of the Ob' River region the Berdsk soil complex at the base contains two soils separated by a loesslike layer. The structural and chemical properties of the lower soil are those of a leached chernozem, formed during the last interglaciation under conditions similar to those of the forest-steppe in the area today. The upper soil dates to an interstadial interval within the Early Zyryanka (Early Valdai) Glaciation. Each soil is modified in its upper part by cryogenic processes. Above a typical loess layer the Iskitim soil complex, which also contains two chernozemic soils separated by a loesslike layer, is dated 33,000 to 19,500 years ago with 10 different samples from wood, charcoal, bone, humus, and alluvium. Covering the Iskitim soil complex is a widespread loess formed from 19,000 to 14,000 years ago, which grades laterally on slopes to colluvial and solifluction deposits. Locally still other thin soils and loess layers occur.

In Central Asia (Lazarenko, chap. 13) the loess thickness reaches 200 m. Key sections in the Tadzhik Depression feature the Olduvai paleomagnetic reversal of 1.6 to 1.8 million years ago near the base. Nine soil complexes occur above the Matuyama/Brunhes reversal of 700,000 years ago. Five short reversed phases occur in layers generally less than 0.2 m thick, of which the Blake double interval (120,000 to 110,000 years ago) and the Laschamp event (about 20,000 years ago) are recognized. The upper loess, which contains the Laschamp reversal, is correlated with the post-Bryansk (Late Valdai) loess of the Russian plain. Thermoluminescence dates suggest correlations for older loesses, e.g., the third soil complex (containing the Blake event) has dates of 115,000 to 130,000 years ago.

The buried soil complexes have at the base a well-developed reddish brown profile with a sharp horizon of carbonate accumulation in disseminated form. Seasonal wetting is indicated by micro-ortsteins, coarse structure, lack of clay migrations, and rodent and molluskan faunas. Upward in each complex the indications of drier conditions increase, e.g., less distinct leaching of carbonate, fewer signs of waterlogging, and increases in soluble salts, but these stratigraphic trends are less prominent than the horizontal changes that occur outward from the mountains.

The loess itself was formed under arid conditions, as indicated by the high content of carbonate and soluble salts, the presence of insect larval capsules (which inhibit desiccation), and a xerophilic molluskan fauna. Pollen content implies birch-pine forests on mountain slopes that are barren of trees today, and a rich herbaceous flora in the foothills.

Loess accumulation, which was initiated in the valleys and thus reduced the relief, amounted to 1.0 to 1.5 m per 1000 yr in the late Pleistocene. Late Quaternary orogenesis deformed the loess cover to make loess hills. Climatic conditions became progressively drier with each soil complex, culminating in the Berdsk complex (Middle Valdai).

The long controversy about the desert versus the periglacial origin of the loess in Central Asia can be considered from mineralogical studies (Minervin, chap. 14). Loess particles consist of silt-size microaggregates, in which a core of a primary mineral (e.g., quartz) is surrounded by calcite and a jacket of clay minerals, iron oxides, amorphous silicic acid, finely dispersed quartz, and carbonates. Experiments with freezing and thawing can reduce wet mineral fragments to silt and fine sand, but heating and cooling of air-dried fragments is ineffective. The process in quartz and feldspar crystals involves crushing the thin walls of dislocation channels by ice pressure. The breakage produces free-radicals, which cluster and hydrate in water. The amorphous surface layer on the quartz fragments reacts with water to form silicic acid, which combines with calcium. Calcite is a further product.

The polymineral jacket on the primary minerals results when clay minerals adsorb $Fe(OH)_3$, and this in turn adsorbs silica gel and organic matter. Dehydration during freezing and sublimation results in bonding the mineral jacket to the primary particles through holes in the calcite envelope. Weathering experiments with cold water saturated with carbonic acid resulted in a 3 to 15% loss of mass and the formation of secondary carbonate and silicate minerals and iron and aluminum oxides.

Observations on loess formation in semi-desert areas showed that silt dust when wetted and then dried produces a hard rock (takyr). Freezing then causes expansion and increases the porosity, which is not lost with subsequent thawing. With further freezing and drying the material resembles typical loess. But generally in this region the silt is dry rather than moist when freezing temperatures occur, so that the full transformation to typical loess does not take place. Apparently the postsedimentation processes of freezing under severe Pleistocene periglacial climatic conditions are critical in loess formation.

Of particular interest in connection with the distribution of loess is the hypothesis that the Arctic Ocean was so thickly frozen that it formed a "climatic dry land," which with the adjacent subarctic portions of Eurasia and north America supported a permanent anticyclone with cloudless skies, high summer solar radiation, very cold winters, and an arctic-steppe vegetation in Beringia (Tomirdiaro, chap. 15). The loess cover was broken by massive polygonal ice veins to produce the edoma complex of Yakutiya north of latitude 72°N, visible especially on the Arctic coast and islands. Ice content may exceed 90% of the mass to a depth of 35 m, with vertical ice veins 9 m wide and earth veins only 3 m wide. Thawing from the surface results in subsidence of the ice veins, leaving a microrelief of loess hillocks.

This arctic or shelf type of edoma represents slowly accumulated loess containing clay but little sand, deposited under the stable air of a permanent anticyclone. Fine dust settled on the exposed arctic shelf and perhaps over the frozen Arctic Ocean. A dust layer of 1 m could be swelled to 20 m by ice segregation. It may have supported the steppe vegetation required for the Arctic mammoth.

The subarctic type of edoma, found in the American sector and in northeastern Siberia south of latitude 72°N, consists of frozen sandy stratified loess with narrow syngenetic ice veins, dated to the Sartan Glaciation (Late Valdai). Beneath this is buried woody peat and then loess with a higher ice content and wider ice veins, of Zyryanka age (Early Valdai).

More than 90 radiocarbon dates are available from the ice complexes (Kaplina and Lozhkin, chap. 16). An older ice complex (Zyryanka = Early Valdai) was partly thawed during two subsequent warm intervals to produce thermokarst depressions in which woody peat or lake sediments were deposited. Renewed ice accumulation during the Late Valdai built up the surface by 20 m. Thawing began once again about 12,000 years ago.

Vegetation History

The large number of Late Quaternary pollen records from the Soviet Union allows examination of the vegetation during the climatic optimum of the last (Mikulino) interglaciation, the glacial maximum of Late Valdai time, and the different phases of the Holocene. The coverage of sites is great, and data are available from all areas except parts of Siberia, Central Asia, the mountainous regions of the south, and the Soviet Far East. The source and quality of these records vary considerably, inasmuch as pollen assemblages are described from such different materials as peat deposits, lake sediments, paleosols, and fluvial sands and gravels. The radiocarbon dates necessary to establish an absolute chronology and thus permit precise correlations among sites are often lacking, especially from Pleistocene sites. Some of the Holocene records are well-dated, having as many as six to 11 radiocarbon dates from wood and organic lake mud *in situ*. At other sites the age is inferred from the geologic or geomorphic position of the polleniferous sediments, from associated paleontologic remains, or from correlation with nearby dated pollen records.

In addition to the pollen data, which provide direct information on the fossil flora, vegetation reconstructions are based on an analysis of phytogeography and community structure within the modern vegetation. Three types of flora are recognized: migration, orthoselection, and relict (Grichuk, chap. 17). Migration floras are typical of the East European Plain, where Pleistocene glaciation was most extensive and plant distributions greatly altered by repeated glacial-interglacial cycles. The composition of the modern migration flora, as well as that of previous interglaciations, is determined by the location of glacial refugia and by the rate and direction of postglacial migration.

An example of an orthoselection flora is the pine-birch forest of western Siberia, where conditions during each ice age were particularly severe. Unlike the region to the west, occupied by migration floras, this landscape does not offer large areas that could have served as glacial refugia for temperate species. As a result, extinction rather than range adjustments has been the main process of floristic change, and plant communities have become progressively depauperate with each successive glaciation. Long-term processes, such as the development of continentality in Late Cenozoic time, have had a greater impact on the development of an orthoselection flora than the repeated glaciations of the Pleistocene.

Relict floras are found in the southern Maritime Territories, part of the Caucasus, Central Asia, and the southern Baikal region. The fact that these floras show little change in Quaternary time suggests a climate of relative stability.

Pollen records are interpreted in terms of the changing temperature and moisture conditions that characterize each glacial-interglacial cycle. Interglaciations begin with a warm-dry (thermoxerotic) stage, followed by a warm-humid (thermohygrotic) phase. The glacial periods experience a cold-humid (cryohygrotic) stage, then a cold-dry (cryoxerotic) stage. Identification of these different stages in pollen records across the country is the primary means of correlation. By use of this method, a map constructed from pollen data at 268 sites shows the vegetation at the thermohygrotic-thermoxerotic transition of the Mikulino Interglaciation, i.e., the climatic optimum. The interglacial patterns bear close resemblance to present vegetation distributions. Unlike today, however, in Mikulino time polar deserts were completely absent, tundra was more restricted, and the northern borders of the boreal forest and broad-leaved forest lay farther north. More subtle comparisons are also noted, e.g., birch featured more prominently in the boreal than today. Broad-leaved trees and cedar (*Pinus sibirica*) were common in the southern boreal forest. White beech (*Carpinus betulus*) was more widespread in the southern forests of the eastern European Plain. Elm and oak were present in the broad-leaved forests of southwestern Siberia, where they do not grow today. Forest-steppe occupied much of the area of present-day steppe, and true steppe was confined to the southeastern East European Plain, Kazakhstan, and the western Altay. Deserts may also have been more restricted, although little information is available from those areas.

The vegetation map for the glacial maximum (cryohygrotic-cryoxerotic transition) of Late Valdai time is based on 187 sites and very few radiocarbon dates. Glacial vegetation seems to have no exact modern analogue: the periglacial areas of the country were covered by a mixture of steppe, forest-steppe, and polar desert. Open forests of larch, pine, and birch in combination with steppe and tundra species extended from eastern Siberia to the periphery of the Scandinavian ice sheet. In the Far East larch and birch were common trees in an open forest-tundra landscape. Broad-leaved and coniferous trees were mainly confined to the southern European USSR, the southwestern Ural and southern Altay Mountains, and the Far East. Most of the southern part of the country was covered by poorly characterized *Artemisia* steppe and semidesert.

The great number of sites of Holocene age lends detail to the postglacial reconstruction, and as a result the vegetation history is more detailed (Khotinskiy, chap. 18). Many sites are radiocarbon-dated (about 700 dates are available for 1000 pollen records), but correlation among sites primarily rests on proper assignment of the Blytt-Sernander scheme to each record. In this volume the Blytt-Sernander classification is used to define both chronologic (time-parallel) and climatic units, thereby making an assumption *a priori* that climatic events were synchronous and that the resulting vegetational response (at least of the local component) was immediate across Europe and Asia.

The scale of vegetation change across the USSR in late-glacial and Holocene time is impressive. The late-glacial vegetation was characterized by very heterogeneous communities of tundra, steppe, and forest, most of which have no modern analogue. Birch, pine, and spruce were the dominant trees west of the Urals; larch was common to the east. In the coastal Far East, tundra and mountain tundra prevailed. The pollen sequence shows three cool, tundra-like intervals and two intervening warm, more forested periods, similar to the climatic oscillations of northern Europe. Khotinskiy suggests that the cold late-glacial periods represent times of increased seasonality and decreased moisture.

After about 10,000 yr B.P. communities began to sort themselves into more modern associations, and distinct latitudinal zones of tundra, forest, and steppe replaced the previous hyperzonal vegetation. During the Early Holocene (the Boreal Period) the coverage of forest was the same as today's, although treeline extended as much as 200 km farther north, evidenced by the occurrence of fossil larch and birch wood dating to this period in the tundra of the Taimyr Peninsula. Birch was the dominant tree from eastern European USSR to southwestern Siberia. Spruce was also widespread, growing in parts of Siberia where it is not found today, thus indicating humid and less continental conditions. In Middle Holocene time (the Atlantic Period), the forest zone expanded dramatically northward at the expense of tundra. In European USSR the northern limit of boreal forest shifted north as much as 400 to 500 km from its present position. At the same time broad-leaved trees, including oak, elm, lime, and hazel, typical today of the Baltic, Belorussian, and western Ukranian regions, migrated far into the forest zones of middle latitudes, forming a belt of temperate forest in the European USSR that was three to four times wider than that of today and in places was as much as 1200 to 1300 km wide. Surprisingly, the southern limit of forest was relatively unchanged and similar to that of the present day. The vegetational reconstructions of the northern and southern limits of forest suggest that Middle Holocene temperatures were considerably warmer in the north than they are today, but that there was little or no change in the south.

A series of short-term climatic fluctuations is inferred between 6400 and 3200 yr B.P., bringing to an end the Middle Holocene thermal maximum in the Soviet Union. By Late Holocene time (late Subboreal Period) the forest/tundra border had shifted southward 400 to 500 kilometers in some areas. The modern spruce taiga, which includes pine and birch in the western and central parts and pine, larch, and birch in the east, became the dominant vegetation type. Broad-leaved trees were only minor elements of the coniferous forests, much restricted from their Middle Holocene distribution. Particularly noteworthy is the decline of elm in Soviet forests at 4600 yr B.P., coinciding with its demise in northern Europe. The elm fall is attributed to climatic cooling at the end of the Atlantic Period, rather than to forest clearance by humans or to widespread disease, which are offered as alternative hypotheses farther west. The establishment of cooler conditions during the "Little Ice Age" is inferred in Subatlantic time by the expansion of birch-pine forest at the expense of spruce.

Peatlands developed during the Holocene over vast parts of northern European USSR, western and central Siberia, and the Far East (Neustadt, chap. 19). In western Siberia several stages are distinguished in peatland formation, beginning with the formation of numerous lake basins in late-glacial time. These basins were filled with peat and merged into several large peatlands in Early and Middle Holocene time, and in Late Holocene time they coalesced into the vast peat areas of today, some of which cover 500,000 km^2. The development of peatlands seems to be independent of the many climatic fluctuations inferred from the pollen data in Chapter 18.

Fauna

Most of the Late Quaternary vertebrate remains are associated with archaeological sites (Markova, chap. 20). Interglacial faunas, dated by their association with late Acheulian artifacts, are reported from the Central Russian Plain and the Caucasus. Much of the fauna, especially the small mammals, bears close affinity with modern types, and fossil mammals indicative of forest, steppe, and forest-steppe are recovered in sites within those environments today. Thus, the biogeographic distribution of interglacial and postglacial mammals is probably very similar. In Early and Middle Valdai time the mammalian faunal complex, consisting of mammoth, woolly rhinoceros, reindeer, polar fox, and pied lemming, was widespread over the Russian Plain and extended into the Crimea and Caucasus. The southern Russian Plain was inhabited by steppe species (including horse, bison, saiga antelope) as well as mammoth and reindeer. Woodland mammals are found at most fossil sites, suggesting localized but widespread pockets of forest. In Late Valdai time the fauna was more heterogeneous. Tundra and steppe species occurred together on the Central Plain north of about latitude 50°N. South of that, the remains of periglacial, steppe, and woodland mammals are found, but the woodland types (for example, brown bear, boar, beaver, wood vole, wood mouse) are usually confined to riparian sites. Northern animals apparently ranged as far south as the Crimea, and southern forms penetrated farther north than they do today. Reindeer is the most common fossil of this period, and its remains are abundant across the entire Central Plain, the Black Sea, and the Crimea. The early disappearance of large animals like the mammoth and woolly rhinoceros in Crimean faunas is attributed to Paleolithic hunters.

Markova (chap. 20) uses the modern distribution of small mammals with limited range and distinct ecology to make inferences about the past climate of the Central Russian Plain. The mammal data suggest that the full-glacial climate was cooler, drier, and more seasonal than that of today, which corroborates other evidence for enhanced continentality in Late Valdai time. An analysis of contemporaneous Siberian faunas corroborates these climatic interpretations, although faunal differences were maintained between eastern and western Siberia despite the widespread cold conditions (Vereshchagin and Kuz'mina, chap. 21). The most distinctive glacial-age assemblages are reported in the Far East and include raccoon dog, badger, otter, tiger, horse, wild boar, and Manchurian hare. The mammalian diversity of the region probably resulted from faunal exchange with the Indo-Malayan region to the south.

Insect faunas from Late Pleistocene sediments of Siberia (Kiselev and Nazarov, chap. 22) yield paleoclimatic information similar to that derived from the mammal records. The interglacial faunas are very similar to those of today and seem to reflect an environment like the present. During Valdai time the insect assemblages suggest widespread open landscapes, but as with the interpretation of the small-mammal data, provincial differences between eastern and western Siberian insects were maintained to some degree.

Inland Sea Basins

Bordering the Soviet Union in basins or on marine shelves are marginal seas that today have wide connections to the oceans, but during the times of eustatic sea-level depression some of them were much more constricted, e.g., Sea of Japan (Chepalyga, chap. 23). In addition, semi-isolated basins like the Black Sea and the Baltic and White Seas have limited connections to the ocean but in the past may have been completely isolated. Finally, fully isolated basins like the Caspian Sea fluctuated in level and in salinity in response to climatic change and the input of glacial meltwaters. The records of these changes consist primarily of former strandlines and of fossils sensitive to ecological conditions.

During the Mikulino Interglaciation the Black Sea was a marine basin connected to the Mediterranean Sea, and its level was 8 to 12 m higher than it is today. It was slightly larger and of similar salinity. Its surface waters were warmer, and the deep waters were charged with hydrogen sulfide. At the same time the Caspian Sea was a closed basin with brackish water. Its level was above the present level, because the Mikulino climate was more humid than that of the present even though it was warmer, as indicated by pollen studies in the region.

In the north during the Mikulino Interglaciation the Arctic seas flooded the previously glaciated areas around the White Sea and up the valleys, providing connections to the Baltic Sea and the North Sea. Farther east in northern Siberia it covered an area 1400 km from north to south and 1900 km from west to east. A large and varied marine fauna of boreal species indicates that the Gulf Stream penetrated far to the east; in addition, two species came in from the Bering Sea. The fauna suggests that the water temperatures were 2°C to 4°C warmer than today and that most of the sea did not freeze.

In the Early Valdai the level of the Black Sea was lowered to more than 100 m below modern sea level, in response to eustatic depression of sea level. The Straits of Bosphorus were dry, and the Black Sea became isolated, as is the Caspian Sea today. Salinity is estimated as 5 to 10‰ under conditions of a cold climate marked by periglacial steppes and coniferous forest.

The Caspian Sea at this time (about 60,000 years ago), on the other hand, rose to 76 m above its present level, and it was 2.5 times its modern size. It overflowed through the Manych Strait into the Black Sea. Subsequently it lowered through several stillstands to 4 to 6 m above the present level by the end of the Early Valdai. This major transgression is attributed to reduced evaporation throughout the drainage basin, resulting in increased input from the Volga River. The salinity at this time was 12 to 13 ‰, as indicated by mollusk remains, and the water temperature was low, perhaps because of the influx of glacial meltwaters, as suggested by $\delta\ ^{18}0$ values of -12 to -14.5‰.

In the Middle Valdai interstade the Black Sea transgressed to near its present level, and its salinity was similar. The Caspian Sea dropped to a low level because of warmer climatic conditions. The Arctic seas transgressed the North Siberian Lowland, and the fauna indicates warmer water temperatures than today.

In the Late Valdai, the lowered sea level in the world ocean caused the Black Sea level to drop once again, isolating it from the Mediterranean marine waters. The water was freshened by the inflow of glacial meltwater from the north and by the reduced evaporation under conditions of a cool periglacial climate, and the basin apparently had an outlet through the Bosphorus, which has since been partially filled with alluvium. The water was cool and not chemically stratified.

The Caspian Sea transgressed in the Late Valdai to be 1.5 times its present size and 28 m deeper, but with a salinity similar to that of the present and a cooler temperature, with ^{18}O values in mollusks suggesting inflow of glacial meltwater.

With the lower ocean level during the Late Valdai Glaciation, the Sea of Japan was transformed to a semi-isolated basin with a restricted outlet and water temperature 8°C to 10°C lower than today's. Decreased salinity and temperature and increased incidence of berg-rafted particles in the north are attributed to inflow from the Amur and other rivers.

During the Holocene sea level rose again and the Black Sea once more received Mediterranean waters. Its level transgressed to 3 to 5 m above the present in the Middle Holocene, but it has lowered in the last 3500 years, with minor fluctuations. The Caspian Sea level dropped in the Early Holocene as the climate became drier, but it rose in the Middle Holocene to its present level. The Sea of Japan expanded with the postglacial rise in ocean level, and in the Middle Holocene the surface-water temperatures were as much as 3°C to 4°C warmer than they are today.

In the Baltic basin a series of disconnected proglacial lakes was joined to make the Baltic Ice Lake about 12,000 years ago, draining to the North Sea across Sweden. As the ice retreated and sea level rose, the basin became flooded to make the Yoldia Sea 10,500 to 9000 years ago. With isostatic uplift of the outlet the sea was excluded and Lake Ancylus succeeded. With further rise in sea level, Denmark Straits were opened and the marine waters once again came into the Baltic basin to make the Littorina Sea, whose strandline has subsequently been tilted by continued isostatic uplift. A final regression to the Limnaea Sea involved further freshening of the water.

Paleoclimatic Reconstructions

A variety of methods that yield paleoclimatic estimates and information on past circulation systems are presented in the next part of the volume. One method considers the modern distributions of specific plants (or groups of plants) that occur together in the fossil record and relates them to the climatic parameters that delimit their distribution today (Grichuk et al., chap. 24). Climagrams provide a graphic means of displaying the climatic values that circumscribe the modern ranges of the fossil taxa. The values where the ranges overlap are inferred to be the conditions responsible for sympatry of species in the past. This method of paleoclimatic reconstruction works best in areas of migration floras, where most of the species are present in successive interglacial floras. Orthoselection floras are less suitable because the modern flora is usually more depauperate than that of previous interglaciations, and the climagrams therefore utilize fewer taxa. The area of climatic overlap (shown by the climagram) for an orthoselection flora is usually greater and less precise than it would be if the flora were more diverse. The method does not work for relict floras, for they occur in areas of relative climatic stability that have changed little in Quaternary time.

Climagrams were used to reconstruct climate at 25 sites of the Mikulino Interglaciation and seven of the Late Valdai Glaciation. The number of sites in this analysis as well as the number of radiocarbon dates used to establish the chronology is small. However, the sites typify a wide geographic region and are correlated on the basis of their geologic position and pollen stratigraphy. The reconstruction of full-glacial conditions between 20,000 and 18,000 yr B.P. is very tentative, for it depends on various kinds of sites and almost no radiometric dates. One of the records, the Pucha section, lies within the limits of the last glaciation and is dated to 21,410 yr B.P.; several other sites are inferred to be full-glacial on the basis of their pollen record; and still another site (Khotylevo II) is associated with archaeological remains, and the pollen may reflect local disturbance rather than regional vegetation.

Velichko (chap. 25) uses the climatic estimates presented in Chapter 24 and earlier chapters to develop spatial climatic reconstructions for interglacial and full-glacial time. Today the climate of the European USSR, which is the area of migration floras, is dominated by low-pressure systems in the North Atlantic Ocean. Western Siberia, which is marked by an orthoselection flora, is governed by continental air masses. The relict flora of the Soviet Far East is under the influence of monsoonal systems, which result from an interplay between maritime Pacific and continental Siberian air masses.

During the Mikulino Interglaciation, circulation was latitudinal, much as it is today, but seasonality was decreased as a result of warmer winters, especially in the eastern Arctic. Temperatures south of latitude 50°N were the same or slightly cooler than those of today; precipitation was apparently higher and more uniformly distributed across the country. The climatic record implies that westerly air masses from the Atlantic brought storms to the European USSR, both in summer and in winter. Greater activity in the Atlantic, produced by a northward shift in the Gulf Stream, apparently allowed moisture to penetrate farther inland and to higher latitudes.

The reconstruction of full-glacial conditions in Late Valdai time is based on limited pollen data, but paleocryologic data, fossil insects, and fossil mammals are also considered. The lack of modern analogues for many of the biological assemblages owes its explanation to the fact that meridional circulation patterns replaced the present latitudinal or zonal condition. This situation arose with the southward shift of the Gulf Stream, which diminished the influence of the Icelandic Low and reduced the penetration of westerly storms in winter. Over much of the Soviet Union the winter climate was controlled by a large high-pressure system, which formed from a coalescence of Polar, Siberian, Central Asian, and Scandinavian anticyclones. The effect was to bring very low temperatures, dry conditions, and little cloud cover to most of the country. The greatest decrease in winter temperatures is registered in the Northwest near the Scandinavian ice sheet and in the southwestern zone of permafrost. Winters in Siberia were also more severe than today, with temperatures dropping to −70°C or −80°C in the arctic regions. The greatest change in summer temperature occurred in the western and southwestern USSR as well, contributing to the formation of the Scandinavian ice sheet and the vast periglacial region there. By contrast, summer temperatures in southern and central Siberia changed little from present values. Central Asia was cold and particularly dry at the glacial maximum as a result of the blocking effect of high-pressure systems on westerly sources of moisture. In the Far East the climate continued to be monsoonal, but less precipitation was brought in from the Pacific Ocean with the reduced contrast between land and sea-surface temperatures.

An interesting feature of the summer climate in the western USSR was the creation of the "glacial monsoon" from the juxtaposition of a cold high-pressure area over the ice sheet and a warmer low-pressure system to the south. Cold air descending southward off the ice sheet was adiabatically warmed and dried over the periglacial region. During winter, dust transported by the glacial monsoons in summer settled out of the atmosphere and led to the formation of loess deposits on the Russian Plain.

A second approach used to reconstruct Holocene climatic history utilizes the relationship of vegetation zones and climatic parameters (Savina and Khotinskiy, chap. 26). This approach assumes that the modern vegetation of the Soviet Union is delimited by specific climatic parameters (of extreme and average conditions) and that this relationship was unchanged through Holocene time. It also assumes that these vegetation zones have a long history, which can be identified by pollen analysis, and that reasonably good analogues exist for the fossil reconstructions. By use of this method, the sums of air temperatures, mean January and July temperatures, humidity, and evaporation factors for different periods of the Holocene are estimated. The results for the Boreal Period support the qualitative reconstruction from pollen data: compared with present conditions, January temperatures were lower in the European USSR and higher in the Asian part. July temperatures show almost no change.

Klimanov (chap. 27) uses multivariate statistical techniques to make climatic inferences directly from pollen data, thus bypassing uncertainties associated with the vegetational reconstructions. The statistical information method, which establishes an empirical relationship between contemporary pollen and climate variables, is applied to the fossil pollen data to determine the best estimate of past climate. The inferred climate during the Atlantic Period indicates that the greatest rise in July temperatures occurred at high latitudes, whereas at low latitudes temperature changes were negligible. January temperatures decreased from west to east. Synoptic reconstructions of the Atlantic Period suggest that the Gulf Stream was more active in the North Atlantic and that the Siberian High was less active. Khotinskiy (chap. 28) suggests that the climate of the Holocene is no different from that of other interglaciations because it shows a cyclicity of warm-moist and warm-dry events. The Early and Middle Holocene represent the thermohygrotic and thermoxerotic stages, whereas the Late Holocene may be the beginning of the cryohygrotic stage and the onset of the next glaciation.

Human Cultures

The migration of early peoples into the Soviet Union was sporadic, and the level of cultural development depended largely on the available resources (Praslov, chap. 29). The oldest evidence of early Paleolithic people is found in cave and open-air sites in the Caucasus, southern Russian Plain, Transcarpathia, Central Asia, and Kazakhstan. This distribution of sites suggests that humans did not inhabit areas north of latitude 48°N in either the Early or the Middle Pleistocene. With the warmer conditions of the Mikulino Interglaciation, the area of human occupation was extended as far north as latitude 54°N, and by the last glaciation it occupied much of northern Asia up to latitude 71°N. Late Paleolithic peoples were apparently able to adapt to the cold, dry glacial landscapes as a result of the development of hearth and dwelling structures and new techniques in splitting stone. Unfortunately, many of the late Paleolithic sites in northeastern USSR, which would provide information on the migration of early humans into western Beringia, are still controversial in both their age and stratigraphic context.

The Early Holocene was a time of great environmental change, and, with diversified resources, human cultures also changed. Mesolithic cultures date between 10,000 and 6000 yr B.P. in the forest zone and between 10,000 and 8000 yr B.P. in the southern part (Dolukhanov and Khotinskiy, chap. 30). The sites are dated largely by typology and distributed unevenly across the entire country. In most regions hunting, fishing, and gathering were the primary economies.

Neolithic sites, ranging in age from 7000 to 4000 yr B.P., are abundant, a fact that must reflect the increased population density of that time. The diversification of the landscape is evidenced by the great difference in the economies of different regions. Domestication of plants and development of agriculture are first recorded at this time, with the center of cultivation located in the foothills of Central Asia, Transcaucasis, and the Carpathian Mountains. In the forest zones foraging economies continued to dominate.

The Bronze Age developed in the Late Holocene about 3000 yr B.P. It was recorded by a decline in agriculture and an increase in nomadic pastoralism, although the correlation with vegetational changes inferred from the pollen records of this time is uncertain. Deforestation of the landscape by human activity occurred relatively late in the USSR. The Central Russian Plain was first deforested about 200 to 300 years ago, and other northerly regions show the effects of humans more recently. While environmental change had a strong impact in shaping human cultures, human impact on the natural environment has been negligible until very recently.

Introduction

A. A. Velichko

The preparation of monographs dealing with the evolution of the natural environment of the USSR and USA reflects an important stage in joint Soviet-American research on problems in paleoclimatology, carried out in accordance with the plans for the implementation of the Agreement between the USSR and USA on Cooperation in the Area of Environmental Changes on Climate. Director of the research being done in the Soviet Union is Academician I. P. Gerasimov.

The Soviet Union occupies a considerable portion of extratropical Eurasia, and its territory is traversed by practically all natural extratropical climatic belts. This accounts for the development of a wide range of zonal landscape types, with a complex system of local differences, and it opens up extensive possibilities for studying large-scale natural changes as well as the reactions of the natural components in different geographic zones to marked fluctuations of climate.

This monograph examines the natural changes that have taken place during the last 100,000 to 125,000 years. Such a chronologic range is of fundamental importance, for it covers completely the last climatic macrocycle and the start of a new one, of which the present is a part. The last climatic macrocycle of the Late Pleistocene in the USSR, as in other extratropical regions, includes two principal components: a warm interglacial stage and a cold glacial stage. Macrocycles similar to the last constitute the principal stable component of macroscale global fluctuations of climate, at least they have during the last million years. Interpretation of the characteristics of natural climatic changes in the course of such a macrocycle will undoubtedly aid the investigation of the general principles of long-term climatic fluctuation on the Earth, and the study of the present climate from an evolutionary viewpoint will permit a clearer prediction of future natural trends.

One of the main problems that arises in interregional paleogeographic reconstructions is the correlation of natural events in time. It is well known that such correlations are made on the basis of several methods (lithostratigraphic, paleontologic, and radioisotopic). For the last Late Pleistocene macrocycle (at least its second half), as well as for the Holocene macrocycle, such interregional correlations are the most reliable because of the accuracy of the radiocarbon method of absolute-age determination, although the probability of errors must be considered. Nevertheless, the availability of data and the reliability of correlations are very much greater for the last stage than for older stages, and the paleogeographic reconstructions and paleoclimatic models are more reliable.

In general outline it is possible to compare the events in most of the USSR, as is evident from Table I-1, although a good many problems exist in the correlation of natural events as well as in the paleogeographic reconstructions themselves. I will briefly describe the relationship of the main phases of development in the Late Pleistocene and then turn to some of the unsolved problems.

Most agreement among the investigators exists for the interglacial stage that marks the beginning of the last macrocycle. In the European part of the USSR, this interglaciation is most often referred to as the Mikulino; in Siberia, it is referred to as the Kazantsevskoye Interglaciation. Characteristic of that time was a general increase in forest cover and in availability of heat and moisture, especially in the West. Moisture supply was also greater in areas occupied by present steppe. On the whole, however, the zonal structure was similar to the present one.

The subdivision of the subsequent glacial epoch is not uniform in different regions. In the European part of the USSR, this epoch is called the Valdai, which is subdivided into the Early Valdai and Late Valdai Glaciations, separated by the Middle Valdai interval. Most researchers consider the latter interval to be primarily a cool but climatically nonuniform megainterstade, although some believe it to be of interglacial character. In the view of most investigators, the Late Valdai Glaciation was the more widespread, whereas the Early Valdai ice sheet did not extend

Table I-1.

		European USSR			Sea Basins		
		Glacial Regions (N. S. Chebotareva, I. A. Danilova-Makarycheva, M. A. Faustova, L. N. Voznyachuk, and others)	Periglacial Regions—Loesses and Soils, Cryogenic Horizons (A. A. Velichko)		Arctic Basin (V. I. Gudina and S. L. Troitskiy)	Black Sea (A. L. Chepalyga)	Caspian Sea (G. I. Rychagov)
Glacial Epoch	Late Valdai Glaciation	Salpausselkä Palivere Alleröd Middle Dryas Bölling Luga and Neva advances Raunis warming Vepsovo advance Warming Maximum stage and beginning of degradation (Yedrovo stage)	Loess III (Altynovo) Gleying level Loess II (Desna)	Yaroslavl' Vladimir	Regression	Novoevksin regression semifreshwater basin, −100 to −40 m	Late Khvalyn transgression brackish-water basin ±0 m
	Nonglacial Interval	Dunayevo warming Warming and cooling Upper Volga warming Cooling Krutitsa warming	Bryansk interval (soil) 30,000 to 25,000 yr B.P. Loess I (Khotyleva) Krutitsa interval (soil)	Smolensk phase "b"	Karginskiy transgression sea basin (only in Siberia)	Surozh transgression semimarine basin, −10 m	Yenotayevka regression, −50 to −60 m
	Early Valdai Glaciation	Kurgolovo cooling and glaciation within the confines of the Baltic Shield	Mezin complex: Intra-Mezin loess	Smolensk phase "a"	Regression	Pre-Surozh regression semifreshwater basin, −100 m	Early Khvalyn transgression brackish-water basin, +48 m
Interglacial Epoch		Mikulino	Mikulino Interglacial soil	Interglacial	Boreal transgression	Karangat transgression sea basin +8 to +10 m	Late Khazar brackish-water basin −10 to −15 m

much beyond Scandinavia. However, there are schemes in which the dimensions of the two are considered equivalent.

Somewhat different interpretations have been proposed for the glacial regions of Siberia, although there as well two glacial epochs separated by a nonglacial one are distinguished within the Late Pleistocene. Each glacial epoch is considered independent in this case. The earlier one is called the Zyryanka, and the later, the Sartan (although in some of the latest schemes the entire Late Pleistocene is termed the Zyryanka). There are also independent regional names for individual glacial stages, as described in the corresponding sections of this monograph.

The interval separating the Zyryanka and Sartan Glaciations has been regarded for a long time as the climatically complex Karga Interglaciation, approximately 50,000 to 25,000 years ago. Such an approach offers better prospects for correlation with the European part, where the Bryansk (Dunayevo) warming is also distinguished for the period 30,000 to 25,000 years ago. It is also possible that the warming of about 50,000 years ago corresponds to the Krutitska (Upper Volga) Interstade in the European part of the USSR.

In contrast to the European part of the USSR, for Siberia, according to the opinion of most investigators, the maximum ice extent was during the Late Pleistocene (Zyryanka) glaciation, and that during the Sartan Glaciation the ice occupied a more modest area. Even smaller was the Late Pleistocene glaciation in northeastern Asia, where it was restricted to mountainous areas.

It is possible that the indicated differences in the degree of contemporaneous glacial development in Siberia and eastern Europe are real. On the whole, the differences are explained by the theory of metachronicity of glaciation (Gerasimov and Markov, 1939). This theory calls for reduction in glacial extent from west to east in the USSR because

Correlation of Late Quaternary Glacial and Interglacial Events in the USSR

Western Siberia			Central Siberia		Northeast USSR		Central Asia
	Glacial Regions (S. A. Arkhipov)	Periglacial Regions (I. A. Volkov)		Glacial Regions (L. L. Isayeva)	Glaciations (V. G. Bespalyy)	Loessial-Glacial Complex (T. N. Kaplina and A. V. Lozhkin)	Loesses and Soils (A. A. Lazarenko)
Late Zyryanka (Sartan) Glaciation	Polar Ural phase Sopkeyskiy phase Tyuteyskiy warming (?) Salekhard-Uval phase	Bagan loess Suma subcomplex Yel'tsovka loess	Sartan Glaciation	Noril'sk (north Taimyr Melkolamskiy) phase N'yapan (Mokoritto, Upper Taimyr) phase Karaul (Ezhangodosyntabul'skiy-North Kokora) phase	Sartan (Bokhapcha II, Iskaten', Khaymikin) mountain-valley glaciation	Mus-Khaya cooling Formation of loessial-glacial complex	
Middle Zyryanka complex interstade, 50,000 (55,000) to 24,000 yr B.P.	Karginskiy warming 29,000 to 25,000 yr B.P. Lokhpodgort Glaciation Zolotoy Mys warming Cooling, 45,000-44,000 yr B.P. Shuryshkary warming 50,000 (55,000) to 45,000 yr B.P.	Iskitim soil complex, 32,000 to 24,000 yr B.P. Tula loess Upper soil of Berdsk soil complex	Karginskiy Interglaciation	Lipovsko-Novoselovskiy warming, 30,000 to 24,000 yr B.P. Konoshchel'skiy cooling (Zhigansk glacial complex, 33,000 to 30,000 yr B.P.) Malaya Kheta warming Early cooling Early warming	Karginskiy (Penzhina) Interglaciation (interstade?)	Kuranakh-Salin warming (30,000 to 24,000 yr B.P.) Formation of loessial-glacial complex (34,000 to 30,000 yr B.P.) Khomus-Yurmkhskiy warming	Soil complex 1
Early Zyryanka Glaciation	Khoshgort stage Interstade Kormuzhikhantskiy stage	Loess Cryogenesis phase	Zyryanka Glaciation	Murukta stage Interstade Lowever Tunguska stage	Zyryanka (Bokhapcha I, Vankarem, Tylkhoy) mountain-valley and submontane glaciation	Formation of loessial-glacial complex	Loess Soil complex 2 Loess
	Kazantsevo	Lower soil of Berdsk soil complex		Kazantsevo	?	?	

of the blocking influence of anticyclonic air masses expanding from Siberia toward Europe.

A second, more differentiated explanation has been offered for the out-of-phase character of earlier and later glaciation in Siberia and eastern Europe. This explanation is based on the hypothesis that there was asymmetry in the glacial development in the Northern Hemisphere (Velichko, 1980). According to this hypothesis, at the beginning of the Late Pleistocene, primarily in Siberia, cooling reached a level such that snowfall increased. In northern Europe, however, because of the longer influence of warm Atlantic air masses, an increase in snowfall came later. Thus, the Early Valdai Glaciation of eastern Europe was less extensive than the Zyryanka Glaciation in Siberia. During the Late Pleistocene, as the cooling increased, the anticyclonic masses in Siberia prevented the formation of vast ice sheets, whereas in eastern Europe, still subject to the intrusion of relatively moist air masses from the Atlantic, the ice sheet was able to expand for a longer period of time. The stable anticyclonic masses in northeastern Asia generally prevented the formation of ice sheets there and instead promoted the development of permafrost.

During the Late Pleistocene, permafrost spread widely over the entire USSR, including the European part. There, thanks to detailed studies in the periglacial region, it has been possible to distinguish cryogenic horizons (Smolensk, Vladimir, Yaroslavl') and to use them for correlation purposes for the first time. Similar horizons may be expected in periglacial regions of western Siberia as well. In the loessial periglacial regions both in the European part of the USSR and in western Siberia, a great similarity was observed in the sequence of loess formation and fossil-soil development. The correlations between the Mezin soil complex of eastern Europe and the Berdsk soil complex of western Siberia and between the Bryansk and the Iskitim complexes indicate a common character in the development of the principal natural processes over vast areas of the extraglacial part of the temperate belt during the cold epochs of the Pleistocene.

A common trend is also observed in certain features of climatic change in periglacial regions. Thus, judging from a study of loess strata in both regions, the first half of the Late Pleistocene glacial epoch was more humid than the second. The same relations exist for the loess-ice complex in northeastern Asia. An extremely severe continental climate developed over the entire USSR during the period of maximum cooling 20,000 to 18,000 years ago, when the permafrost region moved farthest to the south, the forest belt was completely destroyed, and the system of natural zoning underwent a complete rearrangement, in comparison with the interglacial, to produce a special hyperzonal type of natural cover.

Recent studies have demonstrated the substantial role of cryogenic processes in the formation of loess, not only of the so-called periglacial belt but also in Central Asia. It should be noted, however, that the correlation of Late Pleistocene events of Central Asia with events in regions farther to the north continues to be a problem. This situation hinders a definitive solution of not only special problems but also general fundamental ones. Chief among them is the correlation of arid and pluvial periods with events in the glacial regions. In the glacial stage, there occurred a marked aridization not only of the steppe areas but also of areas farther north, which also were converted from forest into steppe and forest-steppe. The section of the monograph discussing loesses of Central Asia also shows the correlation of the loess-accumulation epochs with the glaciation epochs. On the whole, aridization took place during the glacial epochs over vast areas south of the glacier margin.

This does not mean, however, that the correlation of glaciation with aridity is a rule without exceptions. The hypothesis of asymmetry in the development of the glacial epochs shows that, whereas in the Eastern Hemisphere correlation of glaciation with aridity is a general rule, in the Western Hemisphere a correlation of glaciation with pluvial conditions is characteristic. The out-of-phase character of these relations is accounted for by the special characteristics of the atmospheric circulation model that has been proposed.

A good many problems exist in studies of glaciation of individual mountain areas, such as the Caucasus, Pamir, and Altay, for the dynamics of mountain glaciers are an important indicator of climatic fluctuations. The importance of such data is already being revealed by paleoclimatic constructions based on data of paleoglaciologic modeling.

Finally, the material presented here makes it possible to compare natural events on dry land with the state of adjoining sea basins. Analysis of the history of sea basins shows that their reaction to climatic changes was not the same and depended on the type of sea basin. Thus, the Black Sea, which was linked to an ocean, underwent a drop in level during the glacial epochs, whereas for the Caspian Sea, in an inland basin, transgressions corresponded to glaciations according to most investigators. However, one must not overlook the fact that these data are not entirely consistent with radiocarbon dates (as will be evident from the corresponding section of this monograph). Obviously, this question requires further consideration.

Very important is the material characterizing the Holocene, for it directs the reader toward an analysis of the present state of the environment. This material shows that the course of natural processes in different regions was not uniform. At the same time, data on all the regions of the country clearly indicate that the optimum of the recent interglaciation has already passed.

A general cooling trend is also attested to by paleoclimatic reconstructions based on methods developed essentially by Soviet authors. Different chronologic sections impose different requirements on the quality of existing paleobotanic data. Whereas Holocene vegetational formations were similar to recent ones, those of the Mikulino Interglaciation differed appreciably. Therefore, methods based on the analysis of modern vegetation can be applied to the Holocene, whereas for the last interglaciation only methods based on an analysis of the species composition are applicable. All paleoclimatic reconstructions are based on concrete facts; therein lies the fundamental characteristic of the selected approach.

Comprehensive data on the dynamics of change in natural components during the Late Pleistocene and Holocene are supplemented with sections describing the characteristics of dispersal of primeval people in the Late Paleolithic, Mesolithic, and Neolithic. From this, the reader gets an idea of one of the initial stages of interaction between humans and the environment and the success of human cultures in mastering the landscape.

Such are the basic characteristics of this monograph. One should also note that the authors and editors of this book did not intend to produce a continuous spatial and chronologic description of the dynamics of all natural components. This type of treatment would be difficult to fit even into several volumes. This book mainly contains correlations based on results of recent studies and new reconstructions and models defining the most-advanced trends in paleogeographic research.

At the same time, particular emphasis should be placed on the fact that the proposed paleogeographic constructions are based on the work of many investigators, primarily the fundamental work of such Russian and Soviet scientists as L. S. Berg, I. P. Gerasimov, P. A. Kropotkin, K. K. Markov, G. F. Mirchink, V. A. Obruchev, V. N. Saks, N. A. Strelkov, and V. N. Sukachev. Investigators of different regions and specialists in different methods participated in the preparation of this monograph. Their opinions and scientific conclusions do not always agree. However, it is my view that an attempt to select a group of authors primarily for the purpose of agreement of points of view is not always helpful in reflecting the actual state of affairs in the study of a given area. For example, it would be much more effective to propose only one definite reconstruction of the ice sheets of northern Eurasia. However, such a reconstruction cannot at the present time be supported by the evidence. Therefore, the book presents various concepts so as to enable the reader to obtain more objective information.

The monograph was prepared at the Institute of Geography of the Academy of Sciences of the USSR. Its preparation involved the participation of leading experts from a number of scientific and scientific-industrial institutions as well as educational institutions in the USSR: the Institute of Geography of the Academy of Sciences of the USSR (Moscow), the Zoological Institute of the Academy of Sciences of the USSR (Leningrad), the Institute of Archaeology of the Academy of Sciences of the USSR (Leningrad Division), the Institute of Geology and Geophysics of the Siberian Division of the Academy of the Sciences of the USSR (Novosibirsk), the Northeastern Combined Scientific Research Institute of the Far Eastern Scientific Center of the Academy of Sciences of the USSR (Vladivostock), the Geography and Geology Departments of Moscow State University named after M. V. Lomonosov (Moscow), the Polytechnic Institute (L'vov), and the Industrial Scientific Research Institute of Construction Engineering of the State Committee for Construction of the USSR (Moscow).

Considerable assistance in the preparation of the illustrative materials for the monograph was extended by the Cartography Section of the Institute of Geography of the Academy of Sciences of the USSR. Editing of the maps was done by I. N. Chuklenkova.

The authors hope that the reader will gain a reasonably complete understanding of the characteristics of the natural development of the USSR during the Late Pleistocene and Holocene. They wish to express their most sincere thanks to their American colleagues for the considerable amount of work involved in the urgent translation and editing of this monograph.

Late Pleistocene Glaciation of the Northern USSR

CHAPTER 1

Late Pleistocene Glaciation of European USSR

M. A. Faustova

According to radiometric, geologic, glaciomorphologic, and paleobotanic data, the Valdai Ice Age was characterized by two cold stages in the European USSR. To a large degree, the problem of the age of these stages, and particularly of the early stage, is connected with the relationship of glaciations and marine transgressions in the Late Pleistocene in the extreme north of the USSR.

The marine series represented in sections of the Kola Peninsula was differentiated by Lavrova (1960) into sediments representing the boreal (correlated with the Eemian) and White Sea transgressions. However, the age of these transgressions as well as the relationship of marine deposits to moraines has since become the subject of protracted discussion. As a result of detailed studies using faunal, diatom, chemical, and palynologic methods, it has been established that the lower band of marine deposits, that is, the Ponoy layers (the names of the layers are those of rivers on the Kola Peninsula), correspond to the warm-water boreal transgression (with water temperatures 4° to 6°C higher than present ones) (Gudina and Yevzerov, 1973; Yevzerov, Lebedeva, and Kagan, 1976). However, radiocarbon dates of shells from Ponoy layers in the interval 46,000 to 33,000 yr B.P. (minimum and maximum dates: 33,650±4000 yr B.P. [TA-271] and 46,540±1770 yr B.P. [LU-1373], with two dates that were beyond the limit) suggested correlation with the Middle Valdai Interglaciation (Yevzerov, 1970; Gudina and Yevzerov, 1973), although a final date for shells exceeding 30,000 yr B.P. was interpreted by Arslanov (Arslanov et al., 1975) as the minimum age. It was not until the end of the 1970s that new dating by the uranium-thorium method finally confirmed the Mikulino age of the Ponoy layers. Three sections of Svyatonosskiy Bay and the Malaya, Kachkovka, and Chapoma Rivers gave the following ages: 97,000±4000 yr B.P. (LU-455B) for deposits in the first section, 102,000±4000 yr B.P. (LU-452A) and 114,000±4000 yr B.P. (LU-452B) for the second, and 85,500±3200 yr B.P. (LU-464) and 86,000±3900 yr B.P. (LU-464-B) for the third (Arslanov, Yevzerov, et al., 1981).

The upper band of marine deposits (Strel'ninskiye layers), separated from the Ponoy layers by an erosion surface locally marked by gravel and, according to Apukhtin and others (1977), by a basal till, came to be treated as sediments of the independent Middle Valdai transgression. Climatic conditions at this time were colder than the present ones. Thus, in sections of the Kola Peninsula, two separate marine members have now been identified that precede the glacial horizons of the first and late stages of the Late Pleistocene Valdai Glaciation.

Early Valdai Glaciation

On the slopes of the Baltic Shield in Karelia, Early Valdai glacial deposits are separated from the Late Valdai moraine by Middle Valdai peat dated as 46,700 yr B.P. (start of warming) and 43,900 yr B.P. (climatic optimum) (Ekman et al., 1979). The age and interstadial or interglacial rank of the warm interval are still under discussion. In the Russian Plain, there are no obvious sections with Early Valdai till overlying Mikulino deposits and overlain by plant-bearing Middle and Late Valdai sediments and Late Valdai till. In the north of the Russian Plain (in the Onega and Ladoga Basins and the Severnaya Dvina Basin), the major portion of the Early Valdai section consists of lacustrine-glacial and lacustrine deposits, indicating limited continental glaciation (Devyatova, 1980). In the basin of Severnaya Dvina River and its tributary Vaga, Early Valdai aquatic sediments consistently replace Mikulino sediments upward, and only locally are they separated by cover loams that include an older till redeposited by solifluction.

The lack of Early Valdai glacial deposits in the sections, as well as the absence of Late Valdai radiometric dates for the outermost moraine on the Russian Plain, have enabled a majority of investigators to conclude that the Early Valdai ice did not extend beyond the limits of the Baltic Sea basin (Chebotareva et al., 1971; Voznyachuk, 1973; Gerasimov, 1973; Chebotareva and Makarycheva, 1974; "Structure and Dynamics of Europe's Last Ice Sheet," 1977; Devyatova, 1980). A similar conclusion has been reached for extraglacial regions of the Russian Plain (Velichko, 1975; Ivanova, 1977). However, in addition to concepts of the limited dimensions of the Early Valdai Glaciation, there persists the idea of a maximum spread of glaciation in the Early Valdai (Raukas and Serebryanny, 1971; Serebryannyy, 1978; Arslanov, Breslav, et al., 1981).

The period separating the end of the Mikulino Interglaciation from the emergence of the glacier on the European Plain has been referred to by certain authors as the ice-free Valdai (Chebotareva et al., 1971). In the view of other investigators, a considerable portion of this period was taken up by the Middle Valdai Megainterstade (Voznyachuk, 1973). The term "ice-free Valdai" means that during the first half of the Valdai the glacier did not reach the Russian Plain but remained within the confines of the Baltic Shield and that the climate was fairly cold, with alternating warm and cool intervals of different intensity and duration. During the relatively warm intervals, the vegetation was represented by periglacial tundra and arctic-alpine associations.

In the deposits of the initial stages of the Valdai, two coolings separated by two warmings were recorded—the Upper Volga warming (Tarasovo warming according to L. N. Voznyachuk) and the Kruglitsa warming (first identified in Belorussia) (Voznyachuk, 1961, 1973). The stratotypic sections are located on the Minsk Upland at the village of Tarasovo and in the Kruglitsa area. The name Upper Volga Interstade for the first post-Mikulino warming was proposed by A. I. Moskvitin in 1950.

The absolute ages of the first post-Mikulino warm and cool intervals are not certain, for the final radiocarbon dates (in the intervals 64,000 to 58,000 and 55,000 to 52,000 yr B.P.) are minimal. During the warm phases, the forest cover increased, with an expansion of pine and spruce (Chebotareva and Makarycheva, 1974).

Middle Valdai Nonglacial Interval

The middle of the ice-free interval is marked by sediments with radiocarbon dates of approximately 50,000 to 30,000 yr B.P. (Figure 1-1). This period was also characterized by short cool and warm intervals, with tundra-steppe associations and thin birch forests alternating with increased conifer-birch forest cover. The most significant warming within this interval (39,000 to 38,000 yr B.P.) is reflected most

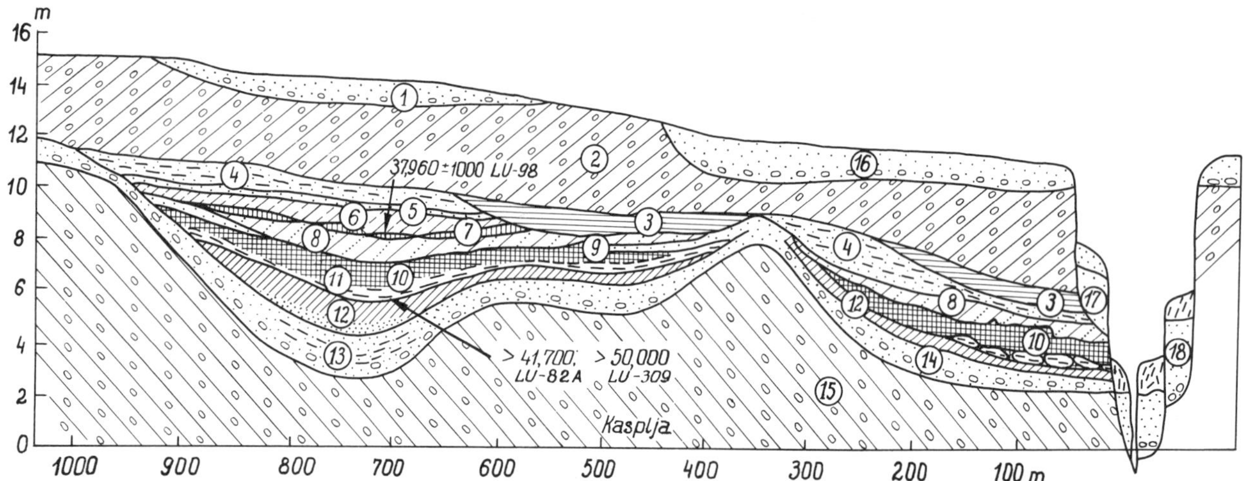

Figure 1-1. Geologic profile of the left bank of the Kasplya River above the mouth of the Nevorozhka River, opposite the village of Sloboda, (after L. N. Voznyachuk [Arslanov et al., 1973].) (Radiocarbon dates in years B.P.)

Numbers:

1. Sand with basal gravel and boulders of fluvioglacial origin (late-glacial deposits)
2. Till, moraine of maximum stage of Valdai Glaciation
3. Varied clays
4. Fine sand with reddish brown ortsands, occasionally changing to silt
5. Inequigranular sand, with gravel, pebbles, and boulders
6. Loam and sandy loam, both altered by solifluction (layers 2 through 6 formed in the first half and culmination phase of the maximum stage of Valdai Glaciation)
7. Loam with an interlayer of muddy peat (Middle Valdai Interstade deposits)
8. Loam with sand lenses, heavily cryoturbated
9. Inequigranular sand with pockets of loam (cryoturbation) and an admixture of gravelly material
10. Humified loam
11. Muddy peat with vegetal remains (interstadial)
12. Sandy loam (8 through 12 are of Lower Valdai age according to L. N. Voznyachuk or from the end of the Mikulino and the beginning of the Valdai according to F. Yu. Velichkevich)
13. Alluvial (?) sand, inequigranular, horizontally and obliquely laminated, Late Pleistocene
14. Sandy-gravelly deposits, with boulders (basal level)
15. Boulder loam, Dnepr moraine
16. Alluvial sand of the 10- to 12-m terrace of the Kasplya River
17. Alluvial sand of the 8- to 9-m terrace (16 and 17 are of late glacial age)
18. Alluvial deposits of the 6- and 3- to 4-m floodplain of the Kasplya River and its tributary (Holocene-age deposits).

Table 1-1. Radiocarbon Dates for Organic Matter from Submorainic Deposits in the Interval 30,000 to 17,000 yr B.P., Fixing the Late Valdai Age of the Boundary of the Latest Scandinavian Ice Sheet on the Russian Plain

Section	Date (yr B.P.)	Laboratory Number
Drechaluki near Surazh, BSSR	17,770±170	LU-95A
	17,900±160	LU-95A
	18,370±180	LU-96A
	23,630±370	LU-97A
Pokrovskoye on the Puchka River near Kubenskoye Lake, Vologda Province, RSFSR	21,410±150	LU-18B
	21,880±110	LU-18A
Gozha River near Grodno, BSSR	18,730±1230	LU-76A
	22,740±1870	LU-76B
	23,200±520	LU-76C
	22,950±440	LU-89
	25,100±240	LU-90A
	24,860±230	LU-90B
Shapurovo, vicinity of Surazh, BSSR	22,500±210	LU-91
	29,150±850	LU-78A
10 km north of Shenkursk, Vaga River valley	24,900±470	V-40
Dunayevo, Lovat' River, Novgorod Province	27,500±1500	LU-28A
	25,600±360	LU-28C
	25,440±270	LU-28B
Borisova gora, vicinity of Surazh, BSSR	28,170±750	LU-105

Sources: Data from Kh. A. Arslanov, L. N. Voznyachuk et al., 1972. 1972.

fully in the pollen diagrams for the Grazhdanskiy Prospect borehole (Leningrad) and in the quarry near the town of Kashin (Kalinin Province).

Sediments with a radiocarbon age of 30,000 to 17,000 yr B.P. represent the final stage of the ice-free interval (Table 1-1). This period includes the fairly distinct Dunayevo warming (25,000 to 22,000 yr B.P.), which is correlated with the Bryansk interval in the extraglacial zone. The marked cooling that followed was associated with an expansion of the ice sheet and the appearance of arctic and arctic-alpine flora in the sediments, with radiocarbon dates of 24,000 to 18,000 yr B.P. Such flora is known in sections from the village of Drechaluki near the town of Surazh (Vitebsk Province, BSSR), on the Puchka River in the village of Pokrovskoye (Vologda Province), and in the village of Gozha near the town of Grodno (BSSR) (Dorofeyev, 1957, 1963; Kolesnikova and Khomutova, 1971). Also found in the sediments were reindeer antlers (Voznyachuk, 1973) and traces of contemporaneous cryogenic deformation.

The periglacial flora persisted during the entire interval from 50,000 to 24,000 yr B.P., even during periods of relative warming; therefore, this interval is not an interglaciation. Some investigators who previously had distinguished a cool interglaciation for this interval now define it as the Middle Valdai Megainterstade, which has a complex paleoclimatic sequence: three warm phases separated by relatively short cool phases (Arslanov, Breslav, et al., 1981). This interpretation was strengthened when a Mikulino age was determined for the stratotype of the Mologa-Sheksna Interglaciation in the Upper Volga Basin (Arslanov et al., 1967). In addition, sediments initially assumed to be Middle Valdai are now thought to represent an older interglaciation (Arslanov et al., 1974; Chebotareva and Makarycheva, 1974). At the present time, the concept of an interglacial rank for the Middle Valdai is held by Raukas (1976) and Serebraynny (1978), who distinguish the Karukyula Interglaciation within this interval. However, for deposits in the Karukyula section, the concept of an older (Likhvinskiy) age is being advanced on the basis of radiocarbon dates (Arslanov, 1975; Shotton and Williams, 1973), along with a detailed plant-macrofossil study (Velichkevich and Liyvrand, 1976).

Deposits immediately preceding the time of maximum spread of the Late Valdai glacier are most common in western European USSR. In the Belorussian sections, where they have been studied best, they are represented by submorainic lacustrine-alluvial and alluvial deposits containing layers of vegetal detritus. Their age at the village of Gozha is 25,000 to 18,000 yr B.P. (Figure 1-2) and at the village of Drechaluki is 23,000 to 17,000 yr B.P. (Arslanov, Voznyachuk, Velichkevich, Kur'yerova, and Petrov, 1971; Arslanov, Voznyachuk, Velichkevich, Zubkov, et al., 1971). In the Kubenskoye Lake basin in Vologda Province, submorainic lacustrine sediments with peat interlayers have an age of 21,410±150 (LU-18B) and 21,800±110 yr B.P. (LU-18A) (Arslanov et al., 1970). Still farther east, in the Vaga River valley (Onega River basin) north of the town of Shenkursk, submorainic deposits have a date of 24,900±470 yr B.P. (V-40) (Atlasov et al., 1978). All the indicated sections are located near the boundary of the maximum spread of Late Valdai ice. Thus, the age of this boundary is confirmed by radiometric and palynologic data (Figure 1-3).

The marginal formations of the Late Valdai ice at its maximum east of the Onega and Severnaya Dvina Rivers are correlated with the outer, morphologically well-defined marginal relief complex developed in the Mezen' and Pechora Basins, on the basis of radiocarbon dates of 47,000 to 33,000 yr B.P. on submorainic deposits chiefly in the Pechora River basin (Arslanov et al., 1977; Lavrov and Arslanov, 1977; Arslanov, Lazrov, and Nikiforova, 1981). The maximum Valdai Glaciation in the extreme northeastern Russian Plain, where no Early Valdai moraines have been found, is regarded as quasi-synchronous with the Late Valdai maximum farther west.

The Middle Valdai fluctuations of climate and vegetation are comparable to those recorded in the sediments of the northwestern Russian Plain, with the exception of the warming of 47,000 to 45,000 yr B.P., when the climatic conditions are thought to have been close to the present ones for these regions (Loseva and Arslanov, 1975; Berdovskaya and Loseva, 1975; Loseva, 1978; Arslanov, Lavrov, and Nikiforova, 1981). This fact, which is of interest in itself, requires further confirmation, for the dates obtained for the corresponding deposits are at the limit of resolution for the method.

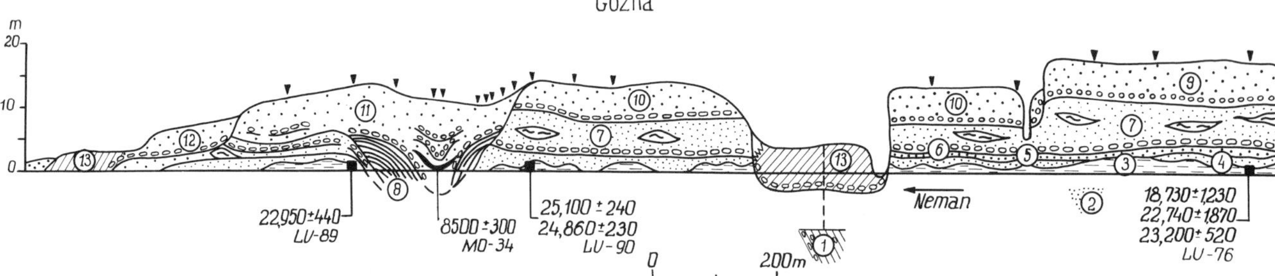

Figure 1-2. Geologic profile of the right bank of Neman River near the village of Gozha, north of the town of Grodno (after L. N. Voznyachuk [Arslanov et al., 1972].) Locations of sections are shown by solid points. (Radiocarbon dates in years B.P.)

Numbers:

1. Moraine of Middle Pleistocene age
2. Finely granular alluvial sand
3. Lacustrine-alluvial silt, with interlayers of vegetal detritus in its upper part, formed before the maximum of the last glaciation
4. Fine-grained alluvial sand
5. Sandy-gravelly deposits changing into sands
6. Sandy loam with boulders and glaciodynamic flow texture, moraine of the maximum stage of Valdai Glaciation
7. Alluvial sand with lenses of sandy silt loam
8. Alluvial clay of late-glacial age
9. Sands and gravel of terrace IV
10. Sands and gravel of terrace III
11. Alluvial sand with inclusions of lenses of oxbow-lake gyttjas of terrace II
12. Alluvial sand of terrace I
13. Floodplain deposits.

Late Valdai Glaciation

On the Russian Plain, the expansion of the Late Valdai ice sheet took place under cold and fairly arid conditions. Paleobotanic data (such as a high proportion of pollen and spores of tundra and subarctic plants and of plants of open habitats and dry, stony slopes) indicate the development of tundra-steppe landscapes at that time. According to entomological data, the climatic conditions during maximum glaciation were similar to those of typical arctic tundras of West Taymyr (Nazarov, 1980).

The expanded ice sheet was characterized by a radial arrangement of ice streams—from west to east, the Baltic, Chudsk, Ladoga, Onega-Kareli, White Sea, Kola-Mezen', Barents Sea-Pechora, and Novaya Zemlya-Kolva ice streams—occupying lowlands in the preglacial relief. In front of the glacier margin were proglacial lake basins, especially in the extreme northeastern Russian Plain. The landforms of the maximum ice margin are less distinct to the east, apparently because of the more continental climate. East of the Onega and Severnaya Dvina Rivers, however, the marginal glacial landforms are again morphologically well defined, with many forms produced by active ice. These more distinct features were probably formed not by Scandinavian but by Novaya Zemlya ice, spread over dry land by the Kola-Mezen', Barents Sea-Pechora, and Novaya Zemlya-Kolva lobes.

As noted above, the maximum spread of Novaya Zemlya ice was not completely contemporaneous with that of the Scandinavian glacier, as confirmed by radiometric dates that place the minimum age of submorainic deposits at the terminus as 18,000 yr B.P. in the western Russian Plain, 24,000 to 21,000 yr B.P. in the center, and 33,000 yr B.P. in the east (Table 1-2). Boulder distributions show that in the western sector the glacier streams and lobes came from Fennoscandia, including the Kola Peninsula and Karelia, and that the ice streams spreading into the basins of the Mezen' and Pechora Rivers came from the Kola Peninsula region as well as Novaya Zemlya (Lavrov, 1977; Velichko, 1979). Ice from Novaya Zemlya forced Scandinavian ice westward, and ice from Scandinavia sometimes extended east of Timan. Apparently, the convergence zone of the Scandinavian and Novaya Zemlya ice sheets was fairly wide.

Initial Stage of Deglaciation (16,000 to 15,000 Years Ago)

During deglaciation, belts of marginal formations were formed, and in the radial ice-shed zones there was formed a complex relief of interlobate uplands and angular masses whose structure is described in detail in several studies ("Structure and Dynamics of the Latest Ice Sheet of Europe," 1977; Gerasimov, 1973).

One can distinguish three major stages in the deglaciation of Scandinavian ice on the Russian Plain. They were controlled in a large measure by the continentality of the climate and by variations in temperature. The initial stage (up to 16,000 to 15,000 yr B.P.) was characterized by the preservation of the glaciodynamic outline of the ice sheet. Slight changes occurred in the Atlantic sector beyond the confines of the Russian Plain. Marginal formations somewhat different in morphology and structure were formed on the southwestern, southern (central), and southeastern slopes of the Scandinavian ice sheet. Whereas at the maximum (Brandenburg) stage on the southwestern (Atlantic) slope in western Europe the ice was more active and left linear marginal zones of stronger relief with a large proportion of push moraines, on the southern and southeastern slopes on the Russian Plain the marginal formations of that time (Bologovo stage) are represented by discontinuous

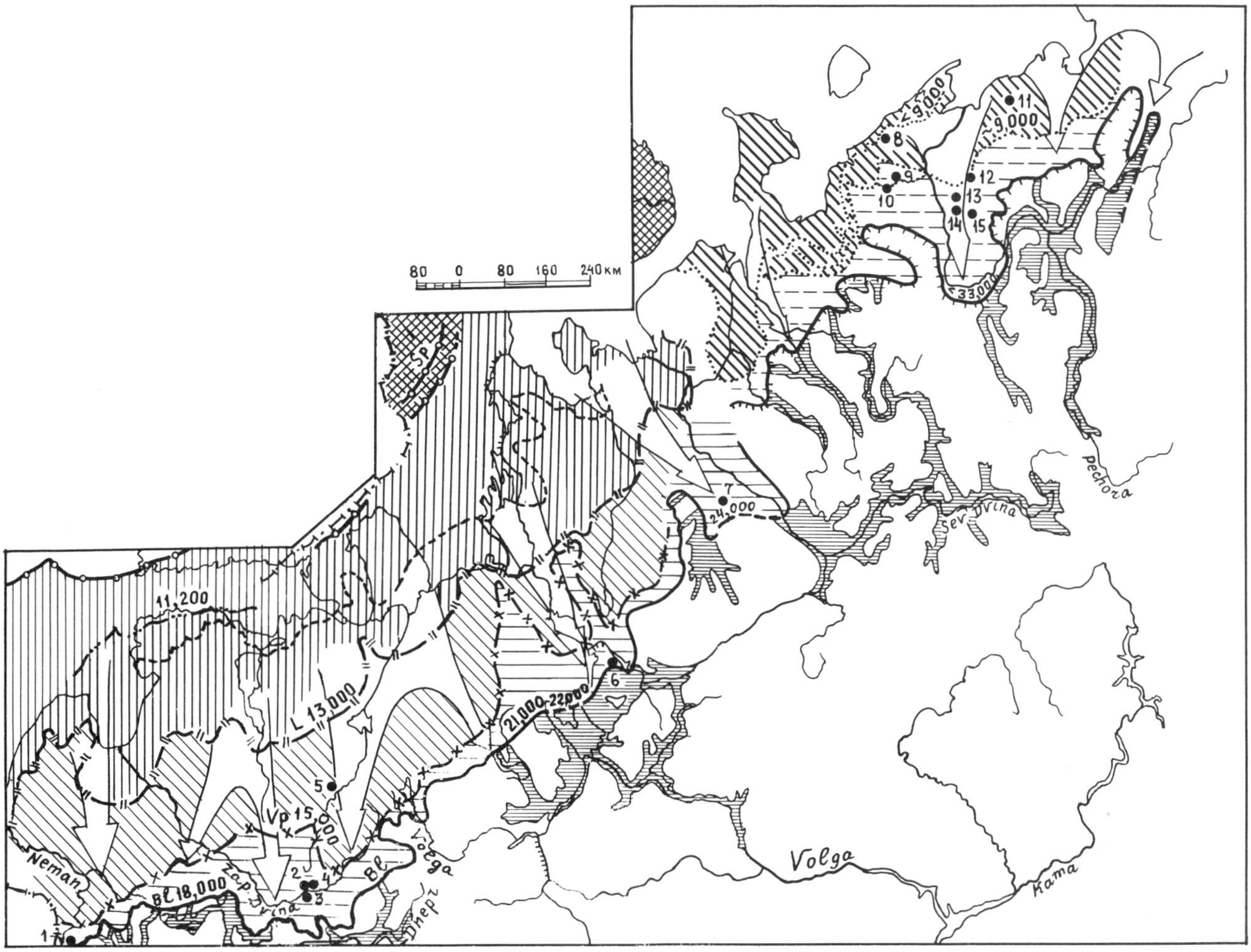

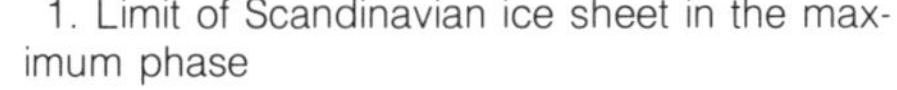

1. Limit of Scandinavian ice sheet in the maximum phase

2. Limit of the Scandinavian ice sheet in the Vepsovo phase

3. Limit of the Scandinavian ice sheet in the Luga phase

4. Limit of the Scandinavian ice sheet in the Neva phase (Pandivere)

5. Limit of the Scandinavian ice sheet during the Palivere advance

6. Limit of the Scandinavian ice sheet in the Salpausselkä phase

7. Limit of Novaya Zemlya ice sheet in the maximum phase

8. Limit of the Scandinavian ice sheet in the final (according to A. S. Lavrov) phase of deglaciation

9. Directions of main glacier flow at the time of maximum ice extent

10. Proglacial water bodies connected with the maximum ice extent

● 13 11. Sections fixing the age of the limit of the Scandinavian and Novaya Zemlya ice sheets (site numbers: 1, Gozha; 2, Drechaluki; 3, Shapurovo; 4, Borisova Gora; 5, Dunayevo; 6, Pokrovskoye (Puchka); 7, Vaga River, north of Shenkursk; 8, Vel't I; 9, Tyrybey; 10, Urdyuga; 11, Chernaya; 12, Shapkina I; 13, Shapkina II; 14, Soz'va I; 15, Soz'va II

12. Areas freed from Scandinavian ice in the Vepsovo phase

13. Areas freed from Scandinavian ice in the Luga phase

14. Areas freed from Scandinavian ice in the Salpausselkä phase

15. Areas freed from Scandinavian ice in a phase later than 10,000 years ago

16. Areas freed from Scandinavian ice in the final phase of glaciation

17. Areas freed from Novaya Zemlya ice starting 9000 years ago.

Figure 1-3. Late Pleistocene glaciation of European USSR. (Compiled by M. A. Faustova from materials in the book, "Structure and Dynamics of the Last Ice Sheet" [1977], and new glaciomorphologic and radiometric data [Arslanov et al., 1975; Il' in et al., 1978; Arslanov, Lavrov, and Nikiforova, 1981; and others].)

Table 1-2. Radiocarbon Dates of Submorainic Alluvial and Lacustrine Deposits in the Interval 37,000 to 33,000 yr B.P., Fixing the Late Valdai Age of the Maximum Stage of the Ice Sheet in the Northeastern Russian Plain

Section	Date (yr B.P.)	Laboratory Number
Soz'va I, Soz'va River, right tributary of Pechora River	33,520±470 35,540±1570	LU-513 A LU-513 B
Tyrybey I, basin of Soyma River, left tributary of Pechora River, Malozemel'skaya Tundra	38,670±870 39,840±570	LU-512 B LU-512 A
Shapkina II, Shapkina River, right tributary of Pechora River	40,650±790 40,860±1260 38,400±640	LU-550 LU-517 B LU-517 A
Shapkina I, Shapkina River, right tributary of Pechora River	42,660±970 43,240±1090 45,280±1200	LU-519 LU-394 C LU-515 A
Urdyuga I, basin of Soyma River, left tributary of Pechora River, Malozemel'skaya Tundra	42,810±1200	LU-533
Vel't I, Vel't River, Timan bank, Malozemel'skaya Tundra	43,250±1110	LU-677
Chernaya I, Chernaya River, Bol'shezemel'skaya Tundra	47,210±1270	LU-674
Soz'va II, Soz'va River, right tributary of Pechora River	>51,200	LU-514 B

Sources: Arslanov, Lavrov, and Nikiforova, 1981; Lavrov and Arslenov, 1977.

forms of small area and low relief, composed primarily of a sandy-gravelly material of dead-ice origin.

During the period of initial deglaciation (Yedrovo stage on the Russian Plain, Frankfurt or Poznan stage in western Europe), larger topographic features were formed but were mainly superimposed on older, pre-Valdai relief. Whereas in the Atlantic sector small oscillations occurred, the deglaciation of the southeastern and eastern slopes was regressive, with preservation of the lobate form. Various portions lost their mobility, changing into dead ice, and primarily accumulation ridges of terminal moraines were formed at the active margin. On the southeastern slope of the Scandinavian ice sheet, such dead-ice masses were particularly large, probably because ice is less active in a continental climate.

Large masses of dead ice in combination with favorable topographic conditions promoted the formation of particularly large proglacial lakes during the initial stages of deglaciation. Small proglacial lakes (Verkhne-Nemanskoye, Verkhne-Vileyskoye, Verkhne-Berezinskoye, Verkhne-Dneprovskoye, etc.) were formed during the maximum stage. In the Upper Volga Basin these were replaced by small meltwater plains and farther east by larger proglacial lakes in the Mologa-Sheksna and Sukhona Lowlands and especially in the basins of the Vaga and Severnaya Dvina Rivers. During retreat from the maximum stage, large proglacial lakes appeared in the basins of the Neman and Vilya Rivers and the Zapadnaya Dvina River and its tributary the Mezha. Large lakes continued to exist in the basins of the Mologa, Sheksna, Sukhona, Vaga, and Severnaya Dvina Rivers.

The formation of proglacial lakes was fostered by the lobate form of the ice sheet, by a general surface slope toward the Baltic Sea (Isachenkov, 1979), and by ice retreat under conditions of continental climate. Large stagnant ice masses persisted primarily in glacial depressions. In this first stage of deglaciation, the time intervals between the formation of recessional belts were short and cold.

Middle Stages of Deglaciation (15,000 to 12,000 Years Ago)

The turning point in the character and rate of deglaciation was about 15,000 yr B.P. This date marked the beginning of the second deglaciation stage of the Scandinavian ice sheet (which lasted to approximately 12,000 yr B.P.). The beginning coincides with the Vepsovo cold stage (Pomeranian in western Europe). The preceding Yedrovo-Vepsovo interval was longer than the previous ones, but it remained just as cold, as indicated by data from glacial depressions, for example, the Dvina-Kasplya glacial depression (Faustova, 1972). Paleobotanic and certain lithologic data (cryogenic differentiation of clastic grains in moraines) attest to the continental climate of the second stage of deglaciation. This is also indicated by the Late Valdai cryogenesis in the extraglacial zone (Velichko, 1973).

By Vepsovo time, the ice sheet had shrunk considerably in size, especially in its eastern part, owing to insufficient nourishment and to the westward displacement of the center of outflow (Chebotareva, 1977). The Vepsovo cold stage is identified on both the European and the Russian Plains as a time of ice advance. Because of the pronounced continental climate of the Russian Plain, however, the ice margin did not reach the boundaries of the preceding stages of deglaciation, as was the case in Denmark, for example. Aseyev (1974) notes a divergence of the boundaries of the maximum and Maritime stages toward the northeast as the Pomeranian features become less distinct. Previously formed marginal formations were destroyed or buried only on the flanks of glacial lobes, where they were close together.

After the Vepsovo advance came the Kresttsy fluctuation in the northwestern Russian Plain. During these two events over the entire Russian Plain, the active ice formed well-defined push moraines, erratic masses, and glacial dislocations in a complex of several marginal zones, beginning in the west as the Baltic Ridge and continuing as the Velikiye Luki-Toropets, Vepsovo, Valdai, Kirillov-Belozersk, and Konosha Ridges. They include the Latgal'skiy, Bezhaniskiy, and Andogskiy Uplands, as well as other interlobate uplands. This so-called main moraine belt (Sokolov, 1949) is characterized by continuity and exceptional freshness and distinctness of glacial landforms; this attests to the transgressive-regressive character (with small oscillations) of the second stage of deglaciation (Basalikas, 1965, 1969; Kudaba, 1969; Faustova, 1972). As the glacier receded, ice-dammed lakes were formed at its margin; the largest were Lake Nizhne-Nyamunskiy and lakes in the Volga Basin.

In the second stage of deglaciation, the ice sheet kept its glaciodynamic outline, with only local changes in structure caused by the occasional appearance of new lobes or ice tongues and the disappearance or shrinking of old ones. The deglaciation rates increased because of rapid ice stagnation and frequent contact between the ice and the adjacent water bodies.

Starting with Vepsovo time, in comparison with the preceding stage of deglaciation, dead-ice formation accelerated, and specific glacial landforms—the so-called "zvontsy," that is, various types of glaciolacustrine plateaus—began to form, caused by deep thawed patches on the ice surface.

The most pronounced warming within the second stage of deglaciation—the Raunis warming—occurred before the Luga advance of the glacier front, which marked the beginning of the late-glacial. The onset of the Raunis warming is dated by submorainic deposits making up the lake terrace on the western edge of the Kubenskoye Lake basin (Vologda Province)—14,300±200 yr B.P. (LU-45B) (Kabaylene, 1970). The end of the Raunis warming is fixed by deposits on the Raunis River northeast of the city of Riga—13,000±500 yr B.P. (Mo-296), 13,250±160 yr B.P. (TA-177), and 13,320±250 yr B.P. (Ri-39) (Stelle et al., 1975). The climatic conditions of the interval were fairly harsh, as indicated by the presence of the following macroremains (sections on the Raunis River and sections on the southern margin of Latgal'skiy Upland): *Betula nana*, *Salix* sp., *S.* cf. *reticulata*, *S. polaris*, *Cerastium* sp., and, especially, *Dryas octopetala*, *Selaginella selaginoides*, and Bryales (Savvaitov et al., 1964; Punning et al., 1968). This warming lasted less than a millennium.

The Luga moraine overlies the sediments of the Raunis warming and underlies the intermorainic sediments in the Kurenurme section in southeastern Estonia, north of the Luga marginal zone. Its age is put at 13,000 yr B.P. on the basis of dates of 12,650±500 yr B.P. (TA-57) and 12,420 ±90 yr B.P. (Tln-35) on the intermorainic sediments (Punning et al., 1968).

Stages of Deglaciation (12,000 to 10,000 Years Ago)

The third and last stage of the deglaciation of the Russian Plain begins at the end of Luga time (about 12,000 yr B.P.). This last turning point was followed by irreversible deglaciation and breakup of the ice sheet under continental climatic conditions, and its rates accelerated markedly as a result of a worldwide warming trend. During the Luga stage, the boundary of the ice sheet on the Russian Plain was already located near the Baltic Sea and White Sea basins (marginal formations of the same age in Poland and East Germany are located within the Baltic Sea and its islands). The Luga marginal formations extend from the Linkuva terminal moraine ridge in Latvia through the Vidzemskiy, Aluksnenskiy, and Khaan'yaskiy Uplands; the hilly ridges north of the town of Ostrov in the Chudsk Lowland; ridges and areas of hilly moraines and kames in the Luga Basin, the Volkhov Depression, and the intervening Shapki-Kirsinskiy kame massif; thence, to the east through ridges adjoining the Olonets Upland and in the Onega River basin.

However, formations recording the Luga glacial recession and the second phase of activation of the glacier front during the late-glacial—the Neva advance—were produced by thin ice. Larger isolated uplands emerged from underneath the ice as nunataks. Marginal zones narrowed and constituent deposits thinned. The ice flow was transformed into glacier tongues, which retained considerable activity only in the first stages. Push moraines with glacial dislocations were formed under the favorable conditions of preglacial relief.

The increased role of meltwater and surface melting led to a major development of supraglacial kames and ablation moraines. Intricate polygenetic complexes with terminal moraines and drumlins are characteristic, along with a varied topography of ablation moraines and kames. Dead-ice forms predominate when the marginal zones are examined in space as well as in section.

Regional differences in the processes of glacier morphogenesis caused by conditions of nourishment lose their importance, for the deglaciation was regulated mainly by a rise in temperature. Nevertheless, in the western part of the sheet the first phase (activation of ice) and the second phase (surface melting) were more distinct.

During the third stage of deglaciation, sharp climatic oscillations occurred, including the Bölling and Alleröd: the Bölling interval dates from 12,750 to 12,250 yr B.P. (from a series of sections in Lithuania on the Ula and Myarkis Rivers, in the Priil'menskiy Lowland on the Lovat' River, in an area south of the town of Velikiye Luki in Smolensk Province near the settlement of Ponizov'ye on the Kasplya River, and in other places). According to some authors, the Luga and Neva cold phases preceded the Bölling and belong to the Oldest Dryas (Gerasimov, 1969; Chebotareva and Makarycheva, 1974; Chebotareva et al., 1978). According to others (Punning et al., 1967; Serebryanny, 1978), the Bölling interval separated these two cold phases, of which the latter is correlated with the Middle Dryas. During the Bölling, the Russian Plain became almost completely free of ice (with the exception of the northern and northwestern parts of Estonia), and arboreal vegetation (mainly tree birch) began to appear.

The Alleröd warming, which on the Russian Plain is dated in the interval 11,950 to 10,800 yr B.P. (Punning et al., 1967; Zobens et al., 1969; Stelle et al., 1975), was studied in many sections of the Baltic Republics (especially Lithuania and Latvia), Karelia, Smolensk Province, and other locations. Unlike other late-glacial deposits, they are mainly represented by peaty sediments. At the beginning of the Alleröd and during the so-called Palivere advance (11,200 yr B.P.), the glacier was still active in the islands of the western Estonian Archipelago and southwestern Estonia south of the Gulf of Finland, but thereafter it quickly receded ("Structure and Dynamics of Europe's Latest Ice Sheet," 1977). Thin forests of pine, birch, and spruce were already growing on the Russian Plain.

The later advance of the glacier margin during the

Younger Dryas (between 11,000 and 10,000 yr B.P.) was formerly known only in Finland (the Salpausselkä moraines), but recent studies in Karelia and the Kola Peninsula reveal the eastern extension of the Salpausselkä. Marginal glacial formations continue northward into the Kuito Lake area in northern Karelia and the Knyazh'ya Bay area in southern Murmansk Province (Il'in et al., 1978), consistent with an early hypothesis (Rosberg, 1899). The same system continues to the western Kola Peninsula ("Structure and Dynamics of Europe's Latest Ice Sheet," 1977). Western Karelia was finally freed from the active glacier over 9500 to 9400 years ago.

Thus, on the Russian Plain, except for a small portion of the Gulf of Finland coast, Scandinavian ice existed from 24,000 to 18,000 yr B.P. until the Alleröd, and, within the confines of the Baltic Shield, until the middle of the Preboreal.

The paleogeographic situation on the Russian Plain during the second half of the late-glacial was related to the deglaciation history of the Baltic Sea and White Sea basins, some of the aspects of which are yet to be definitively elucidated.

The northward retreat of ice during the Bölling warming was accompanied by the emptying of local ice-dammed lakes and the expansion of the southern Baltic proglacial basin, which was preceded by a brackish-water basin (arctic Lomma Sea) formed in the southwestern part of the Baltic Sea basin during the deglaciation of southern Sweden. An oceanic connection with the proglacial basin probably existed at the start of the Alleröd (Lavrova, 1968), and the Baltic Ice Lake itself was formed at the boundary between the Alleröd and Younger Dryas. This lake initially had a very limited connection to the ocean, and only at the boundary of late- and postglacial time (10,200 yr B.P.) was a marine regime reestablished there. (See Serebryanny, 1978.) The lake marks the beginning of the postglacial history of the Baltic region, which is related to the existence of the Yoldia, Ekheney, and Ancylus Seas, as well as others.

The problem of marine incursion into the White Sea is of fundamental importance, for it is associated with the deglaciation history of the adjacent portion of the Barents Shelf and the northeastern Russian Plain. The existing data are ambiguous. Finds of marine fauna enabled several investigators (e.g., Medvedev et al., 1970; Pleshivtseva, 1970) to infer that seawater entered the White Sea and Dvina Bay during the Alleröd and the Younger Dryas. There is also the view that a proglacial basin isolated from the ocean existed in the White Sea for a long time until the Preboreal (10,000 yr B.P.) because of the presence of an ice barrier in the entrance of the White Sea (Yevzerov, Kagan, et al., 1976; Kaligina et al., 1979). According to these concepts, the marine fauna found was not in situ but was redeposited from older interglacial deposits. These concepts are consistent with some of the geomorphic data on a deglaciation of the eastern part of the Baltic Shield and with palynologic data (Malyasova, 1971, 1976) indicating the presence of ice masses in the White Sea basin during late-glacial time.

Deglaciation of Novaya Zemlya Ice Sheet

The deglaciation of the Novaya Zemlya ice sheet, which extended to the northeastern part of the Russian Plain and was not a direct chronologic equivalent of the Scandinavian ice sheet, differed in a number of characteristics. During the maximum stage, the Malozemel'skaya and Bol'shezemel'skaya Tundras and the lower course of the Pechora River were covered over with ice of the Kola-Mezen', Barents Sea-Pechora, and Novaya Zemlya-Kolva lobes ("Structure and Dynamics of Europe's Latest Ice Sheet," 1977). These lobes ended in outlet tongues that formed drumlinoid and fluted forms, which do not occur in the Russian Plain farther west. At the ice margins in the basins of the Mezen', Pechora, and Usa Rivers and their tributaries, huge ice-dammed lakes formed as a result of the flooding of the river valleys.

During the deglaciation, three belts of marginal formations were produced: an outer belt, a transgressive main belt located in the central part of the Bol'shezemel'skaya Tundra (this belt was morphologically the most distinct), and an inner belt. According to the radiocarbon dates of 9990 ± 100 (LU-391) and 9900 ± 110 yr B.P. (MGU-276) obtained from peat and wood in submorainic lacustrine deposits in the area of the mouth of the Pechora River near the village of Markhida, the main belt of marginal formations and the one farther north may have been very young. Those data require thorough investigation, however.

According to Lavrov and Arslanov (1977), the breakup of the marginal zone of the Novaya Zemlya glacier's maximum stage occurred at the beginning of the Bölling. Alluvium of this age, 12,260 ± 180 yr B.P. (LU-364B) and 12,360 ± 170 yr B.P. (LU-390), is overlain with Alleröd lacustrine sediments dated at 11,830 ± 220 yr B.P. (LU-516). A complete breakup of the glacier front occurred during the Preboreal, after which an alluvial terrace began to form in the valleys. Deglaciation involved the formation of large proglacial lakes in the Bol'shezemel'skaya Tundra.

References

Apukhtin, N. I., Klyunin, S. F., and Tkachenko, L. I. (1977). Paleogeography of the eastern part of the Kola Peninsula in the Upper Pleistocene (according to data from 1972-1974). *In* "Stratigraphy and Paleogeography of the Quaternary of the Northern European USSR." (G. S. Biske, ed.), pp. 16-21. Karelian Branch, USSR Academy of Sciences, Petrozavodsk.

Armand, N. N., and Romanova, V. P. (1977). The White Sea glacier flow. *In* "Structure and Dynamics of Europe's Latest Ice Sheet." (N. S. Chebotareva, ed.), pp. 72-80. Nauka Press, Moscow.

Arslanov, Kh. A. (1975). Radiocarbon geochronology of the upper Pleistocene of the European USSR (glacial and periglacial zones). *Bulletin of the Commission for the Study of the Quaternary Period* 43, 3-25.

Arslanov, Kh. A., Auslender, V. G., Gromova, L. I., Zubakov, A. I., and Khomutova, V. I. (1970). Paleogeographic characteristics and absolute age of the maximum stage of Valdai Glaciation in the region of Kubenskoye Lake. *USSR Academy of Sciences Doklady, seriya geologicheskaya* 195, 1395-99.

Arslanov, Kh. A., Berdovskaya, G. N., Zaytseva, G. Ya., Lavrov, A. S., and Nikiforova, L. D. (1977). Stratigraphy, geochronology and paleogeography of the Middle Valdai interval in the northeast of the Russian Plain. *USSR Academy of Sciences Doklady* 233, 188-91.

Arslanov, Kh. A., Breslav, S. L., Zarrina, Ye. P., Znamenskaya, O. M., Krasnov, I. I., Malakhovskiy, D. B., and Spiridonova, S. A. (1981). Climatostratigraphy and chronology of the Middle Valdai of the northwest and center of the Russian Plain. *In* "Pleistocene Glaciations of the Eastern European Plain" (A. A. Velichko and M. A. Faustova, eds.), pp. 12-27. Nauka Press, Moscow.

Arslanov, Kh. A., Filonov, B. A., Chernov, S. B., and Makarova, V. V. (1975). Radiocarbon dates of the Geochronology Laboratory, NIGEI, Leningrad University. *Bulletin of the Commission for the Study of the Quaternary Period* 43, 212-26.

Arslanov, Kh. A., Gromova, L. I., Zarrina, Ye. P., Krasnov, I. I., Novskiy, V. A., Rudnev, Yu. P., and Spiridonova, Ye. A. (1967). Geological age of sediments of old Mologa-Sheksna Lake. *USSR Academy of Sciences Doklady seriya geologicheskaya* 172, 161-64.

Arslanov, Kh. A., Kozlov, V. B., Kalesnikova, T. D., and Semenenko, L. T. (1974). New sections with Mikulino interglacial deposits and their significance in paleogeography. *USSR Academy of Sciences Izvestiya, seriya geograficheskaya* 1, 74-81.

Arslanov, Kh. A., Lavrov, A. S., and Nikiforova, L. D. (1981). On the stratigraphy, geochronology and climate changes of the middle and late Pleistocene in the northeast of the Russian Plain. *In* "Pleistocene Glaciations of the Eastern European Plain" (A. A. Velichko and M. A. Faustova, eds.), pp. 37-52. Nauka Press, Moscow.

Arslanov, Kh. A., Voznyachuk, L. N., Kadatskiy, V. B., and Zimenkov, D. I. (1973). New data on the paleogeography of the Middle Valdai Megainterstade in Belorussia. *USSR Academy of Sciences Doklady* 213, 901-3.

Arslanov, Kh. A., Voznyachuk, L. N., Velichkevich, F. Yu., Kur'yerova, L. V., and Petrov, G. S. (1971). Age of the maximum stage of the last glaciation in the interfluve of the western Dvina and Dnepr Rivers. *USSR Academy of Sciences Doklady* 196, 901-9.

Arslanov, Kh. A., Voznyachuk, L. N., Velichkevich, F. Yu., Zubkov, A. I., and Kalechits, Ye. G. (1972). Age of the maximum stage of the latest glaciation in the region of Grodno. *USSR Academy of Sciences Doklady* 202, 155-59.

Arslanov, Kh. A., Voznyachuk, L. N., Velichkevich, F. Yu., Zubkov, A. I., Kalechits, Ye. G., and Makhnach, N. A. (1971). Paleogeography and geochronology of the Middle Valdai Interstade in the territory of Belorussia's lake region. *USSR Academy of Sciences Doklady* 201, 661-64.

Arslanov, Kh. A., Yevzerov, V. Ya., Tertychnyy, N. I., Gerasimova, S. A., and Lokshin, N. V. (1981). Concerning the age of deposits of the boreal transgression (of Ponoy layers) on the Kola Peninsula. *In* "Pleistocene Glaciations of the Eastern European Plain" (A. A. Velichko and M. A. Faustova, eds.), pp. 28-37. Nauka Press, Moscow.

Aseyev, A. A. (1974). "Ancient Continental Glaciations of Europe." Nauka Press, Moscow.

Atlasov, R. R., Bukreyev, V. A., Levina, N. B., and Ostanin, V. Ye. (1978). Topographical characteristics of the marginal zone of Valdai Glaciation in the Onega-Vaga interfluve and in the Vaga River valley. *In* "Marginal Formations of Continental Glaciations" (V. G. Bondarchuk, ed.), pp. 30-38. Naukova Dumka Press, Kiev.

Basalikas, A. B. (1965). Certain problems in glaciomorphology (in light of new data from a geomorphological study of Lithuania). *In* "Marginal Formations of Continental Glaciation," pp. 14-17. Mintis Press, Vil'nyus.

Basalikas, A. B. (1969). Topographical variety of the glacial-accumulative region. *In* "Continental Glaciation and Glacial Morphogenesis," pp. 65-154. Vil'nyus.

Berdovskaya, G. N., and Loseva, E. I. (1975). Pleistocene lakes in the Shapkina River basin (Bol'shezemel'skaya Tundra). *In* "History of Lakes in the Pleistocene" (O. D. Kvasov, ed.), Vol. 2, pp. 114-19. Institute of Limnology, Leningrad.

Chebotareva, N. S., Grichuk, V. P., Faustova, M. A., Danilova-Makarycheva, I. A., and Guzman, A. A. (1971). Problems of the age of the maximum stage of Valdai Glaciation. *USSR Academy of Sciences Izvestiya, seriya geograficheskaya* 6, 31-46.

Chebotareva, N. S., and Makarycheva, I. A. (1974). "The Latest Glaciation of Europe and Its Geochronology." Nauka Press, Moscow.

Chebotareva, N. S., Makarycheva, I. A., and Faustova, M. A. (1978). Rhythmicity of changes in natural conditions on the Russian Plain during the Valdai Ice Age. *USSR Academy of Sciences Izvestiya, seriya geograficheskaya* 3, 15-25.

Devyatova, E. I. (1980). Periglacial of the Valdai age in the north of the Russian Plain. *In* "Periglacial Formations of the Pleistocene" (V. G. Bondarchuk, ed.), pp. 11-13. Contribution to the Sixth All-Union Conference on the Marginal Formations of Continental Glaciations. Institute of Geological Sciences of the Ukrainian Academy of Sciences, Preprint 80-16, Kiev.

Dorofeyev, P. I. (1957). On the Upper Pleistocene flora of the village of Drechaluki in Belorussia. *USSR Academy of Sciences Doklady* 117, 303-9.

Dorofeyev, P. I. (1963). New data on Pleistocene floras of Belorussia and Smolensk Province. *In* "Materials on the History of the Flora and Vegetation of the USSR." (V. N. Sukachev, ed.), Issue 4, USSR Academy of Sciences Press, Moscow.

Ekman, I. M., Levkin, Yu. M., and Morozov, G. K. (1979). Glaciotectonics of the Late Pleistocene of the Petrozavodsk area. *In* "Paleogeography of the Region of Scandinavian Continential Glaciations" (G. S. Biske, ed.), pp. 35-46. USSR Geophysical Society, Leningrad.

Faustova, M. A. (1972). Relief and Deposits of the Lovat' Lobe of the Latest Ice Sheet. Dissertation, Institute of Geography, Moscow State University.

Gerasimov, I. P. (ed.) (1969). "The Latest Ice Sheet in the Northwestern USSR." Nauka Press, Moscow.

Gerasimov, I. P. (ed.) (1973). "Paleogeography of Europe in the Late Pleistocene." Contributions to the Ninth Congress of the International Quaternary Union." VINITI, Moscow.

Gudina, V. I., and Yevzerov, V. Ya. (1973). "Stratigraphy and Foraminifers of the Upper Pleistocene of the Kola Peninsula." Nauka Press, Novosibirsk.

Il'in, V. A., Lukashov, A. D., and Ekman, I. M. (1978). Marginal glacial formations of western Karelia and their correlation with ridges of the Finnish Salpausselkä. *In* "Marginal Formations of Continental Glaciations" (V. G. Bondarchuk, ed.), pp. 96-108. Transactions of the Fifth All-Union Conference. Naukova Dumka Press, Kiev.

Isachenkov, V. A. (1979). Pleistocene glaciations and the problem of formation of proglacial water bodies in the northwestern Russian Plain. *In* "History of the Lakes of the USSR in the Late Cenozoic" (N. A. Florensov, ed.), pp. 31-33. Proceedings of the Fifth All-Union Symposium. Part 1. Institute of Limnology. Irkutsk.

Ivanova, I. K. (1977). Early interstadials of the Würm Glaciation. *In* "The Late Cenozoic of Northern Eurasia" (K. V. Nikiforova, ed.), pp. 59-69. Contribution to the Tenth Congress of the International Quaternary Union. Geological Institute, Moscow.

Kabaylene, M. V. (ed.) (1970). "History of Lakes," Vol. 2. Proceedings of an All-Union Symposium, Vilna University.

Kaligina, L. V., Rybalko, A. Ye., Spiridonova, Ye. A., and Spiridonov, M. A. (1979). Palynological study of bottom deposits of the northern White Sea as the basis of their stratigraphic differentiation. Leningrad State University, Vestnik 2, 63-71.

Kolesnikova, T. D., and Khomutova, V. I. (1971). New data on the history of the vegetation development of the Valdai Ice Age in Vologda Province. *USSR Academy of Sciences Doklady seriya geologicheskaya* 196, 413-17.

Kudaba, Ch. P. (1969). Marginal glacial formations of the Baltic ridge and diagnostics of glacier margin dynamics. *In* "Continental Glaciation and Glacial Morphogenesis" (P. Vaytekunas, ed.), pp. 155-226. Vil'nyus State University Press, Vil'nyus.

Lavrov, A. S. (1977). The Kola-Mezen', Barents-Pechora, and Novaya Zemlya-Kolva glacier flows. *In* "Structure and Dynamics of Europe's Latest Ice Sheet" (N. S. Chebotareva, ed.), pp. 83-100. Nauka Press, Moscow.

Lavrov, A. S., and Arslanov, Kh. A. (1977). Age and genesis of terraces of the Pechora Lowland: New geochronological and radiocarbon data. *In* "River Systems and Reclamation" (B. V. Mizerov, ed.), No. 41, pp. 128-32. USSR Academy of Sciences, Novosibirsk.

Lavrova, M. A. (1960). "Quaternary Geology of the Kola Peninsula." USSR Academy of Sciences Press, Moscow and Leningrad.

Lavrova, M. A. (1968). Late-glacial and postglacial history of the White Sea. *In* "Neogene and Quaternary Deposits of Western Siberia" (V. N. Saks, ed.), pp. 140-63. Nauka Press, Moscow.

Loseva, E. I. (1978). Middle Valdai sea lake in the west of Bol'shezemel'skaya Tundra. *Bulletin of the Commission for the Study of the Quaternary Period* 48, 103-12.

Loseva, E. I., and Arslanov, Kh. A. (1975). The Middle Valdai horizon in the western Bol'shezemel'skaya Tundra. *In* "Geology and Useful Minerals of the Northeast of the European USSR," 1974 Yearbook, pp. 82-86. Institute of Geology, Komi Branch, Syktyvkar.

Malyasova, Ye. S. (1971). Palynology of bottom sediments of the White Sea and its stratigraphic importance. *In* "Palynology of the Holocene" (M. I. Neustadt, ed.), pp. 77-91. Institute of Geography, Moscow.

Malyasova, Ye. S. (1976). "Palynology of White Sea Bottom Sediments." Leningrad State University Press, Leningrad.

Medvedev, V. S., Nevesskiy, Ye. N., Gosberg, L. I., and others (1970). Structure and stratigraphic differentiation of bottom sediments of the White Sea. *In* "The Arctic Ocean and Its Coast in the Cenozoic" (A. I. Tolmachev, ed.), pp. 253-67. Gidrometeoizdat Press, Leningrad.

Nazarov, V. I. (1980). The insect fauna of the Late Pleistocene of Belorussia. *In* "Paleontology," pp. 92-93. Transactions of Scientific Meetings of the Paleontology Section of the Moscow Society of Nature Researchers for 1977-1978. Moscow.

Pleshivtseva, E. S. (1970). Main stages in the history of the coastal vegetation of the White Sea's Dvina Bay in the period of the boreal and late-glacial marine transgressions. *In* "The Arctic Ocean and Its Coast in the Cenozoic" (A. I. Tolmachev, ed.), pp. 263-71. Gidrometeoizdat Press, Leningrad.

Punning, Ya. M. K., Raukas, A. V., and Serebryanny, L. R. (1967). Geochronology of the latest glaciation of the Russian Plain in light of new radiocarbon datings of fossils in late-bog deposits of the Baltic region. *In* "Transactions of the Second Symposium on the History of Lakes of the Northwestern USSR" (V. G. Zavres, ed.), pp. 139-47. Byelorussian State University, Minsk.

Punning, Ya. M. K., Raukas, A. V., Serebryanny, L. R., and Stelle, V. Ya. (1968). Paleogeographic characteristics and absolute age of the Luga stage of Valdai Glaciation on the Russian Plain. *USSR Academy of Sciences Doklady* 178, 916-19.

Raukas, A. V. (1976). Correlation of Upper Pleistocene deposits and marginal glacial zones of North America and Europe. Fifth All-Union Conference on the Study of Marginal Continental Glaciations. Abstracts, p. 94. Naukova Dumka Press, Kiev.

Raukas, A. V., and Serebryanny, L. R. (1971). Current problems in geochronological investigations of the Late Pleistocene of the Russian Plain. *In* "Problems of Pleistocene Periodization" (V. A. Zubakov and S. M. Tseytlin, eds.), pp. 197-206. Geographical Society, Leningrad.

Rosberg, J. E. (1899). Ytbildningar i Karelen med särskild hänsyn till ändmoränerna. *Fennia* 14, 134-38.

Savvaitov, A. S., Stelle, V. Ya., and Krukle, M. Ya. (1964). Stratigraphic differentiation of Valdai Glaciation deposits on the territory of the Latvian SSR. *In* "Problems of Quaternary Geology" (I. Ya. Danilans, ed.), vol. 3, pp. 183-203. Latvian Academy of Sciences Press, Riga.

Serebryanny, L. R. (1978). "Dynamics of Cover Glaciation and Glacio-Eustacy in the Post-Quaternary." Nauka Press, Moscow.

Shotton, F. W., and Williams, R. E. G. (1973). Birmingham University radiocarbon dates VI. *Radiocarbon* 15, 1-12.

Sokolov, N. N. (1949). Geological structure and history of topographic development. *In* "The Northwest of the RSFSR." (A. A. Grigor'ev, ed.), pp. 8-60. USSR Academy of Sciences Press, Moscow and Leningrad.

Stelle, V. Ya., Savvaitov, A. S., and Veksler, V. S. (1975). Absolute age of chronostratigraphic stages and boundaries of the late-glacial and postglacial periods in the Middle Baltic region. *In* "Status of Methodological Studies in the Area of Absolute Geochronology" (G. A. Afanas'ev, ed.), pp. 187-91. Nauka Press, Moscow.

"Structure and Dynamics of Europe's Latest Ice Sheet" (1977). Nauka Press, Moscow.

Velichkevich, F. Yu., and Liyvrand, E. D. (1976). New data on the flora and vegetation of Karukyula section in Estonia. *Estonian Academy of Sciences Isvestiya* 25 (Chemistry and Geology 3), 215-21.

Velichko, A. A. (1973). "The Natural Process in the Pleistocene." Nauka Press, Moscow.

Velichko, A. A. (1975). Problems of correlation of Pleistocene events in the glacial, periglacial-loessial, and maritime regions of the eastern European Plain. *In* "Problems in Paleogeography of Loessial and Periglacial Regions" (A. A. Velichko, ed.), pp. 7-25. USSR Academy of Sciences Press, Moscow.

Velichko, A. A. (1979). Problems in the reconstruction of Late Pleistocene ice sheets in the USSR. *USSR Academy of Sciences Izvestiya seriya geograficheskaya* 6, 12-26.

Voznyachuk, L. N. (1961). Late-interglacial deposits in Belorussia. *In* "Materials on Belorussia's Anthropogene" (K. I. Lukashov et al., eds.), pp. 159-217. Byelorussian Academy of Sciences, Minsk.

Voznyachuk, L. N. (1973). Contribution to the stratigraphy and paleogeography of the Neopleistocene of Belorussia and adjoining territories. *In* "Problems in the Paleogeography of Belorussia's Anthropogene" (E. A. Levkov, ed.), pp. 45-75. Nauka i Teknika, Minsk.

Yevzerov, V. Ya. (1970). Concerning the age of interglacial deposits of the Kola Peninsula. *In* "Proceedings of the Scientific Session of the Geological Institute of the Kola Branch of the USSR Academy of Sciences," pp. 88-93. Apatity.

Yevzerov, V. Ya., Kagan, L. Ye., Koshechkin, B. I., and Lebedeva, R. M. (1976). Formation of aqueous deposits of the White Sea in connection with the evolution of the natural situation in the Holocene. *All-Union Geographical Society Izvestiya* 108, 421-29.

Yevzerov, V. Ya., Lebedeva, R. M., and Kagan, L. Ya. (1976). Stage of differentiation of abrasion and accumulation (Middle Valdai Interglaciation). *In* "History of Formation of the Relief and Surficial Deposits of the Northeastern Part of the Baltic Shield" (S. A. Strelkov and M. K. Gravé, eds.), pp. 51-76. Nauka Press, Leningrad.

Zobens, V. Ya., Putans, B. D., and Stelle, V. Ya. (1969). First determinations of the absolute age of specimens performed in the Riga Radiocarbon Laboratory. *In* "Problems in Quaternary Geology" (I. Ya. Danilans, ed.), Vol. 4, pp. 141-45. "Zinatne," Riga.

CHAPTER 2

Late Pleistocene Glaciation of Western Siberia

S. A. Arkhipov

In the northern part of the West Siberian Plain, the latest (Zyryanka or Valdai) glaciation includes deposits occurring stratigraphically above the Kazantsevo (Eemian, Mikulino) horizon. According to traditional concepts, the glaciation had two cold phases separated by the Karginskiy Interstade. In works published during the 1950s, only the final (Sartan) stage of the Zyryanka Glaciation was referred to the later phase. From this came the name Sartan Glaciation (Saks, 1953; Strelkov et al., 1959). At the end of the 1960s, it was determined that the later phase included not only the recessional (Sartan) stage but also the earlier Gydan and N'yapan stages of glaciation. In order to avoid confusion, it has been proposed that the later phase be called the Late Zyryanka Glaciation and the earlier phase, the Early Zyryanka or Yermakovo Glaciation (Troitskiy, 1967; Arkhipov, 1971). The complex interstade between them, including two coolings and three warmings, cannot correctly be called the Karginskiy Interstade, as proposed by Kind (1974), for originally Saks (1953) attributed to the Karginskiy age only the time interval between 30,000 and 20,000 years ago. To this interval corresponds only one of the warm phases within the Middle Zyryanka Interstade, obviously synchronous with the Stillfried B (Plum Point, Bryansk) Interstade.

Subdivisions of the Zyryanka Glaciation

According to currently held concepts, the entire mass of Zyryanka Glaciation deposits is distinguished as a super horizon of the same name, which as a stratigraphic subdivision evidently corresponds to the American term "stage." The superhorizon is subdivided into the Lower, Middle,and Upper Zyryanka horizons (substages), for which the previous names (Yermakovo, Karginskiy, and Sartan) have been retained in geologic practice. The proposed subdivisions are based primarily on the stratigraphy of morainic and related marine, lacustrine, and alluvial strata, as well as on radiocarbon dating reinforced by palynologic data (Table 2-1, and Figure 2-1). Thus, according to radiocarbon determinations and with the widely known data on Europe and North America compiled by Kind (1974) taken into account, the age of the Lower Zyryanka horizon in Siberia is usually estimated at 70,000 to 50,000 yr B.P., the Middle Zyryanka horizon at 50,000 (55,000) to 20,000 yr B.P., and the Upper Zyryanka horizon at 22,000 to 10,000 yr B.P. (Arkhipov, 1977; Arkhipov et al., 1977). In addition, several thermoluminescence dates (100,000 ± 17,000 and 110,000 ± 27,000 yr B.P.) have recently been obtained from the base of Kormuzhikhantskiy moraine; these suggest that the origin of the Zyryanka Glaciation dates back to about 100,000 yr B.P. (Arkhipov et al., 1978; the analyses were performed by A. N. Shelkoplyas at the Institute of Geological Sciences of the Academy of Sciences of the Ukrainian SSR).

The Lower Zyryanka horizon includes the Khashgort and Kormuzhikhantskiy moraines in the Ob' River valley and the Yermakovo moraine on the Yenisey, as well as primarily lacustrine sediments in the periglacial zone. At the mouth of the Ob' in the Salekhard region, the Khashgort moraine lies on marine Kazantsevo layers (Arkhipov et al., 1977). The latter are characterized by a complex of arctoboreal foraminifers very similar to those from the Eemian layers of western Europe (Gudina, 1976). Farther south in the Ob' River valley within the confines of Belogor'ye, the base of the Kormuzhikhantve till was dated by the thermoluminescence method at 100,000 ± 17,000 and 110,000 ± 27,000 yr B.P., and the underlying presumably Kazantsevo alluvial and lacustrine sediments were dated at 130,000 ± 24,000 and 130,000 ± 31,000 yr B.P. Lacustrine sediments with peat dated by thermoluminescence at 70,000 ± 15,000 yr B.P. contain fossil seeds and fruits indicating an interstadial warming that may separate the Khashgort stage from the Kormuzhikhantskiy stage.

The Middle Zyryanka horizon is divided into several subhorizons (sub-stages) and layers, which thus far have been primarily of local stratigraphic significance. The

Table 2-1. Stratigraphic Scheme of Zyryanka Glaciation

Super-horizon	Horizon	Sub-horizon	West Siberian Plain				
			Northern Ob' River		Northern Yenisey River		
Zyryanka	Upper		Polar Ural layers: moraines, varved clays		Noril'sk clays: moraines, varved clays, ^{14}C=11,400-10,700 yr B.P.		
			Sopki layers: moraines, varved clays, fluvioglacial sands		N'yapan layers: moraines, lacustrine-glacial clays, sands		
			?		Tiutey layers: lacustrine clays and sands, ^{14}C=16,000-15,000 yr B.P.		
			Salekhard-Uval layers: moraines, fluvioglacial sands, sandy loams, boulders		Gyda layers: moraines, ^{14}C=20,000-19,500 yr B.P.		
	Middle	Karginskiy	Karginskiy layers: alluvial sands, sandy loams, ^{14}C=29,000-25,000 yr B.P.		Alluvial sands, sandy loams, clays, ^{14}C=26,000-23,000 yr B.P.		Karginskiy strata according to Kind (1974)
		Lokh podgort	Lokhpodgort layers: moraine, varved clays	Kazym layers: lacustrine clays, ^{14}C=39,000-36,000 yr B.P.	Lacustrine-glacial and lacustrine clays, loams, sandy loams, ^{14}C=38,000-35,000 yr B.P.		
		Kharsoim	Kharsoim layers: moraine clays and sands with Foraminifera, ^{14}C=37,000-36,000, 40,000 yr B.P.	Zolotoy Mys layers: alluvial sands, clays, peat, ^{14}C=43,000-39,000, >40,000 yr B.P.	Alluvial sands, sandy loams, peat, ^{14}C=43,000-39,000 yr B.P.	Lagoon and marine sands and sandy loams with Foraminifera ^{14}C=46,000-42,000 yr B.P.	
	Lower		Khashgort layers: moraines, varved clays, lacustrine clays, peat, TL=70,000±15,000 yr B.P.; Kormuzhikhantskiy moraine, TL=100,000±17,000, 110,000±27,000 yr B.P.		Yermakovo layers: moraines, varved clays, fluvioglacial sands		
			Marine and alluvial Kazantsevo deposits		TL=130,000±24,000, 130,000±31,000 yr B.P.		

lower Kharsoim subhorizon in the Lower Ob' River region includes estuarine and marine clays with an arctic complex of foraminifers. At the mouth of the Ob' River east of Salekhrad, peat interlayers in estuarine clays yielded radiocarbon dates of 36,400±80 yr B.P. (SOAN-676) and more than 40,000 yr B.P. Also included in the subhorizon are Zolotoy Mys alluvial sands and loams with lenses of peat extensively developed in the lower Ob' River valley. Radiocarbon dates for the peat are 39,150±1200 yr B.P. (SOAN-978), 39,860±1000 yr B.P. (SOAN-976), and 40,800±1300 yr B.P. (SOAN-682) for the upper interlayers and more than 40,000 yr B.P. for the lower ones (Arkhipov, 1977). In the northern Yenisey Valley, distinct stratigraphic analogues occur in the marine and continental facies. (See Table 2-1.) Thus, the Kharsoim Interstade is reliably dated at 50,000 to 40,000 yr B.P. The sediments belonging to it rest on an eroded surface of glacial deposits that, given their stratigraphic position, can only be of Early Zyranka age.

The Lokhpodgort glacial subhorizon was named after a moraine of the same name heretofore observed only in the mouth of the Ob' near Salekhard, where it rests on marine Kharsoim layers. To the south in the Ob' River valley, glaciolacustrine and lacustrine sediments make up Kazym layers with radiocarbon dates of 37,850±80 yr B.P. (SOAN-658) and 39,900±80 yr B.P. (SOAN-681). They have their analogues in the lower Yenisey Valley. (See Table 2-1.)

The superjacent Karginskiy subhorizon is composed of alluvial sands and clays with lenses of peat. In the Salekhard region, these deposits are heavily eroded and overlain by glacial formations of the Salekhard-Uval stage. Karginskiy layers are radiocarbon-dated at 25,900±240 yr B.P. (SOAN-671) and 29,500±520 yr B.P. (SOAN-974). The Karginskiy alluvium is spread much more widely in the periglacial zone of the lower Ob' region and in the northern Yenisey, where it fills buried valleys cut out of Middle and even Lower Zyryanka sediments. However, it should be noted that, according to Kind (1974), the entire mass of sediments separating the Yermakovo and Gydan (Karaul) moraines is referred to the Karginskiy suite on the Yenisey. (See Table 2-1.)

The Upper Zyryanka horizon is subdivided into several climatostratigraphic units, the rank of which has not yet-been definitively established. Best substantiated are the Salekhard-Uval and Gyda layers, which on the Ob' and Yenisey belong to the maximum stage of glaciation (22,000 to 16,000 yr B.P.). Younger N'yapan (15,000 to 13,000 yr B.P.) and Noril'sk (approximately 11,400 to 10,300 yr B.P.) layers (stages) were first identified in the northern Yenisey. (The latter were referred by Saks to the Sartan Glaciation.) Their correlations in the lower Ob' region are considered to be the Sopkei and Polar Ural layers. Thus, the main criterion for the establishment of the principal stages of the Late Zyryanka Glaciation are terminal and recessional moraines, which are very distinct in relief. Interstadial formations have been studied very little.

Climate and Vegetation

Climatic changes during the Zyryanka Glaciation are determined very clearly from the migration of recent vegetation-landscape zones toward the south or north (Arkhipov, 1971). It is now commonly agreed that the period of Early Zyryanka Glaciation in western Siberia involved a practically complete degradation of the dark coniferous forests that had grown there extensively during the Kazantsevo Interglacial. Southern dry steppes were converted into periglacial cold steppes. At the beginning of the glaciation, under cold but moist climatic conditions, the former forest zone was occupied by forest tundra and, at the end, when the climate became markedly continental and dry, by periglacial tundra-steppes (Giterman et al., 1968).

In the past, Early Zyryanka interstadial formations were not known. Today, they are detected in the periglacial zone in the lower Ob' River valley and on the Belogor'ye Upland as a band of lacustrine clays with peat dated by the thermoluminescence method at 70,000 ± 15,000 yr B.P. Fossil plants from buried peat in this zone are scarce but include nuts and cones of *Picea obovata* Ldb., *Pinus* sp.,

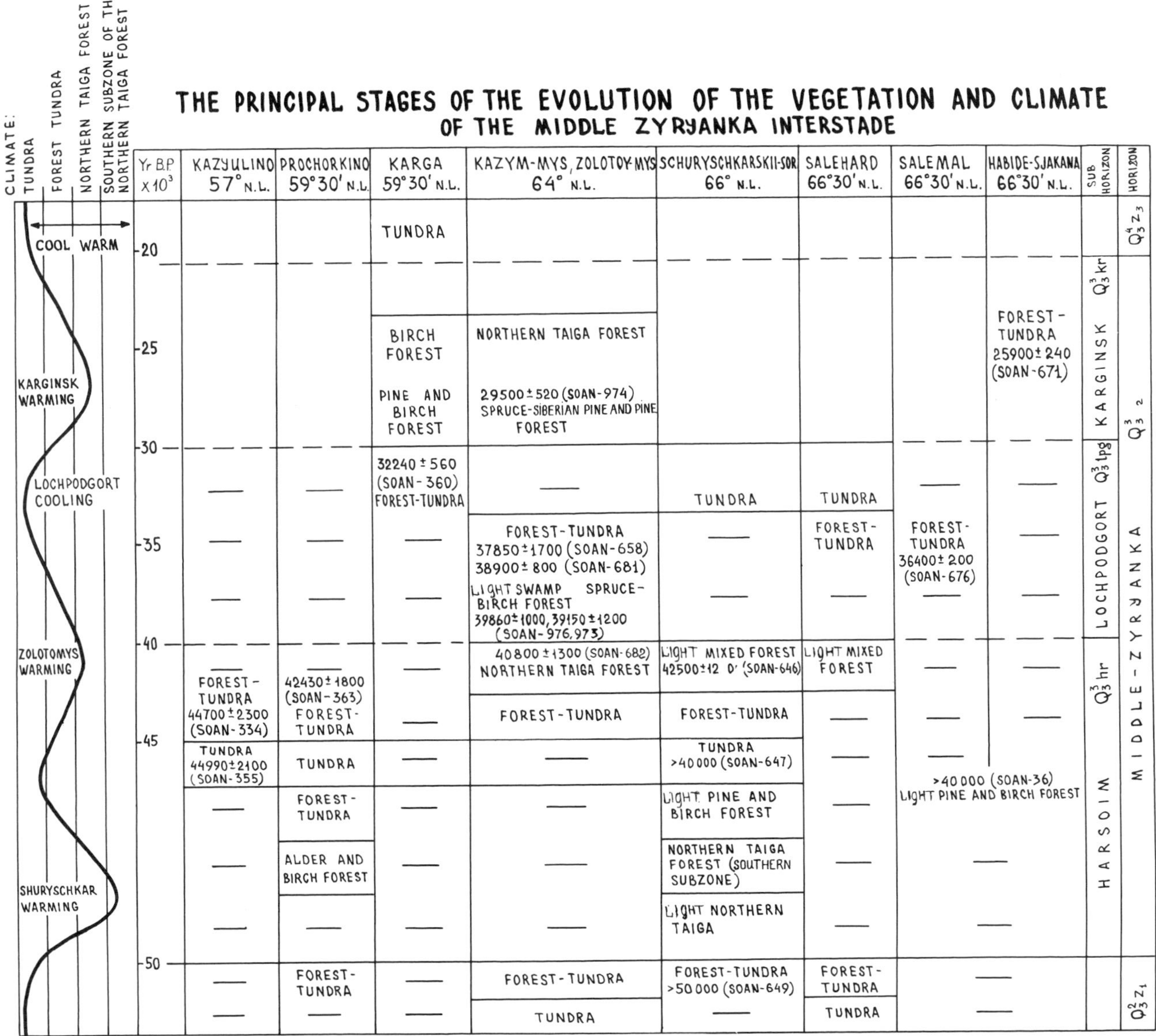

Figure 2-1. Principal stages in the development of vegetation and climate in the Middle Zyryanka Interstadial complex.

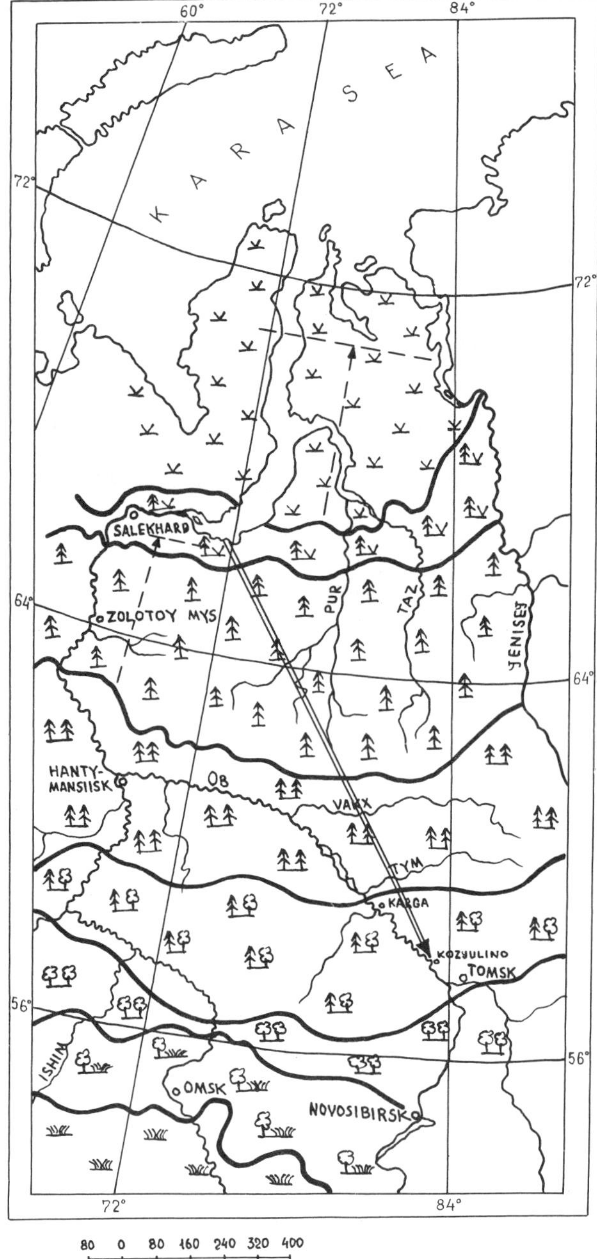

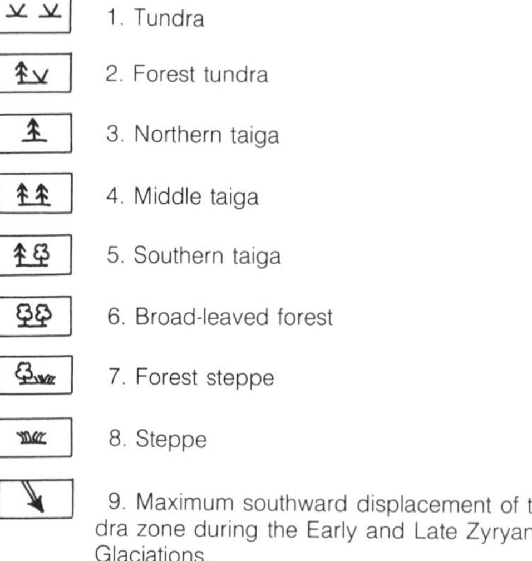

Figure 2-2. Schematic map of zonal types of vegetation in western Siberia. (After Grichuk, 1970.)

Betula sp. (tree), *B. nana* L., *Selaginella* sp., and *Ranunculus flamulla* L. It probably represents a vegetation and climate close to that of the present middle taiga zone.

Middle Zyryanka time cannot be considered as either a cold interval or the only warm interglacial. The complex nature of the paleoclimatic curve is indicated by the Lokhpodgort glacial advance, and the spore-pollen diagrams record repeated changes of vegetation from northern taiga to forest tundra and tundra.

Today, there are several detailed and partially radiocarbon-dated pollen diagrams, mainly for Middle Zyryanka sections located along the Ob' River valley between latitudes 57° and 66°30′N (Figure 2-1). The diagrams are correlated by radiocarbon dates, extrapolation, and palynologic and geologic data. The deposits that had already accumulated by the time of the formation of the Early Zyryanka glacier (more than 50,000 years ago), as well as in the Middle Zyryanka intervals 45,000 to 44,000 and 35,000 to 30,000 years ago and in the maximum stage of Late Zyryanka Glaciation (22,000 to 16,000 years ago), contain spectra indicating that tundra vegetation migrated far south. Vegetation like that of the northern taiga of today is believed to have developed three times within its present limits, and during the short interval between 50,000 (55,000) and 45,000 years ago vegetation like that of the southern subzone of the northern taiga existed in the Salekhard region, which now contains forest tundra.

Several Soviet palynologists (for western Siberia particularly Grichuk [1959, 1970]) have substantiated the hypothesis that each zonal type of present vegetation has a corresponding zonal type of pollen spectrum. It is postulated, therefore, that the main phases on the pollen diagrams summarized in Figure 2-1 correspond to zonal types of vegetation existing in the past. This zonality differed substantially from the present one (Figure 2-2). During the glacial intervals, the tundra migrated southward, almost to the southern boundary of the present taiga (to latitude 57°N). The climate undoubtedly was colder than it is at present. On the other hand, during the Shuryshkary interval (50,000 to 45,000 yr B.P.), the tundra moved far into the north of western Siberia, and the forests of the southern subzone of northern taiga migrated into the Salekhard region. Hence, the climate could have been fairly warm. Finally, the vegetation and climate of the Karginskiy interval (approximately 30,000 to 22,000 yr B.P.) and Zolotoy Mys interval (probably 44,000 to 40,000 yr B.P.) were probably similar to those of the present.

The paleoclimatic curve plotted on the basis of the data cited (Figure 2-1) attests to fluctuations of climate from a cold arctic to a subarctic to the present cold-temperate climate, with probable deviation to temperatures warmer than at present during the Shuryshkary interval. The paleoclimatic curve shows three warm intervals. The longest (about 7000 to 10,000 years) was the last Karginskiy warming, probably characterized by a uniform cold-temperate climate similar to that of the present. The first two, that is, the Shuryshkary and Zolotoy Mys intervals, were of short duration, evidently no longer than two to three millennia, and were separated by a short but very pronounced cooling. The Lokhpodgort cooling was longer, apparently 7000 to 10,000 years in duration.

Still very incomplete is the paleoclimatic characterization of Late Zyryanka time, which on the whole was marked by a more pronounced continentality and dryness. The taiga zone certainly disappeared completely in western Siberia; tundra, tundra-steppe, and periglacial steppe developed during the maximum, that is, the Salekhard-Uval stage, approximately 22,000 to 16,000 (17,000) years ago. The interval between approximately 16,000 (17,000) and 10,000 yr B.P. remains little studied. On the basis of fragmentary data, Kind (1974) distinguishes the Kokorevka warming (13,000 to 12,200 yr B.P.) and the Taymyr warming (11,800 to 11,400 yr B.P.).

Glaciers and Proglacial Lakes

The last few years have seen a revision in the traditional concept that the Zyryanka glaciers were small, thin, and relatively inactive. The revision resulted primarily from a different treatment of the spatial arrangement of marginal glacial formations. Traditionally, they were reconstructed in the form of meridionally oriented belts near the glaciation centers in the northern Urals in the Ob' River area of the West Siberian Plain and near the Putorana and Byrranga Mountains in the northern Yenisey region (Saks, 1953; Strelkov, 1965). The central regions in the northern plain remained beyond the glacier limit, as did the free northward runoff of the great rivers of the West Siberian Plain: the Yenisey, Ob', and Irtysh.

Extensive use of information gained from space explorations (Landsat-type space photographs) combined with geologic studies and radiocarbon dating enabled Astakhov (1976, 1977) to establish the existence of latitude-oriented belts of terminal glacial formations traversing the West Siberian Plain from the Urals to the lower course of the Yenisey River. This finding attests to the movement of glaciers from the north and supports the concept of glaciation of the Kara Sea Shelf (Grosswald, 1977).

This new paleogeographic concept is based on three groups of facts: evidence of a continuous glaciation in northwestern Siberia, existence of vast ice-dammed lakes in the inner regions of the West Siberian Plain, and traces of runoff of ice-dammed waters through the Turgay Hollow to the southwest toward the Aral-Caspian region (Arkhipov et al., 1980).

The margins of the Zyryanka glaciers are marked by hummocky landforms, inclined frontal outwash plains, marginal runoff channels, and terraces of proglacial water bodies. Space and high-altitude photographs enabled V. I. Astakhov to outline a regular system of four belts of marginal formations subconcentric to the coast of the Kara Sea (Figure 2-3). The southernmost (Nadym) belt extending approximately along latitude 65°N generally coincides on the lower Ob' River with the expansion field of the Khashgort and Kormuzhikhantskiy moraines and, thus far, less reliably with the Yermakovo moraine field on the Yenisey River. The next three belts, which are located farther north, are referred by V. I. Astakhov to the maximum stage of the Late Zyryanka Glaciation and to its recessional Tanama and Yamalo-Gyda stages. The first two correlate, respectively, with the Gydan (Salekhard-Uval) and probably the N'yapan (Sopkei) stages of glaciation.

During the Zyryanka glacial epoch, the internal regions of the West Siberian Plain were flooded at least three times as a result of a continuous glacier front in the north, which prevented a free discharge of rivers into the Kara Sea. The last maximum of the lacustrine transgression was recorded by the shoreline at elevations of 125 to 130 m (Volkov and Volkova, 1975). Its lacustrine sediments, radiocarbon-dated at 22,000 to 12,800 yr B.P., form a thin, almost continuous cover over high terraces and lowered interfluves (Arkhipov et al., 1980). The enormous dimensions of Late Zyryanka ice-dammed lakes are surprising (Figure 2-3). It should be kept in mind, however, that the intra-continental lowlands of western Siberia, with elevations of 60 to 130 m, were closed on all sides by uplands, and drainage could take place only through the Turgay Hollow, where the elevation of the "runoff threshold" (Ubogan-Turgay divide) is 126 m. As soon as the level of the ice-dammed basin reached this "threshold" and flooded the vast lowlands of the West Siberian Plain, discharge began along the Turgay Hollow into the Aral-Caspian region, as recorded by the lacustrine and alluvial sediments lining the buried floor of the hollow (Boboyedova, 1966), at least some of which are of Late Zyryanka age (Astakhov and Grosswald, 1978).

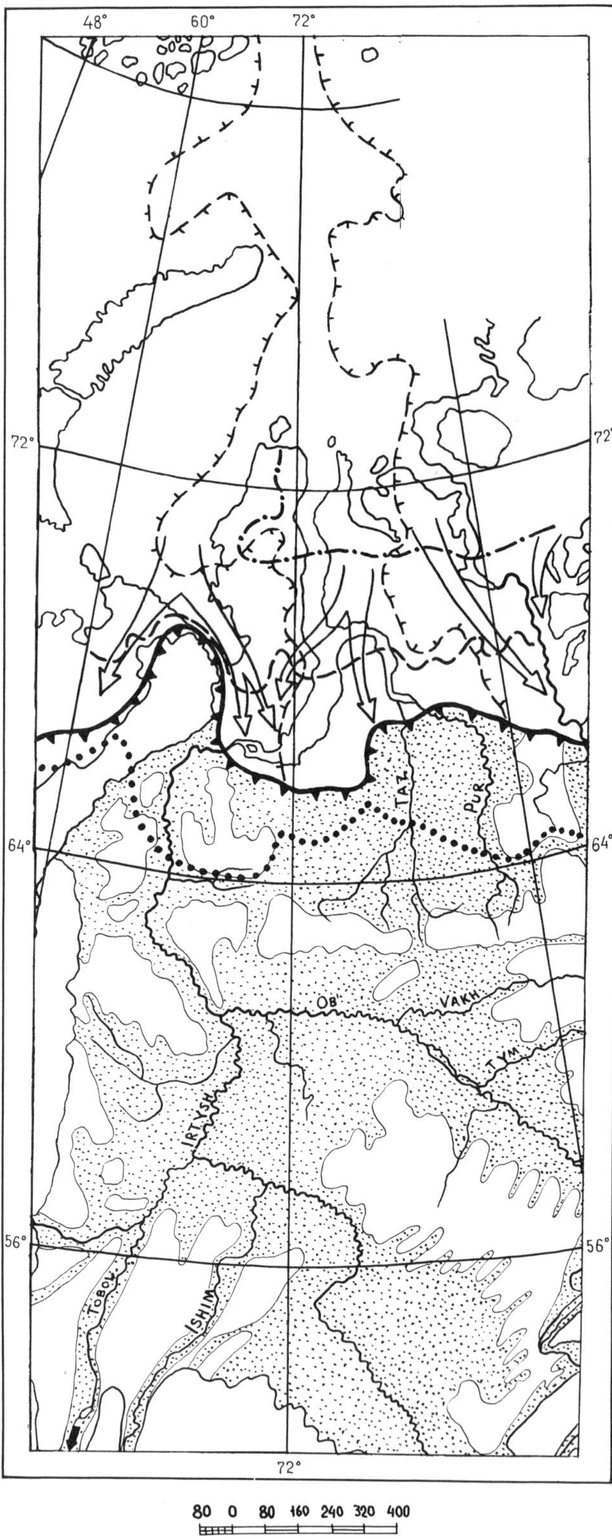

Figure 2-3. Paleogeography of Late Zyryanka Glaciation according to traditional and recent interpretations. (Map compiled by S. A. Arkhipov.)

1. Boundary according to Strelkov et al. (1959) and Dibner (1968)

Marginal zones according to Astakhov (1977):

2. Southernmost Nadym zone, presumably of Early Zyryanka Glaciation

3. Boundary of maximum stage of Late Zyryanka Glaciation

4. Tanama recessional zone

5. Yamal-Gydan recessional zone, one of the last stages of Late Zyryanka Glaciation

6. Presumed direction of flow of glaciers

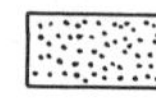

7. Internal low-lying regions of the West Siberian Plain, flooded during the Zyryanka Glaciation, including the maximum Late Zyryanka stage (Volkov and Volkova, 1975)

8. Discharge threshold in the Turgai Hollow, which predetermined the areas of intracontinental flooding

New paleogeographic reconstructions involve certain difficulties concerning the age and correlation of terminal glacial formations combined into the Zyryanka marginal belts. Not all of their elements have yet been dated, and they have not been traced reliably enough everywhere. There are few data on the geology of the Kara Sea Shelf. Traces of recent glaciation thus far have been detected only along its periphery, which adjoins the islands of Novaya Zemlya and Severnaya Zemlya (Dibner, 1968). The shorelines of ice-dammed basins have not been identified everywhere. It is possible that in some cases lacustrine deposits and terminal glacial formations have been wrongly classified as Late Zyryanka. Thus, the paleogeographic concept of Kara Shelf glaciation has not yet been rigorously proved, but it merits serious and comprehensive study.

References

Arkhipov, S. A. (1971). The principal geological events of the latest glaciation and their correlation in western Siberia, Europe, and North America. *In* "VIII International Quaternary Union Congress Proceedings," p. 177.

Arkhipov, S. A. (1977). The Zyryanka Glaciation of the lower Ob' River region of western Siberia (review of new data). International Geological Correlation Program, Project 73/I/24, Report 4. Prague.

Arkhipov, S. A., Astakhov, V. I., Volkov, I. A., Volkova, V. S., and Paychev, V. A. (1980). "Paleogeography of the Western Siberian Plain at the Maximum of Late Zyryanka Glaciation" (V. N. Saks, ed.). Nauka Press, Novosibirsk.

Arkhipov, S. A., Panychev, T. G., Shelekhova, T. G., and Shelkoplyas, V. N. (1978). Glacial Geology of the Belogor'ye Upland. The Western Siberian Plain. The Lower Ob' Region. (In Russian and English.) Guide to the Fifth Session of Project 73/I/24 of the International Geological Correlation Program "Quaternary Glaciations of the Northern Hemisphere." Novosibirsk.

Arkhipov, S. A., Votakh, M. R., Gol'bert, A. V., Gudina, V. I., Dovgal', L. A., and Yudkevich, A. I. (1977). "The Latest Glaciation in the Lower Ob' Region." Nauka Press, Novosibirsk.

Astakhov, V. I. (1976). Geological evidence of the center of Pleistocene glaciation on the Kara Shelf. *USSR Academy of Sciences Doklady* 231, 1178-81.

Astakhov, V. I. (1977). Reconstruction of the Kara center of Pleistocene glaciation from old moraines of western Siberia. *In* "Reports of Glaciological Studies, Chronicles, Discussions" (V. M. Kotlyakov, ed.), no. 30, pp. 60-69. USSR Academy of Sciences, Interdepartmental Geophysical Committee, Moscow.

Astakhov, V. I., and Grosswald, M. G. (1978). New data on the age of sediments of the Turgay Hollow. *USSR Academy of Science Doklady*, seriya geologicheskaya 242, 891-94.

Boboyedova, A. A. (1966). Origin of the Turgay Hollow. *In* "The Quaternary Period of Siberia" (V. N. Saks, ed.), pp. 187-97. Nauka Press, Moscow.

Dibner, V. D. (1968). Old clays and relief of the Barents-Kara Shelf: Direct evidence of its cover glaciation in the Pleistocene. *In* "Problems of Polar Geography" (Ya. Ya. Gekkel' and L. S. Govorukha, eds.), pp. 118-22. Gidrometeoizdat Press, Leningrad.

Giterman, R. E., Golubeva, L. V., Zaklinskaya, E. B., Koreneva, E. V., Matveyeva, O. V., and Skiba, L. A. (1968). "Main Stages of the Development of Central Asia Vegetation in the Anthropogene." USSR Academy of Sciences, Trudy Geologicheskogo Instituta 177. Nauka Press, Moscow.

Grichuk, M. P. (1959). Application of spore-pollen analysis to Siberia. *Nauchnyye doklady vysshey shkoly (Geologo-geograficheskiye nauki)* 1, 113-22.

Grichuk, M. P. (1970). Principles of formation of present spore-pollen spectra as the basis of the interpretation of fossil spore-pollen spectra. *In* "History of the Development of Vegetation of the Extraglacial Zone of the Western Siberian Lowland in the Late Pliocene and Quaternary" (V. N. Saks, ed.), pp. 12-20. Nauka Press, Moscow.

Grosswald, M. G. (1977). The latest Eurasian ice sheet. *In* "Reports on Glaciological Studies, Chronicles, Discussions" (V. M. Kotlyakov, ed.), no. 30, pp. 45-60. USSR Academy of Sciences, Interdepartmental Geophysical Committee, Moscow.

Gudina, V. I. (1976). "Foraminifers, Stratigraphy, and Paleozoogeography of the Marine Pleistocene of the North of the USSR." Nauka Press, Novosibirsk.

Kind, N. V. (1974). "Geochronology of the Late Anthropogene from Isotope Data." No. 257. Nauka Press, Moscow.

Saks, V. N. (1953). "The Quaternary Period in the Soviet Arctic." Transactions of the Scientific Research Institute for Geology of the Arctic 77.

Strelkov, S. A. (1965). "The North of Siberia." Nauka Press, Moscow.

Strelkov, S. A., Dibner, V. D., Zagorskaya, N. G., Sokolov, V. N., Yegorova, I. S., Pol'kin, Ya. I., Kiryushina, M. G., Puminov, A. P., and Yashina, Z. I. (1959). "Quaternary Deposits of the Soviet Arctic." Transactions of the Scientific Research Institute for Geology of the Arctic 91.

Troitskiy, S. L. (1967). New data on the latest cover glaciation of Siberia. *USSR Academy of Sciences Doklady* 174, 1409-12.

Volkov, I. A., and Volkova, V. S. (1975). The great periglacial system of Siberia's runoff. *In* "History of Lakes in the Pleistocene." Abstracts of Papers 2, pp. 133-40. Limnology Institute, USSR Academy of Sciences, Leningrad.

CHAPTER 3

Late Pleistocene Glaciation of North-Central Siberia

L. L. Isayeva

Until the 1960s, deposits of one Late Pleistocene (Valdai) glaciation, which had the local name Zyryanka (Saks, 1953; Strelkov, 1957), were recognized in north-central Siberia. Further studies broadly based on radiocarbon-dating showed that the Valdai in central Siberia included two glaciations—the Zyryanka (Early Valdai) Glaciation and the Sartan (Late Valdai) Glaciation—separated by a long climatically complex sequence (50,000 to 23,000 yr B.P.) referred to as the Karginskiy interval in modern terminology (Troitskiy, 1967; Kind, 1974; Isayeva et al., 1980).

The morphostructural complexity and the vast area of central Siberia account for the presence of a large number of local stratigraphic schemes of Quaternary deposits and the different details of their dissection. Table 3-1 shows the correlation of three morphostructural zones—the North Siberian Lowland, the Central Siberian Highland, and the right bank of the Lena River Lowland (western slope of Verkhoyansk Range)—which were subjected to glaciation during the Late Pleistocene.

Early Zyryanka

According to traditional concepts, the beginning of the Zyryanka (Early Valdai) Glaciation dates back to 70,000 yr B.P. However, thermoluminescence dating in western Siberia suggests an earlier date, now indirectly confirmed by the large dimensions of the reconstructed northern Siberian ice sheet, which required a glacial epoch longer than from 70,000 to 50,000 yr B.P. (See chapter 2.) The glaciation developed from four centers: the North center, located on the Kara Sea Shelf at least during the glaciation maximum, the Putorana center, the Anabar center, and the Verkhoyansk center. This fact is demonstrated by the location of marginal formations on the Central Siberian Highland (Isayeva, 1972), by the petrographic composition of the moraines both on the highland and in the North Siberian Lowland, and by the stone orientation in the moraines (Andreyeva, 1978).

The Zyryanka Glaciation in northwestern central Siberia consisted of two stages represented by two moraine horizons separated in the North Siberian Lowland by barren sandy-aleuritic deposits and in the neotectonic Aganyli and Murukta Basins of the Central Siberian Highland by aleuritic-clayey lacustrine deposits (Figures 3-1, 3-2, 3-3, 3-4, and Table 3-1) (Bardeyeva and Isayeva, 1980). The early-stage till covers almost all of the North Siberian Lowland, where it rests either on a Mesozoic base or on Quaternary marine and continental deposits with microfossils that accumulated under very warm climatic conditions. These deposits are referred to the Kazantsevo interglacial horizon, which most probably corresponds to Europe's Eemian Interglaciation. Deposits of the Zyryanka glacial complex are overlain or bordered by marine and continental sediments of the Karginskiy horizon and have been radiocarbon-dated many times at 46,000 to 23,000 yr B.P. Thus, the Zyryanka Glaciation can be correlated with the beginning of the Late Pleistocene cold stage, that is, the Early Valdai (Isayeva et al., 1980).

On the Central Siberian Highland, deposits of the Zyryanka glacial complex in the Nizhnyaya Tunguska Valley, the Aganyli and Murukta Basins, and the Kotuy River valley rest on alluvial or lacustrine sediments of the Kazantsevo horizon, which are characterized by pollen spectra indicating warm conditions. The Zyryanka moraine, in turn, is overlain by alluvium dated at 35,800 ± 1700 yr B.P. (GIN-493) in the Kotuy Valley and about 37,000 yr B.P. (GIN-61) in the Nizhnyaya Tunguska Valley (Bardeyeva et al., 1980). In addition, alluvium of terraces III and II is covered by proglacial lake deposits associated with the Zyryanka Glaciation moraine and dated at 28,000 yr B.P.

On the right bank of the Lena River (western slope of the Verkhoyansk Range), a ground moraine of the Zyryanka horizon is overlain by alluvium with plant remains dated at 37,000 to 33,000 yr B.P. (Kind et al., 1971; Kind, 1974).

In the early maximum stage of Zyryanka Glaciation,

Table 3.1. Correlation of Stratigraphic Units for Three Morphostructural Zones in North-Central Siberia

Super-horizon	Horizon	Sub-horizon	North Siberian Lowland	Central Siberian Highland	Lena River Plain (Western Slope of Verkhoyansk Range)
Zyryanka	Sartan		Noril'sk glacial strata: moraine, pebble gravel, cobble roundstones Lacustrine aleurites, clays, sands $^{14}C = 10{,}000$-9,000 yr B.P. Alluvium of fluvial terrace II above floodplain, $^{14}C = 15{,}000$-12,000 yr B.P. N'yapan glacial strata: moraine, cobble roundstones, pebble gravel Valvok suite: lacustrine varved clays, aleurites, sands, $^{14}C = 19{,}000$ yr B.P. Alluvium of fluvial terrace III above floodplain, $^{14}C = 17{,}000$ yr B.P. Karaul glacial strata: moraine, cobble roundstones, pebble gravel	Putorana glacial strata: moraine, pebble gravel, cobble roundstones Noril'sk (Melkolamskiy) glacial strata: moraine, cobble roundstones, pebble gravel, sands, aleurites, clays N'yapan glacial stata: moraine, cobble roundstones, sands, aleurites, clays Onega glacial strata: moraine, pebble gravel, sands, aleurites Lacustrine sands, aleurites, clays of individual periglacial basins and a large basin in the Nizhnyaya Tunguska River valley Alluvium of fluvial terrace I above flood-plain, pebble gravel, sands, aleurites; thick, syngenetic repeated-lode ice, ^{14}C (in the upper strata) $= 10{,}000$ yr B.P.	Segemda glacial complex: moraine, loams, pebble gravel, sand Lacustrine-glacial pebble gravel, sands, sandy loams, loams, $^{14}C = 15{,}000$ yr B.P. Sigenekhs glacial complex: moraine, pebble gravel, sand Lacustrine-glacial sands, loams, pebble gravel Ulakhan-Kuyel' glacial complex: moraine, sand, pebble gravel Lacustrine-glacial sands, sandy loams, loams, pebble gravel Eolian, cryogenic-eolian, and loessial cover deposits
	Karginskiy	Upper	Chaykin layers: lacustrine and lacustrine-river sands with alluvial vegetal detritus, $^{14}C = 29{,}000$-23,000 yr B.P. Upper Balakhna layers: lagoon-marine aleurites, sands with mollusk shells, foraminifers, alluvial vegetal detritus, $^{14}C = 33{,}000$-26,000 yr B.P. Baty-Sala band, lacustrine sands, aleurites, clays with vegetal detritus and peat, $^{14}C = 46{,}000$-27,000 yr B.P	Alluvium of fluvial terrace II above floodplain: pebble gravel, sands, aleurites, peat, $^{14}C = 28{,}000$ yr B.P.	Alluvium of fluvial terrace II above floodplain of Lena River: pebble gravel, sands, sandy loams, $^{14}C = 30{,}000$-29,000 yr B.P. Zhigansk glacial complex: moraine, pebble gravel, sands
		Middle	Malaya Romanikha layers: coastal and riverine obliquely laminated sands, sands with pebble and gravel, wood and vegetal detritus, $^{14}C = 46{,}000$-32,000 yr B.P.	Buried alluvial pebble gravel, sands, $^{14}C = 37{,}000$-35,000 yr B.P. Lacustrine and lacustrine-riverine sands, aleurites, clays with vegetal detritus of Murukta and Aganyli Basins Alluvium of fluvial terrace III above floodplain: pebble gravel, rock debris, sands, aleurites Fine cryogenic dislocations, vegetal detritus	Buried alluvial pebble gravel, sands, sandy loams, loams, peat, $^{14}C = 37{,}000$-33,000 yr B.P.
		Lower	Boyarka layers: marine clays, aleurites, sands with mollusk shells, $^{14}C = 46{,}000$-41,000 yr B.P.		
	Murukta	Upper	Northern Kokora layers: moraine, rewashed moraine, cobble roundstones with detritus of marine mollusk shells	Murukta glacial strata: moraine, cobble roundstones, pebble gravel Sands, aleurites, clays of periglacial basins Periglacial alluvium of terrace IV of Nizhnyaya Tunguska (upper part of terrace section); sands, aleurites, cryogenic dislocations	Glacial complex: moraine
		Middle	Yantardakh layers: jointy-kame and presumably marine aleurites, sands, sands with pebbles, and boulders with marine-mollusk-shell detritus	Lacustrine aleurites, sands, clays buried in Murukta and Aganyli Basins and in the Kotuy River valley	
		Lower	North Siberian layers: moraine, varved clays	Lower Tunguska glacial strata: moraine, cobble roundstones, pebble gravel, sands, aleurites, clays Sands, aleurites, clays of periglacial basins	

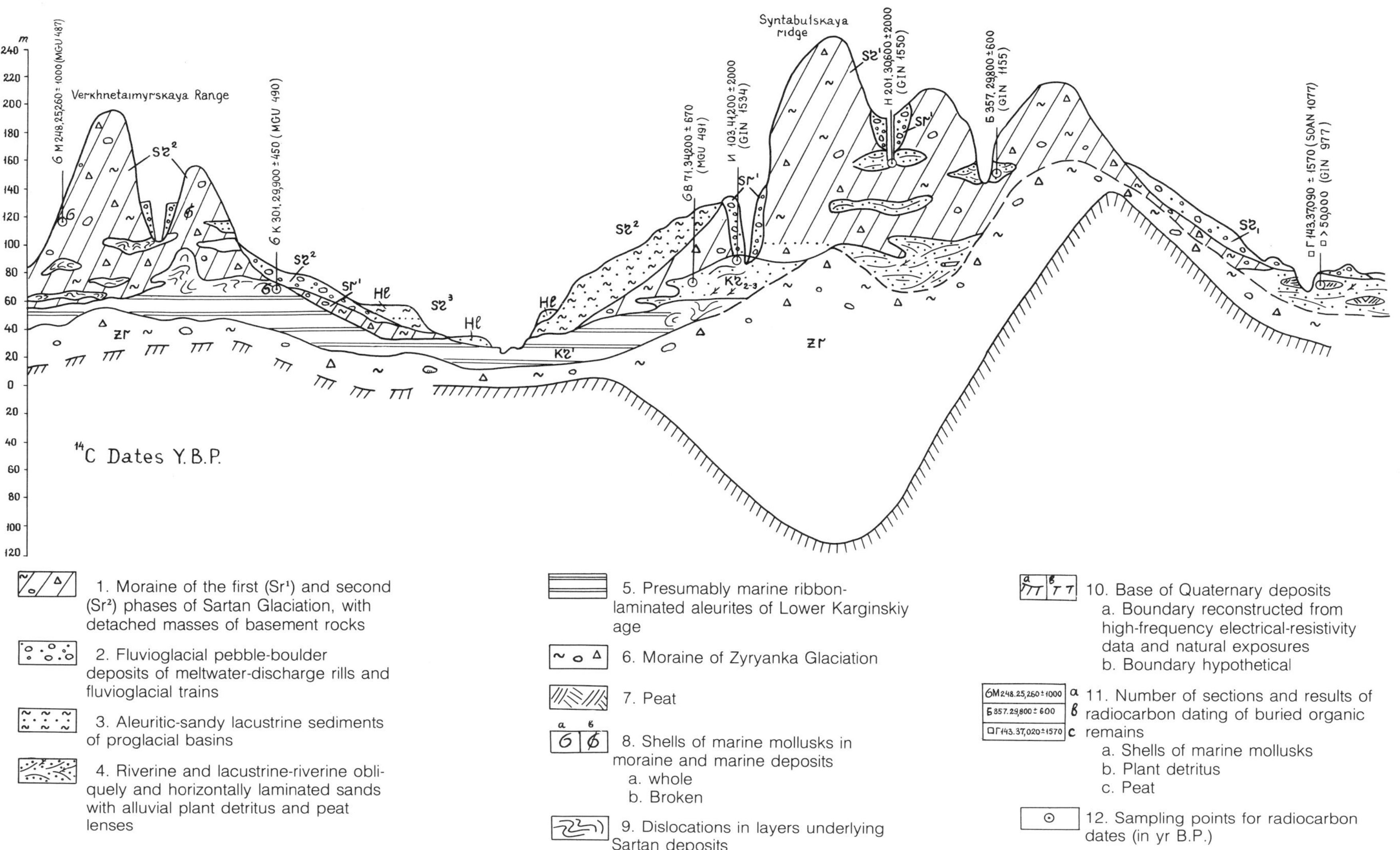

Figure 3-1. Summary geologic-geomorphologic profile through the Upper Taimyr and Syntabul'skiy terminal moraine ridges of Sartan Glaciation, showing locations of specific sections investigated. The following symbols are used: H1, Holocene deposits; Sr^1, Sr^2, Sr^3, deposits of the first, second, and third phases of Sartan Glaciation; Kr^1, Lower Karginskiy deposits; Kr^2, Kr^3, Middle and Upper Karginskiy deposits; Zr, deposits of Zyryanka Glaciation.

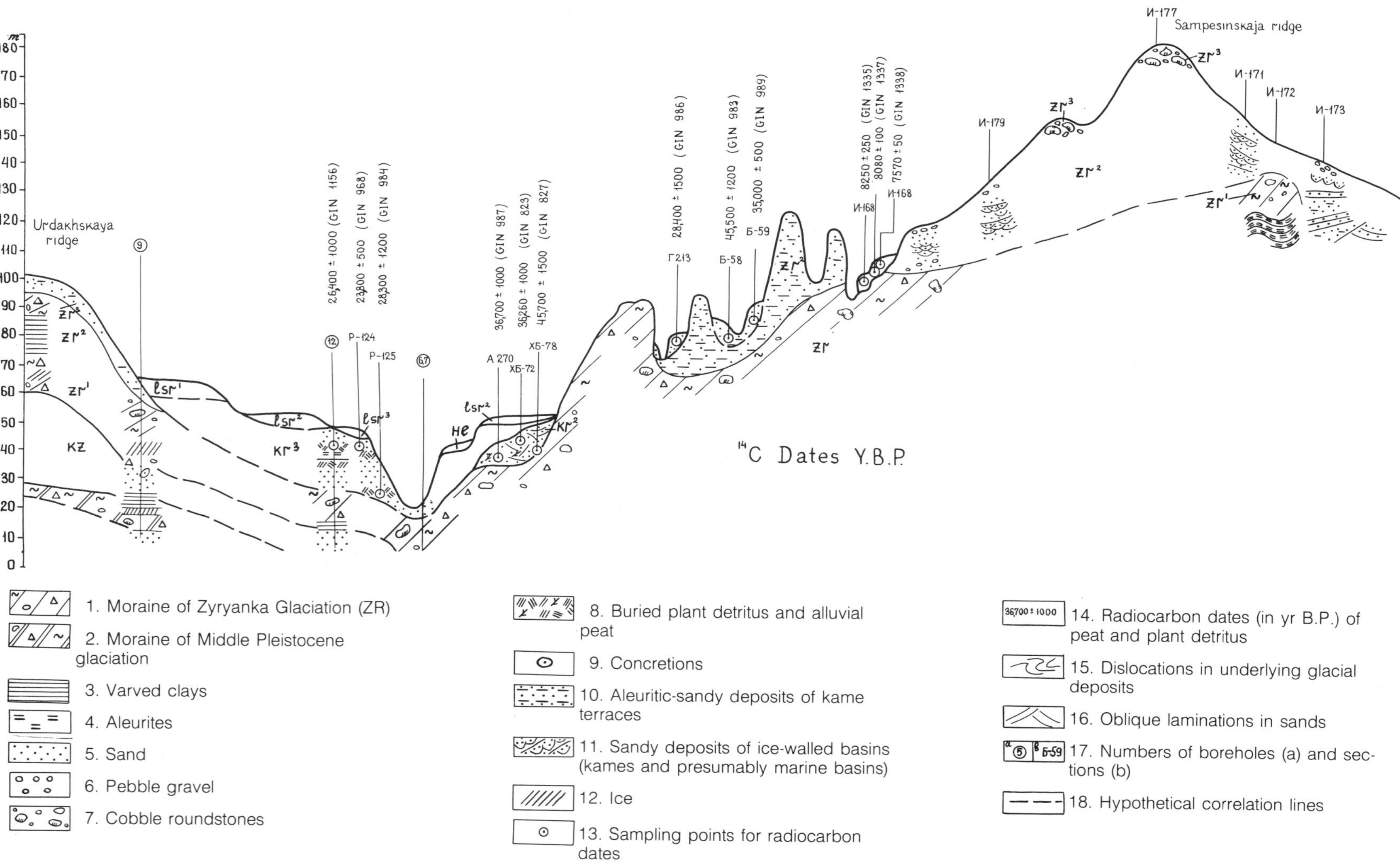

Figure 3-2. Geologic-geomorphologic profile of the southern part of the North Siberian Lowland (Sampesinskiy and Urdakhskiy Ridges). The following symbols are used: H1, Holocene deposits; Sr^1, Sr^2, Sr^3, lacustrine deposits formed during the first, second, and third phases of Sartan Glaciation; Kr^1, Kr^2, Kr^3, Lower, Middle, and Upper Karginskiy sediments; Zr^1, Zr^2, Zr^3, Zyryanka Glaciation deposits: early stage, interstadial, late stage; Kz, sediments of Kazantsevo horizon.

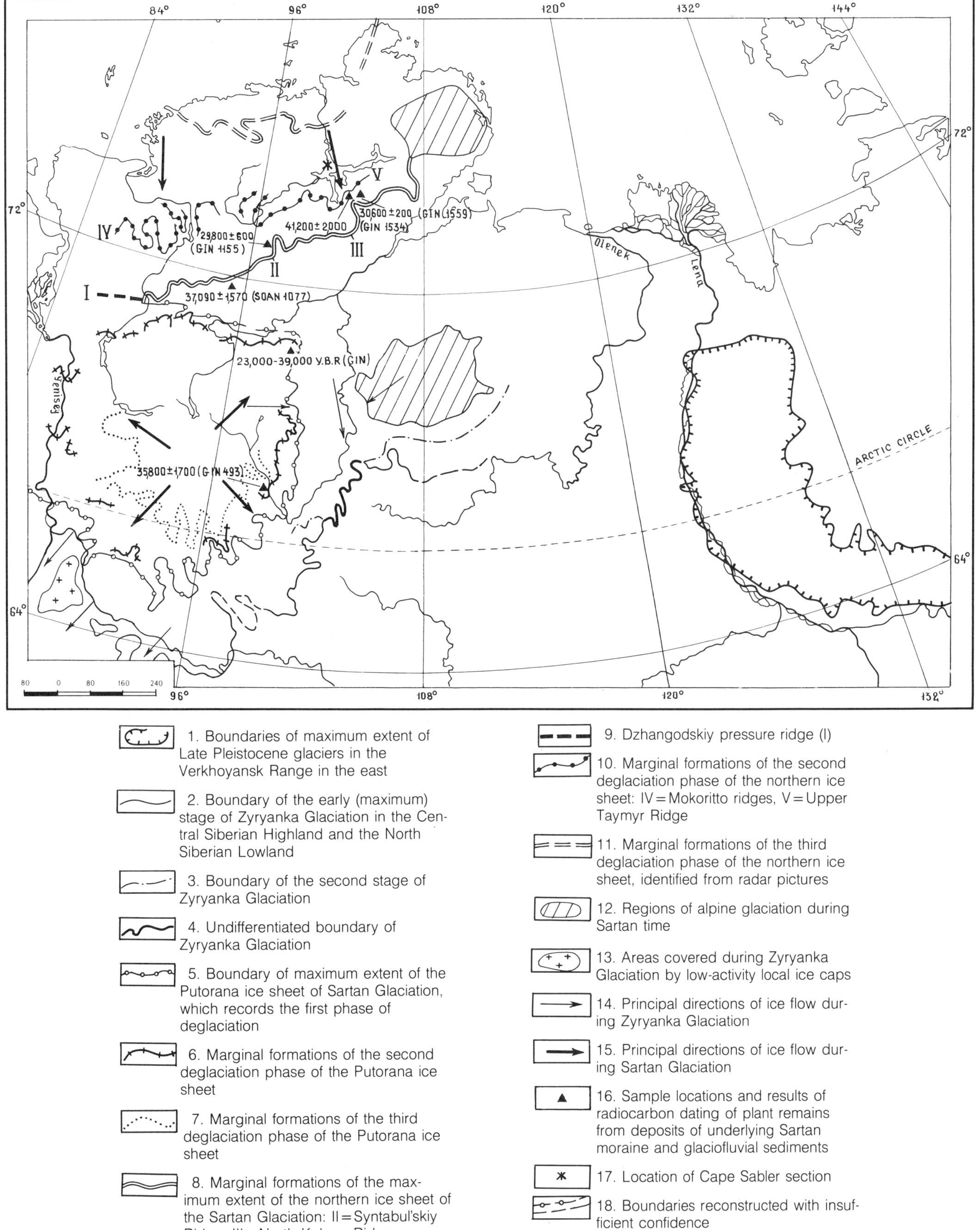

Figure 3-3. Boundary of maximum extent, stages, and deglaciation phases of Zyryanka and Sartan ice sheets in north-central Siberia. (Map compiled by L. L. Isayeva.)

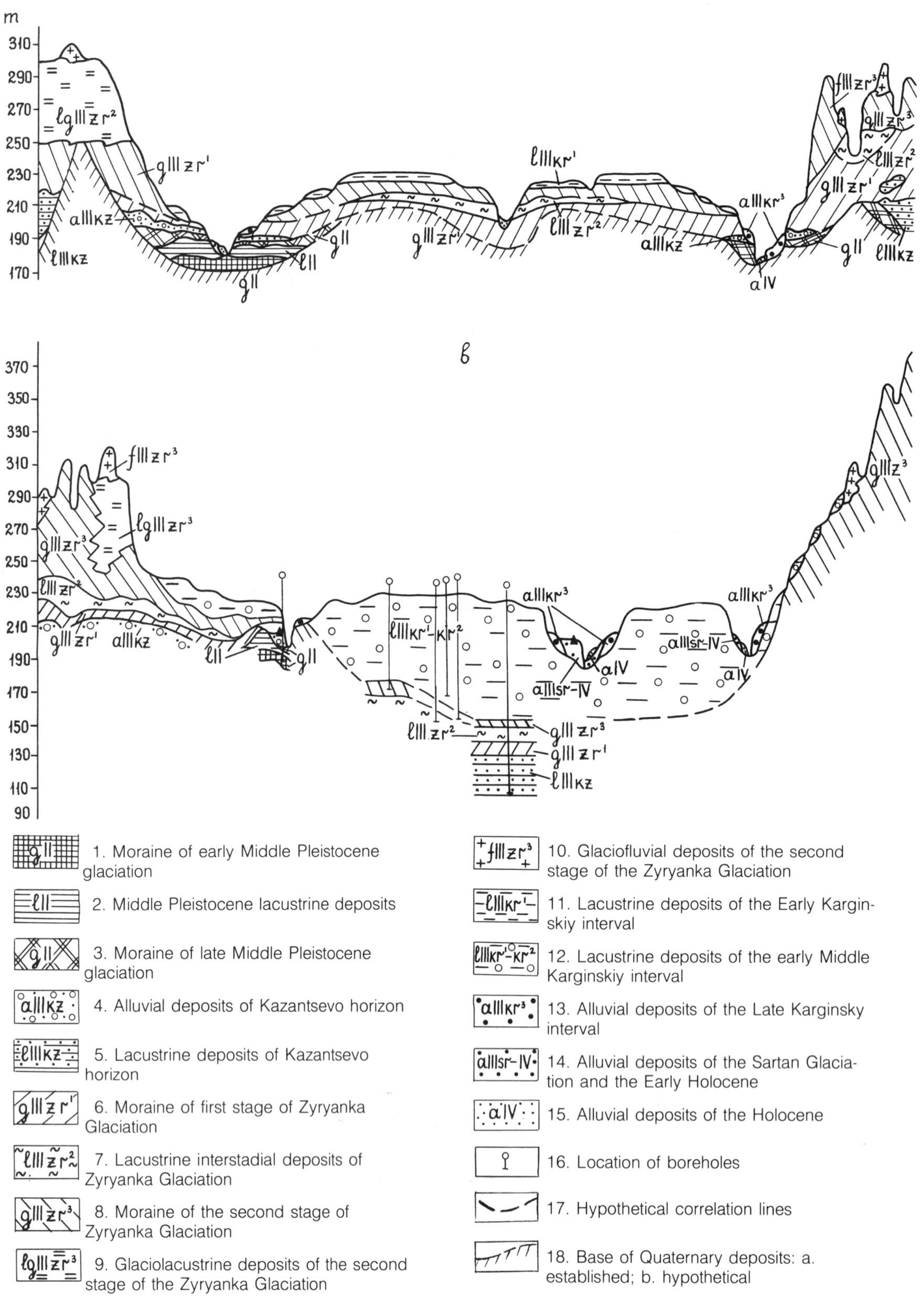

Figure 3-4. Basic geologic-geomorphologic sections through (a) the Aganyli Depression and (b) the Murukta Depression of the Central Siberian Highland.

which corresponds to maximum ice extent, the glaciers of the North, Putorana, and Anabar centers merged into a single cover that occupied the entire northwestern part of the Central Siberian Highland (Figure 3-3) and formed large lobes extending southward along the Yenisey Valley and eastward along Khatanga Bay. It also may have covered the Severnaya Zemlya Islands, for the Zyryanka moraines occur everywhere as well as on the shelf below the present ocean level. There, the moraine rests on Kazantsevo deposits and is overlain by radiocarbon-dated Karginskiy deposits, which also occur on high marine terraces, indicating glacioisostatic uplift of the islands.

In the Verkhoyansk Range, the glaciation was largely restricted to the valleys, although the high parts of the mountains may have been covered by an ice cap. Large piedmont glaciers reached the Lena River valley at the base of the range.

At the maximum of Zyryanka Glaciation, the east-west section of the Nizhnyaya Tunguska Valley was dammed by the ice and was filled by a proglacial lake to an elevation of not less than 280 m. Upstream alluvium is of periglacial type and contains pseudomorphs of ice wedges formed at the same time as the alluvium. This indicates a fairly harsh climate at the glacial maximum (although the size of the pseudomorphs is small, with a maximum depth of 0.7 to 1.0 m). The cold climatic conditions of that time are also indicated by the pollen spectra from the alluvium, which indicate a peculiar periglacial tundra-steppe with grasses, *Artemisia*, chenopods, lichens, and mosses. Woody plants are represented by dwarf birch, alder, and *Alnaster* and possibly by tree birch and larch. Harsh winters with little snow are indicated by numerous mammalian remains (e.g., horse, bison, and mammoth) buried in Zyryanka alluvium. Throughout the Central Yakutya Lowland, pebbles vertically oriented by frost action are found at different geomorphologic levels. It can be assumed that the taiga forests that grew in central Siberia before the glaciation degraded almost completely during the last glaciation, remaining as small islands in the southern part of the region in the Upper Angara Basin.

Middle Zyryanka

During the Middle Zyryanka Interstade, the glaciers on the Central Siberian Highland shrank considerably, remaining only on the Putorana Plateau and the Anabar Highland. The Murukta and Aganyli Basins were occupied by lakes. Pollen spectra from interstadial deposits of the Murukta Basin indicate very thin larch forests, with shrubs and grasses in openings; that is, they indicate a harsher climate than the present one. The ice sheet in the North Siberian Lowland broke up into a series of dead-ice blocks, which did not expand until the next glaciation (Isayeva et al., 1980).

Karginskiy Interval

At the beginning of the Karginskiy interval, the Zyryanka glaciers degraded completely, and the North Siberian Lowland was penetrated by a marine transgression (Andreyeva, 1980), which is represented by clays and by aleurites with shells of bathypelagic marine mollusks (Table 3-1 and Figures 3-1 and 3-2) widely distributed in the basins of the M. Romanikha and Boyarka Rivers. Radiocarbon dates on this part of the section are 43,600±1500 yr B.P. (GIN-673) from wood and 31,000±750 yr B.P. (MGU-486) from shells. The decline of the transgression is represented there by marine and riverine sandy and pebbly sediments containing woody detritus dated at 39,000 to 32,000 yr B.P. In the western (Pyasina River basin) and northeastern parts of the lowland, marine conditions of sedimentation still existed at that time. A further retreat of the sea resulted in the formation of lacustrine and lacustrine-riverine deposits localized in separate topographic depressions (with radiocarbon dates of 29,000 to 23,000 yr B.P.). The aleurites and sands in the upper part of Bolshaya Balakhna Basin, with shallow-water mollusks dated at 33,000 to 26,000 yr B.P., record a lagoon-type basin that remained the longest within the lowland. In the North Siberian Lowland, there also existed closed depressions where transgression waters did not penetrate and where lacustrine aleuritic-argillaceous and sandy deposits were formed during the entire Karginskiy interval.

In the central Siberian Highland, Karginskiy lacustrine and alluvial deposits are locally buried by Sartan glacial sediments. Lacustrine sediments fill up the Aganyli and Murukta depressions, reaching a thickness of 75 m in the latter, and can be traced part way up the valleys of the Kotuy and Moyyero Rivers in these basins. A radiocarbon date of 35,800±1700 yr B.P. (GIN-493) (Bardeyeva et al., 1980) was obtained from alluvial deposits (channel facies) buried under a Sartan moraine in the Kotuy River valley, and a date of about 37,000 yr B.P. was obtained from the pebbly alluvium in the Nizhnyaya Tunguska Valley under sediments of a proglacial Sartan lake. In regions not covered by glaciers and not flooded by Sartan proglacial lakes, Karginskiy deposits are represented by fluvial terraces III and II of large rivers (Nizhnyaya Tunguska, Kotuy, Moyyero). A radiocarbon date of 28,800±500 yr B.P. (GIN 237) (Kind, 1974) originates from an oxbow-lake facies of the Nizhnyaya Tunguska's second terrace.

On the western slope of the Verkhoyansk Range, in the right-bank portion of the Lena River Plain, the Karginskiy horizon is clearly divided into three bands of different ages and of partially different origin. The oldest band is alluvium with radiocarbon dates of 37,000 to 33,000 yr B.P. on plant remains. The alluvium is buried by moraine and fluvioglacial deposits of the Zhigansk glacial complex. The latter, in turn, is overlapped or bordered by alluvium of fluvial terrace II of the Lena River near the village of Zhigansk, with dates of 30,000 to 29,000 yr B.P. (Table 3-1) (Kind et al., 1971; Kolpakov and Belova, 1980).

For Karginskiy time, repeated fluctuations of the climate are recorded by pollen evidence of a change in vegetation from middle taiga forest to forest-tundra and tundra. Three stages of warming, separated by two cold intervals, are indicated by Kind (1974) for the Yenisey Valley. Similar fluctuations for the North Siberian

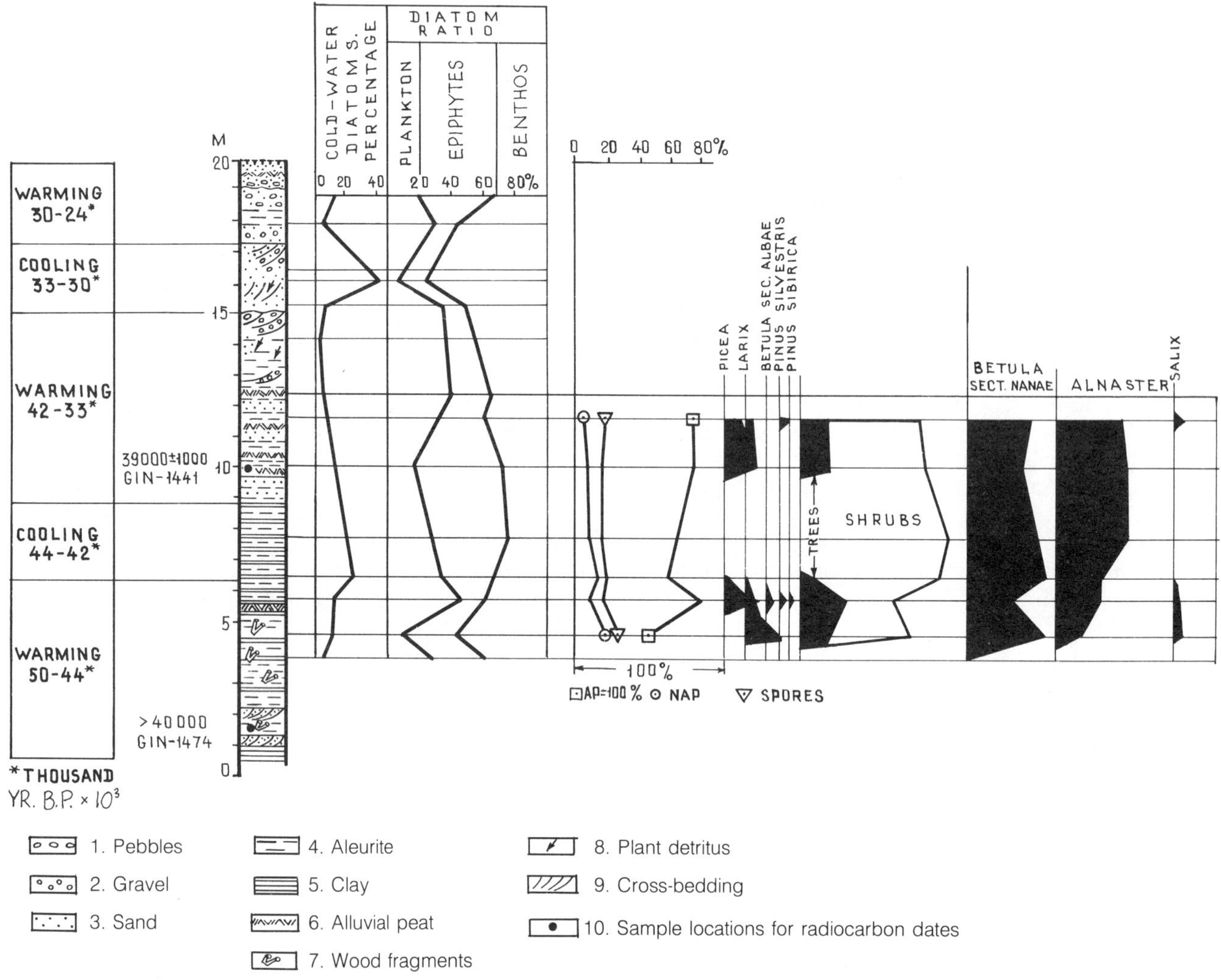

Figure 3-5. Section of lacustrine deposits of Baty-Sala beds (Karginskiy interval), their diatomaceous characteristics (after M. Cherkasova), and their spore-pollen characteristics (after M. Nikol'skaya).

Lowland are based on the pollen and diatom content of lacustrine deposits in the Baty-Sala Basin (Figure 3-5) (Andreyeva, 1980), representative of the entire Karginskiy interval, with three warmings (early: 50,000 to 44,000 yr B.P.; middle: 42,000 to 33,000 yr B.P.; late: 30,000 to 24,000 yr B.P.). Warming epochs are dominated by trees and shrubs (60% to 90%) in the pollen spectra, although trees do not exceed 40% of the sum of trees and shrubs. Of interest is the presence of 20% to 30% spruce pollen. This is remarkable because the nearest spruce at the present time is 500 km south of the Batay-Sala Basin, which is completely in the tundra zone. Macrofossils in the Karginskiy deposits include seeds of taiga elements (duckbean, water milfoil, pondweed), which do not grow in the tundra zone at the present time. Sediments of the Karginskiy warm epochs are characterized by the presence of a small number of cold-water diatoms (up to 12% versus 27% to 43% in sediments of cold epochs). In modern sediments cold-water diatom species (arctic, arctoboreal, and northern alpine) are present in amounts of 10% to 25% (M. N. Cherkasova, personal communication).

Thus, the Karginskiy warm intervals were warmer than the present climate. For the cold epochs, the pollen spectra indicate an almost complete absence of tree pollen and a preponderance of dwarf birch, alder, and *Alnaster*. The sharpest Karginskiy cooling even led to glaciation (Zhigansk glacial complex, 33,000 to 30,000 yr B.P.) in the Verkhoyansk mountain system. The large size of the glaciers (exceeding the boundaries of the mountains), developed in such a short period of time, suggests that the glaciation, which began in the mountains during Zyryanka time, was not completely interrupted, although it did decrease markedly in dimension. Data on the Zhigansk Glaciation are in accord with those on the Lokhpotgort Glaciation in the northwestern Siberia.

Sartan (Late Valdai) Glaciation

The cooling that followed the Late Karginskiy warming led to the formation and development of Sartan (Late Valdai) Glaciation. The Sartan horizon of northern Siberia encompasses glacial deposits and the lacustrine and alluvial deposits associated with it, as well as widely distributed eolian and cryogenic loess-ice formations.

In the North Siberian Lowland, Sartan glacial deposits are represented by ground moraines and a complex series of terminal moraines bordered by fluvioglacial gravel trains. The location of the marginal formations on the lowland, as well as the lithology and long-axis orientation of the stones, indicate two centers of glaciation: a northern one somewhere on the Kara Sea Shelf and a southern one on the Putorana Plateau. The maximum advance of the northern ice sheet is recorded by a frontal ridge, which extends across the entire lowland from the southwest to the northeast (it is known by the name of Dzhangodo-Syntabul-North Kokora Ridge [Figure 3-3]). The deposits beneath this ridge (and possibly also those in detached masses) have dates of 41,000 and 29,000 yr B.P. on plant detritus and 37,000, 29,000, and 25,000 yr B.P. on marine mollusks incorporated in the moraine (Figures 3-1 and 3-2). Peat underlying the fluvioglacial deposits are dated at 37,090±1500 yr B.P. (SOAN-1077) and plant detritus from sediments underlying meltwater deposits yielded a date of 30,600±2000 yr B.P. (GIN-1559). Thus, the Dzhangodo-Syntabul marginal formations are considered to be Sartan (Late Valdai) in age. Farther to the north, another frontal ridge, the Mokoritto-Upper Taimyr frontal moraine ridge, records the advance of two huge glacier lobes from the north across the Byranga Mountains (Figure 3-3), and a large number of separate ridges indicate a partial retreat of the ice sheet. On the northern coast of the Taimyr Peninsula, another frontal moraine ridge identified from aerial photographs may record a third major phase in the development of the northern ice sheet. At the northern base of the Putorana Plateau, Sartan glacial deposits form several marginal arcs that cross each other transgressively. The glacial and fluvioglacial deposits forming the arcs overlie marine and continental Karginskiy deposits, which have several radiocarbon dates from 43,000 to 23,000 yr B.P. (Isayeva et al., 1976).

Widespread Sartan lacustrine deposits in the lowland are represented by aleurites and fine sands with horizontal and wavelike laminations. The deposits form a series of lacustrine terraces 40 to 100 m high, with transition to fluvioglacial trains on the distal side of glacial ridges. The fluvial deposits entrench the lakebeds. In the south-central part of the lowland in the Kheta River basin, the oldest river terrace (enclosed in lacustrine ones) is dated at 15,630±80 yr B.P. (GIN-938) and 13,700±150 yr B.P. (GIN-692). However, in the northern part of the lowland (in the basin of the Upper Taimyra River and Lake Taimyr), lacustrine deposits continued to form in later phases of deglaciation as well, and some lakes have persisted until the present time (e.g., Lake Taimyr, Lake Labaz). Alluvial deposits in the third fluvial terrace of the Khatanga River, whose basin was not occupied by a proglacial lake, have a date of 17,780±200 yr B.P. (GIN-937), and those of the second fluvial terrace of the Kheta and B. Balakhnya Rivers are dated at 14,000 to 10,000 yr B.P. Above the lacustrine deposits or (in the east) on older sediments lie eolian sands forming either single or complex dunes.

On the Central Siberian Highland, the Sartan horizon also includes glacial, fluvioglacial, lacustrine-glacial, and alluvial deposits. The glacial complex is made up of ground moraines, filled terminal formations, and gravel in dead-ice fields contiguous to terminal moraines on the proximal side and characterized by a distinct hill and basin relief. Belts of marginal formations delineate the Putorana Plateau in the east and south (Isayeva, 1972). Traced most clearly and almost continuously (as in the North Siberian Lowland) are three belts, tentatively correlated with the terminal moraines of the northwestern foot of the Putorana Plateau identified by Saks (1953) and Strelkov (1957) as Zyryanka formations. The occurence of Karginskiy deposits (dated from more than 50,000 to 40,000 yr B.P.) (Troitskiy, 1967) under the earliest (Karaul) of these formations now permits one to regard them as of Sartan age.

Lacustrine deposits are represented by varved clays, aleurites, and sands. In the Norilka River basin, at the foot of the northwestern slope of the Putorana Plateau, a radiocarbon date of 19,000 yr B.P. was determined from fine plant detritus from the lower portion of a section of lacustrine clays resting on ground moraine. In the Nizhnyaya Tunguska Valley, covered with Sartan glacial deposits near the estuary, lacustrine deposits of a proglacial lake are represented by aleurites and sands lining the valley to an elevation of 180 m and traced up the river to the settlement of Tura. Farther up the river, they are replaced by alluvial deposits of the first fluvial terrace, which are characterized by the presence of thick syngenetic ice wedges.

In the highland area covered by Sartan glaciers or by waters of glacial lacustrine basins, only the alluvium of fluvial terraces is bordered by Sartan deposits. On the Anabar Highland, small terminal moraines of valley glaciers can be tentatively referred to the Sartan.

At the foot of the western slope of the Verkhoyansk Range, Sartan deposits are represented by ground moraine and an intricately structured complex of marginal deposits forming marginal arcs of piedmont glaciers. As it has already been stated, the alluvial sands underlying them were radiocarbon-dated at 30,000 to 29,000 yr B.P. Three belts of marginal arcs are the most distinct: the Ulakhankyuelskiy, Sigenekhskiy, and Segemdinskiy belts. Dates of lacustrine deposits associated with the youngest belt are 15,850±60 yr B.P. (GIN-333) and 15,100±60 yr B.P. (GIN-332).

Thus, during the Sartan Glaciation the centers were located in the same place as those of the Zyryanka Glaciation, but the dimensions were much smaller (Figure 3-3). Glaciers of the North and Putorana centers came together only in the far west of the North Siberian Lowland, and its central and eastern parts were free of this ice. Minor mountain glaciers developed on the Anabar Highland, but the glaciation of the Verkhoyansk Range was extensive and almost equal to the Zyryanka Glaciation. As in Verkhoyan'ye, on the Putorana Plateau and in the North Siberian Lowland marginal frontal formations recorded three major pauses during the general retreat of the ice sheets.

The retreat of the piedmont glaciers of the Verkhoyansk Range ended around 15,000 yr B.P. However, glaciers continued to exist in the mountains, as recorded by valley terminal moraines. Small glaciers are known in the Verkhoyansk Range at the present time as well. At the maximum glaciation in the central part of the North Siberian Lowland, the vast Pralabaz Lake basin was formed; the lake emptied into the Laptevy Sea around 14,000 yr B.P. As the glaciers retreated, a few other lakes of smaller size were formed in the lowland. During glacial expansion and after the emptying of Pralabaz Lake, winds from the west produced oriented dunes in the central and eastern parts of the lowland (eastern and northeastern directions predominate at present).

Eolian processes were extensively developed at that time in the Lena-Vilyuy Lowland as well, as indicated by the peculiar omnipresent loess-ice cover deposits—loams, sandy loams, fine sands, and buried ice—thought to be of cryogenic-eolian origin. The cover deposits probably started to form during the Karginskiy interval (Zhigansk cooling), but they appear to have accumulated most actively during the Sartan Glaciation. The formation of thick, syngenetic ice-wedge features in Sartan alluvium and the development of eolian processes (formation of a ventifact horizon, sandy dune masses, single eolian forms) in central Siberia indicate very harsh climatic conditions during the Sartan Glaciation. The valleys of shallow rivers were filled with rubbly slope sediments. A layer of cryogenically reworked soils up to 5 m thick was formed on surfaces that were free of vegetation or turf. Tundra or forest-tundra was present even 1000 km from the glacier edges, as evident from pollen spectra from alluvium of terrace I. On this basis, one can postulate that most of central Siberia at the maximum of Sartan Glaciation was a peculiar zone of cold desert and semidesert. Grassy plant associations could have grown along the river valleys and in the zones of periglacial floods, and farther to the south islands of forest vegetation could have occurred. On the whole, the conditions for growth of vegetation were much harsher than during Zyryanka time; practically all of central Siberia was a periglacial zone. The course of climatic changes during the deglaciation stage of Sartan Glaciation cannot be characterized because of an almost total lack of organic remains in deposits of that time.

The Late Pleistocene paleogeography of north-central Siberia described above is only one of the possible interpretations of the data available at the present level of knowledge. Some data contradict this interpretation. On the western shore of Lake Taimyr, a section of a 30-m lake terrace is characterized by a series of successive dates from 30,000 yr B.P. to those of Holocene ages. The section is represented by horizontally stratified monochromatic aleurites and fine sands with plant detritus (moss and grasses). The section has no breaks in sedimentation; its presence in the far north of the North Siberian Lowland and the Byrranga Piedmont contradicts the concept of the spreading of the Sartan ice sheet to the lowland from the north during either the Syntabul'skiy or the Upper Taimyr glaciation phase. In that case, it may be inferred that over the Taimyr in Sartan time only alpine glaciation developed in the highest northeastern part of Byrranga Mountains, where traces of local valley glaciation have been preserved.

References

Andreyeva, S. M. (1978). Zyryanka Glaciation in north-central Siberia. *USSR Academy of Sciences Izvestiya seriya geograficheskaya* 5, 72-78.

Andreyeva, S. M. (1980). The North Siberian Lowland in Karginskiy time: Paleogeography, radiocarbon geochronology. *In* "Geochronology of the Quaternary Period" (I. K. Ivanova, and N. V. Kind, eds.), pp. 183-91. Nauka Press, Moscow.

Bardeyeva, M. A., and Isayeva, L. L. (1980). Identification of the Murukta horizon in Quaternary deposits of northern Siberia. *USSR Academy of Sciences Doklady* 251, 169-73.

Bardeyeva, M. A., Isayeva, L. L., Andreyeva, S. M., Kind, N. V., Nikol'skaya, M. V., Pirumova, L. G., Sulerzhitskiy, L. D., and Cherkasova, M. N. (1980). Stratigraphy, geochronology, and paleogeography of the Late Pleistocene and Holocene of the north of the Central Siberian Highland. *In* "Geochronology of the Quaternary Period" (I. K. Ivanova and N. V. Kind, eds.), pp. 198-208. Nauka Press, Moscow.

Isayeva, L. L. (1972). Marginal glacial formations of the northwestern Central Siberian Highland. *In* "Marginal Formations of Continental Glaciations" (G. I. Goretskiy, D. I. Pogulyayev, and S. M. Shik, eds.), pp. 205-11. Nauka Press, Moscow.

Isayeva, L. L., Kind, N. V., Andreyeva, S. M., Ivanenko, G. V., Nikol'skaya, M. V., Sulerzhitskiy, L. D., and Fisher, E. L. (1980). Geochronology and paleogeography of the Late Pleistocene of the North Siberian Lowland based on radiocarbon data. *In* "Geochronology of the Quaternary Period" (I. K. Ivanova and N. V. Kind, eds.), pp. 191-98. Nauka Press, Moscow.

Isayeva, L. L., Kind, N. V., Kraush, M. A., and Sulerzhitskiy, L. D. (1976). Age and structure of marginal formations at the northern foot of the Putorana Plateau. *Bulletin of the Commission for the Study of the Quaternary Period* 45, 117-23.

Kind, N. V. (1974). "Geochronology of the Late Anthropogene Based on Isotope Data." Nauka Press, Moscow.

Kind, N. V., Kolpakov, V. V., and Sulerzhitskiy, L. D. (1971). Age of the glaciation of Verkhoyan'ye. *USSR Academy of Sciences, Izvestiya seriya geologicheskaya* 10, 135-44.

Kolpakov, V. V., and Belova, A. P. (1980). Radiocarbon dating in the glacial region of Verkhoyan'ye and its framing. *In* "Geochronology of the Quaternary Period" (I. K. Ivanova and N. V. Kind, eds.), pp. 235-38. Nauka Press, Moscow.

Saks, V. N. (1953). "The Quaternary Period in the Soviet Arctic." Proceedings of the Scientific Research Institute of the Geology of the Arctic 77.

Strelkov, S. A. (1957). Stratigraphy of Quaternary deposits of northwestern Siberia and the Taimyr Lowland. *In* "Proceedings of the Interdepartmental Conference on the Study of Unified Stratigraphic Schemes of Siberia" (V. N. Saks, ed.), pp. 373-82. Gostoptekhizdat, Leningrad.

Troitskiy, S. L. (1967). New data on the last cover glaciation of Siberia. *USSR Academy of Science, Doklady* 174, 1409-12.

CHAPTER 4

Late Pleistocene Mountain Glaciation in Northeastern USSR

V. G. Bespalyy

Mountain glaciation in the northeastern USSR is dealt with at length in the works of S. V. Obruchev, D. M. Kolosov, N. Saks, N. A. Shilo, Yu. N. Trushkov, A. P. Vas'kovskiy, O. B. Kashmenskaya, Z. V. Khvorostova, Yu. P. Baranova, S. F. Biske, Yu. I. Gol'dfarb, O. M. Petrov, and many other authors. Despite the enormous volume of research already carried out, there is as yet no consensus on the number and scale of Pleistocene glaciations. During the last decade, radiocarbon dating has shown that well-defined depositional and erosional forms were produced by Late Pleistocene glaciations. This has made it possible to chart glacier forms and to establish the thickness and scale of glaciations. Such work was done by the Cenozoic Stratigraphy and Geomorphology Laboratory of the Joint Northeastern Scientific Research Institute in accordance with studies under the International Geological Correlation Program, Quaternary Glaciations of the Northern Hemisphere.

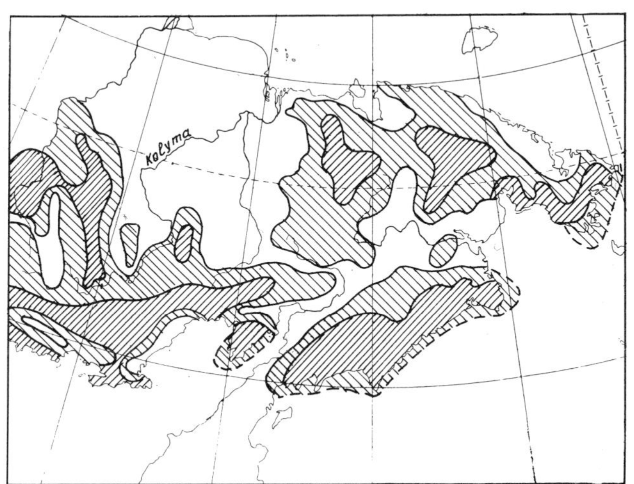

Figure 4-1. Areas affected by Zyryanka Glaciation (light shading) and by Sartan Glaciation (dark shading).

Zyryanka Glaciation

Traces of the first Late Pleistocene glaciation (Zyryanka, according to Saks, 1948) are widely represented in the region (Table 4-1). In the Yano-Kolymskiy Highland they are known as the first Bokhapcha Glaciation (A. P. Vas'kovskiy's term), on Chukotka as the Vankarem Glaciation (Petrov, 1966), and in northeastern Priokhot'ye as the Tylkhoy Glaciation (Bespalyy, 1974). The glaciation encompassed practically all mountain masses above 1000 m (Figure 4-1). Judging from altitudes on cirque floors, the snow line underwent considerable fluctuation. In northern Priokhot'ye, it descended to 400 m (Voskresenskiy and Voskresenskiy, 1977). Even lower was the snow line on the coast of eastern Chukotka, where there are cirques with floor altitudes of 60 to 80 m above sea levels. In the interior of the continent, the snow line gradually rose and reached an elevation of 1450 to 1850 m in the Chersk mountain system (Miller, 1976).

In the southeastern part of Chukotka, in the eastern part of the Koryaki Upland and Kamchatka, and in northeastern Priokhot'ye, the glaciers advanced onto the exposed shelf. Marginal deposits of this glaciation are now submerged. Eroded into them are submerged shorelines created during warm Late Pleistocene intervals, when the sea level was 15 to 45 m below the present level.

The intensity of the first Late Pleistocene glaciation was different in various regions. The heaviest glaciation occurred in Verkhoyan'ye and the Cherskogo mountain system, which acted as orographic barriers to the westward transport of moisture. Most of the valleys were filled with ice. Only narrow ridges, isolated uplands, and the highest and steepest subdivides remained free from ice. From the location of terminal moraines in intermountain areas it is evident that the margins of individual valley glaciers merged, forming thick covers on the piedmonts. Piedmont glaciers were also characteristic of other regions (Bespalyy and Davidovich, 1974).

Table 4-1. Comparison of the Main Events during the Holocene and Late Pleistocene of the Mountain Regions of the Northeastern USSR

Geologic Age	Yana-Kolyma Highland	Chukchi Peninsula	Northeastern Okhotsk Region
Q_{IV}	Recent glaciation; degradation of glaciers	Transgression	Transgression
Q^{4}_{III}	Bokhapcha-II (Sartan) mountain-valley glaciation; $^{14}C = 13{,}860 \pm 690$ yr B.P. (MAG-16), $21{,}700 \pm 800$ yr B.P. (MAG-45)	Iskaten' mountain-valley glaciation; $^{14}C = 17{,}200 \pm 700$ yr B.P. (MAG-44)	Khaymikinskiy mountain-valley glaciation; $^{14}C = 19{,}000 \pm 1100$ yr B.P. (MAG-38), $21{,}020 \pm 400$ yr B.P. (MAG-178)
Q^{3}_{III}	Karginskiy Interglaciation (Interstade?); $^{14}C = 26{,}840 \pm 250$ yr B.P. (MAG-28), $32{,}940 \pm 620$ yr B.P. (MAG-29), $37{,}000 \pm 1800$ yr B.P. (MAG-37)	Karginskiy Interglaciation $^{14}C = 40{,}800 \pm 800$ yr B.P. (GIN-378), $42{,}800 \pm 400$ yr B.P. (GIN-143)	Penzhina Interglaciation; $^{14}C = 26{,}440 \pm 390$ yr B.P. (MAG-130), $30{,}770 \pm 370$ yr B.P. (MAG-210), $31{,}090 \pm 1580$ yr B.P. (MAG-47), $37{,}000 \pm 1500$ yr B.P. (MAG-30), $44{,}600 \pm 2600$ yr B.P. (MAG-41)
Q^{2}_{III}	Bokhapcha-I (Zyryanka) mountain-valley and piedmont glaciation	Vankarem mountain-valley glaciation	Tylkhoy mountain-valley and piedmont glaciation

In the continental part of the region, the intensity of glaciation decreased behind the orographic barriers, and alpine glaciation dominated, with individual glaciers as long as 100 to 120 km.

Karginskiy Interglaciation

The climate warmed about 50,000 years ago, and the glaciers retreated especially on the Maritime plains and in intermontane depressions. In the mountains, however, the glaciers shrank at a slow rate. In the Tobychan River valley, the glacier shrank to half (60 km) its maximum extent over a period of 12,000 to 13,000 years (Miller, 1976). The glaciers did not completely disappear in the mountains of the northeastern USSR during the Karginskiy Interglaciation (Zamoruyev, 1978). The problem of assessing mountain glaciation remains unsolved, for dates for interglacial deposits come mainly from submontane and intermountain areas.

In the Verkhoyansk Range, the Karginskiy Interglaciation was complicated by a brief expansion of the glaciers, lasting only 1000 to 2000 years, called the Zhigansk stage (Kind et al., 1971; Kolpakov, 1979; Kolpakov and Belova, 1980). Farther east, it is not possible to identify moraines of this stage. Moreover, peat bogs up to 8 m thick on the shore of Penzhina Bay have horizons of stumps dated at 32,000 to 31,000 yr B.P., which is close to the age of the Zhigansk stage. About 32,000 years ago, birch forests grew in a region that is now treeless (Alekseyev et al., 1974; Bespalyy and Davidovich, 1974). One can only hypothesize that this discrepancy resulted from different orographic conditions.

Sartan Glaciation

The last, or Sartan (Bokhapcha-II, Iskaten', Khaymikinskiy; see Table 4-1) glaciation was not manifested everywhere in mountain regions. Its age, established by dating periglacial deposits, ranges from 26,000 to 11,000 yr B.P. This glaciation occupied approximately one-half the area of the preceding glaciation and had a nodal character; that is, it spread in isolated sections around the elevated parts of northern Priokhot'ye, the Cherskogo mountain system, the Koryaki Highland, and Chukotka. Sartan glaciers never reached into the foothills. Cirque glaciers and valley glaciers were primarily developed. Only in the Cherskogo mountain system and in the eastern Koryaki Highland was the glaciation of reticulate character. The extent of valley glaciers seldom exceeded 20 to 25 km. Sartan glaciers completely inherited the sculptured forms produced by the Zyryanka Glaciation (Shilo, 1970). In regions not occupied by glaciers, periglacial conditions prevailed.

Holocene

The warming at the start of the Holocene led to the wastage of the glaciers. By the time of the Holocene climatic optimum (7000 to 5000 yr B.P.), the glaciers probably had completely disappeared.

Modern glaciers are associated with the highest mountain masses (Table 4-2). From the relationships of the volume of the sediment produced by the ice masses and from the magnitudes of ablation, one can infer highly active glaciers in the Koryaki Highland and less active glaciers in the Cherskogo mountain system and the Suntar-Khayata Range. In addition, judging from the presence of Holocene moraines possibly related to the "Little Ice Age," a reduction of glaciation is taking place at the present time. In the view of Shilo and Vinogradov (1970), recent glaciers of the Northeast Province are not relicts of an earlier glaciation; instead they developed normally in a favorable orographic-climatic situation.

A study of the character, scale, and intensity of Late Pleistocene glaciations shows that glacier formation took place under the influence of the western and southeastern

Table 4-2. Characteristics of Recent Glaciation in the Mountains of the Northeastern USSR

Region	Number of Glaciers	Total Glaciation Area (km^2)	Height of Firn Line (m)	Glacier Shrinking Rate (mm/year)
Suntar-Khayata Range	208	206	2300-2400	3-7
Ranges of Cherskogo mountain system	223	147	2100-2200	No data
Koryaki Highland	497	194	1600-1700	15-17

Source: Data from Shilo and Vinogradov, 1970.

transport of atmospheric precipitation. The glaciation was manifested most intensively in mountain masses that formed orographic barriers in the path of moist air masses. During the Late Pleistocene, as well in earlier epochs, a general directional and cyclic change of climate in the direction of a cooling and drying took place. These changes were manifested in the periodicity of glaciations, in a successive reduction of their scale and intensity, and in the migration of snow lines.

References

Alekseyev, M. N., Bespalyy, V. G., Geptner, A. R., Lozhkin, A. V., and Chemekov, Yu. F. (1974). The Northeast of the USSR. *In* "Geochronology of the USSR," vol. 3, "The Latest Stage" (V. A. Zubakov, ed.), pp. 238-47. Nedra Press, Leningrad.

Bespalyy, V. G. (1974). Stratigraphic scheme of Pleistocene deposits in Kamchatka. *In* "Aspects of Stratigraphy of Kamchatka's Pleistocene." Proceedings of the Northeastern Joint Scientific Research Institute 59, pp. 109-31.

Bespalyy, V. G., and Davidovich, T. D. (1974). Stratigraphic regions of Kamchatka's Pleistocene. *In* "Aspects of Stratigraphy of Kamchatka's Pleistocene." Proceedings of the Northeastern Joint Scientific Research Institute 59, pp. 26-82.

Kind, N. V., Kolpakov, V. V., and Sulerzhitskiy, L. D. (1971). Age of Verkhoyan'ye's glaciations. *USSR Academy of Sciences Izvestiya, seriya geologicheskaya* 10, 135-44.

Kolpakov, V. V. (1979). Glacial and periglacial relief of the Verkhoyansk glacial region and new radiocarbon datings. *In* "Regional Geomorphology of Newly Developed Areas" (A. I. Muzis, ed.), pp. 83-97. Geographical Society of the USSR, Moscow.

Kolpakov, V. V., and Belova, A. P. (1980). Radiocarbon datings in the glacial region of Verkhoyan'ye and its framing. *In* "Geochronology of the Quaternary Period" (I. K. Ivanova and N. V. Kind, eds.), pp. 230-35. Nauka Press, Moscow.

Miller, V. G. (1976). Two stages of Late Pleistocene glaciation in the upper course of the Indigirka River (El'gi River basin). *Geomorfologiya* 1, 90-94.

Petrov, O. M. (1966). "Stratigraphy and Fauna of Marine Mollusks of the Chukotka Peninsula's Quaternary Deposits." Proceedings of the Geological Institute of the Academy of Sciences of the USSR 155.

Saks, V. N. (1948). "The Quaternary Period in the Soviet Arctic." Proceedings of the Arctic Scientific Research Institute 201.

Shilo, N. A. (1970). Relief and geological structure. *In* "The North of the Far East" (N. A. Shilo, ed.), pp. 21-84. Nauka Press, Moscow.

Shilo, N. A., and Vinogradov, V. V. (1970). Recent glaciation. *In* "The North of the Far East" (N. A. Shilo, ed.), pp. 150-64. Nauka Press, Moscow.

Voskresenskiy, S. S., and Voskresenskiy, Í. S. (1977). Maximum glaciation of mountain structures of the Northeast of the USSR. *In* "Relief and Landscapes" (N. A. Gvozdetskiy and A. I. Spiridonov, eds.), pp. 72-85. Moscow State University Press, Moscow.

Zamoruyev, V. V. (1978). Problems of study of Quaternary glaciation of mountains in the east of the USSR. *Proceedings of the All-Union Geological Institute* (new series) 297, 100-112.

CHAPTER 5

Late Pleistocene Glaciation of the Arctic Shelf, and the Reconstruction of Eurasian Ice Sheets

A. A. Velichko, L. L. Isayeva, V. M. Makeyev, G. G. Matishov, and M. A. Faustova

The problems associated with judging the extent of Late Pleistocene glaciation on Eurasia's Arctic Shelf are closely related to the problems encountered in reconstructing the entire system of ice sheets in northern Eurasia during the Late Pleistocene. Marine-geologic studies at many points of the present Eurasian shelf have resulted in the discovery of deposits and relief forms similar to those formed by glaciers on the continent. Detailed geologic mapping on the European and Asian Arctic Shelves in the 1970s—including echo sounding and photography of the bottom, large-scale profiling, and bathymetric mapping of the surface—has confirmed the hypotheses previously advanced by several authors (V. A. Obruchev, N. N. Urvantsev, S. A. Yakovlev, V. N. Saks, and others) for the existence of past ice sheets within the confines of the present shelf.

Geomorphologic indications of the extent of ice sheets on the shelf include systems of marginal glacial formations as well as marginal longitudinal and transverse trenches and submerged fjords. Marginal trenches and fjords have a stepped longitudinal profile, overdeepened depressions, and U-shaped cross sections, all of which indicate glacial erosion. End moraines are traced as concentric chains over the bottom of trenches and submerged plateaus. Their relief reaches 100 m, and their narrow transverse profile and depths above them range from 50 to 350 m. The ridges are composed of dense boulder clays and loams and are covered with a thick pebble or boulder mantle (Dibner et al., 1971; Blazhchishin and Lin'kova, 1977; Matishov, 1976a, 1976b, 1980). According to G. G. Matishov's studies, the ridges include push moraines, frontal moraines, and coastal morainic ridges. A system of erosional-depositional formations was formed by fluvioglacial flows. In the sea-floor topography, a zone of predominant excavation, a zone of nonuniform glacial accumulation, and a periglacial and periglacial-marine zone can be clearly distinguished.

The dimensions of the Late Pleistocene glaciation of the Arctic Shelf depend on the age determination of the marginal formations. Their dating is based on geomorphologic characteristics and on correlation with dated marginal zones on the coast. Therefore, the reconstructions made thus far are hypothetical.

Barents Sea Shelf

Reconstructions of Late Pleistocene glaciation in the European (Barents Sea) sector of the Arctic are essentially reduced to two variants: (1) an independent Barents Sea ice sheet merged with the Scandinavian ice sheet, forming a dome on the central low-lying shelf zone (Grosswald, 1977, 1980; Schytt et al., 1968), and (2) a more limited Barents Sea glacier on the Arctic Shelf—a concept that has many adherents: V. D. Dibner, G. G. Matishov, A. A. Velichko, V. G. Khodakov, M. A. Faustova, and others. It is assumed that the centers of such glaciation were confined to elevated relief forms—islands and high plateaus (Velichko and Khodakov, 1979). Danilov (1971) and others hold the extreme view that there was no shelf glaciation during the Late Pleistocene.

We briefly discuss the support of modern data for each of the proposed hypotheses. The chief arguments in favor of a single dome in the central low-lying zone of the Barents Sea Shelf spreading out toward its periphery are the saucer-shaped shelf, which is attributed to the weight of the ice sheet, and the pattern of uplift isolines. However, studies made during the 1970s indicate no relationship between the isostatic effect and uplift isolines for the glacial episode of 21,000 to 18,000 years ago (Boulton, 1979). The absence of such a relationship was also indicated earlier by Saks (1948), Strelkov (1968), and Dibner (1970) as well as by Lavrushin (1970) and Lazukov (1972). The hypothesis cited is inconsistent with paleoglaciologic calculations, which indicate that there was not enough time to form a single dome in the central part of the Barents Sea Shelf (Velichko and Khodakov, 1979). In addition, detailed echo sounding has failed to detect traces of glacial activity in the shallows of the Pecheromorskiy Plain (Matishov, 1977), and there are no traces of a linkage of the Scandinavian and Spitzbergen ice sheets. (Data on the limited dimensions of the Spitzbergen ice sheet during the Late Pleistocene are discussed later in this chapter.)

According to the second reconstruction, the Scandinavian and Spitzbergen glaciers in the Late Pleistocene reached the outer edge of the shelf in the west. Between them was a region of shelf glaciation, inside which probably existed a small, independent ice sheet whose center was located on the Medvezh'ye Nadezhdinskiy Upland. The Scandinavian ice sheet was connected to the Lofoten and Ponoy local ice sheets, and it extended northward along marginal trenches in numerous lobes. In the east, the Spitzbergen and Franz Josef Land ice sheets merged with the Novaya Zemlya ice sheet, which in turn merged southward with the Vaygach and Pay-Khoy ice sheets in the Urals, forming lobes that reached the Pechora Lowland in the south. The glacier limits are shown on the shelf by concentric zones of marginal formations linked to the marginal zones of the Scandinavian, Spitzbergen, Novaya Zemlya, and Franz Josef Land ice sheets. On the Barents Sea Shelf, at least four concentric zones of marginal glacier formations can be traced to the periphery of the ice sheets of Scandinavia and Barents Sea islands. These complexes are divided into an outer (pre-Valdai) and an inner (Valdai) complex.

The boundary of Late Pleistocene glaciation on the outer shelf of Norway is represented by the end-moraine complex Egga 1, which is one of five heterochronous glacial complexes of Late Pleistocene glaciation on Scandinavia's outer shelf, referred to 21,000 to 17,000 yr B.P. The complex is up to 50 km wide and is made up of two to four ridges with a configuration reflecting the lobate structure of the ice sheet that descended on the shelf along marginal trenches. The thickness of the deposits forming the ridges is 200 m or more. The Late Pleistocene age is indicated by the fresh relief of the ridges and the presence of exclusively Holocene sediments on their surface (Holtedahl and Sellevoll, 1972). The similarity of the glacial forms on the floor of the Barents Sea and on the coast of northern Norway enabled the adherents of the second hypothesis to draw the boundary of the ice sheet's maximum extent from the Egga 1 end moraine to ridges with depths of 300 to 400 m, located on the Medvezh'ye Lowland and near the western shore of Spitzbergen (Matishov, 1980). The Novaya Zemlya complex of marginal formations at depths of 200 to 300 m in the eastern part of the Central Lowland, and neighboring submarine elevations and marginal trenches, are also referred to the maximum stage of Late Pleistocene glaciation. Such a reconstruction is consistent with currently available geologic material and with the concept that glaciation centers are confined to highlands where snow accumulated (Velichko, 1979). It is also consistent with calculations on the development phases of the Late Valdai Scandinavian glacier (Khodakov, 1973). New data indicating limited glaciation of contiguous sectors of the Arctic (Canada, Greenland) and certain islands of the European and Asian sectors do not yet permit a different reconstruction for the glaciation of the Arctic's European sectors, but already they contradict even the version discussed above.

Thus, radiocarbon dating of bottom sediments of Lake Endlevatn on Andya Island (Vester Aalen Archipelago) on Norway's outer shelf, located on the proximal side of Egga 1 ridges, are 18,000, 18,700, and 19,000 yr B.P. (Vorren, 1978). These dates probably attest to the modest dimensions of the Late Pleistocene glacial maximum, with a date not older than 18,000 yr B.P., as assumed by Grosswald (1977), who regarded the Egga 1 ridges as representing one of the deglaciation stages of the shelf's Late Pleistocene glaciation, as did Andersen (1968, 1976). The data on Lake Endlevatn are consistent with conclusions about the glaciomarine origin of deposits on the Norwegian shelf north of Tromsö, for which a date of 13,350 yr B.P. was obtained from shells. This conclusion was reached on the basis of a detailed lithologic study of sediments (Vorren et al., 1978). Changes in ice volume during the Weichselian, calculated for western Norway by Miller and Mangerud (1980) by means of the amino-acid method, showed that during a major portion of the Weichselian the Norwegian coast was free of ice. Gradually the dimensions of the Late Pleistocene glacier of the Spitzbergen Archipelago shrank. Well known from the literature are sections in marine terraces in which Holocene sediments rest continuously on marine sediments radiocarbon-dated at 40,000 to 26,000 yr B.P. or are separated from them only by a bed of colluvium (Blake, 1961; Salvigssen, 1977; Feyling-Hanssen, 1965; Troitskiy and Punning, 1979). The extent and radiocarbon dating of submarine and continental end moraines led to the conclusion that the ice did not totally cover the coast and that nunataks were very common inside the area occupied by the ice (Salvigssen, 1977). Thus, the Late Pleistocene glacier limit may have been close to the coast.

According to the concept of a limited glacial extent on the Barents Sea Shelf, the continental ice there started to contract with the Pomeranian Vepsovo stage around 15,000 yr B.P. At the same time, the band of shelf glaciers and drifting icebergs increased (Matishov, 1980). Morainic ridges were formed then on intrashelf cuestas, as were ridges on subbathyal plains (Disko and others). Starting at 13,500 yr B.P., the continental ice sheet was within the inner shelf, and during the Younger Dryas it terminated in valley glaciers that scarcely reached beyond the coast islands, as indicated by numerous radiocarbon dates on moraines in the mouths of fjords and on the islands and the coast.

Kara Sea Shelf

Data on the glaciation of the Kara Sea Shelf are scarce. Hypotheses about the formation of the Late Valdai ice sheet on the Kara Sea Shelf and its expansion onto the northern margins of the Asian continent have been actively developed in the last few years by Grosswald (1977, 1980). He reconstructed an ice sheet of up to 2500 km in diameter, with its center located in the northern part of the Yamal Peninsula and adjoining parts of the shelf north of the Tazovskiy, Gydan, and Taimyr Peninsulas. The ice linked up with the Barents Sea ice sheet to form a single ice sheet in the area of the Svyataya Anna Trough. The reconstruction was based mainly on an analysis of marginal glacial formations on the Taimyr Peninsula, the Yenisey Valley, and western Siberia; on data about the transport of detritus from the northern coastal areas of the Taimyr Peninsula across the Byrranga Mountains into the North Siberian Lowland and from the shelf into western Siberia; and on data indicating the development of large proglacial lakes in western Siberia. However, there are facts that argue against such a reconstruction. In the area of the Central Kara Upland and the Svyataya Anna Trough, according to Lastochkin (1977; Lastochkin and Fedorov, 1978), there are no indications of glacial action but only erosional and erosional-tectonic relief forms. According to Dibner (1970), relief forms similar to marginal glacial formations on dry land are known on the floor of the Kara Sea Shelf and east of the Novaya Zemlya Depression along the shores of Novaya Zemlya and farther east on the Yamal-Gydan Shallows. These formations may outline the limits of two different ice sheets—the Novaya Zemlya and the Taimyr-Severnaya Zemlya (Figure 5-1). Moreover, in the reconstruction under consideration, the southern boundary of the Kara ice sheet on the North Siberian Plain is drawn with insufficient reliability. This boundary is drawn along discontinuous remnants of glacial ridges, whose correlation and age have not been adequately substantiated.

Also cautioning against such a reconstruction are the radiocarbon dates of 16,000 to 15,000 yr B.P. from the west coast of the Yamal Peninsula (Figure 5-1), which were obtained from buried peat located in the upper levels of terrace III (Danilov, 1980). The data are still too scarce and the concepts still too insufficiently substantiated to support a long lacustrine transgression in western Siberia, which should exist if a single thick ice sheet existed on the Kara Sea Shelf.

A different reconstruction of Late Valdai ice sheets in northern Asia was proposed by Velichko (1979). Its basis is the hypothesis that ice sheets started in highlands on the northern continental margin (Putorana Plateau, Polar Ural, Byrranga Mountains) and on the islands of Novaya Zemlya and Severnaya Zemlya. At the glacial maximum, the glaciation center was displaced from the Byrranga Mountains to the north coast of the Taimyr Peninsula and adjoining parts of the Kara Sea Shelf, and the North Taimyr and the Severnaya Zemlya ice sheets merged into a single cover. Such reconstructions were made previously by Strelkov (1968) for the Early Valdai Glaciation. Two other ice sheets, formed over the Polar Ural and Novaya Zemlya, also merged into a single cover at the glacial maximum. Their expansion to the east was recorded by submerged glacial forms on the eastern slope of the Novaya Zemlya Depression and by moraines in the southwestern part of the Yamal Peninsula. The boundary of the Taimyr-Severnaya Zemlya cover in the west is also shown by glacial forms indicated by Dibner (1970) in the Yamal-Gyda Shallows of the Kara Sea Shelf. The Dzhangodo-Syntabul'skiy-North Kokora Ridge of the North Siberian Lowland is taken as the southern boundary of the ice sheet. The Polar Ural-Novaya Zemlya and Taimyr-Severnaya Zemlya covers did not link up, leaving the northern part of the Kara Sea Shelf, the Svyataya Anna Trough, and the Central Kara Upland free of ice.

Such a reconstruction is in good agreement with the aforementioned data reported by Lastochkin (1977) and Dibner (1970) on the structure of the Kara Sea Shelf, and it accounts for the absence in western Siberia of any distinct end moraines over a long stretch east of the Ob' River's mouth. It also accounts for the southward transport of detritus from northern Taimyr across the Byrranga Mountains, as well as for the geographic confinement and configuration of the Mokoritto-Upper Taimyr glacial ridges. On the whole, even in this reconstruction, all the known data have not been logically or definitively interpreted. On the one hand, still unexplained is the northernmost end moraine on the north coast of Taimyr; this ridge could have been formed only by an ice sheet located on the Kara Sea Shelf. Without sufficient proof, the glacial origin and Late Valdai age of part of the ridge forms in northwestern Siberia is doubtful, for they are of Late Valdai age. The complexity of these problems is discussed in the chapter dealing with western Siberia (Chapter 2). In such a reconstruction, a major ice advance far to the south up the Yenisey Valley is not very understandable either, although the moraine left by such ice is radiocarbon-dated as Late Valdai in age (Kind, 1974; Troitskiy, 1967).

On the other hand, even such a "minimalist" reconstruction may yield an exaggerated representation of the dimensions of glaciation during the Late Valdai. Thus, a section of Cape Sabler, mentioned in the chapter dealing with central Siberia (Chapter 3), contradicts the hypothesis of an expansion of the Sartan (Late Valdai) ice sheet from the north to the lowland both in the maximum and Upper Taimyr glaciation phases. One can, therefore, postulate that in the entire Taimyr during Sartan time only mountain-valley glaciers developed and that these were confined to the highest northeastern part of the Byrranga Mountains, where traces of local valley glaciation are preserved.

This concept is consistent with data on the emergence of marine Karginskiy deposits locally on the bottom of glacial depressions in the rear part of the Dzhangodo-Syntabul'skiy-North Kokora Ridge, for these areas lie only under a layer of lacustrine deposits. A slight glaciation of Severnaya Zemlya at the Late Valdai maximum (20,000 to 18,000 yr B.P.), similar in size to the recent glaciation, is supported by data obtained there in the last few years. Thus, radiocarbon dating of the mammoth bone observed on many islands of the archipelago at the margins of recent glaciers has shown that the glaciers of 24,000 to 19,000 yr B.P. did not surpass recent ones in size (Makeyev et al., 1979; Arslanov et al., 1980). The glacial maximum of the Late Valdai on Severnaya Zemlya was at 18,000 to 14,000 yr B.P. However, even so, one must not postulate the formation of a continuous cover over the entire shelf, for the glaciers were smaller than recent ones as early as 11,500 yr B.P.

All the contradictions and interpretations of existing data and differences in reconstructions of the entire glacier

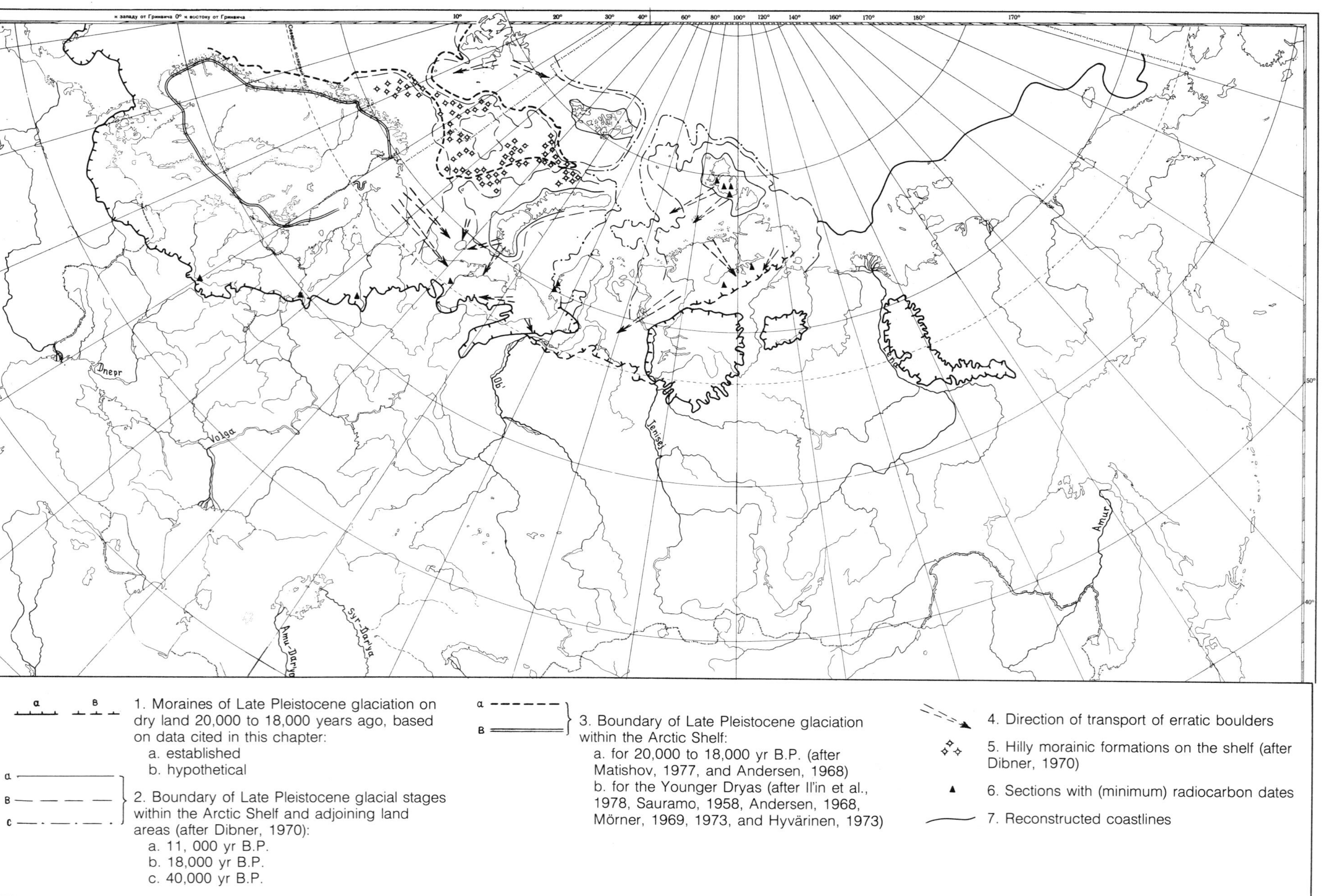

Figure 5-1. Data on the extent of the Late Valdai ice sheet. (Map compiled by A. A. Velichko.)

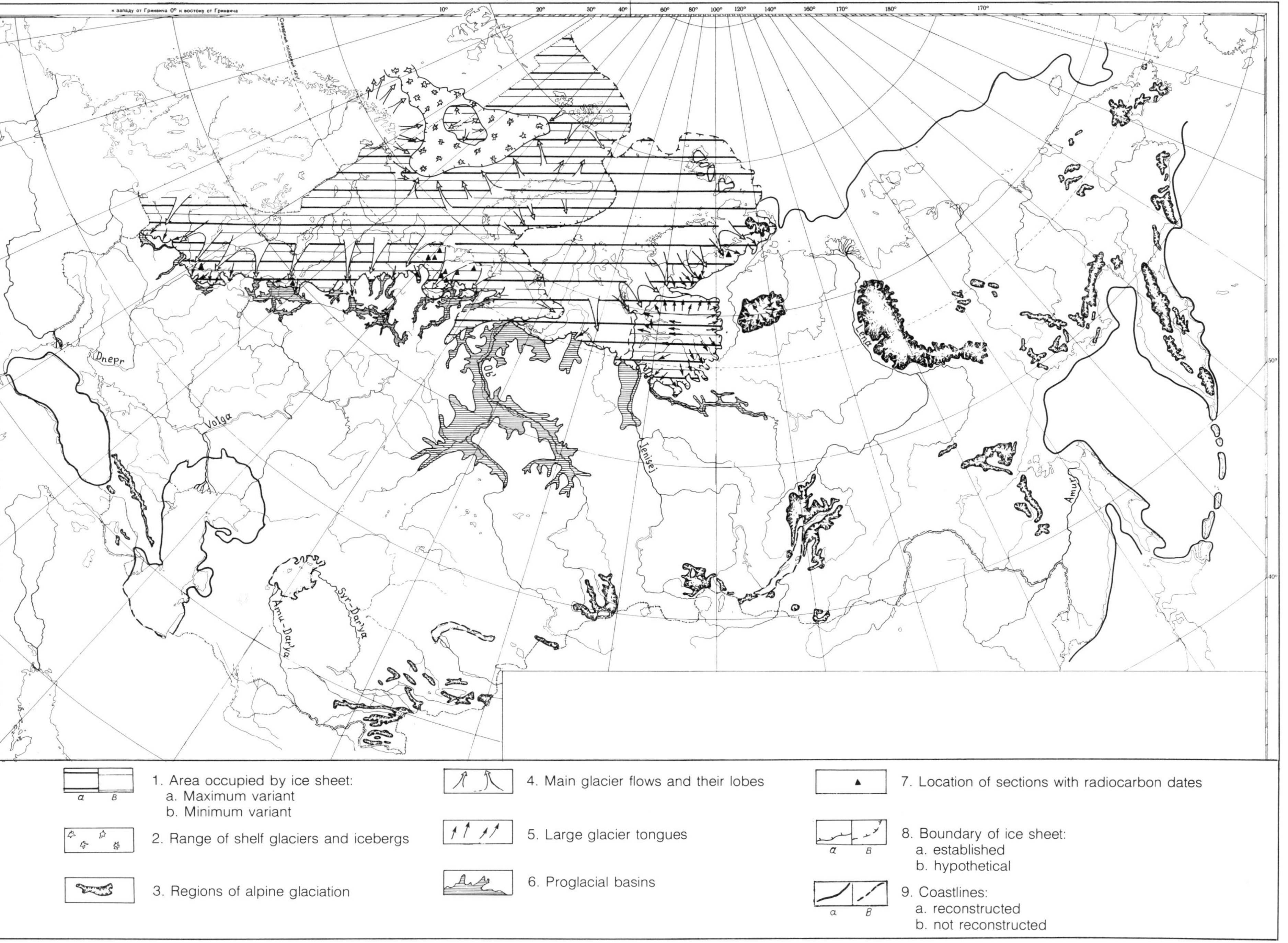

Figure 5-2. Late Valdai Glaciation of Eurasia. (Map compiled by A. A. Velichko, L. L. Isayeva, V. M. Makeyev, G. G. Matishov, and M. A. Faustova.)

system described here result from the insufficient study of the northern regions of western and central Siberia, and in particular the Kara Sea Shelf. A the present time, this insufficiency interferes with the development of an adequately substantiated, unambiguous model of northern Asian ice sheets. That is why the appended map (Figure 5-2) indicates two variants of possible glacial boundaries in the Late Valdai for the north of western and central Siberia and the Kara Sea Shelf. The maximum variant assumes that the glaciers of the Kara Sea Shelf and the northern areas of dry land were almost completely covered by glaciers (Figure 5-2), and the minimum variant specifies the development of a continuous Novaya Zemlya-Polar Ural cover, an ice sheet on the Putorana Plateau, small ice sheets on the Severnaya Zemlya Islands, and mountain-valley glaciers in the mountains of the eastern Taimyr Peninsula and on the Anabar Highland.

References

Andersen, B. G. (1968). "Glacial Geology of Western Troms, North Norway." Norges Geologiske Undersökelse 256, 1-160.

Andersen, B. G. (1976). Map of end moraines and ice-front lines in northern Europe. Unpublished data of CLIMAP Project.

Arslanov, Kh. A., Makeyev, V. M., Baranovskaya, O. F., Malakhovskiy, D. B., and Tertychnaya, T. V. (1980). Geochronology and certain aspects of paleogeography of the second half of the Late Pleistocene of Severnaya Zemlya. *In* "Geochronology of the Quaternary Period" (I. K. Ivanova and N. V. Kind, eds.), pp. 108-76. Nauka Press, Moscow.

Blake, W., Jr. (1961). Radiocarbon dating of raised beaches in Nordaustlandet, Spitsbergen. *In* "Geology of the Arctic" (G. O. Raasch, ed.), pp. 133-45. Proceedings of the First International Symposium on Arctic Geology. University of Toronto Press, Toronto.

Blazhchishin, A. I., and Lin'kova, T. I. (1977). Pliocene glaciation of the Barents Shelf. *USSR Academy of Sciences, Doklady* 236, 696-99.

Boulton, G. S. (1979). A model of Weichselian glacier variation in the North Atlantic region. *Boreas* 8, 373-95.

Danilov, I. D. (1971). Age and principles of stratigraphy of the latest deposits of marine plains of the north of Eurasia. *Moscow State University, Geografiya, Vestnik* 5, 56-61.

Danilov, I. D. (1980). Nature of dislocations in the Pleistocene deposits of the North. *Litologiya i poleznyye iskopayemye* 5, 49-62.

Dibner, V. D. (1970). Geomorphology. *In* "The Soviet Arctic" (L. S. Govorukha and Ya. Ya. Gekkel', eds.), pp. 59-107. Nauka Press, Moscow.

Dibner, V. D., Kotenev, B. M., and Zaderman, M. L. (1971). Geologic-geomorphic observations of the Barents Sea floor from the SEVER-I hydrostat. *In* "Geology of the Sea" (N. N. Lapina, ed.), vol. 1, pp. 140-44. Scientific Research Institute of Arctic Geology Press, Leningrad.

Feyling-Hanssen, R. W. (1965). Shoreline displacement in central Spitzbergen. *Norsk Polarinstitutt, Meddelelser* 93, 1-5.

Grosswald, M. G. (1977). The latest Eurasian ice sheet. *In* "Glaciological Studies: Chronicle and Discussions" (G. A. Avsyuk and V. M. Kotlyakov, eds.), no. 30, pp. 45-60. Moscow.

Grosswald, M. G. (1980). Cover glaciers of continental shelves. Doctoral Dissertation, Institute of Geography, Moscow State University.

Holtedahl, H., and Sellevoll, M. (1972). "Notes on the Influence of Glaciation on the Norwegian Continental Shelf Bordering on the Norwegian Sea." Ambio Special Report 2, pp. 31-38.

Hyvärinen, H. (1973). The deglaciation history of eastern Fennoscandia: Recent data from Finland. *Boreas* 2, 85-102.

Il'in, V. A., Lukashov, A. L., and Ekman, I. M. (1978). Marginal glacial formations of western Karelia and their correlation with ridges of the Finnish Salpausselkä. *In* "Marginal Formations of Continental Glaciations" (V. G. Bondarchuk, ed.), pp. 96-108. Transactions of the Fifth All-Union Conference. Ukrainian Academy of Sciences, Institute of Geological Sciences. Naukova Dumka Press, Kiev.

Khodakov, V. G. (1973). Construction of a model of the European ice sheet, based on an actualistic approach. *In* "Paleogeography of Europe in the Late Pleistocene" (I. P. Gerasimov, ed.), pp. 79-107. USSR Academy of Sciences, Institute of Geography, Moscow.

Kind, N. V. (1974). "Geochronology of the Late Anthropogene from Isotope Data." Nauka Press, Moscow.

Lastochkin, A. N. (1977). Bottom relief of the Kara Sea. *Geomorfologiya* 2, 84-91.

Lastochkin, A. N., and Fedorov, B. G. (1978). Relief and latest history of the development of Eurasia's northern shelf. *Geomorfologiya* 3, 19-27.

Lavrushin, Yu. A. (1970). Problems of stratigraphy and paleogeography of Spitzbergen in the Late Pleistocene. *In* "The Arctic Ocean and Its Shores in the Cenozoic" (E. D. Tulmachev, ed.), pp. 53-56. Gidrometeoizdat Press, Leningrad.

Lazukov, G. I. (1972). "Anthropogene of the Northern Half of Western Siberia: Paleogeography." Moscow State University Press, Moscow.

Makeyev, V. M., Arslanov, Kh. A., and Garutt, V. Ye. (1979). Age of mammoths of Severnaya Zemlya and some aspects of Late Pleistocene paleogeography. *USSR Academy of Sciences, Doklady* 245, 421-24.

Matishov, G. G. (1976a). Certain problems of the geomorphology of glacial shelves. *In* "Geology and Geomorphology of the Quaternary Period of Northern European USSR" (G. S. Biske, ed.), pp. 22-31. Petrozavodsk State University, Petrozavodsk.

Matishov, G. G. (1976b). Glacial morphosculpture and characteristics of glaciation of the Barents Sea Shelf. *Geologiya Morya* 5, 20-26.

Matishov, G. G. (1977). Character of Pleistocene glaciation of the Barents Shelf. *USSR Academy of Sciences, Doklady* 232, 184-87.

Matishov, G. G. (1980). "The Glacial and Periglacial Relief of the Ocean Floor." Moscow State University, Moscow.

Miller, G. H., and Mangerud, J. (1980). Glacial-age temperature in western Norway deduced from amino acid diagenesis and the implications for ice-volume fluctuations. "American Quaternary Association, Sixth Biennial Meeting (Orono, Maine), Abstracts," p. 140. Institute for Quaternary Studies, University of Maine, Orono.

Mörner, N. A. (1969). The Late Quaternary history of the Kattegat Sea and the Swedish west coast: Deglaciation, shorelevel displacement, chronology, isostasy, and eustasy. Sveriges Geologiska Undersökning C-640, 1-487.

Mörner, N. A. (1973). Submarin kvartärgeologi i Östersjön; nagra preliminära resultat och framitidsaspekter. *Svenska Sällskapet för Antropologi och Geografi*, Ymer 1973, 91-96.

Saks, V. N. (1948). "The Quaternary Period in the Soviet Arctic." Trudy Arkticheskogo Instituta, Moscow and Leningrad.

Salvigssen, O. (1977). "Radiocarbon Datings and the Extent of the Weichselian Ice Sheet in Svalbard." Norsk Polarinstitutt, Årbok 1976, pp. 209-24.

Sauramo, M. (1958). Die Geschichte der Ostsee. *Annales Academiae Scientiarum Fennicae AIII (Geologica-Geographica)* 51, 1-522.

Schytt, V., Hoppe, G., Blake, W., and Grosswald, M. G. (1968). "Extent of the Würm Glaciation in the European Arctic." International Association of Scientific Hydrology, Commission of Snow and Ice, Publication 79, pp. 207-16.

Strelkov, S. A. (1968). Glaciation centers in the north of Siberia: Conditions and stages of development. *In* "Neogene and Quaternary Deposits of Western Siberia" (V. N. Saks, ed.), pp. 88-98. Nauka Press, Moscow.

Troitskiy, L. S., and Punning, Ya. M. K. (1979). Pleistocene glaciations on Spitzbergen. *In* "Glaciological Studies: Chronicle and Discussions" (G. A. Avsyuk, ed.), vol. 35, pp. 204-8. Moscow.

Troitskiy, S. L. (1967). New data on the latest cover glaciation of Siberia. *USSR Academy of Sciences, Doklady* 174, 1409-12.

Velichko, A. A. (1979). Problems of reconstruction of Late Pleistocene ice sheets on the territory of the USSR. *USSR Academy of Sciences Izvestiya, seriya geograficheskaya* 6, 12-26.

Velichko, A. A., and Khodakov, V. G. (1979). The problems of reconstructing Late Pleistocene ice sheets of the Arctic Shelf on the basis of specified data and paleoglaciological modelling. *In* "Quaternary Glaciations on the Northern Hemisphere" (V. Sibrava, ed.), pp. 221-35. Report 5 on the session in Novosibirsk (USSR). Geologicheskaya Sluzhba, Prague.

Vorren, K.-D. (1978). Late and Middle Weichselian stratigraphy of Andya, north Norway. *Boreas* 7, 19-38.

Vorren, T. O., Strass, I. F., and Lind-Hanssen, O. W. (1978). Late Quaternary sediments and stratigraphy on the continental shelf off Tromsö and West Finnmark, northern Norway. *Quaternary Research* 10, 340-65.

Mountain Glaciation

CHAPTER 6

Mountain Glaciation in the USSR in the Late Pleistocene and Holocene

L. R. Serebryanny

Mountains account for one-third of the total area of the USSR, and they are concentrated in the southern peripheral regions. According to Meshcheryakov (1972), the surface of the USSR is comparable to an amphitheater open to the north and framed by marginal mountain systems. The latter include the Carpathian Mountains, Crimea, Caucasus, Kopetdag, Gissaro-Alay, Pamir, Tien-Shan, Altay, Sayany, the Baykal mountainous region, mountains of the Far East, Verkhoyano-Kolymskaya mountainous region, and mountains of the extreme Northeast.

The mountains frequently rise several kilometers above the level of the surrounding plains and are capped by glaciers. The area of recent mountain glaciation is significant. As part of the compilation of the "Catalogue of Glaciers of the USSR," it was calculated that, for example, glaciers of the Caucasus occupy 1424 km^2; the Ghissar-Alai, 2175 km^2; the Pamir, 7493 km^2; the Tien-Shan, 7214 km^2; and the Altay, 909 km^2.

The degree of glaciation depends largely on the landforms and on the absolute altitudes of the mountainous regions; in addition, the active influence of climatic factors is apparent. Particular emphasis should be placed on the location of a region in relation to the prevailing moisture-bearing air currents. Under present conditions, the glaciers of the southern mountain belt depend on westerly air transport, which is also manifested in the eastern regions, including the Verkhoyanskiye Range, the Suntar-Khayata Mountains, and the Chersky Range (Koreysha, 1963; Sheynkman, 1978). Thus, the Atlantic Ocean supplies moisture to the glaciers of most of Eurasia, whereas moisture from the Pacific Ocean reaches only the glaciers of extreme northeastern Asia (Kotlyakov, 1968).

The present situation, however, should not automatically be extended to the Pleistocene, when substantial changes may have occurred in the atmospheric circulation system (Velichko, 1980). Special cartographic studies made by N. A. Timofeyeva on data from the Italian National Atlas (Dainelli, 1938) show that in the Alps the thickest Würmian glaciers were located on the southern slopes, not on the western or northern ones as at the present time. This suggests that the moisture-bearing air currents that supplied the Alpine glaciers in the Würm differed sharply from those of the present.

In the study of the Quaternary history of mountain glaciation in the USSR, the geomorphologic approach, as influenced by the classic studies of A. Penck in the Alps (Penck and Brückner, 1901-1909), has predominated. Not only the methods but also the terminology of the Alpine school have frequently been employed in the USSR. On the one hand, this favored interregional correlations, but, on the other hand, this led to an underestimation of the importance of local conditions. As a rule, the appearance of reliable stratigraphic and geochronologic information coincided with the abandonment of Alpine standards and with the development of original concepts for the development of mountain glaciation in the USSR.

At the present time, the literature on early glaciation in the mountains of the USSR includes several hundred titles; most of these studies are geomorphic publications of a regional character. We do not claim to cover all this material in the present chapter, and so readers interested in the most important studies should refer to the review articles of Markov (1939, 1965). Many new books and articles dealing with the glacial history of the mountains have been published in the last 15 years. Because it is impossible to give a clear explanation of these regional data within the scope of this chapter, a discussion of more general problems will be emphasized. The chronologic framework of the discussion is limited to the Late Pleistocene and the Holocene.

Geomorphic Studies

Students of mountainous regions are almost unanimous in acknowledging that traces of Late Pleistocene glaciation expressed in the topography are identified by their better preservation in comparison with older glacial features. It is clear, however, that the degree of topographic preservation is a very indefinite criterion and that it is difficult on this basis to identify clearly the stages of mountain glaciation (Murzayev, 1964; Markov, 1965).

It is also necessary to admit that in mountainous regions there is no direct evidence for multiple glaciation, when one neglects the grouping of moraines in mountain valleys. The few existing finds of interglacial and interstadial deposits are confined to intermontane and submontane basins. At the same time, the development of the tectonics and climate of mountains and basins differed substantially, and, therefore, the question of continuity of mountain glaciers remains difficult to solve.

When the geomorphologic method is used, the moraine complexes identified in mountains are usually connected with river terraces. However, distinct correlations of end moraines and terraces are not observed very frequently, and in unclear situations disagreements arise among the investigators. In addition, the age of terrace cycles has been little studied thus far, and its elucidation makes it necessary to have available a fairly large number of age determinations. The attempt we made to determine the duration of the formation cycle of Abay terraces in the Syr-Dar-ya Basin (Pshenin and Serebryanny, 1981) is possibly the only such experiment for the inland regions of Eurasia.

At the same time, there can be no doubt that the use of geomorphic data is valid for general qualitative estimates of glaciation types. The fresh glacial relief forms are referred by many investigators to Late Quaternary glaciation. When this assumption is taken as the basis and all available cartographic material is analyzed, one finds that glaciation during the Late Quaternary in almost all mountainous regions of the USSR was controlled by the relief and that the glaciers were distributed throughout valleys.

Statements that early glaciation had a mountain-valley character have been most common among Soviet investigators of the Alpine mountain belt. However, even for regenerated mountains—for example, the Altay (Okishev, 1977)—a definite predominance of valley glaciers during the Late Pleistocene is also postulated. Let us note that as long ago as 1939 Markov distinguished mainly traces of valley glaciers among old glacial forms, but he also pointed out traces of large piedmont glaciers in the mountains of Central Asia, where the glaciers descended to altitudes of 900 to 2000 m above sea level. Grosswald was perhaps the only investigator who acknowledged significant ice sheets of Late Pleistocene age on the Sayano-Tuvinskiy Upland and particularly on the Pamir (Grosswald, 1965; Grosswald and Orlyankin, 1979). There are also differences in indicators of the lowering of the snow line below its present position: according to Grosswald, this lowering was no less than 1000 m during the Würm (Valdai), whereas other investigators usually cite figures of about 600 to 800 m, and for Eastern Pamir only 260 to 350 m (Markov, 1965).

At the present time, further use of the geomorphic method in the study of early mountain glaciation should be combined with analytical methods: geochronologic, paleontologic, and lithologic-geochemical. These new approaches are becoming common in the study of a number of mountainous regions of the USSR, and the results obtained are promising.

Lithologic Studies

In the diagnosis of the features of mountain glaciation, it is important to use lithologic-geochemical methods, which make it possible to eliminate sharp discrepancies in many aspects of the glacial history of mountains (Dumitrashko, 1977; Okishev, 1977; Serebryanny et al., 1977). In particular, it is very important to learn how to distinguish clearly glacial formations such as moraines from superficially similar loose deposits and associated relief forms of different origin (mud flow, collapse-slip, avalanche, slope wash, and other forms).

Along with A. V. Orlov and A. S. Medvedev, we conducted studies of different types of moraines—both deposits and relief forms—under the conditions prevailing in the high mountains of the Caucasus. One of the main objectives was to determine the diagnostic criteria of glacial formations. Coarse detritus was subjected to petrographic analysis in four size fractions: 1 to 3, 3 to 5, 5 to 7, and 7 to 10 cm; at the same time, the shape and the surface character of the fragments were studied. The mineralogic analysis of morainic debris involved the use of immersion and X-ray diffraction methods, as well as grain-size analysis. The orientation of elongated pebbles was measured directly in the exposures, and the orientation of sand grains was measured in translucent polished sections. A few geochemical analyses were performed, and color determinations on the till were made with the aid of standard scales.

The lithologic information was considered in close correlation with the study of relief, the large-scale mapping of key areas, and the bio- and chronostratigraphic indicators, all of which ensured comprehensive space-time control. Attention was also given to estimating the climatic background against which the processes of glacial lithomorphogenesis took place. All in all, we were able to follow the process of complex-moraine formation from the initial pickup of the rock fragments to the deposition and subsequent changes in the material (Serebryanny, 1980; Serebryanny and Orlov, 1980; Serebryanny et al., 1981).

One important result of our investigation was the identification of diagnostic differences in the material of surface (medial) moraines and basal moraines in glacial deposits and of associated relief forms in the mountains. It was found that moraines consist basically of a material lithologically similar to a basal till. The surface material was involved only in the uppermost parts of the glacial deposits and in certain horizons of the moraines. This finding completely failed to confirm previous concepts about the formation of glacial deposits as a result of surface-moraine material falling from the front of the glacier (Markov, 1946; Shcherbakova, 1978; Okishev, 1974; Kostenko, 1975; Boulton, 1970). These concepts, which minimized the role of bed erosion by glaciers, were based on visual observations rather than on analyses of the composition and structure of the moraines. The advent of quantitative analysis radically altered previous concepts and helped to confirm the concepts of glacial action in affecting the relief of mountains.

Interesting results were produced by a study of exposures with several tills separated by buried soils. The thickness of the basal till was as much as 15 m, indicating large-scale excavation. On the basis of the interpolation of radiocarbon dates on buried soils, the rate of moraine accumulation was calculated to be about 2 mm per year; this rate can also serve as a measure of the erosion rate under the conditions prevailing in the high Caucasus (Serebryanny et al., 1978). These results are valid for the region under consideration, where moderate-type glaciers are widespread, but they are also confirmed for Spitzbergen and central Tien-Shan, where our lithomorphologic studies were undertaken in 1980 through 1982.

Evolution of Mountain Glaciation during the Late Pleistocene

As noted above, the elucidation of the developmental history of mountain glaciation requires a body of data, that is, geomorphic, lithologic, biostratigraphic, and geochronologic data. Research designed to yield such complete information has been conducted in only a few mountainous regions. In addition to our work in the high Caucasus—work that did not go beyond the scope of the Holocene (Serebryanny et al., 1977, 1978, 1979)—investigations have been carried out in the Issyk-Kul' Basin and other high-altitude basins of the Tien-Shan (Markov, 1971; Shnitnikov, 1979, 1980), as well as in the piedmont of the Fergana Basin (Serebryanny et al., 1980). In the regions of Central Asia studied, the main objective was to work out the history of the latest glaciation, including the ice retreat in the Holocene; there, the most important methods used for supplementing the geomorphic studies were pollen analysis and radiocarbon dating.

In addition, thorough biostratigraphic studies within the entire Pliocene-Pleistocene were undertaken in the mountainous regions of Central Asia. These studies enabled investigators to look at the glacial history from broader points of view. After many years of palynostratigraphic studies of the Pamir and Gissaro-Alay, M. M. Pakhomov identified several glacial epochs separated by interglacials (Nikonov and Pakhomov, 1976; Pakhomov, 1980; Pakhomov and Nikonov, 1977). The same type of data was obtained for the Tien-Shan by Z. V. Aleshinskaya (Markov, 1971), Grigina (1973), and other investigators.

Major disagreements exist over the paleogeographic interpretation of the glacial history of the Central Asian mountains. M. M. Pakhomov and A. A. Nikonov are working on the correlation of glaciation with pluvial intervals and of interglaciations with arid intervals. They have shown that, when the highlands were involved in glacial processes, coniferous and deciduous forest grew at points lower in the mountains. On the other hand, during the interglacial epochs, the mesophytic forest vegetation as well as the meadow vegetation were forced back into the upper belts, giving way to xerophytic associations below. Pakhomov and Nikonov essentially proceed from the assumption that there was greater precipitation during glaciation, in accord with the views of Tolmachev (1944), Zabirov (1955), Agakhanyants (1965), Markov (1965), Sidorov (1979), and other investigators. At the same time, a temperature lowering is also presumed, which was caused not only by global climatic change but also by mountain uplift.

Another view is that the glaciation of high mountains in Central Asia, and in particular in the Pamir, was characterized by arid conditions and that the climate of the interglacial intervals was characterized by pluvial conditions (Velichko and Lebedeva, 1974). This view is shared by investigators of loess regions (A. A. Lazarenko, A. V. Minervin, and A. E. Dodonov) as well as by certain palynologists (Grigina, 1973). Figure 6-1, a pollen diagram of intermorainic deposits of the Altyn-Dara section, is presented here as an example of an interglacial rise of the forest belt. On the whole, Grigina concludes that the glacial epochs were dry and cold and that the interglacial ones were humid and warm. Accordingly, she postulates that elements of broad-leaved forests in the mountains could have existed only during an interglacial period but at the same time notes that spruce forests with some boreal varieties were commonly distributed in the mountains during both glacial and interglacial epochs.

The complicated interpretations of the environmental history of mountainous areas result not only from the succession of glaciations and interglaciations but also from general trends in landscape development. Pakhomov (1980) notes that during the Late Pleistocene and Holocene a slow but irreversible process of desertification and xerophytization took place. Thus, during the last glaciation, extremely dry conditions prevailed in the eastern Pamir and were made more pronounced by the intense upheaval of this region at the end of the Pleistocene (Sidorov, 1979).

Definite evidence of glacial disintegration in the middle of the Late Pleistocene on the Pamir Highland was discovered in beds of blue clays exposed in a mass of buried ice on the eastern shore of Lake Kara-Kul'. These beds, 5 to 10 m thick, underlie pale yellow loams of the lowest lake terraces. The blue clays, which have retained the roots of littoral aquatic macrophytes, have been radiocarbon-dated at 27,700 ± 700 yr B.P. (MGU-257) (Velichko and Lebedeva, 1974). Palynologic data for this section (Pakhomov, 1967) indicate a fairly rich assemblage of tree pollen (willow, pine, elm, alder, and cedar, together amounting to 9% to 11% of the total pollen and spores). This situation is caused by wind transport from the western Pamir, where the slope and floodplain forests probably rose to fairly high levels at that time. Thus, one can postulate that the glacial dimensions were substantially decreased.

The subsequent cooling, which corresponded to the latest Pleistocene glaciation, was associated with a marked depletion of the Pamir mountain flora. Desert-steppe communities typical of an extremely dry climate similar to

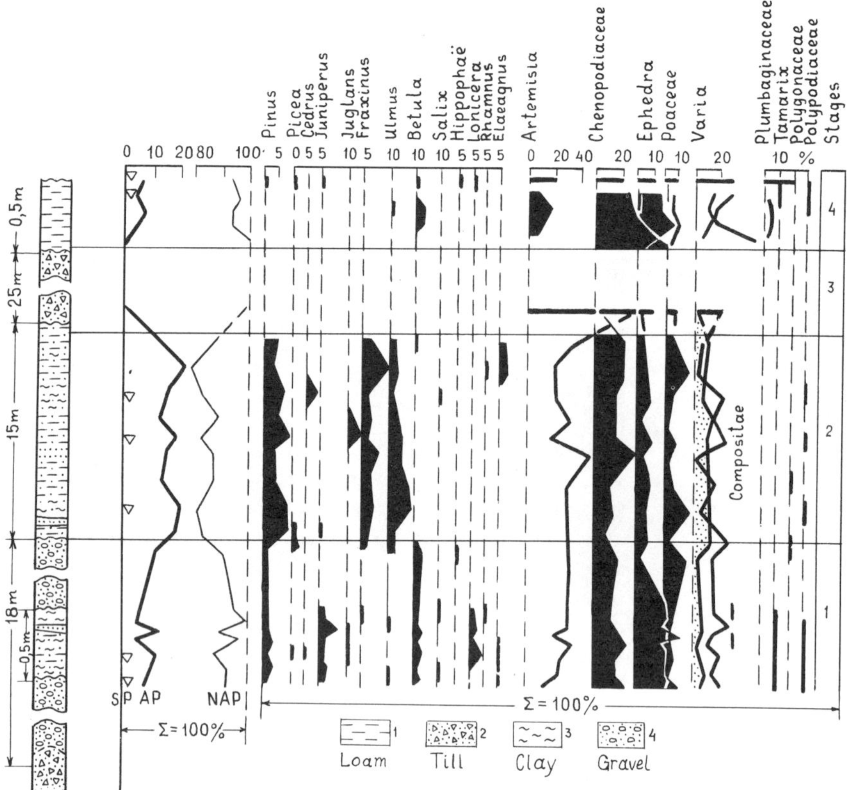

Figure 6-1. Pollen diagram of the Altyndara section. (After Grigina, 1973.)

the present one have been spreading since that time (Pakhomov, 1972). Unfortunately, there is no consensus among investigators concerning the relationships of radiocarbon-dated blue clays on the shore of Lake Kara-Kul' to the end moraines near the lake. Therefore it is difficult to distinguish accurately the limits of the latest Pleistocene glaciation even in this closely studied region of the Pamir. The wastage rate for this glaciation can be estimated from the geographic location of the Oshkhona Mesolithic campsite, radiocarbon-dated at 9500 to 7500 yr B.P. This campsite is located in the Markansu River basin, 12 km from the end of the present Uysu glacier and 7 km from the Karaartskiy moraine, which according to Velichko and Lebedeva (1974) was formed at the maximum of Late Pleistocene glaciation.

Despite lively discussions on the problems of the Pamir's glacial history, of fundamental importance is the conclusion from biostratigraphic data that there have been repeated and substantial variations in the dimensions of the glaciation of this highland. M. M. Pakhomov, A. A. Nikonov, O. K. Chedia, V. A. Ranov, A. K. Trofimov, V. V. Loskutov, A. A. Velichko, I. M. Lebedeva, O. Ye. Agakhanyants, and other investigators are in agreement with this conclusion.

The search for key strata for correlation in the mountains of Central Asia is being carried on intensively. Most agreement has been reached for a section of intermorainic deposits in the Altyn-Dara Gorge on the northern slope of Zaalayskiy Range, discovered in 1961 by K. V. Kurdyukov. At this location, Grigina (1973) identified a Late Pleistocene interglaciation characterized by a marked rise of the forest belt. (See Figure 6-1.) Comparable but not as contrasting were the data obtained by M. M. Pakhomov for the Katta-Karamyk section, located 12 km west of the Altyn-Dara Gorge. The interglaciation in question, called the Altyn-Dara Interglaciation, is placed between moraines of the Middle and Late Pleistocene, which were dated by means of geomorphic characteristics.

In the eastern Issyk-Kul' Basin, traces of an interglaciation of the lower subdivision of the Late Pleistocene have also been observed, confined to the upper part of the key section of lacustrine-alluvial deposits in the Dzhergalan River valley near the village of Orlinoye (Markov, 1971). Our observations have shown that in this region the 36-m thickness of sand and pebble gravel accumulated under conditions of repeated alternation of planar delta sedimentation and ordinary fluvial processes, as controlled by deep and lateral erosion. In this connection, a certain analogy can be drawn to the modern processes on the valley bottom. A regular upward decrease in the amount of pollen of trees and shrubs from 25% to 45% down to 8% along the section was interpreted by E. K. Azykova and Z. V. Aleshinskaya as evidence of climatic change from the cool conditions of the Middle Pleistocene glaciation to the warm conditions of a Late Pleistocene interglaciation. The wood was radiocarbon-dated at more than 48,500 yr B.P.

The solution to the problems associated with the Late Pleistocene history of this region can also be approached by studying lake terraces and recognizing their relationship to glaciations. A Mikhaylovskaya (45- to 50-m) terrace, in whose alluvial-diluvial horizons V. A. Ranov discovered weapons and flakes of a Mousterian culture (Markov, 1971), is known in the basin of the Issyk-Kul'. Considering the present-day concepts of the age of this culture, one should recognize that the incision of the Mikhaylovskaya terrace took place during the first half of the Würm (Valdai). For sediments of the lower Nikolayevskiy (32-m) terrace, Kh. A. Arslanov obtained a radiocarbon date of 26,340 ± 540 yr B.P. for shells taken at a depth about 7 m from the top (Markov, 1971). Hence, one can detect indications of two major stages of glaciation in the north of the Tien-Shan during the Late Pleistocene; the data are closely related to those from other plains and mountainous regions of Eurasia (Serebryanny, 1978). (It is interesting to note that in the view of Shnitnikov [1980] the maximum phase of the latest Pleistocene glaciation of the Tien-Shan occurred 14,000 years ago, that is, somewhat later than the period of maximum expansion of the last ice sheet over the plains of Eurasia.)

Indications of a warmer interval preceding the latest Pleistocene glaciation were found in the center of the Tien-Shan in the basin of Lake Chatyrkel' (Schnitnikov, 1980). For lacustrine sediments extending over the bottom of this basin (3500 m above sea level) above Holocene terraces, radiocarbon dates of 22,000 to 16,000 yr B.P. were obtained. According to G. N. Berdovskaya's pollen data, a rich assemblage of herbs (including meadow mesophytes of the buckwheat, lily, and other families) was represented substantially in the lower portions of the sections against a background of steppe and semidesert types. The greater pollen frequency of coniferous trees—spruce, pine, cedar, juniper—can be interpreted as indicating a higher altitude of the forest belt.

The upper layers of lacustrine sediments accumulated during the latest Pleistocene glaciation. The diversity of herb pollen was reduced, and the wind dispersal of tree pollen nearly came to a halt. Apparently, mountain steppe gave way to semidesert associations adapted to extremely cold conditions. At the same time, the areas of glaciers in the mountains greatly expanded.

As a result of a study of the abundant plant remains in the basal lacustrine clays in the Kerikidon section (440 to 450 m above sea level), 10 km east of Fergana, we were able to document pluvial conditions of a Late Pleistocene epoch of 12,000 years ago (Serebryanny et al., 1980). A curtailment of the role of *Artemisia, Ephedra,* and *Chenopodium* pollen, an increased percentage of pollen of meadow and wetland plants, and a significant proportion (up to 14%) of juniper, birch, and willow pollen were noted in these clays, located in the Abay Terrace. The forest belt was probably located nearby during the Late Pleistocene.

A lowering of the forest belt is also observed in the Bidan area on the Akbura River (1200 m above sea level). Here, in a similar geomorphic situation, M. M. Pakhomov observed a high pollen content of spruce, pine, birch, and other trees.

On the basis of data obtained by studying both sections, we estimate that the lower boundary of the forest belt descended approximately 1000 m below its present position. Similar values exist for the northern slope of Turkestanskiy Range, where during the Late Pleistocene cooling birch groves descended into the valleys to an altitude of 1600 m (Shakhristan); at the present time, birch and juniper grow above 2700 m near glaciers (Pakhomov and Nikonov, 1975). One can postulate that the lower boundary of glacier accumulation on the Tien-Shan also dropped considerably at that time.

Thus, two major intervals of Late Pleistocene glaciation can be distinguished in the mountains of Central Asia. Similar data also exist for other mountain regions of the USSR. In the Northeast, in the region of the Verkhoyano-Kolymskiy Mountains, these intervals are correlated with the Zyryanka and Sartan Glaciations known in Siberia, and the intermediate interval is correlated with the Karginskiy Interglaciation, confirmed by palynologic and radiocarbon data with geomorphic control (Kind et al., 1971; Sheynkman, 1978). During both glaciations, the glaciers emerged through valleys into foothills, but they were larger during the Zyryanka glaciation. The Karginskiy Interglaciation was characterized by a distinct deglaciation up to approximately the present level, where mainly corrie and a few corrie-valley glaciers are preserved in the highest mountain areas.

On the other hand, for the Altay, Okishev (1977) and other investigators admit a practically continuous glaciation of considerable dimensions during the Late Pleistocene. However, in acknowledging the presence of two major phases of glaciation, they postulate that the intermediate interval was characterized by deglaciation of intermontane basins, with the formation of dammed lakes. The age of these water bodies, established by radiocarbon and thermoluminescence dating, was found to correlate with the Karginskiy Interglaciation (Svitoch and Khorev, 1975).

Numerous indications of the two-phase character of Late Pleistocene glaciation have been obtained in the Caucasus, but they are mainly geomorphic in character. The mountain relief shows traces of two levels, separated by a deep incision of about 100 m or more (beyond the confines of highlands), and in the foredeeps and intermontane troughs two terraces stand out. For example, the first and second terraces of the Kuban' Trough (the Cherkessk and Krasnodar Terraces) are referred to two maxima of Late Pleistocene glaciation (Milanovskiy, 1966; Shcherbakova, 1973). The deep erosion separating the glacial troughs of the two levels is regarded by Caucasian investigators as evidence of an interglacial interval, although there is no consensus regarding the extent of glacier recession during this interval (Kovalev, 1967).

Unfortunately, few sections have been found thus far in which one can detect clear-cut relationships between glacial deposits and lacustrine-paludal or alluvial deposits containing macropaleontologic and geochronologic information. The only exception is the data of Goretsky (1962) on sediments of ice-dammed lakes up to 300 m thick, discovered during drilling in the Ullukam and Uchkulan Valleys in the upper reaches of the Kuban' River. Lacustrine sediments including diatomites are interbedded and covered by three horizons of glacial or associated deposits. Appreciable changes in the vegetative cover are established on the basis of pollen analysis. Predominant in certain lacustrine horizons is the pollen of thermophilic deciduous trees (white birch, beech, oak, elm) up to 41%; this finding provides a basis for distinguishing an interglaciation, but Goretsky himself evaluates these results with more restraint and distinguishes only two interstades of Late Pleistocene glaciation. Absolute-age data for these interesting sections are lacking.

On the southern macroslope of the Greater Caucasus are exposures of intermorainic clays (Tsereteli, 1966), 2 km east of the village of Mestia in Svanetiya (1550 m above sea level). The upper moraine of the section is associated with the Kakhurzagyarskiy end moraine, on which part of the village is located (where the Mestiachala River discharges into the Mulkhura River). A Würmian age is attributed to this ridge, and the above-mentioned clays are correspondingly referred to an intra-Würmian (intra-Valdai) interglaciation (Tsereteli, 1966).

The total thickness of clays under the upper moraine exceed 10 m, but no basal moraine at all was observed. A series of samples at 10-cm intervals, analyzed by Ye. S. Malyasova, was practically devoid of pollen. Isolated samples showed coal particles, highly mineralized detritus, plant-tissue fragments of recent appearance (with chlorophyll), and isolated grains of Mesozoic plants—coniferous and cryptogamous. In addition, isolated pollen grains of recent plants were occasionally observed.

In the original work of Tsereteli (1966), the analyst Ye. A. Mal'gina noted that the occurrence of pollen and spores in the clays in question was also extremely sporadic; thus, it provides no basis for identifying an interglacial age. The formation of the end moraine may have been preceded by a period of clay accumulation, most probably under conditions of decelerated river runoff. Very similar clay facies alternating with sandy-pebbly interlayers are accumulating in various parts of the Mulkhura River at the present time. There is no doubt about tectonic activity, which predetermined the deep incision of the Mulkhura River after the formation of the end moraine. Unfortunately, the age of these events remains unknown. Solely on the basis of the good preservation of moraines, one can arbitrarily assume that the glacier descended into the Mestia region during the Late Pleistocene, but there is no basis yet for more accurate dating.

The good preservation of moraines implies that glaciers on the southern macroslope of the Caucasus descended during the Late Pleistocene to elevations of 1200 to 1400 m in the west and 1800 to 2400 m in the east. On the northern macroslope, the glacier margins reached much lower, to the escarpment of the Skalisty Range, 1100 to 1200 m above sea level (Reingard, 1927; Kovalev, 1967). According to Milanovskiy (1966), glaciers may also have emerged in piedmont plains at elevations of 660 to 700 m. Near Sovetskiy, beyond the confines of the Skalisty Range, we observed till 2 to 5 m thick in terrace III at 24 m on the left bank of the Cherek River. In this area, it is possible to apply biostratigraphic methods, for a mass of lacustrine clays is exposed under the till.

In summary, it can be stated that the Late Pleistocene mountain glaciation of the USSR has been insufficiently studied. Most information available is regional, pertaining to the distribution of morphologically expressed traces of early glaciation. However, the lithologic-geochemical confirmation of these data is only beginning, and there is little biostratigraphic or geochronologic information available. Nevertheless, the facts collected thus far make it possible to note the presence of the two major epochs of mountain glaciation during the Late Pleistocene.

Markov (1965) stated that in the mountains of the Caucasus and Central Asia the climate of the glacial epochs was more humid than it is at present. This view still has its adherents and critics today. At the same time, Markov believed that in the mountains of southern Siberia and in the Carpathians the climate was drier because of the location in the periglacial zone of continental glaciation. Thus far, this point of view has not been supported, for data obtained in western Mongolia (Devyatkin et al., 1978) indicate that in this region the glaciations were characterized by pluvial conditions.

A special role in mountain glaciation is played by the tectonic factor, which should also undergo rigorous analytical study. The scope of the present chapter precludes a discussion of highland neotectonics, especially since a stage-by-stage analysis of tectonic movements by different investigators is far from clear. It should merely be noted here that the amplitude of mountain uplift during the Late Pleistocene and the Holocene is usually estimated to have been small compared to the general background of uplift of the latest tectonic stage. Only for the Pamir do certain investigators acknowledge high rates of uplift during the Late Pleistocene and the Holocene (Sidorov, 1979).

Characteristics of Mountain Glaciation in the Holocene

Most Soviet investigators note that Holocene glaciers of the southern mountain belt of the USSR are inherited from the preceding stage. The retreat of mountain glaciers during the Early Holocene is explained by a global improvement of climatic conditions, which reached its fullest expression during the climatic optimum. One can assume with Ahlmann (1953) that at least some of the glaciers, if not all of them, melted at that time and were then regenerated as a result of the subsequent cooling. This idea forms the basis of the concept of the "Little Ice Age," developed for the Caucasus by Tushinskiy (1968), who holds the view that during the epoch of reduced humidity of the first millennium A.D. (the Arkhyz interval) mountain glaciation greatly decreased, glaciers disappeared from mountain passes, and communication across the Alps and the Caucasus improved. During the epoch of increased humidity in the middle and second half of the current millennium (the Fernau stage), an advance of glaciers in highlands occurred, accompanied by intensified avalanche activity.

Shnitnikov (1957) was the first to demonstrate a rhythmic retreat of mountain glaciers in the Holocene, with repeated advances resulting in end moraines of stadial rank.

Our investigations in the high Caucasus emphasize the dynamics of glaciers in the Holocene (Serebryanny et al., 1977, 1978, 1979). For this purpose, use is made of geomorphologic, geochronologic, biostratigraphic, and biologic approaches. Particularly interesting results were obtained by combining lichenometric surveys of moraines with radiocarbon dating and paleobotanic data. The data were correlated with the use of reliable maps.

Special approaches preceded the application of many methods. Samples of humified surface deposits from different vegetational belts were selected to check the pollen data. Analyses of these samples generally confirmed the correlation of pollen counts with the composition of local vegetation. The most significant anomaly was the maximum of forest-plant pollen in the subalpine belt, which resulted from the action of vertical air currents (Klopotovskaya, 1973).

Data obtained from the above studies make it possible to reconstruct the sequence of Holocene changes in climate and vegetation, but our attempts to compare the results with the sequence of end moraines involved some major complications. For example, at the Holocene climatic optimum, vegetational belts were raised, with the upper forest boundary on the northern macroslope of the Greater Caucasus in its central part being approximately 300 m higher than it is today. The Early and Late Holocene were cooler than the Middle Holocene. In the meantime, on both macroslopes of the Greater Caucasus, the glaciers successively retreated over the course of the entire Holocene while undergoing periodic oscillations. We also recorded indications of episodic movements during the Middle Holocene (as confirmed by several radiocarbon dates). The glacier dimensions at that time were greater than in the Late Holocene or at the present time.

As still another example, a paleobotanic study of peat bogs located near glacier margins at 1800 to 2300 m elevation revealed a progressive increase in humidity during the Sub-Atlantic period of the Holocene. Nevertheless, data on the geomorphology, stratigraphy, and lichenometry of moraines very definitely indicate a gradual glacier retreat during this interval.

Attempts to account for these contradictions unfortunately are complicated by a lack of information on the temporal variability of precipitation at different elevations in the Caucasus Mountains. One cannot ignore the possibility that during the Sub-Atlantic a gradual decrease in snow accumulation could have taken place in high firn basins (3500 m or higher) but that at the same time the humidity in the terminal zone (1900 to 2400 m) and in the forest belt not only remained high but even increased. Consequently, opposite tendencies in climatic changes may have occurred in the bottom layers of air and in the free atmosphere.

Another equally possible explanation involves rejecting a direct comparison with present glaciologic conditions in the interpretation of glaciation characteristics in the past. For example, beech was a major tree genus in high-altitude forests of the postglacial climatic optimum, and one can postulate that the seasonal distribution of precipitation at that time differed appreciably from the present distribution. It is well known that beech tolerates prolonged snowy winters (Sukachev and Zonn, 1955), and it may be that the amount of winter snow at times in the Middle Holocene was greater than at the present and that the amount of summer precipitation was somewhat smaller. As a result, the glacier balance could have been determined not only by increased ablation but also by increased accumulation.

Using a combination of different analytical methods, particularly for sections with several moraine horizons and intervening buried soils (Figure 6-2), we determined that glacier advances occurred earlier than 8600 years ago, between 8600 and 6400 years ago, between 6400 and 4200 years ago, around 3000 years ago, and a number of times

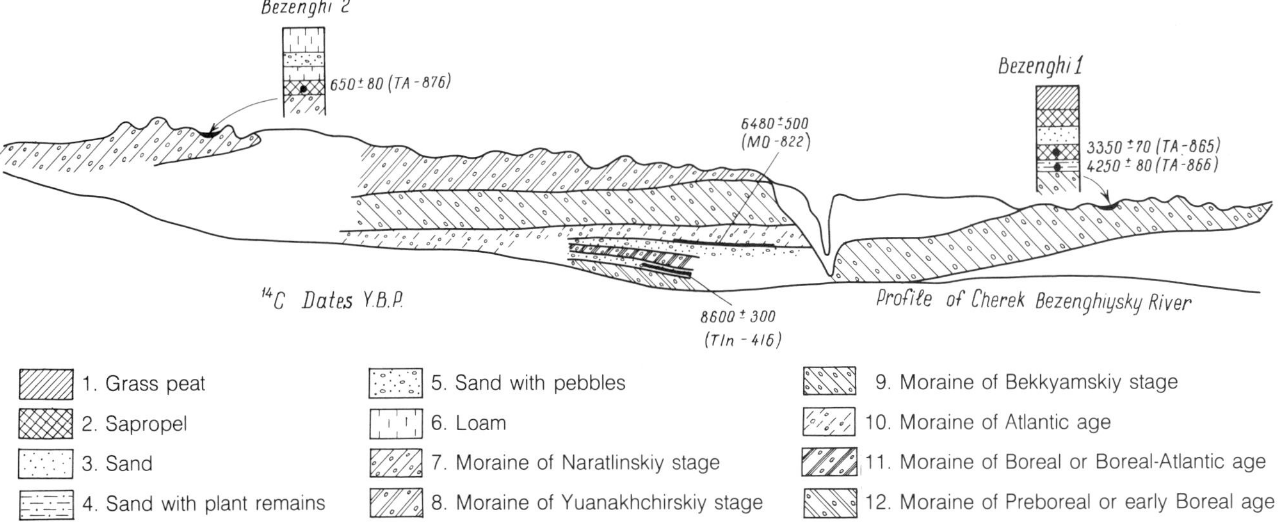

Figure 6-2. Diagram of the structure in exposures of a moraine terrace located 5 to 7.5 km from the snout of Bezenghi glacier.

later against a background of general retreat throughout the Holocene. No data on the duration of these movements are available; we think it entirely possible that, despite the considerable scale of moraine accumulation (the average thickness of each moraine horizon is 12 to 15 m), the duration of the movements could have been short. Nevertheless, it has been established beyond any doubt that at the climatic optimum of the Holocene the glaciers of the high Caucasus did not disappear completely and that each younger moraine was characterized by smaller

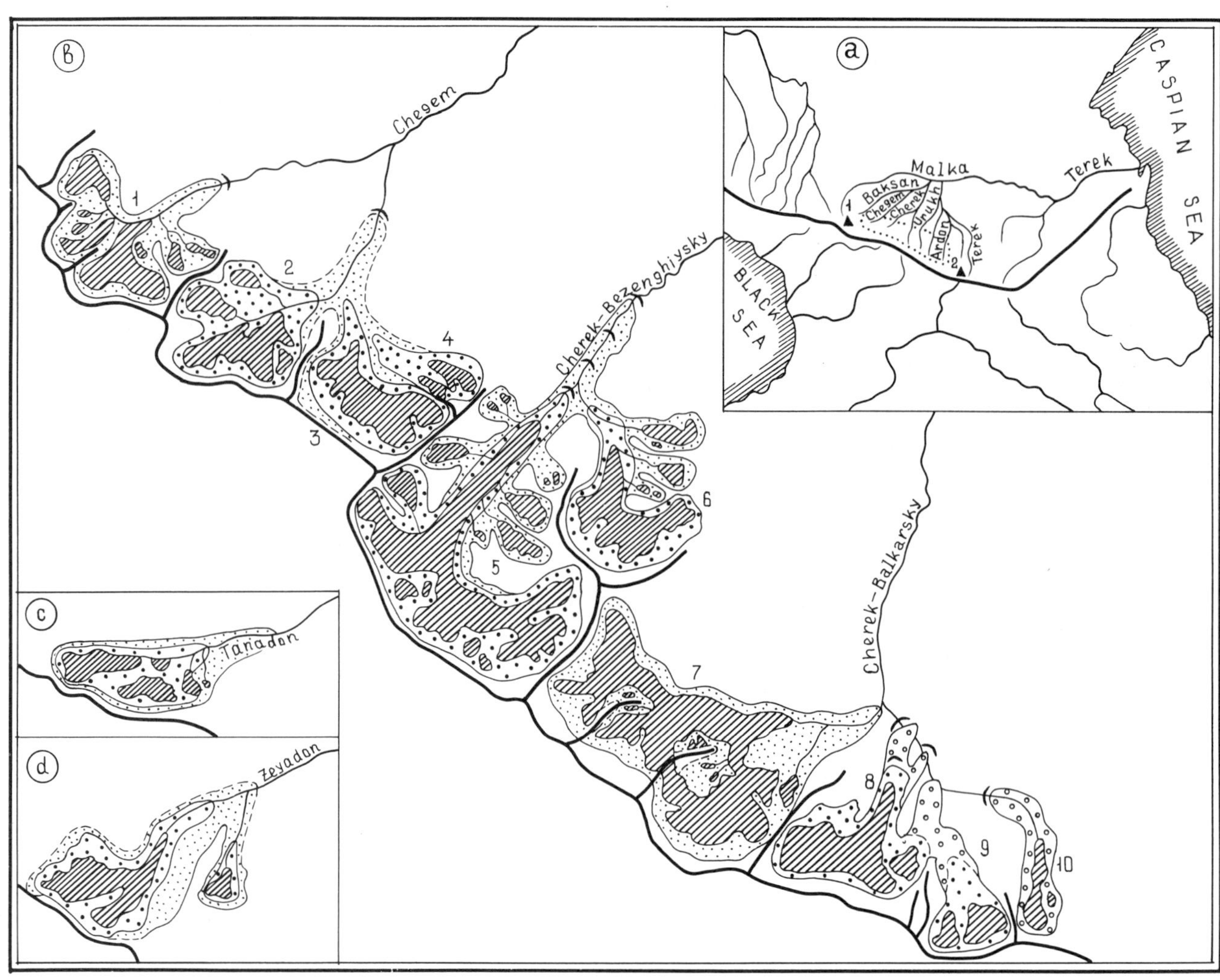

Areas freed from ice:

- From the end of the 13th century to the end of the 19th century
- From the end of the 17th century to the end of the 19th century
- From the end of the 19th century to the beginning of 20th century
- Recent glaciers

a. location of the area in the Greater Caucasus
  1. Town of El'brus
  2. Town of Kazbek (dot-dash line outlines the Chegem-Ardonskiy portion of lichenometric survey

b. Chegem-Cherek group of glaciers:
  1. Bashil
  2. Kulak-Chegemschimar
  3. Shaurtu
  4. Tyutyurgu
  5. Bezengi
  6. Mizhirgi
  7. Dykh-Su
  8. Agashtan
  9. Western Shaurtu
  10. Eastern Shaurtu

c. Tanatsete glacier

d. System of Tseya-Skazka glaciers

Figure 6-3. Changes in glaciers on the northern macroslope of the Greater Caucasus during the last 700 years. (After N. A. Goldkovskaya.)

glacier dimensions than the preceding one. Likewise, we have no evidence to support the view of Tushinskiy (1968) that many Caucasian glaciers disappeared during the Arkhyz interval. Thus, the Holocene and present glaciation of the Caucasus should be considered as having been directly inherited from the Late Pleistocene.

More detailed information on the dynamics of Caucasian glaciers has been obtained through a lichenometric survey on both macroslopes of the highland (Figure 6-3). According to N. A. Golodkovskaya, during the last 700 to 800 years the total decrease of glaciers has been complicated by episodic movements no fewer than 10 times. During these movements, the glacier accumulation limit dropped by a maximum of 150 m below the present level. The distinct rhythmic variability of mountain glaciers (Shnitnikov, 1957) is completely confirmed, and under the conditions prevailing in the Caucasus an 80-year rhythm is recognized as the most pronounced.

The results of our work in the high Caucasus are entirely consistent with the data from Central Asian paleoglaciologic expeditions. According to pollen analyses, two intervals of increased humidity there may have been correlated with major advances of glaciers—at the Pleistocene/Holocene boundary and during one of the intervals of the Middle Holocene (Pakhomov and Nikonov, 1975; Serebryanny et al., 1980; Mikhaylov, 1981). In addition, another whole series of smaller movements can be distinguished, complicating the overall course of glacial retreat. In a majority of cases, as in the Caucasus, the dimensions of the glaciers during each advance were smaller than during the preceding advance. Mikhaylov (1981), using data from a pollen analysis of end moraines in the Fanskiye Mountains, notes that the glacier oscillations occurred in the presence of different relationships of heat to moisture. The use of the dendrochronolgic methods by this investigator has confirmed the validity of classifying the two episodes of mountain-glacier advance under the "Little Ice Age"—in the 12th through 13th and 18th through 19th centuries—this being consistent with data of lichenometric surveys on the Kungey-Alatau (Pomortsev, 1980) and the high Caucasus (Serebryanny et al., 1979). The use of detailed reconstruction methods for mountain-glaciation dynamics opens the way to the development of long-range forecasts, which are of major importance to the Soviet national economy.

References

Agakhanyants, O. Ye. (1965). "Basic Problems of the Physical Geography of Pamir." Part 1. Tadzhik Academy of Sciences, Dushanbe.

Ahlmann, H. W. (1953). "Glacier Variations and Climatic Fluctuations." American Geographical Society, New York.

Boulton, G. S. (1970). On the origin and transport of englacial debris in Svalbard glaciers. *Journal of Glaciology* 9, 213-30.

Dainelli, G. (ed.) (1938). "Atlante Fisico-economica d'Italia." Consociazione Turistica Italiana, Milano.

Devyatkin, Ye. V., Malayeva, Ye. M., Murzayeva, V. E., and Shelkoplyas, V. N. (1978). Pluvial Pleistocene basins of the Great Lakes depression of western Mongolia. *USSR Academy of Sciences Izvestiya, seriya geograficheskaya* 5, 89-99.

Dumitrashko, N. V. (1977). Old glaciation. *In* "General Characterization and History of the Development of Caucasian Relief" (B. A. Antonov and N. S. Shirinovko, eds.), pp. 90-100. Nauka Press, Moscow.

Goretskiy, D. I. (1962). On the age and space relationships of anthropogenic terraces of the Kuban' River. *Proceedings of the Commission for the Study of the Quaternary Period* 19, 33-67. USSR Academy of Sciences, Moscow.

Grigina, O. M. (1973). Forest spore-pollen spectra in extraglacial and periglacial regions of Tien-Shan and their interpretation. *In* "Palynology of the Pleistocene and Pliocene" (V. P. Grichuk, ed.), pp. 145-50. Nauka Press, Moscow.

Grosswald, M. G. (1965). "Development of the Relief of Sayano-Tuvinskiy Upland (Glaciations, Volcanism, Neotectonics)." Nauka Press, Moscow.

Grosswald, M. G., and Orlyankin, V. N. (1979). Late Quaternary ice cap of Pamir. *Transactions of Glaciological Studies, Chronicle, Discussions* 35, 85-97.

Kind, N. V., Kolpakov, V. V., and Sulerzhitskiy, L. D. (1971). Age of the glaciations of Verkhoyan'ye. *USSR Academy of Sciences, Izvestiya, seriya geologicheskaya* 10, 135-44.

Klopotovskaya, N. B. (1973). "Basic Principles of Formation of Spore-Pollen Spectra in Mountainous Regions of the Caucasus." Metsniyereba Press, Tbilisi.

Koreysha, M. M. (1963). "Recent Glaciation of Suntar-Khayata Range." USSR Academy of Sciences, Moscow.

Kostenko, N. N. (1975). "Quaternary Deposits in Highlands." Nedra Press, Moscow.

Kotlyakov, V. M. (1968). "The Earth's Snow Cover and Glaciers." Gidrometeoizdat Press, Leningrad.

Kovalev, P. V. (1967). "Recent and Old Glaciation of the Greater Caucasus." Transactions of the Caucasus Expedition, vol. 8, Khar'kov.

Markov, K. K. (1939). Old glaciation of high-mountain regions of the USSR. *In* "The Ice Age on the Territory of the USSR" (I. P. Gerasimov and K. K. Markov, eds.), Issue 33, pp. 179-233. USSR Academy of Sciences, Institute of Geography, Trudy.

Markov, K. K. (1946). On the form and origin of moraines in mountains. *Moscow State University, Uchenyye Zapiski* 119(2), 59-74.

Markov, K. K. (1965). Region of old glaciation of high mountains in the southern USSR. *In* "The Quaternary Period," vol. 1 (K. K. Markov, G. I. Lazukov, and V. A. Nikolayev, eds.), pp. 302-67. Moscow State University, Moscow.

Markov, K. K. (ed.) (1971). "Section of the Latest Deposits of the Issyk-Kul' Basin." Moscow State University, Moscow.

Meshcheryakov, Yu. A. (1972). "Relief of the USSR (Morphostructure and Morphosculpture)." Mysl' Press, Moscow.

Mikhaylov, N. N. (1981). Dynamics of the natural environment of Fanskiye Mountains in the Holocene. Diploma Dissertation, Institute of Geographical Sciences, University of Leningrad.

Milanovskiy, Ye. Ye. (1966). Fundamental problems in the history of old glaciation of the central Caucasus. *In* "Problems of the Geology and Paleogeography of the Anthropogene" (V. Ye. Khain, ed.), pp. 5-49. Moscow State University, Moscow.

Milanovskiy, Ye. Ye. (1968). "The Latest Tectonics of the Caucasus." Nedra Press, Moscow.

Murzayev, E. M. (1964). Old glaciation of Central Asia. *In* "Development and Transformation of the Geographical Environment" (G. D. Rikhter, ed.), pp. 168-84. USSR Academy of Sciences, Moscow.

Nikonov, A. A., and Pakhomov, M. M. (1976). Stratigraphy and paleogeography of the Anthropogene of mountainous Badakhshan (Tadzhik SSR, Afghanistan). *Bulletin of the Commission for the Study of the Quaternary Period* 46, 73-89.

Okishev, P. A. (1974). Genesis of terraces in the middle course of the Katun' River. *In* "Problems of Altay Glaciology," vol. 2, pp. 46-73. Tomsk State University, Tomsk.

Okishev, P. A. (1977). Dimensions and characteristics of Late Pleistocene glaciation of the Altay Mountains. *Transactions of Glaciological Studies: Chronicle, Discussions* 29, 203-10.

Pakhomov, M. M. (1967). The spore-pollen flora of the first terrace of Lake Kara-Kul'. *USSR Academy of Sciences, Izvestiya, seriya geologicheskaya* 2, 98-100.

Pakhomov, M. M. (1972). Certain features of the history of the vegetation of the mountains of Central Asia in connection with the characteristics of their paleogeography in the Holocene. *In* "Palynology of the Pleistocene" (V. P. Grichuk, ed.), pp. 249-63. Nauka Press, Moscow.

Pakhomov, M. M. (1980). Paleogeographical aspects of the history of the vegetation of the mountains of Central Asia (as illustrated by Pamiro-Alay). *Botanical Journal* 65(8).

Pakhomov, M. M., and Nikonov, A. A. (1975). Contribution to the history of birch groves on the north slope of the Turkestan Range. *Tadzhik Academy of Sciences, Doklady* 18, 52-55.

Pakhomov, M. M., and Nikonov, A. A. (1977). On the Pliocene glaciation and Kokbayskiy Interglaciation of Pamir. *USSR Academy of Sciences, Izvestiya, seriya geologichiskaya* 8, 126-34.

Penck, A., and Brückner, E. (1901-1909). "Die Alpen im Eiszeitalter." Bd. 1-3. Tauchnitz. Leipzig.

Pomortsev, O. A. (1980). Glaciers of the southern slope of the Kungey-Alatau Range as an indicator of the variability of natural conditions. Abstract of a Dissertation Diploma in Geographical Sciences, University of Leningrad.

Pshenin, G. N., and Serebryanny, L. R. (1981). On the polychronicity of river terraces. *USSR Academy of Sciences, Doklady* 257, 1430-32.

Reyngard, A. L. (1927). Concerning the Quaternary glaciation of the Caucasus. *USSR Academy of Sciences, Doklady,* series A, 19, 319-23.

Serebryanny, L. R. (1978). "Dynamics of Cover Glaciation and Glacial Eustacy in the Late Quaternary Period." Nauka Press, Moscow.

Serebryanny, L. R. (1980). "Laboratory Analysis in Geomorphology and Quaternary Paleogeography." Advances in Science and Technology: Geomorphology, vol. 6. VINITI, Moscow.

Serebryanny, L. R., Golodkovskaya, N. A., and Devirts, A. L., et al. (1978). Contribution to the history of glaciation of the high Caucasus in the Holocene. *USSR Academy of Sciences, seriya geograficheskaya* 2, 107-15.

Serebryanny, L. R., Golodkovskaya, N. A., Gey, N. A., et al. (1977). Paleoglaciological studies in the high Caucasus. *Transactions of Glaciological Studies: Chronicle, Discussions* 29, 221-32.

Serebryanny, L. R., Golodkovskaya, N. A., and Il'ves, E. O. (1979). Variations of glaciers of the high Caucasus in historical time (based on lichenometric and radiocarbon data). *All-Union Geographical Society, Izvestiya* 111, 11-18.

Serebryanny, L. R., and Orlov, A. V. (1980). A method of studying the end moraines of mountain glaciers. *Geomorfologiya* 4, 44-53.

Serebryanny, L. R., Orlov, A. V., and Medvedev, A. S. (1981). Study of the shape and surface character of fragments in moraines of Caucasian glaciers. *USSR Academy of Sciences, Izvestiya, seriya geograficheskaya* 4, 117-24.

Serebryanny, L. R., Pshenin, G. N., and Punning, Ya.-M. K. (1980). Glaciations of Tien-Shan and fluctuations of the level of the Aral Sea (stage-by-stage analysis of events in the Late Quaternary history of Central Asia). *USSR Academy of Sciences, seriya geograficheskaya* 2, 52-65.

Shcherbakova, Ye. M. (1978). "Old Glaciation of the Greater Caucasus." Moscow State University, Moscow.

Sheynkman, V. S. (1978). Characteristics of the Upper Pleistocene glaciation of the Verkhoyano-Kolymskiy Highland. *Transactions of Glaciological Studies: Chronicle, Discussions* 34, 93-100.

Shnitnikov, A. V. (1957). Variability of total humidity of continents of the Northern Hemisphere. *USSR Academy of Sciences, Geographical Society Transactions,* New Series 16.

Shnitnikov, A. V. (1979). "Issyk-Kul': Nature, Protection and Prospects for the Use of the Lake." Ilim Press, Frunze.

Shnitnikov, A. V. (ed.) (1980). "The Lakes of Tien-Shan and Their History: Physical Geography and Paleogeography." Nauka Press, Leningrad.

Sidorov, L. F. (1979). "Nature of the Pamir in the Quaternary." Nauka Press, Leningrad.

Sukachev, V. N., and Zonn, S. V. (eds.) (1955). "Broad-Leaved Forests of Northwestern Caucasus." USSR Academy of Sciences Press, Moscow.

Svitoch, A. A., and Khorev, V. S. (1975). Old glacial deposits and glaciations of the Altay Mountains. *USSR Academy of Sciences, Izvestiya, seriya geograficheskaya* 4, 101-8.

Tolmachev, A. I. (1944). The ice age in the history of the development of the vegetation of the Pamiro-Alay. *Tadzhik Branch, USSR Academy of Sciences, Izvestiya* 7, 8-13.

Tsereteli, D. V. (1966). "Pleistocene Deposits of Georgia." Metsniyereba Press, Tbilisi.

Tushinskiy, G. K. (1968). Rhythms in the glaciation and snow abundance in the Caucasus in historical time. *In* "Glaciation of El'brus" (G. K. Tushinskiy, ed.), pp. 256-64. Moscow State University, Moscow.

Velichko, A. A. (1980). Latitudinal asymmetry in the state of natural components of glacial epochs in the Northern Hemisphere. *USSR Academy of Sciences, Izvestiya, seriya geograficheskaya* 5, 109-17.

Velichko, A. A., and Lebedeva, I. M. (1974). Experience in paleoglaciological reconstruction for the eastern Pamir. *Transactions of Glaciological Studies: Chronicle, Discussions* 23, 109-17.

Zabirov, R. D. (1955). "Glaciation of the Pamir." Geografgiz, Moscow.

CHAPTER 7

Late Pleistocene Glacier Regimes and Their Paleoclimatic Significance

I. M. Lebedeva and V. G. Khodakhov

For epochs of major global climatic cooling, glacier advance, and permafrost development, many environmental indicators have decreased effectiveness (Velichko, 1980). However, cryogenic features may be more useful because of their increased areas and because of their direct response to processes of heat and mass exchange with the surroundings according to well-known physical laws. In this respect, glaciers are easier to study than permafrost, because their surface is not separated from the atmosphere by a soil and plant cover, which is a highly variable and capricious heat insulator, and because they exchange with the atmosphere both heat and water, which are key climatological elements in the natural environment.

Calculation of Mass Balance

According to the definition given by Shumskiy (1969), "Glaciers are ice flows of atmospheric origin that have acquired a form and structure determined by motion and that exchange mass and energy with the surroundings." At the same time, the motion and the very existence of glaciers are determined by the heat and mass exchange at their surface. As a rule, on dry-land glaciers, more than 99% of the heat and mass exchange is associated with the upper subaereal surface and so is determined by climate. Only in the central parts of the larger ice sheets, particularly in the case of increased geothermal heat flow, is bottom melt possible at a rate of about 0.5 g per square centimeter per year, which amounts to 15% of the accumulation on the surface. However, in the glacier mass balance as a whole, bottom melt is insignificant in this case as well.

On the glacier surface in the accumulation area, the supply of snow is greater than the discharge of meltwater and evaporation; in the ablation area, the reverse is true. In the extreme, the latter region may be confined to a line, as in many portions of the present Antarctic glacier. These two areas are separated by the glacier equilibrium line. A dry-land glacier (which is the only one considered below) becomes stationary when the supply of snow in the accumulation area is exactly equal to the loss of ice in the ablation area. Otherwise, the glacier advances when the snow accumulation exceeds the wastage, or it retreats when the opposite is true. Because all climatic characteristics have always changed with time, we discuss the stability of climate and glaciers only within certain limits determined by the accuracy of the method of reconstructing glaciation and climate. High-frequency climatic oscillations, even very large ones, are automatically filtered out by the glacier. Mountain glaciers have a short time of reaction to climate changes—on the order of decades, compared to centuries for larger glaciers and millennia for large ice sheets. Therefore, we use a different approach for the reconstruction of mountain glaciers and ice sheets. Accordingly, in the case of mountain glaciers, the paleoclimatic parameters for a stable ice margin are obtained automatically from the reconstruction of their regime, whereas for ice sheets it is necessary to know the basic paleoclimatic parameters for the epochs of advance and retreat.

The first step in both problems is to reconstruct the shape of the former glacier. Most glaciers carry a fair amount of morainic material within ice and on the surface. The process of deposition for this material has been studied by many authors and described analytically by Khodakov (1978). The longer the glacier remains in a stationary state, the larger the end moraine it forms. For over a century, such ramparts have been the main subject of studies of former glaciers. On plains, they outline the position of an ice sheet during the phases of the maximum and the subsequent retreat. In the mountains, lateral moraines and abrasion marks on bedrock may reveal the shape of the glacier in three dimensions (neglecting the convexity of the glacier surface in cross section). The three-dimensional reconstruction of ice sheets can be based either on the solution of problems of dynamic glaciology (Shumskiy, 1969) or on a direct analogy to recent glaciers (Khodakov, 1978). It is our view that insufficient knowledge of the viscoplastic properties of large ice masses makes the second approach preferable.

The second step is the reconstruction of the rate of accumulation and ablation of snow and ice on a glacier surface of a given shape. To date, ablation measurements have been made on dozens of glaciers over the widest range of variations of summer climatic conditions: from the snout of the Franz Josef glacier in the subtropical forests of New Zealand to the polar deserts of Antarctic mountains, where only evaporation is involved. Figure 7-1 shows the results of the most reliable measurements covering the entire period of ablation and accompanied at least by meterologic measurements and in many cases (numbers denoting points) by actinometric measurements as well. This relation is approximated by the formula:

$$A = 0.1(t + 1.3\sqrt{B_k} + 4)^3, \qquad (1)$$

where A is the annual ablation of snow and ice in grams per square centimeter per year, t is the air temperature averaged over three summer months at a height of 2 m above the glacier, and B_k is the total balance of solar radiation for three summer months in kilocalories per square centimeter. Because of the complexity of using radiation in paleoclimatic reconstructions, we hereafter apply a simplified version of equation 1:

$$A = 0.1(t + a)^3. \qquad (2)$$

The term a of equation 2 can range from 8 °C (if heavy cloudiness exists and if the glacier surface is covered mainly with snow) to 12 °C (if sunny weather predominates and the ice surface is contaminated with ablation till), or it may be assumed to be equal to the average value a equals 9.5 °C.

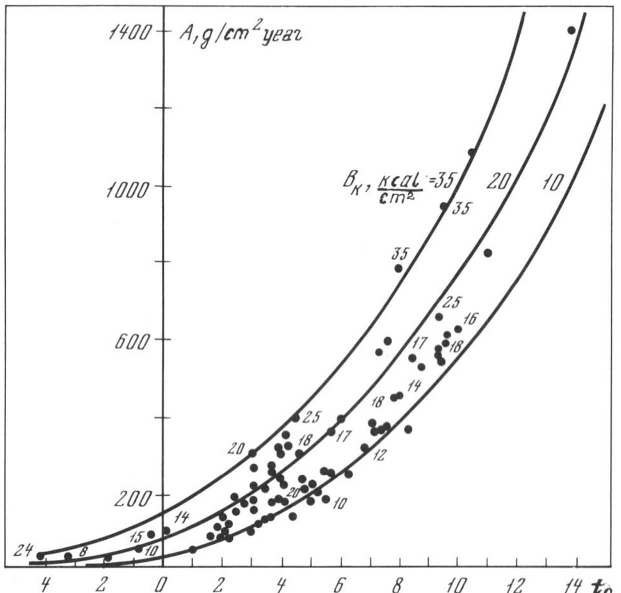

Figure 7-1. Total annual ablation of glacier surface (A) versus mean summer (June-August) air temperature above the glaciers (t_g) and the sum of absorbed solar radiation B_k (numerals near points and curves).

The relation between the summer temperature above the surface of a former glacier, t, and above a nonglacier surface today (relative to the height of the snout of the former glacier H_o t_p in kilometers), is expressed by equation 3:

$$t = t_p + \Delta t_g + g(H\text{-}H_o) + \Delta t_k, \qquad (3)$$

where Δt_g is the cooling influence of the glacier, g is the vertical gradient of the mean summer temperature of air above the glacier's ablation area (most frequently taken to be 7 °C per kilometer but ranging from 3 °C per kilometer at the ocean coast to 10 °C per kilometer in ultracontinental regions), H is the height of the former glacier surface in kilometers and Δt_k is the depression of the summer air temperature in the given region when the glacier was present. The quantity Δt_g for ice sheets and, in many cases, for mountain glaciers can be calculated from the equation (Khodakov, 1978):

$$\Delta t_g = 0.28 \log L - 0.07, \qquad (4)$$

where L is the typical length of the glacier in kilometers. As the typical dimension we use the distance from the ice divide to the edge of the ice sheet or, in a mountain glacier, to the beginning of the taper of the tongue.

Much more complicated is the calculation of the annual total atmospheric precipitation, X, and snow accumulation on the glacier, C, in grams per square centimeter per year. For mountain glaciers, X resulted from the orographic rise of air masses just as today. Furthermore, the decreased atmospheric moisture resulting from the cooling could have been offset, particularly at temperate latitudes, by the intensified circulation, which increased the vertical component of wind velocity in the flow of air around mountains, thus directly influencing the increase in the rate of precipitation (Khodakov, 1978: Chapter 2). Therefore, in the absence of additional evidence, we subsequently assume that, in regions of former mountain glaciation, X during maximum glaciation was equal to the present annual total precipitation X_t. In such an analysis, it is important to determine the dependence of X_t on the altitude H of the locality; therefore, corresponding calculations are possible only for closely studied mountain regions. The fraction of precipitation that falls as snow γ increases with altitude. This effect is reliably taken into account by Lauscher's (1954) formula:

$$\gamma = \alpha - \beta T, \qquad (5)$$

where T is the mean monthly air temperature and the parameters are, on the average $\alpha = 0.5$ and $\beta = 0.05$. This calculation is important for obtaining C on mountain glaciers but not for large ice sheets. In addition, on mountain glaciers, it is also necessary to consider the concentration of snow by wind and avalanches. If total annual snowfall is denoted by X_s, the coefficient of snow concentration on a glacier in a given altitudinal zone (Khodakov, 1978) is:

$$K = C/X_s. \qquad (6)$$

For large mountain-valley glaciers, this coefficient averages 1.1, for medium-sized glaciers 1.4, and for cirque glaciers 2.0.

In principle, the precipitation X_t on plains cannot be applied to the surface of former ice sheets. The latter intensify the effect of rising air currents, as do mountains, although they also block the passage of cyclones and intensify the cooling of air, so that the end result is uncertain. Below, we use for ice sheets the present-day dependence of accumulation C on typical dimension L.

In the accumulation area, the glacier's mass balance also includes the internal accumulation G, that is, the freezing of meltwater and rainwater in the body of the glacier because of the winter cold reserve (Khodakov, 1978: Chapter 7). According to measurements, this value on temperate mountain glaciers is 20 to 60 g per square centimeter per year (the average is 40 g). On the whole, the specific ice balance on the surface of the glacier (grams per square centimeter per year) is:

$$b = C - A + G. \qquad (7)$$

Integration of b over the entire area of the glacier gives the sign and absolute value of the rate of change of its mass; that is, it completely determines the direction of the glacier's development. In the course of such calculations, the altitude of the equilibrium line of the glacier, sometimes called the climatic snow line, is obtained automatically. Moreover, as is done in many paleoglaciologic reconstructions, it is not necessary to establish a direct correlation between the depression of the snow line and climatic change, for the principal components (summer air temperature and annual precipitation) are directly reflected in the quantities C and A in equation 7. Calculations of balance in the manner indicated above were made for various mountain glaciers and for the section through the Greenland ice sheet for which the most direct measurements were available (Khodakov, 1978). In all cases, good agreement was obtained between the calculations and the field data. It goes without saying that such a calculation for former glaciers cannot claim a similar accuracy, primarily because of probable variations in the main parameters of the calculation. Therefore, an attempt is made below to evaluate as objectively as possible the reliability of the final results and the conclusions based on them, often with an appreciable exaggeration of possible variations in the calculation parameters.

Reconstruction of the Regime of the Former Ice Sheet

The balance method of reconstruction was applied in most detail to the northern European Late Pleistocene (Late Valdai, Würmian) ice sheet (Khodakov, 1978) and was subsequently extended to Arctic glaciers (Khodakov, 1979). An essentially nonstationary problem was solved in which the time of change τ of the ice sheet is related to the change of its mass, $\Delta(HL)\varrho$, and to the ice balance on the upper surface b as follows:

$$\tau = \varrho \int_{L_j}^{L_{j+1}} \frac{a(HL)}{bL} \qquad (8)$$

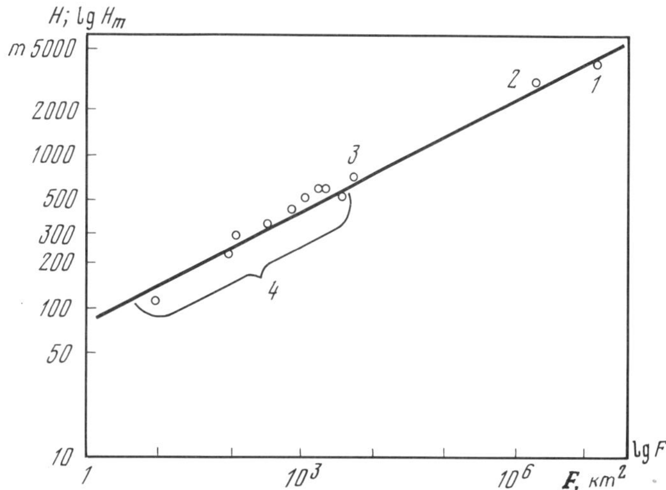

Figure 7-2. Maximum altitude of surface H_m versus area F in recent ice sheets: (1) Antarctic, (2) Greenland, (3) Novaya Zemlya, (4) other glaciers of the Arctic and Antarctic.

For all cases of a glacier spread out on a plain, with a correspondingly small divergence of the streamlines, the quantity L_j in kilometers is simply the length of a segment of the glacier extending from the ice divide to the margin in phase j, with a width of 1 km. The quantity H in kilometers is the average thickness of the glacier, b is the ice balance on its surface in 10^6 tons per square kilometer per year, and ϱ is the average thickness of the glacier, essentially amounting to 0.9×10^9 tons per cubic kilometers.

Statistical treatment of data on the shape of recent ice sheets shows that the maximum height of their surface above the edge H_m is fairly closely related to both the total area of the glacier F (Figure 7-2) and the length of the corresponding glacier segment L (Figure 7-3). The corresponding relationships for the average surface height H are even closer. The ratio H/H_m differs by only 4% from that for a regular ellipse. Using an elliptic approximation of the shape of ice sheets in profile in our further calculations, we obtained by the method of least squares for the best-studied profiles:

$$H = 0.074\sqrt{L} \text{ and } H_m = 0.094\sqrt{L}. \qquad (9)$$

We have already noted the role of L as a thermal predictor (equation 4). For recent ice sheets and possibly for many former ones not too distant from sources of nourishment (ice-free water areas), this predictor also gives the snow accumulation C on the glacier. In Figure 7-4, the quantity C was obtained on glacier transects from direct measurements of the mean annual accumulation averaged over the entire of length L. Basically, such are the initial data of the calculation, the numerical results of which are tabulated by Khodakov (1978). We only briefly describe them here.

Figure 7-5a shows the distribution of average yearly temperature, t_p, over a section through Europe along a straight line from Jan Mayen Island to Moscow. During the last major advance of the ice sheet (lasting for 17,000 years from 35,000 to 18,000 yr B.P.), but outside the zone of

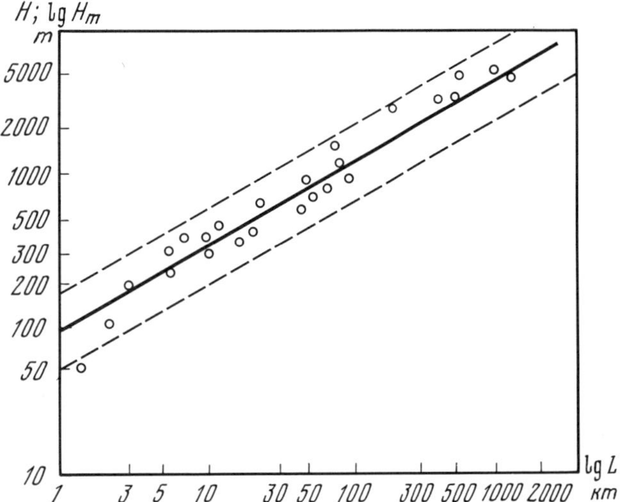

Figure 7-3. Maximum altitude of surface H_m versus typical dimension L of recent ice sheets.

its direct influence (no closer than 100 to 200 km from its margin), this quantity ($t_p + \Delta t_k$) was an average of 8°C lower. It was obtained as an average of various data (Khodakov, 1978) and confirmed by the most recent studies (Kolstrup, 1980). It may be considered demonstrated that the transition from a warm time to a cold one and vice versa did not follow a sine curve but occurred very abruptly and that thereafter the temperature fluctuated around an entirely new main level (Schüpp, 1979). It is evident from Figure 7-5a that, despite a marked cooling, the average yearly temperature in the north of the Central Russian Plain was still fairly high. It is obvious from the calculated curves that an unmelted snow-firn cover was formed only in the Scandinavian mountains and hence that this was the only area from which the glacier could have advanced onto the plains. A somewhat different situation could have existed in the floating sea ice. However, even there the advance of the glaciers undoubtedly came

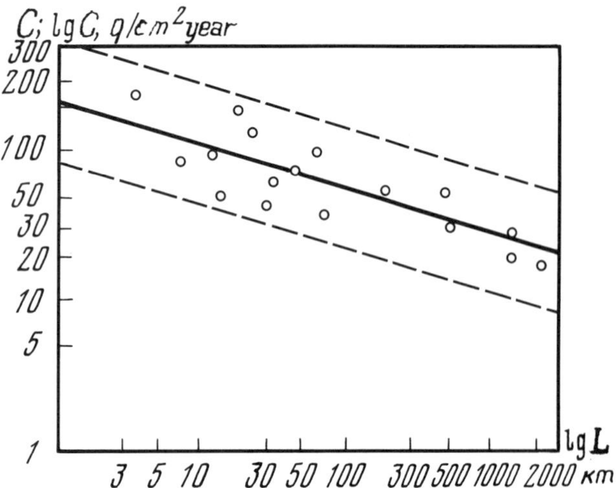

Figure 7-4. Mean annual accumulation C versus typical dimension L of recent ice sheets.

from orographic centers (i.e., recent ice caps), where C was several times greater than the snow accumulation on the plain (Khodakov, 1979).

Figure 7-5b shows advances of the northern European ice sheet in the direction indicated above. A recent small glacier of a Scandinavian type follows the topography in phase 1. In order to reach phases 2, 3, 4, 5, and 6 with all the average calculated parameters, the required time τ equals 2820, 7160, 11,650, 17,000, and 21,300 years, respectively. It is apparent that the control figure (17,000 years) is reached in phase 5, which corresponds to the northern piedmont of the Valdai upland. The unquestionable traces of a former glacier on the upland itself, and even south of it, may be explained either by a long duration of the ordinary advance to phase 6 or by the "degradation advance" effect (Khodakov, 1978), which we take up later.

We also reconstructed the average yearly temperature of air on the surface of the calculated northern European ice sheet. This temperature was found to practically coincide with those on recent ice sheets with a corresponding typical dimension L. However, although both the external mass exchange and the temperature of the surface of the advanced former glacier (and in the stable case the temperature of the entire mass as well) were analogous to the recent ones on the average, its shape should correspond to the shape of recent ice sheets, if only for physical reasons (Shumskiy, 1969). Hence, the calculations for the ice sheet considered here should be regarded as the most probable ones not only statistically but also physically by analogy with the present ice sheets.

The proposed model can consider objectively the most diverse combinations of past conditions, variants that for whatever reasons did not correspond to the conditions known from recent data. Then, the calculation should be made to include not only average but also extreme relationships. Thus, the dashed curves in Figures 7-3 and 7-4 indicates the deviation from the mean with a probability of 5%. The extreme limits in Figure 7-2 correspond to the maximum cloudy and maximum sunny summer weather. Let us merely note that a combination of several sets of low-probability conditions quickly leads to an extremely low, practically unrealistic probability of a deduction.

The paleogeography and geochronology of Europe's latest glaciation have been closely studied, and so an analysis of the behavior of a model glacier is strictly controlled by the experimental data. The total time of retreat for the northern European ice sheet may be assumed to be roughly from 18,000 to 8000 yr B.P., or 10,000 years. This is the main reference figure, and intermediate control of recessional phases is based on dating recessional moraines.

The temperature of the summer period during retreat of the last ice sheet from the European plains was characterized by great instability, but on the average was 2°C to 3°C lower than the present temperature, and toward the end of this period became 2°C to 3°C higher than the present temperature (Holocene optimum). We include in the calculation the constant value Δt_k equals -2.5°C. Thus, it was found that the ice balance of phase 5 of the advance

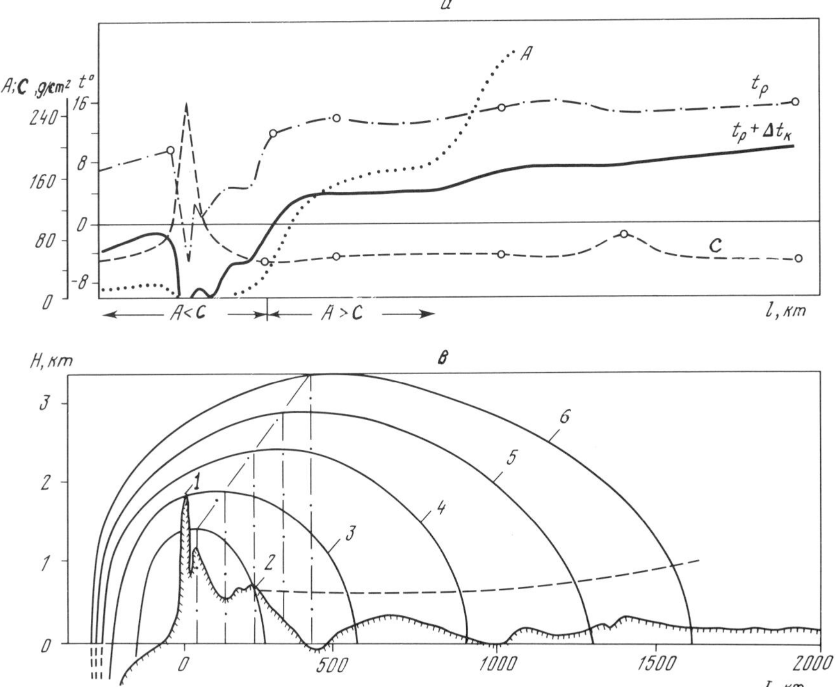

Figure 7-5. (a) Climatic conditions and (b) shape of the last northern European ice sheet in the epoch of its advance. Abbreviations: t_p, recent mean summer air temperature; Δt_k, its depression in the epoch of glacier advance; C, recent total annual precipitation; A, ablation of snow and ice at temperature $t_p + \Delta t_k$; 1, current topographic section, starting at the water divide of the Scandinavian mountains; 2 through 6, numbers of glacier advance phases; dashed line, calculated altitude of its equilibrium line.

(Figure 7-5) was almost exactly equal to zero; that is, the glacier could have remained in place but could not have retreated. In phase 6, the position was similar.

We noted above that the shape of the glacier during the advance was determined by statistical and physical analogy to recent ice sheets. Moreover, the fact that the temperatures of the former glacier ice of different dimensions were identical to recent analogues implies the same ice plasticity and conditions of basal sliding. The rigid ice of the cold epoch gave a fairly sharp form to the surface of the glacier. The marked warming of 5°C to 6°C during the deglaciation must inevitably have led to an appreciable heating of the entire body of the glacier. Moreover, the elevation of the mean temperature of the glacier in the zone covered by the melting of previously accumulated firn was much greater than 5°C to 6°C because of the secondary freezing of the meltwater and possibly because of the appearance of rainwater. We estimate that for the glacier of phase 5 of the advance the ice and firn were heated by as much as 20°C.

This situation must inevitably have resulted in a change in the shape of the glacier. Paterson (1972) estimates a thinning of 35% with a warming of 20°C. Possibly at a certain time and in a certain sector a comparatively slow advance changed to a rapid one, as is seen in recent mountain and valley glaciers (Dolgushin and Osipova, 1971).

Thus, the climatic warming led to (a) a rise of the equilibrium line (dashed curve in Figure 7-6), an increase in the ablation area, a corresponding decrease in the accumulation area, and as a result a decrease of the ice balance (see equation 7); and (b) an increase in flow and hence to an advance of the glacier margin, while the lowering of the surface of the glacier enhanced ablation. At different times and in different places in that epoch, high velocities formed glacier lobes at the ice-sheet margin. This entire set of phenomena is termed a "degradation advance" (Khodakov, 1978), characterized by a rapid decrease in thickness along with an advance and a general loss of glacier mass resulting from the sharply increased ablation of the glacier surface.

Figure 7-6 shows the transition from the advance phase to initial phase 1 of retreat without any change in glacier mass. The glacier margin, in accordance with paleogeographic data, is located on the southern periphery of the Valdai upland. On the basis of elementary geometric considerations, with the elliptic shape being retained, relations (equation 9) for the epoch of glacier degradation becomes

$$H = 0.048\sqrt{L} \text{ and } H_m = 0.061\sqrt{L}. \qquad (10)$$

We see that the decrease in glacier thickness corresponds to a calculation based on completely independent and purely physical considerations (Paterson, 1972). The new shape does not differ from the old one enough to exceed the range of the 5% confidence interval (Figure 7-3).

The calculations made after the introduction of Δt_k equals -2.5°C and with the preceding dependence of C on L (Figure 7-4) give the time of retreat from maximum phase 1 (Figure 7-6) to 2, 3, 4, 5, and 6 as 2740, 6175, 8020, 9065, and 9715 years, respectively. Starting with phase 4, the glacier is devoid of the accumulation area, and its degradation assumes a catastrophic character. Such a situation is known from direct observations in the epoch of recent warming in the Arctic on the Novaya Zemlya and Severnaya Zemlya ice sheets (Khodakov, 1978).

Similarly, Khodakov (1979) has calculated the advance of ice sheets in the eastern sector of the Arctic (Figure 7-7).

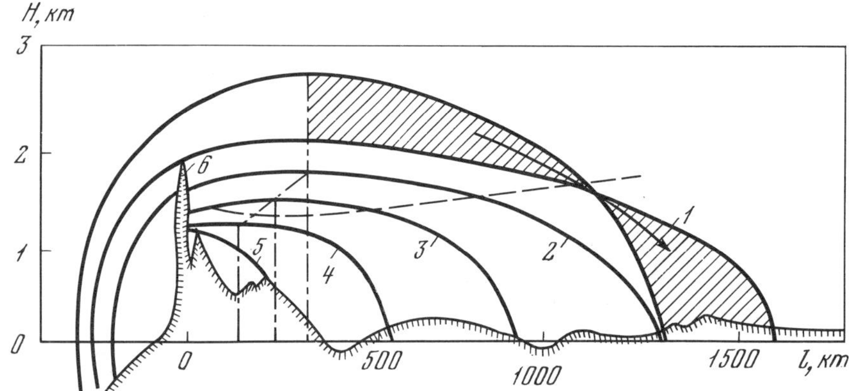

Figure 7-6. Shape of the last northern European ice sheet in the epoch of its retreat (1 through 6 are the numbers of the retreat phases. Lined area shows glacial advance during initial stage of thinning.)

Certain specific aspects of this calculation are related to the conditions of a shallow sea and its ice cover, but they do not play a decisive role. In accordance with existing dates for marine deposits, the control figure for the entire cycle of advance was taken to be 12,000 years, so that phase 5 is absent from Figure 7-7. As follows from calculations based on average data, all the separate ice sheets in the Arctic coalesced at the time of their maximum extent. No analogues exist today. One can only assume that converging ice flows sent active lobes into regions as marked on Figure 7-7. However, the paleogeographic situation of the epoch of advance makes it necessary to postulate that the quantities C used in the calculation (Figure 7-4) were actually smaller, perhaps by a factor of 2. Then, the duration of the advance increases proportionately, and there is not enough time for the glaciers to coalesce.

With all its obvious shortcomings, the model under consideration has the advantage of taking the greatest account of data on recent ice sheets, which reflect in concealed fashion certain effects that cannot yet be adequately described theoretically.

Reconstruction of the Regime of Former Mountain Glaciers

The reconstruction was done for individual glaciers of the Caucasus, Kamchatka, and the Pamir in the USSR and for the Wasatch Range in the USA. It was based on the above-described system of equations 2 through 7, which constitute an actualistic model of mountain glaciation.

The calculation sequence was as follows. First, geomorphic features, palynologic evidence, frost soils, and other characteristics were used to establish the general climatic situation of the period of 20,000 to 18,000 years ago, confirmed insofar as possible by radiocarbon dating (Velichko and Lebedeva, 1974; Khodakov, 1978). Principal attention at this stage was given to the evaluation of snowfall from spore-pollen data and to their distribution with altitude.

Then, calculations of the volumes of accumulation and ablation were carried out at equal-altitude intervals above the glacier margin, as marked by remains of old end-moraine complexes, up to the ice divide, in order to obtain the air temperature for which A equals C plus G. Summer temperature was selected by calculating the increase in volume of snowfall if the total precipitation and volume of ablation remained constant. The reconstruction of the glaciers was made in only four highlands (Table 7-1), but they are exposed to such different climatic conditions, from especially maritime (Kamchatka) to highly continental (Pamir), that the results obtained make it possible to draw certain important global-scale conclusions about the climatic and natural characteristics of the Late Pleistocene glacial maximum.

The majority of glaciers studied are still in existence (with the exception of Little Cottonwood and Bells in the western USA), and they are of the valley type. The altitude of the firn line on them differs very markedly—from 0.5 km on the Koryto Glacier (in maritime Kamchatka) to 4.9 km on the Uysu Glacier (very continental eastern Pamir). Snowfall at these levels amounts, respectively, to 300 g per square centimeter per year—the greatest precipitation on Eurasia's glaciers—and 65 to 75 g per square centimeter per year. In the central part of the Caucasus, where the climate is temperate on the Kolka-Mayli Glacier, the present altitude of the firn line is 3.6 km, and the precipitation at this level amounts to 110 g per square centimeter per year.

The presence of snowfields in the near-crest portion of the valleys of the Wasatch Range, occupied by the Little Cottonwood and Bells Glaciers, made it possible to calculate the summer temperature at their altitude level. The calculation was made for g between a pair of lower-lying weather stations, Cottonwood Dam (1509 m) and Silver Lake (2664 m), by using equations 2 and 5. It was found that the present snowfall at an altitude of 3.5 km in the Wasatch Range amounts to 155 g per square centimeter per year, which is close to the precipitation at approximately the same altitude in the Caucasus (Table 7-1).

We have already substantiated the assumption that at the maximum of Late Pleistocene glaciation the total atmospheric precipitation was approximately the same as it is now and, therefore, that the values used in the reconstructions are those given in Table 7-1. It is shown quantitatively below that Late Pleistocene mountain glaciation could not have been caused solely by an increase of atmospheric precipitation in the range of possible values.

The selection of t at which the volume of ablation on glaciers becomes equal to the accumulation led to the de-

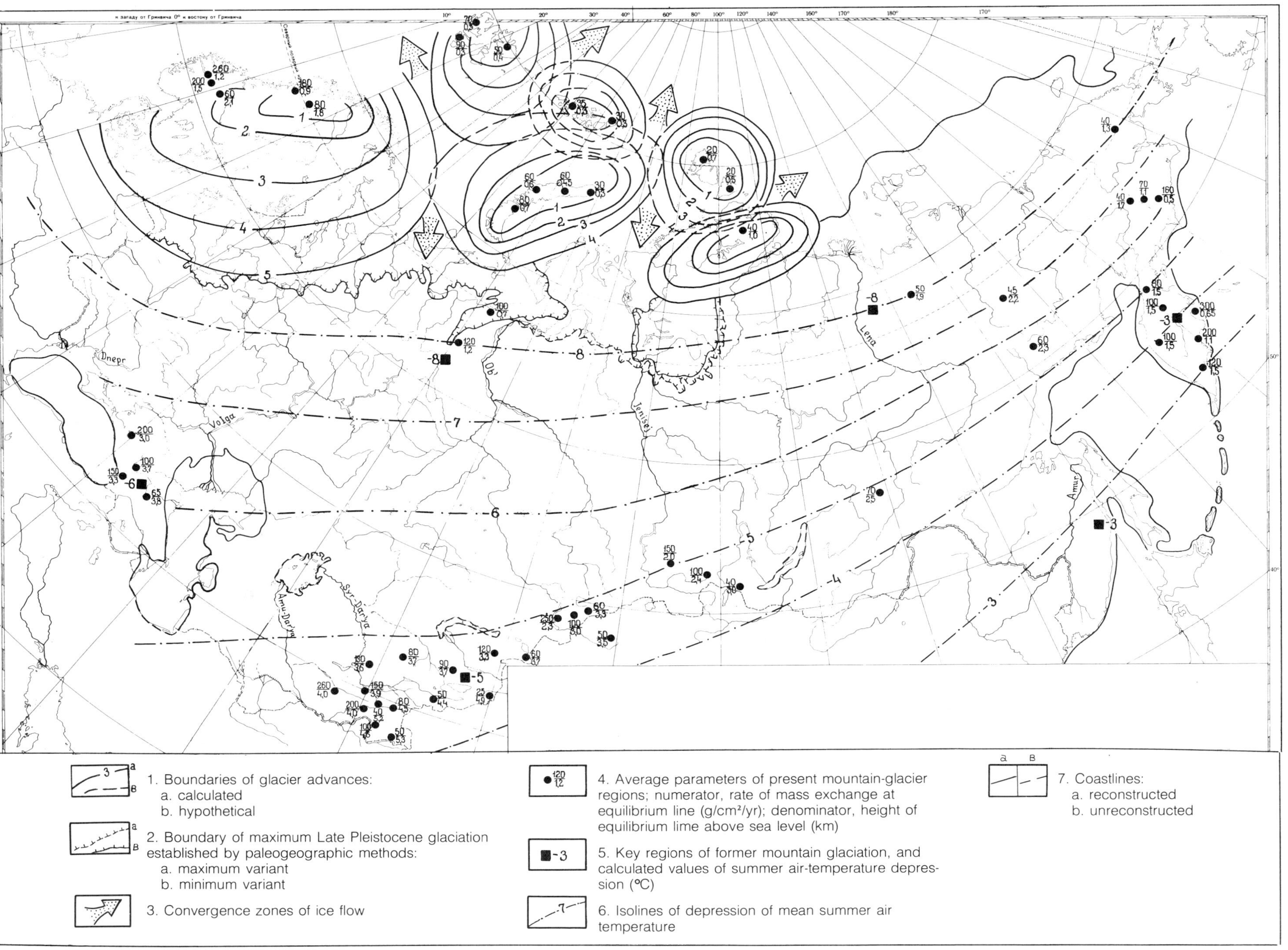

Figure 7-7. Glacioclimatic reconstruction in the regions of ice sheets and mountain glaciers in the USSR at the Late Pleistocene maximum. (Compiled by V. G. Khodakov.)

Table 7-1. Annual Precipitation (X_s), Altitude of Present Firn Line (H), and Firn Line at the Maximum of the Late Pleistocene (H'), $t°$ at H', Firn-Line Depression ($H-H'$), and General Climatic Cooling (Δt_k)

Region	Glacier	X_s (g/cm²/year)	H (km)	$t°$ at H (°C)	H' (km)	$t°$ at H' (°C)	$H-H'$ (m)	Δt_k °C
Caucasus	Kalka-Mayli	110	3.6	2.0	2.7	2.3	900	6
Pamir	Uysu	75	4.9	−0.5	4.3	0.7	600	3
Pamir	Fedchenko	65	4.7	−2.0	4.1	−0.8	600	3
Kamchatka	Koryto	300	0.5	6.5	0.3	4.5	200	3
Wasatch	Little Cottonwood	155	3.5	6.6	2.7	3.2	>900	9
Wasatch	Bells	155	3.5	6.6	2.6	3.9	>900	9

termination of Δt_k, which is the depression of t at 20,000 to 18,000 yr B.P. (Table 7-1). The summer temperature was found to be 9°C below the present one in the Wasatch Range, 6°C lower in the Caucasus, and 3°C lower on the Pamir and Kamchatka.

It is immediately apparent that the difference between these values in the highlands depends on how far they are from the "cold center" located at high latitudes. As the distance from the ice sheet increased, the general climatic cooling became attenuated. It was most pronounced in the Wasatch Range, for the margin of the ice sheet was less than 1000 km away. The Valdai cover was much more distant from the Caucasus, but the Pamir turned out to be the farthest—3000 km away. Therefore, although the Little Cottonwood, Bells, Uysu, and Fedchenko Glaciers were located at almost the same latitude (40°N), the temperature conditions of their existence were quite different.

The fact that it was a lowering of air temperature and not an increase in atmospheric precipitation that played a decisive role in the formation and existence of large mountain glaciers in the Late Pleistocene can be ascertained from the reconstruction of the mass balance of Little Cottonwood Glacier. Its maximum dimensions were recorded by moraines, which were closely associated with dated deposits of Lake Bonneville (Richmond, 1964; Morrison, 1965). The glacier was 18 km long, and it completely occupied a valley that is now empty.

For the same total precipitation as the present 53.7 g per square centimeter per year at the Cottonwood Dam weather station, 109.5 g per square centimeter per year at the Silver Lake weather station, and 201 g per square centimeter per year at the snowfield level, the existence of the glacier was due to the fact that, with cooling, the reduction of ablation that took place was several times greater than the increase in the proportion of snowfall. Under present climatic conditions, 0.42 km³ per year of snow and ice, expressed in terms of water, could emit and evaporate from the area occupied by the glacier; not even a quarter of this amount is supplied by atmospheric precipitation. As the summer temperature dropped by 9°C, this value was reduced fourfold and was compensated by an increase in snowfall of only 30% (Figure 7-8, curves 1, 4, and 5).

Even if the atmospheric precipitation increased to 300 g per square centimeter per year, as is the case only in highlands with a maritime climate, only a mean annual cooling of 8°C would cause glaciation in the Wasatch Range (Figure 7-8, curves 3, 4, and 5).

It is apparent from Table 7-1 that the depression of the firn line was substantially greater in highlands with a temperate climate than in those with a continental one. This results from the fact that as the air temperature decreases the rate of decrease in ablation is greater in the range of high summer temperatures than in low (Figure 7-1). Therefore, at the same Δt_k in regions of temperate climate, where a large amount of precipitation allows the glaciers to descend to low altitudes, the reduction in ablation and the depression of the firn line are more pronounced than in continental regions; there, because of a small amount of precipitation, the glaciers are confined to high altitudes, where the temperatures are cold. For example, a cooling of 1°C will result in a decrease in ablation by 60 g per square centimeter per year if this decrease occurs at a mean annual air temperature of 5°C but only 13 g per square centimeter per year at a summer temperature of −2°C.

According to this logic, one could expect that on Kamchatka, an area with a maritime climate, the depression of the firn line must have been most pronounced. However, it can be seen from the example of the Koryto Glacier that an important role is also played by the orographic characteristics of the glaciation region. In this case, during the cooling period, the Koryto Glacier, which occupied a wide valley with a very low gradient, advanced into the foothills with a large lobe, and the increase in accumulation volume was offset by ablation, with a small depression of the firn line.

Hence, it is clear that the depression of the firn line at the maximum of Late Pleistocene glaciation ranged between very wide limits, depending (1) on how far a given highland was removed from the margin of the ice sheet, (2) on the continentality of its climate, and (3) on the orographic characteristics of the glaciation.

Still another important consequence of the nonuniformity of cooling is evident from the data in Table 7-1. As the altitude of an area increases in highlands, the air temperature decreases by 0.7°C per 100 m. The summer temperature at the firn line in the Late Pleistocene turns out be higher than the present one if the warming associated with the decrease of its altitude is not offset by a general climatic cooling. This factor was important in the mountains of the Caucasus and the Pamir.

In the Wasatch Range, the general climatic cooling was so pronounced that, even when the firn line dropped to 2.7 km, the summer air temperature at this level was substantially below the present temperature in the snowfield belt at an altitude of 3.5 km.

In general, the decline of the firn line caused by cooling results in an increase in the area and volume of accumulation on the glaciers. This also results in an expansion of the

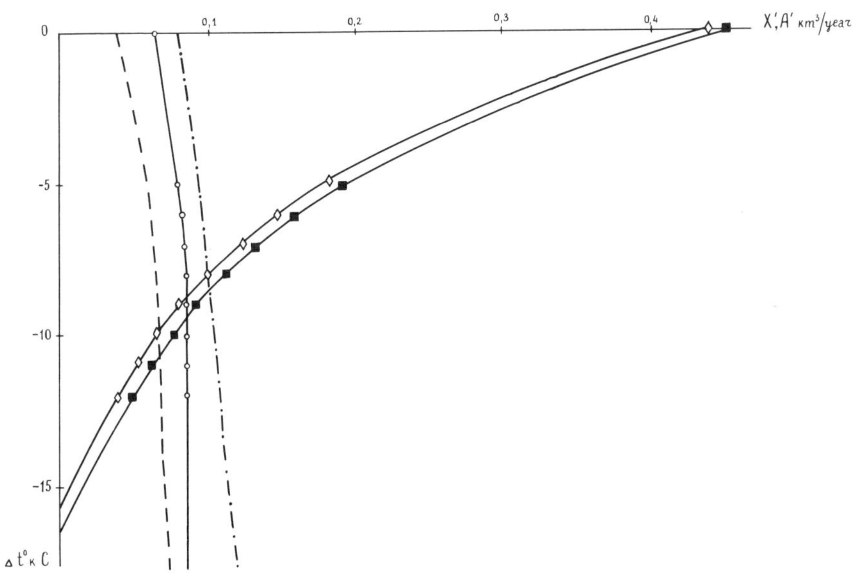

1. X' for total precipitation equal to the present one
2. X' for total precipitation one-third less than the present one
3. X' for total precipitation one-third greater than the present one
4. A' at g=10°C/km and $\Delta t_s = 2$°C
5. A' at g=7°C/km and $\Delta t_s = 4$°C

Figure 7-8. Calculation of volume of accumulation (X^{-}, km³/year of water) and ablation (A^{-}, km³/year of water) on the Little Cottonwood Glacier at the maximum of Late Pleistocene glaciation. Δt_s is for summer temperature, and g is vertical gradient of air temperature.

ablation region, mainly as a result of the elongation of the tongues. However, in the second case, when the air temperature at the altitude of the firn line and in the ablation region was depressed, the glacier tongues must have been particularly large. Therefore, whereas in the Wasatch Range large valley glaciers appeared in areas where there is no glaciation at all at the present time, in the Pamir the increase in glaciation area was approximately equal to that of the present glaciation.

Everywhere at the cooling maximum of the Late Pleistocene the runoff from precipitation in regions of mountain glaciation was much greater than the present one, for the precipitation, as already stated, was approximately the same as now. This results from the fact that the evaporation from the snow and ice surface is much less than from the ground, for the surface does not heat up above 0°C. Therefore, the volume of evaporation from catchment areas occupied by glaciers was appreciably smaller than it is with the present glacier dimensions, because today there are vast areas of exposed ground surface.

However, the surface of the ground also heated up and evaporated less than now because of the cooling. According to our calculations, the runoff from the glacier catchment areas of the Pamir was approximately four times what it is at the present time (Lebedeva, 1977). The same increase in runoff characterized the Little Cottonwood and Bells Rivers in the Wasatch Range.

The increase in runoff caused the transgression of lakes in intermontane troughs of the Great Basin in North America, and the decrease in evaporation from the water surface caused the formation of a huge freshwater lake, Lake Bonneville, which at least in the eastern part almost came into contact with glaciers that had descended to an altitude of approximately 1.5 km.

The flux of solar radiation at the maximum of Late Pleistocene glaciation may be assumed to be equal to the present flux (Gates, 1976). However, because of heavy cloudiness, the total incoming radiation in the region of Lake Bonneville may have been less intense. For this reason, and also because of lesser heating of the surface, the effective outgoing radiation also should have been weaker. It can be postulated, therefore, that the radiation balance was similar to the present one.

At the same time, the heat balance should have undergone appreciable changes because of the decrease in evaporation related to the climatic cooling. With the same total precipitation as now, namely 54 g per square centimeter per year at an altitude of 1.5 km, where the glaciers ended, but with a lower summer temperature (approximately 15°C), the soil was better supplied with water, and there are indications that under these conditions forests grew below the glaciers (Morrison, 1965). Thus, neither lake transgression nor the appearance of forest vegetation in any way signifies a transition to a pluvial epoch, but both are caused only by a cooling of the climate and a rearrangement of the heat balance in the extraglacial regions.

Paleoglaciologic reconstructions make it possible, with a minimum of initial data, to establish a complete set of natural climatic characteristics of the past consistent with its characteristics preserved up to the present. Such reconstructions help identify the errors of other reconstructions in determining the dimensions and ages of glaciations or estimates of paleoclimate, for they can reveal inconsistencies among individual parts in a system of concepts about the nature of past epochs.

A paleoglaciologic reconstruction refuted the hypothesis that the Pamir's large-scale valley glaciation was of Middle Pleistocene age. The prominent elements of the landscape are very "fresh" morainic complexes. It can be shown, in accord with Gerasimov (1964), who first gave a correct explanation of this phenomenon, that "fresh" moraines could have formed only during the maximum of Late Pleistocene valley glaciation. At that time, because of the already nearly modern altitude of the mountains and general climatic cooling, the major portion of the heat supplied to the glaciers was expended in evaporation; therefore, the melting and runoff and their erosive power were low (Lebedeva, 1977).

Another hypothesis that did not stand the paleoglaciologic test was that of a reticulated partial ice-sheet glaciation of the Pamir 20,000 to 18,000 yr B.P. (Grosswald and Orlyankin, 1979). This hypothesis was based on two premises. The first was a strong depression of the July air temperature for Central Asia, −25°C, obtained in the model of Gates (1976), and the second was the adoption of a 900-m depression of the firn line as an average for the entire Northern Hemisphere. The firn line descended into the valleys of the eastern part of the Pamir. There formed an ice reservoir from which glaciers of many hundreds of kilometers ran down in different directions.

A reconstruction of the mass balance of Uysu Glacier, which in this case had a length of 180 km and descended into the Kashgar Depression to an altitude of 2000 m, showed that it should have existed when summer air temperatures were 9°C cooler. However, the annual runoff from this huge but not the largest glacier in the reticular glaciation of the Pamir should have been 40 times the runoff from the recent Uysu Glacier. Thus, one of the smallest rivers of the Pamir, with a mean annual discharge of less than 1 m^3 per second, should have become almost as full as the Syr-Dar'ya River, one of the largest in Central Asia. Such a tremendous increase in mountain runoff should have left the clearest traces in river valleys on the plains and in the Aral Sea region, but no such traces have actually been observed.

According to Gates (1976), the very strong depression of July air temperature for Central Asia indicates that, even in the lowest regions (the Takla Makan Desert, the Tarim Basin), the mean summer temperature should be negative. Such temperature conditions would immediately convert the low areas into ice reservoirs, the level of which would quickly climb to the highest peaks, and all the mountain systems of Asia would be buried under an ice cover.

Paleoclimatic Consequences

The reconstruction of former ice sheets and mountain glaciers in the USSR enables us to give a quantitative estimate of the summer air temperature depression at the time of the stationary phase of maximum glaciation. The annual atmospheric precipitation is everywhere assumed to be equal to the present precipitation. As indicated by an analysis of model results from a reconstruction of the summer paleoclimate of this epoch (Manabe and Hahn, 1977), this hypothesis is entirely admissible for the southern half of the USSR. Only in the Northeast could the decrease in precipitation have been so marked as to appreciably affect the calculated depression of summer air temperature. It is clear from the above discussion that lower total precipitation values would lead to a greater depression in summer air temperature.

Comparison of the values of summer air-temperature depression obtained by a large mathematical model (Manabe and Hahn, 1977) and by our method show that, on the average, the results are practically the same throughout the USSR. The differences are slight in the periglacial zone of Europe (according to both reconstructions, −8°C), somewhat greater in western and eastern Siberia, but major for the Caucasus-Caspian and Central Asian regions. In the Caucasus-Caspian region, the mathematical model outlines a vast area with a slightly positive anomaly, and our model gives an anomaly of −5°C to −7°C. With summers as warm as today's or even warmer, at least a fivefold increase in total annual precipitation would be necessary to force the glaciers to descend to their Late Pleistocene maximum. It is evident from the present data shown in Figure 7-7 that in this case the precipitation at the equilibrium line of the glaciers (3.3 to 3.8 km above sea level) would be as much as 600 g per square centimeter per year, that is, an annual snow layer up to 15 m thick at the equilibrium line and over 20 m thick, on the average, in the accumulation region. Such values are not known anywhere in the world. Even on the glaciers of New Zealand and Iceland, the thickness of the annual snow layer is less than 10 m. Let us note that the mathematical model gives a slightly negative precipitation anomaly in this region, and, hence, this model cannot be used at all to account for the large advance of mountain glaciers.

The situation in the Central Asian mountain mass is more difficult to analyze. On our map (Figure 7-7), the northern half of this mass is characterized by an anomaly of −3°C to −5°C, whereas according to the mathematical model it is −4°C to −8°C and up to −12°C in the Himalayas. The model's precipitation anomaly in this region is close to zero. As a whole, the results given by the mathematical model differ from ours in having a much greater spatial variability, and certain areas have extremely high values of summer air-temperature anomalies that cannot be correlated, even qualitatively, with data on the dimensions of former glaciers.

Despite the obvious inadequacy of the data of Figure 7-7 on former mountain glaciation, as contrasted with an abundance of data on recent glaciation, one can formulate a working hypothesis for further application of the method. The depression of Eurasia's summer air temperature in the epoch of the maximum of Late Pleistocene glaciation had an essentially nonlatitudinal distribution. The most pronounced cooling is attributed to the Northwest and the least to the Southeast, without any marked local differences. Barrier effects were weakly manifested; that is, they were not appreciably intensified as compared to the present effect. The development of the proposed method for reconstructing glacioclimatic conditions of the past requires primarily an increase in the quantity of reliable data and age determinations of moraines, as well as paleoecologic data on atmospheric precipitation.

References

Dolgushin, L. D., and Osipova, G. B. (1971). New data on pulsations of recent glaciers. *In* "Glaciological Studies: Chronicle, Discussions" (V. M. Kotlyakov, ed.), vol. 18, pp. 191-218. Interdepartmental Geophysical Committee, Moscow.

Gates, L. (1976). Modeling the ice-age climate. *Science* 191, 1138-44.

Gerasimov, I. P. (1964). The paleographic paradox of Pamir. *USSR Academy of Sciences, Izvestiya, seriya geograficheskaya* 3, 4-13.

Grosswald, M. G., and Orlyankin, V. N. (1979). The late Quaternary ice cap of Pamir. *In* "Glaciological Studies: Chronicle, Discussions" (V. M. Kotlyakov, ed.), vol. 35, pp. 85-97. Interdepartmental Geophysical Committee, Moscow.

Khodakov, V. G. (1978). "The Water-Ice Balance in Regions of Recent and Old Glaciation of the USSR." Nauka Press, Moscow.

Khodakov, V. G. (1979). Paleoglaciological reconstruction for the epoch of the maximum of Late Pleistocene glaciation of the USSR and certain paleoclimatological consequences. *USSR Academy of Sciences, seriya geograficheskaya* 6, 27-32.

Kolstrup, E. (1980). Climate and stratigraphy in northwestern Europe between 30,000 B.P. and 13,000 B.P. with special reference to the Netherlands. *Mededelingen Rijks Geologische Dienst* 32, 181-253.

Lauscher, F. (1954). Klimatologische Problem des festen Neiderschlags. *Archiv für Meteorologie, Geophysik, und Bioklimatologie, Ser. B.* 6, 37-43.

Lebedeva, I. M. (1977). Role of evaporation in the degradation of Pamir's latest old glaciation. *USSR Academy of Sciences, Izvestiya, seriya geograficheskaya* 1, 71-79.

Manabe, S., and Hahn, D. G. (1977). Simulation of the tropical climate of an ice age. *Journal of Geophysical Research* 82, 3889-911.

Morrison, R. B. (1965). "Lake Bonneville. Quaternary Stratigraphy of the Eastern Jordan Valley, South of Salt Lake City, Utah." U.S. Geological Survey Professional Paper 477.

Paterson, W. (1972). "Physics of Glaciers." Pergamon Press, London.

Richmond, G. M. (1964). "Glaciation of the Little Cottonwood and Bells Canyons, Utah." U.S. Geological Survey Professional Paper 454-D, pp. 1-41.

Schüpp, M. (1979). Das Problem der langfristigen Klimaschwankungen. *Versuchsanstalt für Wasserbau, Hydrologie, und Glaziologie, Mitteilungen* 41, 257-65.

Shumskiy, P. A. (1969). "Dynamic Glaciology." Nauka Press, Moscow.

Velichko, A. A. (1980). Latitudinal asymmetry in the state of natural components of glacial epochs in the Northern Hemisphere. *USSR Academy of Sciences, Izvestiya, seriya geograficheskaya* 5, 5-23.

Velichko, A. A., and Lebedeva, I. M. (1974). Experience with a paleoglaciological reconstruction for eastern Pamir. *In* "Glaciological Studies: Chronicle, Discussions" (V. M. Kotlykov, ed.), vol. 23, pp. 109-17. Interdepartmental Geophysical Committee, Moscow.

Permafrost in the Late Pleistocene and Holocene

CHAPTER 8

Dynamics of Late Quaternary Permafrost in Siberia

V. V. Baulin and N. S. Danilova

The formation of permafrost began in the USSR at the start of the Quaternary (Popov, 1967). The oldest paleontologically dated traces of permafrost were discovered in the northeastern USSR in the valley of the Chukoch'ya River (Arkhangelov and Shaposhnikova, 1974) in the so-called Olerian suite, which is referred to the Early Pleistocene. Depending on climatic fluctuations, the areas occupied by permafrost sometimes increased, covering almost all of the USSR, and at other times greatly decreased. However, in the Far North and Northeast, the permafrost was preserved during the entire Quaternary. It is helpful to start the analysis of permafrost dynamics in the Late Pleistocene with the Kazantsevo (Mikulino) Interglaciation.

The Kazantsevo Interglaciation

In western Siberia, the forest boundary shifted by 5° of latitude (Volkova, 1977) in comparison with the present boundary, and in central Siberia by 2° to 4° (Ravskiy, 1972). In northwestern Siberia, where a transgression of a warm sea took place at that time (Lazukov, 1970), the frost thawed completely under the sea floor, and it was preserved only on islands that at the present time have elevations more than 50 to 60 m above sea level. As shown by models of permafrost formation (Baulin et al., 1967), the thickness of the frozen ground was about 200 to 300 m. The oldest syngenetic frozen strata with ice-wedge polygons preserved up to the present time are found there in coastal deposits consisting of regressive, primarily sandy and sandy-loam facies with a high content of vegetal detritus as well as peat interlayers (Trofimov et al., 1980). The deposits have a comparatively high volumetric ice content for sands and sandy loams, amounting to 40% to 50% or more; their swollen and massive basal, lenticular, horizontally stratified cryogenic structures confirm their syngenetic freezing. The thickness of the syngenetic horizon averages 5 to 6 m and locally increases up to 8 to 10 m. Ice-wedge polygons are more than 12 m thick (as in the region of Neyto Lake on Yamal, etc.) and are fairly widely distributed. On Yamal, they have been traced up to the latitude of the Yuribey River valley, and on Gydan up to latitude 69° N; to the south, they are encountered sporadically. The ice wedges are as much as 2.0 to 2.5 m wide, and the distance between them is about 8 to 10 m. South of latitude 66° to 67°N in western Siberia, permafrost of Kazantsevo time was not preserved, and up to 64°N there are only traces of it in the form of ice-wedge-polygon pseudomorphs and horizons of involutions.

A definite regularity in the stratigraphic distribution can be noted: in the middle portion of the Kazantsevo sediments, there usually are no traces of frost, and typical pseudomorphs are found only at the base and top (exposures near the settlements of Gorki and Kazymskiy Mys,

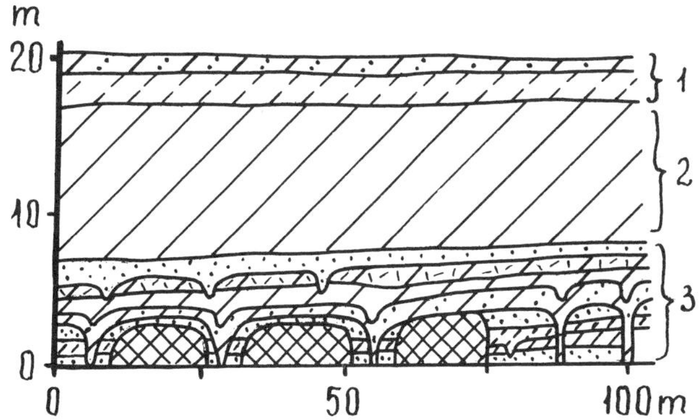

Figure 8-1. Pseudomorphs of polygonal ice veins in two stages at the base of Kazantsevo deposits, Migetl-Posl channel of Ob' River valley. (Sketch by L. M. Shmelev.)

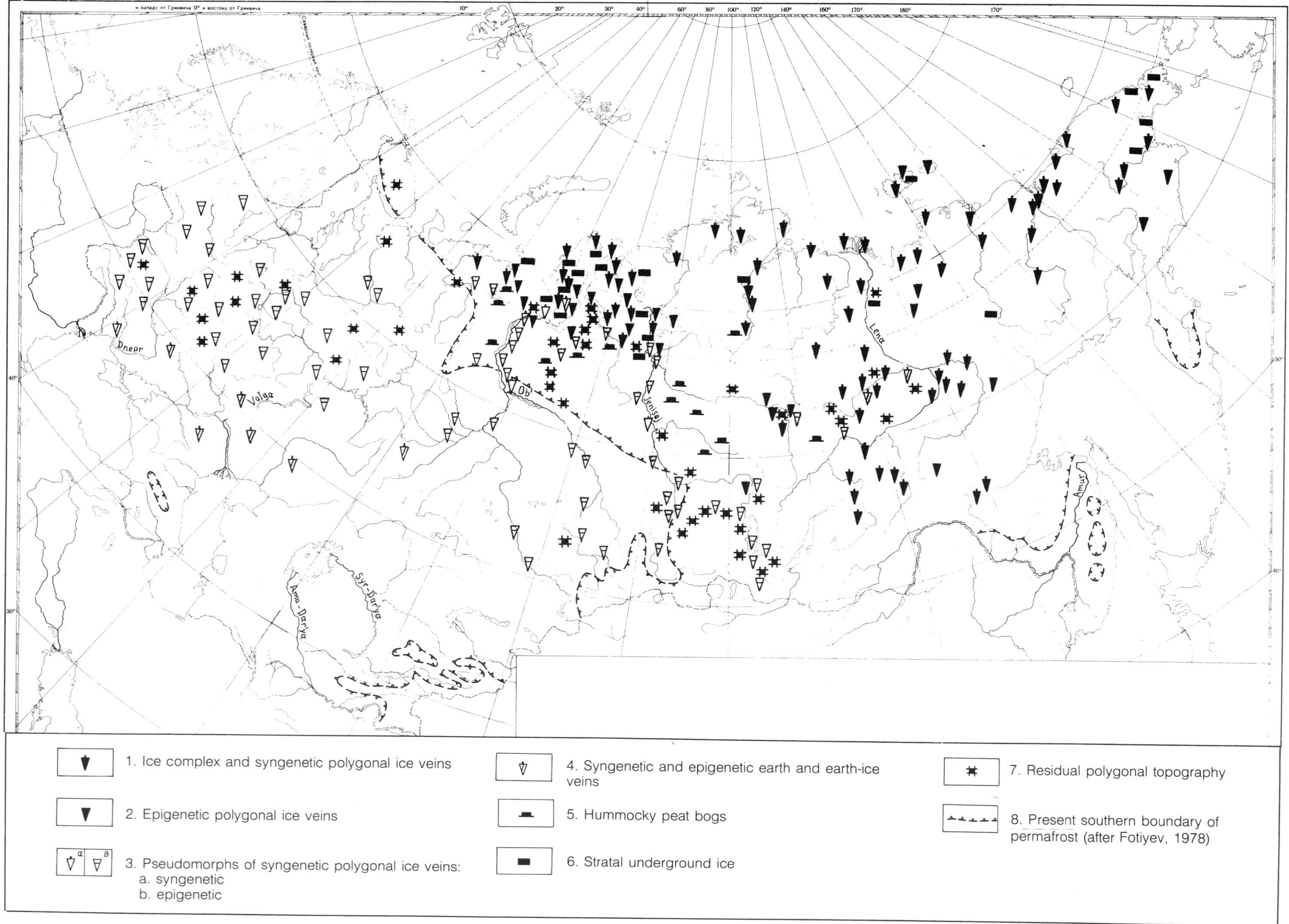

Figure 8-2. Map of Late Pleistocene and Holocene permafrost features in the USSR. (Compiled by V. V. Baulin, Ye. B. Belopukhova, A. A. Velichko, N. S. Danilova, and V. P. Nechayev.)

Mingitnell-posl Channel on the Ob' River, and other places), associated with horizons of involutions and in some cases with indications of syngenesis (Figure 8-1). The size of the polygonal grid is about 8 to 10 m. The features are very widely distributed and have been noted by many investigators. These data indicate much more severe freezing conditions at the beginning and end of Kazantsevo time than at present and a reduction in the middle. In the lower reaches of the Yenisey River, the so-called Innokentinskiy exposure shows evidence for syngenetic freezing of nearshore marine deposits of Kazantsevo age, with frost cracking and ice-wedge-polygon formation (Vtyurin, 1975).

East of the Yenisey River, traces of frost cracking, as an indication of severe freezing conditions, have been found in practically all major regions. In the Angara River basin, pseudomorphs of syngenetic ice-wedge-polygons are present in the Kazantsevo alluvial terraces (Ravskiy, 1972). In the Vilyuy River basin (Katasonova, 1963), in central Yakutiya (Solov'yev, 1959; Katasonov, 1979), and in the entire Northeast, permafrost from that time is preserved with ice-wedge-polygons. The southern boundary during Kazantsevo time in western Siberia is drawn north of latitude 58° to 59°N, and in eastern Siberia permafrost was everywhere.

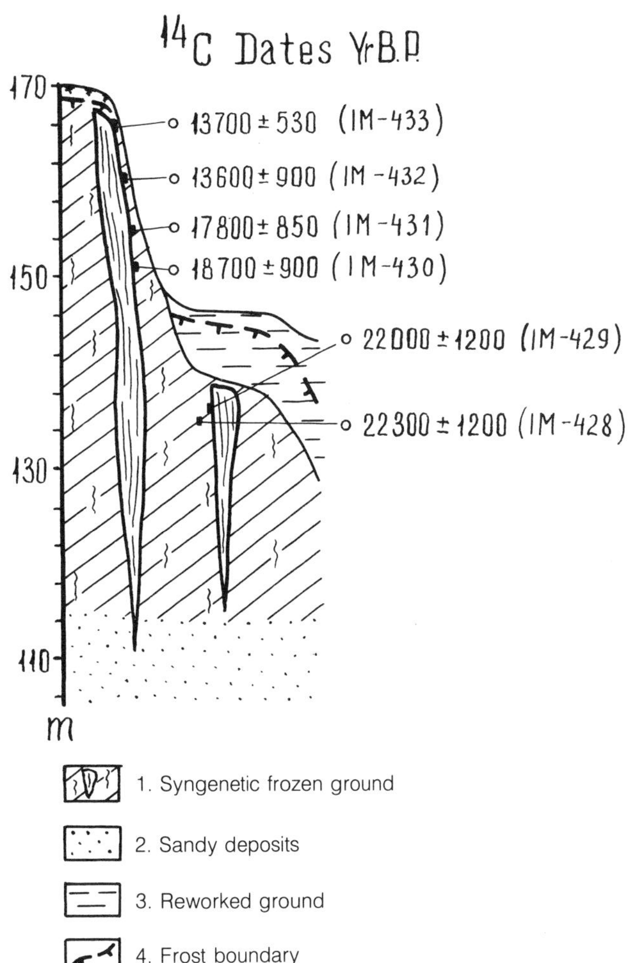

Figure 8-3. Structure and age of syngenetic frozen ground of central Yakutiya. (Sketch by E. M. Katasonov.)

Late Pleistocene Glacial Epoch

The Late Pleistocene was marked by severe climatic conditions (Lazukov, 1970; Velichko, 1973; Kind, 1974). Perennial freezing affected a large part of the USSR, and thick permafrost developed. Severe freezing is indicated by horizons of frozen strata, including thick syngenetic ice-wedge polygons, forming a so-called ice complex (Solov'yev, 1959) of very wide distribution (Figure 8-2) on the North Siberian Lowland, on the Central Siberian Highland (Fotiyev et al., 1974), in central Yakutiya (Katasonov, 1979; Solov'yev, 1959) and in the northern maritime lowlands (Popov, 1967; Vtyurin, 1975). Frozen strata of similar cryogenic structure are present in Transbaykaliya (Kaplina et al., 1975; Nekrasov and Klimovskiy, 1978). In practically all these regions, syngenetic freezing of sediments is indicated by the great thickness (which ranges from 10 to 20 m to as much as 50 to 70 m in certain regions) (Figure 8-3) as well as by the character of the lateral contacts of the ice veins and deposits, the displacement of the ice-vein lattice (Figure 8-4), and the nature of the cryogenic structures between the veins. On the basis of absolute dating, the age of the ice complex is Middle and Late Pleistocene (Markov, 1973; Kaplina, 1978) and the continuity of its formation is established for different regions. In central Yakutiya, syngenetic freezing in underlying sands (Figure 8-3) indicates that permafrost existed there long before the ice complex began to form (according to data from thermoluminescence analysis, over 300,000 years) (Katasonov, 1979).

The paleogeographic analysis of the ice complex is based on its comparison with analogous recent sediments with ice-wedge polygons, which include mainly the floodplain facies of alluvium of major rivers in northern Siberia, although the structures are not exactly the same. As a rule, the old ice complex contains more structure-forming ice, has a more loesslike appearance, and has more organic matter. The old ice-wedge-polygon lattice is approximately one-third to one-half the size of the present one (5 to 10 m versus 20 to 30 m), and the widths of the ice-wedge polygons are larger by the same factor (3 to 5 m versus 1 to 2 m). This relationship holds in almost all regions where the complex is present. For example, in the Vilyuy River valley in mineral soils of the recent flood plain, the ice-wedge polygons are less than 0.5 to 0.8 m wide, with a diameter of about 25 m; they occupy small areas and are present only in regions where the ground temperature drops below −7°C. Old ice-wedge polygons in this region occupy areas of many square kilometers, are present at all terrace levels up to 200 to 250 m above sea level, and are much larger. Much more severe conditions must have prevailed in the Pleistocene than in the Holocene. The differences exceeded the present zonal conditions of the Arctic coast and the interior areas of permafrost south of latitude 60°N.

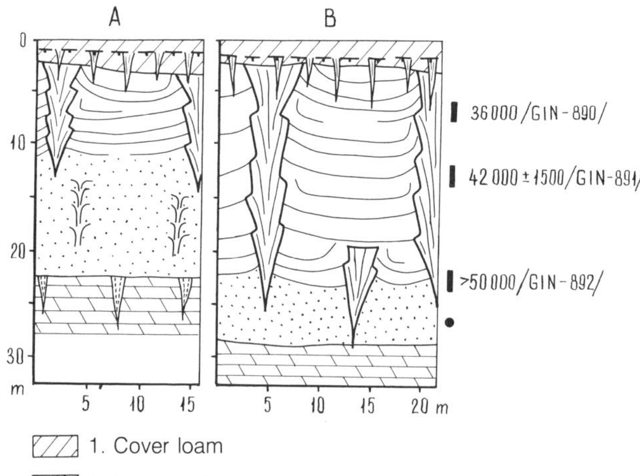

1. Cover loam
2. Ice complex
3. Sand
4. Bedrock
5. Sandy veins
6. Pseudomorphs of polygonal ice veins
7. Permafrost surface
8. Absolute age determinations (yr B.P.)
9. Tiraspol' mammalian faunal data (from Shofman et al., 1977)

Figure 8-4. Cryogenic structure of syngenetic frozen ground on the fourth terrace of the Vilyuy River. (Sketch by N. S. Danilova.) (A) Near Lonkholookh Landing, (B) 8 km above the mouth of Chybyta River.

Kaplina and Kuznetsov (1975) have estimated atmospheric temperatures during the Late Pleistocene for the northern USSR from the widths of ice-wedge polygons and elementary ice veins in the ice complex. They used Dostovalov's formula $\Delta t = m/\alpha \cdot \ell$, where Δt is the annual temperature change (°C), m is the width of an elementary crack, α is the coefficient of thermal expansion of the ground, and ℓ is the dimension of the polygon. As is evident from Table 8-1, for the most typical width of the elementary vein (12 mm) and for the most typical dimension of the polygons (7 m), the mean annual temperature of the rocks is −20°C, and the mean air temperature is still lower (−27°C). The calculated air temperatures differ from the present ones by a minimum of 3°C to 6°C and by a maximum of 15°C to 19°C, that is, by an average of about 10°C. The same value is obtained from a paleogeographic analysis of the conditions of occurrence for recent and Pleistocene ice-wedge polygons in the Far North and in the southern and western regions of the USSR.

The ice complex provides still other data for paleogeographic analysis. The cryogenic structure of the ice complex and especially its saturation with structure-forming ice indicate a high moisture content in the old seasonally melting layer and hence a swampy state of the surface with little evaporation of precipitation in the cool summers. The southernmost point beyond the Yenisey River where Pleistocene syngenetic horizons with ice-wedge polygons are preserved is in the middle course of the Nizhnyaya Tunguska River. There and in sections II and III of terraces near the mouth of the Ilimpei River and near the settlement of Nakanno, the syngenetic alluvial sandy loam with ice-wedge polygons is up to 10 m thick (Tseytlin, 1964). Syngenetic frozen strata have been noted in the Lena River valley near the town of Olekminsk, as well as on all the Late Pleistocene terraces in many valleys of the mountainous Transbaykaliya (Nekrasov and Klimovskiy, 1978; Kaplina et al., 1975).

In western Siberia, syngenetic freezing and growth of ice-wedge polygons also took place during the formation of Late Pleistocene coastal marine and river terraces I, II, and III. North of latitude 67° to 68°N, Late Pleistocene frozen strata have been preserved up to the present time. For example, a syngenetic alluvial sandy loam to sand with ice veins and ice-earth veins (Figure 8-5) has been described from the Yuribey River valley (Yamal Peninsula, latitude 69°N), where its thickness exceeds 10 m (Trofimov et al., 1980). On the Gydan Peninsula (68°N), ice veins have been described for the Messoyakha River valley (Danilov, 1978). North of latitude 69° to 70°N, syngenetic frozen strata with ice-wedge polygons are encountered practically everywhere. The size of the polygons ranges from 4 to 5 m through to 10 to 15 m, and the width of ice-wedge polygons reaches 2.5 m. The vertical thickness is comparable to the thickness of the syngenetic frozen horizon (Baulin et al., 1967).

In the southern regions of the present area of permafrost, there remain only traces of late Pleistocene cooling: ice-wedge-polygon pseudomorphs, horizons of involutions, and disturbances of original stratification. Thus, in south-central Siberia in the Angara River basin and in the upper reaches of the Lena River in the alluvium of terraces I, II, and III, pseudomorphs of syngenetic ice-wedge poly-

Table 8-1. Mean Annual Air Temperature during the Late Pleistocene in the Northeastern USSR

Width of Elementary Crack in Ice-Wedge Polygons (m)	Width of Polygons (m)	Mean Annual Rock Temperature (°C)	Mean Annual Air Temperature (°C)
0.12	7	−20	−27
0.12	10	−15	−20
0.2	10	−25	−32

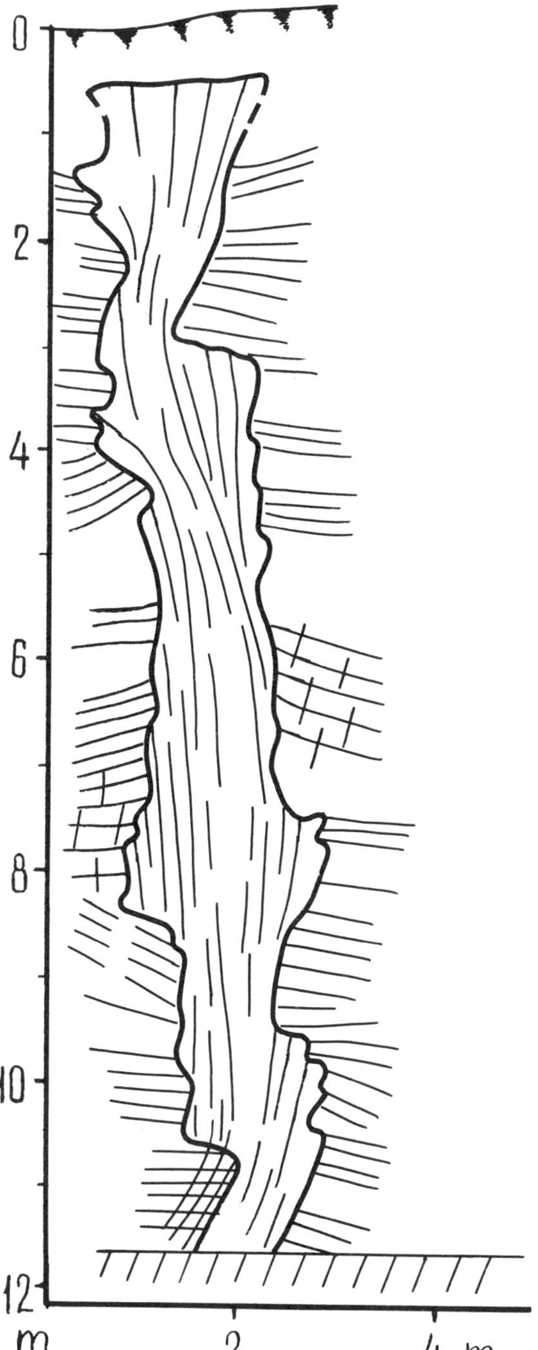

Figure 8-5. Photograph-based sketch of syngenetic ice vein located in the sandy alluvium of third (Zyryanka) terrace of Yuribey River, Yamal Peninsula. (Photo by G. I. Dubikov.)

gons have been described in many sections. As a rule, these pseudomorphs are composed of cover loam and occur only in the lower part of sand; the height of the pseudomorphs usually does not exceed 3 to 4 m (Ravskiy, 1972), and pseudomorphs 10 m high or higher have been described relatively infrequently (Litvinov, 1962). In the same region on terraces and interfluves, a residual polygonal relief that formed during the melting of ice-wedge polygons has an extremely wide distribution. The morphology of this relief and its good preservation indicate its relative youth and the simultaneity of ice-wedge-polygon formation during the Late Pleistocene (Fotiyev et al., 1974).

In western Siberia south of the Arctic Circle in sediments of terraces III and II, ice-wedge-polygon pseudomorphs have been described in many exposures up to latitude 49° to 50°N. North of latitude 55° to 60°N, one continually encounters systems of multistage pseudomorphs, that is, two or three pseudomorph horizons in the same section. The thickness of the pseudomorphs reaches 8 to 10 m, the size of the polygons is less than 10 m (Shmelev, 1966), and isolated pseudomorphs of smaller size are described only south of latitude 55° to 60°N. Thus, in the Nyda River basin (66°30′N) in a section of a 10-m terrace, a ice-wedge-polygon pseudomorph (Figure 8-6) has a vertical thickness exceeding 8 m and a width greater than 2 m. Shown on the right in the figure is a collapse of enclosing sandy loam and sand. On the left, an irregular upward fold of the layers, characteristic of syngenetic ice-wedge polygons, is well preserved. The assignment of a Sartan age to a pseudomorph is based on its geomorphologic location. Analogous but smaller pseudomorphs are described at many points in the southern Taz Peninsula in the Nadym River basin in sandy alluvium of terraces II and III (Danilov, 1978). A large pseudomorph

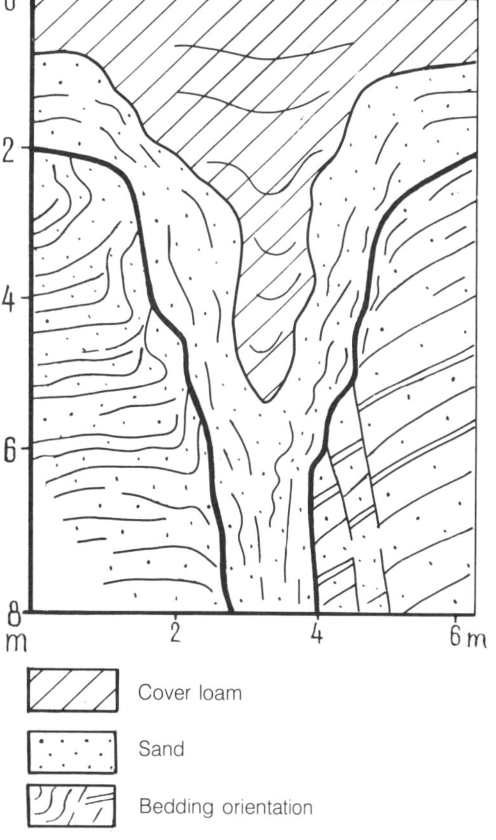

Figure 8-6. Pseudomorph of syngenetic ice vein from the lower Nyda River, western Siberia. (Sketch by N. S. Danilova.)

over 10 m high has been described near the Urals in western Siberia in the basin of the Malaya Sos'va River in sandy alluvium of terrace II (Alekseyev, 1971). South of the Arctic Circle as far as latitude 50° to 52°N, a residual polygonal relief is widely represented.

In the Yenisey River valley, pseudomorphs in alluvium have been described at many points between the settlement of Zyryanovo and the town of Krasnoyarsk (Lavrushin, 1961; Litvinov, 1962; Fotiyev et al., 1974). North of the mouth of the Bol'shoy Pit River, the absolute age of the alluvium including ice-wedge-polygon pseudomorphs is 18,000 to 20,000 years (Zubakov, 1972).

Beyond the confines of the present area of permafrost, ice-wedge-polygon pseudomorphs have been described in the region of the towns of Tyumen', Pavlodar, and Semipalatinsk and others. The height of these pseudomorphs does not exceed 2 to 3 m, but they are fairly widely distributed in the alluvium of terraces II and I (Arkhipov, 1971). According to pseudomorphs occurring in deposits of Zyryanka age (terrace III), the southern boundary of permafrost in western Siberia runs thence along latitude 52°N, in Karginskiy time (terrace II) along latitude 56° to 57°N, and in Sartan time (terraces I and II) along latitude 48° to 49°N (Shmelev, 1966; Baulin et al., 1967).

In addition to the information gained from cryogenic structures, abundant paleogeographic data are provided by the thickness and vertical structure of recent permafrost.

Age of Permafrost

In connection with the study of the freezing-melting process, an attempt was made to estimate quantitatively the age of recent frozen ground of great thickness (Kudryavtsev, 1970; Balobayev, 1973). Account was taken of the fact that thickening depends chiefly on the natural conditions during the entire time of freezing, which determine the heat exchange and temperature amplitude at the surface, the composition (primarily moisture) of the freezing sediments, and the heat flow entering the base of the permafrost from below. The rate of perennial freezing (Table 8-2) was determined by taking into account the harmonic temperature oscillations at the surface but without considering the change of water into ice; for this reason, the calculations show the maximum possible freezing depth. The calculated depths are close to those for rocks; in loose sediments, the thickness may be one-third to one-half as large.

Kudryavtsev (1970) proposes a tentative estimate of the age of permafrost in different present frost-temperature zones of the USSR (Table 8-3). The calculations confirm the old age of permafrost in the northeastern part of the USSR, as established by an analysis of cryogenic structures. Permafrost up to 300 to 500 m thick is dynamic, and permafrost more than 500 to 600 m thick is stable. In addition, the strong dependence of the thickness and age of the permafrost on regional geologic and geographic factors has been confirmed.

In eastern Siberia, thick recent permafrost is in a state of degradation and is melting from below at a rate of 1 to 3 cm per year (Balobayev, 1973).

Table 8-2. Depth of Penetration of Surface-Temperature Oscillations into Frozen Ground

Time (years)	Amplitude of Surface Oscillation (°C)	Depth of Penetration (m)
40	4	78
1000	4	123
2000	4	550
10,000	4	1260

Table 8-3. Age of Permafrost

Present Characteristics		Age (years)	
Temperature (°C)	Thickness (m)	Minimum	Maximum
0 to −1	Up to 120 to 150	11 to 40	100 to 1000 (up to 50,000)
−1 to −3	Up to 200 to 300	100 to 1000 (for rocks)	10,000 to 200,000 (for loose rocks)
−1 to −5	100 to 400	10,000	200,000
−5 to −10	200 to 600	100,000	500,000
−10 to −14	500 to 1000 or more	500,000	

Source: Data from Kudryavtsev, 1970.

In western Siberia (Figure 8-7) and the European Northeast (Figure 8-8), relict permafrost is detached from recent permafrost and occurs at depths from 50 to 160 m up to 200 to 400 m (Baulin et al., 1967). This permafrost dates to the Late Pleistocene and was buried during the warm period of the Holocene. Paleogeographic analysis of relict frozen strata of great thickness existing at the present time has made it possible to reconstruct the thickness of permafrost in the Late Pleistocene and Holocene.

In northwestern Siberia, a regular increase of 100 to 450 m is observed in permafrost thickness from the recent to the older coastal plains of Late Pleistocene age. Through analysis of the data and models of the process of perennial freezing in the Late Pleistocene (Baulin and Chekhovskiy,

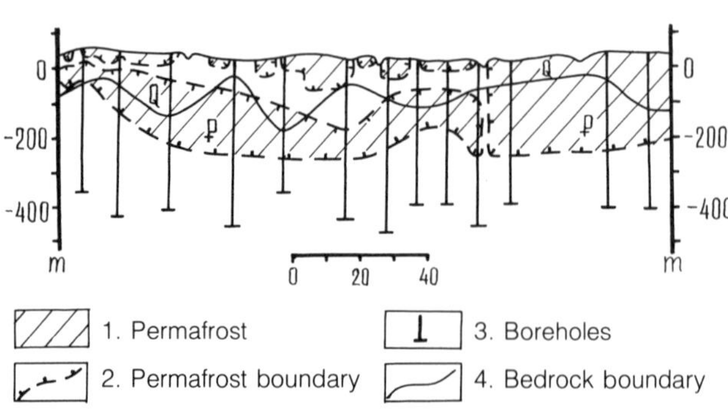

Figure 8-7. Diagram of frost profile, at the Arctic Circle, western Siberia.

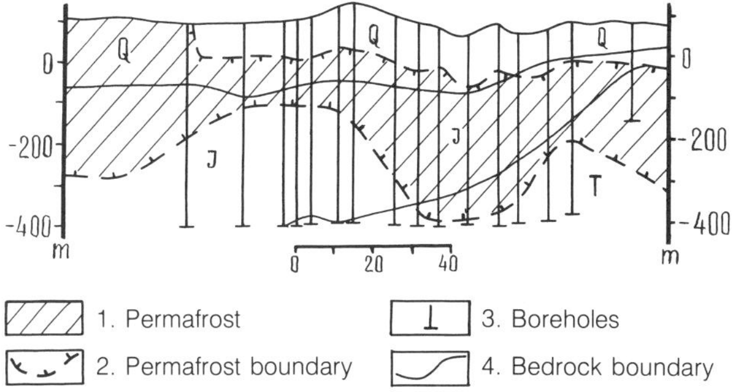

Figure 8-8. Diagram of frost profile from the southern part of Kolva rampart, northern European USSR.

1976), maps of permafrost thickness have been compiled for individual stages of the Late Pleistocene (Figure 8-9). When these maps are compared, an increase in permafrost thickness from Zyryanka time to the end of the Late Pleistocene becomes apparent. In addition, knowledge of this region made it possible, on the basis of permafrost thickness at each geomorphic level and on the basis of freezing time, to calculate the average ground temperatures when each geomorphic level was found (i.e., during the Zyryanka, Karginskiy, and Sartan stages of the Late Pleistocene as well as during the Holocene) (Figure 8-10). The calculation was made for the northern Taz Peninsula, where the present mean annual ground temperature is about −7°C.

Thus, severe freezing conditions continually prevailed during the Late Pleistocene in Siberia. Frost cracking was not interrupted even during the Karginskiy Interglaciation (Shofman et al., 1977). The increased severity of freezing climatic conditions that occurred at that time was recorded by the appearance of ice veins of so-called high generation, which have been described in many exposures of the "ice complex" in the Northeast, for example, in Duvannyy Yar (Kaplina, 1978). The inadequate amount of study and the complexity of a comparative estimate of permafrost facies in the "ice complex" and ice-wedge-polygon pseudomorphs in different regions do not permit one to draw from the cryogenic structure any conclusions about oscillations in the permafrost climate during the Late Pleistocene. For the entire territory, one can say definitively only that the continuous permafrost conditions were substantially more severe during the Late Pleistocene than they are at present and that the permafrost thickened continuously, particularly in the western half of the USSR, that is, in the region where the permafrost dates back to the Late Pleistocene.

At the end of the Late Pleistocene (Figure 8-11), the southern boundary of permafrost ran along latitude 48° to 49°N in the USSR. This boundary was drawn on the basis of the occurrence of ice-wedge-polygon pseudomorphs by analogy with present permafrost conditions. The permafrost region can be broken up into zones of continuous and discontinous occurrence, which in contrast to the present regions assume a distinct latitudinal trend. The southern boundary of the zone of continuous permafrost runs along a massive occurrence of ice-wedge-polygon pseudomorphs

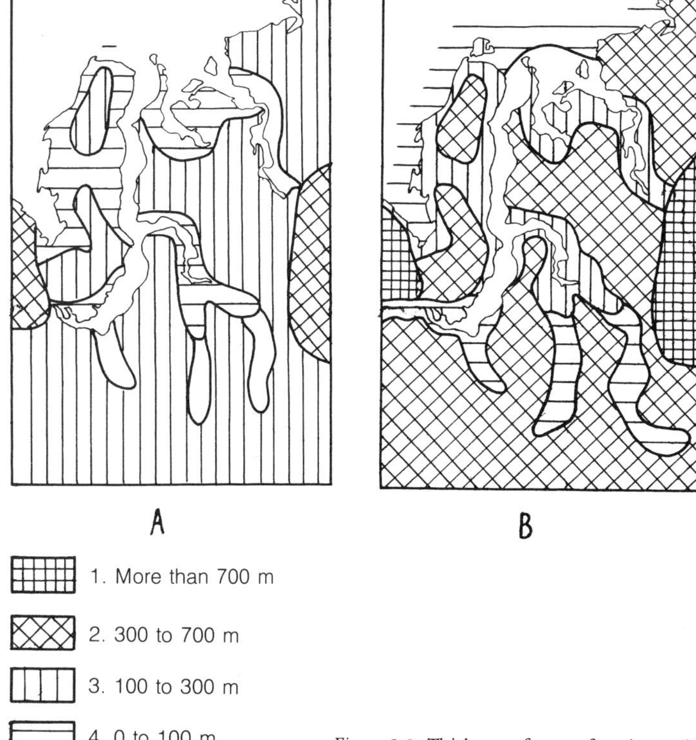

Figure 8-9. Thickness of permafrost in northwestern Siberia during Zyryanka (A) and Sartan (B) time.

over 2 to 3 m deep. The zone of discontinuous occurrence shows up only in the European USSR and Kazakhstan. East of the Yenisey River, the permafrost is continuous, and the southern boundary reaches far into the south, largely because of orographic relief.

The south of the European USSR and Central Asia are regions of deep seasonal freezing. Winter freezing there could have been 3 to 5 m deep or possibly even deeper during the Late Pleistocene.

In the European USSR and southern Siberia, which previously had not been occupied by permafrost, thick permafrost strata were formed. Around the Arctic Circle their thickness reached 600 to 700 m and in the central regions, 300 to 500 m, depending on the local geologic and geothermal conditions.

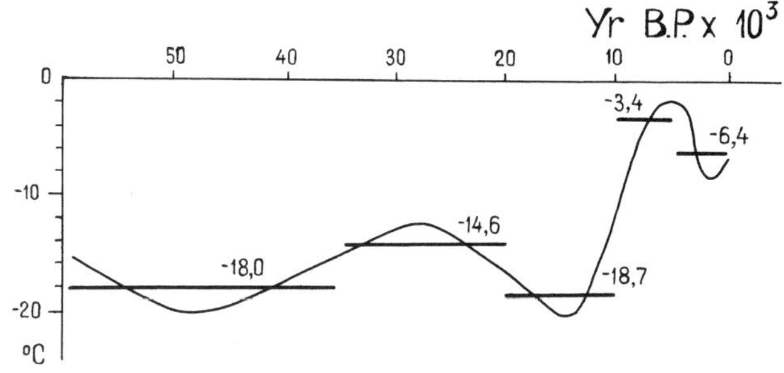

Figure 8-10. Permafrost temperatures (curved line) during the Late Pleistocene. (Averages shown by straight lines.)

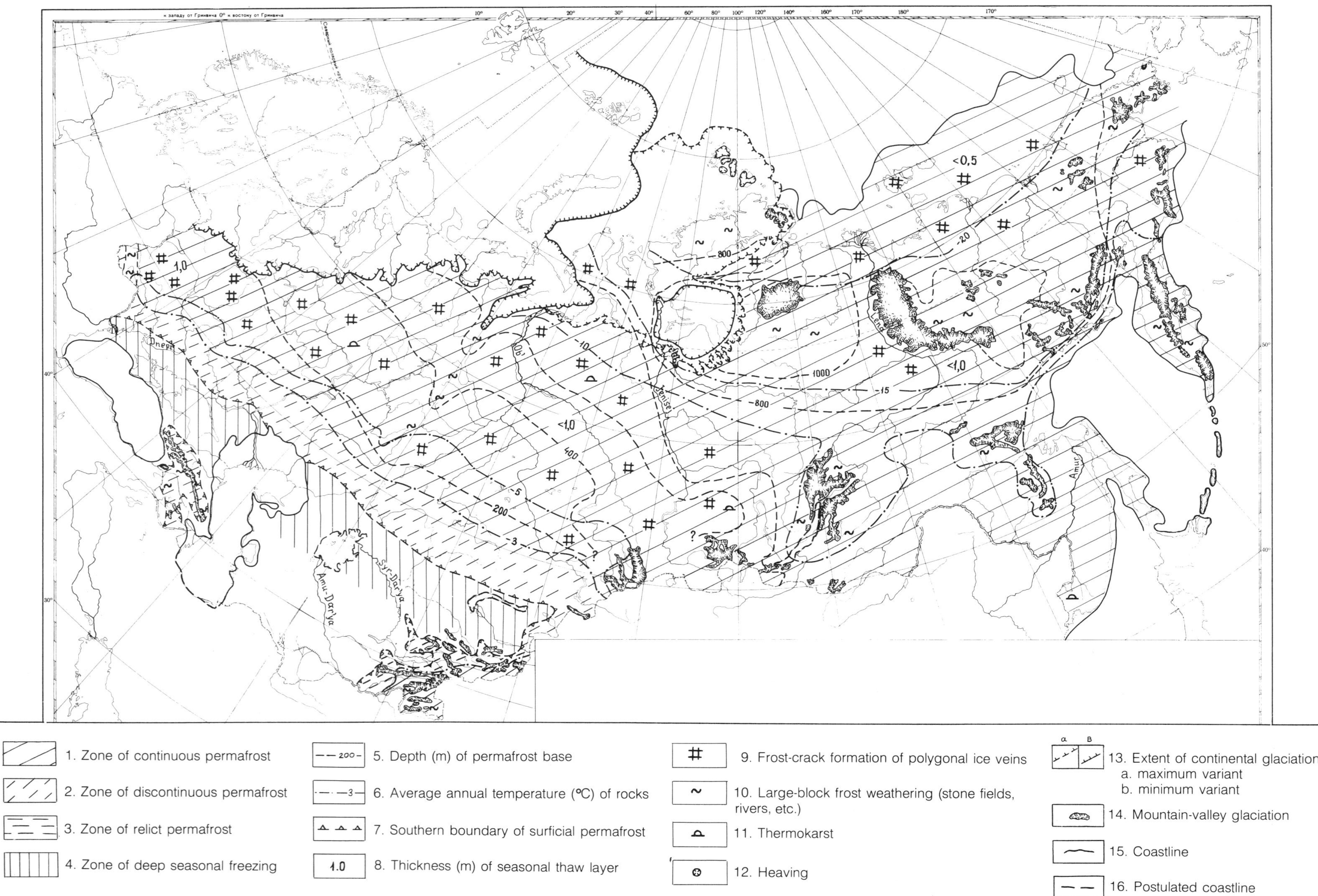

Figure 8.11 Map of permafrost distribution during the Valdai (Sartan) Glaciation. (Compiled by V. V. Baulin, Ye. S. Belopukhova, A. A.

The thickness of permafrost in the northeastern USSR (i.e., regions where it now exceeds 700 to 800 m) was at most only 100 to 250 m greater.

At the depth of zero annual amplitude, the temperature of permafrost rocks, as determined mainly by comparison of recent and Late Pleistocene ice-wedge polygons reached −20°C in the North, whereas at the present time it is close to −10°C to −12°C there. The widespread occurrence of traces of frost cracking in the southern USSR indicates that the permafrost temperatures were −5°C to −15°C over most of the USSR.

Conclusions

Among cryogenic processes most characteristic of the Late Pleistocene is frost cracking, and the ice-wedge-polygon traces are found over the entire USSR down to latitude 50°N. Neogenesis of frozen strata in loose deposits was accompanied by the formation of frost heaves; this process was especially typical in the western sector of the permafrost region. In mountainous areas, frost weathering of large rock fragments predominated, and stone fields and rock streams were formed. Judging from the thickness of involution horizons and the depth of occurrence of the "ice complex," the depth of seasonal thawing did not exceed 1 m, even in the southern regions of permafrost.

References

Alekseyev, V. R. (1971). Cryogenic structures in periglacial deposits of the Kondo-Sosvinskiy Ob' region. *In* "Geocryological Investigations" (I. A. Nekrasov, ed.), pp. 33-53. Yakutskoye Press, Yakutsk.

Arkhangelov, A. A., and Shaposhnikova, Ye. A. (1974). Approximate estimate of paleofrost conditions of formation of Early Pleistocene deposits in the eastern part of the Primor'ye Lowland. *In* "Frost Investigations" (V. A. Kudryavtsev, ed.), Issue 14, pp. 76-80. Moscow State University Press, Moscow.

Arkhipov, S. A. (1971). "The Quaternary Period in Western Siberia." Nauka Press, Novosibirsk.

Balobayev, V. T. (1973). Basic principles of deep freezing of the Earth's crust. *In* "Problems of Geocryology" (S. G. Tsvetkova, ed.), pp. 26-36. Nauka Press, Novosibirsk.

Baulin, V. V., Belopukhova, Ye. B., Dubikov, T. I., and Shmelev, L. M. (1967). "Geocryological Conditions of the West Siberian Lowland." Nauka Press, Moscow.

Baulin, V. V., and Chekhovskiy, A. L. (1976). Thickness of permafrost in western Siberia. *In* "Engineering-Geological and Geocryological Studies in Western Siberia" (G. I. Oubikas and M. M. Koreisha, eds.), pp. 4-31. Stroyizdat, Moscow.

Danilov, I. D. (1978). "Pleistocene of Submarine Subarctic Plains." Moscow State University Press, Moscow.

Fotiyev, S. M., Danilova, N. S., and Sheveleva, N. S. (1974). "Geocryological Conditions of Central Siberia." Nauka Press, Moscow.

Kaplina, T. N. (1978). Rates of accumulation and age of the "ice complex" of Primor'ye lowlands of Yakutiya. *In* "Frost Investigations" (V. A. Kudryavtsev, ed.), Issue 17, pp. 142-48. Moscow State University Press, Moscow.

Kaplina, T. N., and Kuznetsova, I. L. (1975). Geothermal and climatic model of the epoch of accumulation of the edoma suite of Yakutiya's Primor'ye Lowland. *In* "Problems of Paleogeography of Loessial and Periglacial Regions" (A. A. Velichko, ed.), pp. 170-74. Nauka Press, Moscow.

Kaplina, T. N., Pavlova, O. P., Chernyad'yev, V. P., and Kuznetsova, I. L. (1975). "Latest Tectonics and Formation of Permafrost and Groundwaters." Nauka Press, Moscow.

Katasonov, E. M. (ed.) (1979). "Structure and Absolute Geochronology of Alas Deposits of Central Yakutiya." Nauka Press, Novosibirsk.

Katasonova, Ye. G. (1963). Recent permafrost deposits and their older analogues in the northeastern part of the Lena-Vilyuy interfluve. *In* "Conditions and Characteristics of Frozen Ground Development in Siberia and the Northeast" (E. M. Katasonov, ed.), pp. 41-60. USSR Academy of Sciences Press, Moscow.

Kind, N. V. (1974). "Geochronology of the Late Anthropogene from Isotope Data." Nauka Press, Moscow.

Kudryavtsev, V. A. (1978). Formation and development of permafrost strata of rocks. *In* "General Geocryology" (V. A. Kudryavtsev, ed.), pp. 283-332. Moscow State University Press, Moscow.

Lavrushin, Yu. A. (1961). "Types of Quaternary Alluvium of the Lower Yenisey." USSR Academy of Sciences Press, Moscow.

Lazukov, G. I. (1970). "The Anthropogene of the Northern Half of Western Siberia." Moscow State University Press, Moscow.

Litvinov, A. Ya. (1962). Traces of ancient cryogenic processes and phenomena in the vicinity of the city of Krasnoyarsk. USSR Academy of Sciences, Permafrost Institute, *Trudy* 18, 47-63.

Markov, K. K. (ed.) (1973). "Section of Latest Deposits of Mamontova Gora." Moscow State University Press, Moscow.

Nekrasov, I. A., and Klimovskiy, I. V. (1978). "Permafrost of BAM Zone." Nauka Press, Novosibirsk.

Popov, A. I. (1967). "Frost Phenomena in the Earth's Crust (Cryolithology)." Moscow State University Press, Moscow.

Ravskiy, E. I. (1972). "Sedimentation and Climates of Inner Asia in the Anthropogene." Nauka Press, Moscow.

Shmelev, L. M. (1966). Traces of cryogenic phenomena in Quaternary deposits of western Siberia and their paleogeographic significance. *In* "The Quaternary of Siberia" (V. N. Saks, ed.), pp. 429-38. Nauka Press, Moscow.

Shofman, I. L., Kind, N. V., Pakhomov, M. M., et al. (1977). New data on the age of deposits of low terraces in the Vilyuy River basin. *Bulletin of the Commission for the Study of the Quaternary Period* 47, 100-107.

Solov'yev, P. A. (1959). "Cryolithic Zone of the Northern Part of the Lena-Amga Interfluve." USSR Academy of Sciences Press, Moscow.

Trofimov, V. T., Badu, Yu. B., and Dubikov, G. I. (1980). "Cryogenic Structure and Ice Content of Permafrost of the West Siberian Plain." Moscow State University Press, Moscow.

Tseytlin, S. M. (1964). "Comparison of Quaternary Deposits of the Glacial and Extraglacial Zones of Central Siberia (Nizhnyaya Tunguska Basin)." Nauka Press, Moscow.

Velichko, A. A. (1973). "The Natural Process in the Pleistocene." Nauka Press, Moscow.

Volkova, V. S. (1977). "Stratigraphy and History of the Development of Vegetation of Western Siberia in the Late Cenozoic." Nauka Press, Moscow.

Vtyurin, B. I. (1975). "Underground Ice of the USSR." Nauka Press, Moscow.

Zubakov, V. A. (1972). "Paleogeography of the Western Siberan Lowland in the Pleistocene and Late Pliocene." Nauka Press, Leningrad.

CHAPTER 9

Late Pleistocene Permafrost in European USSR

A. A. Velichko and V. P. Nechayev

The European part of the Soviet Union has a special place in the history of the development of permafrost. As is well known, permafrost in this region now occurs only in a narrow band adjacent to the Arctic coast. However, only about 15,000 years ago almost all of eastern Europe was included in the region of permafrost and deep seasonal freezing, as shown by traces in the sediments and landforms.

Evolutionary analysis suggests that the present state of the natural environment may be replaced in a few millennia by an episode of permafrost expansion. Paleogeographic studies in the Russian Plain have shown that epochs of permafrost expansion were frequently replaced during interglaciations by epochs of its degradation over vast areas. In other words, in comparison with regions farther east, the European part of the USSR was characterized by a greater instability of the geothermal regime of the upper part of the lithosphere during the Pleistocene.

In the course of paleocryogenic studies begun during the prewar years, it was suggested that old frost structures were present even in Akchagylian deposits (Moskvitin, 1962), that is, more than 2 million years ago. Reliable indications of permafrost in the form of pseudomorphs of ice-wedge polygons occur as far south as latitude 50°N for the Middle Pleistocene Dnepr Glaciation (Velichko, 1958; Bogucki et al., 1975). The most widespread evidence of paleocryogenic deformation is for the Late Pleistocene Valdai cold epoch.

The Mikulino Interglaciation

The start of the Late Pleistocene—the Mikulino Interglaciation—is the epoch that separated the Middle and Late Pleistocene cryogenic phases through a complete absence of a permafrost region in eastern Europe. According to indirect data, during the Mikulino Interglaciation the permafrost thawed completely over the entire Russian Plain, including the northern band adjacent to the Arctic coast (Sukhodol'skiy, 1978). This conclusion is consistent with paleobotanic reconstructions (Grichuk, 1973), which show that the tundra zone, where permafrost now occurs, was replaced by a forest zone at the optimum of the Mikulino Interglaciation. This is also indicated by paleoclimatic reconstructions given in this volume (Chapters 24 and 25), which show a significant rise of winter temperatures (by 4° to 5°C for January) near the Arctic coast.

The Valdai Ice Age

With the advent of the Late Pleistocene (Valdai) glaciation, the stage of the gradual (although nonuniform) spread of permafrost began, and by the end of the Late Valdai it covered almost all of the European USSR. The method of studying the dynamics of permafrost in the territory where it is now almost completely absent has several specific features. The dynamics of paleocryogenic processes and their character are reconstructed from residual cryogenic structures in stratigraphic sections, especially by macro- and microtextural analyses, which permit identification of old cryogenic structures (Popov, 1960; Kaplina, 1965; Katasonov, 1973; Romanovskiy, 1977). The most reliable indicators of past permafrost are systems of large-crack formations—pseudomorphs of polygonal ice veins. However, as it will be shown subsequently, an important role is also played by fine-crack systems and so-called astructural deformation (solifluction, involutions, etc.). Other important indicators of permafrost development at the end of the Late Pleistocene are complexes of residual frost microrelief (relict cryogenic morphosculpture), identified on the present surface of the East European Plain (Velichko, 1965).

The clearest chronostratigraphic placement of paleocryogenic formations is based on their association with dated horizons in loess and fossil soils in the periglacial zone of the East European Plain. Three separate horizons of paleocryogenic deformation with specific structural features and

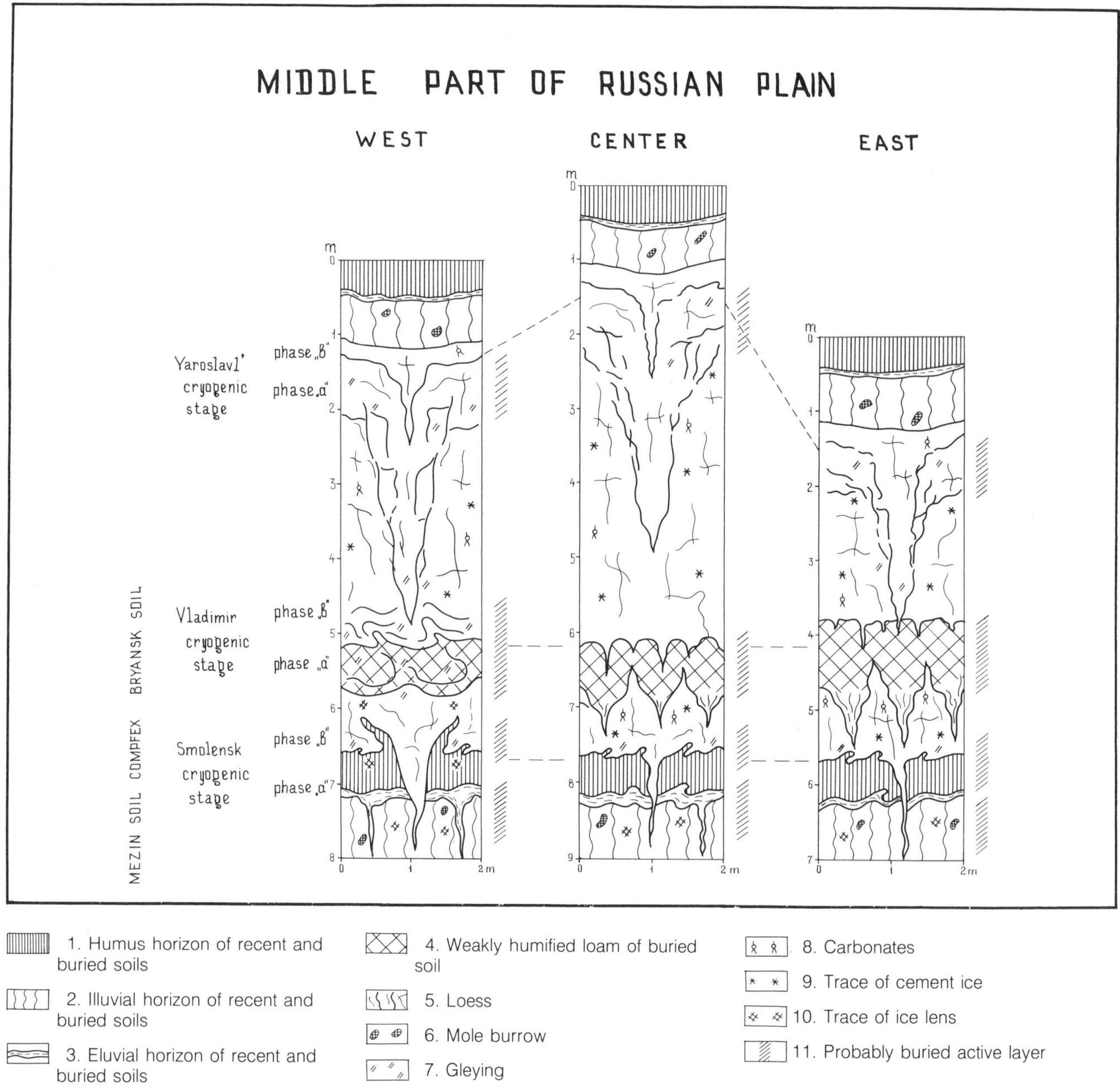

Figure 9-1. Late Pleistocene paleocryogenic deposits in the loess region of the East European Plain.

stratigraphic persistence (Figure 9-1) are identified, called (from oldest to youngest) the Smolensk, Vladimir, and Yaroslavl' horizons (Velichko, 1975), best represented in the middle meridional band of the East European Plain (Dnepr River basin). To the east they are less persistent, and to the west, less discrete—another indication that the dynamics of Valdai permafrost cannot be represented solely as three frost-propagation phases separated in time. The dynamics were more complex.

The earliest stage of cryogenesis is the Smolensk cryogenic stage, which is divided into two phases. The first phase, "a", reflects the earliest (Valdai) cooling after the Mikulino (Eemian) Interglaciation and before the start of a major Early Valdai Interstade (the Krutitsy Interstade), which probably correlates with the Brörup Interstade. (For more detail on the chronostratigraphy of the loessial soils of the East European Plain, see Chapter 11.)

Cryogenic structures of this phase disrupt A_2- and B-horizons of the Mezinskiy soil complex as well as the Middle Pleistocene rocks underlying them, usually without affecting the humus A_1-horizon of the complex. In the western part of the Russian Plain (west of the Dnepr meridian), they are represented mainly by small structures (upper width, a few decimeters; vertical height, up to 1.5 to 2.0 m and seldom larger), with distinct vertical layering of the type of earth veins and ice-earth veins with polygons

3 to 6 m in diameter. But farther east (in the basins of Oka and Don Rivers), they are represented by involutions and cryoturbations (Bogucki et al., 1975; Udartsev and Sycheva, 1975).

Most disturbances described pertain to the layer of seasonal frost, but a few (e.g. wedge-shaped structures and two-layer structures) can attest to the local existence of permafrost. On the basis of the indicated diagnostics of paleocryogenic phenomena for the time before 70,000 years ago, one can postulate that conditions in the central portion of the Russian Plain (latitude 50° to 55°N) were similar to those prevailing in the present regions of deep (at least 1.5- to 2.0-m) seasonal and sporadic perennial ground frost. The present zone of sporadic permafrost is characterized by annual temperature fluctuations of 0°C to -1°C for the ground at the base of the layer and a permafrost thickness of a few tens of meters (Fotiyev, 1978).

In the North, the geocryologic conditions evidently became more and more severe. On the basis of average recent data for the European USSR for the values of the horizontal temperature gradient and the corresponding increment of the thickness of permafrost strata (Kudryavtsev, 1970), it can be postulated that the ground temperature graded from -3°C at latitude 60°N to -5°C near the Arctic Circle and that the permafrost in this region was several meters thick (when there was sufficient freezing time). It must be emphasized that the figures just cited obviously have no absolute significance (in part because the questions about the paleogeography of these areas remain controversial), but they are useful for subsequent comparisons.

The second phase, "b", of the Smolensk cryogenesis is represented by the frost structures that disrupted the whole profile of the Mezinskiy soil complex and the lower part of the Early Valdai (Khotylevo) loess horizon above the fossil soil. The time of its development corresponds to the cooling interval 60,000 to 50,000 years ago. The paleocryogenic phenomena of this phase have been closely studied in the central regions of the Russian Plain. At the latitudes of the cities of Vladimir, Moscow, and Smolensk (57° to 55°N), the profile of interglacial soil is most heavily disrupted by folds and fractures, as well as by solifluction. The extensive development of solifluction and involutions attests to the appreciable moisture content of the active layer (obviously higher than the total moisture capacity). Farther south, in addition to solifluction, there are small polygonal crack forms with earth veins and ice-earth veins 0.2 to 0.3 m wide at the top and up to 1.5 m deep. Such a combination of structures has been recorded at the latitudes of the towns of Ryazan', Bryansk, and Ternopol' (Figure 9-2).

Another indication of high soil-moisture content in the horizons of seasonal frost and the upper permafrost (at least above the lower plasticity limit, i.e., a moisture content in excess of 20% to 25% for sandy loam and sand) is an extensive development of postcryogenic plates and reticulate structure in the lower half of the Early Valdai (Khotylevo) loess and in the entire profile of the Mezin soil complex. Such structures are typical of sections of the Volyno-Podol'skiy Upland. In addition to an appreciable meso- and microjointing of these horizons, the total porosity is less and the volume weight of the soil skeleton more than for loess horizons devoid of distinct traces of postcryogenic structures (often by 15% to 20%). This fact may indicate an appreciable intraaggregate compaction of the rocks under the pressure of segregation ice schlieren formed in the past.

It is possible that the somewhat larger moisture content of the ground in comparison with phase "a" may be explained by an expansion of the area of permafrost and hence by a decline in mean annual temperatures. This hypothesis is supported by the fact that deposits of the first half of the Valdai (the Khotylevo loess and contemporaneous deposits) contain fairly large, wedge-shaped structures penetrating into the underlying deposits. Their width at the top may reach 0.5 to 1.0 m, and the vertical dimensions are 2.0 to 2.5 m. These structures may be diagnosed as pseudomorphs of polygonal ice veins. They have been found in various types of deposit—loess, deluvial loam, and inequigranular sand.

Such structures have been observed by the authors in sections on the Smolensk Upland. They are also known farther east (Kozhevnikov, 1972; Mozzherin and Butakov, 1980), as well as farther south in the western part of the Ukraine (Bogucki et al., 1975). For the latter region, all the structural parameters of this layer (width, height, diameter of polygons) were studied in detail, and the evident ice content attributable to polygonal vein ice was tentatively calculated as very low, no higher than 3% to 5%. Wedge-shaped structures similar in morphology and stratigraphic location have also been observed farther west beyond the confines of the European USSR, for example, in Poland (Jersak, 1973; Maruszczak, 1980). In Siberia, corre-

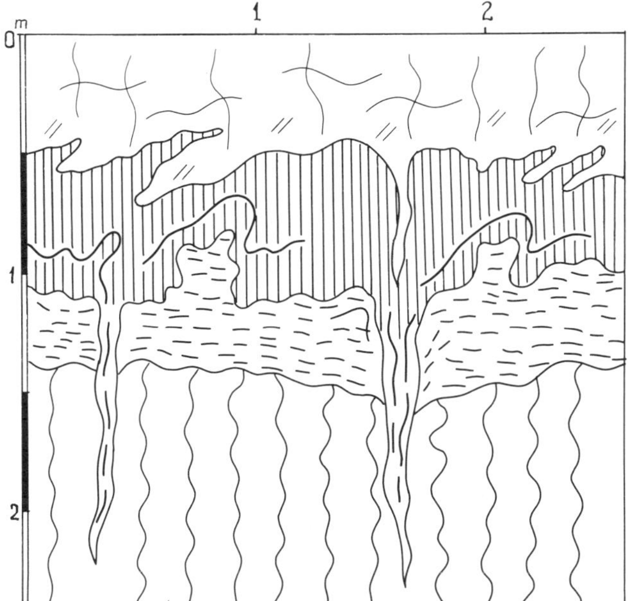

Figure 9-2. Frost deformation in the Mezin fossil soil complex in the Bryansk section. (Sketch by A. A. Velichko.) For symbols, see Figure 9-1.

lative polygonal ice veins and their pseudomorphs are referable to Zyryanka time.

The probably low ice content and relatively small size of the veins but the appreciable width of the polygonal net (up to 30 m or more) suggest that these ice veins formed near their southern limits under less continental climatic conditions, which today correspond to the transitional zone between continuous and discontinuous permafrost (Vtyurin, 1975).

Frost conditions probably became most severe at the final stage of the Smolensk cryogenesis, in contrast to phase "a". The southern boundary of continuous permafrost may have shifted 6° to 8° of latitude south of its present position, reaching the center of the Russian Plain. There, ground temperatures were at least as low as -3°C, and polygonal ice veins formed locally. However, cryogenesis did not undergo its maximum development at that time.

Sections from Volyno-Podoliya contain wedge-shaped structures filled with compact unstratified sediment derived from the upper part of the Early Valdai (Khotylevo) loess horizon. Apparently, the thawing and filling of the cavities occurred only during the deposition of the upper Early Valdai loesses, because this filling was not detected higher in the section. However, in our view, it is unlikely that these processes reflect any appreciable climatic warming. Instead, degradation could have occurred during sharply continental hyperzonal conditions and loess accumulation. At that time, greater dryness and continentality increased the depth of seasonal melting and led to the development of thermokarst over polygonal ice veins. "Dry-frost" facies (Velichko, 1958), which have no distinct natural analogues in the recent cryolithic zone, probably developed later, following Smolensk cryogenesis. The low soil-moisture content at that time (i.e., less than 10% to 15% for sandy loam) prevented the active formation of subterranean ice, except possibly cement ice, even at very low temperatures.

The next phase of active cryogenesis, imprinted in sections of the loessial-periglacial zone, is the deformation of the intra-Valdai Bryansk fossil soil. As is well known, the time of this soil formation is correlated with the Dunayevo

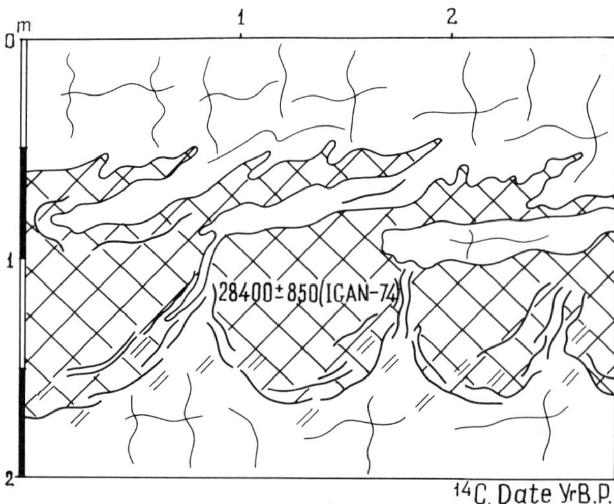

Figure 9-3. Frost deformations in the Dubno fossil soil (correlative of the Bryansk fossil soil) in the Basov-Kut section. (Sketch by V. P. Nechayev.) For symbols, see Figure 9-1.

warming in the glacial region of the Russian Plain (Chebotareva and Makarycheva, 1974). However, paleopedologic, paleontologic, and archaeologic data indicate a very severe climate during the Bryansk interval at 29,000 to 24,000 yr B.P. (Velichko and Morozova, 1972; Markova, 1975; Bader, 1977). This interval probably was "warm" only in comparison with the very cold conditions of most of the Valdai. A marked worsening of the climate immediately following this interval is indicated by the frost structures of the Vladimir cryogenic epoch, which commonly deform the Bryansk soil. A significant indication of this horizon, formed in hyperzonal conditions, is a reduction of zonal differences and a distinct manifestation of provincial characteristics. From the western USSR to the Pridneprovskaya Upland, the chief paleocryogenic processes were solifluction and plastic deformation of the Dubno soil (Bogucki, 1972), a correlative of the Bryansk soil, with weakly developed fine polygonal structures (Figure 9-3). To the east is a relatively narrow (100 to 200-km) submeridional band showing very little deformation (both struc-

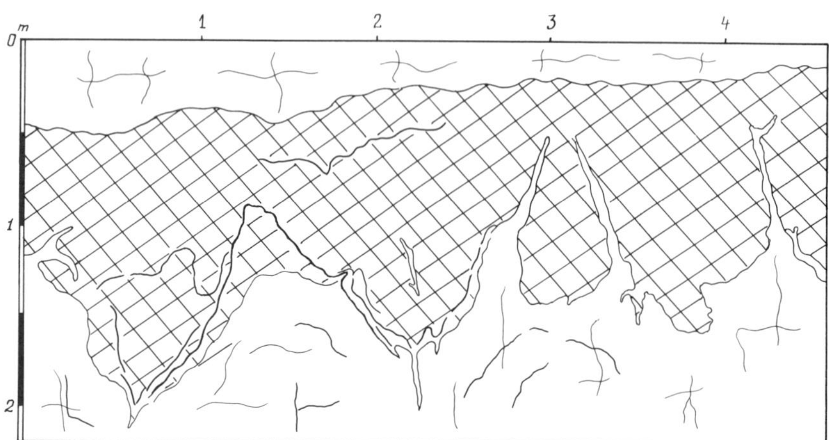

Figure 9-4. Frost deformations in the Bryansk fossil soil in the Schcheptaki section. (Sketch by A. A. Velichko.) For symbols, see Figure 9-1.

tural and nonstructural types) of the intra-Valdai soil. Finely polygonal formations of spot-medallion type (Velichko and Morozova, 1972) are observed east of the Dnepr Valley (Figure 9-4).

Extensive evidence of solifluction and plastic deformation at this chronstratigraphic level indicates a rigid base (i.e., permafrost) below the active layer. The geographic extent of frost features of this stage suggests vast areas of permafrost on the Russian Plain, across which the cryomorphogenic processes varied with longitude. In the West, the relatively thin active layer (no thicker than 0.6 to 0.8 m) was waterlogged, and during seasonal freezing the buried soil was plastically deformed. Then, as the climate became more continental (at the beginning of Late Valdai loess accumulation), the seasonal thaw layer thickened and solifluction became active. However, the intensity of seasonal freezing and the vertical temperature gradients there were insufficient for extensive development of polygonal structures. To the east of the Dnepr meridian, climatic and cryogenic processes were probably most similar to those presently found in south-central Siberia, where today the main cryogenic process is fine polygonal cracking (Fotiyev et al., 1974).

Increased continentality and dryness during the main phase of Late Valdai loess deposition initially retarded cryomorphogenesis and expanded the "dry-frost" facies, similar to that formed prior to the Bryansk interval. Indeed, neither major frost features nor fragments of postcryogenic textures are recorded in the middle part of Late Valdai loesses. However, specific types of micropores formed by the growth of individual crystals of cement ice are widely represented (Velichko and Markova, 1977).

Paleocryogenic characteristics change substantially in the upper part of Late Valdai loesses. In sections from the Russian Plain extending to latitude 48° to 49°N, there are major wedge-shaped structures diagnosed as pseudomorphs of polygonal ice veins (Figure 8-2). These wedge-shaped structures are associated with gray loess 0.5 to 1.2 m thick, gleyed to different degrees (Figure 9-5). Along subvertical blocks, this loess extends into wedge-shaped structures along contact zones. The structural characteristics of the loess horizons—including thickness, degree of gleying and humification, accumulation of iron oxide, and association with pseudomorphs—identify them as buried active layers of the Yaroslavl' cryogenic stage (Nechayev, 1980).

The formation of polygonal ice veins during the Yaroslavl' cryogenesis epoch should correspond to the final stages of Late Valdai loess accumulation. Increased atmospheric humidity (still under severe climatic conditions), characteristic of the final stages of all glacial epochs, permitted the development of polygonal ice veins in rocks subjected to intensive frost cracking. The widespread occurrence of polygonal ice veins in the loess regions and in the morainic and frontal apron plains to the north (Berdnikov, 1976) suggests that climatic conditions during the Pleistocene were most severe on the Russian Plain. At present, such conditions are found in the northern part of the zone of continuous permafrost, where ground temperatures are less than −5°C to −7°C. Polygonal ice veins formed on all relief forms and in different deposits (loess, fluvioglacial and alluvial sands, boulder loams, etc.). In the North, they reached a maximum depth of 4 to 5 m below the buried active layer, but farther south they changed into shallower earth-ice and earth veins 2 to 2.5 m deep. The vein width at the base of the buried active layer was 1.0 to 1.5 m. The average width of the polygons was 15 to 20 m (Figure 9-6). In the Dnepr Basin and the Volyno-Podoliya the ice content represented by ice veins was at most 8% to 12%, considerably less than has been calculated for similar features of Sartan age in Siberia.

Conclusions

Thus, vast areas of the Russian Plain were marked by low-temperature continuous permafrost at the end of the Late

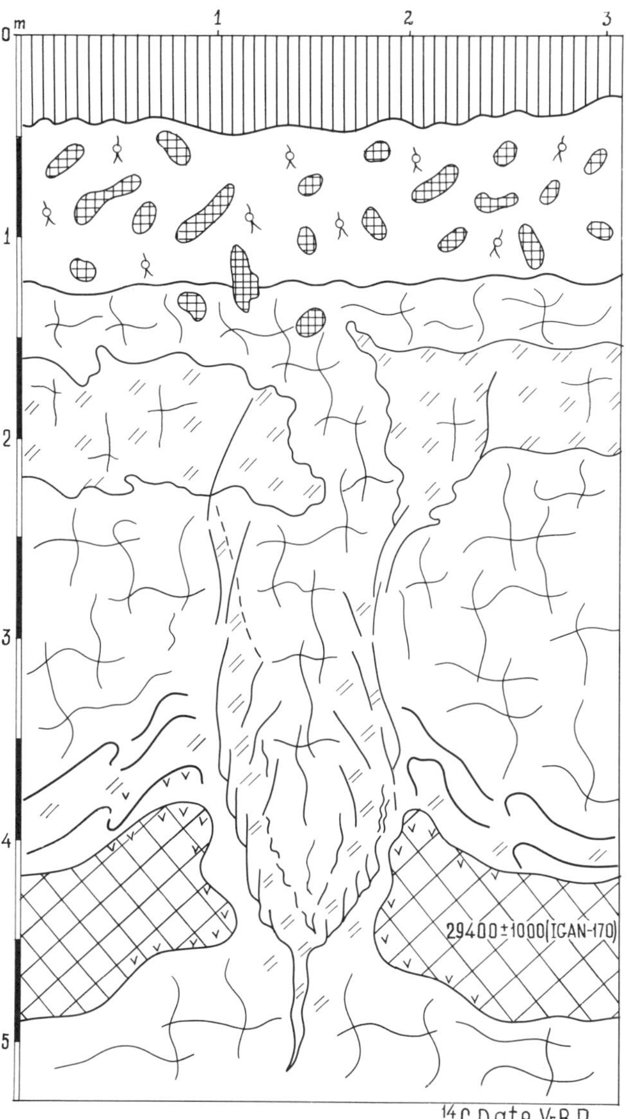

Figure 9-5. Bladed frost structure (pseudomorph of polygonal ice veins) in loess deposits of Krasnoselka section. (Sketch by V. P. Nechayev.) For symbols, see Figure 9-1.

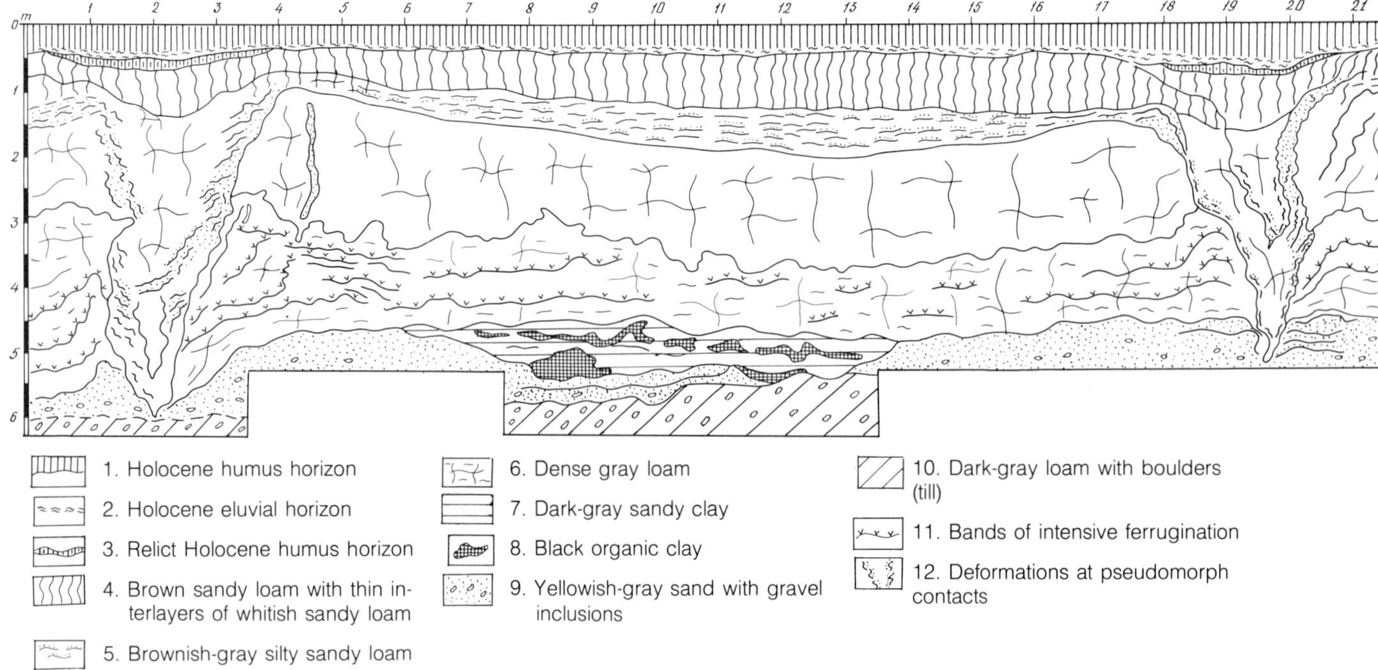

Figure 9-6. Fragment of polygonal network of ice-vein pseudomorphs in the Tuchino section. (Sketch by A. A. Velichko.)

Pleistocene. Permafrost extended 400 to 600 km south of the ice sheet boundary and over 1000 km from the center of the glacier advance. The nature of the cryolithic zone at the ice margin was very complex. The outer belt of the ice sheet was subjected to sharp dynamic movements that had an appreciable influence on the underlying bed and on the development of glacier-dammed basins. This promoted the development of hydrogenic and glaciogenic closed and even open taliks. The cryolithic zone near the ice margin was probably more discontinuous than at a considerable

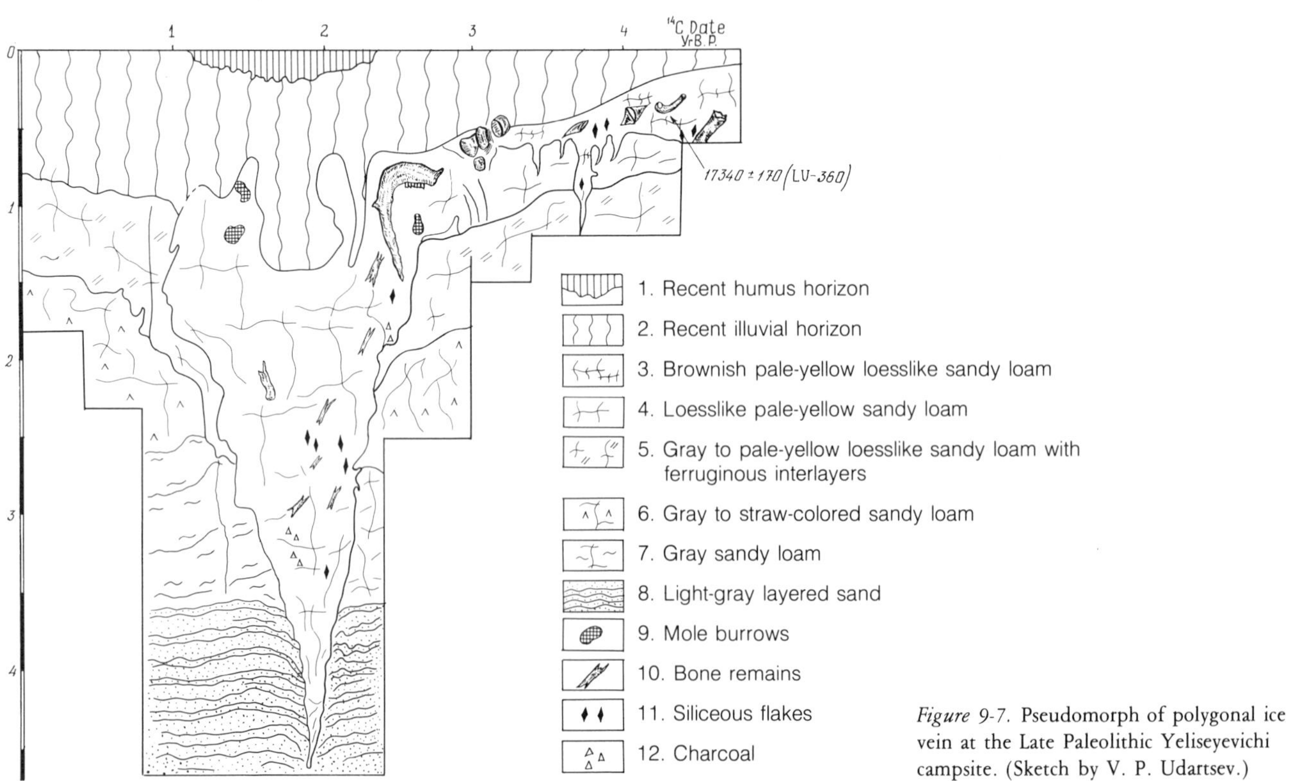

Figure 9-7. Pseudomorph of polygonal ice vein at the Late Paleolithic Yeliseyevichi campsite. (Sketch by V. P. Udartsev.)

distance from the ice. During deglaciation, blocks of dead ice may have acted as "coolers" differentially affecting the temperature regime of the underlying ground. All these questions, which are at the dividing line between glaciology and geocryology, still await definitive answers.

Studies of the Late Paleolithic relicts in the Desna Basin (Velichko et al., 1977; Kurenkova, 1978) suggest that the initial degradation of permafrost and thermokarst in the loess regions of the Russian Plain occurred between 17,000 and 15,000 years ago (Figure 9-7). However, the earliest datings of degradation may reflect local campsite conditions, as shown by the series of wedge-shaped structures of the Yaroslavl' cryogenic epoch exposed in a section near the Timonovka campsite near Bryansk. When the permafrost reached its maximum extent and began to degrade in the South, intensive freezing occurred in the North as the ice cover receded and the proglacial water bodies drained. Frost cracking and wedge-shaped structures developed in the drained areas of the Zapadnaya Dvina, Volga, and Severnaya Dvina Basins (Moskvitin, 1948; Kostyayev, 1960). Polygonal ice veins formed in the southern Baltic region during the Vepsovo and Luga stages (Basalikas and Shvyadas, 1978). Frost cracking and other frost deformation developed north of latitude 58° to 60°N during the Older and Younger Dryas (Meyrons, 1972; Ul'st and Berzin', 1962; Miidel, 1974). The final stages of the cooling at the end of the Pleistocene and the beginning of the Holocene (Preboreal and Boreal periods, are recorded by polygonal vein structures in northern Europe near the Arctic Circle (Rapp and Clark, 1971).

References

Bader, O. N. (1977). Paleoecology and the people of the Sungir' campsite. *In* "Palecology of Early Man" (I. K. Ivanova and N. D. Praslov, eds.), pp. 31-40. Nauka Press Moscow.

Basalikas, A., and Shvyadas, K. (1978). Properties of periglacial sheet formations in southeastern Latvia. *In* "Marginal Formations of Continental Glaciation." Naukova Dumka, Kiev.

Berdnikov, V. V. (1976). "Paleocryogenic Microrelief of the Center of the Russian Plain." Nauka Press, Moscow.

Bogucki, A. B. (1972). Stratigraphy of loesses of the Wolyn Upland. *In* "Lithology and Stratigraphy of Loesses in Poland" (H. Maruszczak, ed.), pp. 59-61. Wydawnictwo geologiczne, Warsaw.

Bogucki, A. B., Velichko, A. A., and Nechayev, V. P. (1975). Paleocryogenic processes in the western Ukraine during the Upper and Middle Pleistocene. *In* "Problems in the Paleogeography of Loessial and Periglacial Regions" (A. A. Velichko, ed.), pp. 80-90. USSR Academy of Sciences, Institute of Geography, Moscow.

Chebotareva, N. A., and Makarycheva, I. A. (1974). "The Latest Glaciation of Europe and Its Geochronology." Nauka Press, Moscow.

Fotiyev, S. M. (1978). "Hydrogeothermal Characteristics of the Cryogenic Region of the USSR." Nauka Press, Moscow.

Fotiyev, S. M., Danilova, N. S., and Sheveleva, N. S. (1974). "Geographical Conditions of Central Siberia." Nauka Press, Moscow.

Grichuk, V. P. (1973). Vegetation. *In* "Paleogeography of Europe in the Late Pleistocene" (I. P. Gerasimov, ed.), pp. 182-216. USSR Academy of Sciences, Institute of Geography, Moscow.

Jersak, J. (1973). Lithology and stratigraphy of loess of highlands in southern Poland. *Acta geographica Lodziensia* 32, 1-139.

Kaplina, T. N. (1965). "Cryogenic Slope Processes." Nauka Press, Moscow.

Katasonov, Ye. M. (1973). Paleofrost investigations, their objectives, methods and some results. *In* "Paleocryology in Quaternary Stratigraphy and Paleography" (V. V. Baulin and S. M. Tseitlin, eds.), pp. 10-22. Nauka Press, Moscow.

Kostyayev, A. G. (1960). Periglacial deposits and structure of low terraces of Valdai age in the Severnaya Dvina River valley. *In* "Periglacial Phenomena on the Territory of the USSR" (K. K. Markov and A. I. Popov, eds.), pp. 188-200. Moscow State University Press, Moscow.

Kozhevnikov, A. V. (1972). Solifluction-deluvial slopes and paleogeography of periglacial zones of plain and mountain territories. *In* "Problems of Study of the Quaternary Period" (A. S. Khomentovsky and S. M. Tseitlin, eds.), pp. 441-54. Nauka Press, Moscow.

Kudryavtsev, V. A. (1970). "Minimum Cryogenic Age of Permafrost in Different Frost-Temperature Zones of the USSR," pp. 117-24. Moscow State University, Vestnik.

Kurenkova, Ye. I. (1978). Radiocarbon datings and paleogeography of certain Late Paleolithic campsites of the middle course of the Desna River. *USSR Academy of Sciences, Izvestiya, seria geograficheskaya* 1, 102-10.

Markova, A. K. (1975). Paleogeography of the upper Pleistocene based on an analysis of small mammals of the upper and middle Dnepr region. *In* "Problems of Paleogeography of Loessial and Periglacial Regions" (A. A. Velichko, ed.), pp. 59-68. USSR Academy of Sciences, Institute of Geography, Moscow.

Maruszczak, H. (1980). Stratigraphy and chronology of loesses in Poland *In* "Stratigraphy and Chronology of Loesses and Glacial Formations of the Lower and Middle Pleistocene in Poland" (H. Maruszczak, ed.), pp. 43-54. Polish Academy of Sciences, Komitet Badan Czwartorzedu, Lublin.

Meyrons, Z. V. (1972). Certain characteristics of the structure of kames of the Kurzemskaya and Latgal'skaya Uplands as an indication of the conditions of their formation. *In* "Marginal Formations of Continental Glaciation" (G. I. Goretsky, D. I. Pogulyaev, and S. M. Shik, eds.), pp. 23-29. Nauka Press, Moscow.

Miidel, A. (1974). Settekiilude leide Pohja Eestist. *Eesti Geografia Selsti Aastaraamat 1971-1972,* 17-23. Tallinn.

Moskvitin, A. I. (1948). Fossil traces of permafrost. *Bulletin of the Commission for the Study of the Quaternary Period* 12, 69-77.

Moskvitin, A. I. (1962). Pleistocene of the lower Volga region. *USSR Academy of Sciences, Institute of Geology, Trudy* 64, 1-264.

Mozzherin, V. I., and Butakov, G. P. (1980). Periglacial phenomena of Middle and Late Pleistocene glaciations in the east of the Russian Plain. *In* Collected Abstracts, "Periglacial Formations of the Pleistocene" (V. G. Bandarchuk, ed.), pp. 24-26. Ukrainian Academy of Sciences, Institute of Geological Sciences, Kiev.

Nechayev, V. P. (1980). Structure of pseudomorphs of polygonal ice wedges in loessial deposits of the southwest of the Russian Plain. *In* Collected Abstracts, "Periglacial Formations of the Pleistocene" (V. G. Bandarchuk, ed.), pp. 30-32. Ukrainian Academy of Sciences, Institute of Geological Sciences, Kiev.

Popov, A. I. (1960). Periglacial formations of northern Eurasia and their genetic types. *In* "Periglacial Phenomena on the Territory of the USSR" (K. K. Markov and A. I. Popov, eds.), pp. 10-36. Moscow State University Press, Moscow.

Rapp, A., and Clark, G. M. (1971). Large nonsorted polygons in Padjelanta National Park, Swedish Lappland. *Geografiska Annaler* 2, 71-85.

Romanovskiy, N. N. (1977). "Formation of Polygonal Vein Structures." Nauka Press, Novosibirsk.

Sukhodol'skiy, S. Ye. (1978). Principles of the spread and formation of the cryolithic zone of the Northeast of the European USSR. *In* "General Geocryology" (P. I. Melnikov, ed.), pp. 5-14. Nauka Press, Novosibirsk.

Udartsev, V. P., and Sycheva, S. A. (1975). Upper Pleistocene loesses and buried soils of the Oka-Don Plain. *In* "Problems of Paleogeography of Loessial and Periglacial Regions" (A. A. Velichko, ed.), pp. 26-43. USSR Academy of Sciences, Institute of Geography, Moscow.

Ul'st, V. G., and Berzin', L. E. (1962). Frost deformations in deposits of the Baltic Glacier Lake and their paleographic significance. *Latvian Academy of Sciences, Institute of Geography, Trudy* 8, 33-41.

Velichko, A. A. (1958). Periglacial structures of the middle Desna River basin and their significance for stratigraphic and paleogeographic constructions. *Biuletyn peryglacjalny* 6, 361-72.

Velichko, A. A. (1965). Cryogenic land forms of the late Pleistocene periglacial zone (cryolithic zone) of eastern Europe. *In* "The Quaternary and Its History" (V. I. Gromos, ed.), pp. 104-20. Nauka Press, Moscow.

Velichko, A. A. (1975). Problems of correlation of Pleistocene events in the glacial, periglacial-loessial and maritime regions of the eastern European Plain. *In* "Problems of Paleogeography of Loessial and Periglacial Regions" (A. A. Velichko, ed.), pp. 7-26. USSR Academy of Sciences, Institute of Geography, Moscow.

Velichko, A. A., Grekhova, L. V., and Gubonina, Z. P. (1977). "The Environment of Early Man of Timonovka Campsites." Nauka Press, Moscow.

Velichko, A. A., and Markova, A. K. (1977). Two main forms of large pores in loesses. *USSR Academy of Sciences, Doklady* 197, 899-902.

Velichko, A. A., and Morozova, T. D. (1972). The Bryansk fossil soil and its stratigraphic significance and natural conditions of formation. *In* "Loesses, Buried Soils and Cryogenic Phenomena on the Russian Plain" (A. A. Velichko, ed.). Nauka Press, Moscow.

Vtyurin, B. I. (1975). "Underground Ice of the USSR." Nauka Press, Moscow.

CHAPTER **10**

Holocene Permafrost in the USSR

V. V. Baulin, Ye. B. Belopukhova, and N. S. Danilova

The general warming trend in the Holocene caused changes in frost conditions. The southern boundary of permafrost shifted north by 20° of latitude, frost cracking was sharply reduced, average annual temperatures rose by 12°C to 13°C, and areas of discontinuity and thermokarst increased. In contrast to present conditions, Late Pleistocene relicts were preserved outside the southern boundary of surface frost, and a wide zone was preserved where the permafrost was deeper and appeared to be buried. The overall thickness of the permafrost was also greater (Figure 10-1).

Because of climatic fluctuations, conditions varied throughout the Holocene, especially west of the Yenisey River. In the European USSR, permafrost persisted only in the Far North and Northeast, where the permafrost dynamics are similar to those described by Belopukhova (1963) for western Siberia and where distinct stages of permafrost development occurred during the Early, Middle, and Late Holocene. During the Early Holocene, the severe freezing conditions that developed during Sartan time persisted, and syngenetic ice veins grew in Early Holocene peat on alluvial terraces of the Taz Peninsula (latitude 67°N) (Sheshina, 1978). At present, syngenetic growth of ice veins is not observed at these latitudes.

Ice-wedge pseudomorphs found in oxbow-lake and floodplain sediments of the Yenisey River that were deposited 9000 years ago (Kind, 1974) also indicate severe climatic conditions. At latitude 64°N (Lavrushin, 1961), these pseudomorphs have been found at depths of over 8 m and are 1.5 m deep. Similar ice wedges form today on the floodplain at latitude 68°N, so apparently the boundary of ice-wedge formation in the Early Holocene was approximately 4° south of the present boundary. Obviously, the boundary of permafrost formation was shifted south by the same amount.

Middle Holocene

Conditions during the Middle Holocene had a major effect on the present-day frost situation. The climate was generally warmer from 9000 to 3300 years ago, according to paleobotanic data (Levkovskaya, 1977; Khotinskiy, 1977) and was characterized by milder frost conditions. This is indicated by the two-layered structure of the permafrost strata in the forest-tundra and the northern part of the northern taiga subzone today, where the partial melting of permafrost during the climatic optimum was followed by the subsequent incomplete freezing of the melted layer during the Late Holocene. The average ground temperature during the entire warm period of the Middle Holocene is calculated from the distribution of two-layered permafrost. In sandy forested areas of northwestern Siberia, it ranged from 0.8°C to 1.1°C in the west to 1.1°C to 1.5°C in the east, or about 1.0°C to 1.5°C higher than present values.

The amount of increase in ground temperature was associated with its transition through 0°C, that is, with the change of state. As the air temperature rose within negative values, changes in average rock temperature were more appreciable. Apparently, during the warmest intervals of the Middle Holocene, ground temperatures in the northern part of the region rose by more than 2°C to 3°C.

Latitudinal shifts in the distribution of cryogenic structures in surface deposits and permafrost formation, such as multiyear frost mounds and ice-wedge pseudomorphs (Figure 8-2), indicate that permafrost during the Late Pleistocene was widespread on the Yamal Peninsula in western Siberia north of latitude 68°N, but to the south only islands of syngenetically frozen Pleistocene deposits and epigenetic ice veins in peat bogs occurred. Ice-wedge pseudomorphs are abundant on the Taz Peninsula and elsewhere between latitude 68° and 66°N, but to the south no ice veins have been preserved, although numerous traces are known. Thus, the southern boundaries of continuous permafrost during the climatic optimum correspond to the limit of widespread Late Pleistocene syngenetic frozen ground, and the southern boundary of discontinuous permafrost is marked by the southern extent of localized Late Pleistocene syngenetic frozen ground. The location of these boundaries is confirmed by an analysis of

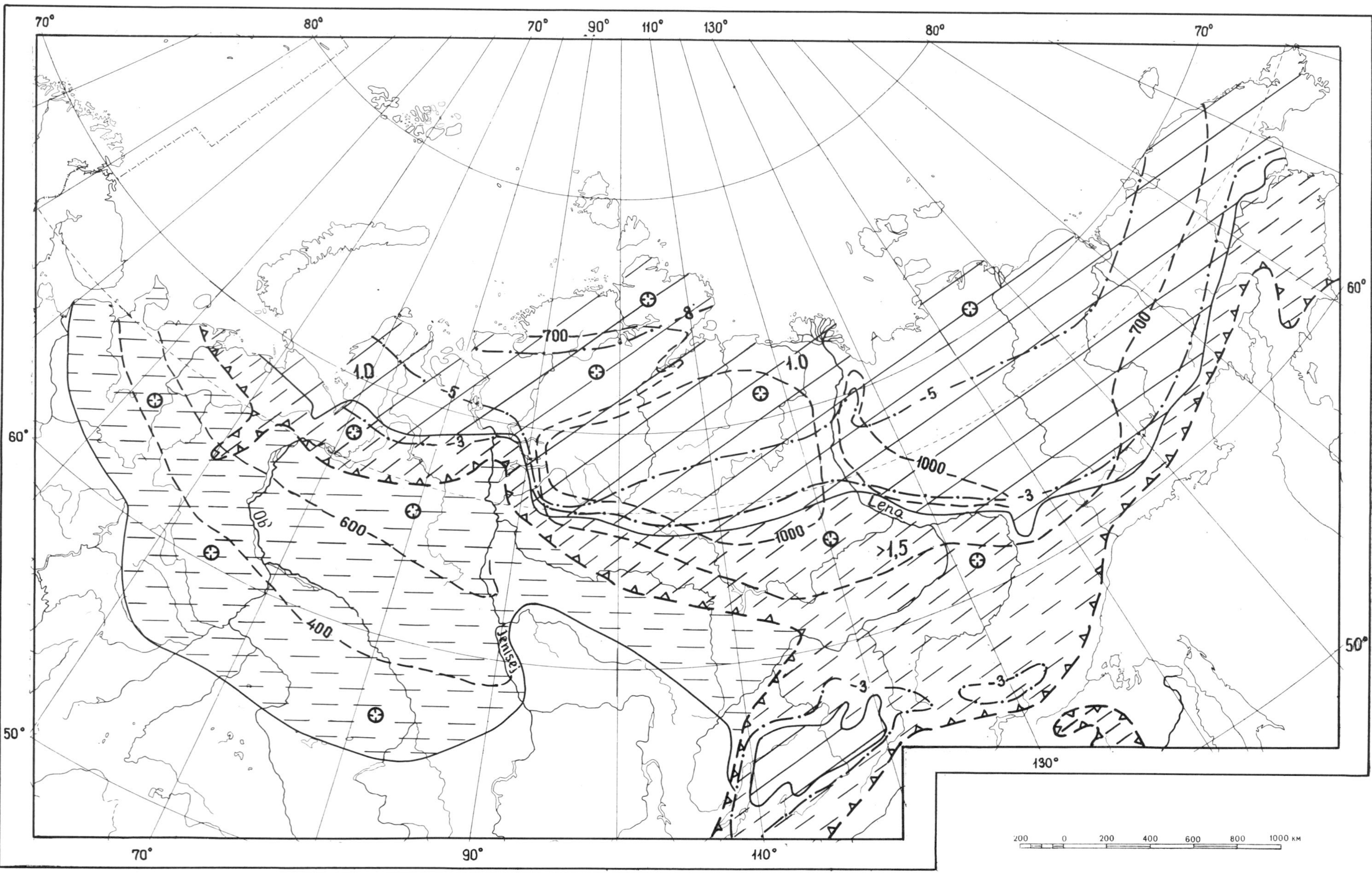

Figure 10-1. Map of permafrost during the Holocene climatic optimum. (Compiled by V. V. Baulin, E. B. Belopukhova, and N. S. Danilova.) (For symbols, see Figure 8-11.)

the occurrence characteristics of mound forms (Figure 10-2).

Three frost zones existing during the period of maximum warmth in the Middle Holocene can be recognized: a zone of continuous permafrost, a zone of discontinuous permafrost, and a zone of deeply buried permafrost. The zone of continuous permafrost was characterized by a widespread occurrence of syngenetic frozen ground with ice veins and by a ground temperature near the southern boundary of about −3°C. Isolated closed taliks formed in thermokarst depressions, in which large frost mounds were subsequently developed.

In the zone of discontinuous permafrost, in which Late Pleistocene syngenetic sediments were almost absent, the ground temperature was between 0°C and −3°C. Peat bogs developed in thawed areas during the climatic optimum of 7500 to 4000 years ago (Levkovskaya, 1977). In the North, the thawing of peat was less than 3 to 4 m deep and limited in area, so that subsequent freezing of taliks with an upper peat layer produced very slightly protruding mounds (up to 1 m) on flat peat bogs. Similar talik dimensions are estimated from the cryogenic structure of surface deposits and from the salinity changes in Paleogene marine sediments at depth. In the South, the taliks exceeded 10 m in depth and frequently developed mounds up to 3 to 5 m high.

The zone of deeply buried permafrost was characterized by the widespread and deep thawing of frozen ground. In the North, near the town of Salehard, the thawed layer was 50 m thick (Baulin et al., 1967). Large hummocky peat bogs, widely distributed along the northern limit, could have been formed only within large, deep taliks. However, by analogy with present conditions, islands of shallow permafrost must have existed not only in the North but also much farther south in this zone. Permafrost may have been preserved under dense, dark coniferous forests on the loamy soils characteristic of the middle taiga subzone, for today perennially frozen ground exists under such conditions at latitude 60°30′N. Old frozen ground has been detected at a depth of only 8 to 10 m. Taliks up to 10 m deep must have formed during the preceding colder epoch, when the middle taiga forests receded southward.

Late Holocene

Late Holocene climatic conditions determined the fairly severe frost conditions. Calculations show that the average ground temperature was 0.4°C to 0.6°C below that of the present. At times, the frost conditions were more severe; at other times, they were milder than at present.

Frost cracking and ice-wedge development are evident in western Siberian peat bogs as far north as latitude 65°N. At present, such features do not form south of 66°30′N, and so the permafrost temperature was approximately 2°C lower than the present one during the coldest period of this stage. Severe frost conditions in the Holocene are also indicated by the occurrence of ice-wedge pseudomorphs at a depth of 2.0 to 2.5 m in the alluvium of the high floodplain of the Yenisey River at latitude 66°N. In the northern part of the northern taiga subzone, widespread hummocky peat bogs formed as a result of thawing ice wedges during the Late Holocene warm interval.

It is logical to assume that the most severe frost conditions indicated by intensive growth of ice wedges occurred early in the Late Holocene. In contrast, milder conditions characterized the second half (the start of the second millennium of the present era).

In contrast to western Siberia, Middle Holocene permafrost was preserved only in the northeastern European USSR along the coast east of the Pechora River (Baranov, 1964; Solomatin, 1965) and as isolated islands extending to the Arctic Circle. Active thermokarst and peat accumulation occurred. During the Late Holocene, peat bogs of the Kola Peninsula, the lower Dvina River, and the middle Pechora River basin (up to latitude 66°N) froze. During the "maximum" cooling, peat bogs underwent frost cracking and formed ice wedges up to 1.0 to 1.5 m deep. The wedges have been preserved only in the eastern Bol'shezemel'skaya Tundra, north of latitude 67°30′N.

In central and eastern Siberia, one can also distinguish three stages of frozen-ground development. In the Nizhnyaya and Podkamennaya Tunguska River basins of west-central Siberia as in western Siberia, widespread hummocky bogs accumulated peat in thermokarst depressions during the second stage (Shumilova, 1963) and developed frost cracking and ice wedges during the third stage. The southern boundary of hummocky peat bogs in central Siberia was latitude 58° to 59°N, or 5° south of its position in western Siberia, as a result of the pronounced continentality of the climate. Farther south in the Angara River basin, a topography of hummocks and flat-bottom steppe depressions formed as the frost and Pleistocene polygonal ice veins thawed. The southern boundary of permafrost in central Siberia runs along the southern limit of relict Pleistocene ice veins or syngenetic frozen strata. (See Figure 8-2.) The zone of discontinuous occurrence is delimited by recent deep taliks (up to 100 to 140 m deep) in sandy deposits of central Yakutiya, which may be regarded as Middle Holocene relicts (Solov'yev, 1959), and includes areas of high-temperature frozen ground in the middle reaches of the Lena and Aldan River basins. However, large masses of frozen ground, primarily the "ice complex" of central Yakutiya, are preserved within this zone. North of the southern permafrost boundary, the Holocene warming is reflected in the depth of the "ice complex" (Danilova, 1967). In the region of permafrost, the "ice complex" is buried except where its active destruction is taking place. The thickness of overlying loams is 1.5 to 2.0 times that of the recent seasonal thaw layer, or about 1.5 to 2.0 m in central Yakutiya and 1 m in the tundra regions. The entire loam layer, including its permafrost portion, is considered to be a Middle Holocene active layer formed when seasonal thawing in loamy soils was 1.5 to 2.0 times greater than the present depth. Thermokarst depressions (alasses) and valleys found throughout the "ice complex" can be regarded as Middle Holocene relicts in the permafrost region.

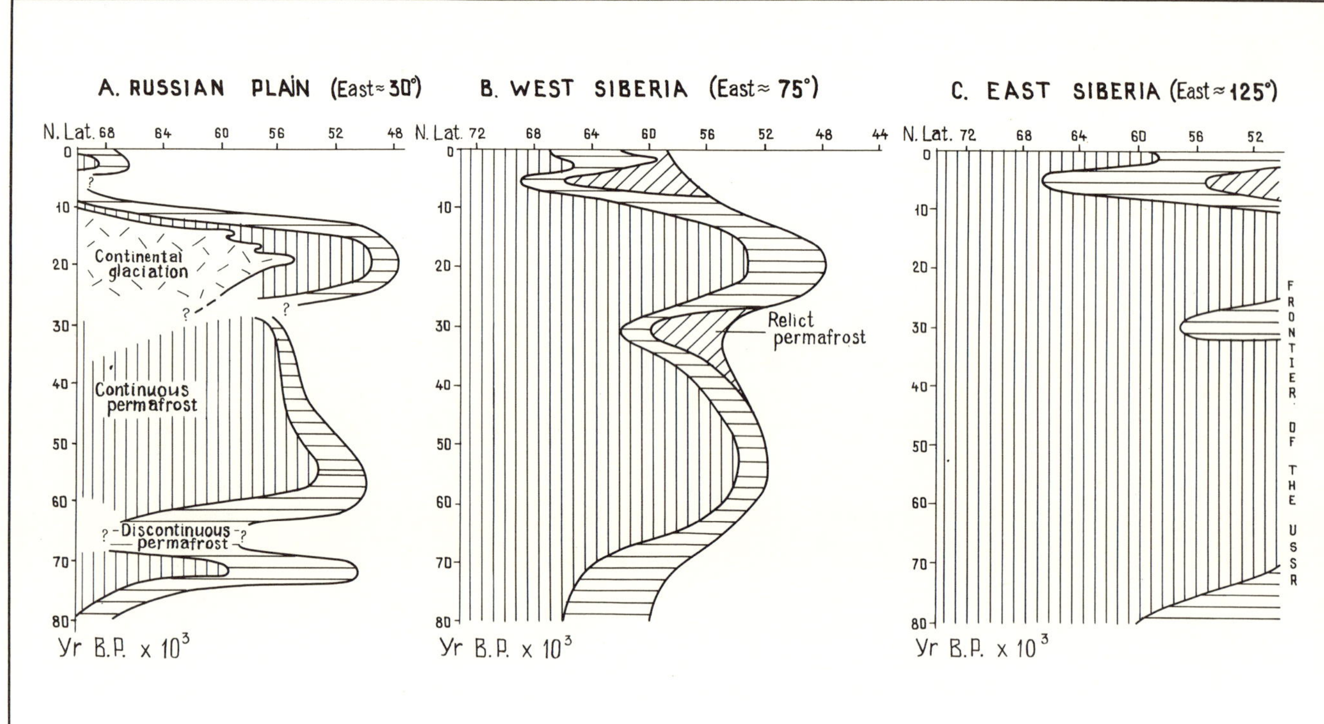

Figure 10-2. Dynamics of Late Pleistocene and Holocene permafrost in the USSR.

Conclusions

During the Holocene, polygonal ice veins were smaller than in the Pleistocene, their distribution was more limited, and the size of the network was greater in floodplain sediments throughout the permafrost region. They were formed in fewer sediment types, and their distributional limits shifted markedly northward by almost 20° of latitude in the European USSR, by 13° in western Siberia, and by 6° in central Siberia. During the Middle Holocene over the entire territory, Pleistocene ice veins were buried as the seasonal thaw layer increased, and polygonal ridges on all the geomorphologic surfaces higher than the second terrace were leveled. The formation of ice veins may have stopped even in the extreme Northeast at that time. In the third stage (Late Holocene), frost cracking substantially increased, and the permafrost boundary shifted 1° to 2° of latitude south of its present position. In the European USSR, western Siberia, and the western half of central Siberia, epigenetic polygonal ice veins up to 1.5 m wide were formed in peat bogs. In the northeastern USSR, syngenetic ice veins were formed in sediments of the upper floodplains as well as on peat-covered surfaces of terraces and interfluves. At present, these ice veins are either buried or have decreased growth rates, and the separation from present seasonal thawing does not exceed 0.5 m. All these polygonal ice veins are important evidence of a sharp although brief Late Holocene cooling.

During the Holocene, a reduction in permafrost thickness occurred throughout the USSR (Figures 10-2 and 10-3). West of the Yenisey River (with the exception of Arctic regions), permafrost became detached from the seasonal freezing layer and descended to a depth of up to 100 m before disappearing altogether. Calculations, modeling, and analysis of recent relict permafrost indicate that during the Middle Holocene up to 100 m of Pleistocene permafrost thawed, and up to the present about 150 to 200 m has thawed (Baulin et al., 1967; Balobayev, 1973). Thus, the thickness of frozen ground was approximately 100 m greater during the Middle Holocene than today (Figure 18-19). These estimates apply to loose and weakly cemented Mesozoic and Cenozoic sandy loams and loams. In bedrock, thawing from below could have exceeded 300 m. Frozen strata at the surface could have thawed to a depth of 100 to 300 m or more, according to the depth of relict frozen ground, the thickness of deep taliks, and so on. Thus, permafrost whose thickness reached 400 to 500 m in Sartan time thawed during the Middle Holocene.

A characteristic frost feature in the first half of the Holocene was the wide zone of relict (buried) frozen ground, which had a width of 1200 km in western Siberia and up to 700 km in the European USSR. This relict frozen ground obviously affected the natural landscapes of that time, especially the vegetation and the occurrence and circulation of groundwater.

The last part of the Late Holocene was characterized by an attenuation in the severity of frost conditions. However, recorded against this background are several stages of increased permafrost severity during medium- and short-period climatic fluctuations.

In conclusion, because of a number of difficulties, paleofrost reconstructions give only a very general pattern of permafrost dynamics. Permafrost features, for example, have different response times and react differently to changes in natural climatic conditions. Thus, there is a considerable lag between fluctuations of heat exchange at the surface and changes in permafrost thickness. Relicts as well as traces of permafrost do not always have sufficient time to equilibrate. Changes in environmental and frost conditions in different regions of the USSR occurred at different times, in different magnitudes, and even in different directions. These facts must be considered in analyzing the map of the Middle Holocene climatic optimum, because this map reflects extreme values of all permafrost characteristics over a long time interval.

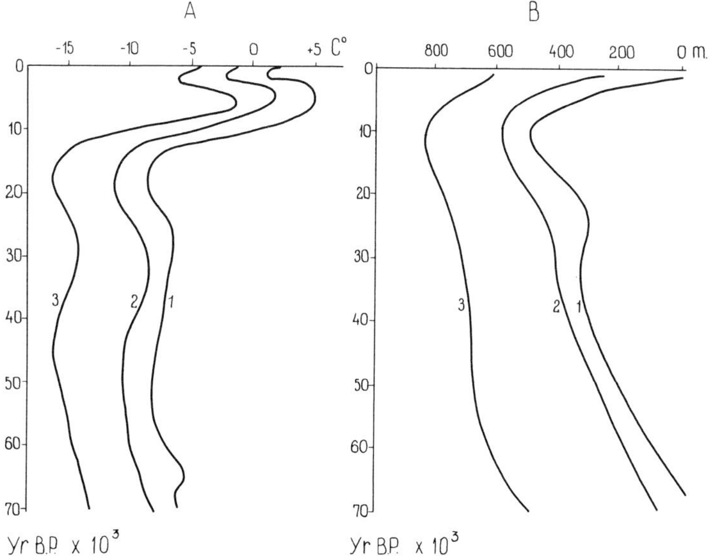

Figure 10-3. Dynamics of Late Pleistocene and Holocene cryolithic zone at latitude 65°N. (A) Variation in temperature. (B) Variation in thickness. (1) Russian Plain (longitude 50°E), (2) western Siberia (longitude 75°E), (3) eastern Siberia (longitude 125°E).

References

Balobayev, V. T. (1973). Basic principles of deep freezing of the Earth's crust. *In* "Problems of Geocryology" (S. G. Tsvetkova, ed.), pp. 26-36. Nauka Press, Novosibirsk.

Baranov, I. Ya. (ed.) (1964). "Geocryological Conditions of the Pechora Coal Basin." Nauka Press, Moscow.

Baulin, V. V., Belopukhova, Ye. B., Dubikov, T. I., and Shmelev, L. M. (1967). "Geocryological Conditions of the West Siberian Lowland." Nauka Press, Moscow.

Belopukhova, Ye. B. (1963). Characteristics of permafrost development in the northwest of western Siberia in the Late Holocene. *In* "Permafrost of Different Regions of the USSR" (A. I. Efimos, ed.), pp. 218-25. USSR Academy of Sciences Press, Moscow.

Danilova, N. S. (1967). Formation of cover loams of central Yakutiya. *USSR Academy of Sciences, Izvestiya, seria geograficheskaya* 4, 82-89.

Khotinskiy, N. A. (1977). "Holocene of Northern Eurasia." Nauka Press, Moscow.

Kind, N. V. (1974). "Geochronology of the Late Anthropogene from Isotope Data." Nauka Press, Moscow.

Lavrushin, Yu. A. (1961). "Types of Quaternary Alluvium of the Lower Yenisey." USSR Academy of Sciences Press, Moscow.

Levkovskaya, G. M. (1977). History of the Holocene forestation of the Arctic in light of radiocarbon data. *In* "Results of Biostratigraphic, Lithologic and Physical Studies of the Pliocene and Pleistocene of the Volga-Ural Region" (V. L. Yakhimovich, ed.), pp. 15-36. USSR Academy of Sciences Press, Ufa.

Sheshina, O. N. (1978). Palynology of the Holocene of the Pur River basin. *In* "Proceedings of a Conference of Young Scientists: Paleontology," pp. 42-48. Moscow State University Press, Moscow.

Shumilova, L. V. (1963). An outline of the region where the Tunguska meteorite fell. *In* "Problems of the Tunguska Meteorite" (G. D. Plekhanov, ed.), pp. 22-34. Nauka Press, Moscow.

Solomatin, V. I. (1965). Syngenetic and epigenetic freezing of marine deposits on the coast of the Pechorskoye Sea. *In* "Underground Ice" (A. I. Popov, ed.), Issue 1, pp. 183-93. Moscow State University Press, Moscow.

Solov'yev, P. A. (1959). "Cryolithic Zone of the Northern Part of the Lena-Amga Interfluve." USSR Academy of Sciences Press, Moscow.

Loesses, Fossil Soils, and Periglacial Formations

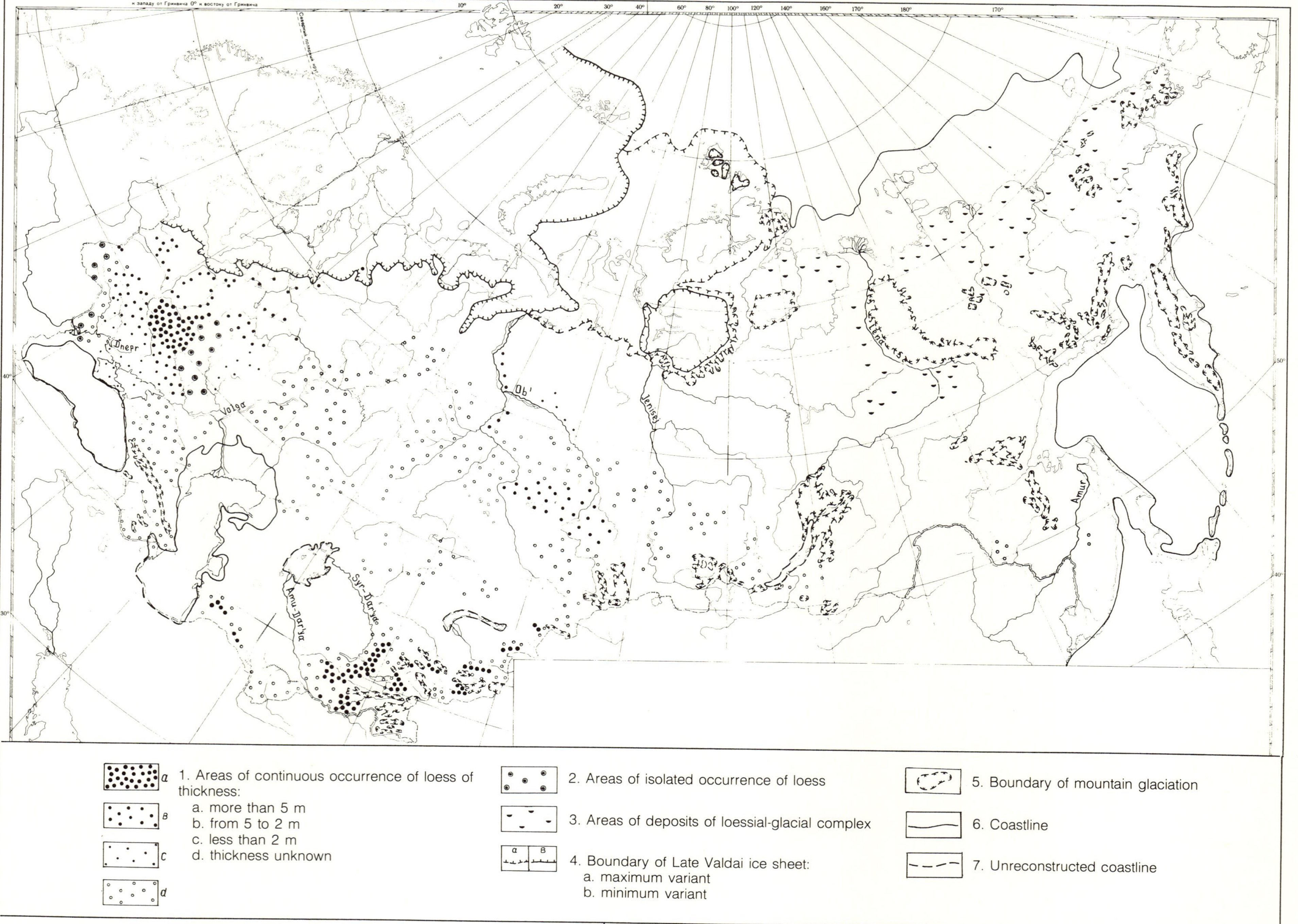

Figure 11-1. Areas of loess of Valdai age. (Map compiled by A. A. Velichko and T. A. Khalcheva on the basis of data from A. B. Bogucki, V. S. Bykova, I. A. Volkov, V. V. Kolpakov, T. D. Morozova, A. A. Lazarenko, V. P. Nechayev, S. V. Tomirdiaro, and V. P. Udartsev.)

CHAPTER 11

Periglacial Landscapes of the East European Plain

A. A. Velichko, A. B. Bogucki, T. D. Morozova, V. P. Udartsev, T. A. Khalcheva, and A. I. Tsatskin

This chapter and the subsequent five concern the development of the natural environment primarily on the basis of loess and loessial soils in various regions of the USSR. Loess and glacial features constitute two of the clearest manifestations of Pleistocene processes. In the USSR, loesses are not uniformly distributed in either Europe or northern Asia (Figure 11-1), nor is their content the same in different regions. Even in such "classic" loess regions as the European USSR and Central Asia, the stratigraphy and lithology differ markedly.

There is no consensus on loess genesis in the USSR. The discussion on the origin of loess, which began about 100 years ago, has not yet been completed. For example, prevalent for a long time was the fluvioglacial hypothesis for loess origin, not only for the European USSR but also for its southern regions. Today, most investigators divide the problem into two parts: (1) mode of sedimentation of homogeneous material, primarily silt and (2) conditions wherein the rock acquired properties characteristic of loess (structure, carbonate content, porosity). With respect to the method of accumulation, many investigators adhere to the polygenetic hypothesis, believing that the loess can accumulate in various ways (wind, water, proluvium, etc.). However, in a given region a single factor is primarily responsible. When the genesis of loess is approached in this manner, one can assume that most investigators regard the main, well-stratified loess with fossil soils (on high plateaus and ancient terraces) as eolian except in the lower parts of valleys, where loessial material may result from river floods, or on high surfaces, where small lakes may have existed in the periglacial zone. It will be shown that loess in not only the European USSR but also in western Siberia is mainly of eolian origin but that in Central Asia the deluvial-proluvial processes may be of major importance. That region may have been more arid than it is at the present time, so the eolian factor may also have been significant. Standing by itself is the co-called "loess-ice complex," widespread on the periglacial plains of the northeastern USSR.

The origin of this group of deposits is highly controversial: the question is, are they eolian or alluvial? It is generally agreed that specific loessial properties, despite definite differences in the depositional conditions in individual regions, are acquired primarily under arid conditions. In such regions as the Russian Plain, western Siberia, the northeastern USSR, and perhaps Central Asia, cryogenic processes participated, as some recently published studies indicate.

The degree and detail of the chronostratigraphic subdivisions of loessial strata are different. In eastern Europe, the study of loesses and intervening soil horizons has a long tradition. In western Siberia, at least for the Late Pleistocene, the loess stratigraphy has many features in common with that of eastern Europe. In particular, for both regions the initial stages of the Late Pleistocene Valdai Glaciation were humid, whereas in the Late Valdai the aridity increased markedly. For the northeastern USSR, a similar sequence is observed. Apparently, the main stages of loess and soil formation during the Late Pleistocene can be correlated all over the periglacial regions of the Northern Hemisphere (Table 11-1). For Central Asia, the correlations are less certain.

Method of Investigation

Fossil soil horizons and paleocryogenic deformations are well preserved in loess. They reflect complex changes in natural conditions ranging from periglacial to interstadial to interglacial. Fossil fauna and flora are also preserved. Obviously, the study of loess requires a comprehensive approach. The methodological problems in the study of cryogenic deformations and fauna and flora are examined in subsequent chapters, so here we briefly discuss methodological aspects in studies of loess and fossil soils.

In modern paleogeography, fossil soils have become one of the most important tools for studying the natural environment of the past, thanks to the work of K. D. Glinka,

Table 11-1. Comparison of Principal Late Pleistocene Fossil Soils in Periglacial Regions of the Northern Hemisphere

Austria	Belgium	USSR				USA
(Fink, 1969)	(Paepe, 1969)	(Velichko, 1977) Russian Plain	(Volkov and Zykina, 1984) Southwestern Siberia	(Lazarenko, 1984) Central Asia	(Kaplina and Lozhkin, 1984) Northeast	(Follmer et al., 1978)
Stillfried complex B	Kesselt soil	Bryansk soil	Iskitim soil complex	?	Deposits corresponding to the time of Kuranakh-Sala warming	Farmdale Soil
Stillfried complex A	Varneton and rocour soil	Mezin polygenetic soil complex	Berdsk soil complex	Soil complex number three	?	Sangamon Soil

V. F. Mirchink, V. I. Krokos, I. P. Gerasimov, M. A. Glazovskaya, R. Ruhe, J. Fink, V. Kubiena, and others. Among analytic methods of importance in the study of fossil soils are the determination of bulk composition of soils and silt, the mechanical composition, the fractional composition of humus, and the optical properties of humic acids. The micromorphological method is widely applied in the USSR and elsewhere. Fossil soils are diagnosed principally on the basis of the structure of the profile, the microstructure of the genetic horizons, the humic state (with allowance for a significant transformation of humus after burial), the type of mineral weathering, and the distribution of stable components. For fossil soils, diagnosis mainly to the level of type and sometimes to subtype is possible. In many cases, because of a lack of modern analogues of fossil soils, only the elementary processes and their combinations in ancient soil formation can be diagnosed.

The study of Late Pleistocene fossil soils has progressed to the point that a higher level of data integration is now possible, namely, the reconstruction of ancient soil mantles and the compilation of paleopedologic maps for particular epochs of soil formation in Europe (Velichko and Morozova, 1973; Morozova, 1981). Soils of this age are the best preserved, and their properties are closest to Holocene soils of the recent surface. The development of a system of diagnostic or morphotypic indexes characteristic of different Late Pleistocene soils provides a sound basis for correlation over extensive areas.

Somewhat more complex is the reconstruction of the environment at the time of loess formation. We have based our arguments on the views of V. A. Obruchev, I. P. Gerasimov, K. K. Markov, and A. A. Velichko in treating loess as a continental and primarily eolian, permanently accumulating soil-like formation with initial geochemical transformation limited mainly to microaggregation, redistribution of secondary carbonates without leaching, and very slight accumulation of humus. Weathering of the mineral mass is revealed by the silt fraction, which predominates in loesses, and by the relationship of groups of minerals of different stability. There are also several coefficients, the most informative of which are listed in Table 11-2.

The weathering characteristics of loesses are also reflected in the surface texture of mineral grains, particularly unstable ones.

The study of loess chemistry involves the determination of humus content, the carbon content of humic and fulvic acids, the optical properties of the humic acids, the carbonate content, and the bulk composition of the loess. Chemical and mineralogic analyses make it possible to refer to the siallite carbonate type of weathering in epochs of loess formation.

There also have been attempts to reconstruct the environmental conditions from data on loess porosity, not so much total porosity as the structure of the pores themselves. In particular, differences have been found in the conditions of loess formation with different ratios of open (channellike phytogenic) and cavernous (closed) macropores.

In the course of our study, we also used a method of studying the shape and surface of quartz sand grains extracted from the loesses. Each type of deposit (alluvial, marine, eolian, etc.) is characterized by its predominant type of sand grain (Cailleux, 1942). In order to obtain data on sand grains from loesses, it was necessary to concentrate them, for typical loesses contain very small amounts of sand (1/5000 to 1/1,000,000 of the total volume of the rock).

Table 11-2. Weathering Coefficients for Minerals in Loessial Soils

Coefficient	Stable Minerals	Unstable Minerals
K-I	Zircon, tourmaline	Amphiboles, pyroxenes
K-II	Zircon, tourmaline, garnet, ore minerals, titanium-containing minerals	Amphiboles, pyroxenes
K-VI	Quartz	Feldspars
K-VII	Zircon, tourmaline, garnet, ore minerals, titanium-containing minerals	Amphiboles, epidote
K-IX	Zircon, tourmaline, titanium-containing minerals	Amphiboles, epidote
K-X	Zircon, tourmaline, titanium-containing minerals	Amphiboles, pyroxenes

Table 11-3. Characteristics of the Surface and Roundness of Sand Grains of Different Types of Deposits

Number	Length at 20×, mm	Width, mm	Radius of Inscribed Circle	Radius of Described Circle	Least Rounding Radius	Grain Surface: Lustrous			Quarter-dull			Half-dull			Dull			Roundness Class					Roundness Coefficient (after A. Cailleux)
						Smooth	Pitted	Shell-like	Smooth	Pitted	Shell-like	Smooth	Pitted	Shell-like	Smooth	Pitted	Shell-like	0	I	II	III	IV	
1	8.0	4.0	2.0	4.0	0.5										1					1			0.125
2	9.0	5.5	2.4	4.5	1.5			1													1		0.334
3	8.5	6.0	2.7	4.3	0.5							1							1				0.116
4	7.0	6.1	2.3	3.5	0.0			1										1					0.0
5	8.0	4.0	1.9	4.0	0.5				1										1				0.120
49	6.8	5.0	2.5	3.4	0.75										1					1			0.125
50	7.0	5.1	2.5	3.6	0.25										1				1				0.103

Roundness and shape of sand grains were classified by the point method (Khabakov, 1946), with our additions concerning the grain surface characteristics (Table 11-3). The data on visual evaluation of grain roundness were compared with morphometric data obtained from the formula worked out by A. Cailleux: $I=r/L$, where r is the radius of least rounding of the grain and L is the length. This made it possible to obtain sufficient agreement of the results of the two methods (Figure 11-2).

Given below are the results of a study of horizons of fossil soils and loesses. The study was based on the methods just described. The last section of this chapter gives reconstructions of changes in environmental conditions for individual regions in which loess-soil series are common, with use also being made of data from paleocryogenic, paleobotanic, and paleofaunistic studies.

Interregional Correlations of Loessial Strata

Loesses are more common in the East European Plain than anywhere else in the Northern Hemisphere. In the North, near the boundary of the latest glaciation (but without a strict correlation with it), so-called cover deposits (derivatives of loesses) are widespread. Farther south, they are replaced by a region of loess proper, especially in the middle and western portions of the East European Plain, where the main loessial regions studied are (from west to east) the Volyno-Podoliya, Dnepr, and Don regions. In the eastern portion of the plain, however, particularly in Transcaucasia, distinct loess deposits are very rare. In the southern part of the plain, loesses are widespread in Stavropol' (northern Caucasus foothills), but the chronologic control of the stratigraphy is poor. The structural characteristics of the loesses in the main regions of the East European Plain enable us to reconstruct environmental changes in the periglacial zone of the European USSR. The following chronostratigraphic scheme is utilized (Velichko, 1977).

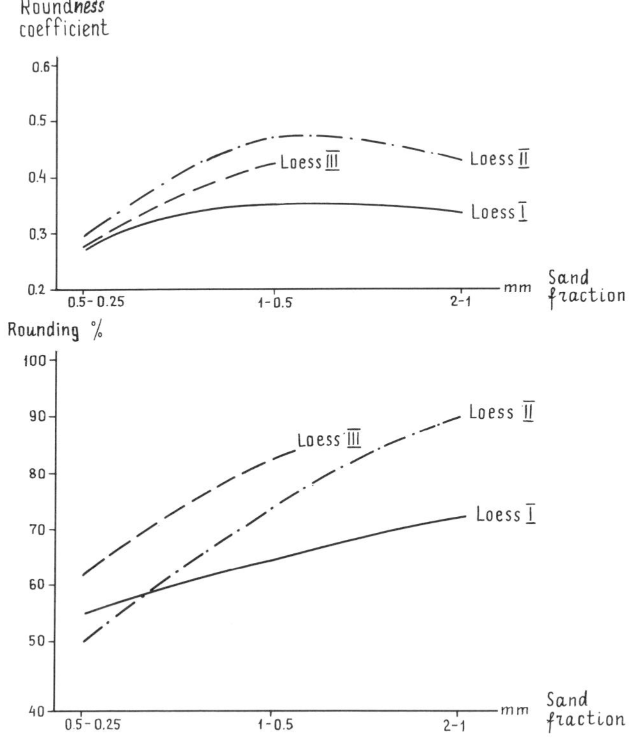

Figure 11-2. Rounding characteristics of sand grains of Valdai loesses according to size fractions.

I. Holocene—Recent soil including relict (Timonovka) soil
II. Late Pleistocene
  A. Valdai Glaciation
    1. Late Valdai
      a. Loess III (Altynovo)
        (1) Yaroslavl' cryogenic horizon
        (2) Gleying level (Trubchevsk horizon), slight oscillation
      b. Loess II (Desna loess)
        (1) Vladimir cryogenic horizon
        (2) Bryansk soil (Bryansk warming interval of 30,000 to 25,000 yr B.P.)
    2. Middle Valdai
      a. Loess I (Khotylevo loess)
        (1) Smolensk cryogenic horizon, phase "b"
    3. Early Valdai
      a. Krutitsa phase of Mezin soil complex (Krutitsa, Early Valdai warm interval)
      b. Intra-Mezin thin loess layer
      c. Smolensk cryogenic horizon, phase "a"
  B. Mikulino Interglaciation
    1. Salyn phase of Mezin soil complex
III. Middle Pleistocene—Loesslike sandy loams corresponding to the Moscow stage; Dnepr glacial deposits; loams with fragments of local rocks corresponding to the Dnepr time; till of Dnepr Glaciation

This scheme was initially worked out for the Dnepr Basin and subsequently confirmed in western Volyno-Podoliya (Bogucki et al., 1975) and in the Oka-Don Plain, farther to the east on the Russian Plain (Udartsev and Sycheva, 1975). These three main loess regions have many common chronostratigraphic features.

VOLYNO-PODOLIYA

Practically all the soil horizons enumerated in the list can be observed in one of the key sections near the town of Gorokhov (Figure 11-3). At the base of the Late Pleistocene strata is the Gorokhov polygenetic soil complex, in which soils of two phases—Salyn and Krutitsa—are distinguished. This complex compares quite well with the soils of the Mezin complex of the summary scheme. In the central regions of Europe, it corresponds to the Stillfried A soil complex (Table 11-1).

In the Novovolynsk section, this complex overlies a moraine of Dnepr Glaciation; this fact confirms the complex's Late Pleistocene age. Its most characteristic feature, particularly in the Gorokhov section, is a combination of dark-colored A-horizon (0.45 m), extensively burrowed by mesofauna, and Bmt-horizon (0.8 m), represented by a rich brown, light, structured loam. This soil complex is disturbed by cryogenic deformations of the Smolensk (Torchin) cryogenic horizon, in which is distinguishable a phase "a," which developed between the Salyn and Krutitsa soil-formation phases, and a phase "b," which completed the Krutitsa soil-formation phase (chernozemlike soil). Both phases of this horizon are characterized by finely polygonal systems and solifluction deformation, a special description of which is given in Chapter 9.

Located above is loess Horizon I, correlated with the Valdai Glaciation. Its thickness is about 0.8 to 1.0 m. It has a marked heterogeneity of texture when compared to younger loess horizons. Developed directly on loess I is an intra-Valdai Dubno fossil soil, which is a stratigraphic analogue of the Bryansk soil dated at 29,000 to 28,000 yr B.P. (Table 11-4). It has the structure of a loamy, homogeneous, gley tundra soil. It is disturbed by solifluction-type cryogenic deformation of the Vladimir cryogenic horizon. As a rule, post-Dubno loesses II and III are much thicker (up to 4 or 5 m) than loess I. These horizons have a more detailed stratification; in particular, loess II can be subdivided into two subhorizons containing cryogenic deformations, including ice-wedge pseudomorphs referred to the Rovno cryogenic stage, according to the local stratigraphic scheme. The upper part of the loess is deformed by ice-wedge pseudomorphs 4 to 5 m thick belonging to the Yaroslavl' (Krasilov) paleocryogenic horizon. The section is capped by Holocene soil.

Table 11-4. Absolute Age of Bryansk Fossil Soil and Its Analogues

Section	Horizon	Material for Dating	Radiocarbon Age in yr B.P.
Krasnoselka	Dubno soil	Total humic acids	29,400 ± 1000 (IGAN-170)
Basov Kut	Dubno soil	Total humic acids	28,400 ± 850 (IGAN-74)
Fat'yanovka	Bryansk soil	Total humic acids	22,300 ± 250 (IGAN-197)
Bryansk	Bryansk soil	Total humic acids	24,920 ± 1800 (M-337)
Arapovichi	Bryansk soil	Total humic acids	24,000 ± 300 (IGAN-46)
Mezin	Bryansk soil	Total humic acids	24,200 ± 1680 (Mo-342)
Novokhopersk	Bryansk soil	Total humic acids	22,840 ± 220 (IGAN-87)
Stillfried	Stillfried B	Charcoal	28,120 ± 200 (GRO-2533)

THE DNEPR BASIN

The most complete sections of the Late Pleistocene are confined to the left bank of the Dnepr Basin, including such Dnepr tributaries as the Desna, Psel, Seym, and others. The Dnepr Basin is the region where numerous studies of loess and fossil soils have been made (Mirchink, Krokos, Veklich, Sirenko, Moskvitin, Velichko, Morozova, and others). The Late Pleistocene series is very fully represented on the left bank of the Dnepr (Velichko and Morozova, 1972) by the key section near the village of Mezin (Figure 11-4).

The Mezin complex at the base extends over moraine deposits of Dnepr age and is overlain by fluvioglacial sands and periglacial deposits of Moscovian age—homogeneous, loesslike sandy loams represented mainly by coarse silt and fine sand, which serve as parent material for the Mezin complex. The Mezin complex includes a well-defined lower (Salyn) forest soil with a light gray to white A2-horizon enriched with siliceous powder and a well-structured Bt-horizon.

An extensive ancient sinkhole near the village of Mezin includes a peat bog of Mikulino age in its central part. A series of sections uncovered facies transitions ranging from

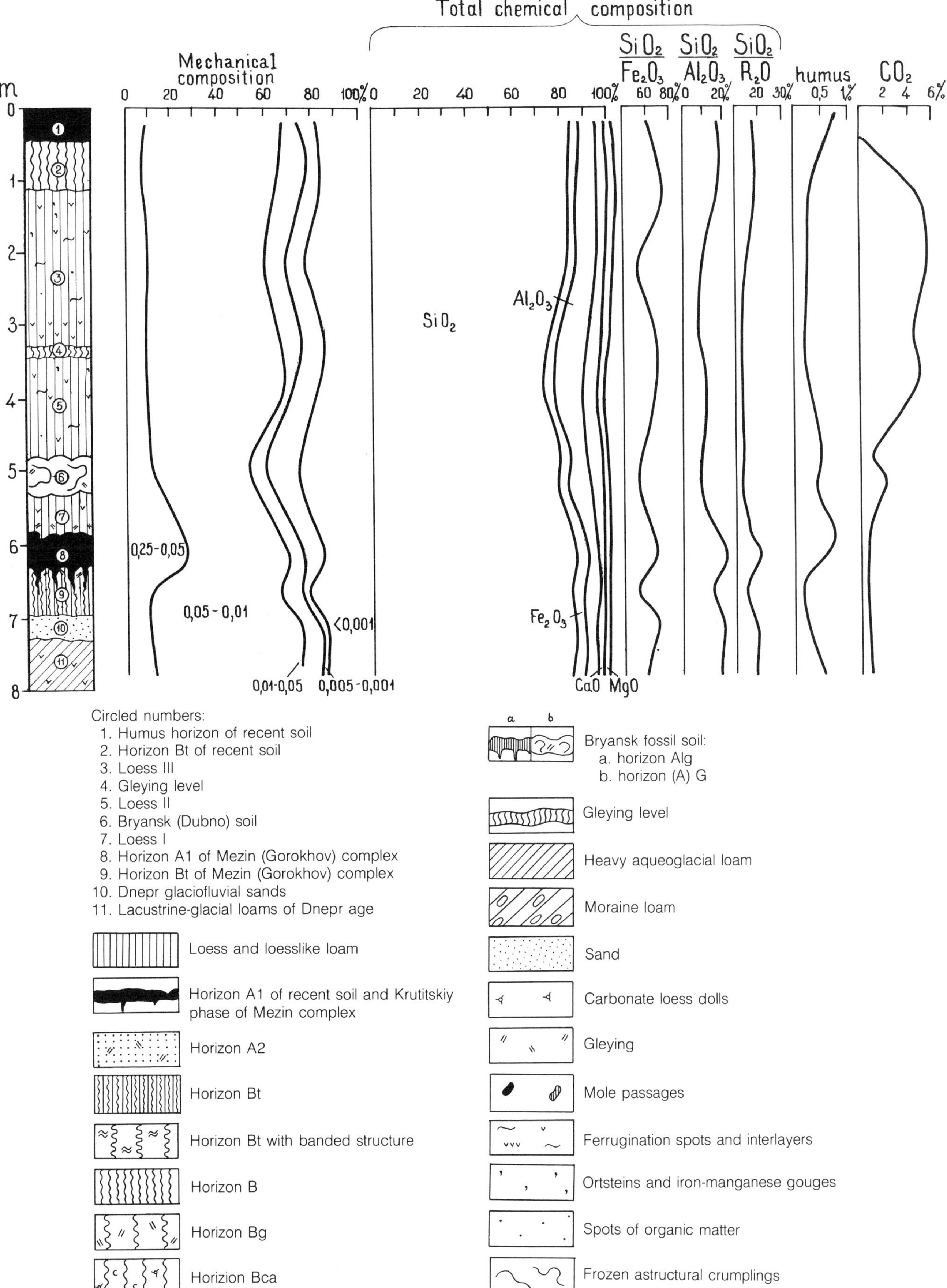

Figure 11-3. Structure of soil-loess strata in the Gorokhov section.

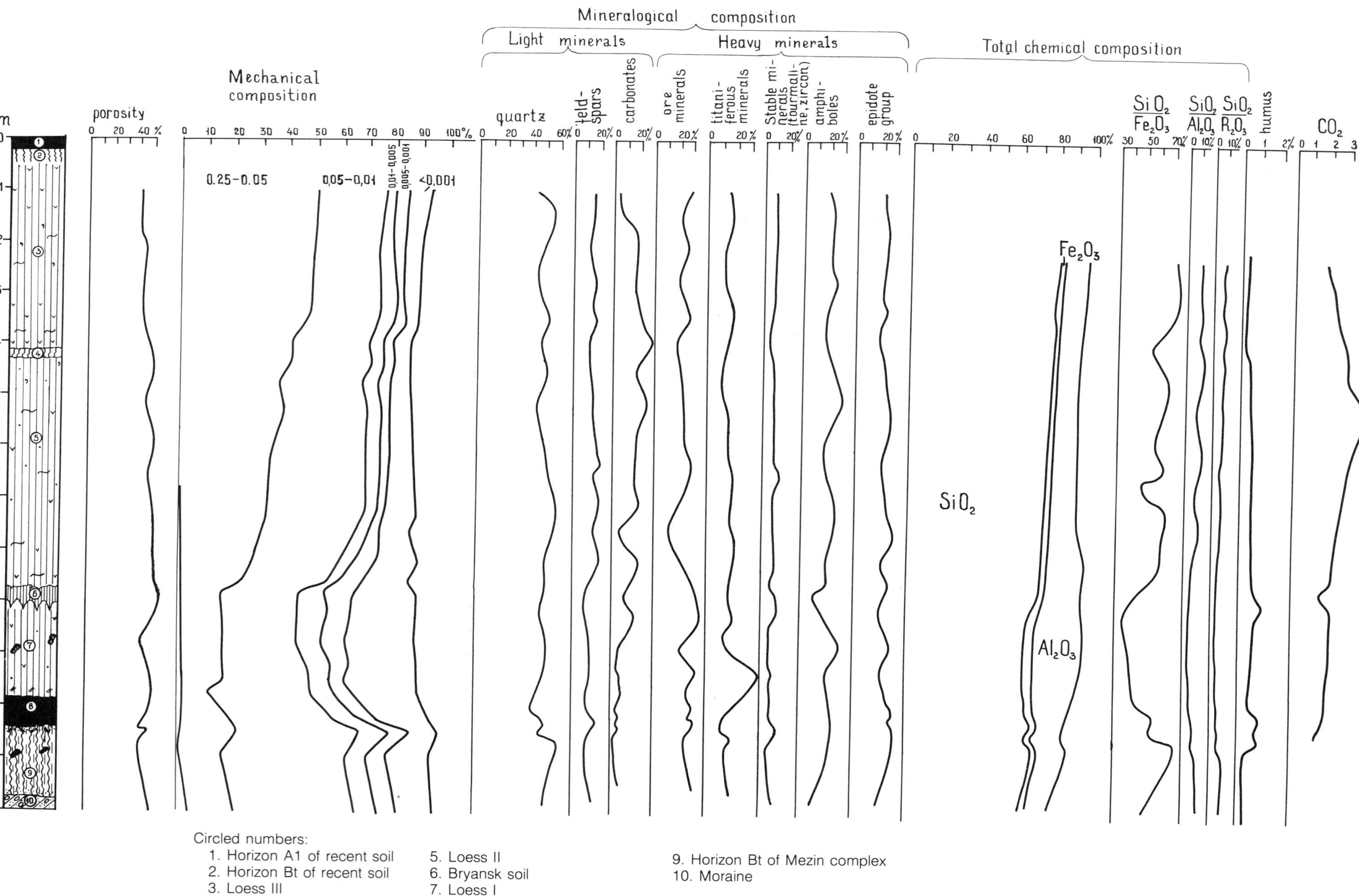

Figure 11-4. Structure of soil-loess strata in the Mezin section. For symbols, see Figure 11-3.

Table 11-5. Isotopic-chronologic Studies of Soils of the Krutitsa Phase of the Mezin Complex and Their Analogues in Central Europe

Section	Soil	Material for Dating	Radiocarbon Age in yr B.P.
Klepaly	Krutitsa soil	Total humic acids	>35,000 (Bratislava)
Platovo	Krutitsa soil	Total humic acids	>37,000 (IG-53)
Melekino	Krutitsa soil	Total humic acids	>35,000 (LGU)
Dol'ni Vestonitse	Upper humified level	Total humic acids	>51,800 (grN-2599)
Czechoslovakia	Lower humified level	Total humic acids	>52,000 (grN-2105) >55,000 (grN-2604)

soils with different degrees of gleying (on the slopes and bottoms of the depressions) to an automorphous soil, with a profile sharply differentiated into genetic horizons—a soil classified as pseudopodzolic. The second phase of soil formation in this section is represented by a soddy-chernozem soil with a thick, dark gray humus horizon containing a well-defined chernozem-type aggregation of humic acids. The Krutitsa soil-formation phase corresponds to the warm (Krutitsa) Interstade, which dates back to the beginning of the Valdai (Table 11-5) and probably correlates with the Brörup Interstade.

The Mezin soil complex was disturbed by frost action in the earliest cryogenetic phase of the Late Pleistocene, that is, the Smolensk phase. The early "a" phase of this complex preceded the Krutitsa phase of soil formation and is represented by wedge-shaped structures in the section; the later "b" phase has mainly solifluction dislocations and fine polygonal fissured structures.

Above the Mezin complex lies the Valdai loess, which is divided into three well-stratified horizons. The lowest horizon consists of loess I (Khotylevo loess), which was deposited slowly and has a heterogeneous granulometric composition and obscure bedding. It served as parent material for the Bryansk fossil soil, which in this region differs structurally from its variants farther west in that it has a light brown to pale yellow humified horizon with relict spots and lenses of a dark gray humus horizon containing coarse humus of fulvate composition and a characteristic ooid type of aggregates. Below it is a thin transition horizon B, represented by a pale yellow loam filled with microcrystalline calcite and numerous iron-manganese nodules. The soil overlies a gleyed loam and is dated at 25,000 to 24,000 yr B.P. (Table 11-4).

The Bryansk fossil soil is marked by a spot-medallion type of deformation in the Vladimir cryogenic horizon, which corresponds to the final stages of the Bryansk interval. Located above is loess II (Desna horizon), which here constitutes the most typical loess, with a thickness of 5 to 6 m, predominance of the 0.01 to 0.05-mm fraction, high porosity, and predominance of quartz in the mineral composition. It is separated from loess III (Altynovka horizon) by a gleying level, which corresponds to the Trubchevsk horizon of the Russian Plain. The gleying horizon is generally apparent as an interlayer of greenish gray gleyed loess with brown ferrugination spots. Loess III is similar in composition to loess II. In its upper part is the Yaroslavl' cryogenic horizon, with ice-wedge pseudomorphs.

Loess III serves as the parent material for Holocene (recent) soils, which in this region are represented by gray forest soil.

THE OKA-DON PLAIN

In the Oka-Don Plain, practically all Late Pleistocene horizons of loesses and fossil soils also overlie the Dnepr moraine, especially in the Middle Oka Basin. For example, the Elat'ma section on the left bank of the Oka Valley within the southeastern margin of Meshchera (Figure 11-5) shows a thick polygenetic fossil soil above the Dnepr moraine and resting directly on thin fluvioglacial sands. On the basis of its stratigraphic position and its similar morphologic indicators, it correlates with soils of the Mezin complex. The clearly differentiated fossil soil profile consists of a humus horizon 0.65 m thick and an eluvial-illuvial horizon of 1.1 m. The nature of the profile implies two independent phases of development: phase I (horizon A2 and horizon Bt) for forest soils formed as a result of eluvial-illuvial processes (Salyn phase), and phase II (horizon A1) for soils with a predominant sod-building process (Krutitsa phase). Separation of the soils into two independent phases is clearly emphasized by the solifluction deformation at the level of horizon A2 (Smolensk cryogenic horizon, phase "a"). In contrast to regions of the Russian Plain farther west, wedge-shaped structures here are isolated and very rare.

The final stage in the formation of the humus horizon of the Krutitsa phase coincides with the next cooling wave (Smolensk cryogenic horizon, phase "b"), which is reflected in the Oka Basin much more broadly in the form of solifluction crumplings, cryoturbations, and ice wedges. Soils of the Mezin complex are overlain by a thin horizon of loesslike loam—loess I, which is often hidden in the profile of the Bryansk soil overlying it. Its thickness in this region does not exceed 1.0 to 1.5 m.

Stratigraphically above loess I lies a soil similar in structure to the Bryansk soil of the Dnepr Basin at a depth of 2 to 5 m, in some cases immediately under the Holocene soil. The humus horizon consists of light gray loam with a dove-colored shading. The soil is dislocated in the Vladimir cryogenic horizon, with wedge-shaped deformations over the entire profile. In the Fat'yanovka section, where

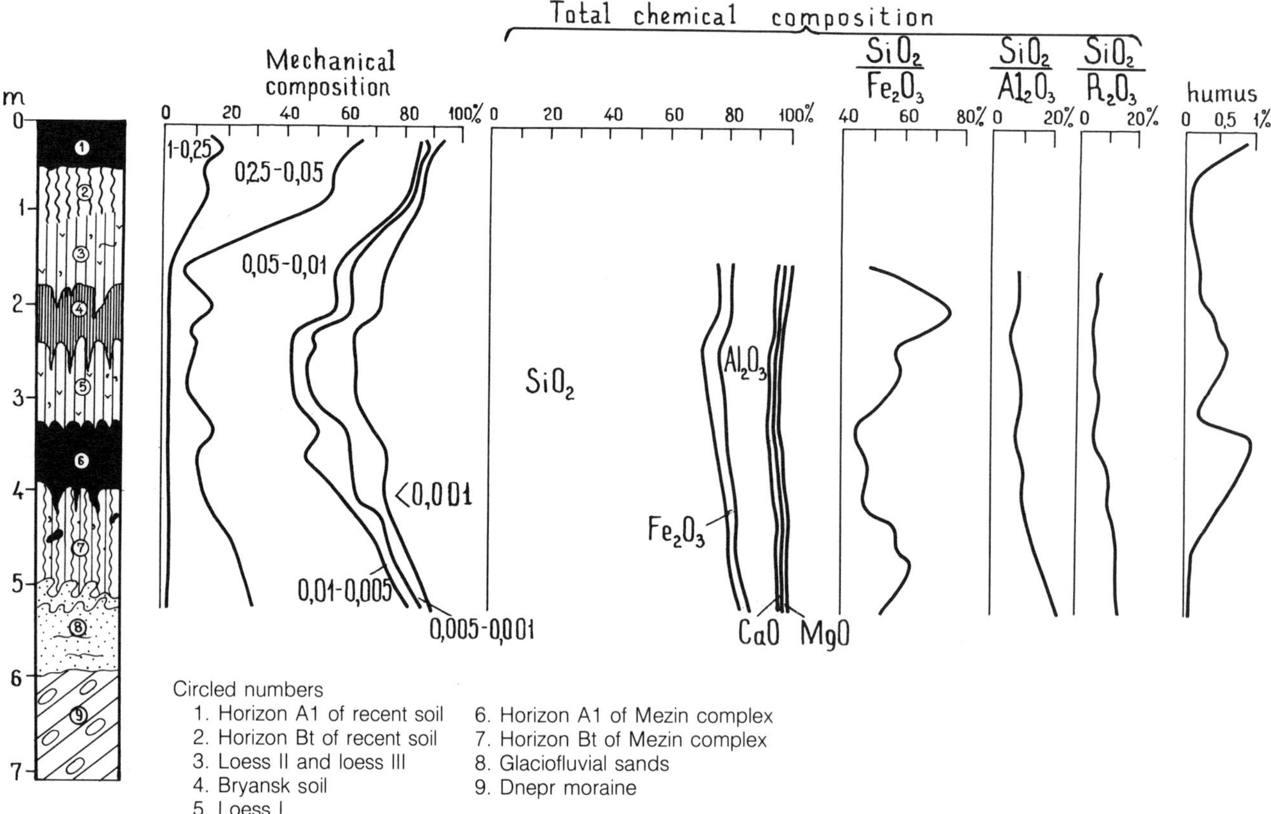

Figure 11-5. Structure of soil-loess strata in the Yelat'ma section. For symbols, see Figure 11-3.

the Bryansk soil lies at a depth of 5 m, a radiocarbon date of 22,300 ± 250 yr B.P. (IGAN-197) has been obtained on humic material (Table 11-4).

The Bryansk soil is covered by a fairly thick horizon of loess II (4 to 5 m), which in individual sections of the Oka Basin as well as in western regions is separated by a rather indistinct soil (Trubchevsk level). The onset of loess formation coincides with the completion of the Vladimir cryogenic stage. Coarse, wedge-shaped structures of the Yaroslavl' cryogenic horizon occur in the upper part of loess III.

Spatial Characterization of Soil and Loess Horizons

THE MEZIN SOIL COMPLEX

Soils of the first (Salynskiy) phase of soil formation (Mikulino Interglaciation) have been closely studied (Morozova, 1963b; Velichko and Morozova, 1963; Morozova, 1972). A map of the Mikulino soil mantle on the Russian Plain shows a great variety, structurally similar to the present mantle. On the basis of soil-profile ȧnalysis, the following types and subtypes are distinguished: (1) brown forest pseudopodzolized, (2) brown forest pseudogley, (3) pseudopodzolic, (4) podzolic sandy (pine-forest sands), (5) podzolic and podzolic-gley, (6) chernozem (leached), (7) meadow-chernozem, (8) surface-gley soils of sinkholes, and (9) chernozem (ordinary).

Within the Russian Plain, a latitudinal zonation is apparent. The entire northern and central part belonged to the zone of forest soils, whose structure and genesis differed for the extreme western and central regions. Three main elementary processes of soil formation—argillization, lessivage, and surface gleying—were involved, usually in some combination, although the relative importance of each remains unclear because of insufficient knowledge of recent soils. Indicators of argillization decreased to the east, and lessivage processes intensified. This makes it possible to consider facies in the distribution of forest soils during the Mikulino Interglaction.

The western margins of the Russian Plain included soils with a distinct manifestation of argillization and surface gleying, along with lessivage processes—brown forest pseudopodzolized and brown forest pseudogley soils, as controlled by the ancient microrelief. Thus, in the Torchin section under flat interfluvial conditions, there was a combination of slightly differentiated, brown pseudopodzolized forest soils (A-Bmt-C) and sharply differentiated brown forest pseudogleyed soils of sinkholes. The former soils were characterized by a clear ferrugination (browning) in the Bmt-horizon, its slight aggregation, clay films over the faces of aggregates, a few illuvial cutans, and the accumulation of silt and iron and aluminum oxides (Figure 11-6A). Soils with a sharply differentiated profile (A-A2g-Bmg-Bmt-BC-C) are characterized by a bleached, desilted horizon A2g with signs of gleying, combined with the hor-

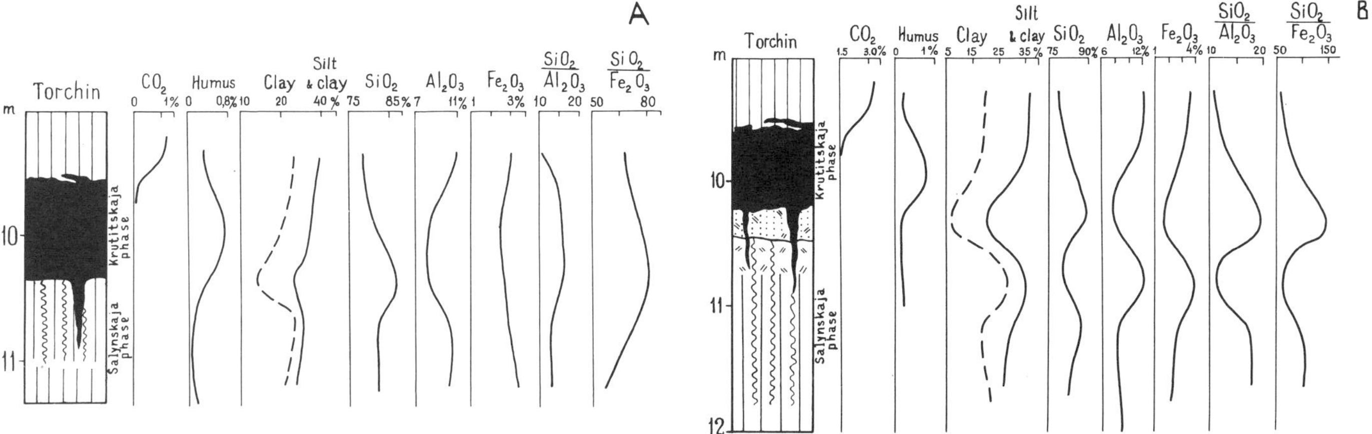

Figure 11-6. Analytical characterizations of soils of the Mezin (Gorokhov) complex of the western margin of the Russian Plain (Volyno-Podoliya). Torchin section, (A) brown pseudopodzolized forest soil of Salyn phase, (B) pseudogley soil of Salyn phase. For symbols, see Figure 11-3.

izon of spotty gleying Bmg and horizon Bmt (Figure 11-6B).

Under the flat interfluvial conditions of the central part of the Russian Plain at that time, forest soils were formed with a sharply differentiated profile, soils that we classify as pseudopodzolic, for in the western regions they were genetically associated with brown pseudopodzolized soils in the soil mantle.

Such soils were widely distributed in the basin of the middle Dnepr, on the Middle Russian Upland, and on the Oka-Don Plain. These soils were distinguished by a thick, sharply differentiated profile formed under the influence of lessivage and surface-gleying processes and consisting of a highly bleached whitish pale-yellow A2-horizon with abundant ferruginous ortsteins and a Bt-horizon consisting of yellowish brown loam with a lancellate-lumpy structure, an abundance of illuvial cutans clay (Figure 11-7), silt segregations, and sesquioxides of iron and aluminum (Figure 11-8).

On the Russian Plain, such soils formed a latitudinal zone, which in the Middle Russian Upland extended beyond the southern boundary of the podzolic soil zone by 350 to 400 km (75 to 150 km on the Oka-Don Plain) and occupied a zone of gray forest soils and, in part, ordinary

Figure 11-7. Humus-clay illuviation cutan in horizon Bt of Salyn phase of the Mezin complex in the Mezin section. (Magnification 5 × 9 with one nicol prism.)

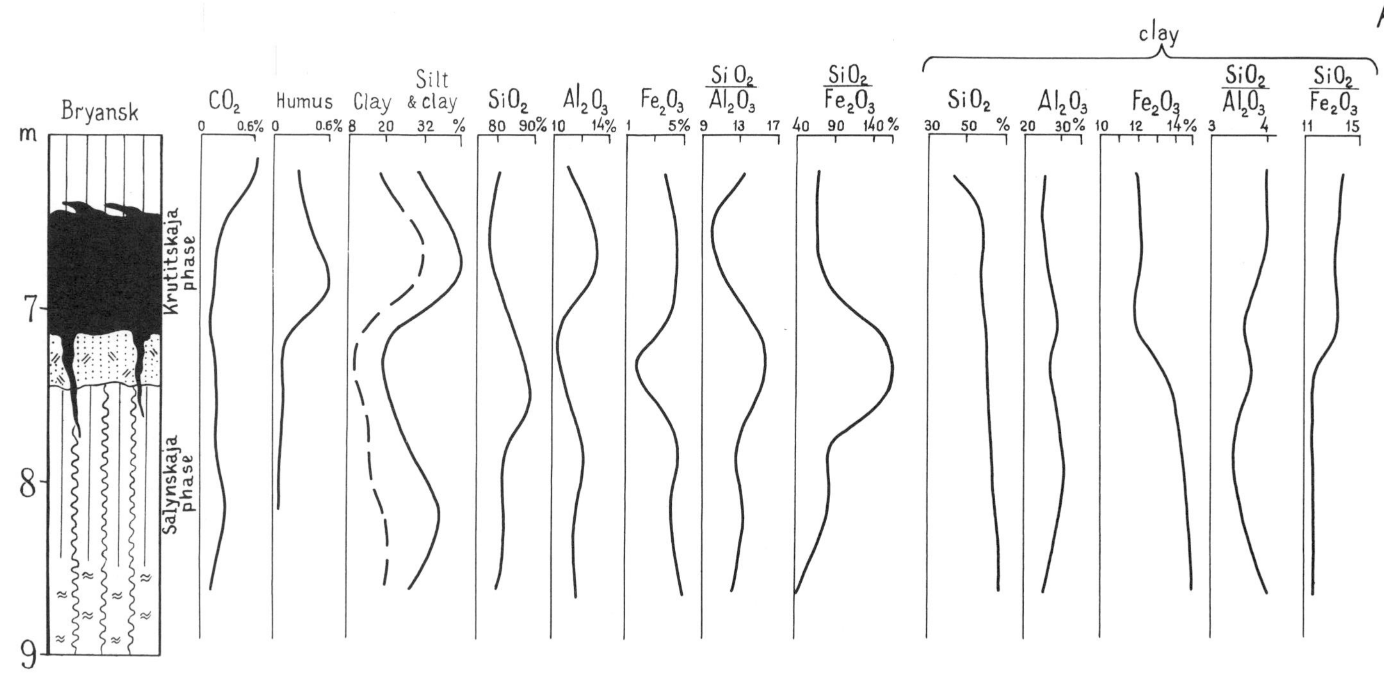

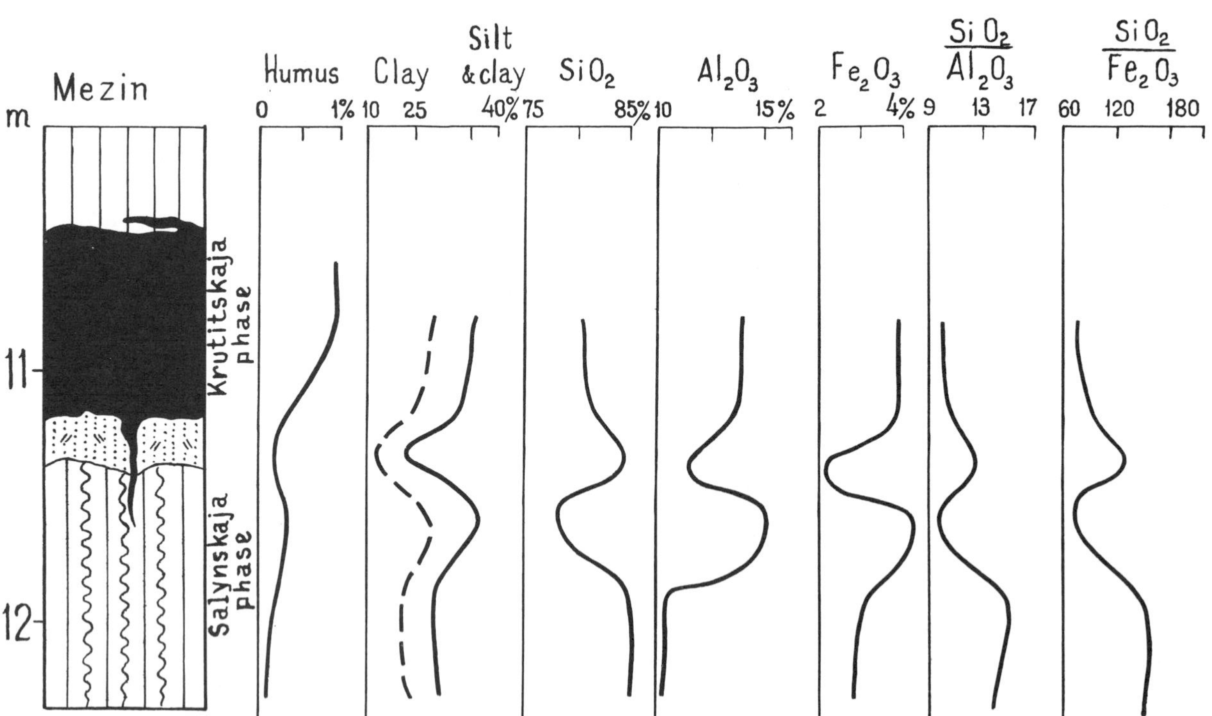

Figure 11-8. Analytical characterization of pseudopodzolic soils of Salyn phase of the Mezin complex of central regions of the Russian Plain. (A) Bryansk section, Dnepr Basin; (B) Mezin section, Dnepr Basin; (C) Mikhaylov section, Oka-Don Plain. For symbols, see Figure 11-3.

chernozems. In the western part of the Russian Plain, however, there was almost no manifestation of this shift. This ancient zone of forest soils changed to a sublatitudinal forest-steppe zone, where forest soils with a differentiated profile were of secondary importance. They were formed in depressions in the microrelief in combination with chernozem and meadow-chernozem soils. In the western part of the Russian Plain within the Podol'skaya Upland, the soil

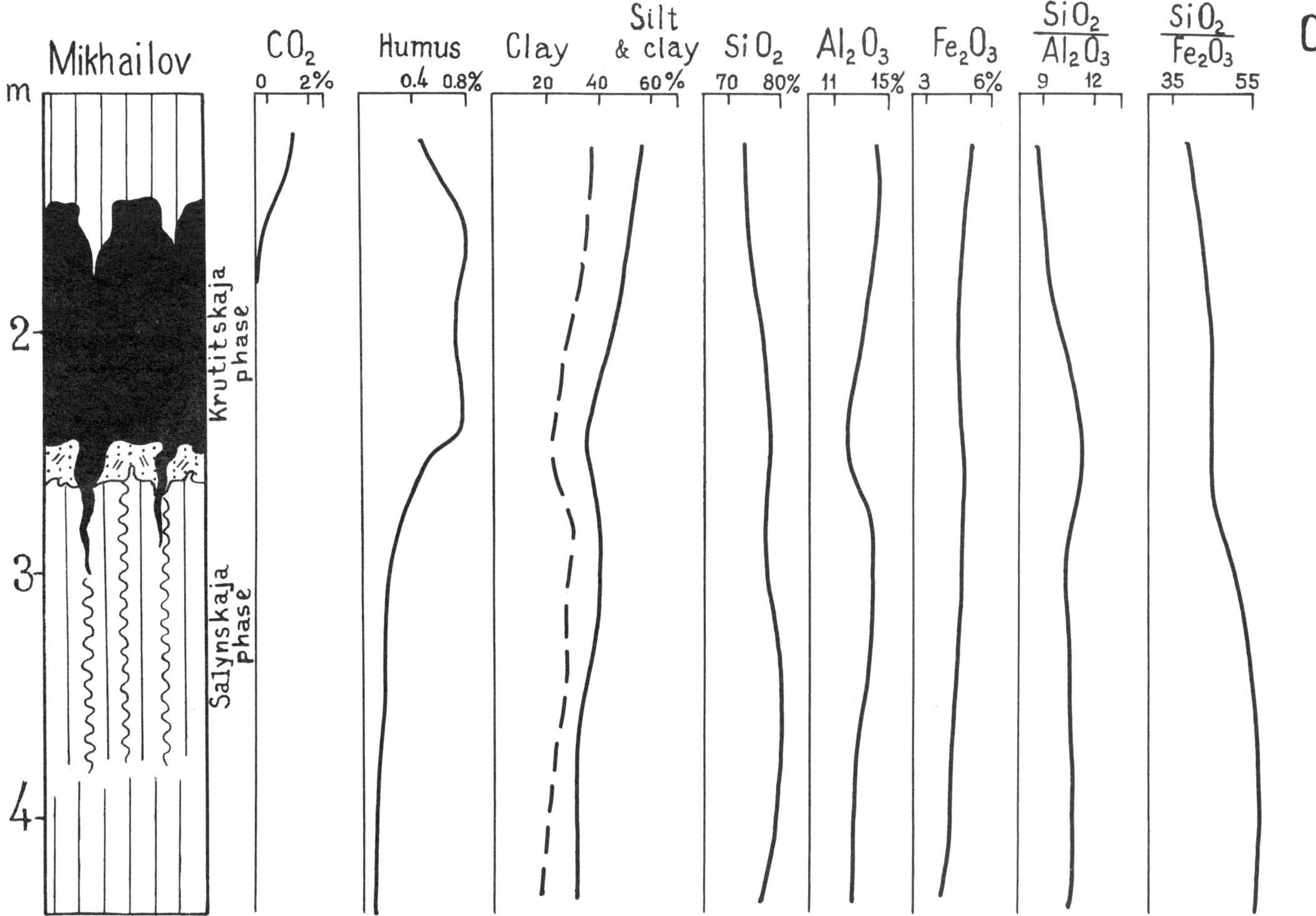

mantle of this zone was made up of two types of soil, which differed in their degree of Bmt-horizon development. Soils of the following profile were developed in areas between padings: A1-horizon, a dark-colored loam heavily disturbed by fauna, with humate-type humus; and B(m)-horizon, a brownish pale yellow loam, structureless and carbonate free. The soils were leached of carbonates and not differentiated with respect to silt or oxides of aluminum and iron.

Ancient padings contained soils having a morphologically distinct brown Bmt-horizon with bleaching in the upper parts, along with abundant clay and humus-clay cutans. Obviously, more argillized and lessivé soils, contrasting with the old chernozems of the above-indicated areas between padings, were formed under increasingly humid conditions. In the forest-steppe zone of the Dnepr and Don Basins, the soil mantle included leached chernozems, meadow chernozems (Figure 11-9), and eluvial gley soils of padings, with distinct signs of surface gleying and lessivage processes.

The southernmost margins of the Russian Plain were in a zone of steppe chernozem soils comparable to recent ordinary chernozems (Figure 11-10), and micellar forms of carbonates prevailed. In the West (Podol'skaya Upland), this zone generally coincided with the present forest-steppe. In the Dnepr Basin it occupied the zone of ordinary chernozems, and in the Oka-Don Plain it extended southward 200 to 250 km beyond the southern boundary of the present forest-steppe.

The distribution of Mikulino soils on the Russian Plain was similar but not identical to the present one. The boundaries of the soil zones were different, reflecting two basic characteristics of natural climatic conditions of the Mikulino Interglaciation in comparison with the present: the forest and forest-steppe soil zones were farther south, and western variants of forest (middle European) soils penetrated into eastern Europe.

SOILS OF THE KRUTITSA INTERSTADE

The Krutitsa Interstade (the second phase of soil formation for the Mezin complex), a paleogeographically unusual time, was characterized by an AC profile uniformly covering the entire Russian Plain, with a humus horizon 0.5 to 1.0 m thick. The parent material consisted of soils of the first (Salynskiy) phase of the Mezin complex (Figures 11-6 through 11-10).

Being the upper member of the Mezin polygenetic complex, soils of the Krutitsa Interstade were strongly influenced both by the transition to cold glacial conditions (cry-

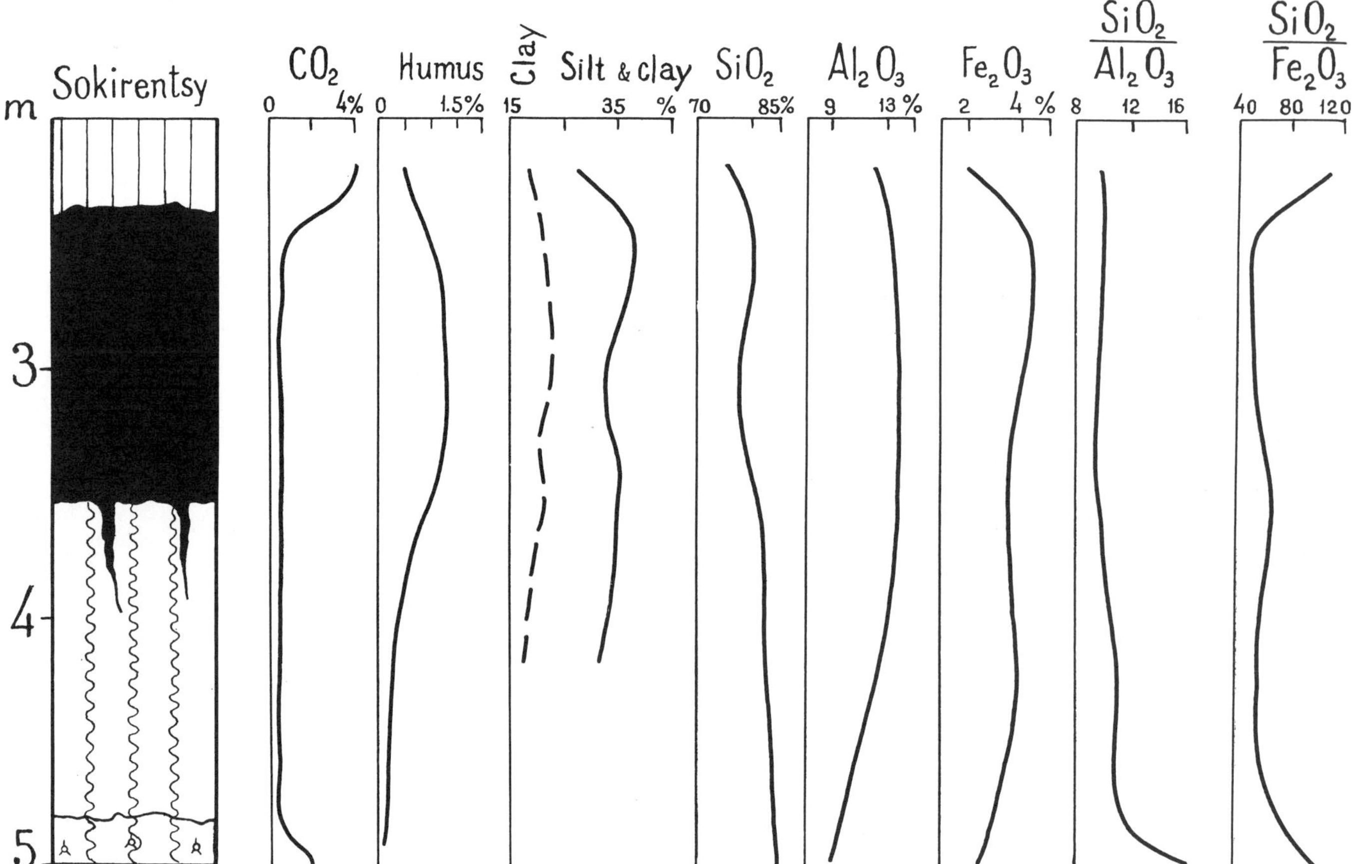

Figure 11-9. Analytical characterization of meadow-chernozem soil of the Mezin complex in the Sokirentsy section. For symbols, see Figure 11-3.

ogenic pedometamorphism of the Early Valdai cryogenic phase, Smolensk phase "b", after Velichko, 1973) and by intensified sedimentation during the transition to the Valdai Glaciation (removal and redeposition). In contrast to soils of the Mikulino Interglaciation and the Bryansk interval (Figure 11-11), the humus in soils of the Krutitsa interval, first described for sections in the Dnepr Lowland (Morozova, 1963b; Morozova and Chichagova, 1968), was found to lack fractions of "free" humic acids (fraction I according to I. V. Tyurin) and fractions of humic acids and their complexes with fulvoacids, combined with relatively stable hydrates of sesquioxides (fraction III according to I. B. Tyurin, as stated by Morozova and Chichagova, 1968). The absence of these two fractions was caused by pedometamorphic processes, that is, dehydration of humic acid molecules and by compaction after prolonged burial. It is assumed that both missing fractions entered into fraction II and were partially combined with calcium, making a nonhydrolyzable residue called humin, which predominates in most of the sections studied. The overwhelming predominance of fraction II (combined with Ca) confirms the hypothesis that soil formation at that time was of the soddy-chernozem type. The humate composition of humus was confirmed by Glushankova (1971) for the Likhvin section, by Sycheva (1978) for sections of the Oka-Don Plain, and by Tsatskin for the Volynskaya Upland (Table 11-6). Studies of the optical properties of humic acids indicate a high degree of condensation of their aromatic nucleus, with values similar to those of recent chernozem analogues (Morozova and Chicagova, 1968; Glushankova, 1971; Sycheva, 1978). The structural uniformity of the soils reflects the uniform natural conditions over a considerable area, that is, the hyperzonality typical of the transition from warm (interglacial) to cold (glacial) epochs (Velichko, 1973).

THE MIDDLE VALDAI HORIZON: LOESS I

Loess I (Khotylevo loess), which separates the Mezin soil complex from the Bryansk fossil soil, is uniform in thickness and in certain physicochemical properties. It has been reworked by soil processes of Bryansk time, so that an evaluation of its properties is difficult. On the Russian Plain, it is 1.0 to 1.5 m thick in the Dnepr Basin, 0.8 to 1.0 m thick in the Volyno-Podoliya, and 0.5 to 1.0 m thick on the Oka-Don Plain.

The stability of the material of loess I in granulometric composition is also typical. In contrast to post-Bryansk loesses, the sand and clay fractions are higher in loess I, along with a predominance of the silt fraction. For example, most representative is the Loess I composition in

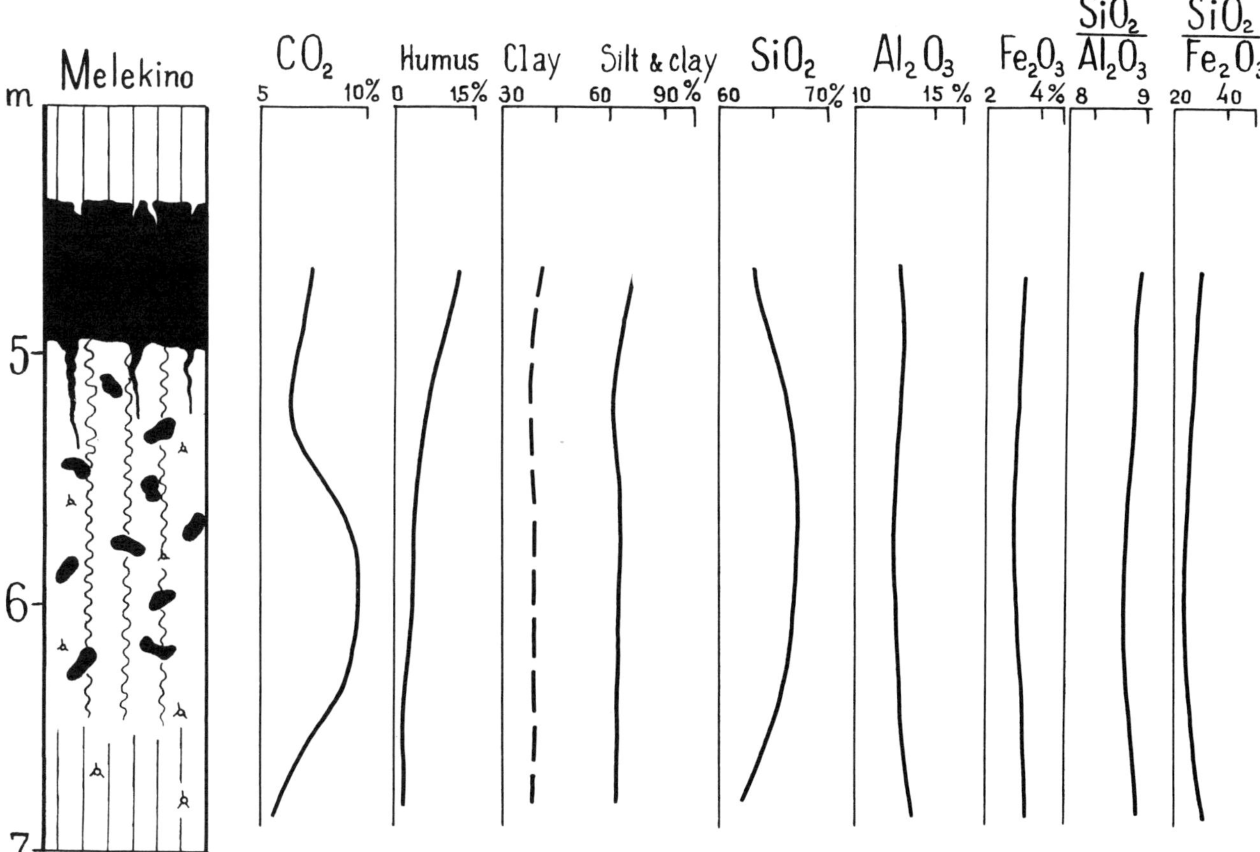

Figure 11-10. Analytical characterization of chernozem soil of ordinary Mezin complex in the Melekino section. For symbols, see Figure 11-3.

the Dnepr Basin, where the sand fraction amounts to 5% to 6% and the clay fraction 10% to 20%. In the northern regions of the Dnepr and Oka Basins, loess I consists of a fairly light sandy loam. To the south and east of these regions, the loam becomes heavier; to the west and east, more sandy (in Volyno-Podoliya the sand is 10% to 20% and the clay 20% to 30%; and in the Oka-Don Plain, 10% to 40%, respectively).

In the northern regions of the Desna and Oka Basins, loess I is free of carbonate. To the south, the carbonate content increases to 4% or 5%. To the west, in Volyno-Podoliya, it is less than 1%. In the Don Basin, however, it increases to 4% or 5%, perhaps because of the carbonate horizon of the Bryansk fossil soil formed on it.

The silicate minerals of this horizon are markedly altered by weathering. Amphiboles have thinned edges, with a coating of pelitized material. Feldspars are also altered. The weathering coefficient K-1 is 0.4 to 0.5.

The rounding and surface texture of quartz sand grains shows little evidence for mechanical (eolian) abrasion during sedimentation. In degree of rounding, quartz grains are fairly heterogenous. In the three fractions analyzed (0.5 to 0.25, 1.0 to 0.5, and 2.0 to 1.0 mm), rounding classes I, II, III, and IV have almost equal proportions, with classes II and III somewhat larger, amounting to 60% to 70% of the total (Figure 11-12). Total rounding ranges from 45% to 65%, and the rounding coefficient (Cailleux, 1942) usually does not exceed 0.3. However, despite the slight rounding, the surface of most sand grains carries traces of wind abrasion, and zero-class grains (those without abrasion) are absent.

The total dulling of sand grains amounts to 70% to 80% (Figure 11-12), but the intensity of this action was low. Thus, the degree of dullness amounts to 25% to 40%. The majority of the grains have a quarter-dull surface. In our view, this indicates that eolian processes of transport and abrasion of silt at this stage were not long or intense.

SOILS OF THE BRYANSK INTERVAL

The paleogeographically complex post-Mikulino time, characterized by several warmings and coolings, ended in a warming of interstadial character. In the periglacial regions of Europe, this warming resulted in the formation of the soils of the Bryansk interval (interstade), radiocarbon dated at 32,000 to 22,000 yr B.P. (Table 11-4) on humic substances from buried soils and on wood charcoal.

Analogues for the Bryansk soils among recent soils are unknown. In addition, the cryogenic pedometamorphism of the Vladimir phase, confined to the final stages of Bry-

Table 11-6. Properties of the Humus of Soils of the Mezin Complex

Section	Age of Soils	Horizon	Humus (%)	Total C (%)	C_{ha}: C_{fa}	0.001% C_{ha} E_{465}
Likhvin	Krutitsa	A	0.99	0.57	1.67	0.193
Likhvin	Mikulino (Salyn)	A1-A2	0.63	0.36	0.62	0.106
Likhvin	Mikulino (Salyn)	A2	0.31	0.18	0.47	0.096
Likhvin	Mikulino (Salyn)	B	0.21	0.12	0.64	—
Mikhaylov	Krutitsa	A	0.71	0.41	1.09	0.146
Bryansk	Krutitsa	A	0.61	0.36	0.7	—
Arapovichi	Krutitsa	A	1.55	0.90	2.1	0.213
Mezin	Krutitsa	A	1.12	0.65	1.9	—
Klepaly	Krutitsa	A	1.43	0.84	3.0	—
Sokirentsy	Krutitsa	A	1.45	0.67	1.54	—
Gun'ki	Krutitsa	A	0.80	0.46	1.5	—
Borisoglebsk	Krutitsa	A	0.93	0.54	1.69	0.179
Klepki	Krutitsa	A	1.22	0.71	2.54	0.185
Platovo	Krutitsa	A	0.71	0.41	1.35	—
Torchin	Krutitsa	A	1.14	0.66	1.8	0.290
4a-73	Mikulino (Salyn)	A1-A2	0.26	0.15	1.6	—
Torchin	Mikulino (Salyn)	B1	0.38	0.22	0.2	—
4-77	Krutitsa	A	1.29	0.75	3.1	0.240

Sixth column: C_{ha}/C_{fa} = carbon content of humic and fulvic acids.

Last column: Index of optical density at 465 μm reduced to C_{ha} = 0.001% (Orlov et al., 1969).

Sources: Data from N. I. Glushankova, T. D. Morozova, O. A. Chicagova, S. A. Sycheva, and A. I. Tsatskina.

ansk time, 25,000 to 22,000 yr B.P., promoted a dislocation of the soil profile, mixing of the soil mass, and secondary gleying. The initial structure of the soils often must be evaluated only from disconnected remnants of soil embedded in a highly gleyed layer. However, the following pattern in the structure and geographical distribution of soils can be outlined (Morozova, 1962; Velichko and Morozova, 1972; Udartsev and Sycheva, 1975; Bogucki and Morozova, 1981).

The northern half of the Dnepr Lowland contained cryogenic soils with a fully developed but highly deformed profile interpreted as above-frost soddy gley soils. Their profile consists of a humified, loamy, light pale yellow-gray or grayish brown A-horizon (0.3 to 0.5 m) with numerous inclusions of fine black and rust-colored iron-manganese concretions; a whitish pale yellow Bca-horizon saturated with silty carbonates, often with indications of gleying; and a Cg-horizon of pale yellow-gray loam with infrequent iron-manganese concretions and bluish and rust-colored, vaguely outlined gley spots. In the northern and northeastern regions (Oka River basin), gleying was intensified, although only a few sections with Bryansk soils are known there, and preservation is poor because of active washout and solifluction. Within the Tambov and Don Lowlands, in the central part of the land along the Dnepr, and on the southwestern margins of the Russian Plain (Podol'skaya Upland), the soils retain the same structural type but without such distinct gleying and with a more developed carbonate illuvial horizon. Tracing sections of Bryansk soil on the plain from north to south (Dnepr lowland region) does not reveal any significant changes in its structure.

The pronounced deformation of the soil horizons by cryogenic processes makes it difficult to identify patterns, particularly their granulometric composition and texture. The soil differs from the overlying loess by its higher content of particles smaller than 0.001 mm and by the molecular relationships of the oxides (narrower than in the loess). At the base of the humified horizons is carbonate accumulation (Figure 11-13). The total humus, despite the soil's relatively light color, is fairly high (0.7% to 1.3%), even with

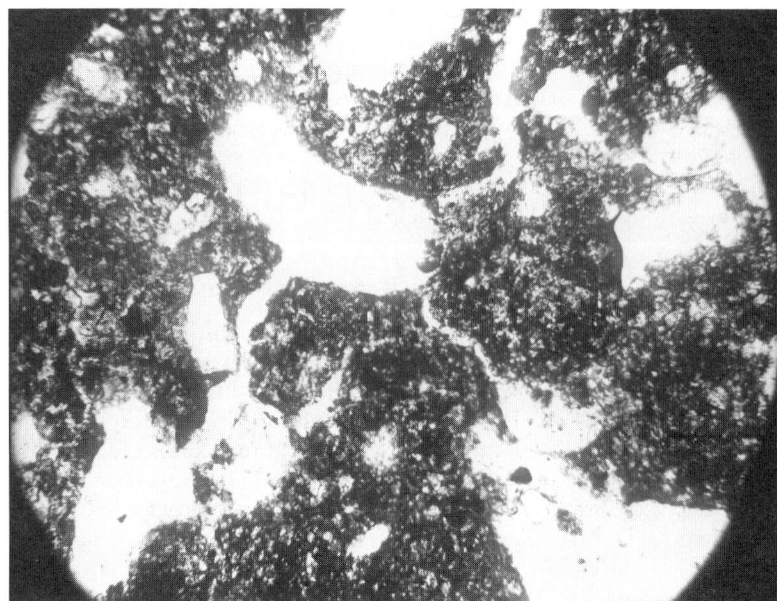

Figure 11-11. Microstructure of humus horizon A1 of soils of Krutitsa phase of the Mezin complex. Gun'ki section, Dnepr Basin. (Magnification 5 × 9 with one nicol prism.)

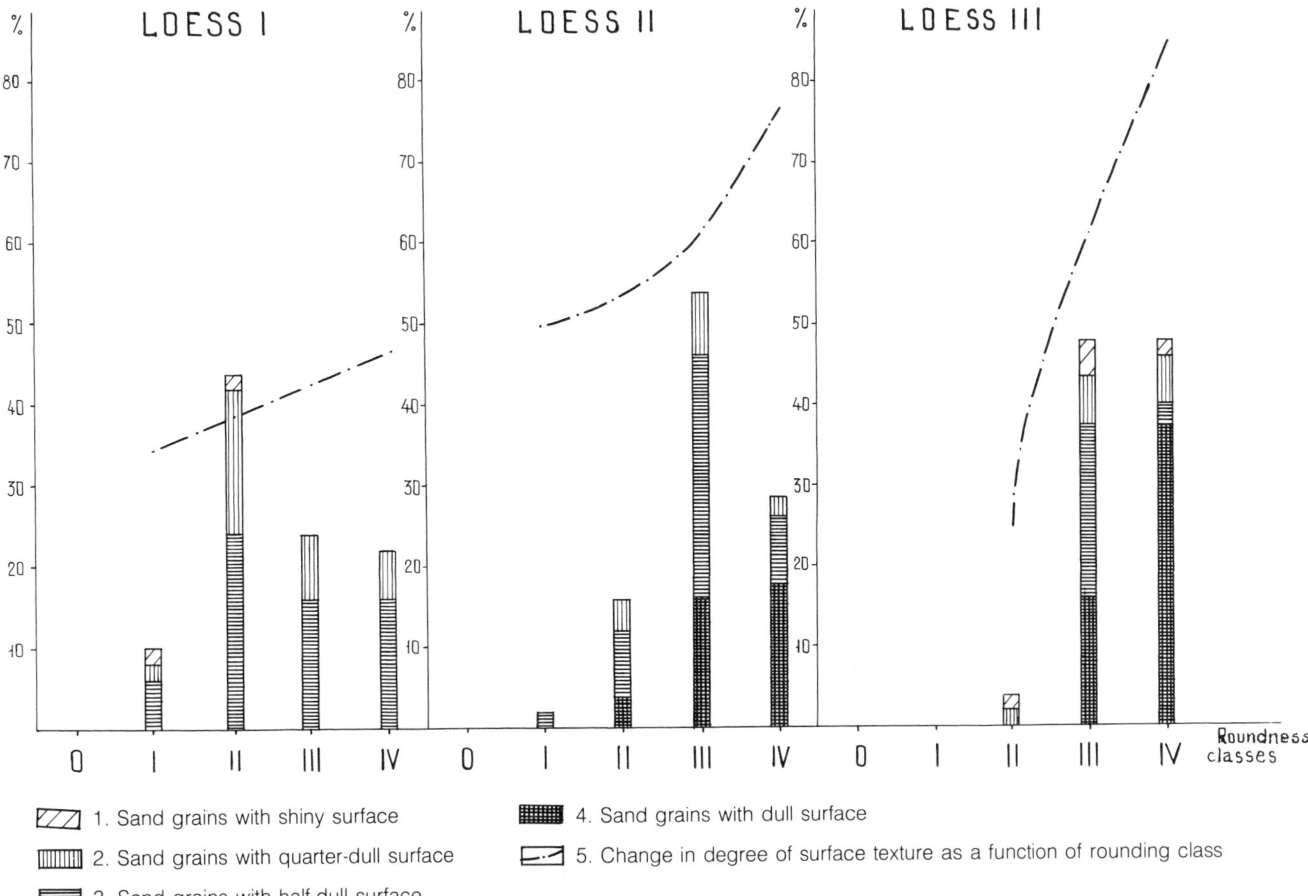

Figure 11-12. Character of rounding and surface treatment of sand grains from different horizons of Valdai loesses.

a correction for its pedometamorphic transformation. Fluvic acids predominate (Morozova, 1963b; Morozova and Chichagova, 1968), and the humic acids are characterized by a lower dispersion coefficient than those of the Krutitsa Interstade (Table 11-7). In the humified horizon, the clay-humus mass is compressed into simple rounded aggregates, frequently delineated by shells of optically oriented clay (Figure 11-14). Inside the aggregates and in the non-aggregated mass, the clayey substance has an imbricate microstructure. No illuvial cutans have been observed. There are many microortsteins and spotty accumulations of iron hydroxide. The peculiar aggregation of the clayey mass in the humus horizon into round, well-defined aggregates of the first order arises during cryogenic coagulation under favorable conditions. This finding is confirmed by experimental data, as is the appearance of a peculiar annular shape of optically oriented clays attributed to the formation of an annular microstructure of segregation ice (Kosheleva, 1958; Konishchev et al., 1973).

As an example of favorable conditions for aggregation, one can cite the conditions prevailing in central Yakutiya, where in the upper horizons of pale yellow, frozen soils one observes well-defined, rounded aggregates of simple structure resembling cluster aggregates of Bryansk soil (Morozova, 1965). Macrofrost deformations are a direct indication of the participation of cryogenic processes in the formation of soils of Bryansk time, soils that, according to the data of A. A. Velichko, formed a fine, hummocky, cryogenous microrelief of spot-medallion type.

We can speak with complete confidence of the uniqueness of soils of this time interval, which is paleogeographically unusual. We do not know of any analogues for these soils, and apparently such analogues did not exist on the Russian Plain during the Holocene. The western portions

Table 11-7. Properties of the Humus of Bryansk Interval Soils

Section	Humus (%)	Total C (%)	(0.001% C_{ha} C_{ha}/C_{fa}	E_{465}
Likhvin	0.731	0.424	0.66	0.085
Likhvin	0.717	0.415	0.19	—
Ivanchino	0.59	0.34	0.28	0.135
Bryansk	0.82	0.476	0.17	0.119
Arapovichi	1.18	0.667	0.65	1.147
Gun'ki	0.97	0.56	0.57	—
Kalach	0.54	0.31	0.29	0.057

Symbols as in Table 11-6.

A

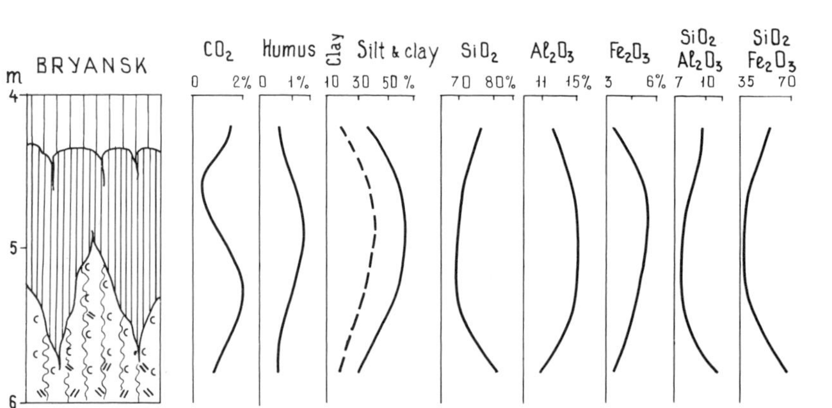

B

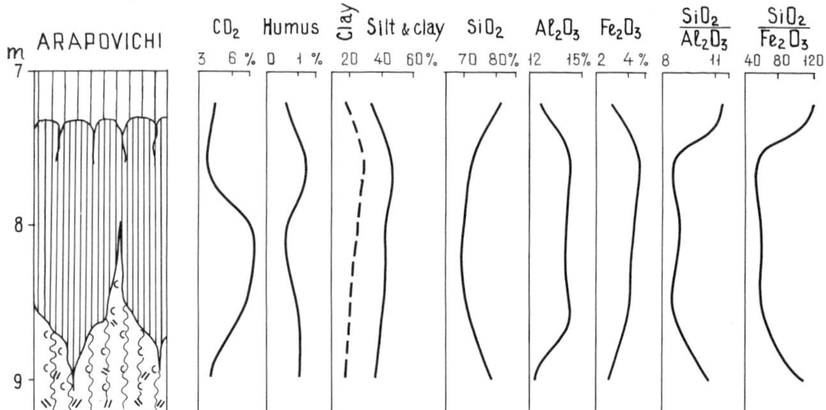

Figure 11-13. Analytical characterization of soils of the Bryansk interval in the Dnepr Basin. (A) Bryansk section, (B) Arapovichi section. For symbols, see Figure 11-3.

of the Russian Plain (Volynskaya Upland) formed a region of soils of fairly uniform structure, arbitrarily called frost-gley soils because of strong indications of gleying, which were reinforced even more due to the later cryogenesis of the Vladimir phase (Figure 11-15). The Bryansk soil mantle generally shows an increase in the rate of soil formation in the southern and eastern regions of the Russian Plain and a decrease in gleying.

Above the Bryansk fossil soil, there are Valdai loess II and III, separated by a weak soil (Trubchevsk gleying level) in the Desna and Oka River basins, where the loess is thickest. These are described only for the Desna Basin, where each of them shows up clearly in the sections.

THE LATE VALDAI HORIZON: LOESS II

The loess II (Desna loess) horizon, which overlies the Bryansk fossil soil, is the thickest of the Late Pleistocene loesses, reaching 3 to 4 m in the Desna Basin and decreasing gradually in all directions. Thus, on the Oka-Don Plain and in Volyno-Podoliya, it is less than 2 to 3 m thick. This horizon in the Desna Basin has the most obvious characteristics of typical loess. It consists primarily of silt (0.05 to 0.01-mm fraction up to 50%), with carbonate content fairly high (4% to 6%) and a porosity of 40%. The weathering coefficient K-1 is low (0.19). Certain mineral grains, particularly feldspars, bear traces of neocrystallization. However, completely fresh grains with distinct crystallographic form are very rare.

The quartz sand grains are different from those of loess I. They are uniform both in degree of rounding and in surface texture. In the Desna Basin, as much as 80% to 90% of quartz grains of the 0.5 to 1-mm fraction from loess belong to classes III and IV. Of these, 30% to 40% are close to an ideally round shape. Grains of class 0 and 1 are encountered singly. The degree of rounding for all samples amounts to 65% to 80%, with a Cailleux rounding coefficient of 0.35 to 0.45. The fairly high degree of rounding is closely related to the type of surface texture (dullness), which reaches 80% to 95%. In some cases, up to 100% of the grains are affected by eolian abrasion, and up to 40% to 50% have a dull surface, which in our view indicates a high degree and duration of eolian action. Most grains are round or elliptical.

THE LATE VALDAI HORIZON: LOESS III

Loess III (Altynovka loess) is the youngest of the Late Pleistocene loesses. It is not much thinner than loess II, amounting to an average of 2.5 to 3.5 m, with its maxi-

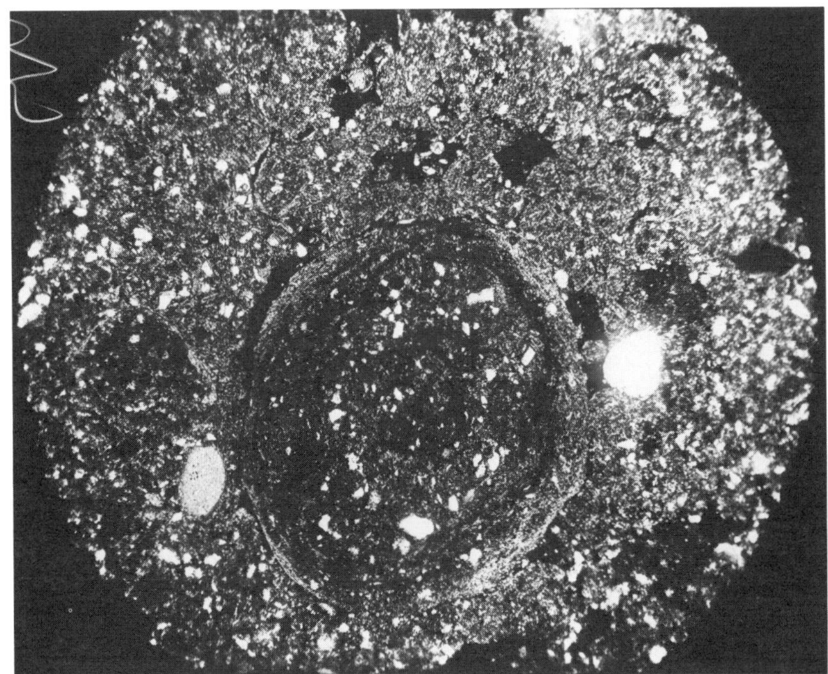

Figure 11-14. Microstructure of humified horizon of Bryansk fossil soil in the Dnepr Basin (Bryansk section). (Ooid microstructure; imbricate and annular microstructure of optically oriented plasma.)

mum thickness also in the Desna Basin. It commonly serves as the parent material for Holocene soils, and this should be taken into account in an interpretation of its properties. Like older Valdai loesses, its sand content increases to the east (the fraction of fine sand amounts to 15% to 20%), as does the amount of clay (up to 30% to 40%). In the West, in the region of Volyno-Podoliya, as compared to the central regions, the composition resembles that of loess II. The silt fraction there amounts to 50% to 60%, the carbon dioxide content is close to 4% to 5%, the porosity is nearly 50%, and the coefficient K-1 is 0.34.

In rounding and surface texture, the sand grains of loess III are also similar to those of loess II but are even more homogeneous in rounding, surface texture, and size. Grains larger than 1 mm are rare; only 10 to 15 grains of this size could be obtained from 100 to 150 kg of loess. Rounding in the 1.0 to 0.5-mm fraction amounts to 75% to 85%, with a rounding coefficient of 0.35 to 0.45, as in loess II. The total dullness is also high, up to 90% to 100%, for an overall dulling of 50% to 70%.

Paleoecology

On the basis of data on the structure of soil, loess, and cryogenic horizons, as well as paleontologic materials contained in them, one can establish the sequence of change in natural conditions from the onset of the Late Pleistocene to its end.

MIKULINO INTERGLACIATION

During the period between about 125,000 and 80,000 yr B.P., in the band of forest soils in the central and southern parts of the East European Plain, the structure of the soil mantle differed appreciably from the present one. At that time there occurred an expansion into eastern Europe of forest soils with a sharply differentiated profile, that is, analogues of pseudopodzolic soils, which at the present time are common in central Europe. An appreciable expansion of the belt of these soils, not only in the latitudinal but in the meridional direction (Figure 11-16), and replacement of a considerable part of the steppe zone by

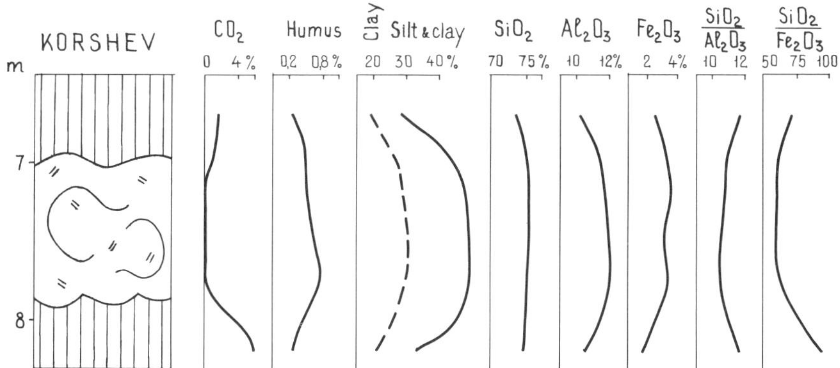

Figure 11-15. Analytical characterization of Bryansk fossil soil on the western margin of the Russian Plain, Korshev section (Volyno-Podoliya). For symbols, see Figure 11-3.

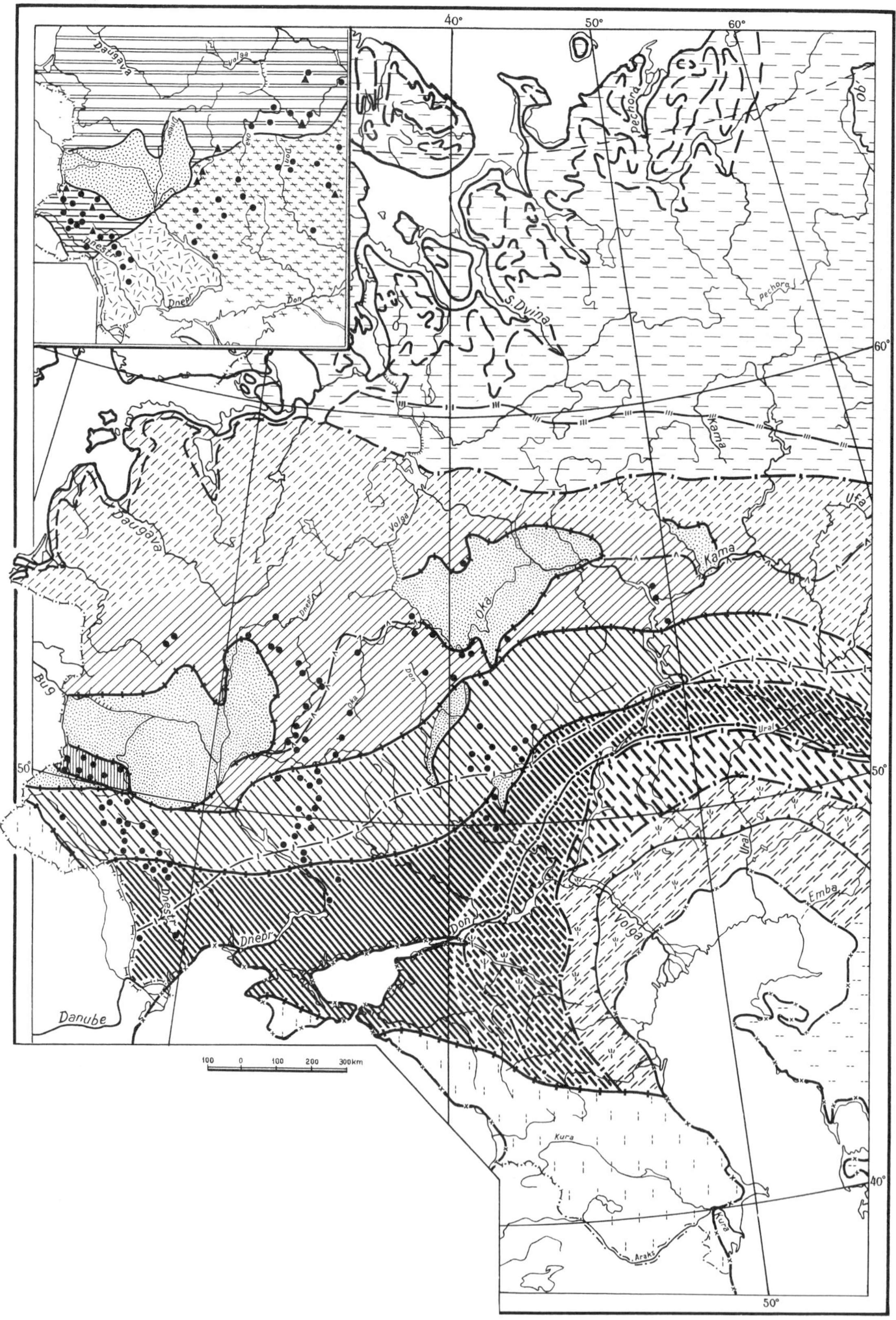

Figure 11-16. Map of the soil mantle of the Mikulino Interglaciation of the Russian Plain. (Compiled by A. A. Velichko, T. D. Morozova, S. A. Sycheva, and N. I. Tsatskin.)

Column *a* = from actual data; column *b*, hypothetical

a б

Soddy-podzolic

Pseudopodzolic

Brown forest soils

Leached chernozems in combination with meadow-chernozemic soils

Soddy-podzolic soils on sands

Typical chernozems in combination with meadow-chernozemic soils

Ordinary chernozems

Chestnut

Saline soils, and areas where soil mantle has not been reconstructed

Mountain soils

Plain soils

Boundaries of paleosol zones

Southern boundaries of recent soil zones:

Podzolic

Soddy-podzolic

Forest-steppe soils (gray forest soils, leached chernozems, podzolized modal soils

Steppe soils (ordinary chernozems)

Dry steppe (chestnut) soils

Key sections

Recent coastline

Coastline of Mikulino Interglaciation after:

N. S. Blagovolin, O. K. Leont'yev, V. M. Muratov, and L. R. Serebryanny

International Map of Quaternary Deposits of Europe, with additions by V. P. Grichuk

A. L. Chepalyga

Inset:

Homogenous gley soils

Above-frost soddy-gley soils

Above-frost soddy-gley illuvial-carbonate soils

Soddy-nongley illuvial-carbonate soils

Frozen soils on sands

Boundaries of paleosol areas

Key sections

Sections with radiocarbon dates

forest-steppe indicate a substantial reduction in the continentality of the climate in eastern Europe and an increase in temperature. The eastward migration of soil associations similar to those of recent central Europe implies a rise mainly in winter temperatures. Moisture changes in the southern plain may not have been very large in absolute values of precipitation (100 to 200 mm), but they did cause a reduction of the steppe zone.

EARLY VALDAI GLACIATION

Early Valdai phase "a" of Smolensk cryogenesis (85,000 to 70,000 yr B.P.), the earliest cooling wave in the periglacial zone of eastern Europe, is recorded by an accumulation of intra-Mezin loess and by cryogenic deformation of phase "a" of the Smolensk cryogenic horizon. It was characterized by conditions favoring the formation of permafrost not only on the northern plain but also on its central part. Although the permafrost in this region resembled that near the southern boundary of the region of present permafrost, where permafrost has an insular distribution, the spread to the latitude of L'vov and Kiev indicates a profound modification of natural conditions far beyond the confines of the expanding glacier.

KRUTITSA INTERSTADE

The loessial periglacial areas show indications of the warmest, most distinct interstade of the entire Valdai Glaciation in the form of soils of the Krutitsa phase (about 70,000 to 60,000 yr B.P.) of the Mezin complex. Steppe conditions are confirmed by fossil finds of small steppe mammals—steppe pika, gray hampster, suslik, narrow-skull fieldmouse, steppe "pestrushka" (Markova, 1975)—in mole burrows in soils of the Krutitsa interval. Soils of chernozemlike appearance were prevalent at that time; they also expanded into areas that had been under broad-leaved forests in the Mikulino Interglaciation. On the East European Plain as well as in central and western Europe, judging from paleopedologic data (Fink, 1969; Velichko and Morozova, 1969), open spaces with chernozemlike soils were displaced in the areas of deciduous forests. Thus, the landscape-climatic conditions were evened out in latitude. Whereas during the Mikulino Interglaciation the Atlantic conditions, with a greater heat and moisture supply, penetrated far to the east (with a simultaneous meridional expansion), during the epoch of the Krutitsa interval there occurred a far-westward expansion of sharply continental conditions associated with a marked expansion (particularly to the east) of open spaces with steppe soil formation. The degradation of the forest zone over a considerable portion of the continent, along with a general reduction in zonality (the phenomenon of hyperzonality, according to A. A. Velichko), are reliable indications of the interstadial character of this chronologic interval. In the central regions one can speak of specific climatic parameters for that time only for a certain interval. Because recent chernozems extend all the way up to the Altay territory, one can infer an appreciable lowering of winter temperatures (possible range of January temperatures from $-4°C$ to $-20°C$).

The July temperatures, however, could have remained close to the present ones (about +20°C), and the precipitation decreased to 300 to 500 mm per year. Thus, on the continent the contrast and continentality of the climate increased.

MIDDLE VALDAI, PHASE "b" OF SMOLENSK CRYOGENESIS

The Krutitsa warm interval, which can best be compared with the Brörup Interstade, was followed by a new wave of cooling and cryogenesis. Whereas cryogenic deformations of phase "a" of the Smolensk cryogenic horizon (from about 60,000 to 55,000 yr B.P.) were heavily veiled by subsequent soil formation during the Krutitsa interval, deformations of phase "b" that disturbed this soil are clearly manifested. Attention is drawn to the latitudinal manifestation of cryogenic deformations. From Volyno-Podoliya to the Don Basin, the deformations indicate the existence of a system of shallow polygons, with sides of 1.5 to 2.5 m formed by earth-ice veins 1.0 to 1.5 m high and 10 to 20 mm wide in the upper part. In the upper 0.2 to 0.4 m, these systems are usually disturbed by solifluction and cryoturbation. The region of permafrost, generally similar in its regime to the conditions of phase "a", extended to a latitude of 51° to 52°N.

On the basis of a calculation on the remains of preserved annual elementary cracks, the duration of the formation of this horizon was short—about 15 to 200 years.

MIDDLE VALDAI COLD EPOCH OF LOESS I (KOTYLEVRO LOESS)

The low-temperature conditions leading to the development of permafrost most probably lasted even longer (about 55,000 to 30,000 yr B.P.) during the formation of loess I. This is indicated not only by loess in the upper part of fissures but also by fissured frozen-ground formations in loess I. The temperature regime corresponded to the formation of permafrost, but the ground moisture permitting the development of cryogenic deformation existed on the western plain. Only the so-called "dry" frozen ground existed in the central and eastern regions. However, in comparison with the younger loess horizons, the loess I interval had less extreme continentality, as indicated by occasional indistinct, thin bedding and by a more clayey composition. The fairly high content of both clayey and sandy fractions indicates that the optimum conditions for loess formation had not yet occurred. This is also supported by the very slight eolian abrasion of the sand grains. Finally, loess I is only 1.5 to 2.0 m thick over a vast area, attesting to low rates of accumulation (about 0.05 mm per year) and homogeneous conditions during a long period of the Middle Valdai.

LATE VALDAI, BRYANSK INTERSTADE

The Bryansk (about 30,000 to 24,000 yr B.P.) fossil soil is the chronologic analogue to the Stillfried B soil in central and western Europe and the Farmdale soil of North America (Table 11-1). The loess-accumulation epochs have no analogues today. The conditions of the Bryansk interval (interstade) also appear to be unusual in many respects and have no modern analogues. Indeed, structurally homogeneous cryogenous soils (Figure 11-16, inset) occurred widely over the entire periglacial region of the East European Plain , starting at least with the latitude of the city of Vladimir, where a Bryansk soil profile was found (56° to 57°N), and farther south as well. Did a warmer phase occur at the start of the Bryansk Interstade? We do not have any such indications thus far. On the contrary, all available data point unequivocally to cold conditions.

The very homogeneous structure of the Bryansk soil profiles from north to south in the basins of the Oka, Don, and Dnepr attests to a distinct hyperzonality for that epoch. Soils similar to the pale yellow cryogenous soils of modern Yakutiya predominated in the central and eastern parts of the plain, and cryogenous gley soils similar to those of Taimyr predominated in the west, where moisture was higher. Very severe continental conditions for that time are indicated also by paleontologic data. Markova (1975) was able to extract from mole burrows of that soil in the Arapovichi section (Desna Basin) the remains of such small mammals as the collared lemming, which inhabited the northern Arctic, and several steppe species: the steppe vole, suslik, steppe marmot, and narrow-skull fieldmouse. The last-named species now inhabits both steppe and tundra.

The results of palynologic studies of profiles of this soil are very significant. Z. P. Guvonina in the same Arapovichi section and E. E. Gurtovaya in the Boyanichi section in Volyno-Podoliya found pollen of *Betula nana* and *Alnaster fruticosus*. Using the climatogram method, these authors obtained estimates of climatic parameters in the following ranges: January temperatures from −19°C to −21°C, July temperatures from +18°C to +14°C, and total annual precipitation from 450 mm to 350 mm.

Consistent with these conclusions, present regions with analogous soil data suggest winter temperatures even lower (down to −30°C). Thus, the Bryansk interval was a very cool, hyperzonal epoch differing from loess-accumulation epochs by the cessation (attenuation) of loess accumulation and a certain increase in moisture.

VLADIMIR CRYOGENIC HORIZON

The Bryansk soil profile was formed against a background of frozen ground (Morozova, 1962). However, a wave of distinct cryogenesis occurred at the end of the interval (about 23,000 to 22,000 yr B.P.), for the already formed soil profile was deformed by cryogenesis. In vast areas of the basins of the Oka, Don, and Dnepr, a microrelief of fissured formations of spot-medallion type was formed, of the type well known in the present permafrost region. On the western East European Plain, solifluction dominated, indicating a higher moisture content. This epoch differed from the Bryansk interval proper through an increase in the amplitude of annual temperatures, primarily as a result of a lowering of winter temperatures, but probably also as a result of a certain decrease in precipitation, especially in the central and eastern parts of the plain.

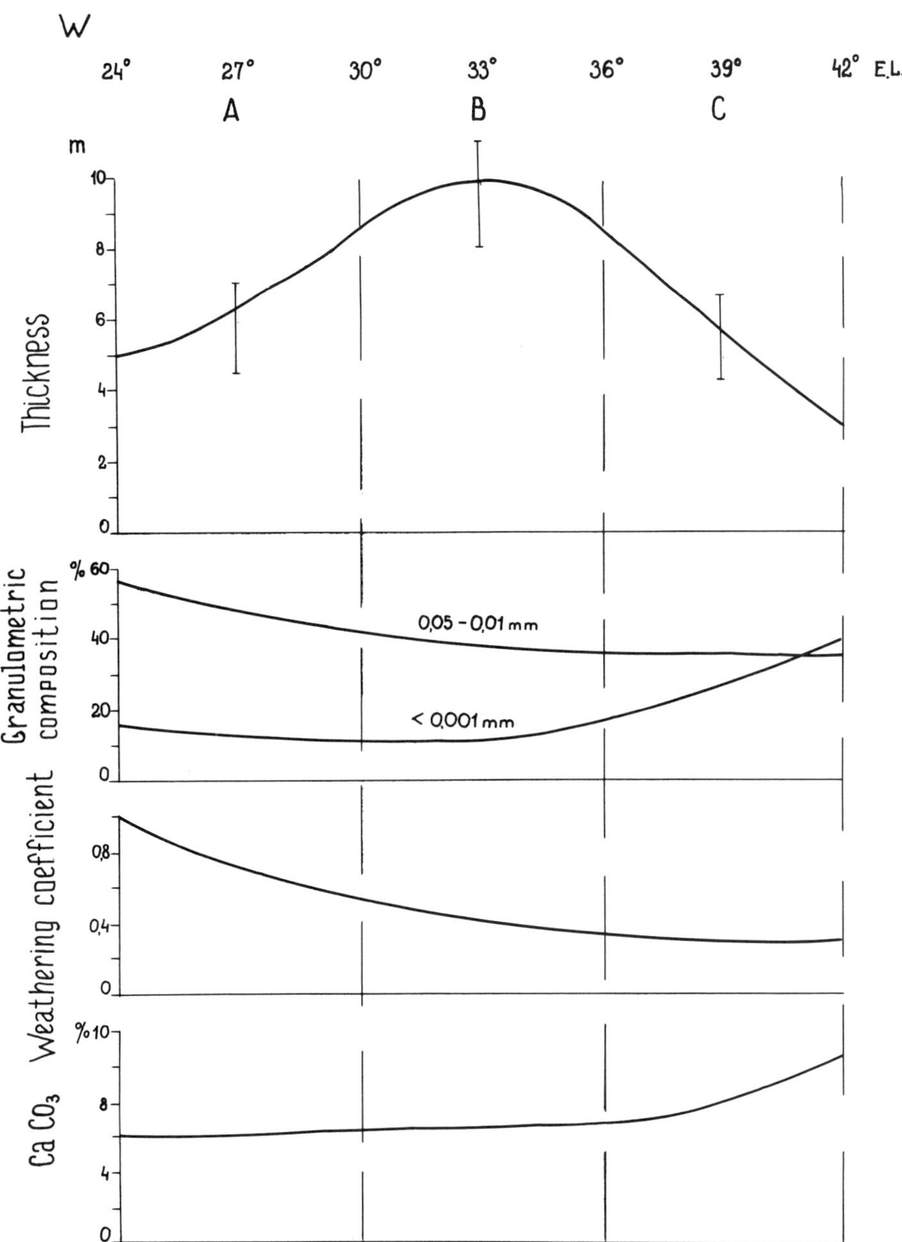

Figure 11-17. Change in the properties of Valdai loesses in the latitudinal direction across the Russian Plain. Loess regions: (A) Volyno-Podoliya, (B) Dnepr Basin, (C) Oka-Don Plain.

LATE VALDAI EPOCH OF LOESS ACCUMULATION

The Late Valdai interval (about 22,000 to 12,000 yr B.P.) is recorded by the Desna horizon of loess II, a gleying level of the Trubchevsk interval (about 16,000 to 15,000 yr B.P.), and the Altynovo horizon of loess III. The post-Bryansk stage of the Late Valdai was characterized by an expansion of loess-forming processes. Within this stage there was a fairly indistinct interval of more moderate climate—the Trubchevsk interval of about 16,000 to 15,000 years ago recorded in the central regions as a gleying level but not elsewhere. The entire post-Bryansk stage of loess formation (horizons of loess II and loess III) is treated as a whole, particularly because the properties of the two loess horizons are similar.

This was the phase of the most active accumulation of the purest carbonate loesses during the entire Pleistocene, and the loesses themselves are the least weathered—their weathering coefficient of 0.19 is close to that of Yakutian loesses (Khalcheva, 1972). The morphoscopy and morphometry of sand grains are similar to those of desert sand grains. A palynologic study of the loess (Grichuk, 1972) also indicates a sharply continental climate under permafrost conditions and open tundra-steppe, also shown by fossil finds of mammoth, reindeer, and arctic fox (Velichko et al., 1977). On the basis of data on concentration centers of flora, Z. P. Gubonina has established that the landscape of that time most closely resembled the Churapcha steppe of central Yakutiya, and extreme dryness (precipitation under 200 mm per year) is also attested by the micro-

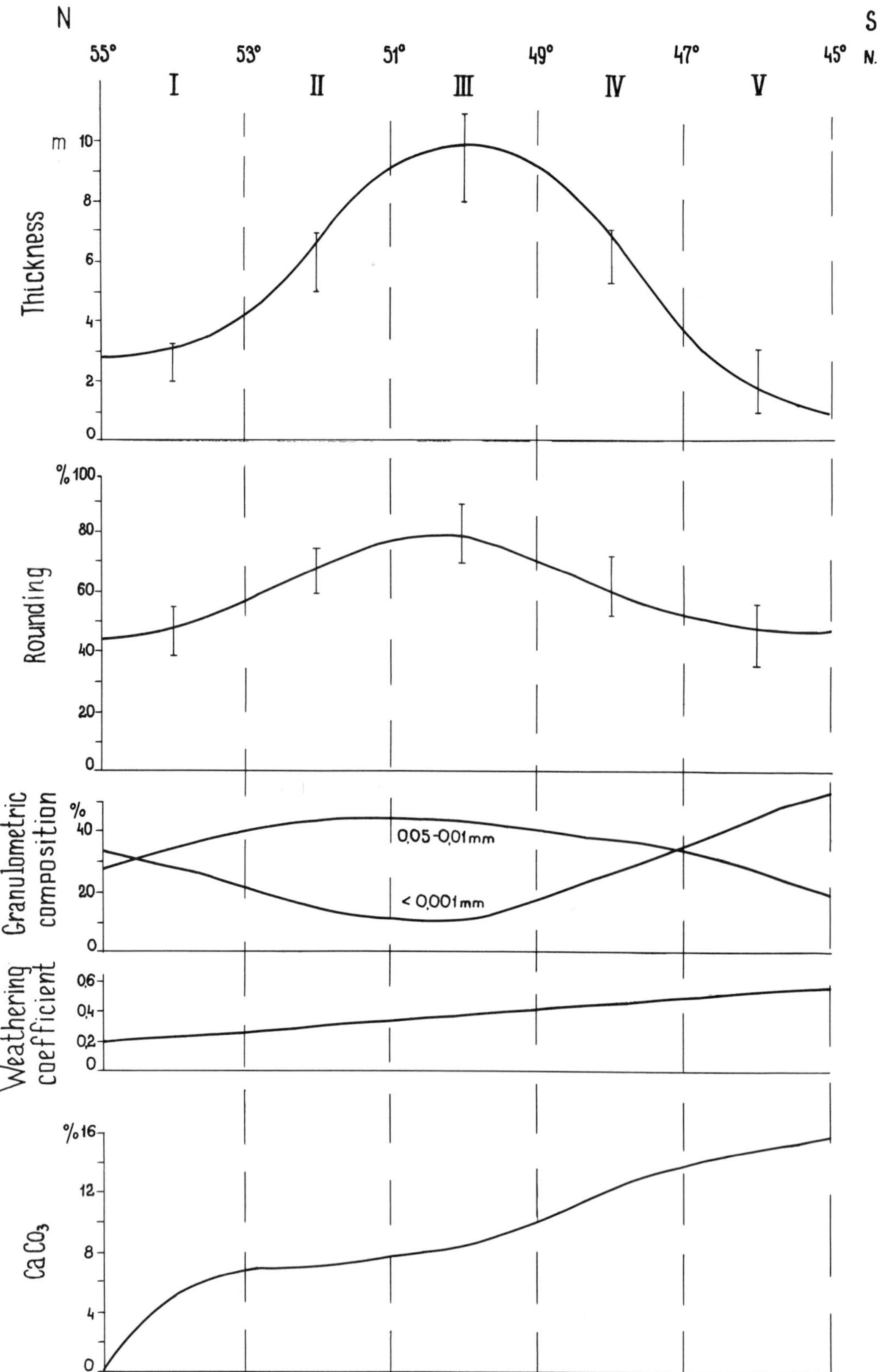

Figure 11-18. Change in the properties of Valdai loesses according to the zones along a meridional profile across the Russian Plain (from I to V).

morphological similarity of the loess to the gray soils of recent deserts (Morozova, 1963a).

On the whole, there were most extremely continental conditions of the entire Late Pleistocene and possibly of the entire Pleistocene. The very low temperatures of the winter months (below −30°C) were combined with very low precipitation (under 200 mm per year). In the West, however, the moisture content might have been somewhat higher; there the facies of dry permafrost was replaced by permafrost containing pseudomorphs of frost wedges inside the loess. An indirect indication of very severe "frost" conditions of loess formation in regions farther east is the high content of closed pores in these loesses, which has been attributed to segregation ice.

Our work, carried out in different regions, makes it possible to identify certain spatial elements in the structure of the loess region of the East European Plain during that epoch. The maximum accumulation of post-Bryansk loesses (Figure 11-17) was 6 to 8 m thick in the central part of the plain at about latitude 52°N, decreasing insigificantly to the west but sharply to the east. The clay fraction increased in the same direction, as if the marginal zone of accumulation were located there. The decrease in accumulation took place both north and south of this latitude (Figure 11-18), but to the north the loess graded into cover deposits with both clayey and sandy fractions, and to the south the finer fractions increased with a fairly high degree of homogeneity.

THE YAROSLAVL' CRYOGENIC PHASE

Radioisotope data on the age of pseudomorphs of ice wedges indicate that the strongest cryogenesis wave began to develop about 18,000 years ago. In the epoch of about 18,000 to 16,000 yr B.P., the permafrost region reached its maximum in eastern Europe both in its southward advance (to latitudes of 50° to 48°N) and in thickness (up to 150 to 200 m). Such an extensive development with ice and ice-earth veins should be correlated with a certain increase in moisture content during the final phases of the Valdai Glaciation.

Paleobotanic analysis of an ice-wedge pseudomorph at the Late Paleolithic campsite of Timonovka II in the Desna River basin (Velichko et al., 1977) suggests that the January temperatures at the center of the East European Plain were under −30°C, that is, more than 20°C lower than the present temperatures in that region. The July temperatures, however, declined much less—by 3°C to 5°C.

After the Alleröd warming, when many ice-wedge systems were degraded, the former polygonal systems were revived in the Younger Dryas, as expressed in the formation of younger generations of pseudomorphs in the tops of already existing fissured structures.

References

Bogucki, A. B., and Morozova, T. D. (1981). Buried soils of the Mezin (Gorokhov) complex of Volynskaya Upland and adjacent regions. *In* "Problems of Paleogeography of the Pleistocene of Glacial and Periglacial Regions" (A. A. Velichko and V. P. Grichuk, eds.), pp. 128-50. Nauka Press, Moscow.

Bogucki, A. B., Velichko, A. A., and Nechayev, V. P. (1975). Paleocryogenic process in western Ukraine in the Upper and Middle Pleistocene. *In* "Problems of Regional and General Paleogeography of Loessial and Periglacial Regions" (A. A. Velichko, ed.), pp. 80-89. USSR Academy of Sciences, Institute of Geography, Moscow.

Cailleux, A. (1942). The eolian periglacial actions in Europe. *Mémoires de la Société Géologique de France, Paris, N. S.* 21 (46).

Fink, J. (1969). Le loess en Autriche. *In* "Stratigraphy of Europe's Loesses." Supplement au Bulletin de l'Association francaise pour l'Etude du Quaternaire, Paris.

Follmer, L. R., Berg, R. C., and Acker, L. L. (1978). Soil geomorphology of northeastern Illinois. *In* "A Guidebook for the Joint Field Conference of the Soil Science Society of America and the Geological Society of America." Illinois Geological Survey, Urbana, Illinois.

Glushankova, N. I. (1971). Characteristics of the group composition of humus of buried soils of the Likhvin key section. *Moscow State University, Vestnik, seria geograficheskaya* 5, 109-13.

Grichuk, V. P. (1972). Results of a paleobotanical study of loesses of the Ukraine and south of the Central Russian Upland. *In* "Loesses, Buried Soils and Cryogenic Phenomena on the Russian Plain" (A. A. Velichko, ed.), pp. 26-48. Nauka Press, Moscow.

Kaplina, T. N., and Lozhkin, A. V. (1984). Age and history of accumulation of the "ice complex" of the maritime lowlands of Yakutiya. *In* "Late Quaternary Environments of the Soviet Union" (A. A. Velichko, ed.), pp. 147-51. University of Minnesota Press, Minneapolis.

Khabakov, A. V. (1946). Rounding indices of coarse gravel. *Sovetskaya geologiya* 10, 98-99.

Khalcheva, T. A. (1972). Variety of the mineralogical composition of loess horizons of the Russian Plain. *In* "Loesses, Buried Soils and Cryogenic Phenomena on the Russian Plain" (A. A. Velichko, ed.), pp. 49-59. Nauka Press, Moscow.

Konishchev, V. N., Faustova, V. N., and Rogov, V. V. (1973). Reflection of cryogenic phenomena in the microstructure of Quaternary deposits. *In* "Micromorphology of Soils and Loess Deposits" (S. V. Zonn, ed.), pp. 61-66. Nauka Press, Moscow.

Kosheleva, I. T. (1958). Micromorphology of tundra soils as a possible indicator of their genesis. *USSR Academy of Sciences, Izvestiya, seria geograficheskaya* 3, 88-92.

Lazarenko, A A. (1984). The loess of Central Asia. *In* "Late Quaternary Environments of the Soviet Union" (A. A. Velichko, ed.), pp. 125-31. University of Minnesota Press, Minneapolis.

Markova, A. K. (1975). Paleogeography of the Upper Pleistocene based on an analysis of fossil small mammals of the upper and middle Dnepr region. *In* "Problems of Regional and General Paleogeography of Loessial and Periglacial Regions" (A. A. Velichko, ed.), pp. 59-68. USSR Academy of Sciences, Institute of Geography, Moscow.

Morozova, T. D. (1962). Fossil soil of the Valdai Interstade. *USSR Academy of Sciences, Doklady* 143, 405-8.

Morozova, T. D. (1963a). Some results of a micromorphological study of loesses. *In* "The Anthropogene of the Russian Plain and Its Stratigraphic Components" (M. I. Neustadt, ed.), pp. 86-99. USSR Academy of Sciences Press, Moscow.

Morozova, T. D. (1963b). Structure of ancient soils and principles of their geographical distribution in different epochs of the Upper Pleistocene. *Pochvovedeniye* 12, 26-37.

Morozova, T. D. (1965). Micromorphological characteristics of pale yellow soils of central Yakutiya in relation to cryogenesis. *Pochvovedeniye* 11, 79-89.

Morozova, T. D. (1972). Micromorphological peculiarities of fossil soils and some problems of the paleogeography of the Mikulino (Eem) Interglacial on the Russian Plain. *In* "Soil Micromorphology" (S. Kowalinsky, ed.), pp. 595-606. Panstwowe Wydawnictwo Naukowe, Warsaw.

Morozova, T. D. (1981). "Development of the Soil Mantle of Europe in the Late Pleistocene." Nauka Press, Moscow.

Morozova, T. D., and Chichagova, O. A. (1968). Study of the humus of fossil soils and their importance in paleogeography. *Pochvovedeniye* 6, 34-43.

Orlov, D. S., Grishina, I. A., and Eroshicheva, N. L. (1969). "Humus Biochemistry." Moscow State University Press, Moscow.

Paepe, R. (1969). Lithostratigraphic units of Belgium's Upper Pleistocene. *In* "Stratigraphy of Europe's Loesses." Supplement au Bulletin de l'Association Francaise pour l'Etude du Quaternaire, Paris.

Sycheva, S. A. (1978). Soils of the Mezin complex of the Oka-Don Plain. *USSR Academy of Sciences, Izvestiya, seria geograficheskaya* 1, 81-92.

Udartsev, V. P., and Sycheva, S. A. (1975). Upper Pleistocene loesses and buried soilş of the Oka-Don Plain. *In* "Problems of Regional and General Paleogeography of Loessial and Periglacial Regions" (A. A. Velichko, ed.), pp. 26-42. USSR Academy of Sciences, Institute of Geography, Moscow.

Velichko, A. A. (1973). "The Natural Process in the Pleistocene." Nauka Press, Moscow.

Velichko, A. A. (1977). Die Erforschung von Lössgebieten und die Paläogeographie der Eiszeitepochen. *Schriftenreihe für geologische Wissenschaften* 9.

Velichko, A. A., Grekhova, L. V., and Gubonina, Z. P. (1977). "Habitat of Primitive Man of Timonovka Campsites." Nauka Press, Moscow.

Velichko, A. A., and Morozova, T. D. (1963). The Mikulino fossil soil and its peculiarities and stratigraphic significance. *In* "The Anthropogene of the Russian Plain and Its Stratigraphic Components." USSR Academy of Sciences Press, Moscow.

Velichko, A. A., and Morozova, T. D. (1969). The loesses of Belgium. *USSR Academy of Sciences, Izvestiya, seria geograficheskaya* 4, 69-76.

Velichko, A. A., and Morozova, T. D. (1972). The main horizons of loesses and fossil soils of the Eastern European Plain: Stratigraphy and paleogeography. *In* "Loesses, Fossil Soils and Cryogenic Phenomena on the Russian Plain" (A. A. Velichko, ed.). Nauka Press, Moscow.

Velichko, A. A., and Morozova, T. D. (1973). The soil cover of the Upper Pleistocene (Mikulino) Interglacial. *In* "Paleogeography of Europe in the Late Pleistocene" (I. P. Gerasimov, ed.), pp. 225-40. VINITI Press, Moscow.

Volkov, I. A., and Zykina, V. S. (1984). Loess stratigraphy in southwestern Siberia. *In* "Late Quaternary Environments of the Soviet Union" (A. A. Velichko, ed.), pp. 119-24. University of Minnesota Press, Minneapolis.

CHAPTER 12

Loess Stratigraphy in Southwestern Siberia

I. A. Volkov and V. S. Zykina

Predominantly subaerial sediments accumulated in the southern part of the West Siberian Plain during the Late Pleistocene on geomorphic levels above the floors of large depressions and river valleys. Extensive studies of these deposits have shown that the principal factor in denudation and transport was wind. The sediments comprise a separate subaerial formation having a characteristic spatial and facies differentiation and a rhythmic structure reflecting climate-caused irregularities in sedimentation (Volkov, 1971b, 1980). Regions marked predominantly by old deflation and accumulation of coarse traction deposits are distinct from regions marked predominantly by atmospheric dust deposits. Examples of the first type are many areas of the interfluves of the Ishim and Irtysh River valleys and of the Ishim and Tobol, as well as of the Barabinskaya Steppe and the western part of the Kulundinskaya Steppe (Volkov, 1968). Regions of the second type include the eastern part of Kulunda and the plain at the Altay foothills (Volkov, 1971b).

The regions where deflation and eolian traction deposits were predominant are characterized by extensive old deflation depressions and eolian depositional landforms. The cover of old subaerial deposits has a highly variable thickness and consists mainly of sand of varying grain size, occasionally even with an admixture of stones. In addition to grains of quartz, feldspar, and other minerals, there are sand-size and pebble-size pellets of dry clay (Figure 12-1), clay sands (Kukla, 1969), sands of clayey debris (Lozek, 1964), or clayey sand (Fedorovich, 1950; Volkov, 1971a, 1971b). When the clayey grains are wetted, they may be deformed, soaked, and changed into a homogeneous clayey mass.

In the Late Pleistocene, there were several active stages of subaerial deflation and accumulation of eolian material. Formations dating to the deglaciation of the Sartan (Late Valdai) continental glaciation are preserved best. The widely distributed, large sublatitudinal dunes (locally referred to as "griva") were formed at that time, as were hummocks and a thin cover of variable thickness. Both the crests and the cover between them consist mainly of eolian traction deposits. The crests are common on all old landforms, including the regionally developed second terrace

Figure 12-1. Surface of a sample of loesslike loam (Petropavlovsk region). (Magnification 20 × .)

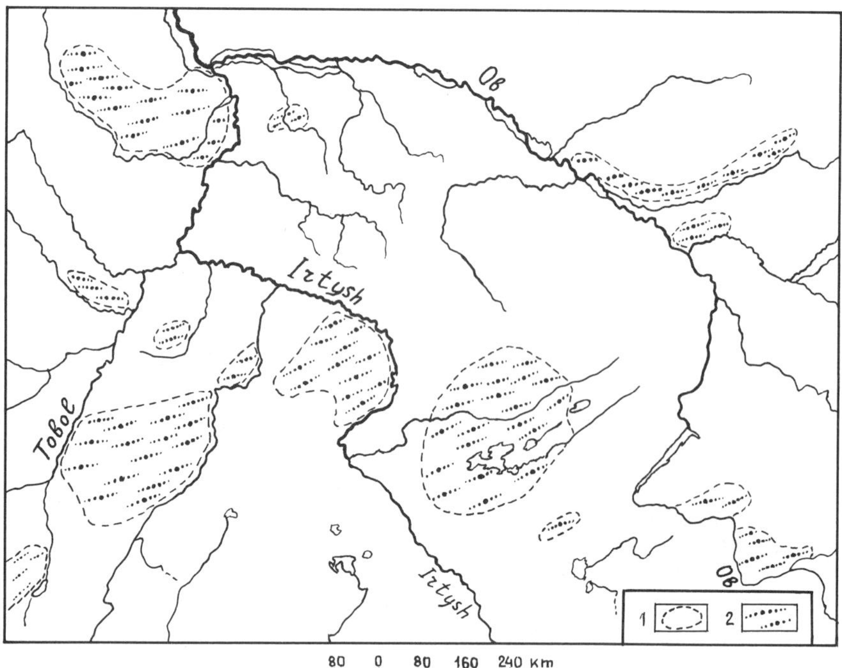

Figure 12-2. Principal regions of traction eolian deposits and Late Quaternary eolian ridge relief on the West Siberian Plain. (1) areas of ridge relief, (2) prevailing trend of crests.

above the floodplain (Figure 12-2), which was completed at the beginning of the Sartan glacial stage (about 20,000 yr B.P.). The formation of the dunes and the cover between them, developed on this terrace and on older formations, began about 20,000 yr B.P. and ended about 14,000 yr B.P. The upper age limit of the formation of the crest relief was determined by radiocarbon dates of 14,800 ± 150 yr B.P. (SOAN-111A), 14,200 ± 150 yr B.P. (SOAN-78), and 13,600 ± 230 yr B.P. (SOAN-111) on mammoth bones from the Volch'ya Griva Paleolithic campsite in the eastern part of the Barabinskaya Steppe (Firsov and Orlova, 1971; Volkov, 1973; Panychev, 1979).

Another active eolian stage (possibly a complex of several phases) occurred after alluvial and lacustrine sediments accumulated on the regionally developed first terrace above the floodplain in the late-glacial (Volkov, 1971a). The dominant small-hummock and small-crest eolian sands developed mainly on late-glacial riverine and lacustrine deposits before the swamping of western Siberia, which began around 9500 yr B.P. Traces of other, minor phases of Late Pleistocene subaerial processes in regions of deflation and accumulation of eolian traction deposits are poorly preserved and have been inadequately studied.

Much more extensive stratigraphic and paleogeographic data were obtained in regions where accumulation of atmospheric dust predominated. In the Ob' River region of Novosibirsk, the main regional subdivisions of the extensive local Late Pleistocene loess were traced, and a stratigraphic scheme of their subdivision was proposed (Volkov, 1971b). Further paleopedologic studies and radiocarbon and paleomagnetic dating substantially expanded and refined this scheme (Volkov and Zykina, 1977a, 1977b; Volkov and Arkhipov, 1978; Zykina, 1980; Panychev, 1979). A stratotypic region near the town of Iskitim was studied in detail, and the occurrence, structural characteristics, and environment of formation of different horizons (covers) of loess and intercalated soil complexes were placed within the chronologic limits of the entire Brunhes paleomagnetic epoch (Volkov and Zykina, 1977a; Volkov and Arkhipov, 1978).

Berd Soil Complex

At the base of the Late Pleistocene loess lies the Berd soil complex (Figure 12-3), which has a total thickness of about 2.5 m and consists of two humified soils separated by an interlayer of loesslike loam 0.5 to 1 m thick (Figure 12-4). The complex is extensively developed on interfluves, but on slopes it is not always preserved. The lower soil is usually distinctly developed on the loesslike loam of the Tazovian Glaciation (Suzun loess). The dark gray A-horizon is up to 60 cm thick. In its lower part, there is a well-developed, dense network of humus streaks down to a depth of 1.5 m. The B-horizon (illuvial) is leached of carbonates. The transitional BC-horizon contains carbonate accumulations in the form of pseudomycelium and fine spots.

The upper soil is separated from the lower by a layer of loesslike loam having humified spots and affected by processes of solifluction and soil formation. The humus horizon (A) is only 30 cm thick in the upper part and is usually light-colored. The illuvial horizon (B) gradually changes into the underlying loam. Carbonate accumulations occur throughout in the form of fine streaks, tubules along the course of plant roots, and indistinct spots.

In the entire soil complex, there are many animal burrows up to 10 to 15 cm in diameter (mole passages). Some are filled with humified loam, some with a light yellowish brown loesslike loam. The top of the soil complex is sharp-

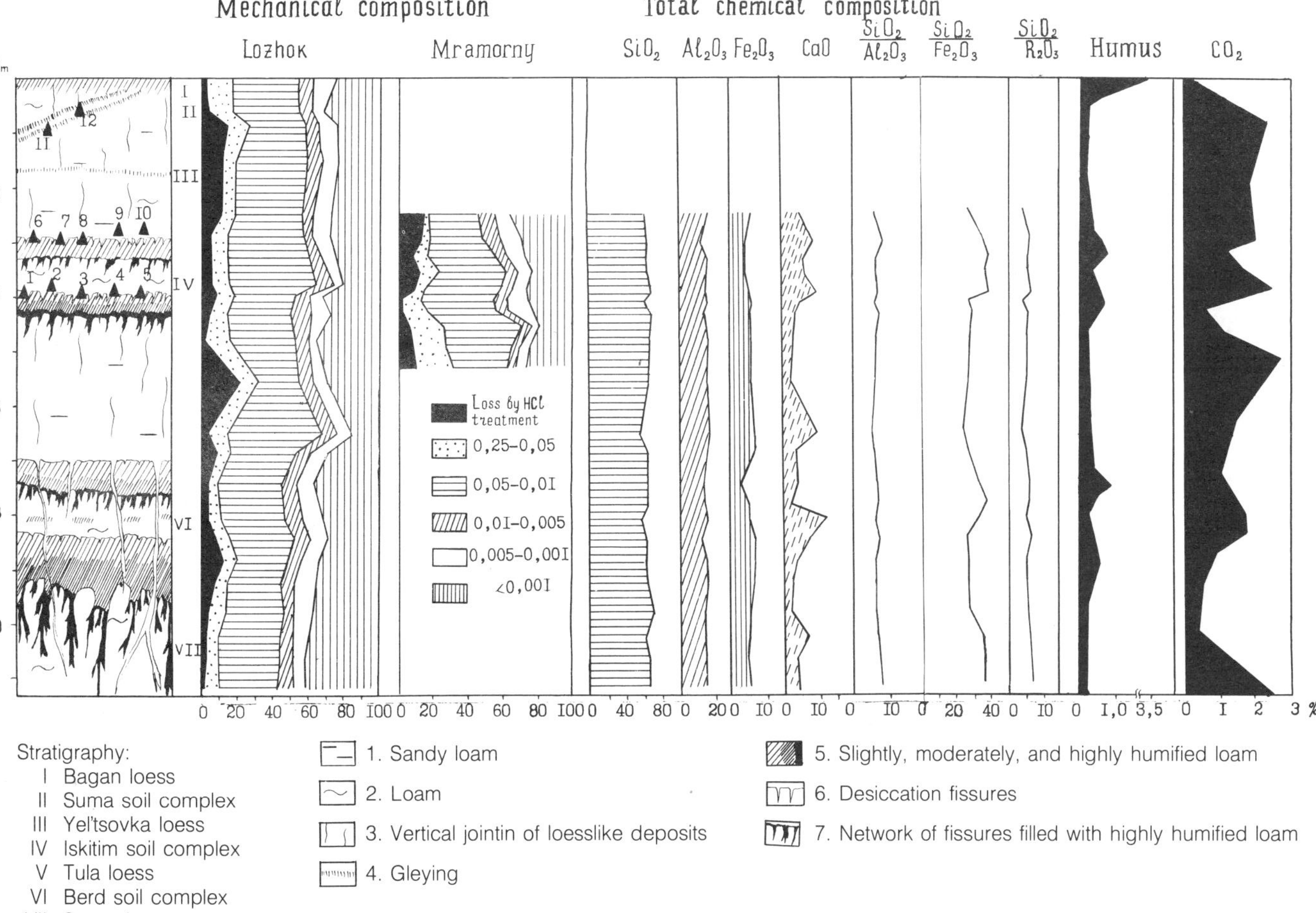

Figure 12-3. Composite of upper Quaternary strata of loess deposits (southeastern part of the West Siberian Plain, Ob' region of Novosibirsk). Triangles with numbers indicate radiocarbon dates: (1) 33,100 ±1600 yr B.P. (SOAN-165), (2) 32,780±670 yr B.P. (SOAN-629), (3) 29,000±450 yr B.P. (IGAN-168), (4) 30,000±1000 yr B.P. (IGAN-169), (5) 30,050±850 yr B.P. (SOAN-1587), (6) 26,300±700 yr B.P. (IGAN-167), (7) 24,300±380 yr B.P. (IGAN-199), (8) 24,490±320 yr B.P. (SOAN-1623), (9) 21,700±900 yr B.P. (SOAN-12), (10) 19,550±900 yr B.P. (SOAN-76), (11) 14,200±150 yr B.P. (SOAN-78), (12) 13,600±230 yr B.P. (SOAN-111).

ly bounded and broken up by a network of desiccation fissures up to 3 m deep and 2 cm wide filled with loesslike loam.

Studies of the soil complex include morphological descriptions, micromorphological analyses, and a detailed study of organic matter (total content of humus, its bulk and fractional composition, amino acid composition, carbohydrate content, and ultimate composition and optical properties of humic acids). The mechanical composition of the soils has been determined, as well as the bulk chemical composition, the carbon dioxide content, the composition of the aqueous extract, and the acidity. The structure of the profile, the macro- and micromorphological characteristics of the soils, their physiochemical properties, the characteristics of the humus profile, and the optical characteristics of humic acids (Dergacheva and Zykina, 1978, 1979) indicate that the soils were originally formed under climatic conditions similar to those of the present.

The thick lower soil developed over a long period of time in the forest-steppe zone under warm and fairly humid climatic conditions during the Kazantsevo Interglaciation. Its structure is similar to that of leached chernozems. The loam above the lower soil was deposited during one of the early stades of the Early Zyryanka Glaciation. The upper soil was formed during the next warm interstade (probably corresponding to the Brörup Interstade) under steppe conditions, as is the case with normal chernozems. Each soil was modified under humid and cold climatic conditions by cryogenic processes, as indicated by the bleached upper part of the humus horizons, the tonguelike lower part, and the peculiar mesostructure of the streaks. Traces of frost processes are more clearly manifested in the lower soil.

Tula Loess

Stratigraphically above the Berdsk soil complex lies the Tula loess, which is 1.5 to 2.5 m thick. Homogeneous, unstratified, light brownish yellow, normal loess dominates,

Figure 12-4. Berd soil complex in the northeastern part of the quarry near Lozhok.

occasionally varying to a light loesslike loam. Thin veinlets of carbonates locally occur along traces of herbaceous roots. Indistinct spots of gleying and iron staining are locally visible in the lower part. Thin interlayers of fine and coarse sand are also present. The cover is separated from subjacent deposits by a sharp boundary, with desiccation fissures formed at the beginning of accumulation. These and other indications point to very dry climatic conditions. The change in loess color to yellowish gray on slopes is probably a result of the secondary rehumification of the cover under moist climatic conditions.

Iskitim Soil Complex

The Iskitim soil complex is located on interfluves and gentle slopes on top of the Tula loess. In places it also extends over the regionally developed second terrace of the Ob' River and its major tributaries. It is about 1.5 m thick and consists of two soils separated by an interlayer of loesslike loam up to 70 cm thick. Three fossil soils are present in some sections.

The lower soil is clearly differentiated into a brownish dark gray to black humus A-horizon up to 40 cm thick; an illuvial B-horizon with a distinct horizon of carbonate accumulation (pseudomycelia); and a C-horizon with many animal burrows 5 to 12 cm in diameter. The upper Iskitim soil consists of a gray humus horizon (A), an illuvial horizon less distinct than in the lower soil (B), and a C-horizon whose lower part contains a lot of carbonate in the form of pseudomycelia, small spots of loess with lime nodules, and occasional small loess dolls. Holes of burrowing animals are up to 12 cm in diameter and occur everywhere.

In morphological and micromorphological characteristics, physicochemical properties, and basic characteristics of organic matter, the lower Iskitim soil can be classified as a soil of the chernozem series. The upper soil may also be regarded as a chernozem-type soil, but one formed under somewhat more humid climatic conditions. Both soils developed under conditions similar to those of the present, but they did not reach the maturity of the recent profile. The lower soil was formed over a longer period of time than the upper soil, for its profile is better differentiated into horizons.

The profiles of both soils show indications of a secondary transformation by cryogenic processes under humid and cold climatic conditions. They include a bleached upper part of the humus horizons, an uneven lower boundary with tongues and streaks, and tunnels of burrowing animals interrupted by wedgelike humic streaks. Locally, the accumulation horizon is substantially modified by frost-solifluction processes and consists of humified folds, spots, and interlayers. The loam separating the soils is also markedly changed.

The age of the Iskitim soil complex has been determined by 10 radiocarbon dates from different sections and in different laboratories (Volkov, 1973; Volkov and Arkhipov, 1978; Panychev, 1979; Gerasimov et al., 1980). From the lower soil in the Mramornyy Quarry near the village of Shipunovo, the following dates were obtained: 33,100 ± 1600 yr B.P. (SOAN-165) from wood charcoal, 32,780 ± 670 yr B.P. (SOAN-629) from a skull bone of a woolly rhinoceros found on the surface of the lower soil, and 29,000 ± 450 yr B.P. (IGAN-168) from the sum of the humic acid fractions of the humus horizon. Still another date, 30,000 ± 1000 yr B.P. (IGAN-169), was obtained in the Lozhok Quarry from the humus horizon of the lower soil, also on the basis of the sum of humic acid fractions. It is possible that the latter two dates, obtained from organic matter of the humus soil horizon, are somewhat young. A date of 30,050 ± 850 yr B.P. (SOAN-1587) was obtained in the Ogurtsovo section on the left side of the Ob' River from wood charcoal in the lower humified interlayer. The Karginskiy age of the Iskitim soil complex is indicated by radiocarbon dates from the middle part of the Karsnyy Yar section on the right bank of the Ob' River below Novosibirsk. These dates were obtained from waterborne sediments, which change into the Iskitim soil complex along the strike (Volkov and Arkhipov, 1978).

From the A-horizon of the upper soil in the Mramornyy Quarry section, a radiocarbon date of 26,300 ± 700 yr B.P. (IGAN-167) was obtained from the first cold fraction of humic acids. Another date, 24,900 ± 380 yr B.P.

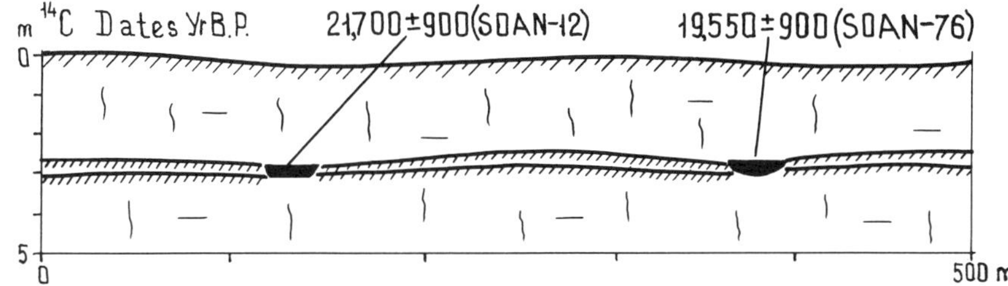

Figure 12-5. Radiocarbon dates from humified lenses of upper Iskitim soil. (1) Present-day surface, (2) level of fossil soils, (3) loess, (4) humified lens.

(IGAN-199) (based on the sum of humic acid fractions), was obtained from the humus horizon of the upper soil in the middle part of the side of the In' Valley near the village of Baryshevo. In the Ogurtsovo section, where three humified interlayers correspond to the Iskitim soil complex, a date of 24,490 ± 320 yr B.P. (SOAN-1623) was obtained from wood charcoal found in the middle interlayer.

Of importance are dates determined in the interfluve of the Shipunikhi and Koynikhi Rivers (left tributaries of the Berda River). They were obtained from the upper part of upper Iskitim soil, already altered by secondary processes, on the bottom of buried discharge rills from humus lenses (Figure 12-5). These lenses have a radiocarbon age of 21,700 ± 900 yr B.P. (SOAN-12) and 19,550 ± 900 yr B.P. (SOAN-76). The dates fix the time of maximum moistening of climate at the end of formation of the Iskitim soil complex; they show that accumulation of the superjacent loess cover began no earlier than 19,000 years ago.

Taken together, the radiocarbon dates definitely indicate that the Iskitim soil complex was formed during the Karginskiy Interstade. There is reason to believe that the lower soil was formed before the Konoshchel'ye cooling (35,000 to 31,000 yr B.P.) (Kind, 1974), if one considers that the dates on the lower soil may be too young because of contamination from organic matter from the humus horizon and that the fragments of wood charcoal used for dating could have entered the soil not during its formation but much later, when the soil underwent secondary transformation under cryogenic conditions. As indicated by the majority of the dates, the upper soil was formed during the Late Karginskiy Interstade, and its secondary transformations occurred at the beginning of Sartan Glaciation.

Yel'tsovka Loess

On interfluves and their slopes, the Yel'tsovka loess is stratigraphically superposed on the Iskitim soil complex. This loess cover usually serves as the parent material for recent soil. It was traced in the Ob' River region near Novosibirsk; in the Kulundinskaya, Barabinskaya, and Ishimskaya Steppes; and in the Kuznetskaya Depression. The cover usually rests on the Iskitim soil complex and other formations synchronous with it. It is widespread over ancient landforms, including the regionally developed second terrace. Such a stratigraphic position is also held by a surface mass of relatively coarse-grained sediments forming crests and located on the surface between them (crest mass).

In the Barabinskaya Steppe, the predominantly sandy crest mass grades eastward to the cover of Yel'tsovka loess. A normal, porous, lightly packed, brownish yellow, unstratified carbonate loess threaded with shallow root passages and vertical fissures predominates. A vertical wall is well preserved in the sections. On gentle slopes, the loess may gradually change to a light loam.

Many sections in the middle part of the loess show a horizon of slight ferrugination or gleying, with a thickness of 3 to 7 cm, and in such sections the lower part of the loess may be somewhat lighter than the upper part. This may be caused by the water logging of the lower part of the loess during the formation of the gleying horizon.

As indicated by radiocarbon dates of subjacent and superjacent sediments, the Yel'tsovskiy loess was formed from 19,000 to 18,000 through 15,000 to 14,000 yr B.P. under very dry climatic conditions, which reached a maximum at about 16,000 yr B.P. in the temperate belts of the Northern and Southern Hemispheres (Volkov, 1971b, 1976; Bowler, 1973, 1975, 1976).

A characteristic feature of Yel'tsovka loess is the fact that it continuously covers the slopes and bottoms of small erosion channels in the upper soil of the Iskitim soil complex. Particularly on valley slopes, the composition of the loess cover changes. Bedding occurs, and on the whole the material is coarser. At the base of ancient erosion scarps separating the second terrace from older landforms, a "periglacial alluvium" was developed that may be gradational to the Yel'tsovskiy loess cover on interfluves. Actually, these are not riverine deposits but deluvial-proluvial and surface-flow solifluction deposits. This fact is indicated by the slope of the bedding in this mass relative to the axial part of the valley. Thus, during the accumulation of Yel'tsovka loess, colluvial sediments accumulated rapidly in the valleys under conditions of very slight erosion.

Suma Soil Complex

In some areas, especially on slopes and crests, the Yel'tsovka loess is covered by younger sediments preserving the Suma soil complex. The lower soil of this complex is present in a stratotypic section near the Volch'ya Griva Paleolitic campsite (Kargatskiy region of Novosibirsk Province). An illuvial horizon up to 30 cm thick has an indistinctly lumpy prismatic structure and an uneven upper boundary broken up by a network of fissures. This horizon shows a greater accumulation of silt and iron and aluminum in comparison with the subjacent and superjacent strata. The microstructure of the horizon involves substantial argillization, optical orientation of clay minerals, and films on grains of the mineral skeleton. Different amounts of slightly developed late-glacial soils are included in the Suminskiy soil complex. Three fossil soils should probably be present in full sections. The Suminskiy soil complex was formed during the late-glacial. It has been little studied.

Bagan Loess

The youngest subaerial loess is the Bagan loess. Like the Suma paleocomplex, it occurs only locally. Its thickness almost never exceeds 1.0 to 1.5 m. Poorly sorted, dark, inhomogeneously colored brownish yellow and brownish gray loam predominates and is locally sandy. Bagan loess rests on the Suma soil complex (or on the weathering horizon on Yel'tsovskiy loess) and older formations. It was formed during the late-glacial and the Early Holocene. In the valleys, age analogues of Bagan loess are fine-hummock and fine-crest eolian sands on a regionally developed first terrace above the floodplain (Kudryashevskaya Terrace).

Conclusions

Thus, the structure of the Late Quaternary subaerial formation of the southern part of the West Siberian Plain indicates that as it accumulated there was a frequent alternation of three different types of environment. (1) Under climatic conditions similar to the present ones, in the natural setting of an interglaciation or interstade, fossil soils were formed. (2) Under the humid and cold climatic conditions of the first half of glacial stades, a secondary transformation of these soils occurred. (3) Under the dry climatic conditions of the deglacial phase, the loess covers accumulated. In the valleys, the climatic changes were reflected in the abundance of runoff. During the interstades and interglaciations, the runoff was moderate. It reached its maximum abundance, which was many times the present value, at the end of each interglaciation (or interstade) and the beginning of the next glacial stage. During the deglaciation, on the contrary, the runoff became scarcer than it is at the present time, and it almost ceased in shallow valleys. For Karginskiy-Sartan time, such a sequence in the change of natural environment can be traced fairly distinctly in the absolute radiocarbon chronology.

References

Bowler, J. M. (1973). Late Pleistocene environments in the Southern Hemisphere. Evidence from playa lakes in southern Australia. Ninth Congress of the International Union for Quaternary Research, Abstracts, pp. 37-38, Christchurch, New Zealand.

Bowler, J. M. (1975). Deglacial events in southern Australia: Their age, nature, and palaeoclimatic significance. *In* "Quaternary Studies" (R. P. Suggate and M. M. Creswell, eds.) pp. 75-82. Royal Society of New Zealand Bulletin 13, Wellington.

Bowler, J. M. (1976). Aridity in Australia: Origins and expression in aeolian landforms and sediments. *Earth-Science Reviews* 12, 279-310.

Dergacheva, M. I., and Zykina, V. S. (1978). Composition of humus of Pleistocene fossil soils of the Ob' region of Novosibirsk. *Geologiy I Geofizka* 12, 81-92.

Dergacheva, M. I., Zykina, V. S. (1979). Amino acid composition of humic acids of Late Pleistocene fossil soils of the Ob' region of Novosibirsk. *Geologiya I Geofizka* 6, 115-18.

Fedorovich, B. A. (1950). "The Face of a Desert." Moskul'tprosvetizdat, Moscow.

Firsov, L. V., and Orlova, L. A. (1971). Radiocarbon dating of bones of the Volch'ya Griva campsite. *In* "Reports of Field Studies by a Far East Archeological Expedition" (A. P. Okladnikov, ed.) Issue 2, pp. 132-34. Institute of History, Philology and Philosophy, USSR Academy of Sciences, Siberian Branch, Novosibirsk.

Gerasimov, I. P., Zavel'skiy, F. S., Chichagova, O. A., Voroshenko, V. V., Cherkinskiy, A. Ye., Aleksandrovskiy, A. L., and Lykhin, V. L. (1980). Radiocarbon studies conducted by the radiometric laboratory of the Geography Institute of the Academy of Sciences of the USSR. *Byulleten' komissii po izucheniyu chetvertichnogo perioda* 50, 200-213.

Kind, N. V. (1974). Geochronology of the Late Anthropogene from isotope data. *USSR Academy of Sciences, Institute of Geology, Trudy* 257.

Kukla, I. (1969). Clay sands and marker. *In* "Loess-Periglacial-Paleolith on the Territory of Central and Eastern Europe" (I. P. Gerasimov, ed.), pp. 154-57. VINITI Press, Moscow.

Lozek, V. (1964). Eine Lösserie mit roten fossilen Bodenbildungen bei Milanovce im Mitral-Tal. Sb. geol. ved. A, N 2, 27-38.

Panychev, V. A. (1979). "Radiocarbon Chronology of Alluvial Deposits of the Cisaltay Plain." Nauka Press, Novosibirsk.

Volkov, I. A. (1968). Climate fluctuations in the extraglacial belt (western Siberia, Turan, Caspian region). *In* "The Cenozoic in Western Siberia" (V. A. Mikolaev, ed.), pp. 48-58. Nauka Press, Novosibirsk.

Volkov, I. A. (1971a). Climate fluctuations of the late-glacial and Early Holocene in the south of the West Siberian Plain. *Geologiya I Geofizka* 8.

Volkov, I. A. (1971b). "The Late Quaternary Subaerial Formation." Nauka Press, Moscow.

Volkov, I. A. (1973). Paleogeographical significance of certain radiocarbon datings in the south of western Siberia. *Geologiya I Geofizka* 2, 3-8.

Volkov, I. A. (1976). Role of the eolian factor in the evolution of relief. *In* "Problems of Exogenic Relief Formation" (N. A. Florensov, ed.), pp. 264-89. Nauka Press, Moscow.

Volkov, I. A. (1980). Cyclicity of formation of Quaternary subaerial sediments of the temperate belt and climatic fluctuations. *In* "Cyclicity of Formation of Subaerial Rocks" (I. A. Volkov, ed.), pp. 25-34. Nauka Press, Novosibirsk.

Volkov, I. A., and Arkhipov, S. A. (1978). Quaternary deposits of the Novosibirsk region (operational data). Institute of Geology and Geophysics, Siberian Branch, Academy of Sciences of the USSR, Novosibirsk. (Rotaprint.)

Volkov, I. A., and Zykina, V. S. (1977a). Fossil soils in a key section of cover deposits of the Ob' Region of Novosibirsk. *Geologiya I Geofizika* 7, 83-94.

Volkov, I. A., and Zykina, V. S. (1977b). Rhythm of loessial deposits in the area of the town of Iskitim in Novosibirsk Province. *In* "Paleogeographical Principles of Efficient Utilization of Natural Resources" (M. J. Veklich, ed.), Part 2, pp. 125-27. Naukova Dumka Press, Kiev.

Zykina, V. S. (1980). Cyclicity of the structure of the Quaternary mass of subaerial sediments based on a study of fossil soils in the Iskitim region. *In* "Cyclicity of Formation of Subaerial Rocks" (I. A. Volkov, ed.), pp. 139-43. Nauka Press, Novosibirsk.

CHAPTER 13

The Loess of Central Asia

A. A. Lazarenko

The loess deposits of Central Asia are up to 200 m thick and have a chronologic range of up to 2 million years, as well as a complex facies structure and geomorphic expression. Loess genesis can be considered in the framework of a unified eluvial-deluvial-proluvial concept (Vasil'kovskiy, 1952; Lazarenko, 1967, 1980a, 1980b; Mavlyanov, 1949, 1950, 1958, 1960; Turbin and Aleksandrova, 1970). The role of the eolian factor is not denied in principle, but it is thought to be much smaller than believed by a number of other investigators (Kes', 1963; Kriger, 1951, 1965; Lomonovich, 1953; Fedorovich, 1960).

The loess stratigraphy is best studied in a series of key sections in the Tadzhik Depression (Lazarenko et al., 1977; Dodonov et al., 1977). There, in the Karatau 1 and Chashmanigar stratotypes, the visible loess thickness amounts to 90 m and 180 m, respectively, and in the second case its base is located below the Olduvai paleomagnetic event (1.6 to 1.8 million years ago), that is, in the Pliocene. I consider this entire stratum as a series (Kayrubak series) with two suites: the Chashmanigar Suite (Upper Pliocene to Eopleistocene) and the Utogan Suite (Pleistocene). The Pleistocene includes 10 regionally manifested, major upper "levels" of soil formation in the rank of soil complex (pedocomplex). Of these, soil complexes 1, 2, and 3 belong to the Late Pleistocene; soil complexes 4, 5, and 6 to the Middle Pleistocene; and soil complexes 8, 9, and 10 to the Early Pleistocene. The Matuyama-Brunhes reversal (690,000 years ago) is confined to the base of soil complex 9, and the bottom of soil complex 10 usually shows a regional angular unconformity (Figure 13-1). In the upper part of the sections, referred to the Brunhes epoch, five "short" episodes occur in thin (usually up to 0.1 to 0.2 m) intervals of reversed magnetization. Of these intervals, the most important for further correlation within the framework of the Late Pleistocene are the Laschamp (about 20,000 years ago, in the upper loess horizon) and the Blake, a double interval (about 120,000 to 110,000 years ago, at the level of soil complex 3 and beneath it).

I should mention a certain age "slip" of the boundaries of the identified stratigraphic horizons in relation to isochronous magnetostratigraphic levels. The Laschamp episode occurs in the upper part of the upper loess horizon in the Fakhrabad section, in the middle part in the Kayrubak section, or in the lower part in the Karatau 1 section. In its upper part, the Blake double episode is found in different sections in different parts of soil complex 3: in its middle (Kayrubak section), in its base (Karatau 1 section), or even below the latter (Lakhuti section). This fact leads to the conclusion that the correlated horizons of loesses and buried soils are not strictly synchronous in different sections, but their age "slip" usually does not go beyond a few thousand years, which on the scale of the Pleistocene is insignificant.

Correlation of Loesses

Comparison of the finely dissected loess of the Tashkent region (Lazarenko et al., 1980) and the Alma-Ata region with previously published similar data on the Tadzhik Depression (Lazarenko et al., 1977) shows them to be practically identical. This finding reflects common features in the development of the loess and offers a real opportunity for the creation of unified stratigraphic subdivisions of this formation for Central Asia and southern Kazakhstan.

There are still many difficulties in making long-range correlations of the loess of Central Asia and other regions (for example, eastern Europe, where its stratigraphy has been worked out in detail by Velichko, 1975, among others). At the present time, because of a lack of reliable geochronologic data, I can mention only the following approximate correlation of Late Pleistocene loesses. The uppermost loess horizons (the Altynovka and Desna horizons in the center of the Russian Plain and the Sanglak horizon in Central Asia) are unequivocally synchronous, because the Altynovka and Desna horizons lie above the Bryansk

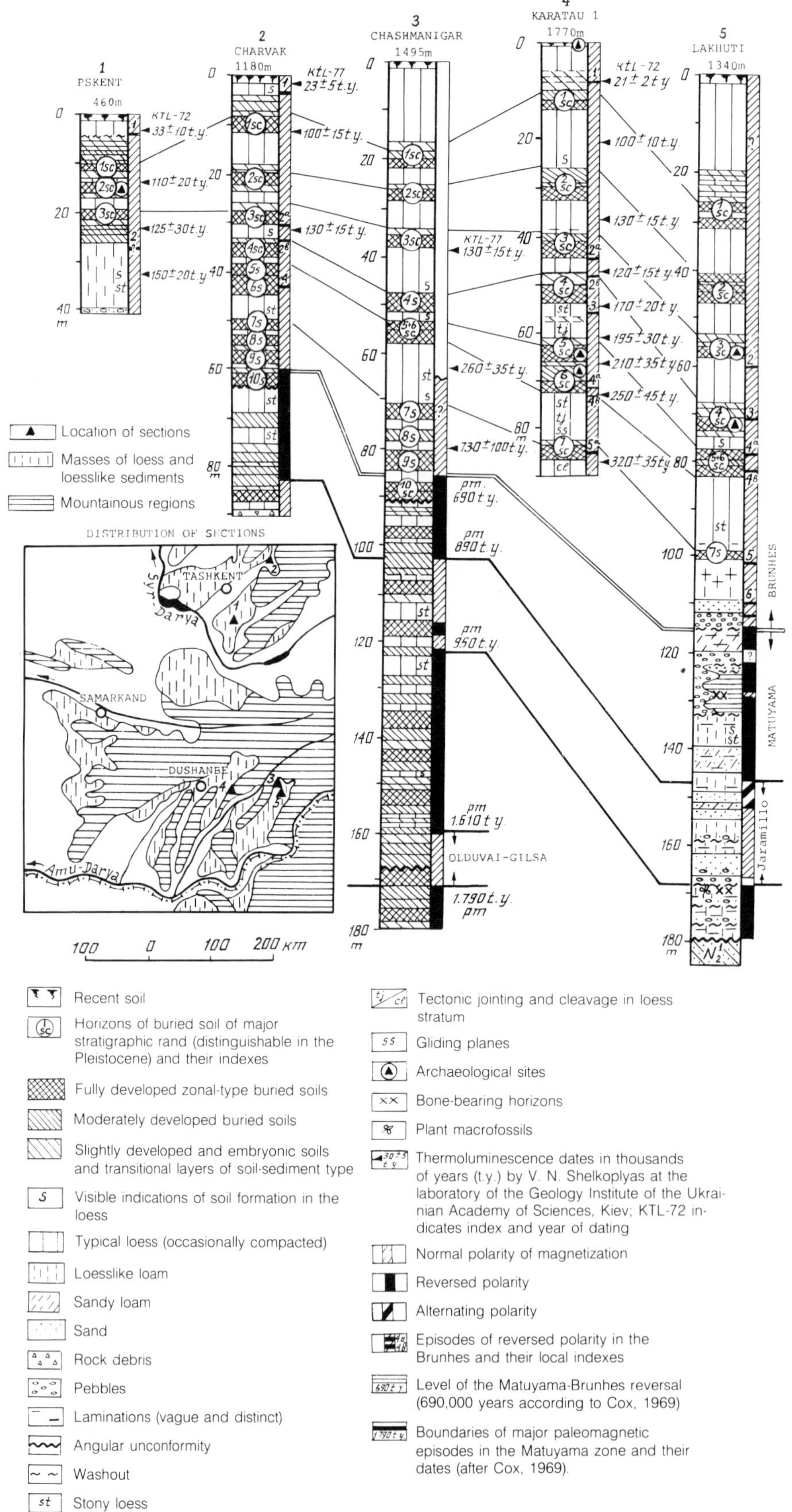

Figure 13-1. Correlation of key sections of loess formation in the Tashkent region and the Tadzhik Depression. (Compiled by Lazarenko [1977], who used the data of A. V. Pen'kov, V. V. Semenov, and V. N. Shelkoplyas.)

Table 13-1. Correlation of Loesses of Central Asia with Those of the European USSR

European USSR			Central Asia (Tadzhik Depression)	
Soil Complex	Climatic Phase	TL Dates (yr B.P.)	Stratigraphic Horizon	TL Dates (yr B.P.)
1	Late Valdai (glacial loess)	20,000-35,000	Sanglak (first loess horizon)	17,000-33,000 (47,000)
2	Moscovian (glacial loess)	From 170,000-180,000 to 210,000-220,000	Tiana (fifth loess horizon)	150,000-210,000
3	Dneprovian (glacial loess)	From 230,000-250,000 to 270,000-320,000	Zimistans (sixth loess horizon)	250,000-300,000

Notes: The thermoluminescence dates are from the data of V. N. Shelkoplyas (Shelkoplyas, 1973; Lazarenko et al., 1977). The radiocarbon dates for the Late Valdai are available for this interval for the European USSR.

buried soil (about 30,000 to 25,000 yr B.P.) and because the Laschamp paleomagnetic episode is confined to the upper loess horizon in Central Asia (about 20,000 yr B.P.), as confirmed by a series of thermoluminescence datings performed by B. N. Shelkoplyas (Lazarenko et al., 1977). As far as older loess horizons are concerned, it is necessary to refer mainly to thermoluminescence dates, which thus far should be regarded as tentative. The closest chronologic relationships can be noted to a first approximation, as in Table 13-1.

From these comparisons one gets the impression that the loess horizons of Central Asia correspond to cold climatic intervals of the European USSR, usually associated with glaciations. Hence, buried soil complexes should generally be correlated with warm climates of interglacial intervals. (Apparently, the indicated correspondences can be discussed only in the most general terms, for maximum cooling, glaciation, and loess accumulation could have occurred at different times.) The third soil complex in Central Asia evidently corresponds to the Mikulino Interglaciation (its optimum phase). As already mentioned, the Blake paleomagnetic episode belongs in this interglacial, as confirmed by a series of thermoluminescence datings in the interval of 115,000 to 130,000 years. Evidently, the alternation of loess horizons and buried soil complexes resulted from the rhythmicity and general trend of the climatic changes.

Buried Soils

The greatest structural complexity is characteristic of the soil-stratigraphic horizons (particularly the upper two horizons) usually represented by soil complexes and less often by individual soils. At the base of each complex lies a dark soil with a fully developed profile corresponding to the climatic optimum of at least regional scale (Lazarenko, 1980a). The soil's color is brown or light brown, usually with a noticeable reddish shade (Munsell 7.5 YR 5/4, less frequently 10 YR 5/4 or 5 YR 5/4, 4/5, etc., for the A1-horizon). It is generally 2.0 to 2.5 or even 3.0 m thick, with the following genetic horizons: CA, 0.4 to 0.6 m; A1 and B1, 0.8 to 1.1 m; B2, 0.3 to 0.5 m; BCca, 0.7 to 1.0 m; with gradual transitions between them (except for a very sharp boundary between the B2- and BCca-horizons). (I propose index CA for the specific upper horizon of buried soils, intermediate between humus horizon A1 and the superjacent rock [usually loess or the intermediate sediment type leading up to it]. The CA-horizon is absent from recent soils.) Clay accumulation is considerable (content of the fraction: less than 0.001 mm, 10%-15%; fraction less than 0.005 mm, 35%-40%). The carbonate profile is sharply differentiated in cases with complete leaching in the A1- and B-horizons and concentration in the BCca-horizon. Humus and readily soluble salts are present (up to 0.1% each) (Figure 13-2). The predominant soil structure (A1- and B-horizons) is coarsely granular (up to 3 to 5 mm, but often not very distinct), with finely nutty structure.

The macro- and micromorphology indicates formerly increased surface moistening, evidently of a seasonal character, especially the fine ferruginous features (microortsteins) up to 0.1 mm, diffuse rings, intense black-brown and black pigmentation of phytopores, and so on. Substantial seasonal moistening (in meadows?) is indicated by the increased thickness of the soil profile as a whole and of the A1- and B1-horizons in particular, very slight differentiation of the latter two, predominance of coarse-grained structure attributed to a developed network of fibrous grass roots, predominance of calcium in the absorbing soil complex, and marked predominance of finely dispersed calcium carbonate in the BCca-horizon, with few distinct carbonate concretions and pseudomorphs along roots. The clayey fraction of less than 0.001 mm has a relatively weak optical orientation. Its active displacement through the soil profile (lessivage) is detected infrequently and in weak form, which also suggests a seasonal wetting regime (it results, however, in the leaching of carbonates out of a major portion of the soil profile). This is also indicated by a very slight change in the mineral composition of the clays

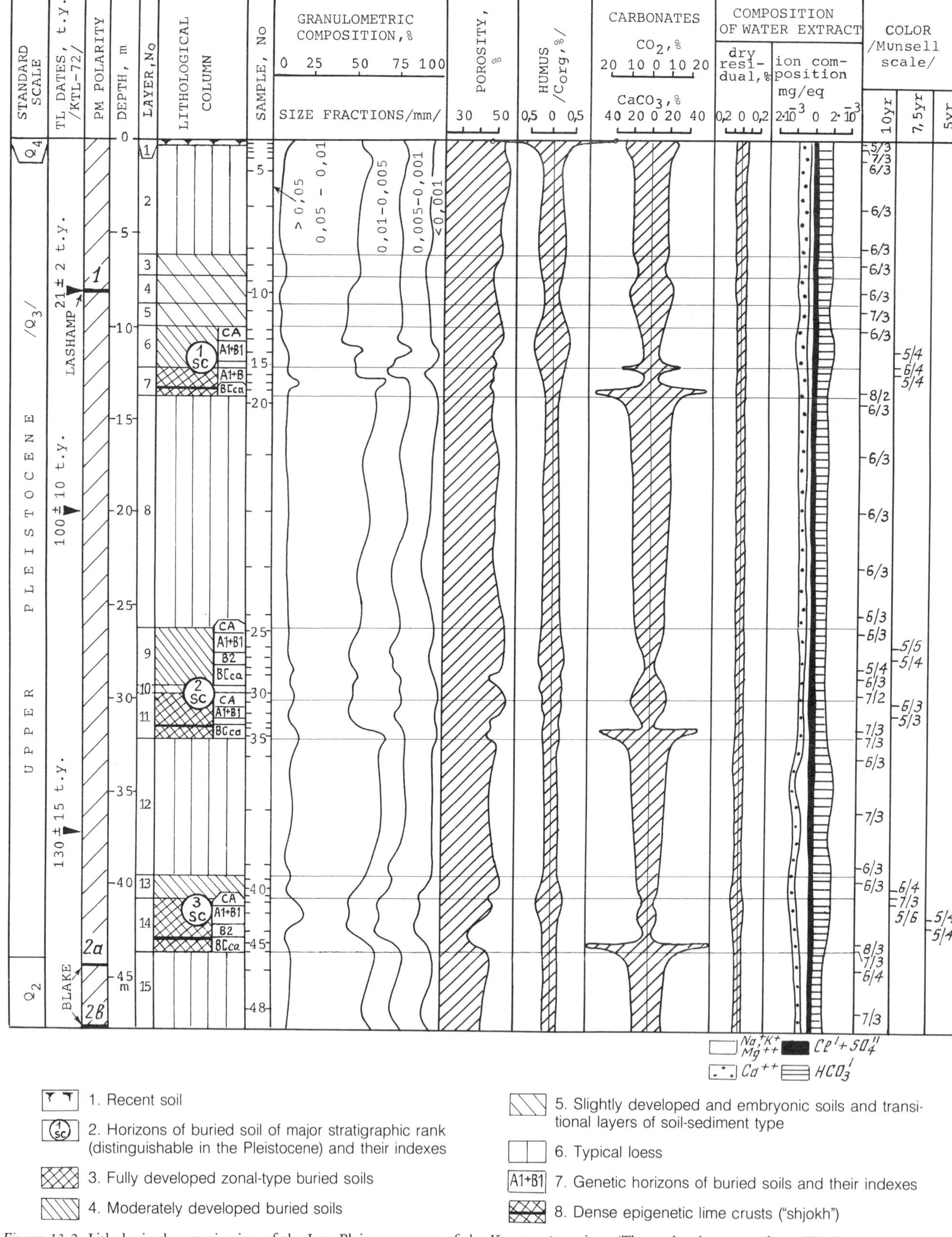

1. Recent soil
2. Horizons of buried soil of major stratigraphic rank (distinguishable in the Pleistocene) and their indexes
3. Fully developed zonal-type buried soils
4. Moderately developed buried soils
5. Slightly developed and embryonic soils and transitional layers of soil-sediment type
6. Typical loess
7. Genetic horizons of buried soils and their indexes
8. Dense epigenetic lime crusts ("shjokh")

Figure 13-2. Lithologic characterization of the Late Pleistocene part of the Karatau 1 section. (Thermoluminescence dates [TL] in thousands of years [t.y.].)

through the soil profile, particularly by more mixed-layer mica montmorillonite in the lower soil horizons compared to hydromica.

Moistening of these soils was characterized not only by periodicity (seasonal fluctuation) but also by an appreciable contrast. The A1- and B1-horizons are noticeably reddish brown. Films of manganese oxides with a characteristic "metallic" luster occur occasionally along structural joints. Globules of clayey material occur up to 1 to 2 cm in size with a radial structure. The rodent fauna, found mainly in the upper (CA) horizons transitional from soil to overlying loess, includes such representatives of open spaces as the northern mole vole (*Ellobius* sp.), the hamster (*Cricetulus* sp.), and the "nekornezubaya" fieldmouse (Microtinae) (as determined by V. S. Zazhigin). Very warm climatic conditions with a periodic but not acute moisture deficit are also indicated by the composition of the mollusk fauna, which is found fairly often in these upper soil horizons as well as in the BCca-horizons (*Angiomphalia regaliana, Leucozonella angulata, Pseudonapaeus albiplicatus,* etc., as determined by A. A. Shileyko). Ecologically this fauna also requires thorough drainage of the substrate and fairly tall and thick herbage (most likely together with shrubs and open forest). This finding is not contradicted by the pollen data, as I will show later in this chapter. During the dry season, the surface of the soils in question was often exposed to steppe fires, as indicated by finds of very fine charcoal fragments of uniform size (about 1 to 2 mm) in the upper part of the A1-horizon, as well as isolated burned lenses.

On the whole, the several fully developed Pleistocene soils under consideration are fairly similar, but each has characteristic features. For example, in soil complex 3 (of Mikulino age), the soils are usually the reddest (Munsell 5 YR 5/4) and have maximum values of magnetic susceptibility ($\varkappa$ up to $200 \cdot 10^{-6}$ cgs or even more), evidently because of the dehydration of iron hydroxides. This soil complex in western, southern, and part of central Tien-Shan (regions of Tashkent and Samarkand and the Fergana and Tadzhik Depressions) is represented by soils of dry subtropics, similar in type to cinnamon soils, and in eastern Tien-Shan (region of Alma-Ata) by moderately warm soils of chernozem and/or chernozemlike type. Toward the periphery of the piedmont loess region, this zonal type of soil increasingly assumes the features of aridity and changes into soils similar to gray cinnamon, gray desert, and possibly gray-brown semidesert soils. Included in soil complex 3 were also Paleolithic finds (the richest site was Obimazar), which, according to V. A. Ranov, belonged to the same group of old Karatau-type pebble cultures as the known finds in soil complex 4 and soil complex 5 especially (Lazarenko et al., 1977).

Generally toward the top of the section in each succeeding soil complex, a gradual increase of indications of drier conditions is noted (Lazarenko et al., 1977, 1980). This increase is also definitely manifested in a comparison of the "main" soil of soil complex 3 with that in soil complex 2 and particularly in soil complex 1: a less distinct differentiation of the carbonate profile and degree of leaching of carbonates; the attenuation of waterlogging indications; the appearance of a characteristic fine, crumbly structure and occasionally of larval capsules that protect insects from desiccation (this being characteristic of certain arid soils, including gray desert soils); and an increase in readily soluble salts in soil complex 1. For synchronous "main" soils and soil complexes as a whole, the increase in indications of aridity as one moves away from the mountains is much more distinct than upward in a vertical section. As a result, the soils in the Tashkent region and the Tadzhik Depression usually change from nearly cinnamon (the Charvak, Khumsan, Karatau 1, Chashmanigar sections, etc.) to gray cinnamon and gray desert soils (the Pskent, Keles, Karakchi, Farkhabad sections, etc.).

Medium-developed (and, in part, slightly developed or, more accurately, slightly differentiated) brown and light brown soils, usually present in the upper parts of soil complexes, evidently were formed under intermediate conditions of moisture (between those of loess and of fully developed soils) and possibly of temperature. These soils may reflect finer phases of climatic fluctuation against a background of major pluvials. Finally, the least-developed embryonic soils and sediment types transitional to loess are of no particular stratigraphic importance; they are local and mainly reflect delays and interruptions in loess accumulation.

Loess Horizons

The loess horizons themselves were formed under more arid conditions, as has been very definitely substantiated by lithologic, geochemical, and paleontologic data. The content of carbonates and readily soluble salts (Figure 13-2) increases, as does the number of insect larval capsules; indications of temporary waterlogging are less common, and the mollusk fauna is more xerophilic and uniform, with *Ponsadenia* (*Bradybaena*) *semenovi, Pseudonapaeus potaninianus,* and similar species. Such trends also show up in individual loess horizons both vertically upward and horizontally with increasing distance from the mountains.

The pollen spectra obtained by N. S. Bolikhovskaya, R. E. Giterman, M. P. Grichuk, and M. M. Pakhomov at all eight key sections of the Tadzhik Depression and Tashkent region (Lazarenko et al., 1977, 1980) have a complex mixed composition and, unfortunately, do not usually show distinct or sharp differences between the loess horizons and the buried soils. Nevertheless, they do indicate a vertical vegetational zonation in the mountains as well as a much more moist climate than at present. Thus, on the mountain slopes during the Late Pleistocene, the birch-pine forests, which today are completely absent, were extensively developed in these regions and included alder, spruce, fir, oak, linden, and other tree species. The grassland vegetation of the foothills included herbs that are indicative of greater moisture than is present today.

Stages in Loess Accumulation

On the whole, a series of stages is noted in the accumulation of the loess formation in Central Asia. One of the most distinct stages corresponds to the middle and main part of the Late Pleistocene. At this time thick loess accumulation first exceeded the boundaries of narrow accumulation zones in valleys and began to extend on a wide regional scale to adjacent slopes and planation surfaces, occupying vast areas and obscuring minor landforms. As a result, the topography of the loess foothills became much flatter. The rate of accumulation generally amounted to about 0.3 to 0.5 m per 1000 years, practically equal to the accumulation rate in the Middle Pleistocene and nearly four times faster than in the Early Pleistocene. The rate of accumulation of loess in the Late (and Middle) Pleistocene was at least 1.0 to 1.5 m per 1000 years.

At the beginning of the Late Pleistocene, the tectonic and climatic situation in Central Asia as a whole and in the Tadzhik Depression in particular generally resembled the Middle Pleistocene situation, but it subsequently changed considerably, particularly at the end of the stage. This was manifested in an intensive Late Pleistocene/Holocene phase of lastest orogenesis (Tadzhik phase), which produced the main features of the present medium-sized mountain topography of the Tadzhik Depression and which deformed the loess cover, producing the "loessial" ranges of Karatau, Kugitek, Tian, and Rangon (Lazarenko, 1980b).

As previously noted, the climatic optimum of the Late Pleistocene corresponded to the time of formation of the third soil complex (about 100,000 years ago). Further change in climatic conditions was expressed in a rhythmic alternation of the phases of relative moistening (of the second and first soil complexes) and desiccation (of the loess horizons separating them) against a general background of progressive drying of climate. On the whole, forest communities on mountain slopes contracted, and grasslands in the foothills gradually changed to desert-steppe. The role of xerophiles in the mollusk fauna increased upward in the section, particularly in the loess horizons. Although the type of soil formation generally remained the same, indications of more arid conditions increased in each succeeding soil complex. On the whole, the soil and plant zonation had more contrast. Indications of drier climatic conditions are strongest in the uppermost loess horizon (Sanglak horizon, about 30,000 to 20,000 yr B.P.), where cooling is suggested as well by the appearance of Ericales pollen and an increase in spruce and juniper pollen and in spores, which represent the vegetation of a substantially expanded alpine belt.

Shortly before the Holocene, loess accumulation in Central Asia ceased almost everywhere, with the exception of local redeposition on slopes and a reduced formation of eluvial-deluvial loess in certain mountain regions.

The above data were obtained from a detailed study of the most typical and thickest loess masses of Central Asia and southern Kazakhstan: the Tadzhik Depression and the Tashkent and Alma-Ata regions. Along with additional data on a number of adjacent regions, they reveal features of great similarity in the paleoclimatic history of the loessial foothills of the vast territory of the southern, western, and part of the central Tien-Shan and western spurs of the Pamir.

References

Cox, A. (1969). Geomagnetic reversals, *Science* 163, 237-45.

Dodonov, A. E., Melamed, Ya. R., and Nikiforova, K. V. (eds.) (1977). "Excursion Guide for an International Symposium on the Problem 'Boundary of the Neogene and Quaternary System' " (3-13 October 1977, Dushanbe, USSR). Nauka Press, Moscow.

Fedorovich, B. A. (1960). Problems of the origin of loess in relation to the conditions of its ocurrence in Eurasia. *USSR Academy of Sciences, Institute of Geography, Trudy* 80, 96-117.

Kes', A. A. (1963). Basic structural features of loessial topography. *In* "The Ideas of Academician V. A. Obruchev Concerning the Geological Structure of North and Central Asia and Their Further Development" (S. V. Obruchev, ed.), pp. 19-41. USSR Academy of Sciences Press, Moscow and Leningrad.

Kriger, N. I. (1951). The loesses of the Kirgiz Range. *USSR Academy of Sciences, Doklady* 78, 355-57.

Kriger, N. I. (1965). "Loess, Its Properties and Relationship to the Geographical Environment." Nauka Press, Moscow.

Lazarenko, A. A. (1967). On the problem of Central Asian loesses. *USSR Academy of Sciences, Doklady* 174, 913-16.

Lazarenko, A. A. (1980a). Buried soils of the loess formation of central Asia and their paleogeographic significance. *USSR Academy of Sciences, Doklady* 252, 181-85.

Lazarenko, A. A. (1980b). Main features of the structure and neotectonics of the loess formation of the Tadzhik Depression in connection with problems of its genesis. *In* "Processes of Continental Lithogenesis" (E. V. Shantsev, ed.), pp. 186-94. Nauka Press, Moscow.

Lazarenko, A. A., Bolikhovskaya, N. S., and Semenov, V. V. (1980). Experience with a detailed stratigraphic subdivision of the loess formation of the Tashkent region. *USSR Academy of Sciences, Izvetiya, seria geologicheskaya* 5, 53-66.

Lazarenko, A. A., Pakhomov, M. M., Pen'kov, A. V. Shelkoplyas, V. N., Giterman, R. Ye., Minina, Ye. A., and Ranov, V. A. (1977). Possibility of climatostratigraphic subdivision of the loess formation of Central Asia. *In* "The Late Cenozoic of Northern Eurasia" (K. V. Nikiforova, ed.), Vol. 1, pp. 70-133. USSR Academy of Sciences, Institute of Geology, Moscow.

Lomonovich, M. I. (1953). Origin of the loess of the Trans-Ili Alatau. *Kazakhstan Academy of Sciences, Izvestiya, seria geologicheskaya* 17, 48-76.

Mavlyanov, G. A. (1949). Genesis of loess and loesslike rocks as the main factor in the evaluation of their physical properties. *Uzbak Academy of Sciences, Institute of Geology, Trudy* 3, 64-85.

Mavlyanov, G. A. (1950). Origin of the loess and loesslike rocks of southern regions of Central Asia. Data from a study of the Quaternary of the USSR. Issue 2. USSR Academy of Sciences Press, Moscow and Leningrad.

Mavlyanov, G. A. (1958). "Genetic Types of Loesses and Loesslike Rocks of the Central and Southern Parts of Central Asia and Their Geological Engineering Properties." Uzbek Academy of Sciences Press, Tashkent.

Mavlyanov, G. A. (1960). Loesses and loesslike rocks of Turkestan. Scientific notes of the Central Asia Scientific Research Institute of Geology and Mineral Raw Materials (SAIGIMS). Issue 4. Tashkent.

Shelkoplyas, V. N. (1973). Application of the thermoluminescent (TL) method to the dating of Pleistocene formations. *In* "Chronology and Climatic Stratigraphy" (V. A. Zubakos, ed.), pp. 121-27. All-Union Geographical Society, Leningrad.

Turbin, L. I., and Aleksandrova, N. V. (1970). The loess rocks of Tien-Shan. *In* "Data on the Geology of the Cenozoic and the Latest Tectonics of Tien-Shan" (O. K. Cheolia, ed.), pp. 89-105. Ilim Press, Frunze.

Vasil'kovskiy, N. P. (1952). On the origin of loess. *Uzbek Academy of Sciences, Institute of Geology, Trudy* 8, 47-62.

Velichko, A. A. (1975). Problems of correlation of Pleistocene events in the glacial, periglacial-loessial and maritime regions of the East European Plain. *In* "Problems of Regional and General Paleogeography of Loessial and Periglacial Regions" (A. A. Velichko, ed.), pp. 7-25. USSR Academy of Sciences, Institute of Geography, Moscow.

CHAPTER 14

Cryogenic Processes in Loess Formation in Central Asia

A. V. Minervin

The origin of loess in the hot dry (interglacial) or the severe cold (periglacial, glacial) climate of Central Asia has been controversial. The object of this report is to examine the paleogeographic conditions of formation of the specific chemical/mineral composition, structure, texture, and modification of loess in the Pleistocene history of Central Asia.

The following two stages of continental lithogenesis should be distinguished with particular rigor in this problem: (1) sediment genesis by different modes under diverse environmental conditions and (2) transformation of sediments into loess as a result of the complex set of physical, chemical, and cryogenic processes in epigenesis (Strakhov, 1960), subaerial diagenesis (Shantser, 1966), and cryolithogenesis (Popov, 1979). The second stage is of decisive importance in engineering geology, for it provides a key to the determination of the nature of the subsidence character of loess. In this chapter two fundamental assumptions are important: (1) In the Pleistocene of Central Asia, the epochs of loess formation are associated with periglacial zones and with a general planetary cooling of climate (Kostenko, 1962; Turbin and Aleksandrova, 1970; Nesmeyanov, 1977). (2) Loess has frequently been frozen for one or more years (Velichko, 1973; Popov, 1968; Fedorovich, 1962; Fotiyev, 1978).

Mineralogy

In different structural-tectonic regions of Central Asia, 95% to 98% of loess in the natural state consists of structural elements stable in water, that is, microaggregates or globules 0.1 to 0.01 mm in diameter. The core of a globule is composed of primary minerals such as quartz (less commonly feldspar), which have a regular crystal form. The core is surrounded by a thin, perforated calcite envelope covered by a jacket of clay minerals (hydromicas, mixed-layer minerals, montmorillonite, kaolinite, chlorite), iron oxides, amorphous silicic acid, finely dispersed quartz, and carbonates (Figures 14-1, and 14-2). In order to separate the "jacket," the loess is saturated with sodium ion (Gedroyts, 1955), and particles smaller than 5 μm are elutriated from the dispersed samples. The grains of loess from which the "jacket" is thus removed are then analyzed under a scanning microscope, which reveals the perforated envelopes of finely dispersed carbonates (Figure 14-3).

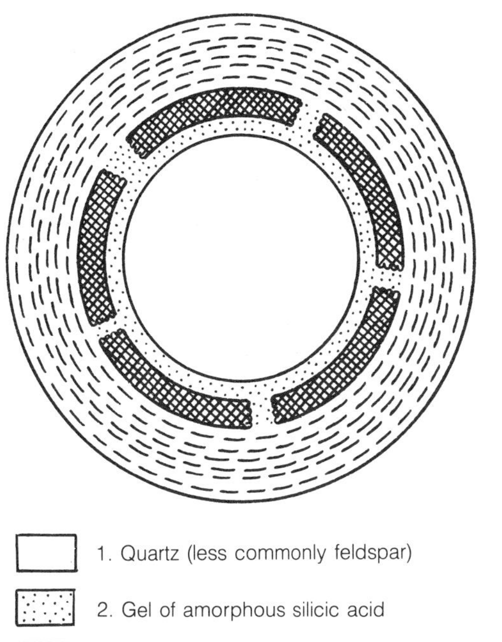

Figure 14-1. Structure of globular aggregates in the loess of Central Asia.

Experiments with 1000 cycles of freezing and thawing of

Figure 14-2. Globular aggregate from loesses of the Tashkent complex (Q_2), Chirchik of the Akhangaran alpine intermontane trough of western Tien-Shan.

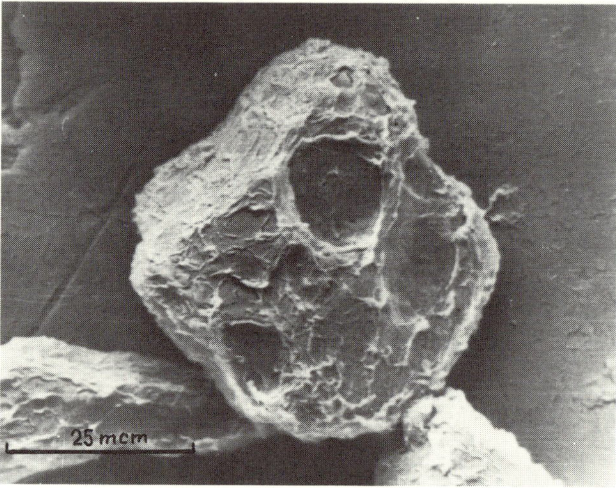

Figure 14-3. Perforated calcite envelope on the surface of quartz from loesses of the Chuyskiy alpine intermontane trough of northern Tien-Shan.

different primary minerals (diameter 0.25 to 0.1 mm) show that up to 90% of water-saturated loess becomes intensively fragmented to the size of coarse silt and fine sand (Minervin, 1980). Quartz and microcline were fragmented most uniformly and effectively. In alternate freezing and intensive heating of air-dried mineral particles (modeled after recent desert landscapes of Central Asia), no fragmentation was observed. The initial dispersion of the material was completely preserved in experiments on alternate wetting and drying at above-zero temperatures (Table 14-1). The experiment emphasizes the decisive role of cryogenic weathering in the formation of the main loess fractions in orogenic source regions of Central Asia. The effective breakage can be explained by the microstructure of

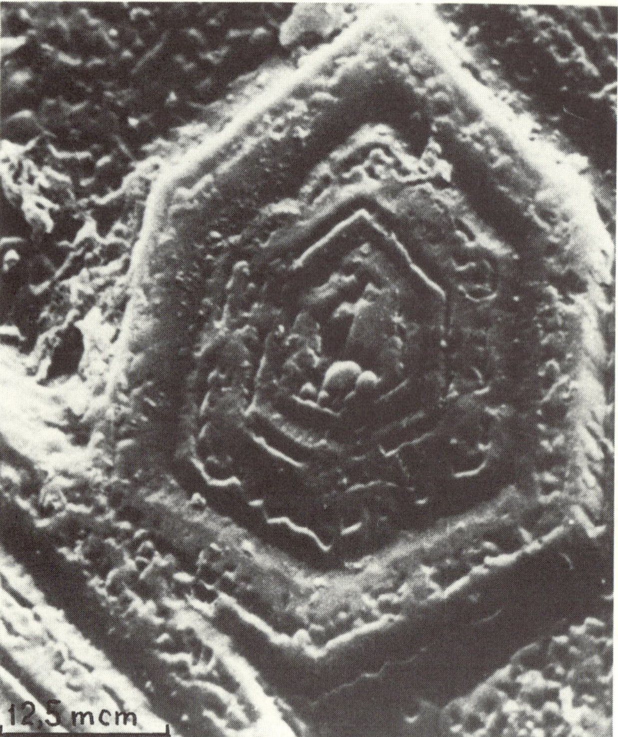

Figure 14-4. Imperfect block structure of a natural quartz crystal from a vein of the Talasskiy Ala-Tau Range of Tien-Shan.

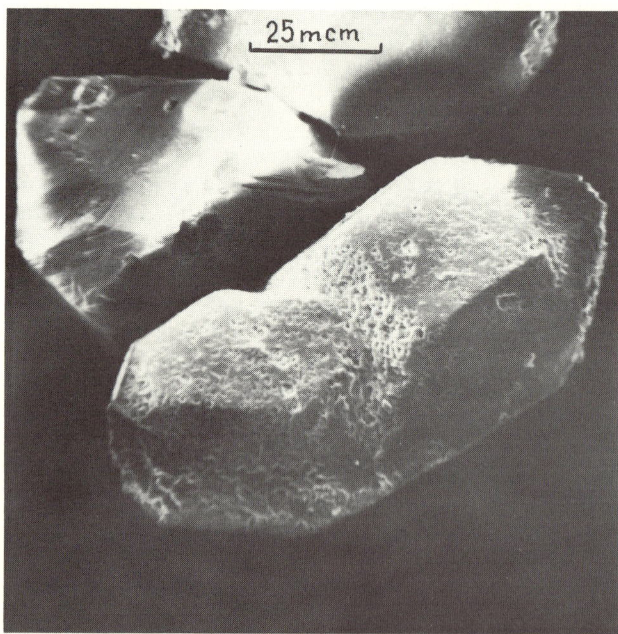

Figure 14-5. Core of globular aggregate-microblock of quartz from loess of Ilyak complex of the Kulyab intermontane trough of the Tadzhik depression.

quartz crystals (Figure 14-4), which consist of individual units of a ditrigonal prism separated by linear defects in the form of dislocation channels 2 to 3 μm in diameter and numbering 2×10^7 per square centimeter (Chepizhnyy et al., 1973; Minervin, 1980). The elementary units of natural quartz and feldspar are 100 to 10 μm in diameter. The disintegration of quartz and feldspar crystals as a result of cryogenic weathering is caused by the crushing of thin walls of dislocation channels less than a micron in diameter by the pressure of freezing water. The fragmentation process reaches the level of microunits—the stablest elements of the crystal structure (with a strength of 5000 to 8000 kg per square centimeter), which did not break up under the pressure of ice (2115 kg per square centimeter) (Shumskiy, 1955). Such is the mechanism governing the formation of the initial coarse-silt and fine-sand fractions of loess under the severe climatic conditions of Pleistocene cold epochs in the mountains of Central Asia (Figure 14-5).

The formation of the structural elements of loess (concentric globular aggregates) is caused by cryogenic processes in alluvial regions during transport and deposition. The splitting of a quartz crystal in the course of cryogenic fragmentation results in the formation of a free-radical surface (Kiselev, 1970).

Some of the bonds link up with each other.

```
       O⁻                                          x
       |                                          /
 x—O—Si—O—x        x—Si⁺—x                       O
       |               |                        /
       x               x                    ⁺Si
                                            /  \
                                           O    O
                                          /      \
                                         x        x

    O⁻      O⁻
      \    /
       Si                      Si⁺⁺
      /   \                   /    \
     O     O                 x      x
    /       \
   x         x
```

The free radicals cluster together, forming a kind of active center. On contact with water, the active centers hydrate to form silanol groups.

```
          O⁻                     OH     OH
          |                      |      |
 x—Si⁺—O—Si—x   + H₂O  ——→   x—Si—O—Si—x
    |     |                      |      |
    x     x                      x      x
```

Cryogenic fragmentation involves amorphization of the surface of quartz particles. The amorphized layer thus formed exhibits increased solubility; when it reacts with water, silicic acid is formed. Molecules of this acid react with the OH groups on the quartz surface to form siloxane bonds as in the following diagram:

```
                                      OH
                                      |
                                   O—Si—OH
                                      |
                                      O
          OH + H₄SiO₄                 |
 quartz   OH + H₄SiO₄   ——→  quartz  O—Si—OH + 5H₂O
          OH + H₄SiO₄                 |
                                      O
                                      |
                                   O—Si—OH
                                      |
                                      OH
```

In glacial and periglacial regions, the solubility of carbon dioxide in natural waters at low temperatures is appreciable, and the weak carbonic acid dissolved the carbonate rocks, thereby becoming saturated with calcium ions and HCO_3^-. The calcium ions became adsorbed on the hydrated surface of quartz and feldspar particles and formed surface calcium hydrosilicates. The carbonic acid displaced the weaker silicic acid from the surface calcium hydrosilicates and formed calcite ($K_1 = 4 \times 10^{-10}$; $K_2 = 5 \times 10^{-17}$).

The patchy distribution of active centers on the surface of a quartz particle accounts for the perforated honeycomb character of the carbonate envelope on the surface of quartz particles in loess.

In the accumulation regions, natural waters containing amorphous SiO_2, $Fe(OH)_3$, Ca^{++}, HCO_3^-, and polymineral clay material in suspension form coagels of mutual precipitation (Popov and Kudryavtseva, 1940). In this connection, the formation of a complex polymineral "jacket" on the surface of silt-sized loess particles takes place as follows. Positively charged micelles of $Fe(OH)_3$ sol are adsorbed on the surface of clay particles. It is well known that the adsorption of $Fe(OH)_3$ on clay minerals is limited. Maximum adsorption is observed when the concentration of $Fe(OH)_3$ sol is 23.12 mg per gram. This can account for the specific content of Fe_2O_3 (4% to 5%) in loess, as well as for the dependence of its content on the amount of clayey substance.

Micelles of $Fe(OH)_3$ sol in turn are capable of adsorbing negatively charged particles of silica gel and organic matter (Tsekhomskiy, 1960). A significant role in the formation of coagels of mutual precipitation is also played by calcium bicarbonates present in natural waters. During the transport of quartz and feldspar particles and their deposition in regions where the natural waters are warmed or frozen, a drop in carbon dioxide partial pressure occurs, causing one molecule of carbon dioxide to escape into the atmosphere. Calcium bicarbonate is then converted into finely dispersed calcite.

The coagel particles are attracted by long-range molecular forces (Deryagin, 1967; Zimon, 1976) to the surface of primary quartz and feldspar particles; this forms on the latter a kind of polymineralic "jacket." The substance of the "jacket" interacts with the amorphized layer of a quartz particle through the "holes" in the calcite envelope.

Table 14-1. Results of Modeling of the Formation of Coarse-Dust (Loess) Fraction under Various Conditions

Experiment Number	Experimental Conditions		Mineral	% Content of Fractions, Particle Diameter in mm					
				0.25-0.1	0.1-0.05	0.05-0.01	0.01-0.005	0.005-0.001	Less Than 0.001
1	Before the experiment		Quartz	100	—	—	—	—	—
			Microcline	100	—	—	—	—	—
			Calcite	100	—	—	—	—	—
			Biotite	100	—	—	—	—	—
2	Freezing (−10°C) and thawing (+15°C to +20°C)	In water-saturated state	Quartz	11	20	68	1	—	—
			Microcline	7	44	48	1	—	—
			Calcite	6	15	20	30	29	—
			Biotite	98	1	1	—	—	—
3	Freezing (−10°C) and thawing (+50°C)	In air-dried state	Quartz	98.5	1	0.5	—	—	—
			Microcline	98	1.5	0.5	—	—	—
			Calcite	93.5	5	1	0.5	—	—
			Biotite	100	—	—	—	—	—
4	Wetting and drying (under laboratory conditions: +18°C to +20°C)		Quartz	100	—	—	—	—	—
			Microcline	100	—	—	—	—	—
			Calcite	100	—	—	—	—	—
			Biotite	100	—	—	—	—	—

The aging of colloids in the course of dehydration during freezing, drying, and sublimation and on subsequent transformation of the silt into a loess causes strong bonding of finely dispersed "jackets" to the surface of primary particles.

Model Studies

The physiochemical characteristics of weathering in glacial and periglacial loess regions were modeled by selecting igneous, sedimentary, and metamorphic rocks and primary minerals of Precambrian to Cenozoic age in the source regions for Central Asian loess (the ranges of Tien-Shan, Pamir, Kopet-Dag). An attempt was made to determine the role of carbonic acid in weathering at normal atmospheric pressure and low temperature (+0.5°C to +1.0°C).

The weathering processes were modeled for mountainous regions, where the conditions are as similar as possible to periglacial and glacial conditions, with a 550-day constant filtration of low-temperature water saturated with 800 to 900 mg per liter carbonic acid at pH 3.8 to 4.0. The experiments were carried out with Soxhlet-type devices (Pedro, 1971). A total of 2800 liters of low-temperature carbon dioxide-saturated water was filtered through each sample at a rate of 5 liters per day.

Results show that 3% to 15% of the mass of most rocks and minerals is lost by leaching, although quartz undergoes practically no chemical change (Table 14-2). Hydromicas, mixed-layered minerals, minerals of the kaolinite and chlorite group, calcite, aragonite, dolomite, siderite, and amorphous silica were formed in the weathering filters. Rinds of weathering products consisting of 80% to 90% oxides of iron (hydrohematite, hydrogoethite, limonite) and aluminum (böhmite, hydrargillite) were formed on rock and mineral fragments. The experiments explain the formation of the polymineralic composition of loess in the specific periglacial paleogeographic situation of Pleistocene cold epochs in the mountain structures and depositional plains of Central Asia.

Subsidence is the basic geologic engineering characteristic of Late Pleistocene loess that was formed under the influence of cryogenic processes. The high-porosity and subsidence properties of loess were studied by modeling under natural and laboratory conditions for different structural-tectonic regions with different supplies of heat and moisture (Central Asia, western Siberia, northern Kazakstan, Steppe Altay, Minusinskiy intermontane trough), and the results were applied to the loesses now developed in the thawed zone of Central Asia.

The Eolian Hypothesis

A study of the natural process of formation of recent loess from the time of deposition of eolian silt to its transformation into a subsidence loess in the hot, dry climate of Central Asia was conducted in the city of Ashkhabad and its environs, where as a result of heavy dust storms a considerable accumulation of silt (10 to 30 tons per hectare) was noted. Natural moistening occurred under different conditions, as follows.

Direct wetting of dust with rain occurred in December 1975. Immediately after sedimentation the silt porosity was 51% and the moisture content about 2%. Subsequently, the dust sediment was uniformly drenched with rain; this resulted in a moisture content of 30%, with the intial porosity unchanged (air temperatures were +10°C

Table 14-2. Chemical-Mineral Composition of Weathering Products Extracted by Low-Temperature CO_2-Containing Water from Rocks and Minerals in a 550-Day Filtration

			Minerals and Oxides in Weathering Filters (%)							
Experiment Number	Rock Mineral	Extracted Products (%)	Amorphous SiO_2	Mixed-Layered Minerals, 2:1	Hydromica	Chlorite + Kaolinite	Calcite with an Admixture of Aragonite	Calcite	Dolomite	Siderite
1	Granite	2.5	39	20	16	–	–	25	Trace	Trace
2	Gneiss	2.8	36	20	15	Trace	–	29	–	–
3	Basalt	8.3	38	20	20	Trace	22	–	Trace	–
4	Gabbro-diabase	8.5	35	29	10	Trace	20	5	–	Trace
5	Mica schist	7.9	5	25	45	10	–	–	–	10
6	Amphibolite	8.2	33	29	15	Trace	23	–	Trace	Trace
7	Porphyrite	8.0	36	19	20	–	20	Trace	2	Trace
8	Syenite	7.4	35	25	25	Trace	–	10	Trace	–
9	Aleurolite	2.3	22	18	30	5	–	20	–	5
10	Quartz-feldspar sandstone	2.5	38	22	15	–	–	25	Trace	Trace
11	Argillite	2.5	3	42	50	5	–	Trace	Trace	Trace
12	Limestone	14.4	–	5	5	–	5	80	5	Trace
13	Anorthite	8.5	35	15	25	Trace	Trace	25	–	–
14	Biotite	8.0	5	35	40	5	–	–	–	15
15	Quartz	0.003	100	–	–	–	–	–	–	–

to +14°C during the day, +2°C to +4°C at night). The deposits then dried to a moisture content of 5%; this resulted in shrinkage to a porosity of 35%. The dried deposits were transformed into a strong, tough, monolithic takyr-type rock. With renewed rains the moisture content rose to 29% and the porosity to 37%. After light frosts (the air temperature during a 12-hour period ranged from −0.5°C to −5.2°C), the material froze and swelled to 50% porosity at a moisture content (ice content) of 27% (the temperature at the lower boundary of the deposits reached −0.4°C). With warmer weather (+12°C to +14°C in the daytime, +2°C to +6°C at night), the expanded high-porosity deposits dried, the moisture content declined to 5%, and the porosity remained fairly high (50%). After freezing, the dried deposits acquired the features of a typical loess: a pale-yellow color, macroporosity, homogeneous structure, absence of laminations, and visible micelles of carbonates.

The subsidence properties of eolian deposits were studied in two states. Wetted silt that is dried to dense takyr did not exhibit subsidence properties even under a load of 20 kg per square centimeter, probably because of the compacted state (porosity 35%) caused by drying, uncharacteristic of loess in general. When wetted, frozen, and dried, silty takyr deposits changed into a typical loess in their morphological characteristics and subsidence character (Figure 14-6) (Minervin, 1979).

Meteorologic observations in Central Asia show that over 100 years powerful dust cyclones with a significant accumulation of loess occurred 12 to 14 times, and in only one case (at the end of December 1975) did the light frost in southern Turkmenia coincide with the moistened state of the windblown loess. In all other cases, the subzero winter temperatures acted either on dust or on dried takyrlike silts and did not form loess from them. Therefore, the above example of conversion of takyr into a subsidence rock is not characteristic and cannot be used as the basis for asserting that typical loess was formed in a hot, arid climate of Central Asia. At the same time, this single case is the key to understanding the formation of thick subsidence loess in Central Asia under conditions of seasonal and multiyear cryolithogenesis.

In the second test area (in February 1968), a study was made of the redeposition of windblown dust by rainwater from small hills to the foot of the slopes in the Kopet-Dag Plain. The study can be used as an example of the eolian-deluvial hypothesis. When dust dried at above-zero temperatures without cryogenic treatment, the deposits were converted into nonsubsidence takyr at very high pressures (15 to 20 kg per square centimeter). Observations of the porosity and moisture content of the deposits in this area were conducted for 4 years. In winter, cryogenesis acted on the dry sediment, which contained only strongly bound water that did not freeze at subzero temperatures. During the 4-year period, the moistening and freezing never coincided. The processes of swelling and high-porosity formation did not occur, and the takyr rocks retained their nonsubsidence character.

The Proluvial Hypothesis

Formation of recent proluvial deposits during intensive mud flows into a shallow lake on the Kopet-Dag Plain, 14 km east of the city of Ashkhabad, resulted from erosion of loess in the Pervomayskiy Ravine. As the lake dried, a high-strength, low-porosity, polygonally fissured takyr

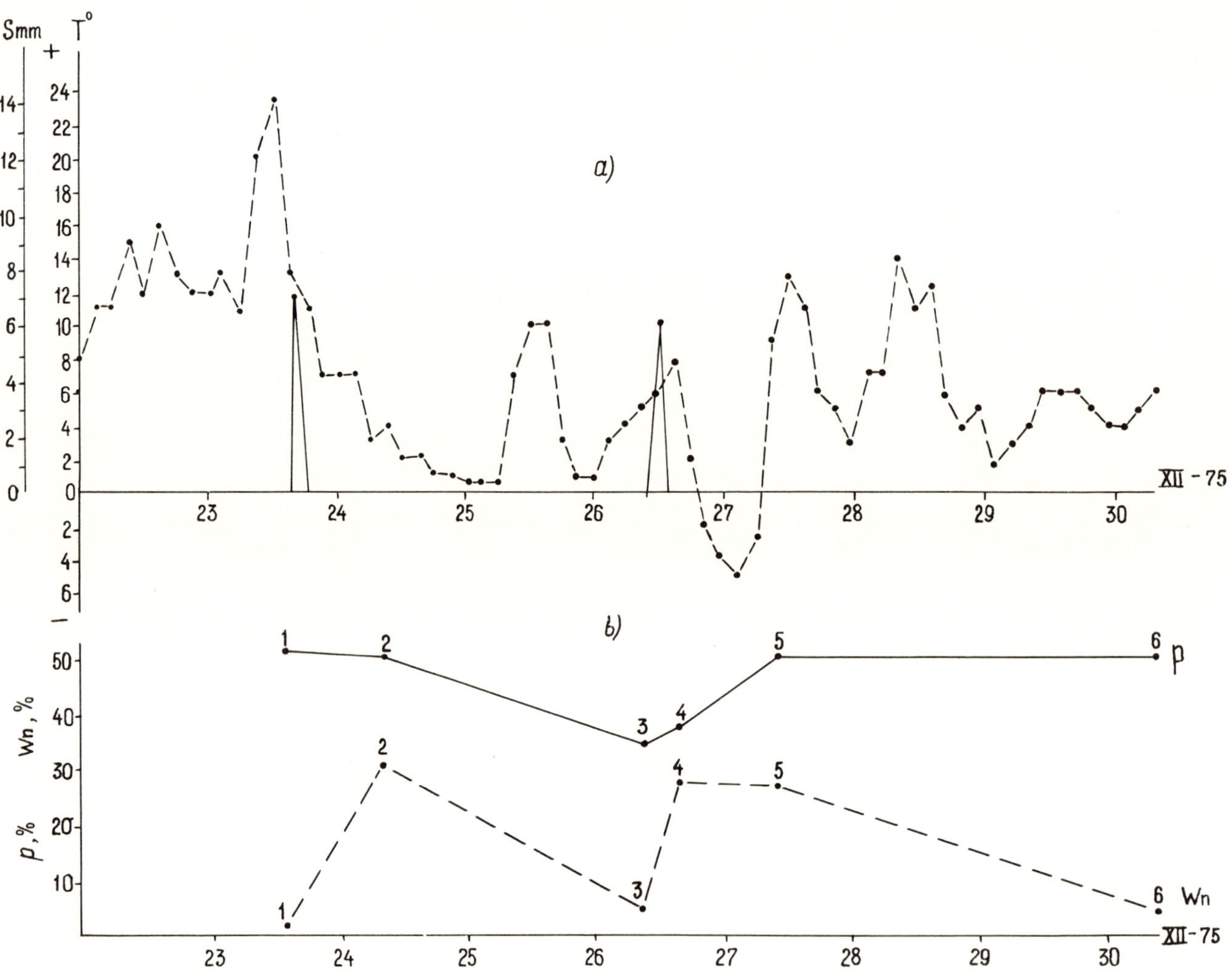

Figure 14-6. Transformation of loose dust into subsidence loess: (a) air temperature (T,C°) and precipitation (S, mm) during the period from 23 to 30 December 1975 (according to the data of the Ashkhabad weather station); (b) dynamics of formation of porosity (p) and natural moisture content (W_n in %) of recent eolian formations during their transformation into subsidence loess: (1) time of dust deposition; (2) wetting with rain; (3) drying and shrinkage (transformation into a dense, nonsubsidence aleurite); (4) new wetting of aleurite by rain; (5) freezing, swelling (loosening); (6) drying after freezing (conversion into subsidence loess).

rock was formed without any subsidence character, although its petrographic characteristics corresponded to a typical loess. The behavior of the porosity and moisture content of such a takyr formation was regularly studied over the course of 3 years. Even under subzero winter temperatures, the takyr did not assume the characteristic features and subsidence properties of loess, because during the winter months of light frost on the air-dry rock the strongly bound water did not freeze. Experiments carried out under natural conditions suggest that in the present hot, arid climate of Central Asia, where subsidence loess is most common, the subsidence properties of eolian and proluvial sediments that went only through the water stage are not formed and not manifested at very high pressures.

The Soil Science and Engineering Geology Department of Moscow State University modeled the formation of subsidence properties in samples of recent takyrlike silts of proluvial loess. Nonsubsidence takyr with a moisture content of 25% to 27% was subjected to unidirectional freezing (from top to bottom) at −5°C for 12 hours under a closed system. After freezing, the samples were dried unidirectionally under laboratory conditions at +18°C to +20°C for 4 days. After several freezing-thawing-drying cycles, the deposits acquired high porosity, morphological features of loess and distinct subsidence properties (Figure 14-7).

Natural modeling of subsidence formation in proluvial loess was also performed in a natural situation different from that of Central Asia, with deep seasonal freezing of surface rocks (southern West Siberia, northern Kazakhstan, Steppe Altay, Minusinsk depression). The subsidence processes were modeled in an elementary layer 1 m thick and in a thick mass of artificial loess.

In both cases during the autumn, silts were deposited

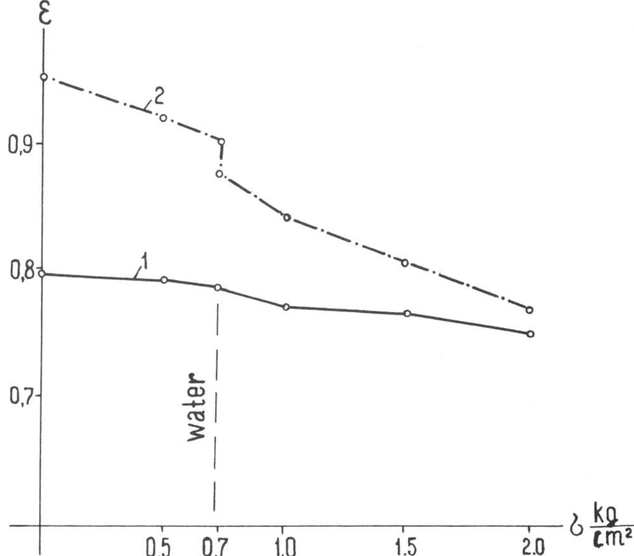

Figure 14-7. Development of a subsidence character in a sample of recent proluvial takyr of loessial composition during wetting-freezing-drying: (1) before wetting-freezing-drying cycle (natural sample); (2) after one cycle of wetting, freezing, and drying.

from natural loess in trenches 10 to 2 m^3 in volume; they froze at the start of winter and swelled to a porosity of 50% to 52% (with an ice content of about 30%). An intensive sublimation of ice from loosened silts took place during a powerful Asian anticyclone in the winter. In April and May, the silts assumed the typical characteristics of natural loess, with distinct subsidence properties (Minervin, 1975).

The Alluvial Hypothesis

Modeling of the subsidence character of alluvial loess was performed on the high floodplain of the Amu Darya River during a 5-month pump test with the groundwater level dropping by 8 m. After drainage and drying in the hot summer, the alluvial floodplain deposits were compacted during shrinkage to a porosity of 35%; subsidence character was absent both at atmospheric pressure and at a pressure of 15 kg per square centimeter.

A second experiment was carried out on the high floodplain of the Ob' River in Novosibirsk Province during a 9-month pump test with a 6-m drop in groundwater level in cryptostratified alluvial floodplain loess. Drainage of water from the alluvium, deep seasonal freezing, and thawing and drying during the hot summer months converted the water-saturated floodplain alluvium into a typical subsidence loess.

The Soil Hypothesis

Berg (1947) and numerous adherents of the soil hypothesis attribute a considerable role in the formation of loess to macropores and branching canals (passages of dead plant roots). Also, numerous investigators consider these macropores to be the cause of the subsidence properties of loess. In order to check this hypothesis, the following experiment was set up in the field. Silt from loess was deposited in trenches in water and dried to a porosity of 35% to 38%. Steppe and semidesert vegetation was sown on the silt surface and left for 2 years. The vegetation was then removed, and the area was left without plants for 2 years. After the trenches were opened, the artificial loess was found to be threaded with macropores from dead roots numbering 15 to 20 per square centimeter) but the total porosity did not increase, and the loess retained its nonsubsidence character.

On penetrating into the loess, roots of steppe and semidesert plants developed a pressure up to 45 kg per square centimeter and compacted only the walls of root passages without increasing the total porosity. As a result of biochemical activity, the roots encrusted the walls of the passages with carbonate salts and substantially increased the structural porosity of the loess. Phytogenic macropores did not close up during subsidence. In field models and in loess of the active layer, the walls were not encrusted with salts in macropores caused by ice; it was precisely such macropores that were responsible for the subsidence character of loess, and they closed up during subsidence. The experiment confirms the results previously obtained by Velichko and Markova (1971). It was also found that weathering crusts (from laterite to kaolinite) do not have subsidence properties. Many years of field studies conducted by the Engineering Geology Department of Moscow State University showed that the subsidence character of loess is formed in zones with the cryoeluvial type of weathering (northern Kazakhstan, southern West Siberia, Steppe Altay, Minusinskiy intermontane trough, and the southwestern Siberian Platform).

In conclusion, the composition, structure, and subsidence character of loess in Central Asia resulted from a set of postsedimentation and hypergenic physiochemical processes under severe Pleistocene periglacial climatic conditions, a decisive role having been played by seasonal and multiyear cryolithogenesis.

References

Berg, L. S. (1947). "Climate and Life." State Publishing House of Geographical Literature, Moscow.

Chepizhnyy, K. I., Sergeyeva, N. E., and Barsanov, G. P. (1973). Symmetry and structure of dissolution forms on the pinacoid of quartz crystals. *A. Ye. Fersman Mineralogical Museum, Vypod* 22, 96-106. Nauka Press, Moscow.

Deryagin, B. V. (1967). Inertia-less deposition of particles from a fluid flow on a sphere under the influence of van der Waals attraction forces. *In* "Studies in the Area of Surface Forces" (D. V. Deryagin, ed.), pp. 69-81. Nauka Press, Moscow.

Fedorovich, B. A. (1962). Frost formations in the steppes and deserts of Eurasia. *In* "Problems of Stratigraphy and Paleogeography of the Quaternary (Anthropogene)" (V. I. Gromov, ed.). *Commission on the Study of the Quaternary, Trudy* 19. USSR Academy of Sciences, Moscow.

Fotiyev, S. M. (1978). "Hydrogeothermal Characteristics of the Cryogenic Region of the USSR." Nauka Press, Moscow.

Gedroyts, K. K. (1955). "Selected Works." Vol. 1. State Publishing House of Agricultural Literature, Moscow.

Kiselev, V. F. (1970). "Surface Phenomena in Semiconductors and Dielectrics." Nauka Press, Moscow.

Kostenko, N. P. (1962). Loessial rocks in the mountains of southern Central Asia. *In* "The Latest Stage of Geological Development of Tadzhikistan's Territory" (V. I. Popov, ed.), pp. 119-41. Donish Press, Dushanbe.

Minervin, A. V. (1975). Role of hypergenesis processes in the formation of the subsidence character of loessial rocks in southern Siberia: Genetic principles of geological engineering study of rocks. "Proceedings of International Conference" (E. M. Sergeev, ed.), pp. 305-14. Moscow State University, Moscow.

Minervin, A. V. (1979). Formation of the subsidence properties of loesses from windblown dust under present conditions of central Asia. *Inzhenernaya Geologiya* 3, 78-85.

Minervin, A. V. (1980). Modeling of the conditions of formation of coarse dust particles of loessial rocks. *Inzhenernaya Geologiya* 1, 51-60.

Nesmeyanov, S. A. (1977). "Correlation of Continental Rock Masses." Nedra Press, Moscow.

Pedro, Zh. (1971). "Experimental Studies of Geochemical Weathering of Crystalline Rocks." Mir Press, Moscow.

Popov, A. I. (1968). The cryogenic factor in the formation of loessial and loesslike rocks. *In* "Recent Exogenic Processes" (V. G. Bodnarchuk, ed.), Part 2, pp. 67-79. Naukova Dumka, Kiev.

Popov, A. I. (1979). Cryolithogenesis and its place in the lithogenesis system. *Problemy Kriolitologii* 8, 7-26.

Popov, A. I., and Kudryavtseva, M. M. (1940). Study of coagels of mutual precipitation. *Pochvovedeniye* 8, 54-67.

Shantser, Ye. V. (1966). Outline of a study of genetic types of continental sedimentary formations. *USSR Academy of Sciences, Institute of Geology, Trudy* 161.

Shumskiy, P. A. (1955). "Principles of Structural Glaciology." USSR Academy of Sciences, Moscow.

Strakhov, N. M. (1960). "Principles of Lithogenesis Theory." Vol. 1. Academy of Sciences of the USSR, Moscow.

Tsekhomskiy, A. M. (1960). Structure and composition of the film on quartz sand grains. *In* "The Weathering Crust" (I. I. Ginzburg, ed.), Issue 3, pp. 293-312. USSR Academy of Sciences, Moscow.

Turbin, L. I., and Aleksandrova, N. V. (1970). Formation of the loessial rocks of Tien-Shan. "Proceedings of International Symposium on the Lithology and Genesis of Loessial Rocks" (G. A. Mavlyakov, ed.), Vol. 1, pp. 345-54. FAN Press, Tashkent.

Velichko, A. A. (1973). "The Natural Process in the Pleistocene." Nauka Press, Moscow.

Velichko, A. A., and Markova, A. K. (1971). Two main forms of large pores in loesses. *USSR Academy of Sciences, Doklady* 197, 899-902.

Zimon, A. D. (1976). "Adhesion of Dust and Powder." Khimiya Press, Moscow.

CHAPTER 15

Periglacial Landscapes and Loess Accumulation in the Late Pleistocene Arctic and Subarctic

S. V. Tomirdiaro

In connection with the extensive Late Pleistocene oceanic ice cover in the more humid regions (Velichko, 1973), a pronounced cryoxerophytization of landscapes took place even in western Eurasia. At that time, the eastern part of the Arctic and Sub-Arctic (so-called Beringia) was located in the interior of a huge continent, for the thickly frozen Arctic Ocean, unopened for millennia, formed a true "climatic dry land" in this area. This "dry land," together with the adjacent continents of North America and Eurasia, produced in the climatic sense a single "supercontinent" with a huge arctic-subarctic landscape hyperzone (a term coined by Velichko, 1973). Here, as within the deep interior of the supercontinent, a special cryoarid zone was formed, along with a permanent arctic global anticyclone that no longer exists (Tomirdiaro, 1980). This resulted in an almost constant cloudless sky and consequently a sharp increase of solar insolation and summer temperatures, along with an even greater drop of winter temperatures (Kaplina and Kuznetsova, 1975). This in turn led to the recently proposed concept of arctic steppe or even arctic prairie (Tomirdiaro, 1972) on the Bering dry land. In 1979, this problem was dealt with at a special international symposium at Burg Wartenstein (Hopkins et al., 1982). Arctic frost-steppe landscapes were local variants of the huge periglacial frost steppes of the entire transcontinental superzone of that time.

In this entire superzone, the main geologic processes were the formation of huge covers of loess and the development of a stable permafrost with polygonal ice veins (Velichko, 1973). Analogous processes should also have prevailed within Beringia. Indeed, there are preserved peculiar frozen loess masses, called edoma complex in Yakutiya and Goldstream Formation in Alaska (Péwé, 1975). ("Edoma" is a local Russian word that apparently means earth "eaten up" by lakes. It refers to high residual masses of Pleistocene loess plains, which the population watched being "consumed" by a very active process of Holocene lacustrine thermokarst formation [Tomirdiaro, 1978].) Deposits of an edoma complex in northeastern USSR were long regarded mainly as a variety of floodplain alluvium (Popov, 1967). Shilo (1971) criticized this concept, and the present author in northern Yakutiya and Chukchi Peninsula established an eolian genesis for the edoma deposits and attributed them and other frozen loess strata of Pleistocene Eurasia to a single time of formation of loess and ice in the past (Tomirdiaro, 1980). These deposits have now begun to be distinguished as a Siberian type of loess (Trush and Kondrat'yeva, 1980). On arctic maritime lowlands of Yakutiya and on islands of the Novosibirsk Archipelago north of latitude 72°N, the material differed so sharply from ordinary loess edomas that for a long time it had been mistaken for huge buried glaciers (Toll', 1897). The chief constituent is ice preserved to the present time only in the major islands of the Novosibirsk Archipelago, and on the coast of northern Yakutiya approximately between the Indigirka River and the mouth of the Anabar River. In the south, this region is bounded on the plains by approximate latitude 72°N, and near mountain structures by the foothills.

Until recently, it was assumed that deposits of all the edomas studied could be unified into a single edoma suite; the edoma outcrop of Duvanny Yar on the Kolyma River, composed of loess with a very low ice content and fine ice veins, was designated as the stratotype for the entire "edoma suite" (Sher, 1971). In the meantime, the author's studies distinguished a separate regional arctic or glacial edoma, that is, the shelf type (Figure 15-1).

The arctic type of frozen-loess deposits (shelf-type edomas or arctic-type loess) is characterized by the following features: (1) The total average content of Pleistocene subterranean ice amounts to 85% to 93% of the volume of the entire mass of deposits to a depth of 30 to 35 m or more. (2) The width of extremely hypertrophied syngenetic ice veins averages 8 to 9 m, and the mineral mass of frozen-loess columns, or so-called "earth veins," is only 2 to 3 m in diameter. (3) The soil mantle covering this ice-filled

Figure 15-1. Formation of arctic or ice edomas of northern Yakutiya (shore of Dmitriya Lapteva Straits). (Photograph by the author.)

stratum is only 0.5 to 0.6 m; it thaws every year, and the entire ice stratum correspondingly sinks continuously and irreversibly (is thermally leveled), for it is not in thermal equilibrium with the present climate even in the interior of the Arctic. (4) As a result of these subsidences, there is formed over the entire surface in this area a special microrelief (not known anywhere else) in the form of a continuous network of grass-covered, rounded hillocks of loessial blocks protruding above the ice. (Such hillocks must not be confused with the well-known "baidzharakhi," which also are remnants of loessial blocks. They develop only on disrupted slopes and not on horizontal surfaces of usually loessial edomas without an artificial disturbance of the thick soil layer present there.) This is observed even on such northern islands of the Novosibirsk Archipelago as Kotel'nyy and Novaya Sibir'. (5) On the whole, this type of deposit is rapidly disappearing from the Arctic.

The best-known outcrops on the continental coast are the Khaptashinskiy edoma on the coast of the East Siberian Sea, three outcrops at Oyagosskiy Yar (Tomirdiaro, 1980), and Cape Bykovskiy on the coast of the Laptev Sea (Are, 1980). Also known are large ice outcrops of edomas of this type on the islands of Bol'shoy Lyakhovskiy, Stolbovyy, Novaya Sibir', and Kotel'nyy.

Photographic documents dating back to the beginning of the 20th century indicate that the Vasil'yevskiy and Semenovskiy Islands in the Laptev Sea, which thawed in the middle of the 20th century, also were a glacial edoma that conformed to the characteristics indicated above (Tomirdiaro, 1980). The Diomida and Merkuriya Islands, which were located near the Bol'shoy Lyakhovskiy Island and rapidly disappeared, probably were also of the same type. Radiocarbon dating of organic matter from an edoma of the Oyagosskiy Yar gave either a Karginskiy age or an infinite date. A sample from a layer at a height of 18 m above sea level was dated at more than 41,000 yr B.P., one at 26 m was dated at 37,700 ± 200 yr B.P. (MAG-543), and one at 30 m was dated at 34,200 ± 200 yr B.P. (MAG-544).

The latest datings are doubtful, however, because of the extensive development of ice veins at this level, which in the Karginskiy layers in other edomas are characterized by a sharp diminution and frequently even by interruptions along the vertical. In addition, the pollen diagram of glacial edoma in the Oyagosskiy Yar is characterized by a tundra-steppe spectrum typical of the cold epochs of the Pleistocene (Tomirdiaro, 1980) but not by a forest or a forest-tundra typical of a Karginskiy spectrum (Figure 15-2). Probably only the Zyryanka base of the glacial edoma has been preserved here up to the present time. And its upper Sartan layers, with an equally high ice con-

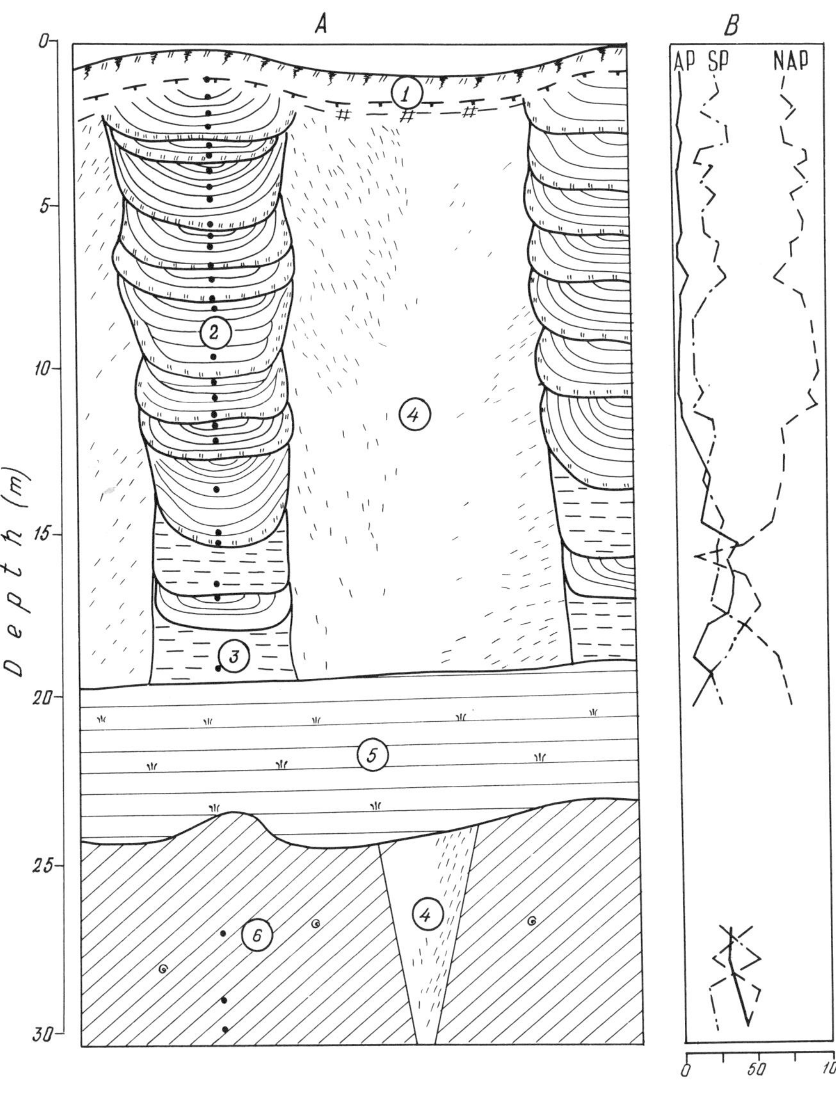

Figure 15-2. Cryolithologic structure and pollen diagram of shelf- (Arctic) type edoma complex. Stratotype is the Oyagosskiy Yar exposure.
(A)
1. soil cover layer
2. bands of nonconformable ice and ice-earth intercalations in earth columns, delineated by an interlayer of heavily peat-laden loam
3. bands of very fine, conformable, interrupted ice layers
4. ice in ice veins
5. talus
6. marine sediments.
(B)
Overall composition of pollen spectra. (Abbreviations: AP, arboreal pollen; SP, spores; NAP, nonarboreal pollen.)

tent, has apparently already thawed (been thermally leveled) during the past 10,000 years of Holocene time. Thus, all the last remnants of this formation have so far continued to thaw out and to disappear gradually from the surface.

The loess in the earth blocks is a very silty loam; its cryotexture has thick schlieren and is fully stratified, with nonconformable bands of cryogenic laminations. Thin intercalations of peat are found at the base of such bands.

This type of loess-ice stratum formation probably predominated up to the beginning of the Holocene on the Arctic Shelf of the northeastern USSR and is termed the arctic type of loess. In the American and Canadian sectors of the Arctic, such deposits are unknown. All the other edoma deposits of the northeastern USSR and Alaska, which are known to consist of thick strata of frozen loess and comparatively thin syngenetic ice veins, are termed the subarctic type of loess, as described below.

The topmost thick section of subarctic edomas, up to 25 to 30 m thick, is composed of loess classified as very silty sandy loam, with fine and very fine syngenetic ice veins less than 2 to 3 m wide and with continuous incompletely stratified microschlieren cryotexture or even a massive cryotexture. This means that interlayers or schlieren of ice in excess of capillary size are absent in frozen loess. This also means that as sedimentation continued only very fine streaks of loess extended into the permafrost, so that even on the moistened bottom of the active layer thick schlieren of seasonally forming segregation ice were not detached or incorporated in the permafrost. This is the only way that the combination of masses of microschlieren cryotexture and sufficient moisture content of the rock itself can be explained. Essentially, these are huge masses of windblown soil of eastern Siberia's periglacial plains, accumulated at a rate of about 1 mm per year according to absolute datings (Tomirdiaro, 1980). Such a sharply increased accumulation was also characteristic of Late Pleistocene frozen loess of the Russian Plain (Velichko, 1973). Hence, the thickness of ice inclusions is less than 1 mm. The microschlieren and biogenic bedding are usually conformable and are very slightly deformed in areas of contact with ice veins. On the whole, such a stratum should be considered as a special cryopedolite and a relict of the original Pleistocene rock from which the European periglacial cold loess was formed during subsequent diagenesis.

Edoma strata of this type are characterized by very xeric, typically steppe or tundra-steppe pollen spectra (Kaplina, 1979; Kaplina et al., 1978; Tomirdiaro, 1980) and have been dated back to Sartan time on the basis of remains of buried roots and stems of herbaceous vegetation. In the northeast, the radiocarbon datings were as follows: A sample from a 30-m outcrop of the Mus-Khaya edoma on the Yana River at a depth of 2 m from the top was dated at 11,500 ± 210 yr B.P. (MAG-137); a sample from the same location at a depth of 15.5 m was dated at 23,360 ± 720 yr B.P. (MAG-175) (Lozhkin, 1977); a sample from the loess mass of the high terrace of the Lena River near the town of Yakutsk was dated at 19,600 ± 500 yr B.P. (IM-360) (Kostyukevich et al., 1978); samples from an aleuritic edoma mass with thick ice veins at the base of a section of the Aleshkin Terrace of the Kolyma River were dated at 14,980 ± 100 yr B.P. (MAG-470) and 15,000 ± 200 yr B.P. (MAG-468); and a sample from a depth of 3 m from a 33- to 35-m key loess outcrop of Omolono-Anyuy edoma of the Duvanny Yar on the Kolyma River (Sher and Kaplina, 1979) was dated at about 18,000 yr B.P., according to a preliminary report by A. V. Lozhkin.

The lower section of all the investigated edomas of this subarctic type was composed of loess with a much higher ice content and thick schlieren cryotexture and with much wider ice veins, up to 4 to 5 m (Figure 15-3). It was separated from the upper sections by thick layers of buried peat and soils with numerous woody macroremains dating back to the time of the Karginskiy Interstade, so we conclude that it is loess of Zyryanka age. Thus, the subarctic type of frozen loess has a microschlieren cryotexture and fine ice veins in the Sartan (Late Valdai) loess and a thick schlieren cryotexture and wide ice veins in the Zyryanka (Early Valdai) loess. This structure probably reflects a higher moisture content and a slower rate of loess accumulation in the Early Valdai than in the Late Valdai. A slowing down of loess accumulation by another few orders of magnitude characterizes the structure of arctic-type ice edomas. In fact, after the thawing of such edomas up to 30 m high, there remains a layer of mineral substance of only 1.5 to 2 m. The finest suspended matter in the upper layers of the atmosphere probably was deposited here—an atmosphere that during these loessial epochs had a particularly high global dust content. The shelf-type edomas are distinguished by a very low content of sand and an appreciable content of clay (Tomirdiaro, 1980). South of latitude 72°N, all the edomas become not only low in ice content, but their granulometric composition changes sharply; that is, a considerable amount of sand appears, and the content of clay sharply decreases (Tomirdiaro, 1980). It is evident that the front of the great Arctic anticyclone passed somewhere at this latitude. A shelf-type edoma stratum covered the entire shelf exposed at that time, that is, all of Arctic Beringia. However, there was nothing in principle to interfere with its formation there or even farther to the north over the Arctic Ocean, which was then deeply frozen. The ice of this water area was also covered with a light sediment of eolian dust produced by drift over long distances. It was sufficient for this layer to reach a thickness of 0.6 to 1 m in order to enable it to be "swelled" by ice to a 20-m layer of shelf-type edoma. In any event, the author believes that a layer of fine dust existed on the ice in the Arctic Ocean at that time. A steppe-and-meadow vegetation thrived on this layer as well as on the ice edomas. In principle, the Arctic mammoth loessial steppe could have covered not only the shelves but also the deep Arctic Ocean of that time. This huge territory is referred to as the loess-ice Arctic. The epochs of accumulation of edoma loess are clearly correlated with the global epochs of loess accumulation. They accumulated in Zyryanka time (mainly the thick schlieren strata) and in Sartan time (microschlieren strata). Accumulation of soils and peat proceeded during Kazantsevo and Karginskiy time, as well as during the Holocene.

Figure 15-3. Formation of subarctic or the northeastern frozen loess edomas in USSR. (Photograph by the author.)

References

Are, F. E. (1980). "Thermoabrasion of Seashores." Nauka Press, Moscow.

Hopkins, D. M., Matthews, J. V., Jr., Schweger, C. E., and Young, S. B. (1982). "Paleoecology of Beringia." Academic Press, New York.

Kaplina, T. N. (1979). Spore-pollen spectra of ice-complex sediments of Yakutiya's Maritime Lowlands. *USSR Academy of Sciences, seriya geograficheskaya* 2, 85-93.

Kaplina, T. N., Giterman, R. Ye., Lakhtina, O. V., Abramov, B. A., Kiselev, S. V., and Sher, A. V. (1978). Dubannyy Yar: A key section of upper Pleistocene deposits of the Kolyma Lowland. *Bulletin of the Commission on the Study of the Quaternary* 48, 170-74.

Kaplina, T. N., and Kuznetsova, I. L. (1975). Geotemperature and climatic model of the epoch of accumulation of the edoma suite of Yakutiya's Maritime Lowlands. *In* "Problems of Paleogeography of Loessial and Periglacial Regions" (A. A. Velichko, ed.), pp. 170-74. Nauka Press, Moscow.

Kostyukevich, V. V., Ivanov, I. Ye., and Nesterenko, S. V. (1978). Radiocarbon dates of the Geochemistry Laboratory of the Geocryology Institute, Siberian Branch, Academy of Sciences of the USSR. *Bulletin of the Commission on the Study of the Quaternary* 48, 213-20.

Lozhkin, A. V. (1977). Radiocarbon datings of Upper Pleistocene deposits of Novosibirsk Islands and the age of the edoma suite of northeastern USSR. *USSR Academy of Sciences, Doklady* 235, 435-37.

Péwé, T. L. (1975). "Quaternary Stratigraphic Nomenclature in Unglaciated Central Alaska." U.S. Geological Survey Professional Paper 862.

Popov, A. I. (1967). "Permafrost Phenomena in the Earth's Crust Cryolithology." Moscow State University Press, Moscow.

Sher, A. V. (1971). "Mammals and Pleistocene Stratigraphy of the Far Northeast of the USSR." Nauka Press, Moscow.

Sher, A. V., and Kaplina, T. N. (eds.) (1979). "Excursion Guide XI, Stage XIV, Pacific Ocean Scientific Congress (1979)." VINITI Press, Moscow.

Shilo, N. A. (1971). Periglacial lithogenesis in the general scheme of the process of continental rock formation. *USSR Academy of Sciences, Severo-Vostochnogo Kompleksnogo Nauchnoissledovatel' skogo Instituta DVNTs, Trudy* 38, 3-56.

Toll', E. V. (1897). Fossil glaciers of Novosibirsk Islands and their relation to mammoth carcasses and the ice age. *Zapiski Russkogo Geograficheskogo Obshchestva* 32, 150-52.

Tomirdiaro, S. V. (1972). "Permafrost of Highlands and Lowlands." Book Publishers, Magadan.

Tomirdiaro, S. V. (1978). "Natural Processes and Development of Territories of the Permafrost Zone." Nedra Press, Moscow.

Tomirdiaro, S. V. (1980). "The Loess-Ice Formation of Eastern Siberia in the Late Pleistocene and Holocene." Nauka Press, Moscow.

Trush, N. I, and Kondrat'yeva, K. A. (1980). Ice complex of the permafrost region Siberian type loess. *In* "Collected Works of the All-Union Conference on Problems of Loessial Rocks in Seismic Regions." FAN Press, Tashkent.

Velichko, A. A. (1973). "The Natural Process in the Pleistocene." Nauka Press, Moscow.

CHAPTER 16

Age and History of Accumulation of the "Ice Complex" of the Maritime Lowlands of Yakutiya

T. N. Kaplina and A. V. Lozhkin

Widely distributed over the Maritime Lowlands of Yakutiya on the upper part of interfluves are deposits called the edoma suite, which are 10 to 60 m thick and consist of ice-filled aleurites and include thick syngenetic polygonal ice veins. Because of their unusually high ice content, these deposits are frequently referred to as an "ice complex" (IC). A. I. Popov's hypothesis of an alluvial genesis of this stratum, advanced during the 1950s and developed later on, is contrasted with S. V. Tomirdiaros' (1980) hypothesis of an eolian origin. Age estimates range from Middle or Upper Pleistocene to Zyryanka (Sher, 1971) or Zyryanka-Sartan (Tomirdiaro, 1980). In the last few years, a large number of additional radiocarbon dates for various sections of the Maritime Lowlands (Figure 16-1) permit a systematic evaluation of available data to determine the age of the IC.

The IC almost everywhere overlies strata of completely different type: lacustrine and paludal sediments with an abundance of wood, including a full-grown birch (which does not grow in this area at the present time), and with pollen spectra reflecting a more temperate vegetation than the present one. In many sections, IC-type deposits are divided into two strata separated by lacustrine-paludal sediments with an abundance of wood (Figure 16-2, sections 10, 15, 23) (Sher and Kaplina, 1979; Kaplina et al., 1980). Thus, "edoma suite" and "ice complex" are not synonyms, and it is necessary to distinguish the "lower" IC from the "upper" IC, which is an edoma suite, and to date them separately.

More than 90 radiocarbon dates now available (Figure 16-3) can be divided into three groups (Figure 16-3, I-III): (I) dates obtained from deposits underlying the IC; (II) dates obtained directly from the IC, and (III) dates obtained from deposits overlying the IC.

(I) Dates from deposits underlying the IC cover a considerable time range (Figure 16-3, I). They include a series of infinite dates but also a series of finite dates in the range of 45,000 to 34,000 yr B.P. and a smaller group in the range of 30,000 to 24,000 yr B.P. As already stated, these deposits contain tree pollen and wood. Many dates were obtained from wood, including birch wood.

(II) Dates from the glacial complex (Figure 16-3, II) are numerous, but very few of them are in vertical stratigraphic sections (sections 5, 6, 23). Some of the dates exceeding 40,000 to 45,000 yr B.P. are either from the "lower" IC, from lower layers of the continuous stratum of IC, or from the lowest strata of the "upper" IC. They indicate that the IC preserved today began to accumulate at least as early as the Zyryanka cooling.

The finite dates from 41,000 to 11,500 yr B.P. range mostly from 41,000 to 28,000 yr B.P. The latter group comes from the lower layers of a continuous IC or (in sections with two bands) from the "lower" IC or from the lower part of the "upper" IC. Thus, an IC of considerable thickness is younger than 28,000 yr B.P.

In the 24,000 to 11,500 yr B.P. interval, the number of dates is small: sections of Mus-Khaya (Figures 16-1 and 16-2, section 5), Khaptashinskiy Yar (10), Shandrin River (16), and Duvanny Yar (23). The dates are for the middle and upper layers of the IC; the material used consisted of grass stems and thin bands of peat. The pollen spectra show a predominance of grass pollen or spores; this implies open landscapes for that time.

(III) The dates from sediments overlying the IC (Figure 16-3, III) are somewhat different in geologic position. They include not only those from sediments overlying the "upper" IC, which are grouped in the 11,500 to 8000 yr B.P. interval, but also dates from sediments overlying the "lower" IC; this accounts for the duplication of certain dates in groups I and III. In addition, the group of dates from two very interesting sections on the Mali Anyuy River—Molotkovskiy Kamen' (24) and Stanchikovskiy Yar (25)—are also included here. They reveal sections of alassy (thermokarst depressions), which make it possible to reconstruct the history of IC accumulation.

Three older dates in the Stanchikovskiy Yar were ob-

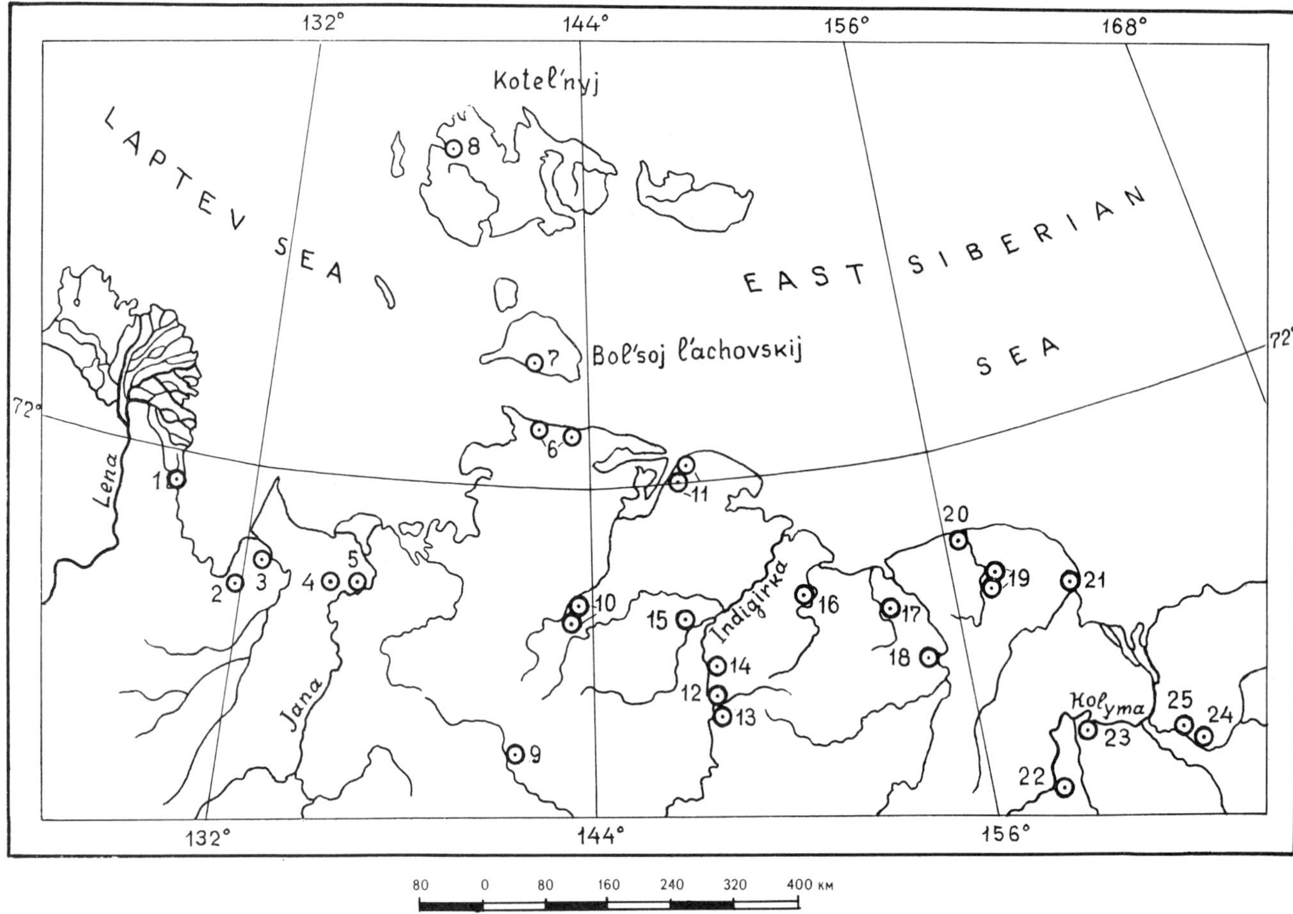

Figure 16-1. Location of radiocarbon-dated sections within the ice complex in the Maritime Lowlands of Yakutiya.

tained from a peat bog covering an IC fragment exposed in the base of the alass and from taberal formations (thawed and compacted IC) overlying this fragment. Taberal formations also overlie the peat bog. Dates from three peat bogs separated by taberal formations were obtained in Molotkovskiy Kamen' (Figure 16-2, section 24). An IC fragment emerges under the lower layer of the taberal formations. The latter contain pollen spectra with a predominance of grass pollen, whereas the peat bogs contain spectra dominated by tree and shrub pollen.

On the whole, the group III dates confirm the conclusion reached above on the basis of group II dates, that is, that IC accumulation began before 45,000 to 50,000 yr B.P. Further, these dates clearly show a break in the accumulation of the IC and its thawing in the interval between 47,000 and 34,000 yr B.P., a resumption of IC accumulation between 34,000 and 30,000 yr B.P., another break from 30,000 to 24,000 yr B.P., and further accumulation between 24,000 and 12,000 yr B.P.

The series of most numerous dates in group III covers the interval of 11,500 to 8000 yr B.P. All these dates pertain to layers located at the surface of edoma masses and outliers or overlying taberal and lacustrine deposits in the alasses. They clearly record the completion of IC accumulation.

Thus, the accumulation of the IC on Yakutiya's Maritime Lowlands began before 50,000 yr B.P., that is, earlier than the Karginskiy warming. The lower layers of the IC probably formed during the Zyryanka cooling. IC layers may date back to the Middle Pleistocene (section 15; Kaplina et al., 1980; section 6: Kolesnikov, 1980).

Radiocarbon dates combined with paleobotanic data make it possible to state that the period from 50,000 to 34,000 yr B.P. in the Maritime Lowlands was characterized by an appreciable warming, for the vegetation zones were shifted northward. For the Maritime Lowlands, the optimum of this warming was first established and dated by Lozhkin (1976) in the valley of the Khomus-Yuryakh River (section 17), so that the name of this river can be proposed for this warming. The latter corresponds to the Igarka warming of the Yenisey region of Siberia and the Port Talbot Interstade of North America (Zubakov, 1974).

At the start of the Khomus-Yuryakh warming, the previously accumulated IC thawed, and pseudomorphs of ice veins and thermokarst depressions (alassy) were formed. During Khomus-Yuryakh time, these were then filled

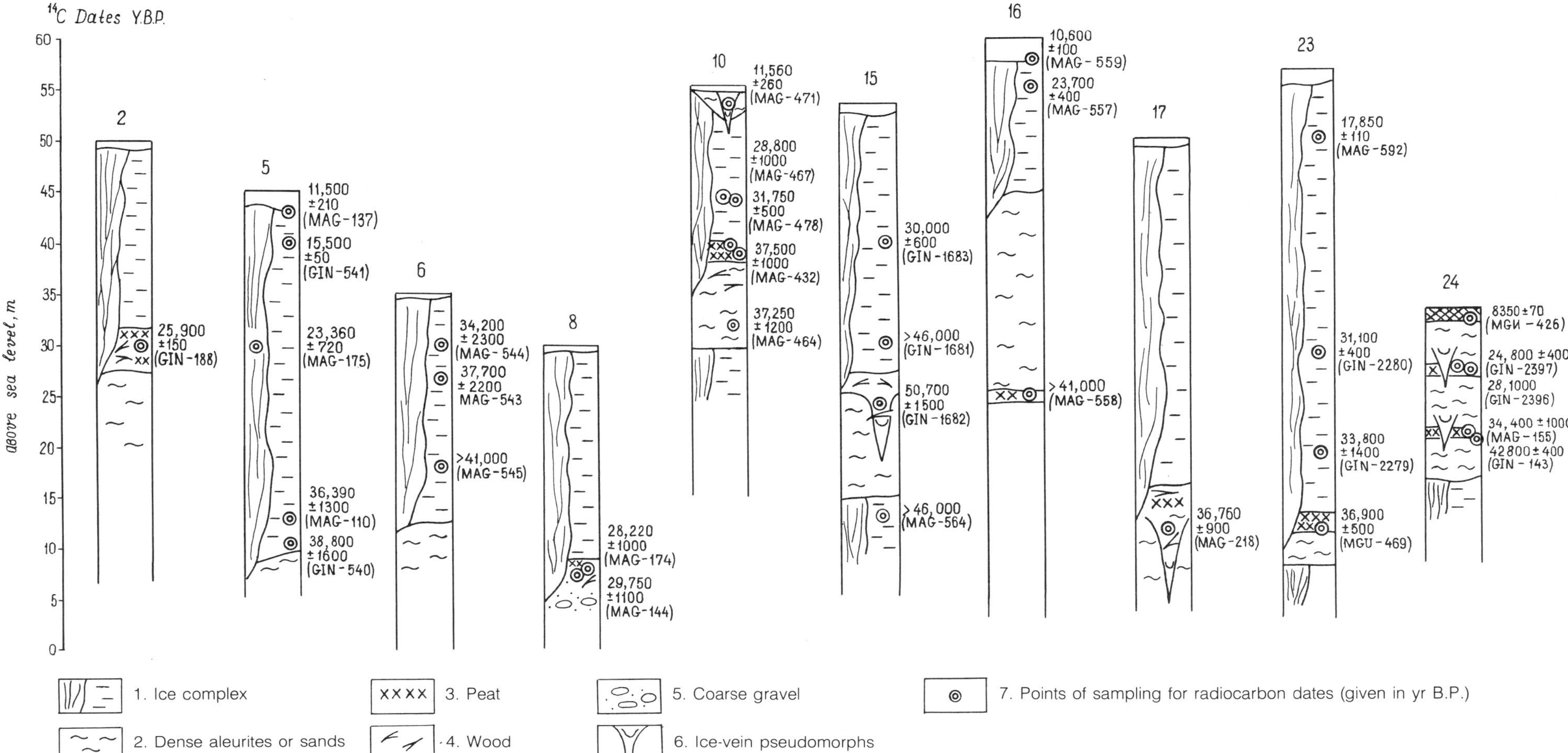

Figure 16-2. Typical sections revealing the ice complex. Numbers of sections (at the top) correspond to their numbers on the map (Figure 16-1).

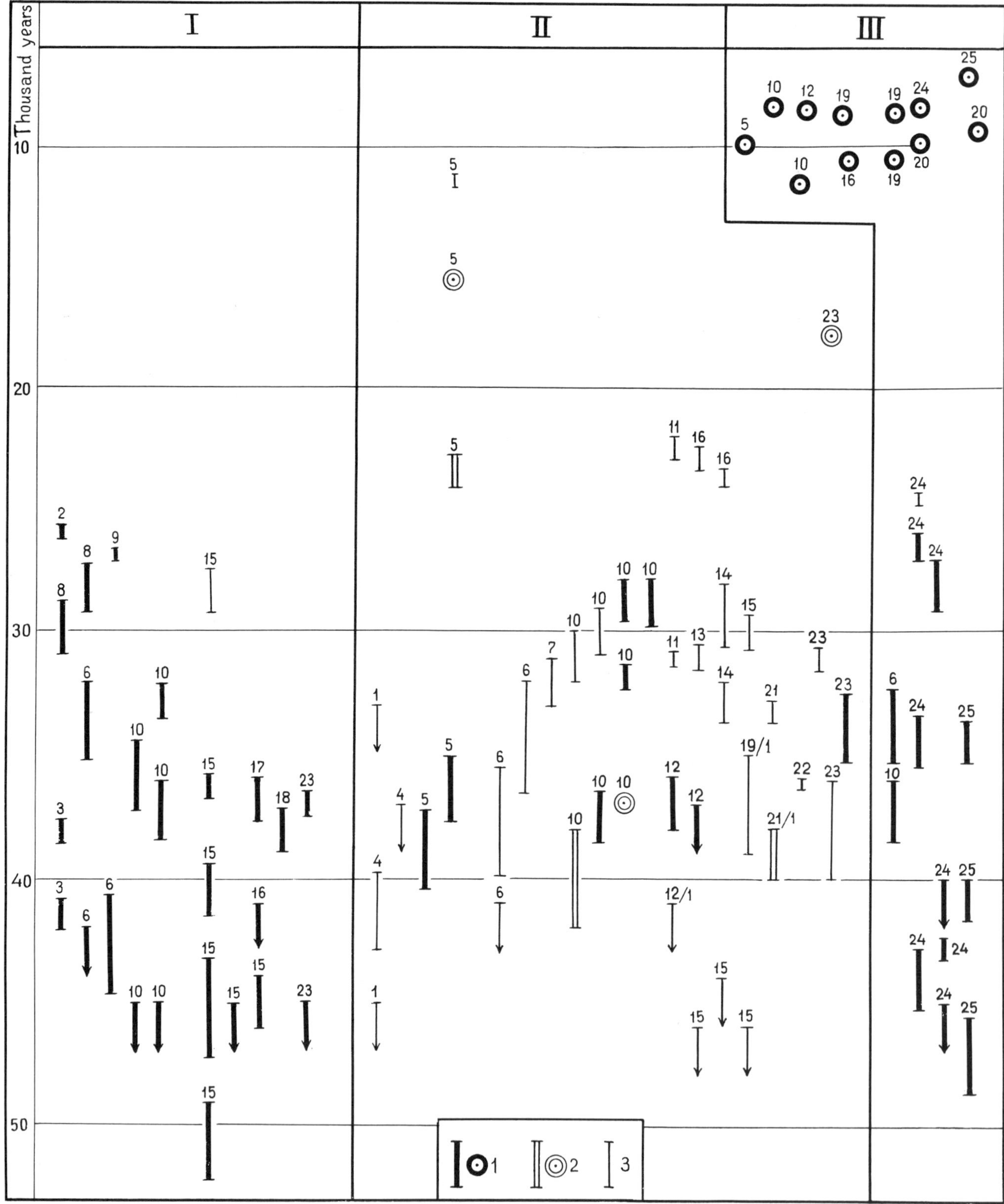

Figure 16-3. Radiocarbon dates giving the age of the ice complex. The numbers of the dates correspond to the numbers of the sections from which they were obtained (Figure 16-1). (1) Dates accompanied by pollen spectra with a predominance of tree and shrub pollen, (2) dates accompanied by pollen spectra with a predominance of grass pollen or spores, (3) dates for which no pollen spectra are available. (I) Dates from sediments underlying the ice complex, (II) dates from the ice complex, (III) dates from sediments overlying the ice complex.

with lacustrine, paludal, and locally alluvial sediments containing peat and wood, from which numerous dates of group I and parts of group III were obtained. The preservation of IC fragments under the lacustrine sediments indicates that the thawing of frozen ground was not continuous and did not result in their complete degradation. The mean annual temperature of the ground rose only to the range of subzero values.

At the end of Khomus-Yuryakh time, IC accumulation resumed, as indicated by the dates of group II, obtained mainly from the lower layers of the "upper" IC, which are also characterized by tree pollen. The dates of 34,000 to 30,000 yr B.P. come not only from the IC but also from the dated taberal layers in sections 24 and 25.

Renewed warming from 30,000 to 24,000 yr B.P. caused further thermokarst activities, as confirmed by the dates of groups I and III. Sediments of that time include lacustrine and paludal deposits containing wood and extending to the northern margins of the lowlands (sections 6 and 8). Palynologic data also indicate an increased forest cover in the region.

Deposits of this interval were first described and radiocarbon dated for the lowlands by Timashev (1972) in the Kuranakh-Sala River valley (section 2), so the local name of Kuranakh-Sala warming is proposed, correlating with the Bryansk Interstade in the European USSR (Velichko, 1973), the Novonazimovskiy warm interval in the Yenisey region, and the Plum Point Interstade of North America (Zubakov, 1974). Sediments of the Kuranakh-Sala warming in certain sections (10, 14) include buried soils, which may be dated in the future.

The geologic location of radiocarbon dates in the 24,000 to 11,500 yr B.P. time interval, combined with the dates from group I, indicated very intensive accumulation of IC, resulting in strata up to 20 m thick. According to palynologic data, this period was very cold and arid.

Deposits of that time interval were studied in most detail palynologically and radiocarbon-dated in the Mus-Khaya section on the Yana (section 5), where they comprise only a part of the IC section and are located at a height of 16 to 18 m above the water (Kondrat'yeva et al., 1976; Lozhkin, 1977; Kolpakov, 1979). The local name of Mus-Khaya can be proposed for the cold interval corresponding to the Sartan cold interval and for the formation of loess II and III in the European USSR (Velichko, 1973). In the sections of the Khroma (10), Allaikha (15), and Shandrin (16) Rivers, the Mus-Khaya sediments are over 25 to 30 m high; and, in the sections of the Duvanny Yar on the Kolyma River (23), they are more than 18 m above the river. But in several sections (11, 14) they are located at only 2 to 5 m above sea level.

The dates of group III indicate that the accumulation of IC ended around 12,000 yr B.P., that is, in a period corresponding to the Alleröd. From that time on, there began an active fragmentation of the edoma surface, composed of the IC and thermokarst, and it was followed by the filling of thermokarst depressions by lacustrine and paludal deposits of the Holocene (Kaplina and Lozhkin, 1979).

Such is the history of accumulation of the IC, as inferred from radiocarbon dates. Obviously, it is yet to be refined. Of particular importance in this respect will be layer-by-layer dating of the thickest key sections. The dating of soil horizons buried in the IC, which thus far have not been given sufficient attention, looks very promising.

References

Kaplina, T. N., and Lozhkin, A. V. (1979). The age of alass deposits of Yakutiya's Maritime Lowlands (based on radiocarbon data). *USSR Academy of Sciences, seriya geologicheskaya* 2.

Kaplina, T. N., Sher, A. V., Giterman, R. E., Zazhigin, V. S., Kiselev, S. V., Lozhkin, A. V., and Nikitin, V. P. (1980). Key section of Pleistocene deposits on the Allaikha River (lower reaches of the Indigirka). *Bulletin of the Commission on the Study of the Quaternary* 50.

Kolesnikov, S. F. (1980). Cryogenic structure of the Kuchchuguy Suite of the Yana-Indigirka Lowland. Scientific and Technical Collection of Abstracts of the USSR State Committee for Construction, series 15, "Engineering Research in Construction, Geocryological Studies," No. 1. Moscow.

Kolpakov, V. V. (1979). Glacial and periglacial topography of the Verkhoyansk glacial region and new radiocarbon datings. *In* "Regional Geomorphology of Newly Developed Areas" (A. I. Muzis, ed.), pp. 83-98. Nedra Press, Moscow.

Kondrat'yeva, K. A., Trush, N. I., Chizhova, N. I, and Rybakova, N. O. (1976). Characteristics of Pleistocene deposits in the Mus-Khaya Outcrop on the Yana River. *In* "Cryological Research" (V. A. Kudryavtsev, ed.), No. 15, pp. 60-93. Moscow State University Press, Moscow.

Lozhkin, A. V. (1976). Vegetation of western Beringia in the Late Pleistocene and Holocene. *In* "Beringia in the Cenozoic" (V. L. Kontrimavichus, ed.), pp. 72-77. USSR Academy of Sciences, Far Eastern Scientific Center. Vladivostok.

Lozhkin, A. V. (1977). Radiocarbon datings of Upper Pleistocene deposits of Novosibirsk Islands and age of the edoma suite of the northeastern USSR. *USSR Academy of Sciences, Doklady* 235, 435-37.

Sher, A. V. (1971). "Mammals and Stratigraphy of the Pleistocene of the Far East of the USSR and North America." Nauka Press, Moscow.

Sher, A. V., and Kaplina, T. N. (eds.) (1979). "Guidebook of Scientific Excursion on the Problem 'Late Cenozoic Deposits of the Kolyma Lowland'." Fourteenth Pacific Ocean Scientific Congress, Stage 11. All-Union Institute of Scientific

Timashev, I. Ye. (1972). Stratigraphy of the Pleistocene of the western margin of the Yana-Indigirka Lowland. *Isvestiya Vysshykh Uchebnykh Zavedeniy, Geologia y Razvedka* 10, 21-25.

Tomirdiaro, S. V. (1980). "The Loess-Ice Formation of Eastern Siberia in the Late Pleistocene and Holocene." Nauka Press, Moscow.

Velichko, A. A. (1973). "The Natural Processes in the Pleistocene." Nauka Press, Moscow.

Zubakov, A. V. (ed.) (1974). "Geochronology of the USSR," Vol. 3. Nedra Press, Leningrad.

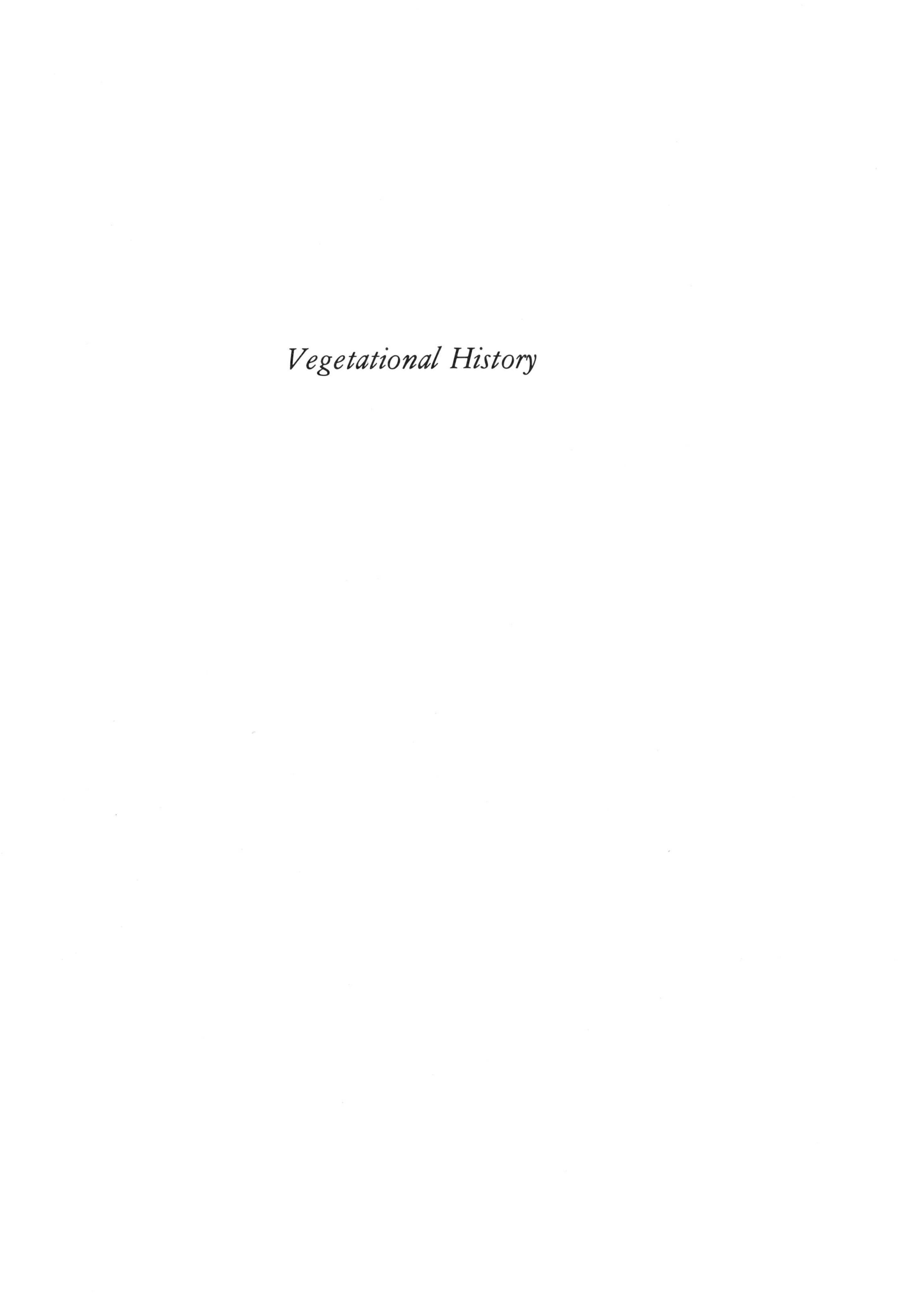

Vegetational History

CHAPTER 17

Late Pleistocene Vegetation History

V. P. Grichuk

A comparatively thorough study of the Late Pleistocene vegetational history of the USSR has been made possible by the more comprehensive paleobotanic information provided by over 900 papers dealing with pollen analysis published from 1962 through 1977. Detailed palynologic studies of a great many sections are available for all the principal regions of the country, except for parts of Siberia and the deserts of Central Asia. The extent of Late Pleistocene studies varies for different parts of the USSR, but palynologic data are available for large regions of the country. Data on recent (subfossil) pollen spectra from most parts of the USSR ensure the reliability of phytocenotypic interpretations. Few studies discuss the vegetational history over a large area, but mention should be made of monographs by Giterman and others (1968), Saks (1970), and Makhnach (1971).

A systematic examination of Late Pleistocene vegetational history is a difficult task. The Late Pleistocene is subdivided into four formal units, which are presented in Table 17-1.

The extent of Early Valdai (Zyryanka) Glaciation and the paleogeographic rank of the Middle Valdai (Karginskiy) interval, as either interglaciation or interstade, is discussed among Quaternary scientists, but the amount of paleobotanic data from these horizons is comparatively limited. Therefore, this survey will examine only the initial and final stages of the Late Pleistocene: the Mikulino Interglaciation (Eemian of western Europe) and the Late Valdai Glaciation (Late Würm of western Europe), the maximum phase of which occurred between 20,000 and 18,000 years ago.

Sources of Paleobotanic Data

In floristic and vegetational reconstructions, the palynologic and plant-macrofossil analyses and the determinations of plant tissues and leaf imprints are of greatest importance. The most versatile tool is pollen analysis, which makes it possible to obtain relevant data at any stratigraphic level from practically any deposit. The separation method of preparing pollen samples (Grichuk, 1937) permits study not only of peat and organic lacustrine sediments but also clay, loam, sand, and coarse gravels. The cavitation method (Grichuk et al., 1967) is more effective for loess and buried soils.

Negative results in pollen analysis (if carefully conducted) can be informative, because a consistent absence of a taxon in the spectrum is direct evidence of its real absence or very insignificant occurrence in a region during a particular time interval. Paleobotanic data provide only selective information on past vegetation. Krasilov (1972) noted that, in addition to loss of information as a result of fossilization, considerable loss derives from the impossibility of identifying all the vegetal remains.

A second and extremely important source of information necessary in reconstructing past vegetation is provided by florocenogenetic analysis of modern vegetation and in particular of relict flora. Before extensive pollen and macrofossil (carpological) analyses, florocenogenetic analysis was the chief source of information on Late Pliocene and Quaternary flora and vegetation (Maleyev 1941, 1948; Kleopov, 1941; Sochava, 1946; Krasheninnikov, 1951). The studies of Krasheninnikov (1951) on the Pleistocene

Table 17-1. Subdivision of the Late Pleistocene

Index	European USSR	Siberia
Q_{III}^4	Late Valdai (glacial)	Sartan (glacial)
Q_{III}^3	Middle Valdai	Karginskiy
Q_{III}^2	Early Valdai (glacial)	Zyryanka (glacial)
Q_{III}^1	Mikulino	Kazantsevo

Note: Units adapted by the Ministry of Geology of the USSR.

floristic complex are confirmed by newly available dated paleobotanic materials. Florocenogenetic analysis provides basic information on past plant cover in certain regions, including the Arctic (Tolmachev and Yurtsev 1970), certain mountainous regions of Siberia, and desert plains of Central Asia and southern Kazakhstan (Korovin, 1961). The results reveal the most significant stages in vegetational history and provide phytocenotic information for extreme conditions.

It is not possible here to discuss the method of florocenogenetic analysis, but we cite some comparatively simple examples using relict species. Figure 17-1 (A-D) shows the distribution of certain species that are common in the broad-leaved and coniferous/broad-leaved forests of Europe and the Far East. These species are also fairly common in isolated locations thousands of kilometers away in the so-called "dark taiga" of the Altay and south of Krasnoyarsk along the Yenisey River.

Figure 17-1A shows the occurrence of *Polystichum braunii* Fée in Europe and eastern Asia. A number of species, including *Actaea spicata* L., *Geranium robertianum* L., and *Asperula odorata* L., have similar distributions. Other species, for example *Circaea lutetiana* L., *Asarum europaeum* L., *Brachypodium silvaticum* (Huds.), and *Bromus beneckeni* (Lge) Trimen, are common only in Europe and in isolated locations in the Altay (Figure 17-1B). The third group of species, such as *Festuca extremiorientalis* Ohwi, *Menispermum dahuricum* D.C., *Viola dactilioides* Roem, and *Osmorhiza aristata* (Thunb.) Mak. and Jabe (Figure 17-1C), has its principal range in eastern Asia and in isolated locations in the southern Baikal region, along the Yenisey River, and in the Altay.

The isolated locations of these species in the Altay and in other regions of Siberia constitute surviving remnants of a once-continuous belt of broad-leaved forests extending across all of Eurasia. Floristic materials are available that permit the age of this belt to be estimated. A fairly large number of endemic species, including *Tilia sibirica* Bayer (Figure 17-1D), also grow in these same forests in the Altay. *T. sibirica* is morphologically very close to both the European *T. cordata* Mill. and the Far Eastern *T. amurensis* Rupr. and is considered by many systematists to be a subspecies of the former. Other Altay endemic species, such as *Galium krylovii* Iljin, *Brunnera sibirica* Stev., and *Dentaria sibirica* (Schulz) N. Busch, differ little from closely related European species; this fact suggests that the Altay nemoral species became isolated fairly recently and that the broad-leaved forest belt extended across Siberia during the Mikulino (Kazantsevo) Interglaciation. Analysis of the ecologic connections among the mentioned species within their present-day range permits a detailed description of the composition and cenotic features of Mikulino-age formations in the Altay and other regions that feature nemoral species.

As a second example we can cite data on glacial relicts in the flora of the southeastern Middle Russian Upland, graphically referred to by the well-known botanist Kozo-Polyanskiy (1931) as the "land of living fossils." As an example, Figure 17-2 (A-D) shows the occurrence of four species found in relict habitats within the upper forest and alpine belts of the Altay, the Sayan, and other mountains farther east in Siberia. The ecology of these and other relict species within their principal ranges makes it possible to characterize past plant formations. The main formations in this region on solonetz soils during the last glaciation were open pine, larch, and birch forest; grassy and herbaceous meadow-steppe; and grassland. Paleobotanic data from the upper loess horizon in this part of the Middle Russian Upland (which is reliably correlated with the last glaciation) made it possible to date this epoch and confirmed the ecologic and phytocenotic conclusions (Grichuk, 1972).

Special information is provided by data on endemic species. The abundance of paleoendemics, particularly those with a narrow, localized distribution, is a definite indication that the environmental conditions of a region were unchanged over a long period. In cases where paleoendemics of genus rank are found, one can be confident of stable conditions over a very long period (Tolmachev, 1974a). Such regions are found in the southern part of the USSR in the Caucasus (Kolkhida and northeastern Dagestan), Central Asia (the Karatau Range, Kendaktas Mountains, and Kungey-Ala-Too Range), the southern Baikal region, and the southernmost Soviet Far East. This fact must be considered, of course, when interglacial and glacial vegetation is to be reconstructed. In conclusion, a vegetational reconstruction must combine paleobotanic information, which permits reliable geologic correlation and interpretation, with the results of florocentric analysis, which specifies phytocenotic phenomena. A good example of such an approach is the well-known monograph by Vul'f (1941).

The botanical history of the USSR provides support for three basic types of flora (Lavrenko, 1938): (1) *relict floras*, characterized by an abundance of old, relatively unchanging elements; (2) *orthoselection floras*, which have been changing in the same direction for a long time (primarily in the direction of increased aridity or continentality; and (3) *migration floras*, which migrated to a given territory after the preceding flora was eliminated during periods of glacial or periglacial conditions. These types of flora require a differentiated approach to the reconstruction of past vegetation.

Methods and Procedures for Dating Materials

Determining the age of materials used in reconstructing vegetation becomes complex when one deals with the entire Late Pleistocene over the entire USSR. Radiocarbon dating is limited to the last glaciation and the preceding Middle Valdai (Karginskiy) interval. For the entire USSR, only a limited number of dated sections spans the maximum phase of glaciation, 20,000 to 18,000 years ago, and it is difficult to reconstruct the vegetation of the last glaciation.

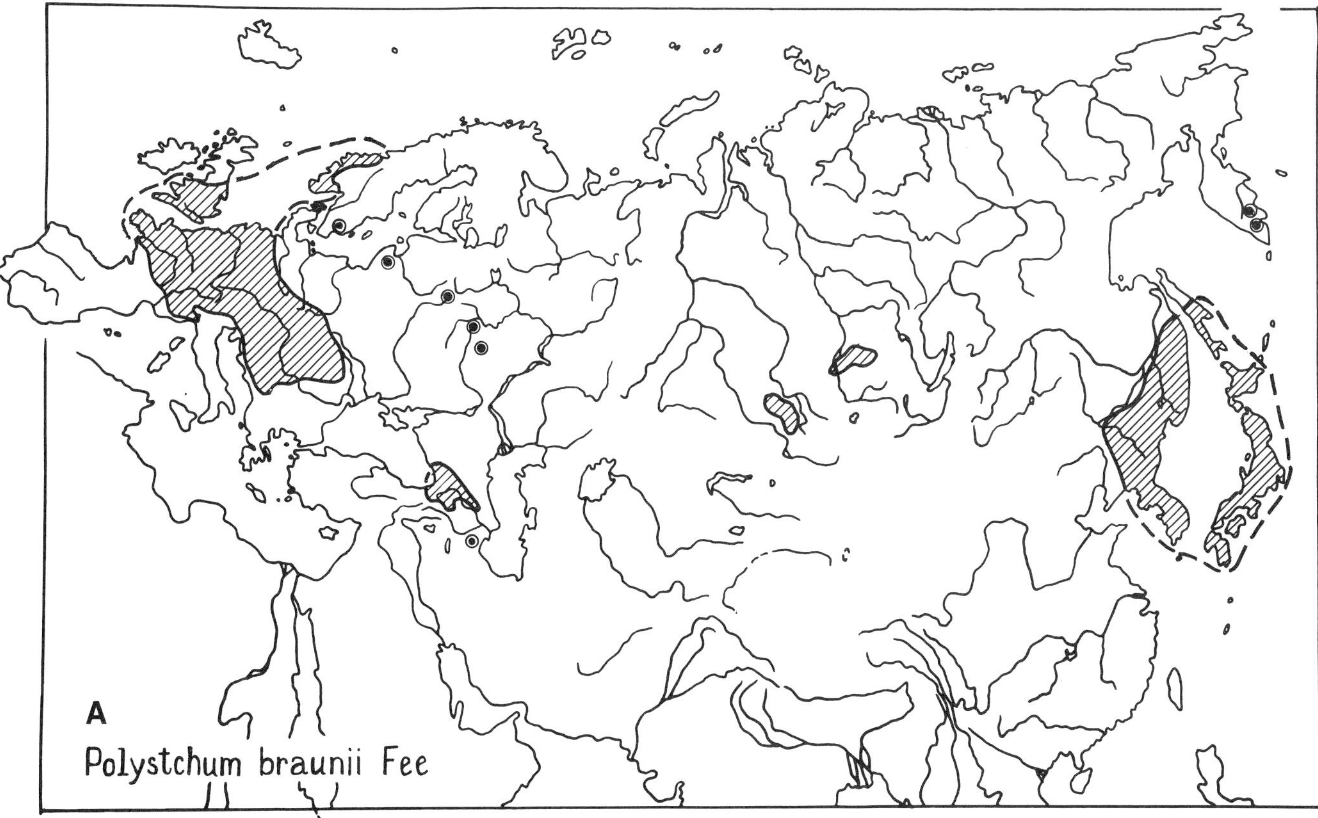

Figure 17-1. Ranges of certain species typical of broad-leaved forests, showing disjunct populations in the Altay and in central Siberia beyond their European and Far Eastern ranges.

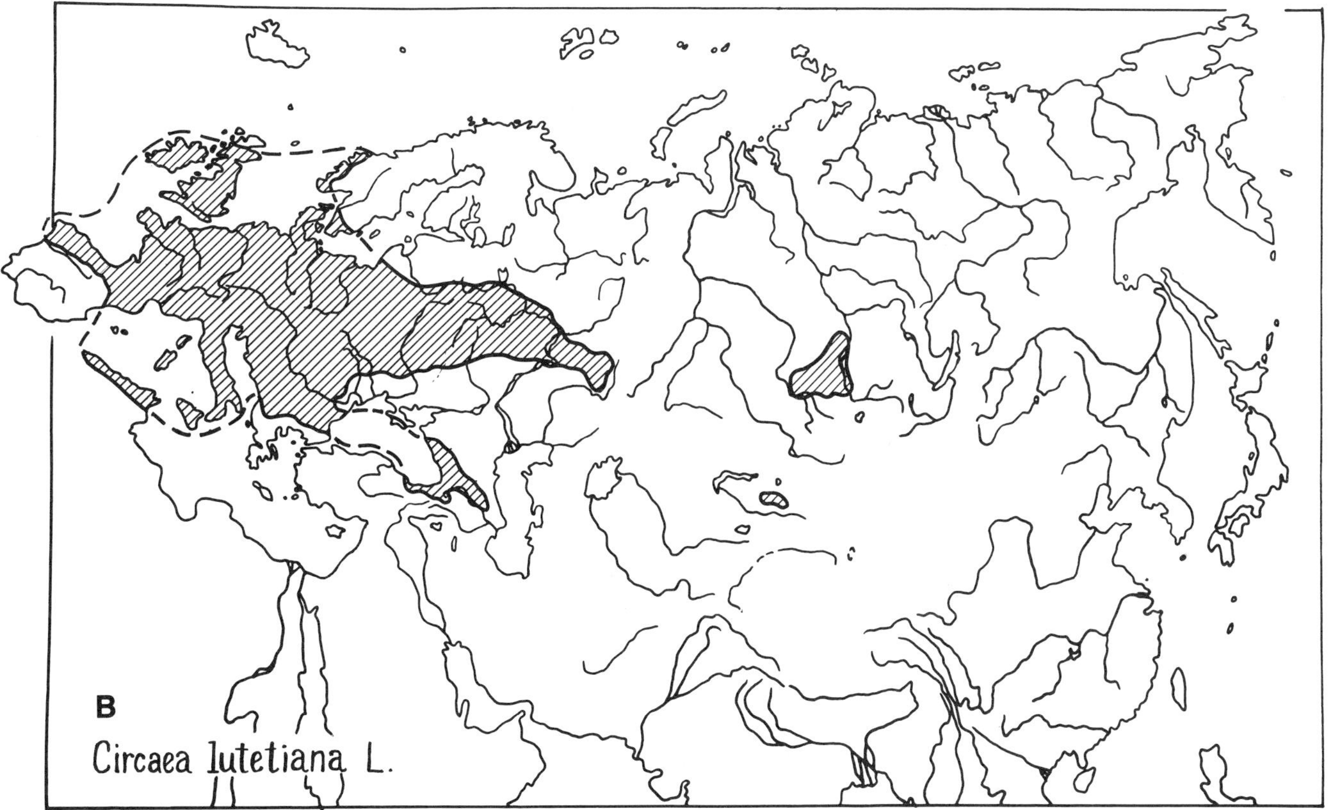

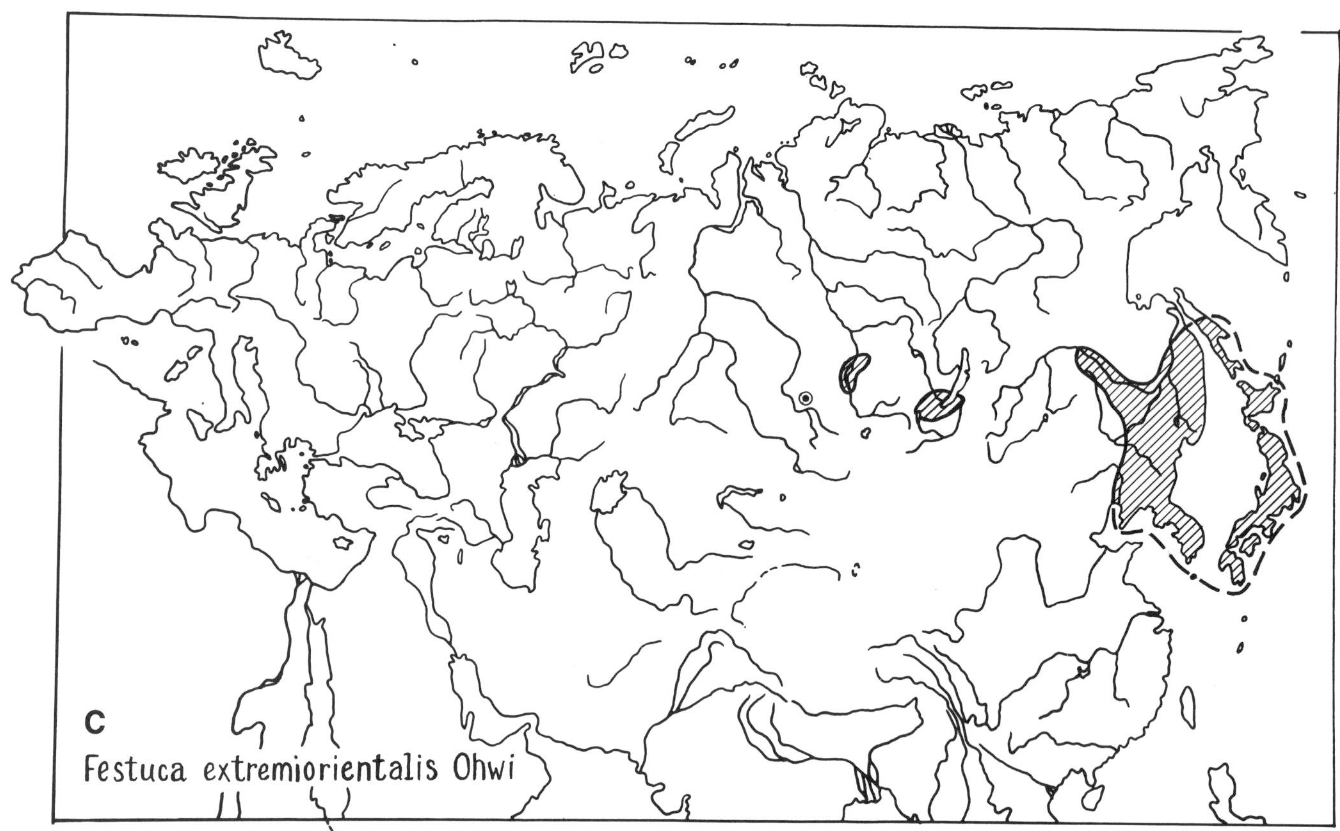
C
Festuca extremiorientalis Ohwi

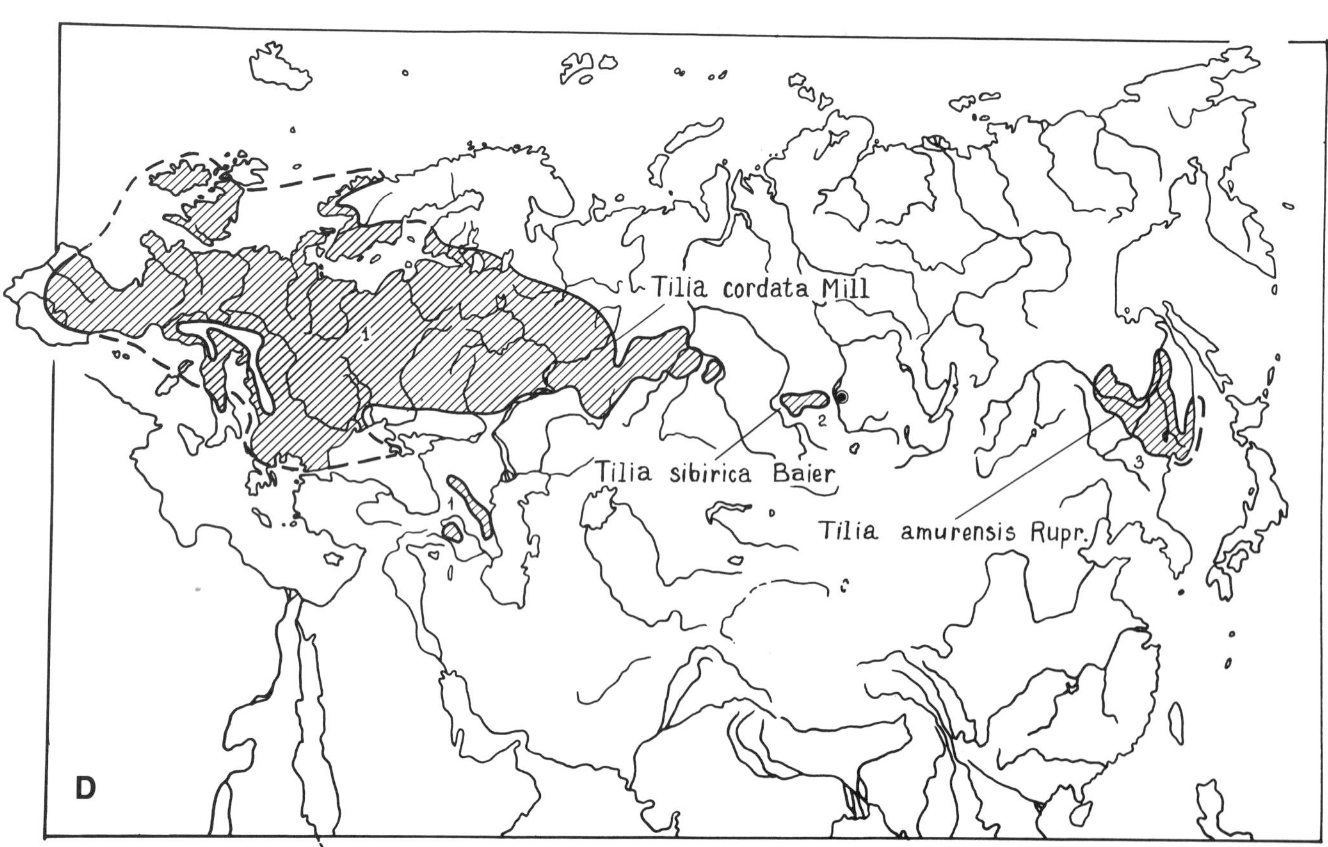
Tilia cordata Mill
1
1
Tilia sibirica Baier
2
Tilia amurensis Rupr.
3
D

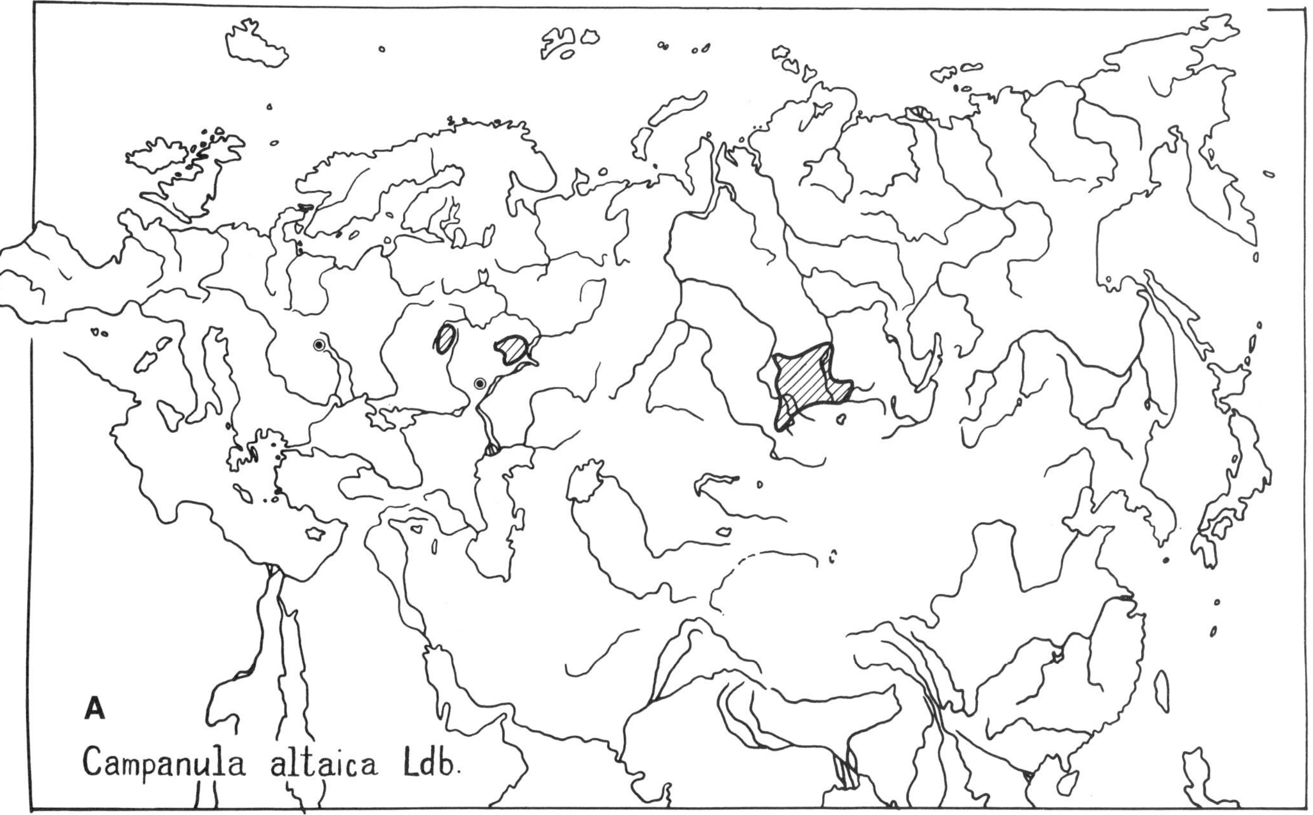

Figure 17-2. Ranges of certain Siberian species found in isolated habitats in the East European Plain (relicts of the ice age).

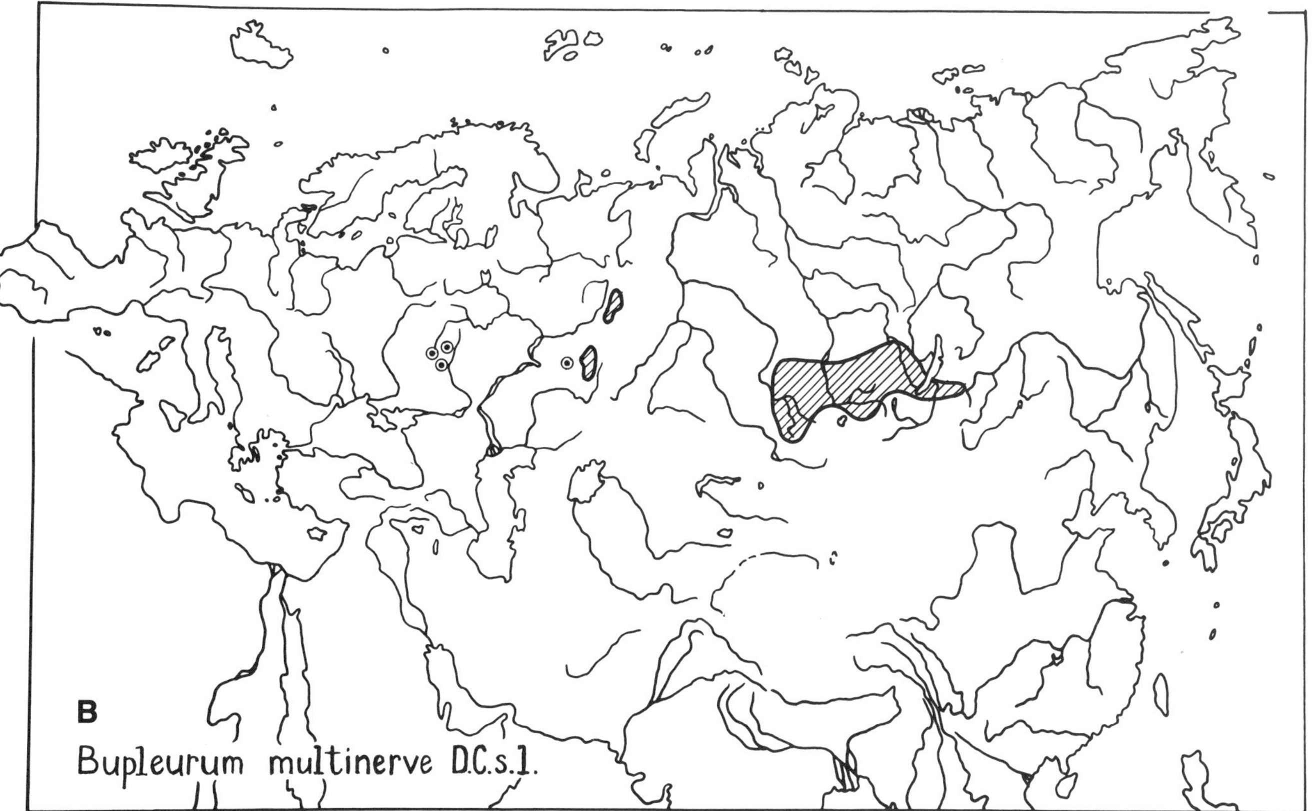

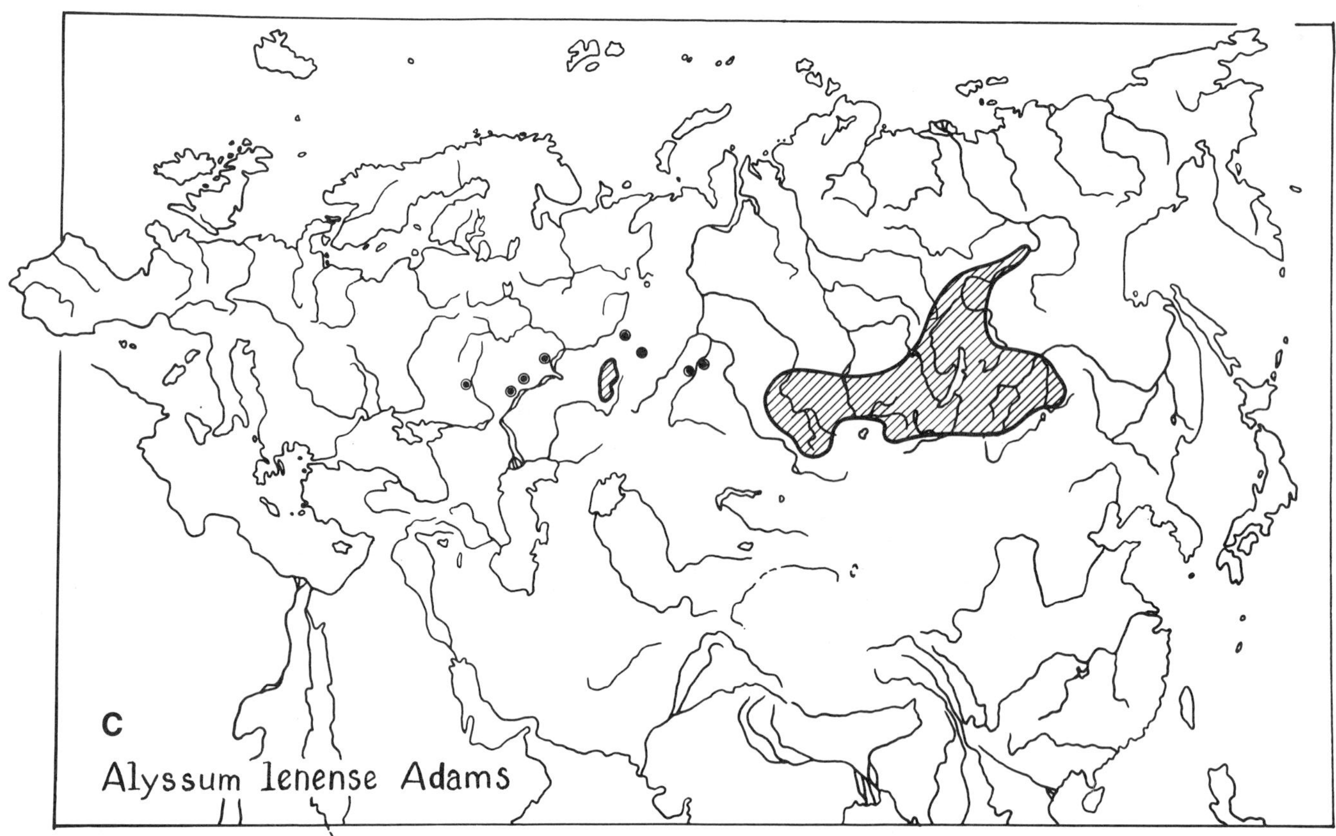
C
Alyssum lenense Adams

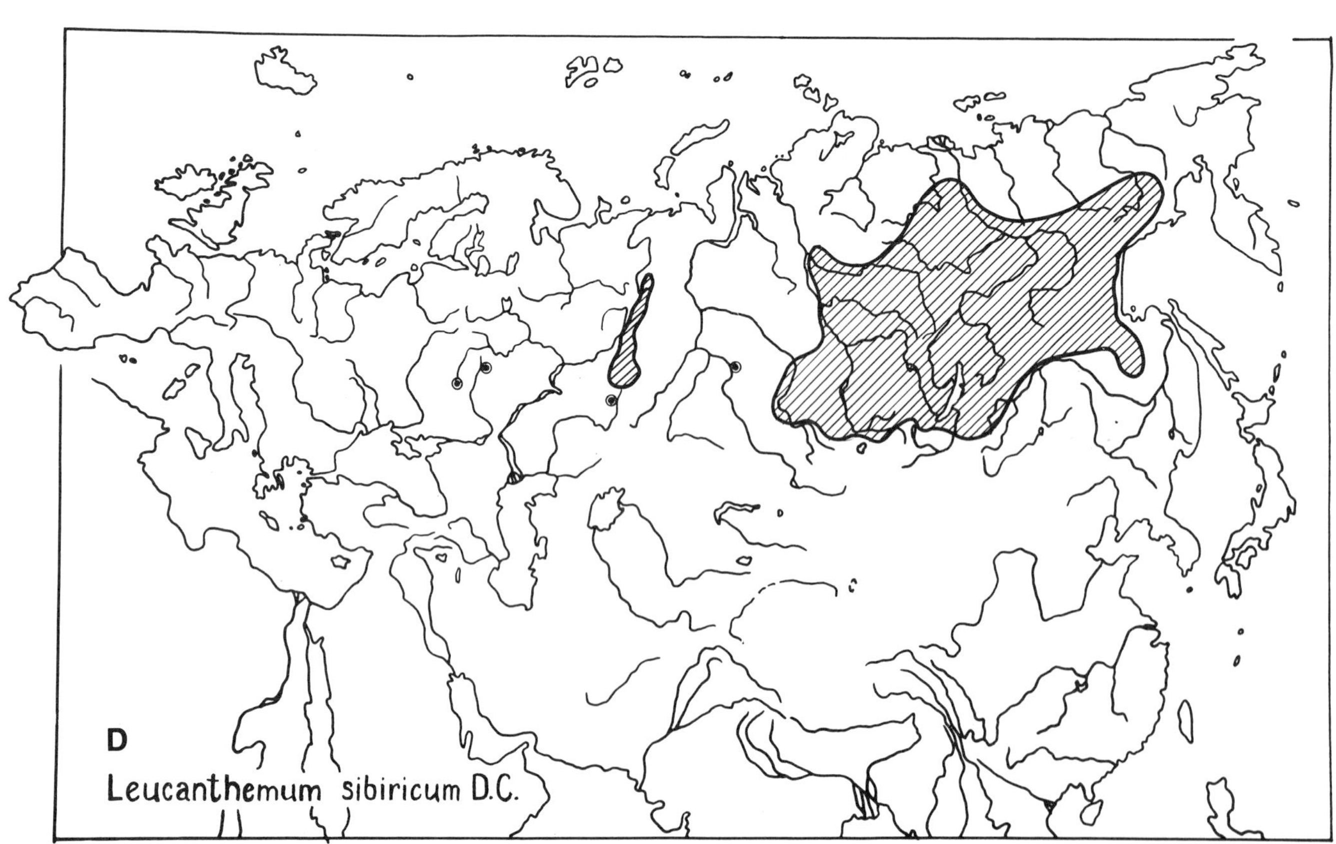
D
Leucanthemum sibiricum D.C.

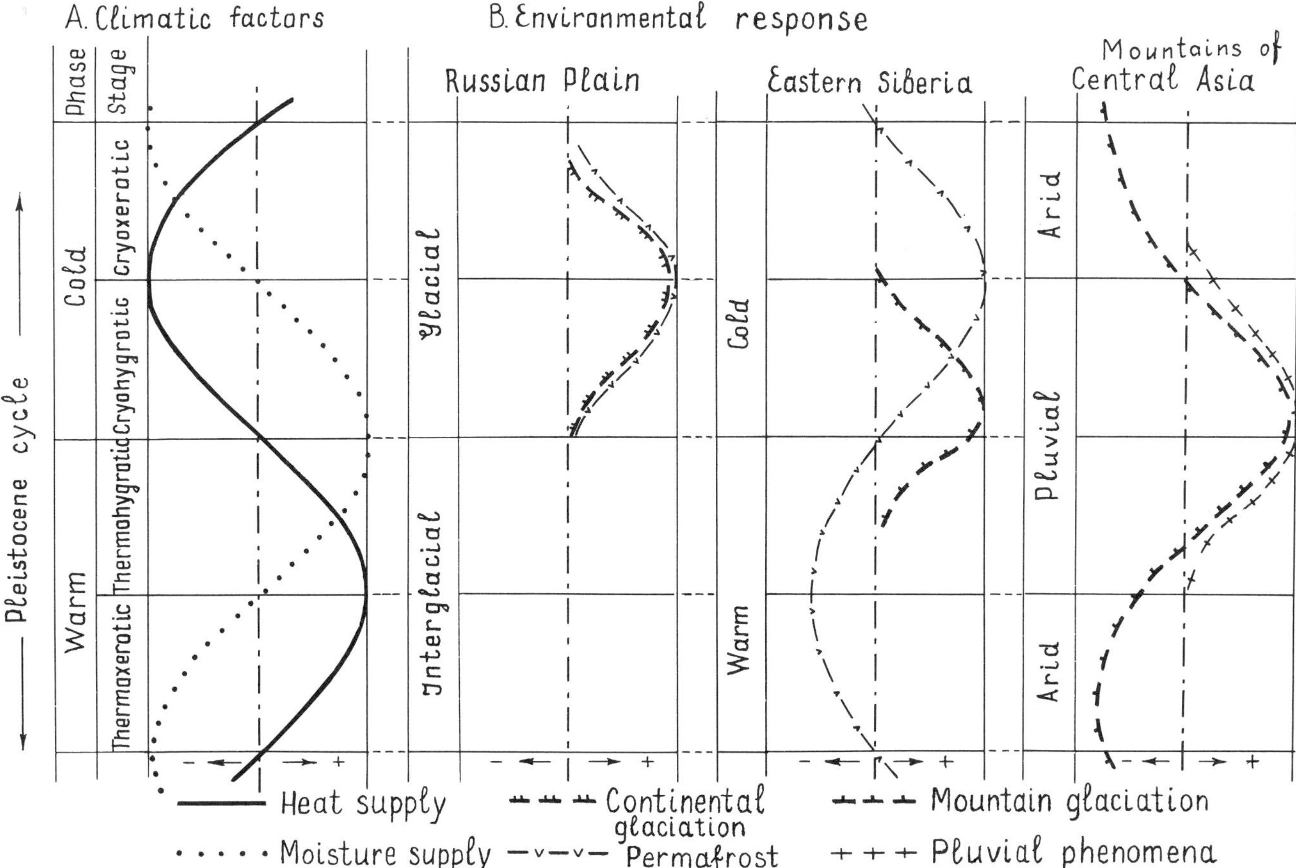

Figure 17-3. Correlation of environmental changes within one Pleistocene climatic cycle. (After Grichuk, 1971.)

Paleobotanic data provide a reliable means of dating interglacial deposits in the European USSR, but in the Asian USSR such data are useful only in dating the Kazantsevo Interglaciation in the southern West Siberian Lowland, the southern Baikal region, and the Indigira-Kolyma mountain region. In the mountains of Central Asia, these interglacial deposits are distinguished only in those regions where the entire Quaternary section has been studied and the corresponding layers identified. Late Valdai (Sartan) deposits cannot be identified solely on paleobotanic information. At present, a versatile dating approach should combine geologic, geomorphologic, and paleobotanic data. Particularly in river-terrace deposits, layers of a given epoch are identified by their stratigraphic position and by age determinations based on fossil plant remains. Horizons are established that correspond to the interglacial climatic optimum or the maximum phase of glaciation.

A regional reconstruction of the vegetation must be based on short time intervals. The chronologic intervals shown on a map must not only be of short duration but also be representative of an entire glaciation or interglaciation. These requirements are best met through the use of those time intervals that characterize extreme climatic conditions. Because radiometric dating of Late Pleistocene sediments is quite limited, narrow chronologic intervals can be identified only from paleoclimatic criteria. Analysis of extensive paleobotanic and geologic material makes it possible to recognize a time interval spanning an interglaciation and the succeeding glacial epoch—the so-called Pleistocene cycles of Markov (1955). This concept can be summarized as follows: Periodic changes in vegetation and geologic phenomena can be attributed to a nonsynchronous course of two main climatic factors—heat supply and moisture supply. Combinations of these determining factors make it possible to identify four climatic stages over the course of one cycle: the thermoxerotic, thermohygrotic, cryohygrotic, and cryoxerotic stages (Figure 17-3A). These stages are determined by atmospheric circulation and thus allow correlation of climate-caused natural phenomena in remote territories. On the geologic time scale the indicated boundaries may be considered synchronous. The climatic stages result in environmental changes that show the development of continental and mountain glaciation, permafrost, and pluvial phenomena on the Russian Plain, in central and eastern Siberia, and in Central Asia (Grichuk, 1971) (Figure 17-3B).

The stages and the transitions between them are determined with a fair amount of confidence from paleobotanic data. The duration of the transitions must, by definition, be short. Within an interglacial epoch the transition from

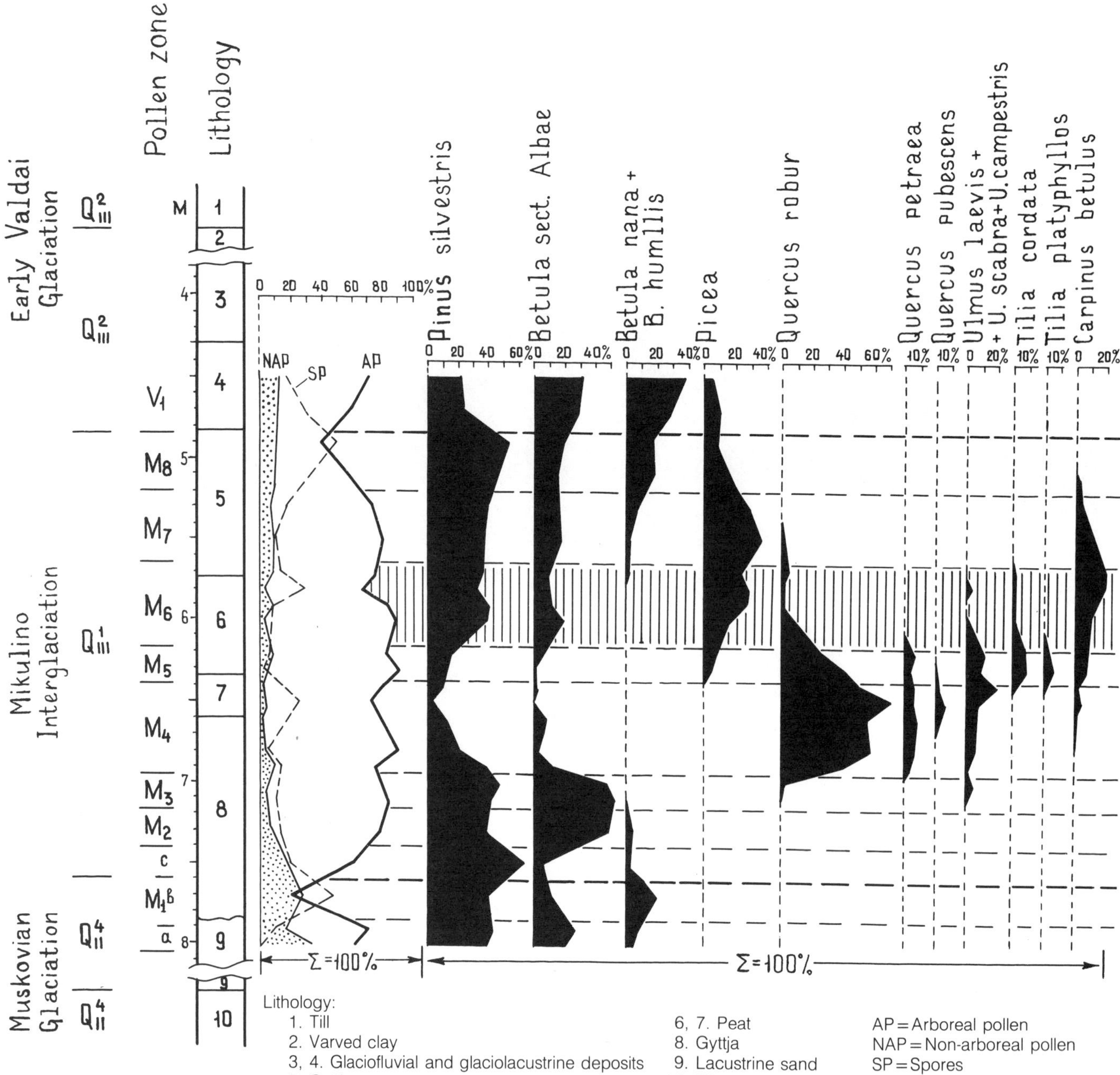

Figure 17-4. Pollen diagram of lacustrine-paludal deposits of the Mikulino Interglaciation near the village of Nizhnyaya Boyarshchina.

the thermoxerotic to the thermohygrotic stage is associated with the maximum heat supply during the climatic optimum. The transition from the cryohygrotic to the cryoxerotic stage during a glacial epoch is associated with the minimum heat supply during the maximum development of ice sheets in Europe. Data from 456 sections in which the horizons corresponding to the indicated intervals could be identified accurately were used in compiling the maps. Because of the small scale of the maps, the vegetation zones were reconstructed and in some cases supplemented with provincial demarcations for both the Mikulino (Kazantsevo) Interglaciation and the Late Valdai (Sartan) Glaciation.

Vegetation during the Mikulino (Kazantsevo) Interglaciation

In order to characterize the vegetation during the Mikulino (Kazantsevo) Interglaciation, it is essential first to examine vegetational changes over time. Naturally, it is not possible in a brief outline to discuss the vegetational history of all regions, so I discuss zones that have a significant nemoral flora of broad-leaved deciduous trees. Historical changes reached their maximum amplitude within those zones.

To characterize vegetational (phytocenotic) rearrange-

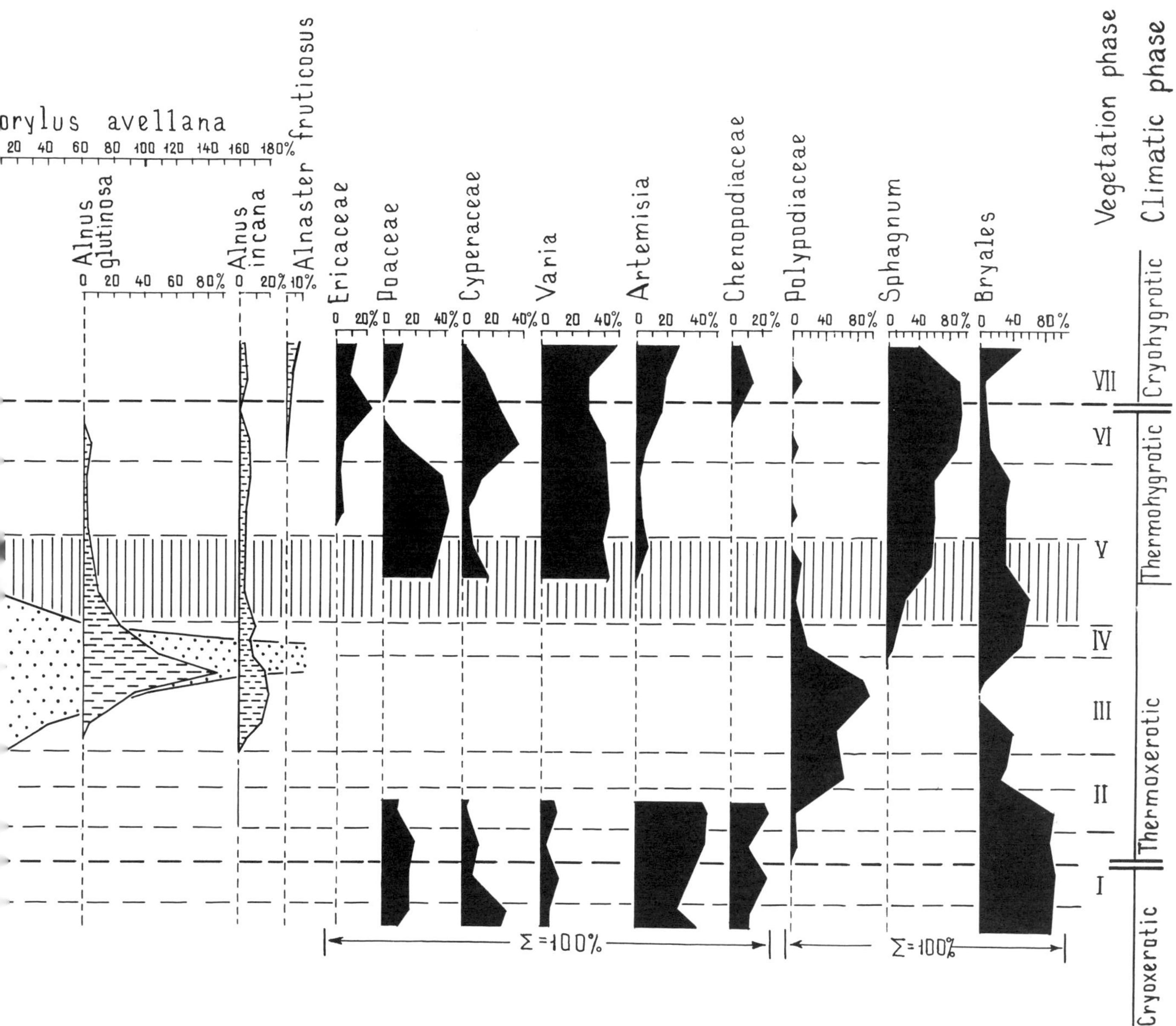

ments in the Central Russian Plain caused by the succession of interglacial climatic stages, I discuss a section near the village of Nizhnyaya Boyarshchina, one of the few sections in the Russian Plain that spans the entire interglacial epoch (Figure 17-4). It lies within a vast old lacustrine basin on the interfluve of the upper Dnepr and Zapadnaya Dvina Rivers. The more than 150 species identified in these deposits by pollen and plant-macrofossil analyses permitted the following characterization of interglacial vegetation, starting with the oldest phase.

A. *Cryoxerotic stage* (its final part) of the Moscovian Glaciation.
   I. Insular forests of *Picea obovata* and elements of periglacial vegetation.

B. *Thermoxerotic stage* of the Mikulino (Kazantsevo) Interglaciation.
   II. Monodominant pine and birch forests with minor participation of spruce, oak, and elm.
   III. Polydominant oak forests with *Quercus robur, Q. petraea, Q. pubescens, Corylus colurna,* and *Ulmus campestris* and an abundant understory of *Corylus avellana.*
   IV. Polydominant broad-leaved forests that included the three oak species mentioned above,

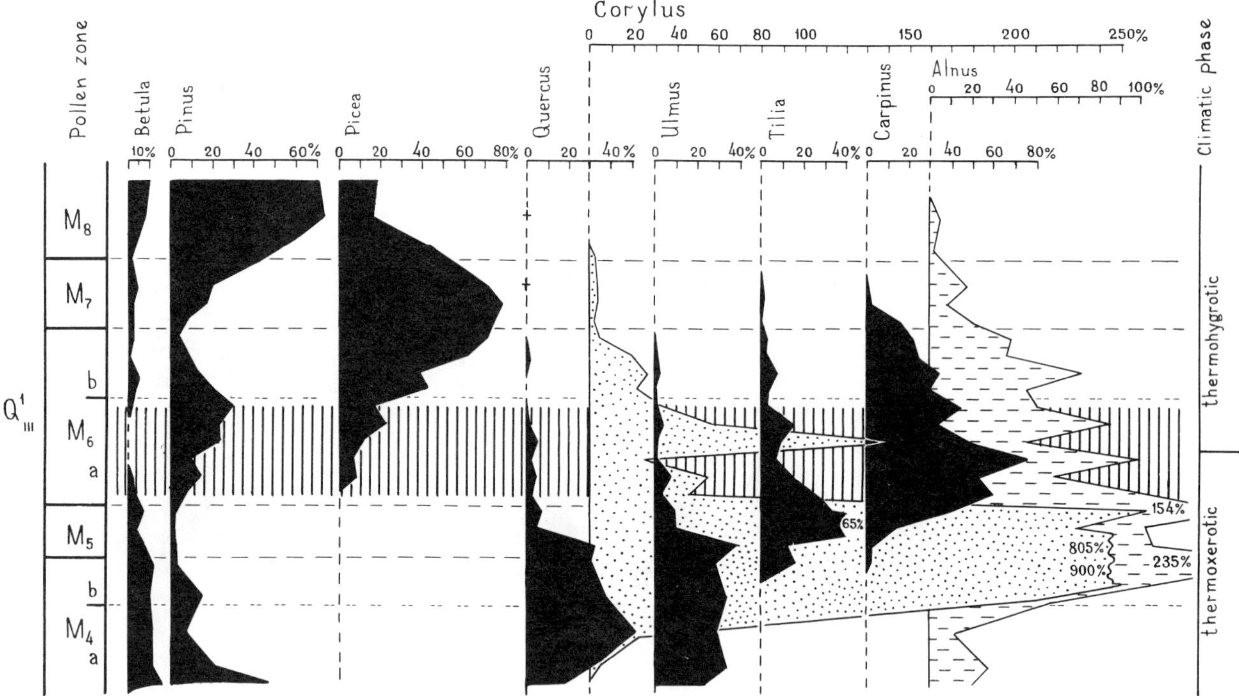

Figure 17-5. Pollen diagram of lacustrine-paludal deposits near the settlement of Mikulino. (Data from Grichuk, 1961.)

along with *Tilia cordata, T. latyphyllos,* and *Carpinus betulus,* as well as *Osmunda cinnamomea* in the ground cover.

C. *Transition* from the thermoxerotic to the thermohygrotic stage.

   V. Polydominant coniferous/broad-leaved forests of complex composition that included *Carpinus betulus, Quercus robur, Q. petraea, Tilia cordata,* several species of *Ulmus,* and *Picea excelsa.*

D. *Thermohygrotic stage.*

   VI. Coniferous/broad-leaved forests with a predominance of *Carpinus betulus* and *Picea excelsa.*

   VII. Monodominant spruce and pine forests, limited participation of coniferous/broad-leaved formations, and shrub associations including *Alnaster fruticosus, Betula humilis,* and *B. nana.*

E. *Cryohygrotic stage* (its initial part) of the Early Valdai (Zyryanka) Glaciation.

   VIII. Monodominant coniferous and birch forests, with wide distribution of paludal shrub formations.

These phases divide the pollen diagram into segments reflecting the successive changes in regional vegetational formations during the interglacial epoch. Pollen diagrams are commonly divided into pollen zones, that is, segments characterized by the predominance of pollen of a given woody plant (or a collection of them). Although such pollen zones contain little vegetational information on phytocenoses, they are convenient units for detailed stratigraphic subdivision and correlation even of distant sections. The left-hand side of the diagram from Nizhnyaya Boyarshchina (Figure 17-4) shows the zonation of Grichuk (1961), widely accepted in palynologic studies of the East European Plain. These zones are similar to those assigned to western Europe by Jessen and Milthers (Grichuk, 1961).

Pine and birch forests grew in the Nizhnyaya Boyarshchina Basin during the entire interglacial epoch as a result of edaphic conditions, and thus many changes in the nemoral flora are not expressed very clearly in the pollen diagram. For example, in phase V (marked by vertical ruling on the diagram) the pollen of *Carpinus betulus,* a plant with the highest requirements for heat and moisture, does not exceed 20%, whereas under different soil conditions these changes are much more apparent as shown by a pollen diagram of interglacial deposits near the village of Mikulino (Figure 17-5). There, the horizon corresponding to the transition from the thermoxerotic to the thermohygrotic stage (zone M_6 of Grichuk, 1961) reflects a distinct maximum of *C. betulus* pollen, up to 80%.

The sequence and characteristics of vegetational formations shown in the diagrams are typical of the entire East European Plain from latitude 47°N to 62°N. Outside the upper Dnepr and Zapadnaya Dvina River Basins, where the described sections are located, the vegetational characteristics of the identified phases naturally change according to the composition of the nemoral flora. Forest formations during the extreme climatic optimum have no close analogues in the modern European vegetation because the flora of the entire forested region in eastern Europe belongs to the migrational type (as defined above) and results from migration after each glaciation.

The formation of nemoral forests in the East European Plain has been the subject of numerous investigations, which have been summarized by Grichuk (1949) and are illustrated in Figure 17-6. The formation of boreal-type forests, however, has thus far been treated only in the most general outline.

Detailed paleobotanic studies of the Mikulino (Kazantsevo) Interglaciation in the Urals and most of Siberia have been made at only a limited number of locations, in part because this region has almost no organic lacustrine-paludal deposits, which in western and eastern Europe are the chief source of paleobotanic materials. Except in the extreme North, the Mikulino (Kazantsevo) Interglaciation is represented by alluvial sediments exposed along low terrace scarps (channel facies). Because of the imbrication of the alluvial layers, each alluvial section has unconformities that result in a somewhat fragmented vegetational record.

Deposits of the second terrace of the Ob' River and its tributaries in the middle latitudes of western Siberia are believed by a majority of investigators to represent the Mikulino (Kazantzevo) Interglaciation. In many cases, the palynologic data span only part of the interglaciation. One of the most representative sections is from the Chulym River second terrace near the village of Tegul'det, about 200 km east-northeast of the city of Tomsk in the northeastern foothills of the Kuznetskiy Alatau Range (Grichuk, 1970). The limited paleobotanic data from this section (Figure 17-7) permit only general or tentative vegetational interpretations, as shown by the following vegetational sequence:

A. *Cryoxerotic stage* (its final part) of the Taz Glaciation.
   - Ia. Insular spruce forests and periglacial steppes, with *Artemisia* and species of Polemoniaceae, which in Siberia are common in the alpine belt and tundra zone.
   - I. Dark coniferous forests and birch forest-steppe, with *Artemisia* and species of Plumbaginaceae, which in Siberia are confined to steppe vegetation on slightly saline soils.

B. *Thermoxerotic and thermohygrotic stages* of the Mikulino (Kazantsevo) Interglaciation.
   - II. Birch forests and limited dark coniferous forests of *Picea* and *Pinus sibirica*.
   - III. Birch and dark coniferous forests consisting of *Abies, Picea, Pinus sibirica*, and minor broad-leaved species (*Quercus, Ulmus, Tilia*), and *Alnus*.
   - IV. Birch forests with Siberian cedar (*Pinus sibirica*) and Scotch pine (*P. silvestris*) forests, as well as alder formations.

The third phase reflects a transition from the thermoxerotic to the thermohygrotic stage of the interglaciation, providing there are no major discontinuities in the alluvial section. The final part of the interglacial epoch is not represented in this section. The presence of *Lycopodium annotinum* spores in the uppermost layers indicates the existence of coniferous forests near the study site. This species is uncommon in the southern forest-tundra today; its occurrence is a definite indication that phase IV is not the final interval of the interglaciation.

There are definite similarities between the character of the vegetation during the climatic optimum of the Mikulino (Kazantsevo) Interglaciation (phase III) and the Holocene vegetation in adjacent regions of western Siberia (Grichuk, 1970; Khotinskiy, 1977; Neustadt, 1957). However, the pollen assemblages are rather different, and the importance of spruce forests in this region during the Holocene was considerably diminished. The Chulym section shows that migrational processes played a very minor role in the formation of the flora during the interglacial climatic optimum. All the vegetation zones were formed from local floristic elements and record one stage in one development of an *orthoselection flora*. This type of flora is characteristic of southeastern western Siberia, almost all of eastern Siberia, and Kazakhstan and Central Asia.

The easternmost deposits of the Mikulino (Kazantsevo) Interglaciation are found in Zolotoy Rog Bay, near Vladivostok. Marine sediments (borehole 1005) at a depth of 7 to 10 m (Korotkiy et al., 1980) were assigned to the Mikulino (Kazantsevo) Interglaciation on the basis of the palynologic data and an altitudinal estimate of contemporaneous sea level. (This age estimate is contradicted by the radiocarbon date of 29,000±250 yr B.P. [MGU-325] obtained from a layer of lagoon peat covering sediments with *Ostrea* shells. The authors of the study believe that this date is too young because it does not correspond to the position of the sea level [Korotkiy et al., 1980]. It should be added that dates of 31,820±750 [MAG-340] and 24,750±400 yr B.P. [SOAN-289] were obtained nearby in the Partizanskaya River Basin from deposits having a cold-tolerant flora that did not contain even single pollen grains of broad-leaved species. There is no doubt that the very thermophilic flora from borehole 1005, in which the total pollen of broad-leaved species was 75%, cannot occupy an intermediate position between these dates.)

Pollen analysis (Figure 17-8) shows that this section spans only part of the interglacial epoch, and the following vegetational changes are recognized:

- I *Quercus* forests with a minor participation of *Juglans, Tilia, Ulmus*, and *Phellodendron* and limited coniferous/broad-leaved forests of *Pinus funebris* Kom., *P. koraiensis* Sieb. and Zucc., *Abies*, etc.
- II Polydominant broad-leaved forests that included predominantly *Quercus* but also *Juglans, Tilia, Ulmus, Carpinus, Betula* sect. Costatae, and a number of coniferous species.
- III Mixed coniferous/broad-leaved forests of *Quercus*, other deciduous species, *Abies, Picea* sect. Eupicea and Omorica, *Pinus koraiensis* Sieb. and Zucc., and *P. funebris* Kom.

Despite limited floristic information, the thermoxerotic and thermohygrotic stages of the interglacial epoch are reflected in this section, and the intervening transition is

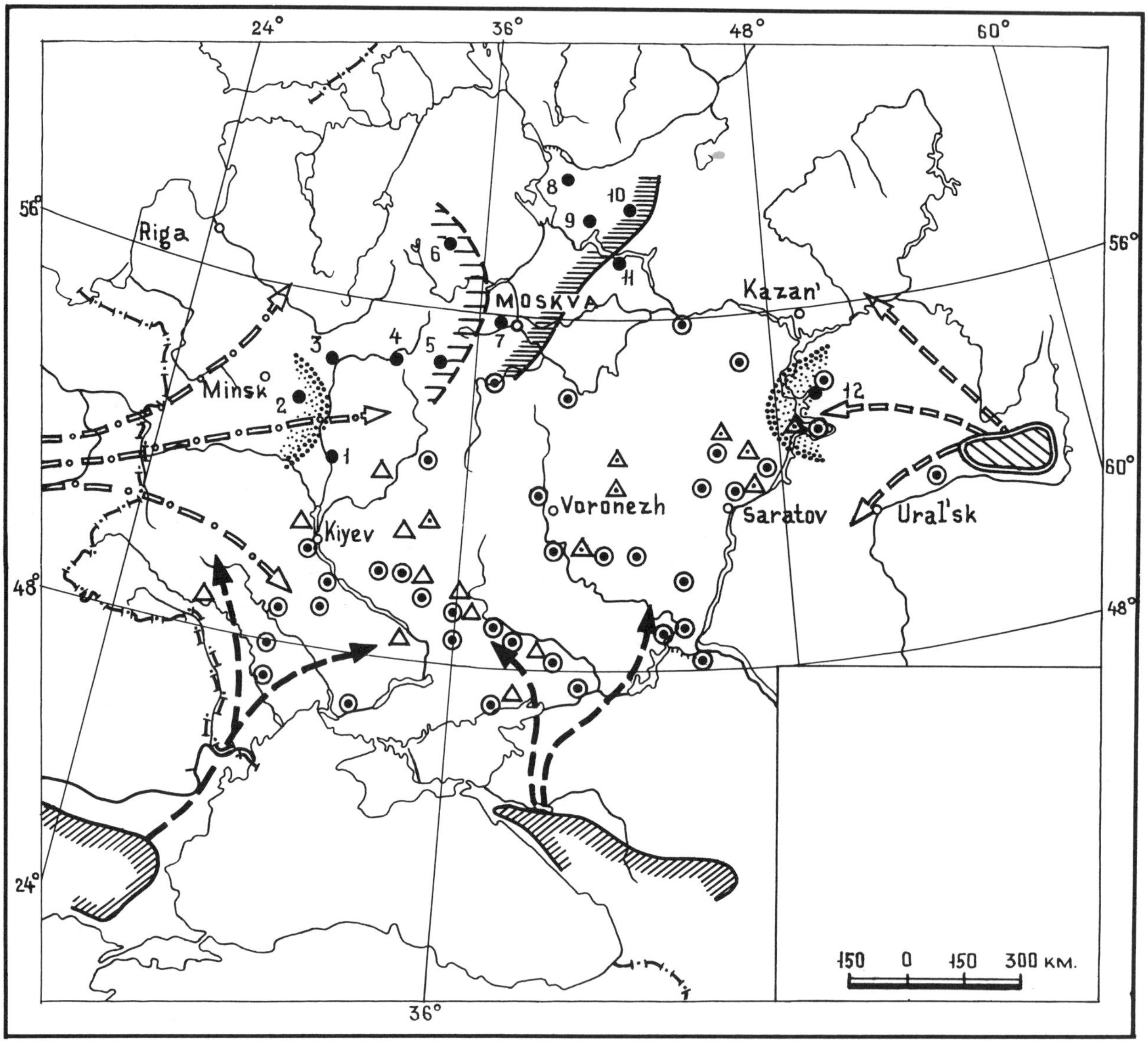

Refugia of broad-leaved trees during the Moscovian Glaciation:

Balkan and northern Caucasus

Southern Ural Mountains

Relicts of Mikulino Interglaciation in the present vegetation (Kleopov, 1941, and others):

Balkan and northern Caucasus

Central Europe

Southern Ural Mountains

Migration directions:

Of Balkan and northern Caucasus relicts

Of European mesophilic elements

Of southern Ural mesophilic elements

Distribution limits:

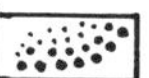
Eastern and western limits of abundant *Tilia* at start of climatic optimum

Eastern limit of abundant *Tilia* at culmination of climatic optimum

Southeastern limit of abundant *Tilia* at end of climatic optimum

Sites with Mikulino plant fossils:

1. Loyev
2. Murava
3. Mikulino
4. Novyye Nemkari
5. Drozhzhino
6. Vyshniy Volochek
7. Potylikha
8. Irkhino
9. Chernaya Sluda
10. Balchuk
11. Ples
12. Cheremshan

Figure 17-6. Migrational history of the broad-leaved forests of the East European Plain during the Mikulino Interglaciation. (Modified from Grichuk, 1949.)

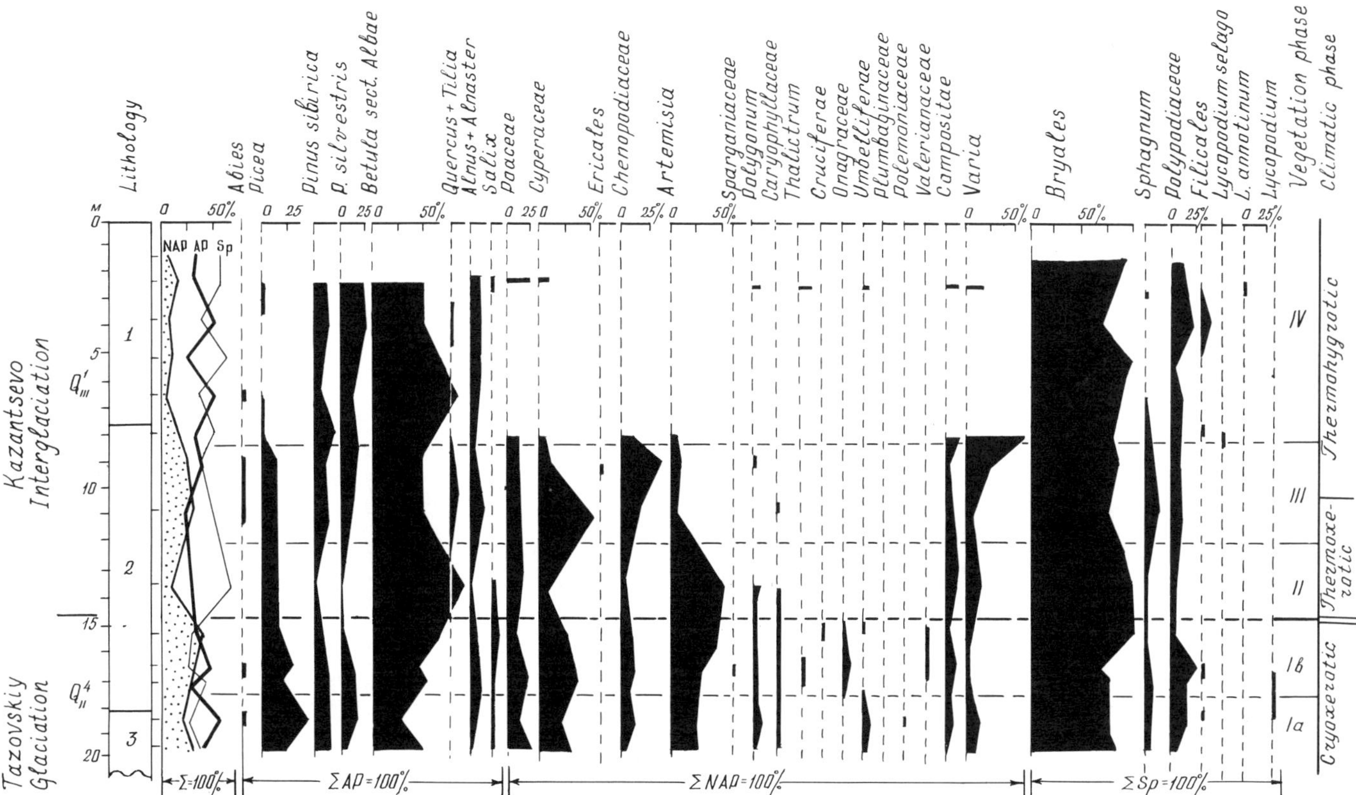

Figure 17-7. Pollen diagram of a section from the second terrace of Chulym River near the settlement of Tegul'det. (Analyses by M. P. Grichuk and T. G. Svirdova.) Lithology: (1) Loam with interbedded sand (floodplain facies). (2) Sand and sandy loam (fluvial facies). (3) Pebbly sand (basal horizon of fluvial facies).

seen in the second phase by the noticeable presence of mesic elements. As indicated by L. P. Karaulova, the oldest pollen spectra at 7.5 to 10 m depth are very similar to those of the Holocene climatic optimum from the southern Maritime Territory. They are also very similar to recent pollen spectra from polydominant forests distributed over maritime mountain slopes just south of Vladivostok. This would indicate that the flora of the southern Maritime Territory is of *relict type.*

Vegetation during the Climatic Optimum of the Mikulino (Kazantsevo) Interglaciation

The vegetation during the Mikulino (Kazantsevo) Interglaciation is reconstructed for the epoch of maximum warmth, that is, the culmination of the climatic optimum. The basic unit used is the vegetation zone, which is the same unit used in small-scale general maps of modern vegetation. The legend of the map contains 12 basic subdivisions, but, because of appreciable meridional differentiation of natural conditions over the vast expanse of northern Eurasia, a provincial distinction was made for a number of zones. As a result, 23 units are given on the map.

The reconstruction is based primarily on paleobotanic data obtained at 268 sites (Figure 17-9). In some sections, the climatic optimum is only partially represented, and identification of this interval is more or less arbitrary. The presence of closely spaced sections in many regions reveals crude errors that easily occur. Full use was also made of florocenogenetic analysis of modern vegetation. The map (Figure 17-10) shows that during the interglacial climatic optimum the vegetation zones were similar to the present ones. Among the chief differences are the absence of polar deserts and a distribution of tundra more limited than today's. Typical tundra appeared only in northeastern Siberia and on the northern islands that formed during the Kazantsevo marine transgression.

Boreal and broad-leaved forests were important, and a marked northward expansion of the forest zone is recorded. Boreal forests are divided into northern and southern subzones based on the coenose-forming species. The southern subzone is distinguished by the occurrence of broad-leaved species on the plains and by cedar (*Pinus sibirica*) in the mountains. Unlike modern taiga forests, birch was a dominant species in both subzones. Similar vegetation is found today as a narrow band along the southern forest boundary in western Siberia (the so-called "white taiga") and in central and southern Kamchatka. Dark coniferous forests of *Picea, Pinus sibirica,* and *Abies* appeared only in mountainous regions of the Urals, the Altay-Sayan Highland, and the southern Okhotsk region. In addition, in northern central Siberia and in the moun-

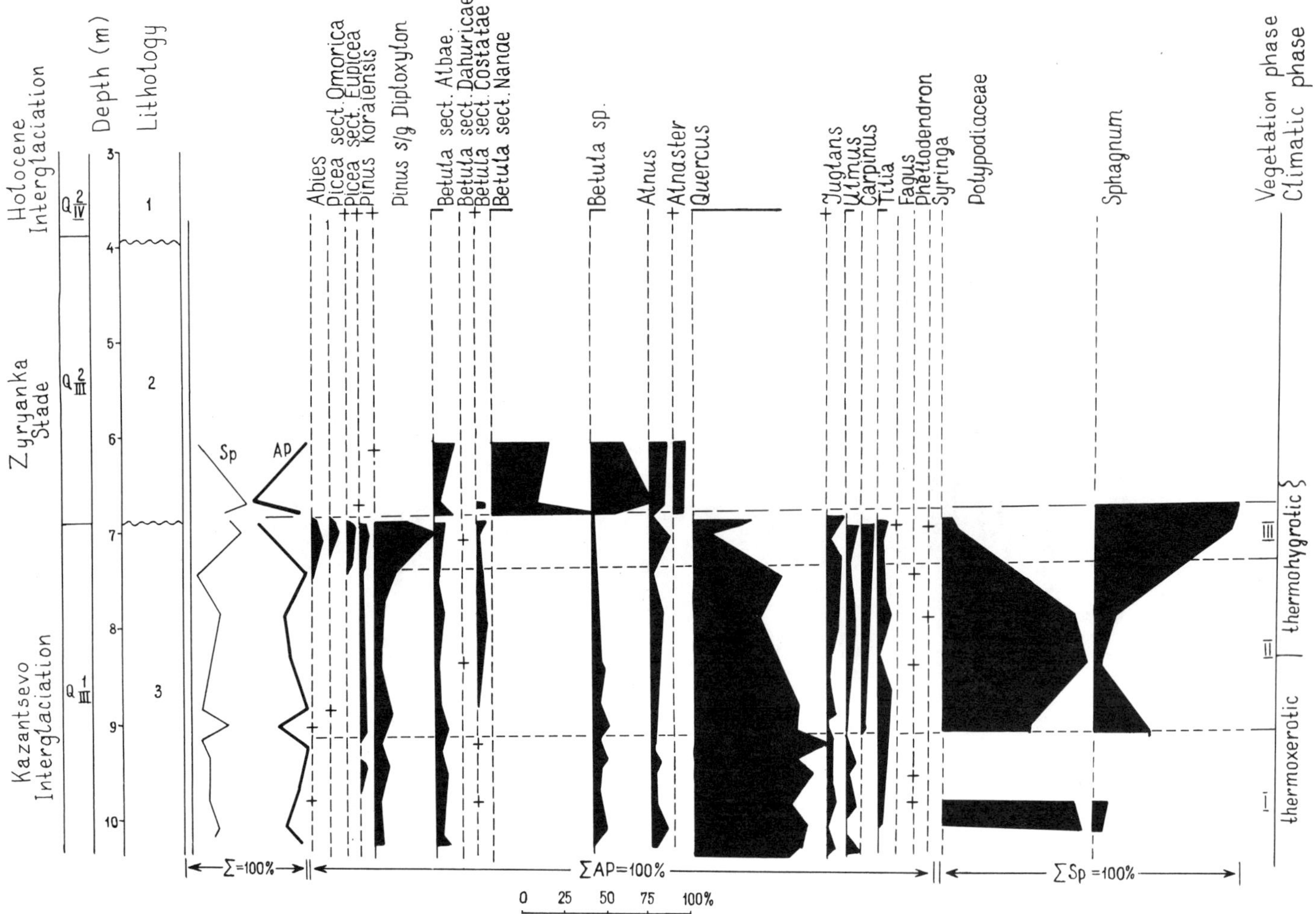

Figure 17-8. Pollen diagram of Late Pleistocene deposits from Zolotoy Rog Bay (bore hole 1005). (Analyses by I. P. Karaulova.) Lithology: (1) Holocene sandy argillaceous marine deposits. (2) Alluvial deltaic detrital loam. (3) Organic lagoonal and argillaceous marine deposits.

tains of eastern Siberia, pine and larch forests were widely distributed, just as they are today.

Broad-leaved forests on the East European Plain were more widely distributed than they are today. The northern boundary up to the Severnaya Dvina Basin was markedly displaced, but the southern boundary shifted insignificantly to the southern boundary of the present European forest-steppe. In the western part of the East European Plain, broad-leaved forests were dominated by white beech (*Carpinus betulus*) and have no close modern analogues in either eastern or western Europe. In the Far East, broad-leaved forests were only slightly more extensive and were very similar in composition to those existing at the present time. The same applies to broad-leaved forests of the southern mountains (Crimea, Greater and Lesser Caucasus, northern Tien-Shan).

Almost the entire present steppe area in the southern East European Plain was occupied by forest-steppe. The forest component consisted of broad-leaved species—in the west primarily white beech (*Carpinus betulus* L.) and in the east oak and other taxa. In southwestern Siberia, *Quercus* and *Ulmus* grew in forests, but birch was the dominant species. Typical herb-grass and grass steppes are reconstructed only in the southeastern East European Plain and in southern Kazakhstan to the foothills of the western Altay. Farther to the east, steppe formed as isolated "islands" in the Minusinskiy Basin and along the Selenga, the upper Angara, and the middle Shilka and Argun' Rivers. This reconstruction is inferred from paleobotanic data and confirmed by historicofloristic materials.

The vegetational history of the semideserts and deserts of Central Asia is based almost entirely on historicofloristic data because paleobotanic data are available at only six localities. The scanty data suggest that typical semishrub and ephemeral steppe and semidesert existed during the entire interglaciation. Palynologic data from southern Kazakhstan indicate that during the interglacial climatic optimum the northern limit of steppe shifted southward into the region now occupied by desert.

The reconstruction of the mountain vegetation in the Caucasus and Central Asia is also based mainly on historicofloristic data because of the limited amount of paleobotanic data. There was an appreciable expansion of broad-leaved and coniferous/broad-leaved forests and a restriction of

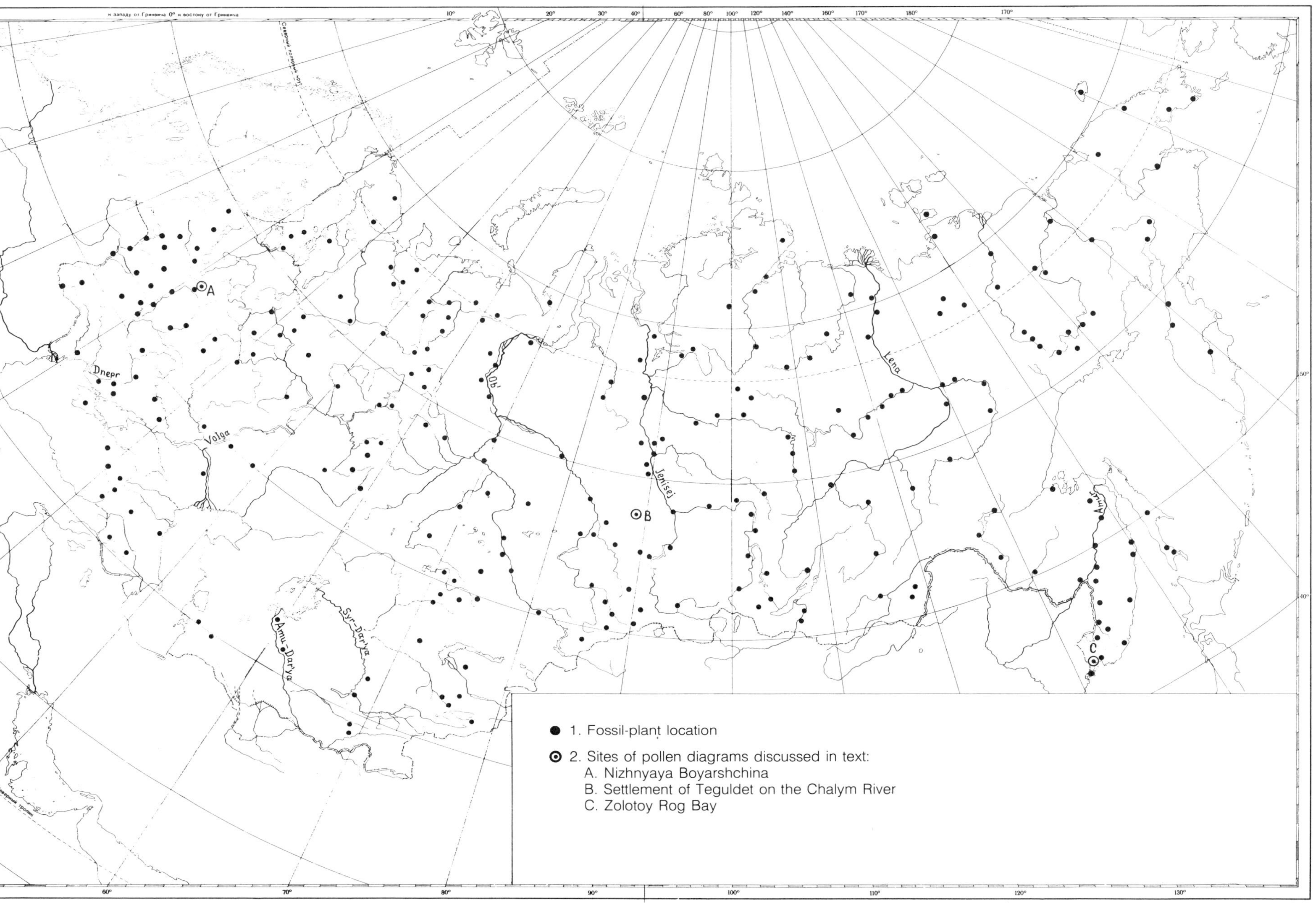

Figure 17-9. Location of sites used to construct the vegetation map of the Mikulino Interglaciation in Figure 17-10.

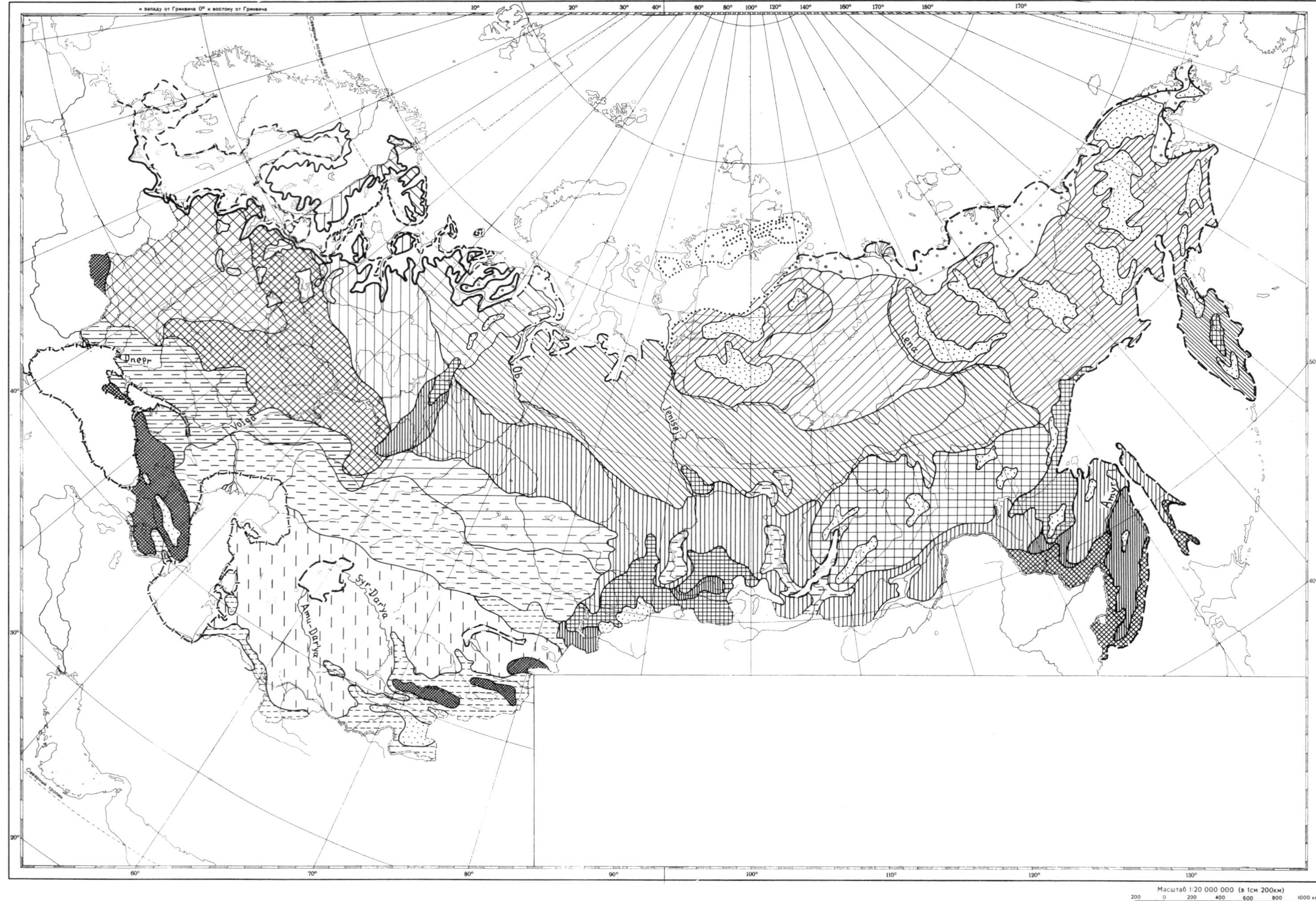
к западу от Гринвича 0° к востоку от Гринвича
Dnepr
Volga
Ob
Jenisej
Lena
Амур
Syr-Dar'ya
Amu-Dar'ya
Масштаб 1:20 000 000 (в 1см 200км)
200 0 200 400 600 800 1000 км

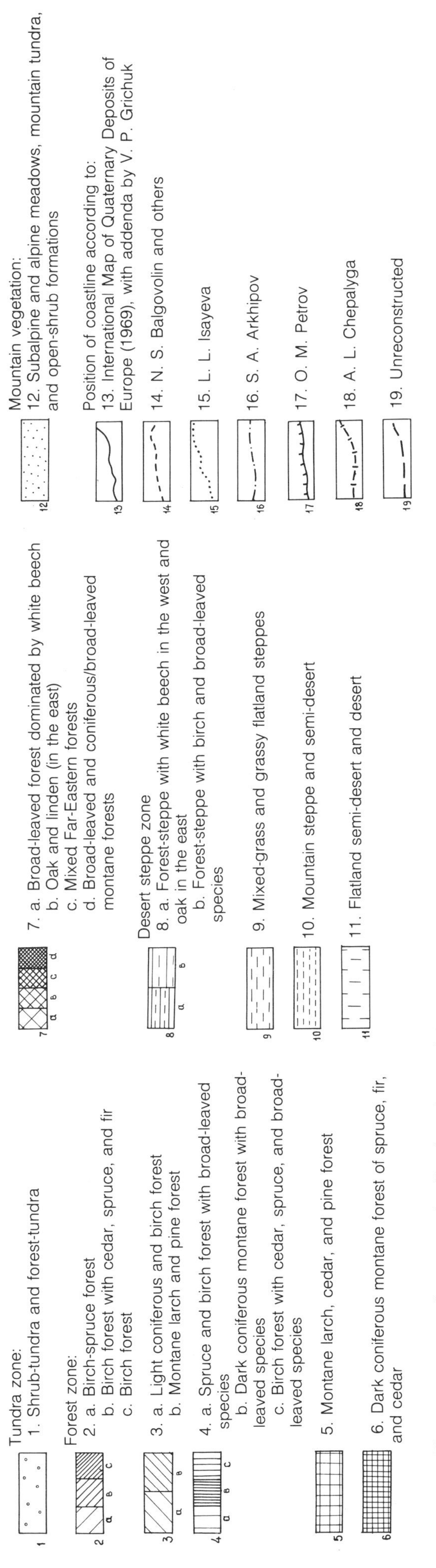

Figure 17-10. Vegetation map at the climatic optimum of the Mikulino (Kazantsevo) Interglaciation (time of maximum warmth). (Compiled by V. P. Grichuk.)

alpine vegetation in these areas. The latter phenomenon is confirmed by paleobotanic data from only one locality in the Caucasus (Svanetiya) and one in Central Asia (Tien-Shan), but many localities show an appreciable altitudinal rise in the forest boundaries.

Alpine vegetation is completely reconstructed from historicofloristic data, but this does not detract from the reliability of the reconstruction. Endemic alpine species (paleoendemics) with narrow ranges provide evidence that alpine vegetation existed in certain mountains for a long time (Tolmachev, 1974b).

Vegetation during the Late Valdai (Sartan) Glaciation

The possibility of examining the successive vegetational changes during the last glaciation is made very difficult by the absence of representative and complete sections. Comparison of composite sections has thus far been impractical in view of the poor correlation. Most radiocarbon determinations are from layers that either have no paleobotanic material or have been studied in little detail (for example, Kind, 1974). Therefore, I discuss only two sections, which span long periods of time close to the maximum phase of the last glaciation.

The first section lies in a vast old lake depression in the upper Volga Basin near the town of Ivanovo, approximately 250 km northeast of Moscow. There, under a Holocene peat bog, a borehole revealed a thick section of loams and sands interbedded with gyttja and peat of the last glaciation. Detailed pollen analysis (Figure 17-11) fully confirms this date and indicates that deposition began during the Dunayevo interval and continued into the Bologovo and Krestsy stages of Valdai Glaciation, approximately 28,000 to 24,000 years ago to 14,000 years ago (Gerasimov, 1969; Chebotareva and Makarycheva, 1974). This section reveals successive vegetational changes characteristic of the periglacial region of the central East European Plain.

A. *Cryohygrotic stage.*
   I. Monodominant spruce and pine-birch forests with limited oak and linden.
   II. Moderately mesophilic pine and mixed conifer-birch forests and *Sphagnum* bogs with *Betula nana.*
   III. Insular spruce-birch and pine forest with *Lycopodium clavatum* and meadow communities with *Botrychium boreale.*

B. *Transition stage* between the cryohygrotic and cryoxerotic stages.
   IV. Meadow steppes with spruce-birch and pine forests and solonetzic meadows with *Atriplex tatarica.*

C. *Cryoxerotic stage.*
   V. Complex cryoxerophilic vegetation—a combination of steppe associations with *Ephedra distachya* and *Eurotia ceratoides,* tundra-shrub communities with *Lycopodium alpinum* and *Selagi-*

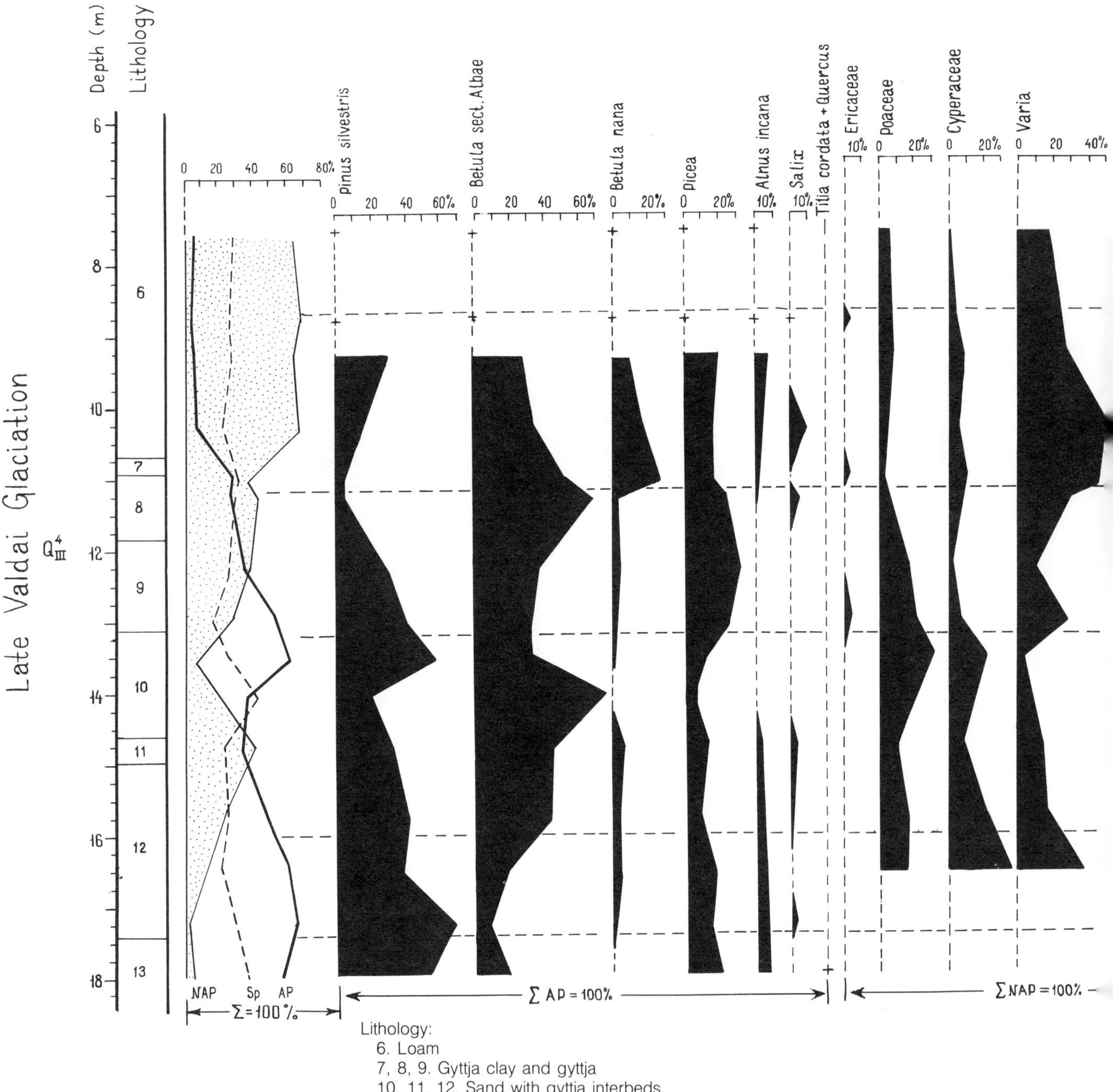

Figure 17-11. Pollen diagram from Late Valdai lacustrine deposits near the town of Ivanovo (hole 63). (Analyses by L. V. Kalugina and M. Kh. Monoszon.)

nella selaginoides and solonchak vegetation with *Echinopsilon* sp.

VI. Forest-free cryoxerophilic vegetation—predominance of steppe associations with *Artemisia laciniata* and *Kochia prostrata,* solonchak vegetation with *Salicornia herbacea, Salsola foliosa,* etc., and tundra elements (*Lycopodium alpinum*).

Pollen representations of these plants are absent today not only on the East European Plain but also in Siberia. These formations and the occurrence of northern boreal and Turanian floristic elements illustrate one more stage in the development of a migration flora. Glacial-age floras from the Russian Plain show the transition from the cryohygrotic to the cryoxerotic stage corresponding to the maximum stage of glaciation (Grichuk, 1969). In this sequence of vegetational changes, phase IV of meadow-steppe is confined to that time. Phase VI represents the final stage of Valdai Glaciation, when the southeastern boundary of

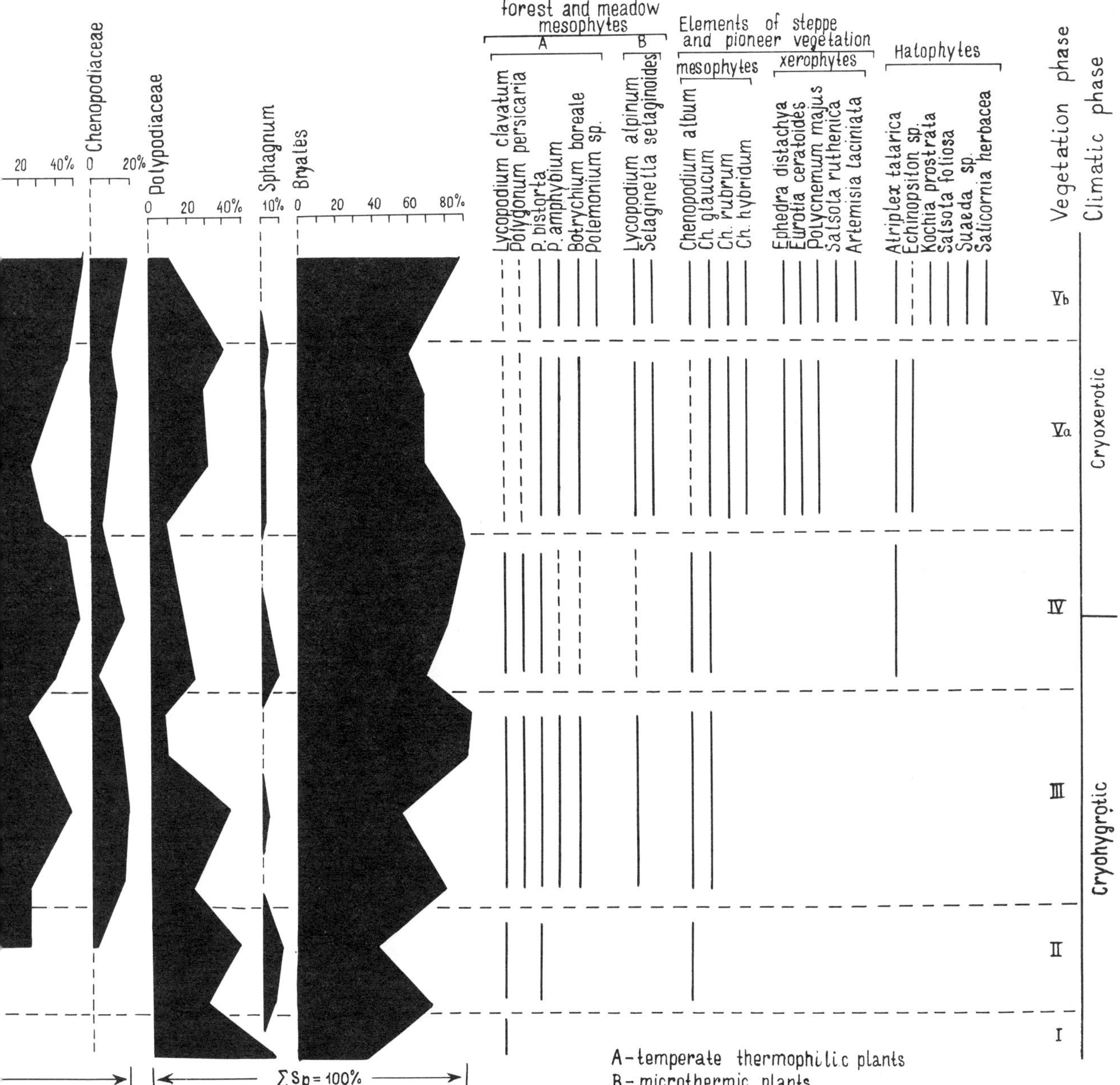

the ice sheet was near the Baltic Sea and the Gulf of Finland (Gerasimov, 1969).

In the middle latitudes of Siberia, the most representative and well-dated section of the glacial maximum stage is found in the cis-Altay Plain of southwestern Siberia. It is exposed in a 20-m terrace of the Chumysh River near the village of Kytmanovo in the upper Ob' River Basin, about 240 km southeast of Novosibirsk. There, the Chumysh River valley comes right up to the foothills of the Salairskiy Range—a mountainous area 500 to 600 m above sea level. At the base of the terrace, under a mass of channel alluvium, sediments of an oxbow-lake facies are exposed. A date of 24,240 ± 2700 yr B.P. (SOAN-31) was obtained on wood fragments from this facies (Panychev, 1979). Pollen analysis of the alluvium and overlying loesslike loams (Figure 17-12) shows large amounts of Chenopodiaceae pollen (Vdovin et al., 1969), probably from plants on the local sandy alluvium and thus of no climatic significance. Also,

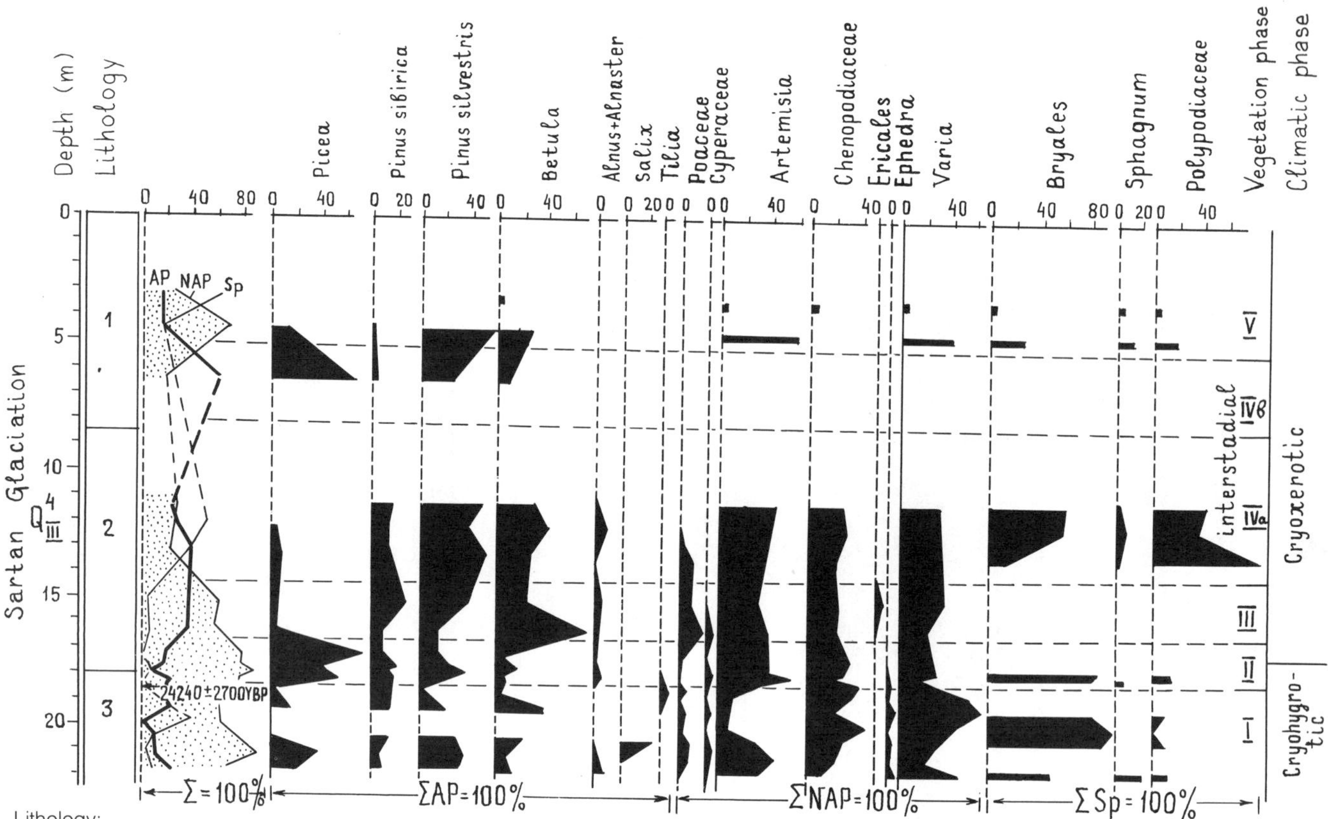

Lithology:
1. Loessial loam with a soil horizon
2. Interbedded sand and loam (fluvial facies)
3. Clay with organic detritus (oxbow facies)

Figure 17-12. Pollen diagram from a 20-m alluvial section of the Chumysh River near the settlement of Kytmanovo. (Analyses by M. R. Potakh.)

the Chumysh River valley lies in the Altay Mountains, where the pollen production of the montane vegetation is abundantly represented in alluvial pollen spectra.

A phytocenotic interpretation of the data makes it possible to distinguish the following two vegetational changes.

A. *Cryohygrotic stage.*
   I. Herb-grass steppe and mixed coniferous and birch forests. A date of 24,240 ± 2700 yr B.P. is assigned to the end of this phase.
   II. Periglacial *Artemisia*/herb steppe, insular spruce forests, and pine-cedar (*Pinus silvestris* and *P. sibirica*) forests in the mountains.

B. *Cryoxerotic stage.*
   III. Birch forest-steppe and forests of cedar (*Pinus sibirica*) and pine in the mountains.
   IVa. Herb-grass steppe and mixed forests dominated by *P. silvestris* in the foothills (first half of the interstadial epoch).
   IVb. Increased importance of spruce formations (second half of the interstadial epoch).
   V. Steppe formations with *Artemisia* and pine-birch forests in the foothills.

The pollen spectra from this section are very similar to modern spectra from floodplain deposits to the south along the upper Ob' River within the forest-steppe and steppe (Grichuk, 1970). One can conclude that the plant formations existing in the Salair foothills during the last glaciation have analogues in the present vegetation of southeastern western Siberia. Hence, the section provides us with information on one more stage in the formation of the orthoselection flora in Siberia. The phytocenotic interpretation of the Chumysh River section is unquestionably tentative and based on limited paleobotanic data, but it does show that coniferous forests were widespread in the middle-latitude mountains of western Siberia before, during, and after the glacial maximum. A similar pattern occurs in middle-latitude central and eastern Siberia, but, unfortunately, no detailed studies of the last glaciation have been published as yet for these areas or for the Far East.

The vegetation during the transition from the cryohygrotic to the cryoxerotic stage represents the coldest phase of the Valdai (Sartan) Glaciation. Paleobotanic, geologic, and radiocarbon data from the East European Plain show that this period also corresponds to the maximum advance of the Scandinavian ice sheet about 20,000 years ago and to the maximum ice advance on the Middle Siberian Pla-

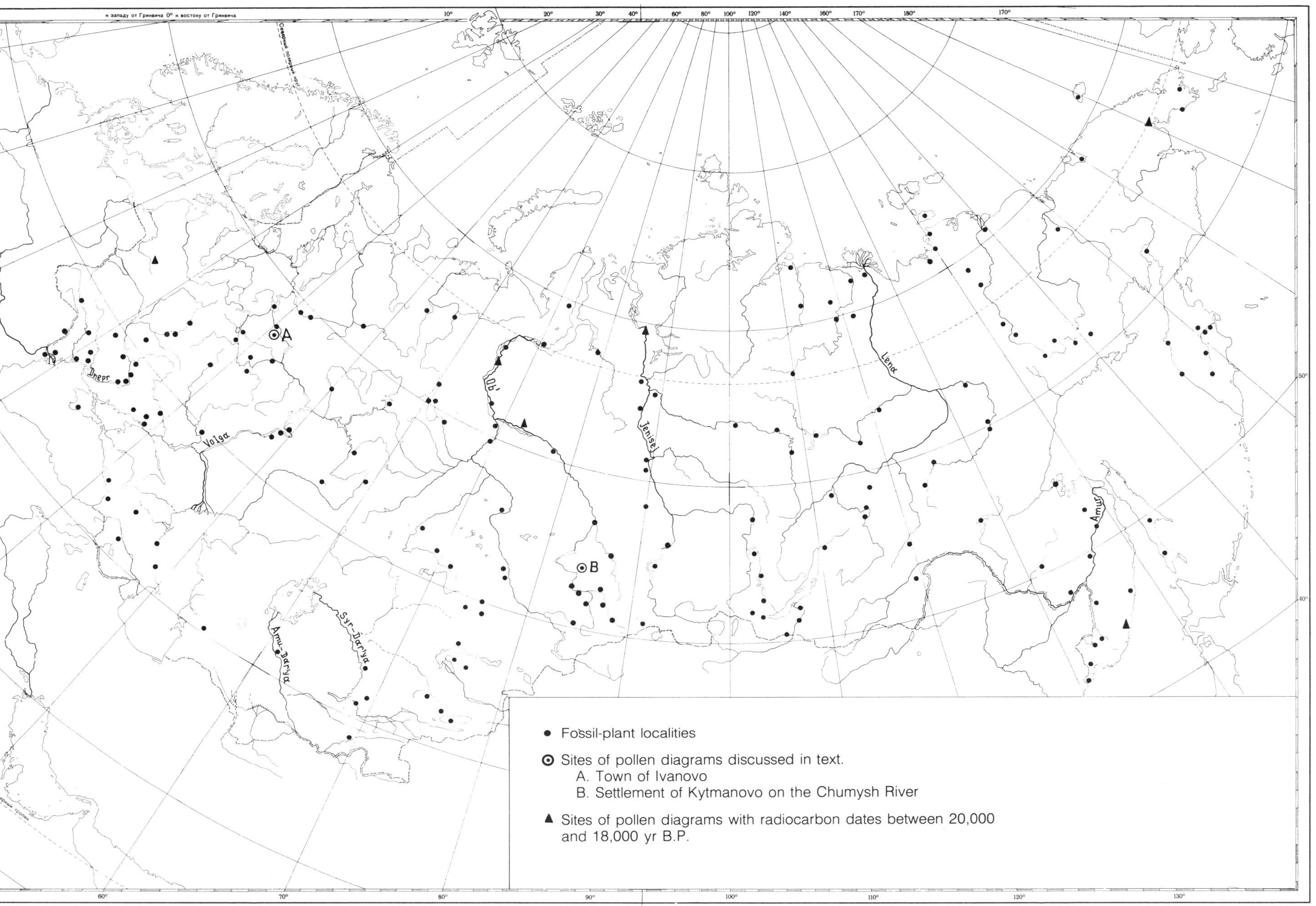

Figure 17-13. Location of sites used to construct vegetation map of the Late Valdai Glaciation.

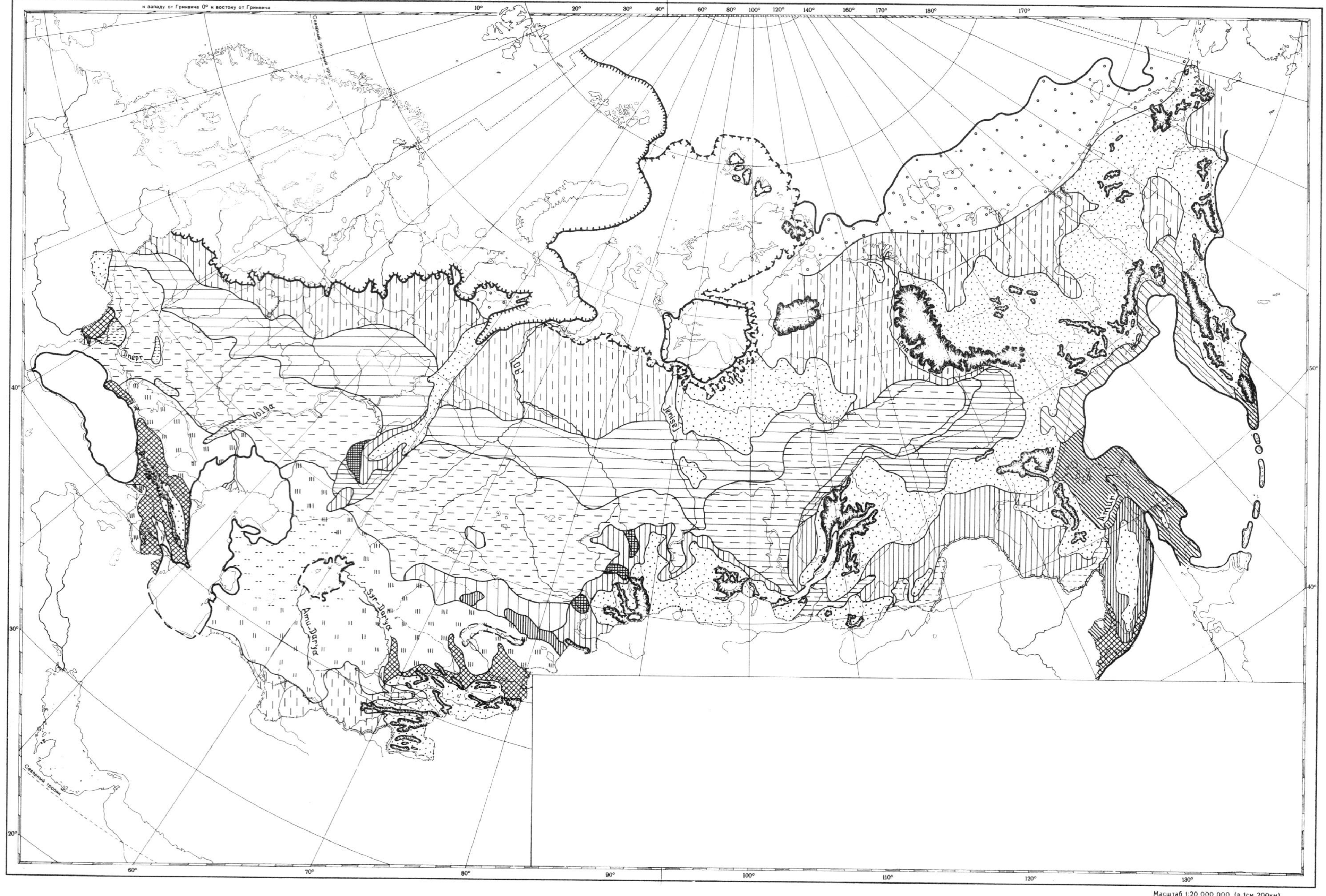
к западу от Гринвича 0° к востоку от Гринвича
Северный полярный круг
Северный тропик
Dnepr
Volga
Ob
Jenisej
Lena
Amur
Syr-Dar'ya
Amu-Dar'ya
Масштаб 1:20 000 000 (в 1см 200км)
200 0 200 400 600 800 1000 км

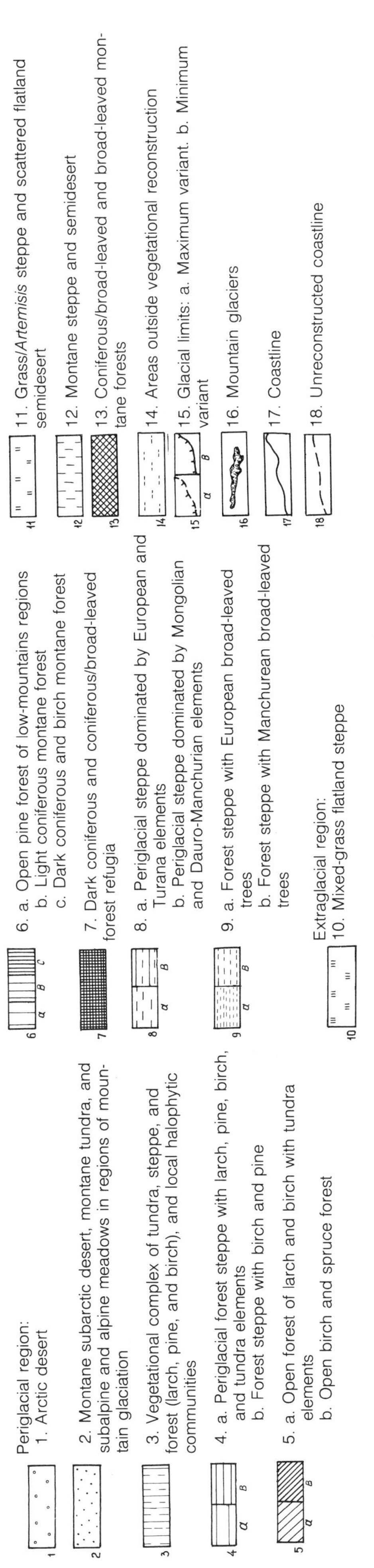

Figure 17-14. Vegetation map during the maximum of Late Valdai (Sartan) Glaciation (time of minimum warmth). (Prepared by V. P. Grichuk.)

teau (Kind, 1974). There are as yet no reliable estimates of the time of maximum alpine glaciation in eastern Siberia.

Geologic and geomorphic criteria are most important in selecting sites for paleobotanic studies, but strata deposited during the coldest phase can be identified only paleobotanically. In practice, this is possible only in thick continuous sections in which pollen analysis can reveal vegetational changes. Unfortunately, there are fewer than 10 radiocarbon dates from paleobotanic sites of this interval, and the inevitable unconformities in these deposits prohibit extrapolation from older or younger sites.

Only 187 sites were used for reconstructing full-glacial vegetation (Figure 17-13), but at the same time published historicofloristic data were particularly useful, if only because relict glacial floras are commonly found in the present vegetation. As was the case for the Mikulino (Kazantsevo) Interglaciation, the basic map unit employed in the reconstruction of glacial vegetation is the vegetation zone. Thirteen zones are distinguished: the first nine encompass the *periglacial region*, and the remaining four types span the *extraglacial region*, where the diminished heat supply never caused appreciable restructuring of the flora or vegetation. The extraglacial region includes low and middle altitudes in the Caucasus, southernmost Kazakhstan, Central Asia, and the southernmost part of the Far East. Provincial floristic differences are given for certain zones (Figure 17-14), resulting in 19 mappable units.

The zones in the periglacial region have limited and rather distant analogues in the modern vegetation. This region was characterized by widespread steppe and forest-steppe vegetation and polar deserts and by the near absence of forest vegetation. Polar deserts are reconstructed in parts of eastern Arctic Siberia and on islands in the East Siberian Sea; this suggests that the exposed Arctic Ocean Shelf was entirely covered by this type of vegetation. High-mountain deserts and tundras are reconstructed for the mountains of central and eastern Siberia.

Vast areas along the periphery of the Scandinavian ice sheet were covered by a very unique and complex vegetation consisting of tundra, steppe, forests of *Larix, Pinus,* and *Betula*, and scattered halophytic communities. Although referred to in eastern Europe as "periglacial vegetation," these formations extended far into eastern Siberia, well beyond the glaciated plains. Farther south and in Europe, as well as in Siberia, forest-steppe vegetation was dominated by steppe-type herb communities, *Larix, Pinus,* and *Betula*. Tundra elements occurred sporadically in topographic depressions. Because of the uniqueness of these plant communities, this vegetation zone is referred to as "periglacial forest-steppe." At the same latitude but in the more oceanic climate, near the Okhotsk Sea and on Kamchatka, an open forest with *Larix* and *Betula* prevailed along with tundra.

Farther south, on the plains of Europe and western Siberia, the vegetation is viewed as periglacial steppe because of its floristic composition. Isolated forest communities dominated by broad-leaved species grew in the European steppe. In the southwestern Urals and in the western Altay Mountains of Siberia, coniferous/broad-leaved forests existed in isolated refuges.

The southernmost USSR represents the extraglacial region, because both paleobotanic and historicofloristic data consistently show that vegetation of the glacial maximum has no analogues. *Artemisia*/grass steppe and semideserts are reconstructed for the plains of the extraglacial region, and mountain steppe and coniferous/broad-leaved and broad-leaved forests are reconstructed for the mountains. Similar forests undoubtedly existed in the extreme south of the Far East as well.

References

Chebotareva, N. S., and Makarycheva, I. A. (1974). "The Last Glaciation of Europe and Its Geochronology." Nauka Press, Moscow.

Gerasimov, A. I. (ed.) (1969). "Endemic Mountainous Plants of Central Asia." Nauka Press, Novosibirsk.

Giterman, R. E., Golubeva, L. V., Zaklinskaya, E. D., Koreneva, E. V., Matveyeva, O. V., and Skiba, L. A. (1968). "Main Stages of the Development of Central Asia's Vegetation in the Anthropogene." Nauka Press, Moscow.

Grichuk, V. P. (1937). A new method of treating sedimentary rocks for purposes of pollen analysis. *Soviet Section of the International Society for the Study of the Quaternary, Trudy* 3, 159-65.

Grichuk, V. P. (1949). Exploration of the process of formation of broad-leaved forests in the Eastern European Plain in the Quaternary. *Voprosy geografii* 12, 79-96.

Grichuk, V. P. (1961). Fossil floras as a paleontological basis of the stratigraphy of Quaternary deposits. *In* "Relief and Stratigraphy of Quaternary Deposits of the Northwestern Russian Plain" (K. K. Markov, ed.), pp. 25-71. USSR Academy of Sciences Press, Moscow.

Grichuk, V. P. (1969). Glacial floras and their classification. *In* "The Last Ice Sheet in the Northwestern European USSR" (I. P. Gerasimov, ed.), pp. 57-70. Nauka Press, Moscow.

Grichuk, M. P. (1970). Principles of formation of recent spore-pollen spectra as the basis for interpreting fossil spore-pollen spectra. *In* "History of the Development of Vegetation of the Extraglacial Zone of the West Siberian Lowland in the Late Pliocene and Quaternary" (V. N. Saks, ed.). Nauka Press, Moscow.

Grichuk, V. P. (1971). Analysis of the structure of the Pleistocene vegetational cover across the USSR. *Pollen et Spores* 8, 101-16.

Grichuk, V. P. (1972). Principal stages of the history of vegetation in southwest of the Russian Plain in the Late Pleistocene. *In* "Palynology of the Pleistocene" (V. P. Grichuk, ed.), pp. 9-53. USSR Academy of Sciences, Institute of Geography, Moscow.

Grichuk, M. P., Shumova, G. M., and Shiporina, I. A. (1967). Application of a new method of isolating pollen from Pleistocene loesslike and clayey deposits. *Moscow State University, Vestnik* 3, 56-59.

Khotinskiy, N. A. (1977). "The Holocene of Northern Eurasia." Nauka Press, Moscow.

Kind, N. V. (1974). "Geochronology of the Late Anthropogene Based on Isotope Data." Nauka Press, Moscow.

Kleopov, Yu. D. (1941). Main developmental characteristics of the flora of broad-leaved forests of the European USSR. *In* "Materials on the History of the Flora and Vegetation of the USSR" (V. L. Komarov, ed.), pp. 183-256. USSR Academy of Sciences Press, Moscow and Leningrad.

Korotkiy, A. M., Karaulova, L. P., and Toritskaya, T. S. (1980). "The Quaternary Deposits of the Maritime Territory." Nauka Press, Novosibirsk.

Korovin, E. P. (1961). "Vegetation of Central Asia and Southern Kazakhstan," Book I. Uzbek Academy of Sciences Press, Tashkent.

Kozo-Polyanskiy, B. M. (1931). "In the Land of Living Fossils: An Outline of the History of Mountain Pine Forests on the Steppe Plain of the Central Black Earth Region." Uchebno-pedagogicheskoye Press, Moscow.

Krasheninnikov, I. M. (1951). Principal modes of development of southern Urals' vegetation in relation to the paleogeography of northern Eurasia in the Pleistocene and Holocene. *In* "Geographical Studies" (A. I. Solovyev, ed.), pp. 170-217. Geografgiz Press, Moscow.

Krasilov, V. A. (1972). "Paleoecology of Terrestial Plants." Far East Research Century, Vladivostok.

Kurentsova, G. A. (1973). "Natural and Anthropogenic Alternations of the Vegetation of the Maritime Territory and Southern Amur Region." Nauka Press, Novosibirsk.

Lavrenko, Ye. M. (1938). History of the flora and vegetation of the USSR based on data of the present distribution of plants. *In* "The Vegetation of the USSR" (Yu. D. Tsinserling, ed.), Vol. 1, pp. 235-96. USSR Academy of Sciences Press, Moscow and Leningrad.

Makhnach, N. A. (1971). "Stages of Development of Belorussia's Vegetation in the Anthropogene." Nauka i Tekhnika Press, Minsk.

Maleyev, V. P. (1941). Tertiary relicts in the flora of western Caucasus and principal stages of the Quaternary history of its flora and vegetation. *In* "Materials on the History of the Flora and Vegetation of the USSR" (V. L. Komorav, ed.), Vol. 1, pp. 61-144. USSR Academy of Sciences Press, Moscow and Leningrad.

Maleyev, V. P. (1948). Principal stages in the development of vegetation of the Mediterranean and mountain regions of the south of the USSR (Caucasus and Crimea) in the Quaternary period. *Gosudarstvennogo Nikitskogo Botanicheskogo Sada, Trudy* 25, 1-251.

Markov, K. K. (1955). Geography of USSR territory in the Quaternary-Anthropogene (basic concepts). *In* "Outlines of the Geography of the Quaternary" (K. K. Markov, ed.), pp. 5-24. Geografgiz Press, Moscow.

Neustadt, M. I. (1957). "History of the Forests and Paleogeography of the USSR in the Holocene." USSR Academy of Sciences Press, Moscow.

Panychev, V. A. (1979). "Radiocarbon Chronology of Alluvial Deposits of the Cis-Altay Plain." Nauka Press, Novosibirsk.

Saks, V. N. (ed.) (1970). "History of the Development of Vegetation in the Extraglacial Zone of the West Siberian Lowland in the Late Pliocene and Quaternary." Nauka Press, Moscow.

Sochava, V. B. (1946). Some aspects of the florogenesis and phytocenogenesis of the Manchurian mixed forest. *In* "Materials on the History of the Flora and Vegetation of the USSR" (V. L. Komarov, ed.), Vol. 2, pp. 283-320. USSR Academy of Sciences Press, Moscow and Leningrad.

Tolmachev, A. I. (1974a). "Introduction to Plant Geography." Leningrad State University, Leningrad.

Tolmachev, A. I. (ed.) (1974b). "Endemic Mountain Plants of Central Asia." Nauka Press, Novosibirsk.

Tolmachev, A. I., and Yurtsev, V. A. (1970). History of the Arctic flora in relation to the history of the Arctic Ocean. *In* "The Arctic Ocean and Its Littoral in the Cenozoic" (A. I. Tolmachev, ed.), pp. 87-100. Gidrometizdat Press, Leningrad.

Vdovin, V. V., Votakh, M. R., and Zudin, A. N. (1969). Materials pertaining to the stratigraphy of the Pleistocene of the Chumysh Salair region. *In* "The Quaternary Geology and Geomorphology of Siberia" (V. N. Saks, ed.), pp. 58-67. Nauka Press, Novosibirsk.

Vul'f, E. V. (1941). Historical plant geography. *In* "History of Floras of the Globe" (S. Yu. Lipshits, ed.), pp. 3-546. USSR Academy of Sciences Press, Moscow and Leningrad.

CHAPTER 18

Holocene Vegetation History

N. A. Khotinskiy

Studies of the Holocene vegetational history in the USSR have been based mainly on pollen analyses. Early analyses of lacustrine-paludal deposits were made by Sukachev (1906), Dokturovskiy (1918), and Dokturovskiy and Kudryashov (1923). The study of arboreal pollen provided a comprehensive picture of Holocene forest history in the USSR (Neustadt, 1957). New information has since been obtained by isolating nonarboreal pollen and spores and by closer taxonomic identifications. Advances in the study of the Holocene vegetational history have been described (Gudelis and Neustadt, 1961; Neustadt, 1965, 1969, 1971; Kotinskiy and Koreneva, 1973, Savina, 1975). Very recently, studies have focused on the history of the northern and eastern USSR, where numerous palynologic and radiocarbon data have already been obtained. Regions that remain poorly studied include the mountainous regions of the southern USSR, the plains of Central Asia, and part of Siberia and the Far East. Radiocarbon dating has made it possible to correlate vegetational records across northern Eurasia from western Europe to the Pacific coast (Khotinskiy, 1977).

Methodological Premises

The duration of the Holocene, or more accurately the age of its lower boundary, ranges between 16,000 and 8000 yr B.P. in the USSR. Despite existing discrepancies, many investigators (Markov, 1965; Neustadt, 1957; Kind, 1974; Khotinskiy, 1977) have come to the conclusion that the lower boundary should be synchronous and time parallel, rather than time transgressive, for all regions. The difficulties in determining the lower boundary of the Holocene are due to the fact that at the end of the last glaciation several sharp natural changes occurred that reflected the fluctuating character of the climatic transition.

The paleogeographic data obtained show that the most significant environmental changes in the USSR occurred around 10,300 yr B.P. At that time, hyperzonal vegetation was quickly replaced by zonal vegetation, atmospheric circulation was transformed from a meridional to a zonal type, and the Paleolithic stage was succeeded by the Mesolithic stage. In accordance with the opinions of most Soviet investigators, this dividing line will be taken as the beginning of the Holocene.

A modified Blytt-Sernander scheme (Figure 18-1), which adequately reflects the character of global climatic fluctuations, was adopted as the chronologic-paleogeographic standard of the Holocene. Use of this scheme as an international standard could significantly facilitate correlation and comparison of Holocene paleogeography throughout the world.

The most complete and reliable information on Holocene vegetational history was obtained from palynologic studies of well-dated peat sections. In regions where bogs are absent, such steppe and desert areas, palynologic data on alluvial and other inorganic deposits have been used. About 1000 pollen diagrams and about 700 radiocarbon dates are used to reconstruct the Holocene vegetation, and vegetational maps for the Boreal period (9000 to 8000 yr B.P.) (Figure 18-2) and the Late Atlantic time (6000 to 4600 yr B.P.) are presented (Figure 18-3). All these data were used to compile the vegetation maps.

The Holocene vegetational history can be traced through pollen diagrams from the several sites. The Polovetsko-Kupanskoye mire (Figure 18-4), located 150 km north of Moscow, lies in a subzone of mixed broad-leaved/coniferous subtaiga forest. The diagram covers the late-glacial Alleröd and Younger Dryas periods and the Holocene from the Preboreal period to the present time.

The Melent'yevo bog (Figure 18-5) is located in northeastern European USSR in the northern dark coniferous taiga. The diagram characterizes vegetation during the Alleröd period and the Holocene from the Boreal period to the present time (with some gaps).

The Medvyanka River valley mire (Figure 18-6) lies near the town of Tula in the broad-leaved forests of the south-

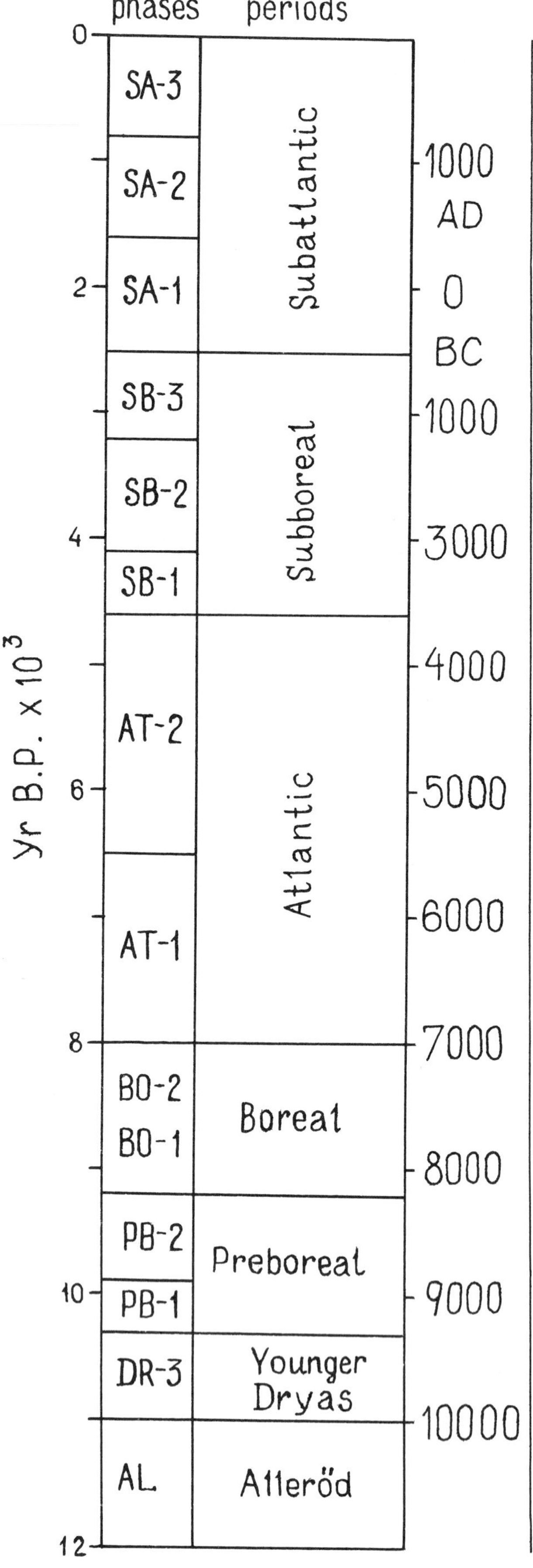

Figure 18-1. Chronologic subdivision of the Holocene.

ern forest zone of the European USSR. The diagram spans the Younger Dryas period and the Holocene from the Preboreal period to the Sub-Atlantic period.

The Chernovskoye mire (Figure 18-7) is located 30 km northwest of the city of Sverdlovsk, on the eastern slope of the Urals in the southern taiga. The diagram covers almost the entire Holocene from the late Preboreal to the present.

The Nizhnevartovskoye mire (Figure 18-8) is located in the southern taiga of western Siberia in the Ob' River valley. The diagram characterizes vegetation during the entire Holocene.

The Uandi mire (Figure 18-9) is found on Sakhalin Island, in the Tatar Straits, within the dark coniferous montane forest. The diagram spans the Younger Dryas and the entire Holocene.

The Medvyanka River valley mire was the only site where the pollen sum was based on the total arboreal and nonarboreal pollen, excluding *Alnus*. At the remaining sites, the pollen sum includes just arboreal pollen, excluding *Corylus*.

In the vegetational reconstruction, the Holocene pollen spectra seem to reflect adequately the character of past vegetation. This assumption is supported by numerous data comparing the pollen spectra of surface samples with modern vegetation. Overall, modern pollen spectra in the USSR seem to delineate vegetation zones and subzones (Neustadt, 1957), but detailed studies have been confined generally to the European USSR and western and central Siberia.

Correlation of Holocene deposits by palynologic and radiocarbon data is based on the following assumptions. Any pollen diagram integrates widespread vegetational changes, because pollen spectra combine local and regional features (Khotinskiy, 1977). Synchronous changes in pollen diagrams reflect a simultaneous and regional change in vegetation caused by a sharp climatic change. Thus, regional subdivisions of the Holocene can be discovered from palynologic data, as can local vegetational patterns. Regional climatic phases identified from paleobotanic and other paleogeographic data are subsequently correlated to form a regional Holocene standard. Account is taken of the fact that vegetational change does not always coincide exactly with climatic change. The lag in vegetational response following secondary or brief climatic fluctuations is well known, and vegetational boundaries related to migration are metachronous. Therefore, it is useful to be guided by the nonmigratory (autochthonous) component of the vegetation, because it reflects almost "instantaneously" to major climatic changes. Holocene climatic fluctuations effected a more or less simultaneous change in *local* vegetation over vast areas, suppressing some elements but favoring the development of others. The pollen levels reflecting changes in local vegetation should be synchronous over the entire region affected by the climatic change. Many plants (for example, spruce, birch) have the ability to switch to vegetative reproduction, during which the pollen productivity sharply decreases. Under favorable conditions, pollen production can increase as the plants switch back to seed reproduction.

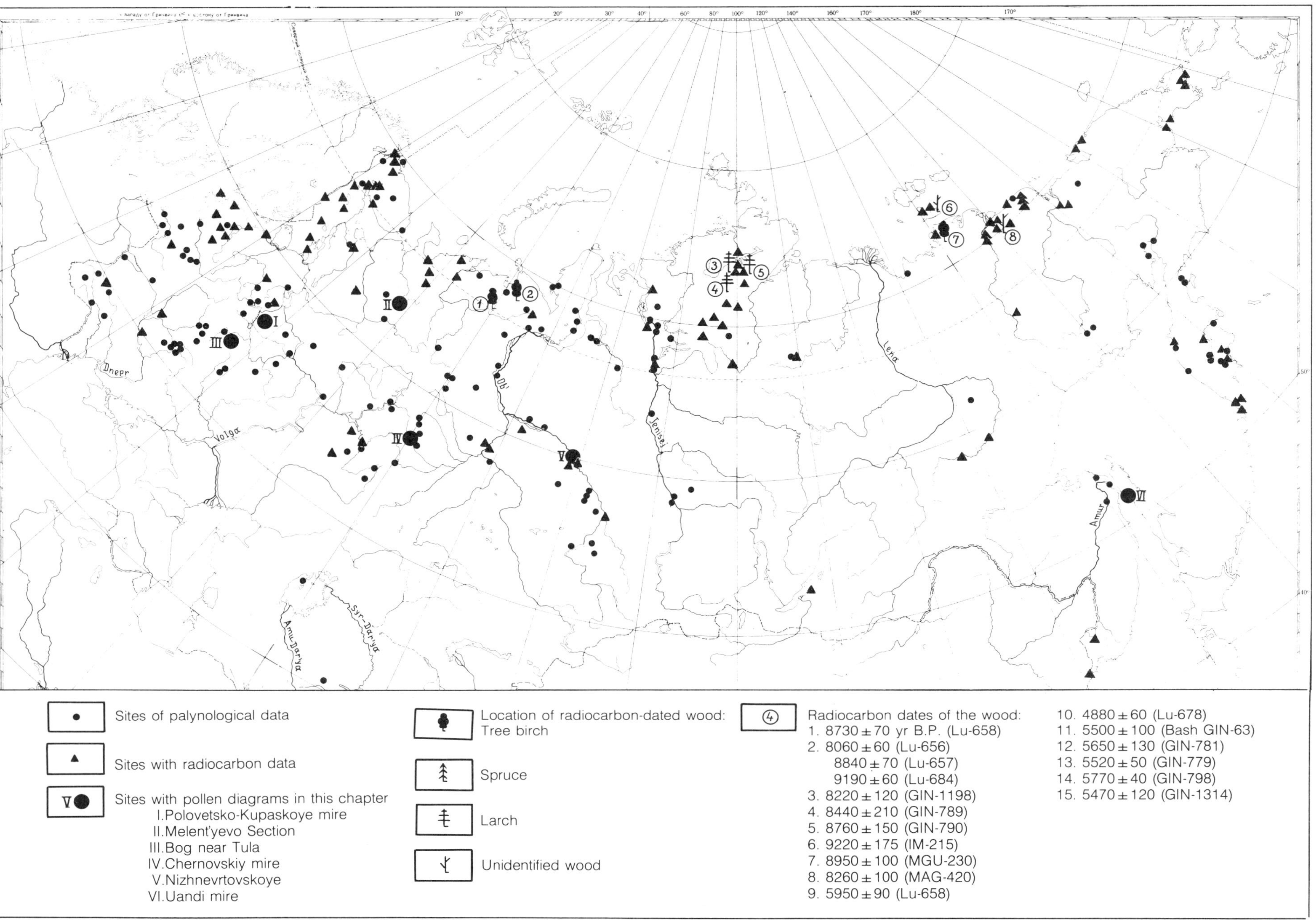

Figure 18-2. Map of data for the Boreal period (9000-8000 yr B.P.). (Compiled by T. A. Serebryannaya.) Sites of radiocarbon-dated wood are based on materials provided by S. M. Andreyeva, Kh. A. Arslanov, G. F. Gravis, G. V. Ivanenko, T. N. Kaplina, Yu. A. Lavrushin, A. B. Lozhkin, L. D. Nikiforova, M. V. Nikolyskaya, L. D. Sulerzhitskiy, L. V. Tarakanov, N. S. Shuliy, A. S. Lavrov, and V. M. Makeyev.

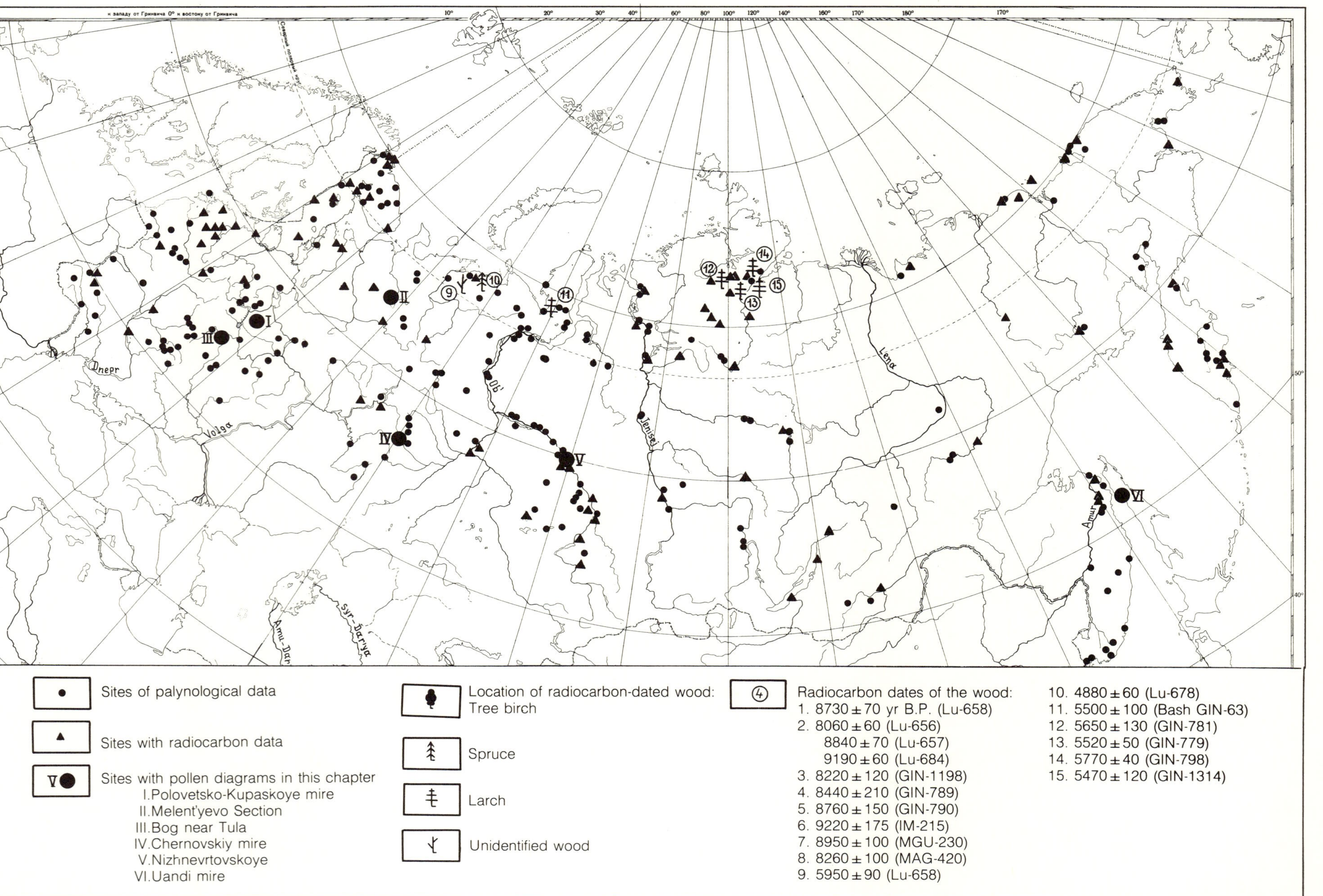

Figure 18-3. Map of data for the second half of the Atlantic period between 6000 and 4600 yr B.P. (Compiled by T. A. Serebryannaya.)

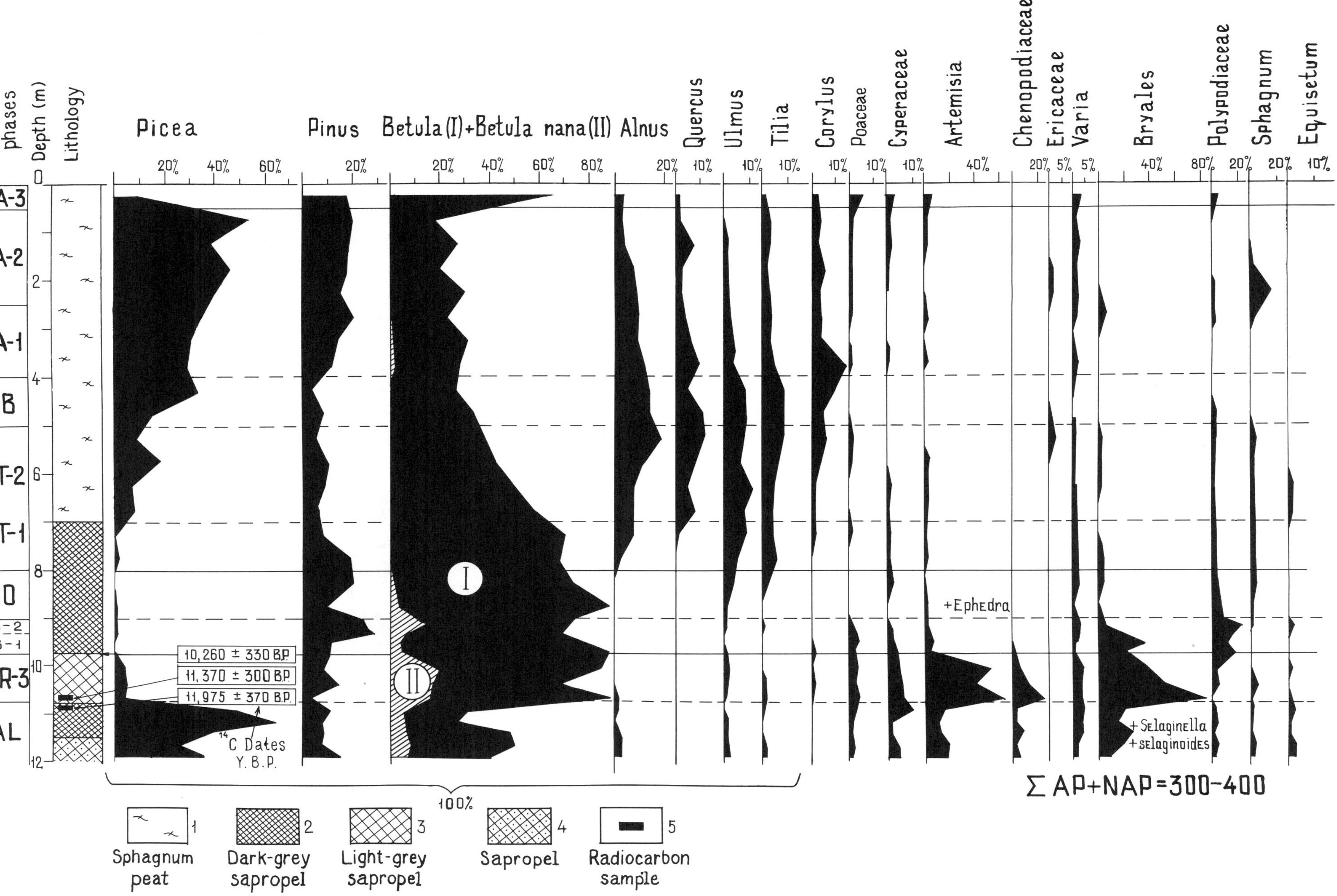

Figure 18-4. Pollen diagram of Polovetsko-Kupanskoye mire. (Analyst: N. A. Khotinskiy.) (The radiocarbon date at the DR-3/PB boundary, 10,260 ± 330 yr B.P. (Mo-268), was obtained from a section of Somino Lake, 1 km south of Melekhovo Bog.)

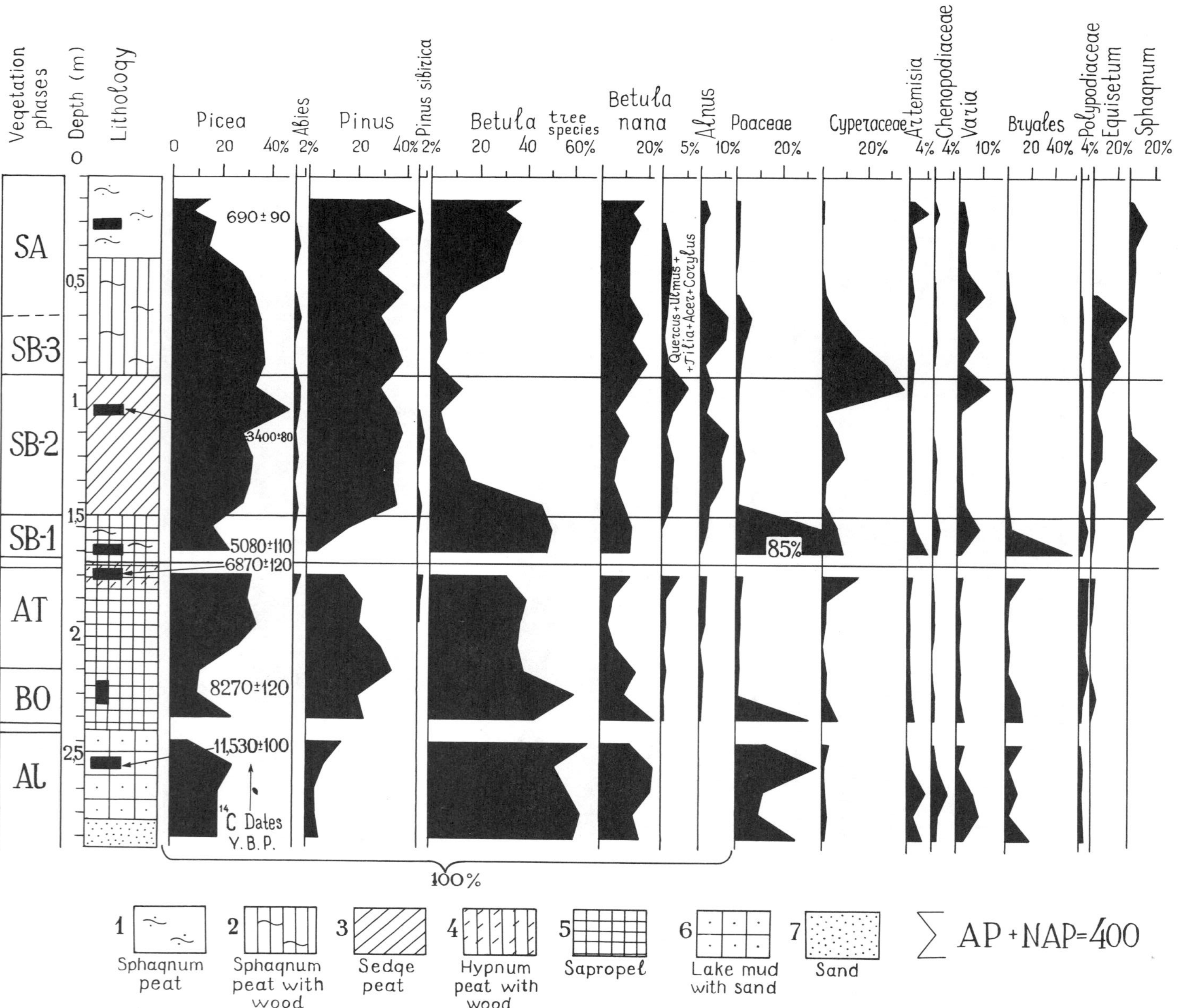

Figure 18-5. Pollen diagram of Melent'yevo mire. (Analyst: L. D. Nikiforova.)

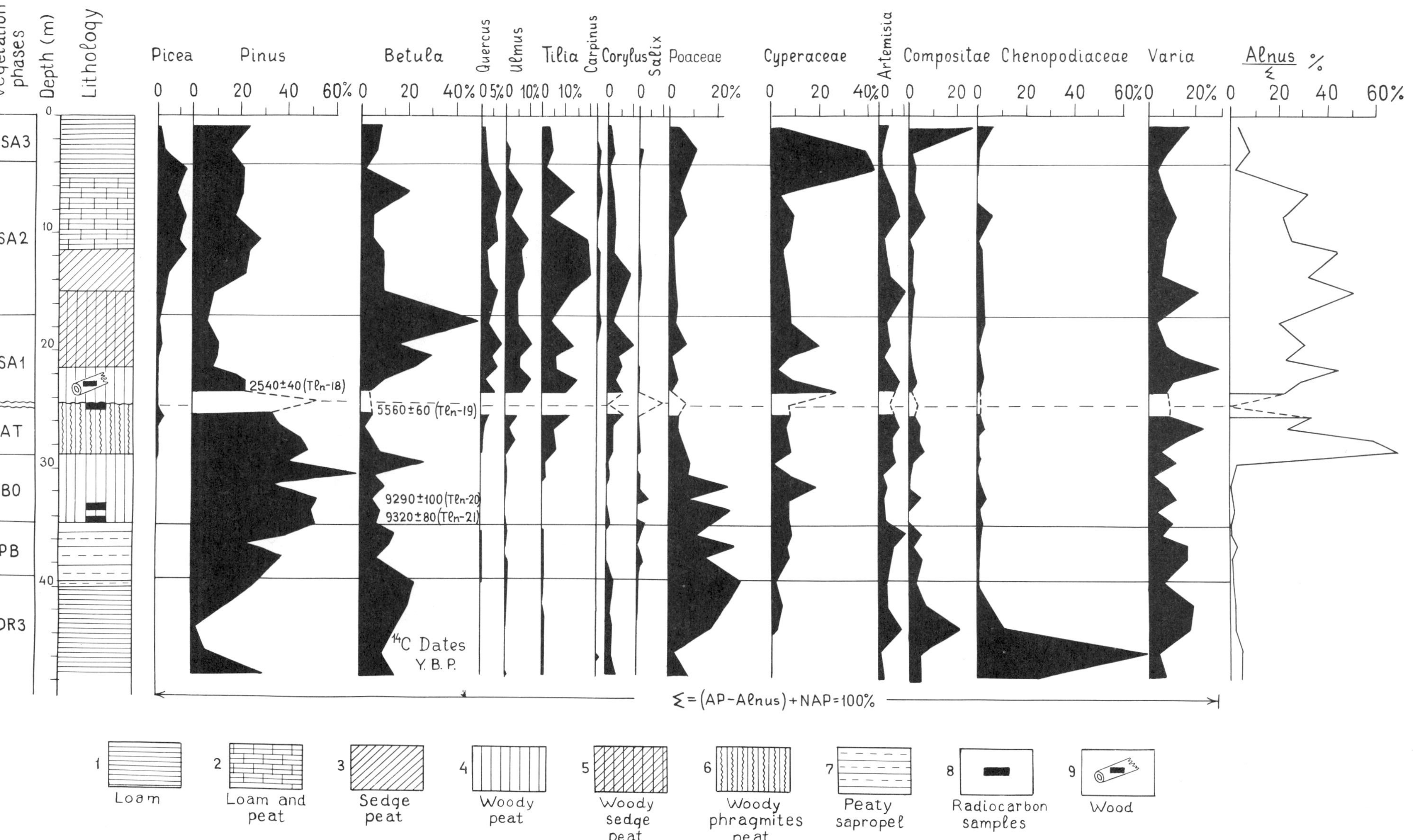

Figure 18-6. Pollen diagram of bog near the town of Tula. (Analyst: T. A. Serebryannaya.)

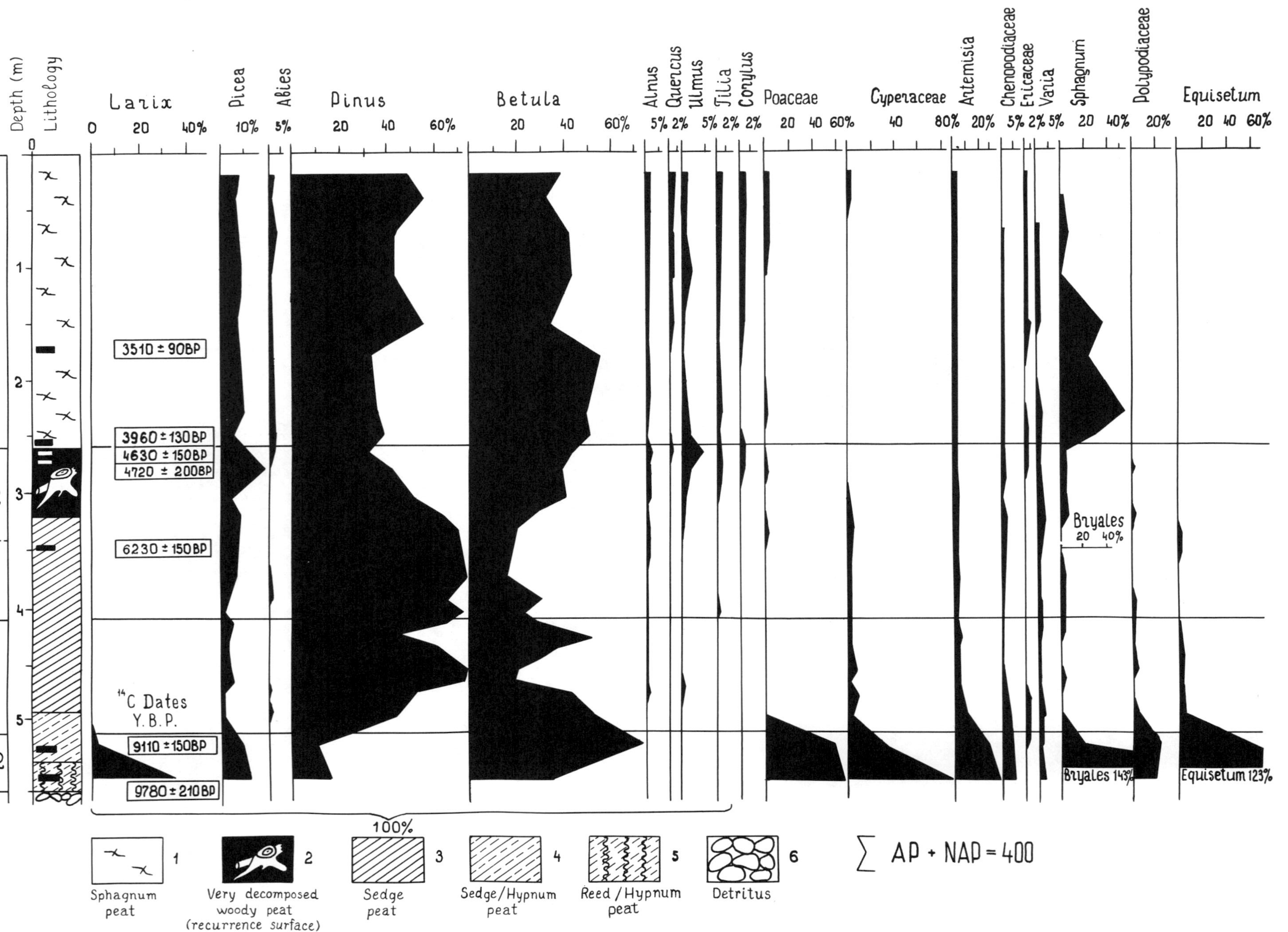

Figure 18-7. Pollen diagram of Chernovskoye mire. (Analyst: N. A. Khotinskiy.)

Thus, changes in pollen spectra may be erroneously attributed to plant migrations when, in fact, they reflect only a change in the reproductive mode of local vegetation.

The Holocene record of northern Eurasia contains a number of bio- and climatostratigraphic boundaries that reflect synchronous vegetational and climatic change over enormous areas. The following boundaries are delimited by numerous radiocarbon dates: the late-glacial/postglacial boundary between the Younger Dryas and Preboreal periods (DR-3/PB) is at about 10,300 yr B.P., the Preboreal/Boreal (PB/BO) at 9200 yr B.P., the Boreal/Atlantic (BO/AT) at 8000 yr B.P., the Atlantic/Subboreal (AT/SB) at 4600 yr B.P., and the Subboreal/Sub-Atlantic boundary (SB/SA) at about 2500 yr B.P. These boundaries considerably facilitate long-range correlation of Holocene deposits.

Principal Stages and Boundaries in the Vegetational History

Late-glacial time is notable as having a severely continental climate, when a large part of the USSR was covered by a combination of tundra, forest, and steppe elements—the phenomenon of zone mixing (Grosset, 1961) or hyperzonality (Velichko, 1973). The late-glacial pollen spectra contain not only forest and tundra elements but also taxa that today are found in the steppe and desert zones. The widespread occurrence of steppe is suggested by the significant amount of *Artemisia* and Chenopodiaceae pollen. Tundra communities are recognized by pollen of dwarf birches (*Betula nana*, etc.) and spores of tundra species of *Lycopodium* and *Selaginella*. (See late-glacial segments of the Polovetsko-Kupanskoye, Melent'yevo, and Uandi diagrams, Figures 18-4, 18-5, and 18-9.)

In late-glacial time, there occurred an unusual meeting of the North with the South, during which present-day vegetation zones shifted and formed unusual assemblages. Within a degraded forest zone, tundra communities penetrated far south, and steppe communities far north, over vast areas of northern Eurasia. Birch, pine, and spruce were part of the forest vegetation west of the Urals, and larch stands grew to the east. In the coastal regions of the Far East (Kamchatka, Sakhalin), there are no indications of extensive "cold" steppe development. Instead, late-glacial pollen assemblages are dominated by tundra and mountain-tundra taxa (Figure 18-9). Thus, in both the latitudinal and meridional profiles one notes certain provincial differences in the late-glacial vegetation of northern Eurasia.

Late-glacial pollen data from western Europe and the European USSR indicate fluctuations in the vegetation and climate. Three cold Dryas epochs (DR-1, DR-2, DR-3) are separated by the relatively warm Bölling (BO) and Alleröd (AL) interstades. Cold stages were characterized by increased continentality and widespread treeless landscapes covered by "cold" steppe and tundra communities. During the interstades, continentality became somewhat attenuated, and the role of woody formations increased. Results reported by Kind (1974) and other investigators suggest that similar fluctuations also occurred in Siberia.

The unique environmental conditions during late-glacial time were chiefly the result of a climate that featured short, hot summers; long, cold winters; large temperature differences; and little precipitation. Analogues of such conditions have not been found anywhere. Only the present climates of Yakutiya and the continental regions of the Northeast remotely resemble the past climate, because remnants of the late-glacial vegetation complex of the tundra, forest, and steppe, in particular the "cold" steppe, are found in central Yakutiya and in the Yana, Indigirka, and Kolyma River basins, growing adjacent to forest and tundra communities (Karavayev, 1965; Yurtsev, 1974).

The Northeast provides an incomplete but close analogue of the late-glacial tundra and steppe landscape that covered vast areas of northern Eurasia. Tundra and steppe plants did not form mixed associations: Steppe communities usually developed on the southern, well-heated slopes of water divides, river terraces, and similar areas, whereas tundra communities usually grew in moist areas of low relief.

Steppe and tundra communities could have coexisted in a severely continental climate, because the summers were short but fairly warm. The short vegetative period of many plants growing both in modern tundra and steppe environments is a result of insufficient heat in the tundra and insufficient moisture in the steppe. Four types of climatic regimes are observed in the modern steppe zone of the USSR (Budyko, 1977): an arctic regime in the winter, a tundra regime in the early spring, a forest regime in the late spring, and a steppe regime in the summer. If the tundra zone is evaluated in the same fashion, similar climatic stages should be recognized. It is this similarity that permitted tundra and steppe vegetation to form such an unusual complex.

The boundary between late-glacial and postglacial time or between the Younger Dryas and the Preboreal period (DR-3/PB) delimits a period of appreciable change in the physiographic conditions throughout northern Eurasia. In pollen diagrams, this boundary at approximately 10,300 yr B.P. separates the late-glacial period of almost treeless landscapes from the postglacial period of forest vegetation (Figures 18-4, 18-5, and 18-9).

The transition to the postglacial epoch involved a rapid restructuring of landscapes from hyperzonal to zonal type. The unique late-glacial vegetation separated into tundra elements in the North, steppe in the South, and forests of birch, pine, spruce, and larch over a large part of northern Eurasia. The major vegetation zones of northern Eurasia (tundra, forest, and steppe) developed within a few centuries—almost instantaneously in geologic time. Such rapid vegetational changes are the result of a sharp climatic warming and a shift from meridional to zonal circulation patterns.

PREBOREAL PERIOD

The Preboreal period (PB) was a transitional stage in which there was a partial return to late-glacial vegetation during

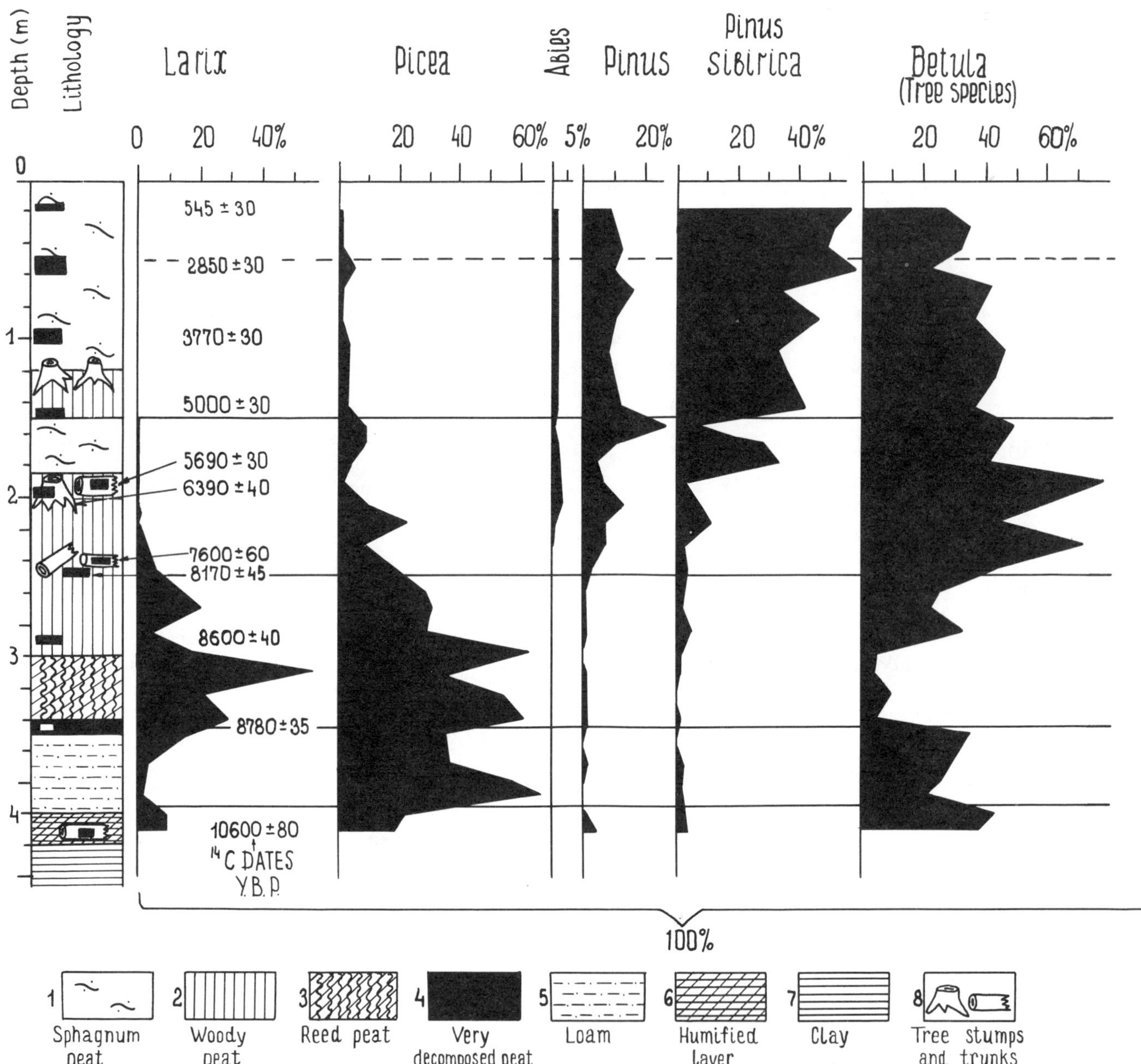

Figure 18-8. Pollen diagram of Nizhnevartovskoye Bog. (Analysts: M. I. Neustadt and E. M. Zelikson.)

the so-called Pereslavl' cooling between 10,000 and 9500 yr B.P., an event synchronous with the Piottino stage in western Europe (Behre, 1967). Evidence of the Pereslavl' cooling was first detected in a section near the town of Pereslavl'-Zalesskiy (150 km north of Moscow) by the increase in *Betula nana, Artemisia,* and Chenopodiaceae pollen (Khotinskiy, 1964). Early Preboreal deposits dominated by tree pollen are assigned to the Polovetsian warming, named from the Polovetsko-Kupanskoye bog (Figure 18-2). The Polovetsian warming (10,300 to 10,000 yr B.P.) is synchronous with the Friesland Interstade in western Europe.

Subsequently, similar changes in Preboreal vegetation were identified in the Urals (Khotinskiy, 1970) and Siberia (Kind, 1974). A general pattern is detected throughout this region. The forest formations, dominant during the Polovetsian warming, were later crowded by glacial-type grass-shrub communities during the Pereslavl' cooling. The partial restoration of late-glacial vegetation during the Pereslavl' climatic deterioration was undoubtedly caused by a new widespread cooling.

BOREAL PERIOD

The Boreal period (BO) was characterized by the complete formation of zonal vegetation in the USSR, that is, by the

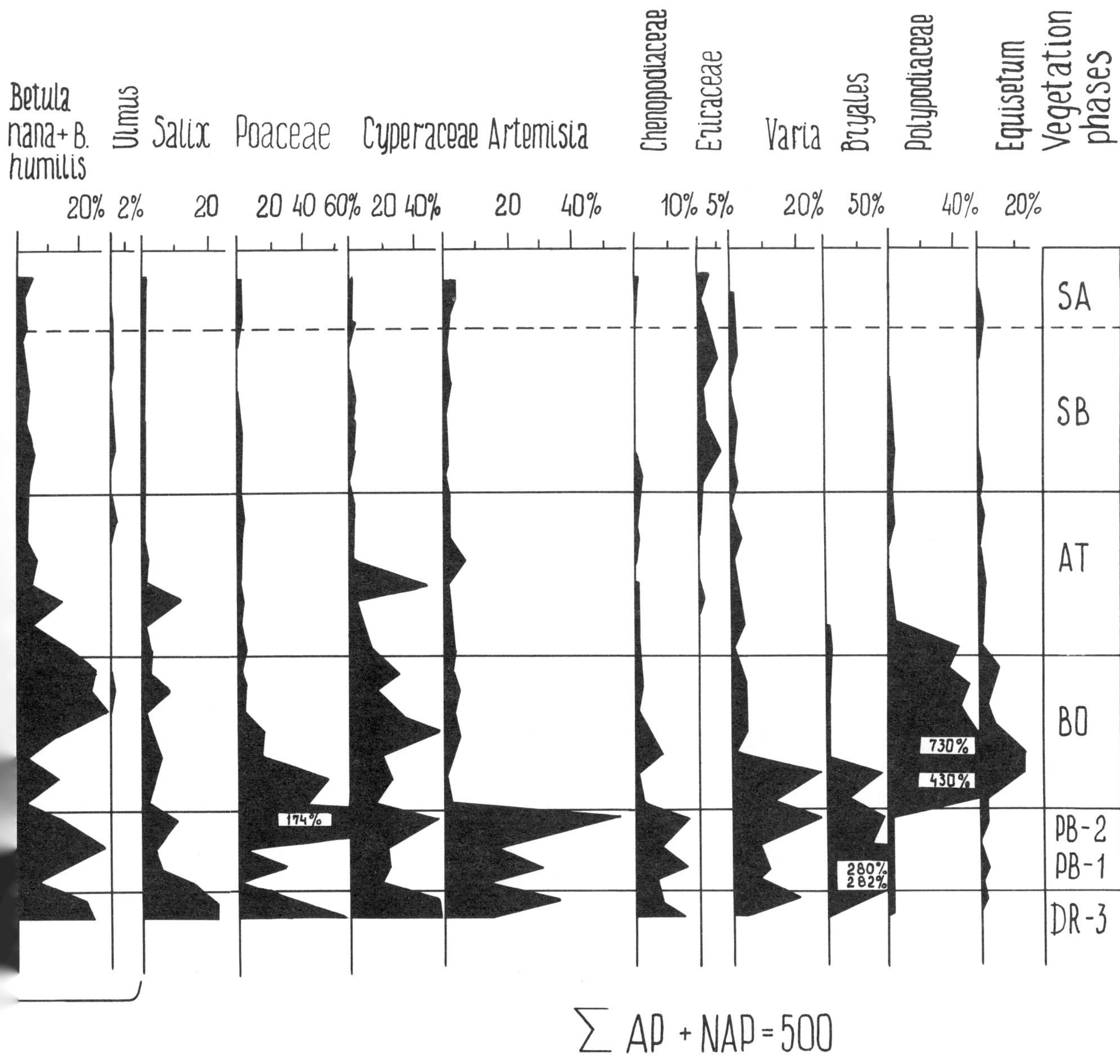

formation of tundra, forest, and steppe zones (Figure 18-10). Late-glacial vegetation had disappeared in most regions, although vegetation differed substantially from the present.

The tundra zone in the European USSR occupied only a narrow maritime strip along the Kola and Yugorskiy Peninsulas. A northward movement of treeline during the Boreal period is indicated by the remains of tree birch found in the modern tundra zone, dated at 8060 ± 60 yr B.P. (LU-6565), 8730 ± 70 yr B.P. (LU-658), 8840 ± 70 yr B.P. (LU-657), and 9190 ± 60 yr B.P. (LU-684) (Figure 18-2). During the Boreal period, tundra expanded to the east, but the southern boundary was 100 to 200 km farther north than at present. On the Taimyr Peninsula, larch wood found within the present tundra is dated 8760 ± 150 yr B.P. (GIN-790) and 8440 ± 210 yr B.P. (GIN-789) near the Vol'shaya Balakhnya River and at 8220 ± 120 yr B.P. (GIN-1198) near Lake Taimyr.

In the northeastern USSR (east of the Lena River), the coastline lay north of its present location, and the Novosibirsk Islands were part of mainland Asia. It is evident that the forest vegetation in these regions, consisting of sparse birch groves, moved at least 100 to 200 km north of its present limit because numerous remains of Boreal-age birch wood have been found in the modern tundra zone. The northernmost site of tree birch, dated at 8950 ± 100

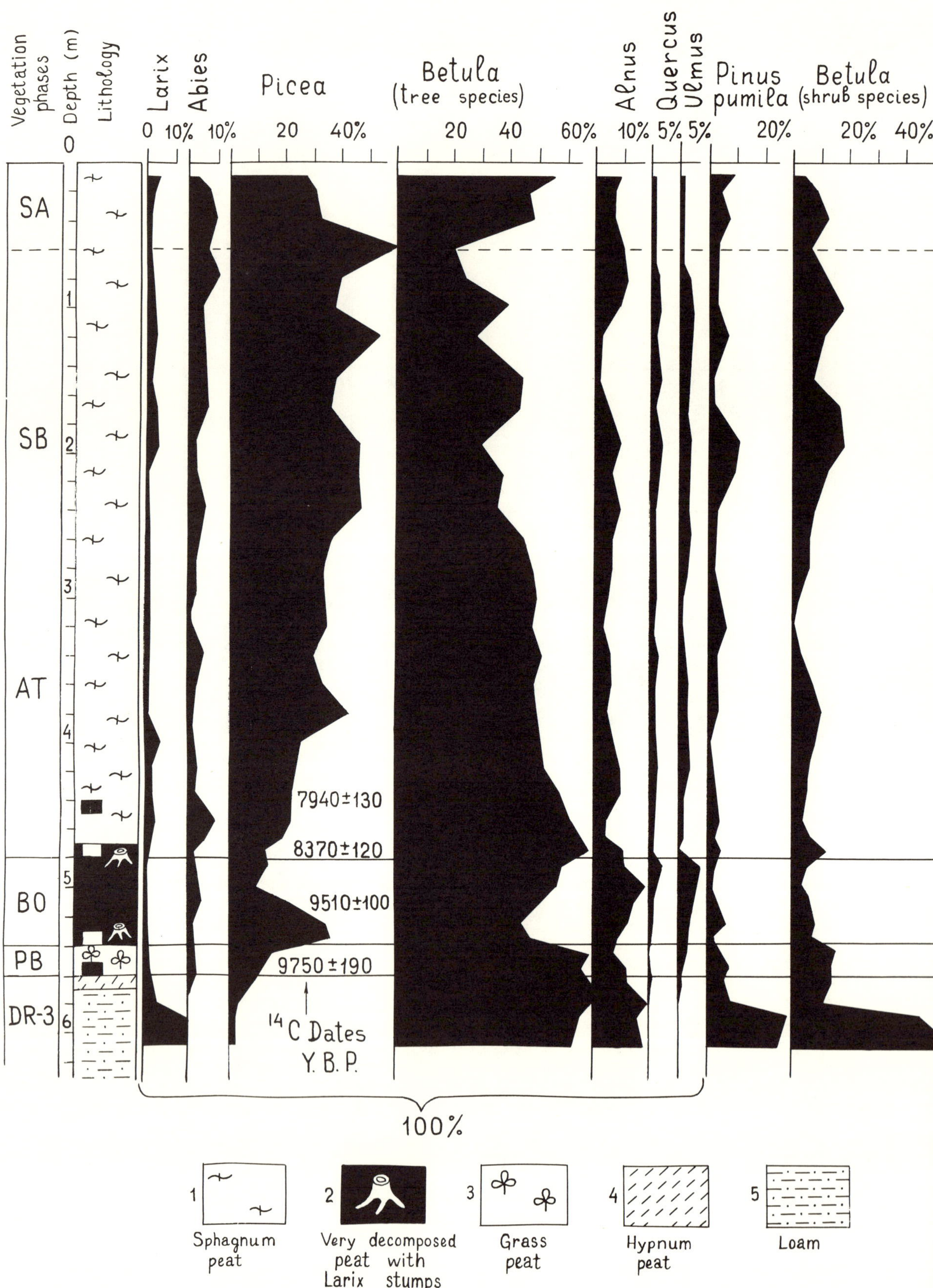

Figure 18-9. Pollen diagram of Uandi Bog. (Analyst: N. A. Khotinskiy.)

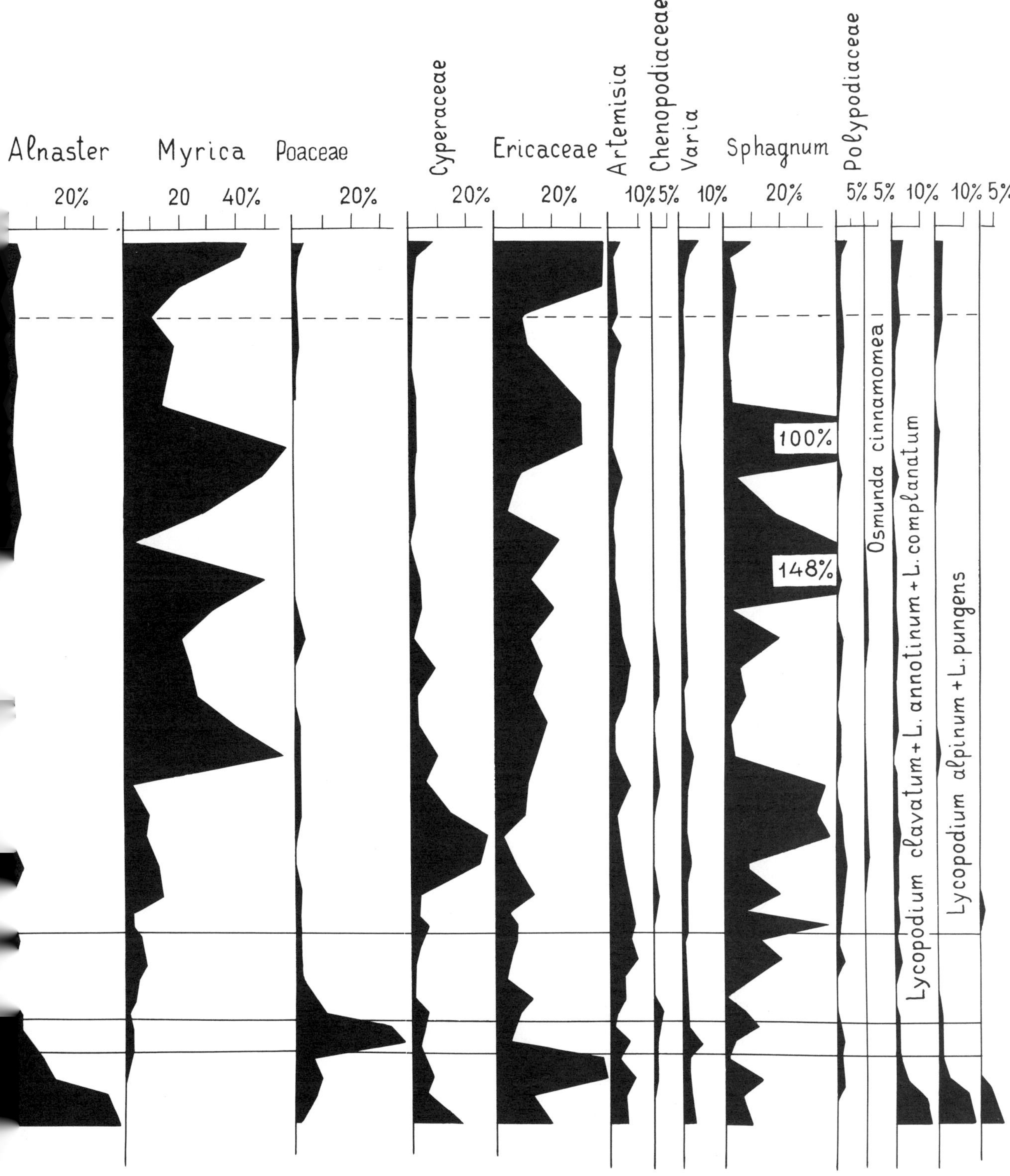

∑ AP + NAP = 500
excluding Sphagnum

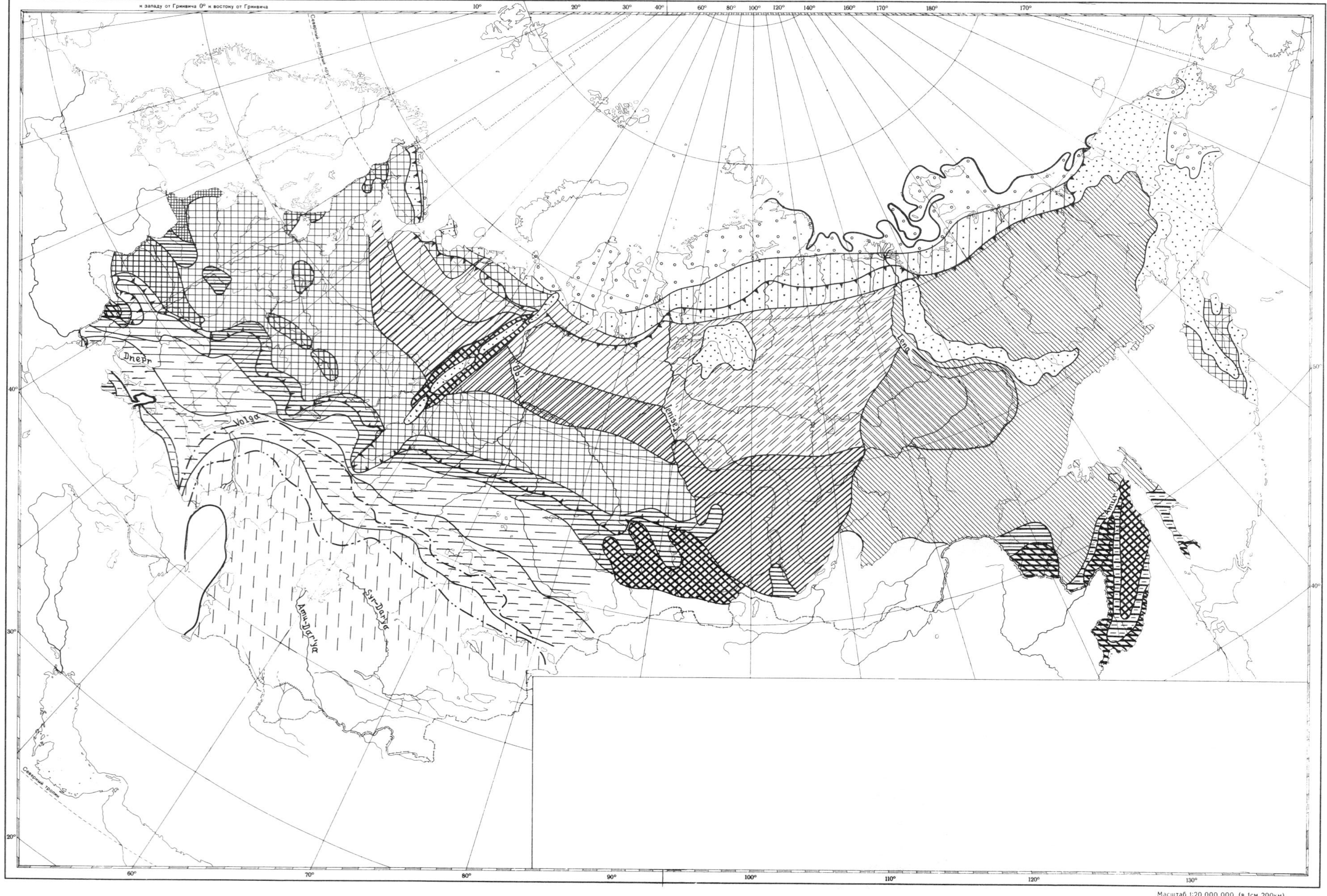

к западу от Гринвича 0° к востоку от Гринвича
Северный полярный круг
Северный тропик
Dnepr
Volga
Ob'
Jenisej
Lena
Amur
Syr-Dar'ya
Amu-Dar'ya
Масштаб 1:20 000 000 (в 1см 200км)
200 0 200 400 600 800 1000 км

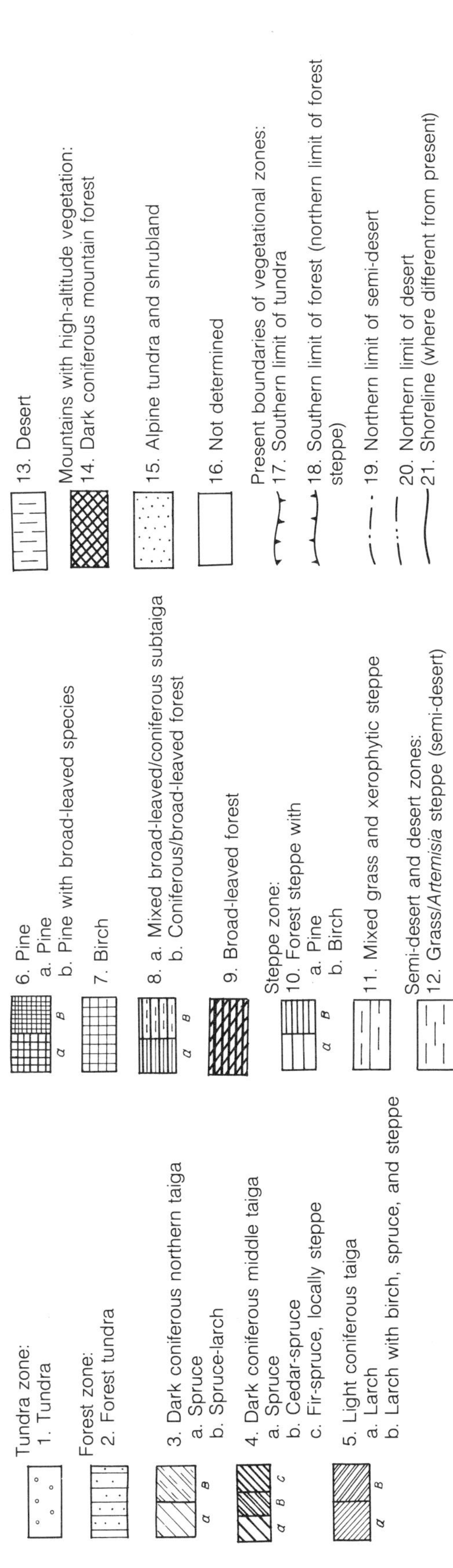

Figure 18-10. Vegetation map for the Boreal period, 9000-8000 yr B.P. (Compiled by N. A. Khotinskiy, with the use of data by M. I. Neustadt, T. A. Serebryannaya, A. T. Artyushenko, A. A. Seybutis, O. F. Yakushko, G. A. Yelina, L. D. Nikiforova, V. S. Volkova, G. M. Levkovskaya, M. V. Nikol'skaya, V. A Belova, A. I. Tomskaya, T. D. Boyarskaya, T. N. Kaplina, and A. V. Lozhkin.

yr B.P. (MGU-230), is near Van'kina Bay (Figure 18-2). (T. N. Kaplina and A. V. Lozhkin report that wood from the Bol'shoy Lyakhovskiy and Kotel'nyy Islands is also from the Boreal period, but, until this report can be verified, these areas are assigned to the tundra zone.) The general advance of the forest vegetation into the tundra during the Boreal period must indicate an appreciable warming across the northern USSR.

The forest zone in the Boreal period occupied an area similar to the present, although in places the forests were different. A subzone of dark spruce taiga extended from northeastern European USSR across the Urals and western Siberia to central Siberia (Figure 18-10). The abundance of spruce pollen in the Boreal sediments (Figure 18-8) indicates that spruce forests were an important part of the vegetation in many areas of Siberia, whereas now they are a minor component or are completely absent. In central Siberia, vast areas of the Yenisey and Lena River basins were occupied by spruce-larch forests. Farther south, they were replaced by cedar (*Pinus sibirica*) and spruce forests and fir and spruce forests. Steppe occurred along the river valleys and within the larch forests of central Yakutiya, where it still survives today. The widespread occurrence of spruce throughout Siberia during the Boreal period indicates a warmer, moister, less continental climate than at present. The intensive development of dark spruce taiga requires a temperate climate, with considerable precipitation, especially during the vegetational period (Tolmachev, 1954).

In the southern Far East along the Amur River and in the Ussuri Territory and Sakhalin, thermophilous broad-leaved and coniferous/broad-leaved forests spread considerably during the Boreal period. At the Uandi bog on Sakhalin (Figure 18-9), the highest pollen percentages of broad-leaved species (*Quercus* and *Ulmus*) are dated at approximately 8500 yr B.P. and assigned to the Boreal period.

Pollen spectra from the Boreal period in the European USSR are characterized by high percentages of birch and pine pollen (Figures 18-4 and 18-6), which suggests that pine-birch forests predominated over large areas of the forest zone. This was indeed the era of birch forests, for they extended from northwestern Europe across the European USSR into the southern Urals and southwestern Siberia. Modern analogues of these birch forests have been preserved only in the southern forest zone of western Siberia, where a belt of aspen-birch forest is located. In the European USSR, pine forests were important on sandy soils and changed little throughout the Holocene. At the same time, coniferous/broad-leaved forests (with *Ulmus, Tilia, Quercus, Corylus*) appeared in the western, southwestern, and other regions of European USSR during the Boreal period. However, extensive development of these thermophilous species was prevented by the relatively dry and cool climate found in most parts of the European USSR. The northern boundary of the forest steppe on the plains of the European USSR was somewhat north of its present position. Steppe communities grew in the southern forest zone as remnants of their former late-glacial distribution.

In the Ukraine south of the forest-steppe, steppe vegeta-

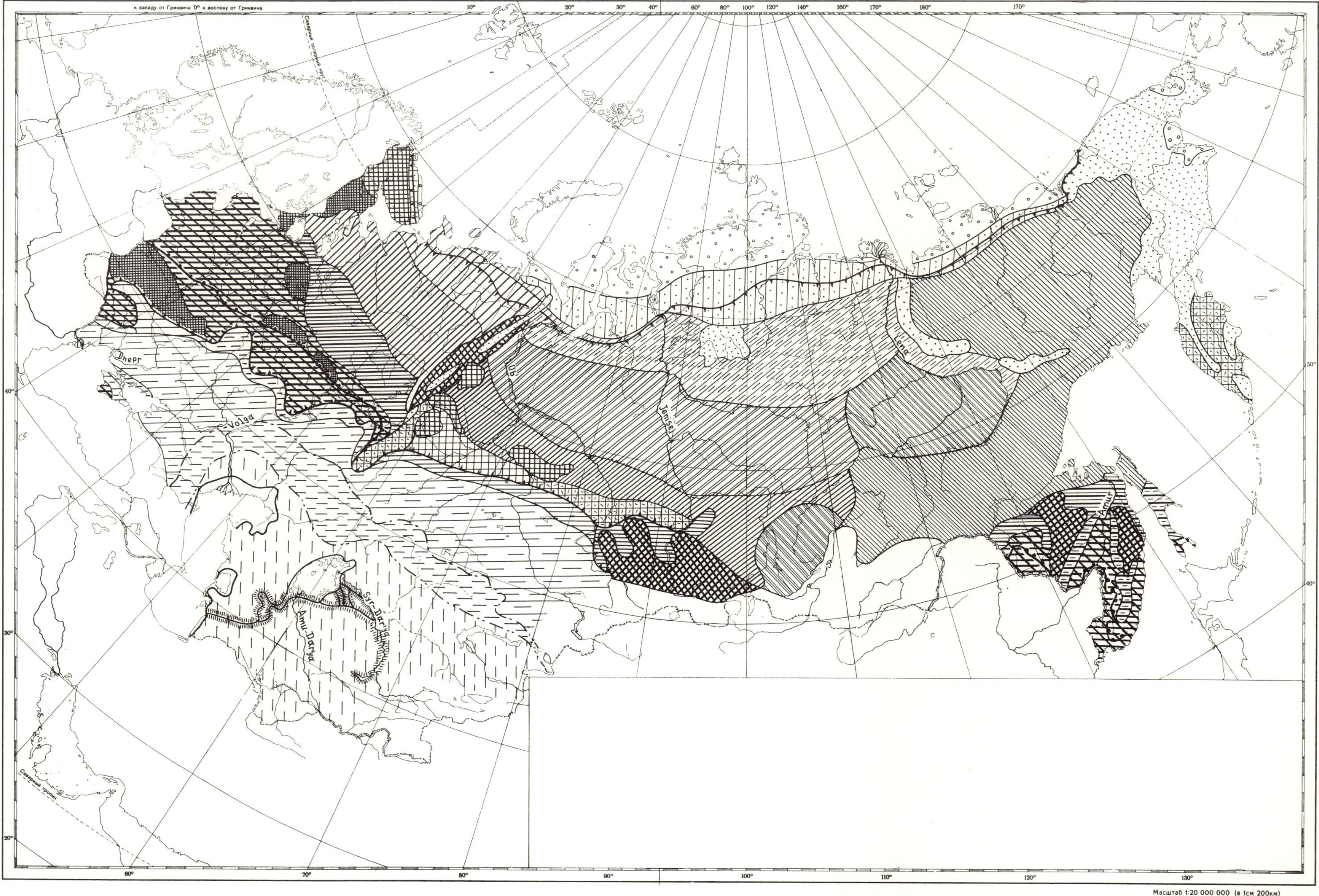
к западу от Гринвича 0° к востоку от Гринвича
Северный полярный круг
Dnepr
Volga
Ob
Jenisej
Lena
Amur
Syr-Darya
Amu-Darya
Северный тропик
Масштаб 1:20 000 000 (в 1см 200км)
200 0 200 400 600 800 1000 км

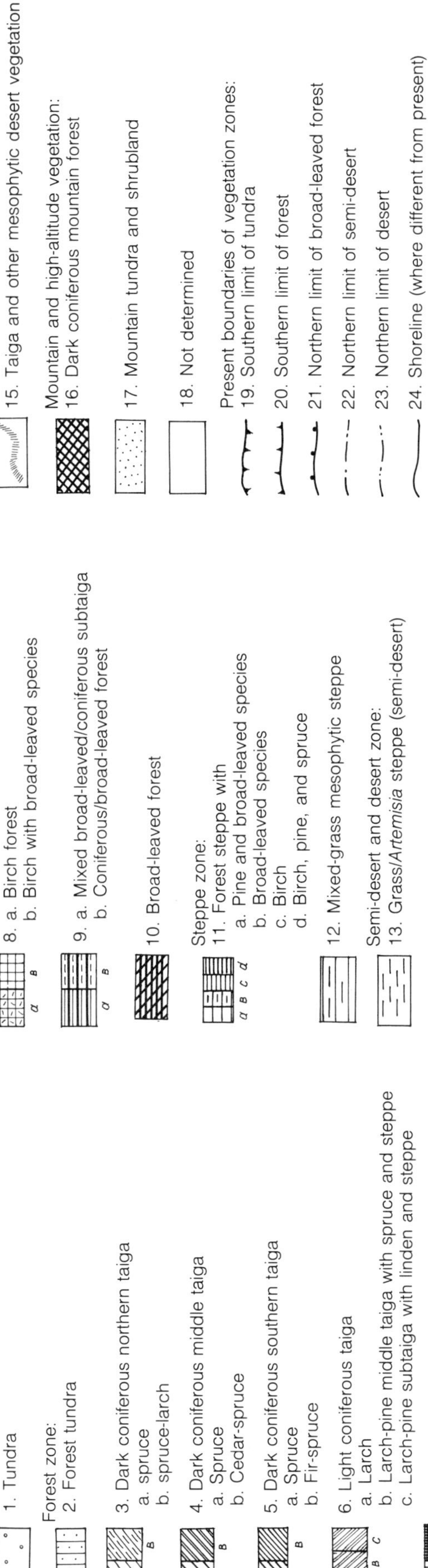

Figure 18-11. Vegetation map for the second half of the Atlantic period, 6000-4600 yr B.P. (Compiled by N. A. Khotinskiy with the use of data by M. I. Neustadt, T. A. Serebryannaya, A. T. Artyushenko, A. A. Seybutis, O. F. Yakushko, G. A. Yelina, L. D. Nikiforova, V. S. Volkova, G. M. Levkovskaya, M. V. Nikol'skaya, V. A. Belova, A. I. Tomskaya, T. D. Boyarskaya, T. N. Kaplina, and A. V. Lozhkin.)

tion was dominated by *Artemisia* and had a more xerophytic character than the semidesert steppe that is now found in the maritime regions of the Black Sea and the Sea of Azov. Apparently, xerophytic steppe and semidesert formations were widespread in the Ukraine considerably north of their present position during the Boreal period. A similar shift to the north probably occurred into the eastern European USSR and western Siberia as well if continentality increased to the east.

The Boreal period of northern Eurasia ended in a cooling around 8000 yr B.P. A marked southward shift of the tundra-forest boundary occurred in the northeastern European USSR (Nikiforova, 1980). In Siberia, the Novosanchugovskiy cooling between 8000 and 7900 yr B.P. is characterized by a replacement of taiga vegetation by forest tundra in the north and by a degradation of dark coniferous taiga in the south (Kind, 1974). A correlative cooling caused a reduction of thermophilous broad-leaved species on Sakhalin. The pollen diagram from the Uandi bog (Figure 18-9) shows a marked decline of broad-leaved species at the Boreal-Atlantic boundary, 8370 ± 120 yr B.P. (Vs-33). Analogous climatic changes in western Europe and North America are the Viskan cooling in Sweden, the Venediger glacial advance in the Alps, and the Cochrane-Cockburn glacial advance in Canada between 8300 and 8000 yr B.P. (Mörner, 1973).

ATLANTIC PERIOD

The Atlantic period (AT) is characterized by the increased differentiation of the vegetation, the expansion of the forest zone to the north and south, and the widespread occurrence of thermophilous taxa in the forest vegetation. In the European USSR and in the southern Far East, broad-leaved forests underwent maximum development, and upper treeline reached its highest altitude in many mountainous regions. These trends are most clearly recorded in the second half of the Atlantic period, between 6000 and 4600 yr B.P. (Figure 18-11).

The tundra zone in the northern European USSR disappeared almost completely. Spruce wood found near the coast of Pechoraya Bay (Figure 18-3) is dated at 4880 ± 60 yr B.P. (LU-678) and 5950 ± 90 yr B.P. (LU-658). In western Siberia, forests advanced 300 to 400 km north of their present position. Larch wood found on the Yamal Peninsula near the Yuribey River has an age of 5500 ± 160 yr B.P. (BASh GIN-63). The central-Siberian forest advanced 100 to 150 km to the north, as indicated by finds of larch wood collected in the tundra of the Taimyr Peninsula and dated at 5470 ± 120 yr B.P. (GIN-1314), 5770 ± 40 (GIN 798), 5650 ± 130 (GIN-178), and 5520 ± 50 (GIN-779) (Figure 18-3). In the maritime regions of the northeastern USSR, the tundra/forest border shifted only several tens of kilometers north of its present position, indicating a climatic warming less than that of the Boreal period.

The forest zone in the Atlantic period had a more complex structure than in the Boreal period. The subzone of dark coniferous taiga was subdivided into northern, middle, and southern belts, whereas during the Boreal period only a northern and middle taiga existed. In the European USSR, dark spruce taiga was located 400 to 500 km north of its present position. In the southern and middle taiga, the appearance of broad-leaved species is noted.

A northward shift of taiga forests occurred on a small scale in western Siberia. Spruce-larch forests grew in the northern part of the forest zone. Farther south, they were replaced by middle-taiga cedar (*Pinus sibirica*) and spruce forests and southern-taiga fir and spruce forests. In the southern taiga, a wide belt of birch forest with a mixture of broad-leaved species (*Ulmus*, *Tilia*, *Quercus*) west of the Irtysh River reached its Holocene maximum.

On the Central Siberian Plateau, northern-taiga spruce-larch forests were also common, although spruce was less important than during the Boreal period. Middle-taiga cedar-spruce forests and southern-taiga fir-spruce forests grew farther south. *Tilia*, *Ulmus*, and *Quercus* grew in the southern forests near Lake Baikal, although those trees are absent there at the present time.

Over a large area of the northeastern USSR, as in the Boreal period, there were light coniferous forests of larch during both the Atlantic and Boreal periods. Their Holocene history, however, has not been determined adequately.

In the European USSR, a belt of mixed forests including broad-leaved species as well as birch, pine, and spruce was found south of the spruce taiga. To a certain extent, these mixed forests resembled the modern coniferous/broad-leaved (subtaiga) forests that are now found approximately 500 km farther south.

Farther south extended a wide belt of broad-leaved forests composed of *Quercus*, *Ulmus*, *Tilia*, and *Corylus*. During the Atlantic period, broad-leaved species spread eastward into the forest zone of the European USSR. These species migrated from the Baltic region, Belorussia, and the western Ukraine, rather than from the south as was previously believed. An opposing westerly and northwesterly migration of broad-leaved species occurred from the southern Urals, where they had survived the last glaciation. On many Holocene diagrams from the forest zone of the European USSR and the Urals, the pollen of broad-leaved species is most abundant at the end of the Atlantic period (Figures 18-4 and 18-7). The belt of broad-leaved forests attained a width of 1200 to 1300 km in the western European USSR and a width of 300 km in the east. Broad-leaved forests also reached their maximum extent in the southern Far East at that time. At present, this forest belt has a width of only 200 to 400 km. The northern boundary in the Atlantic period shifted more than 800 km, but the southern boundary almost coincided with the present one.

Recent palynologic data from deposits dated by radiocarbon within the southern forest zone and forest-steppe of the European USSR and western Siberia (Serebryannaya, 1980; Khotinskiy, 1977) show that the forest/steppe boundary, which earlier had shifted from north to south, became stationary in the second half of the Atlantic period at roughly its present position. The stability of this boundary is surprising, particularly when one considers the major shift of up to 400 to 500 km that took place in the forest/tundra border to the north at that same time. Grichuk (1969) noted that around 5400 yr B.P. regions of Eurasia and North America located north of latitudes 45°N to 55°N experienced winter and summer temperatures greater than those recorded at present. To the south a kind of "neutral" belt existed, where the temperature regime differed little from the present one, and still farther south, beyond latitude 40°N, winter and summer temperatures were lower than today's. The forest-steppe coincided with this belt and accounted for the stability of the forest/steppe border in the southern European USSR and western Siberia. The great shifts in the tundra/forest boundary in the North during the Late Holocene indicate the great variability in climate in maritime territories compared to the relative climatic stability of interior northern Eurasia.

The character of vegetation in steppe and desert zones during the Atlantic period can be determined only tentatively, for representative Holocene sections in these areas are lacking. It can be assumed that these zones were established during the Atlantic period close to their present position and that they have not shifted since that time. This does not mean, however, that the climatic conditions within these zones remained absolutely stable; in fact, data indicate that arid periods alternated with moist periods. In Central Asia, for example, the so-called Lyavlyakanskiy pluvial epoch between 8000 and 4000 yr B.P. is characterized by the formation of soils, considerable flooding of the Karakumy and Kyzylkumy Deserts, and the appearance of lakes (along which numerous Neolithic tribes settled) (Mamedov, 1978). On the basis of these data, Mamedov (1978) concluded that precipitation during the Atlantic period was twice as great as at present in this region, and that vast areas of Central Asia were covered with steppe. The pollen data from the steppe and forest-steppe zones of western Siberia, however, do not indicate a penetration of steppe into Central Asia, where a pluvial epoch undoubtedly occurred. Rather, they show an extensive spread of local mesophytic vegetation but not in any radical zonal rearrangements.

During the transition from the Atlantic to the Subboreal period (AT/SB), the tundra advanced southward, broad-leaved species and other thermophilous taxa partially disappeared, and other changes in the vegetation occurred that indicate a climatic cooling. Palynologic data indicate that this transition encompassed two cool intervals at about 4900 and 6400 yr B.P. These climatic events were expressed with different intensity across northern Eurasia and show up clearly in the Camp Century paleotemperature curve from Greenland (Dansgaard et al., 1970) (Figure 18-12).

SUBBOREAL PERIOD

The Subboreal period (SB) is identified as a complex stage in the vegetational and climatic history of northern Eurasia. Pollen data obtained do not support the traditional concepts of xerothermic conditions throughout the period or of an expansion of steppe into the present-day forest zones. Palynologic data permit the Subboreal period to be subdivided into three phases: SB-1, early-Subboreal cool-

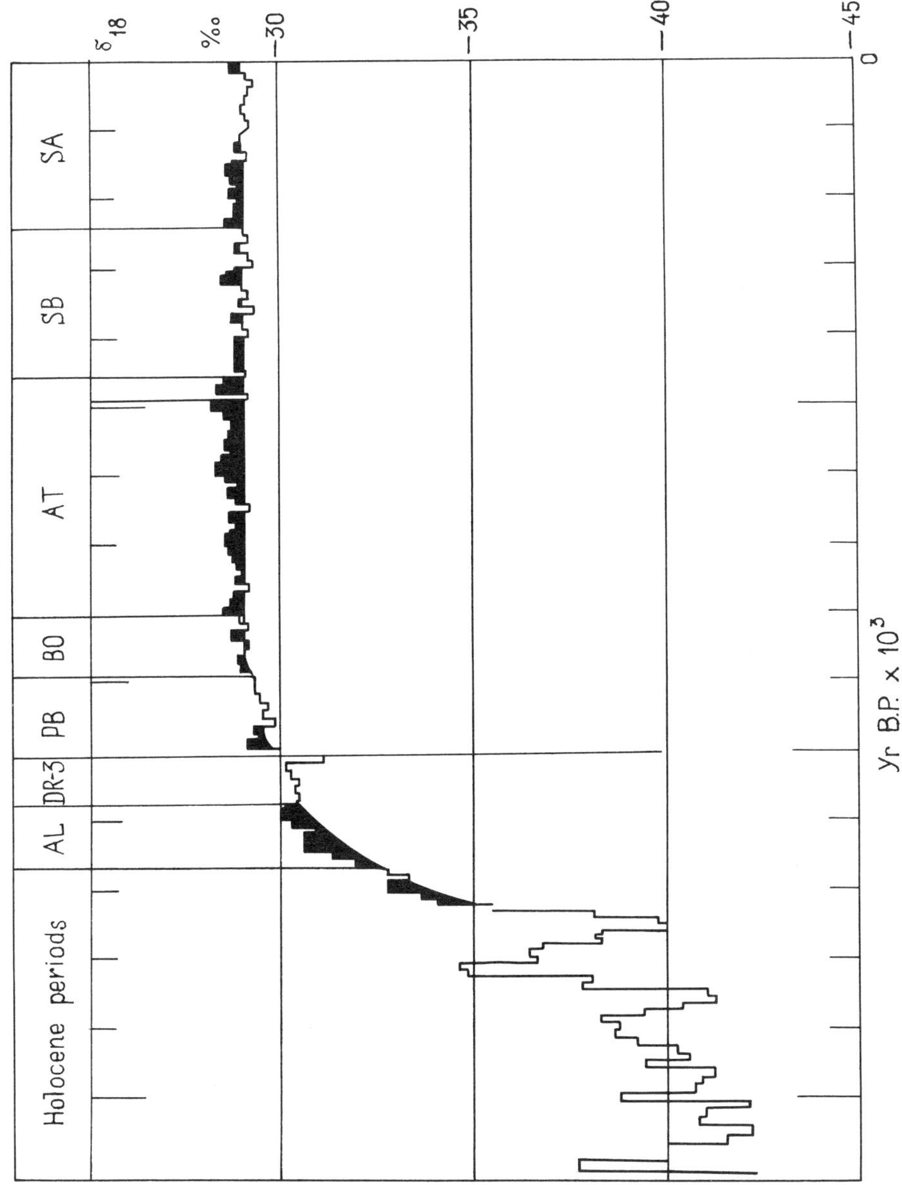

Figure 18-12. $^{18}O/^{16}O$ ratios from the Camp Century core, Greenland. (Dansgaard et al., 1970.)

ing (4600 to 4100 yr B.P.); SB-2, middle-Subboreal warming (4100 to 3200 yr B.P.); and SB-3, late-Subboreal cooling (3200 to 2500 yr B.P.).

During the early-Subboreal cooling (SB-1), tundra in the European USSR expanded southward 200 to 300 km from its position during the Atlantic period. The northern boundary of broad-leaved forests moved southward as well by 400 to 500 km. This cooling restricted the northward spread of broad-leaved species and caused their partial disappearance over vast areas.

A puzzling phenomenon—a universal decline in *Ulmus*—took place in European forests at that time. In the Holocene deposits of northwestern Europe, the amount of *Ulmus* pollen decreases sharply above the Atlantic-Subboreal contact, and pollen of weeds related to various aspects of human farming activity appears consistently. Some western European investigators have suggested that changes in the broad-leaved forests were caused by anthropogenic factors (Berglund, 1969) and that the *Ulmus* decline was the result of Neolithic tribes using the leaves and young shoots for cattle feed. However, the decrease in *Ulmus* occurs not only in western Europe but simultaneously across the entire forest zone of the Russian Plain and Urals, where Neolithic hunters and fishermen did not raise cattle (Khotinskiy, 1977). Thus, the synchronous decrease in *Ulmus* over vast areas cannot be a result of human activity; rather, it is the result of a sharp climatic cooling at the start of the Subboreal period. This cooling caused the decline of thermophilous tree species and also an appreciable rearrangement of the vegetation throughout northern Eurasia.

In western Siberia, the tundra expanded into the present forest zone. The dark coniferous forests declined and broad-leaved species disappeared. Paludification began on a large scale, resulting in the formation of huge peatlands on the West Siberian Plain. In central Siberia and the northeastern USSR, degradation of forest vegetation occurred near its northern limit at the beginning of the Subboreal period. This trend is seen most distinctly north of Yakutiya, where the northern limit of spruce and tree birch was shifted to the south and southwest around 4600 yr B.P. On Kamchatka and Sakhalin, pollen spectra of early Subboreal time indicate an altitudinal lowering of the treeline in the mountains and a widespread occurrence of cold-tolerant shrubs. Partial degradation of broad-leaved and coniferous/broad-leaved forests occurred in the southern Far East. These examples illustrate the synchronous and unidirectional reaction of vegetation across a vast area to a global climatic cooling.

The middle-Subboreal warming (SB-2) is recorded most distinctly in the northern European USSR, where the northward shift of zones was no less significant than that at the end of the Atlantic period (Nikiforova, 1980). During this interval, the tundra zone occupied only the northernmost Kola and Yugorskiy Peninsulas. The northern boundary of the forest in western and central Siberia lay 150 to 250 km north of its present position. Buried larch wood dated at 3790 ± 150 yr B.P. (GIN-1154) and 3430 ± 50 yr B.P. (GIN-1278) was found in the present tundra zone on the Taimyr Peninsula. In the northeastern USSR, the advance of forest into the tundra was less significant and amounted to only a few tens of kilometers.

The forest zone in the middle-Subboreal period retained a complex, differentiated structure characteristic of the Atlantic period. Dark spruce taiga in the European USSR occupied approximately the same area as during Atlantic time. Forests of the southern and middle taiga contained an appreciable admixture of broad-leaved species. The northern limit of *Ulmus*, *Tilia*, *Quercus*, and *Corylus* reached latitudes 64°N to 65°30′N, some 200 to 400 km north of its present position. Thermophilous plants reached their northernmost Holocene position during this period.

In Siberia, the Subboreal and Sub-Atlantic periods combined into a more or less unique stage of vegetational and climatic development. In the taiga of western Siberia, spruce and fir forests declined and cedar (*Pinus sibirica*), pine, and birch became important. The widespread occurrence of cedar is indicated by the high percentages of cedar pollen found in the Subboreal and Sub-Atlantic periods in diagrams for Siberia (Figure 18-8). In areas previously occupied by spruce taiga, pine forests became widespread, indicating increased aridity and continentality in western Siberia during the Subboreal period. Similar climatic trends are seen in central Siberia, where dark coniferous taiga (which had predominated up to that time) was replaced by light coniferous larch and larch-pine taiga.

In the northeastern USSR, the vegetation was dominated by larch forests, much as it is today. Mountain pine (*P. pumila*), widespread in the mountains, and Siberian cedar (*P. sibirica*) in regions of western Siberia were at their optimum biologic development during the Subboreal and Sub-Atlantic periods.

In the Far East, an intensive spread of dark spruce and spruce-fir taiga occurred during the Subboreal and Sub-Atlantic periods. *Carpinus*, *Juglans*, and *Phellodendron* appeared in the broad-leaved forests of the southern Far East. In many maritime regions (Kamchatka, Sakhalin), *Myrica*, an indicator of oceanic climate, became widespread. At the Uandi bog (Sakhalin), the content of *Myrica* pollen reaches 70% in Subboreal and Sub-Atlantic deposits (Figure 18-9). All these data indicate increased moisture in the Far East, in contrast to the increased aridity noted in interior Siberia.

In the European USSR south of the dark coniferous taiga, the belt of mixed coniferous/broad-leaved subtaiga forests considerably expanded southward from its position during the Atlantic period. The boundary between this and broad-leaved forests shifted southward about 400 to 600 km in the western regions. Thus, during the middle-Subboreal period, broad-leaved species reached their maximum northward position in most areas at a time when they were somewhat restricted in the central European USSR. Apparently, the middle-Subboreal warming was expressed more strongly in northeastern Europe than in the middle belt of the forest zone.

The northwestern part of the Ukraine was occupied by mixed pine and broad-leaved forests with hornbeam (*Carpinus*). Hornbeam also appears in the forests of Belorussia, indicating a moisture increase in these regions.

The forest/steppe border in the European USSR and western Siberia did not undergo any appreciable shifts during the Subboreal period. The widely held hypothesis that, during the Subboreal period, steppe penetrated northward into the forest zone seems unfounded. Instead, recent palynologic data consistently indicate that a northward migration of steppe occurred not in the Subboreal period but at the end of the last glaciation, when the climate of northern Eurasia sharply increased in continentality. Subsequently, the steppe, which was crowded by forest, retreated southward, moving a greater distance in oceanic than in continental regions.

The late-Subboreal cooling (SB-3) is most clearly seen in the European USSR, where the tundra/forest border shifted southward, broad-leaved species were less abundant in many areas, and spruce taiga reached its maximum extent. A similar expansion of spruce forests is noted in the Far East (Kamchatka, Sakhalin) but not in interior Siberia, where spruce forests were restricted during the Subboreal period. Apparently, moisture regimes were different in continental and oceanic regions of the USSR.

The boundary between the Subboreal and Sub-Atlantic periods (SB/SA) is very difficult to establish. In Europe it was determined to be at the so-called Boundary Horizon of raised peat bogs, which formed around 2500 yr B.P. during the transition from a xerothermic Subboreal period to a moist Sub-Atlantic period. However, the Subboreal period was a highly heterogeneous stage and was not xerothermic overall. Pollen data show that a moister climate began not in the Sub-Atlantic but in the last third of the Subboreal period (SB-3 phase) around 3200 yr B.P. In addition, the Boundary Horizon is not a rigidly fixed chronologic level (Khotinskiy, 1971); rather, it developed during the Subboreal, the Atlantic, and even the Boreal periods (Figures 18-7 and 18-9).

SUB-ATLANTIC PERIOD

The Sub-Atlantic period (SA) has been poorly studied because until recently most attention has been given to reconstructing the vegetation of older phases and because in northern and eastern USSR organic sediments of that time are frequently absent. In the northern European USSR, broad-leaved species were restricted as dark spruce forests spread during a period of increased moisture beginning as early as the Subboreal period, around 3200 yr B.P. At 700 to 800 yr B.P., the spruce forests were superseded by birch and pine forests because of climatic cooling (Little Ice Age), not because of the onset of agriculture as some investigators have suggested. This issue is discussed in more detail in Chapter 30, which deals with the Holocene dispersal of early man in the USSR.

In many regions of Siberia, dark spruce forests, which had degraded as far back as the beginning of the Subboreal period, did not recover during the Sub-Atlantic period. Instead, forests of pine and Siberian cedar, which developed under dry and cold climatic conditions, became increasingly important. East of the Urals, larch forests were widespread, although their Holocene history is still inadequately known because of the poor preservation of larch pollen. Thus, the climatic differences established between relatively oceanic and continental regions of northern Eurasia in the Boreal period continued into Sub-Atlantic time.

In closing, the following conclusions can be drawn about vegetational development in the Holocene. The Holocene vegetational history reveals an alternation between relatively stable stages and periods of rapid change. During the transition from the last glaciation to the Holocene around 10,300 yr B.P., there was an almost instantaneous rearrangement from hyperzonal to zonal vegetation as a result of global warming and decreased continentality across northern Eurasia. Rapid and unidirectional changes in vegetation over vast areas occurred during cool intervals at the end of the Boreal period, at the Atlantic/Sub-Atlantic boundary, and at other levels. These changes seem to reflect a rapid transformation of local vegetation rather than prolonged plant migrations.

The development of Holocene vegetation followed an interglacial cycle from cold-tolerant to warm-loving and then back to cold-tolerant vegetation. With increased warming, increased complexity and differentiation of the vegetational structure occurred. In the southern Far East, thermophilous broad-leaved forests began to develop as early as the Boreal period, whereas in the European USSR, their expansion occurred during the Atlantic period. Appreciable regional differences show up in the history of the dark spruce forests. During the Boreal period, the maximum spread of spruce forests occurred in interior Siberia; during the Subboreal and Sub-Atlantic periods, it occurred in relatively oceanic regions, such as the European USSR and the Far East.

The degree of change in the vegetation and in the position of zonal boundaries was not uniform across the USSR. Changes were more dynamic in the North as well as in oceanic regions. The most mobile boundary was the tundra/forest border, which during the Holocene repeatedly shifted hundreds of kilometers. In addition, the forest zone had a relatively stable southern boundary, which became established in the Atlantic period and subsequently did not undergo any appreciable displacement.

Steppe vegetation penetrated into the present forest zone during the period of severe continentality at the end of the last glaciation and not during the Subboreal period as previously believed. In the Holocene, steppe shifted southward, to a greater extent in oceanic than in continental regions. The chief trends in the development of the present vegetation are a deterioration of thermophilous vegetation and an increase in the role of cold-tolerant components.

References

Behre, K. (1967). The late glacial and early postglacial history of vegetation and climate in northwestern Germany. *Review of Palaeobotany and Palynology* 4, 149-61.

Berglund, B. E. (1969). Vegetation and human influence in south Scandinavia during Prehistoric time. *Oikos*, Supplement 12, 9-28.

Budyko, M. I. (1977). "Global Ecology." Mysl' Press, Moscow.

Dansgaard, W., Johnsen, S. J., Clausen, H. B., and Langway, C. C., Jr. (1970). Ice cores and paleoclimatology. *In* Nobel Symposium 12, "Radiocarbon Variations and Absolute Chronology" (I. U. Olsson, ed.), pp. 337-54. Wiley, New York.

Dokturovskiy, V. S. (1918). Data from a study of bogs. *Vestnik torfyanogo dela* 4, 10-37.

Dokturovskiy, V. S., and Kudryashov, V. V. (1923). Pollen in peat. *Nauchnoeksperimentalnogo Toryfyanogo Instituta, Izvestiya* 5, 33-44.

Grichuk, V. P. (1969). Experience in reconstructing certain climatic elements of the Northern Hemisphere in the Atlantic period of the Holocene. *In* "The Holocene" (M. I. Neustadt, ed.), pp. 41-57. Nauka Press, Moscow.

Grosset, G. E. (1961). Fluctuation of the boundary between forest and steppe in the Holocene in light of the study of zone mixing. *Byulleten Moskovskogo obstichestva ispytateley prirody* 66, 65-84.

Gudelis, V. K., and Neustadt, M. I. (eds.) (1961). "Problems of the Holocene." Institute of Geology and Geography, Lithuanian Academy of Sciences Press, Vil'nyus.

Karavayev, M. N. (1965). The vegetative cover. *In* "Yakutia (Natural Conditions and Resources of the USSR)" (S. S. Korzhuev, ed.), pp. 247-92. Nauka Press, Moscow.

Khotinskiy, N. A. (1964). Comparison of schemes for zonal division of late-glacial and postglacial time by means of synchronizing levels. *USSR Academy of Sciences, Doklady* 150, 74-77.

Khotinskiy, N. A. (1970). Change in vegetation and climate at the start of postglacial time. *USSR Academy of Sciences, Doklady, seriya geologicheskaya* 6, 112-117.

Khotinskiy, N. A. (1971). The problem of the boundary horizon. Appendix to Field Guide for Itinerary number 1-B. Third International Palynological Conference, Moscow.

Khotinskiy, N. A. (1977). "The Holocene of North Eurasia." Nauka Press, Moscow.

Khotinskiy, N. A. (1978). "Section through the Latest Deposits of the Lower Amur Region" (K. K. Markov, ed.). Moscow University Press, Moscow.

Khotinskiy, N. A., and Koreneva, E. V. (eds.) (1973). "Palynology of the Holocene and Marine Palynology." Nauka Press, Moscow.

Kind, N. V. (1974). "Geochronology of the Late Anthropogene from Isotope Data." Nauka Press, Moscow.

Mamedov, E. (1978). On the problem of pluvial paleoclimates of the USSR's deserts. *In* "Problems of the Physical Geography and Agroclimatology of Central Asia" (N. A. Koga, ed.), pp. 40-49. Tashkent.

Markov, K. K. (1965). Principal changes in the character of the earth's surface in the Holocene. *In* "Paleogeography of the Quaternary" (G. I. Lazukov, ed.), pp. 5-18. Moscow State University Press, Moscow.

Mörner, N. A. (1973). Climatic changes during the last 35,000 years as indicated by land, sea and air data. *Boreas* 2, 33-53.

Neustadt, M. I. (1957). "History of Forests and Paleogeography of the USSR in the Holocene." USSR Academy of Sciences Press, Moscow.

Neustadt, M. I. (1965). "Paleogeography and Chronology of Upper Pleistocene and Holocene Based on Data of the Radiocarbon Method." USSR Academy of Sciences, Institute of Geography. Nauka Press, Moscow.

Neustadt, M. I. (1969). "The Holocene." Nauka Press, Moscow.

Neustadt, M. I. (ed.) (1971). "Palynology of the Holocene." USSR Academy of Sciences, Institute of Geography, Moscow.

Nikiforova, L. D. (1980). Change in the natural environment in the Holocene in the northeast of the European USSR. Ph.D. dissertation, Moscow State University, Institute of Geography, Moscow.

Serebryannaya, T. A. (1980). Contribution to the history of the west of the Central Russian Upland. *Bulletin of the Commission on the Study of the Quaternary* 50, 178-85.

Sukachev, V. N. (1906). Data from a study of bogs and peat bogs of a lacustrine region. *Proceedings of the Freshwater Biological Station of the St. Petersburg Society of Naturalists* 2, 161-262.

Tolmachev, A. I. (1954). "Contribution to the History of Formation and Development of Dark Coniferous Taiga." USSR Academy of Sciences, Moscow and Leningrad.

Velichko, A. A. (1973). "The Natural Process in the Pleistocene." Nauka Press, Moscow.

Yurtsev, B. A. (1974). "Problems of Botanical Geography of Asia's Northeast." Nauka Press, Leningrad.

CHAPTER 19

Holocene Peatland Development

M. I. Neustadt

The Holocene can be regarded as a time of great peat formation, resembling in many ways the older epochs of extensive peat development related to coal accumulation. The Holocene peatlands with the greatest peat thickness are about 10,000 to 12,000 years old, but some peatlands started later and expanded into new areas. The rate of peat formation generally ranges between 0.2 and 0.8 mm per year in the USSR, mainly as a function of climatic and geomorphic conditions, water regime, vegetation, etc.

D. A. Gerasimov introduced the concept of "peat basins," and later Nikonov (1946, 1948, 1955) defined them as vast, interconnected deposits forming in bedrock depressions. Lands that underwent particularly intensive peat development in the Holocene form peat basins. It is impossible to give a map of peat formation, so I will present instead a map of peat basins (Figure 19-1). Twenty-two peat basins, containing 70% of the peat reserves, have been identified, of which the largest are the western Siberian, Pripyat', and Volkhov basins.

Detailed descriptions of some basins can be found in the literature (Neustadt, 1936). From the list of coal, shale, and peat basins presented in Table 19-1, it can be seen that recent peat basins are equal in area to coal and shale basins.

Peatlands are chiefly confined to the temperate forested zone. Kats (1948, 1971) in a description of peatlands throughout the world distinguished 140 provinces. Markov and Khoroshev (1975) estimated the probable area of peatlands in the USSR at 83 million ha; but Sabo (1980) calculated that in the USSR alone the surface area of boggy lands, including peat bogs, is about 245 million ha, excluding tundra and forest-tundra regions, where boggy areas are very large.

Peat formation occurs in the northern regions of European USSR and western Siberia (Neustadt et al., 1977; Pyavchenko, 1980), as well as in central Siberia and the Far East, where it is related to partial thawing of permafrost. Peatlands have also developed in the tundra and forest-tundra zones, although often without peat deposition (Table 19-2). These processes are also observed in other

Table 19-1. Selected Coal, Shale, and Peat Basins

Coal basins	Area (km²)	Shale Basins	Area (km²)	Peat Basins	Area (km²)
Karaganda	3000				
Suchan	6000			Polistovskiy	6000
				Meshchera	16,000
Kuznetskiy	26,000			Balakhnino-Vetluzhskiy	18,000
Irkutsk	35,000			Volkhov	32,000
Donetskiy	60,000	Baltic region	60,000		
Near Moscow	120,000	Olenekskiy	135,000	Pripyat'	100,000
Lena	750,000			Western Siberian	500,000[a]
Tunguska	1,000,000				

Sources: Coal basins after Matveyev, 1960; shale basins after Nikonov, 1946, 1948.
a. The basin area is much greater in studies made after 1948.

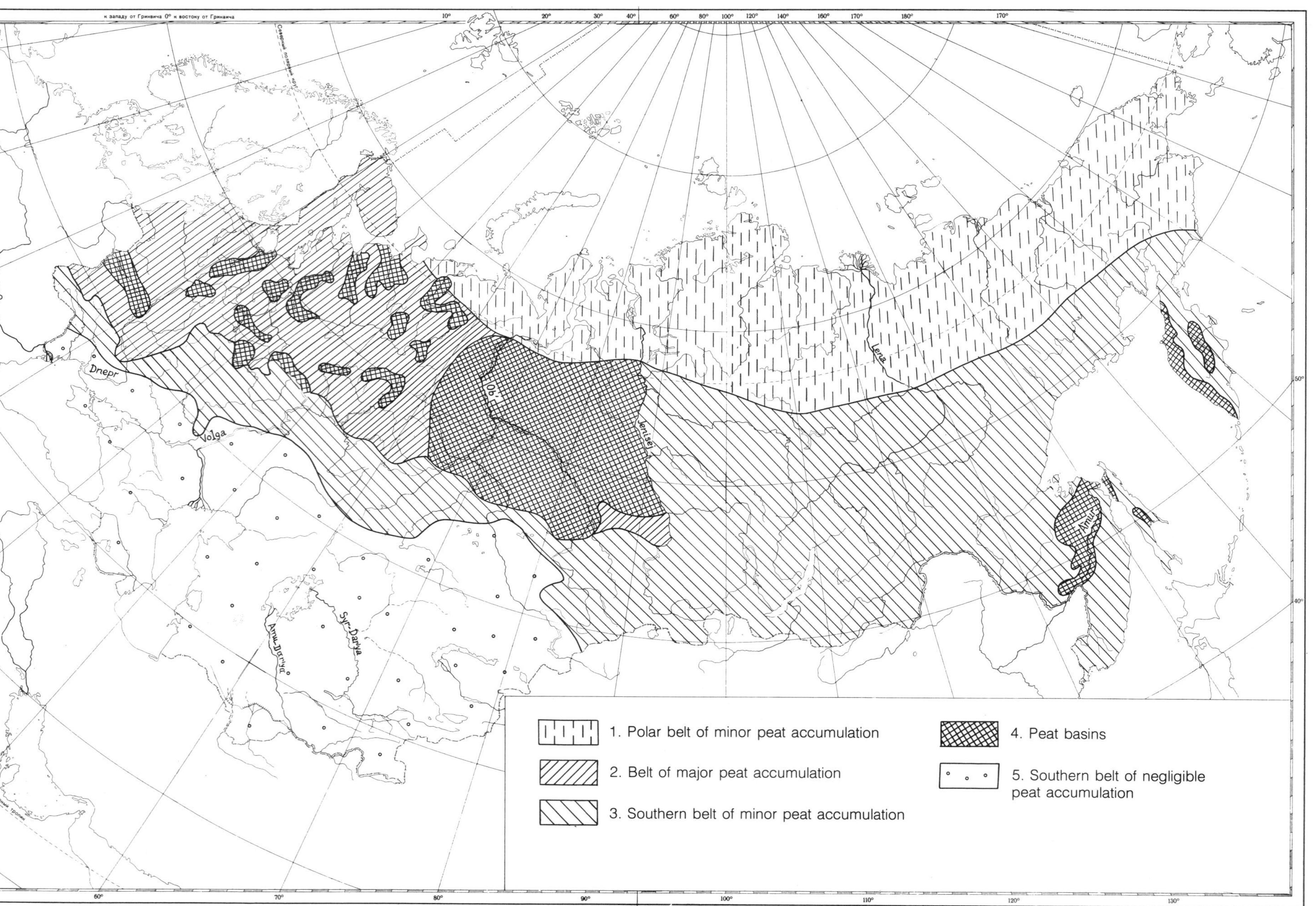

Figure 19-1. Areas of peat accumulation in the USSR.

Table 19-2. Peat Cover and Thickness in Accumulation Belts of the USSR

Peat Accumulation Belt	Peat Coverage of Surface	Average Thickness of Peat	Reserve of Absolutely Dry Peat per 1 ha of Surface
Polar belt of slight peat accumulation	Less than 1%	1.0 m	Less than 15 tons
Belt of intensive peat accumulation	1% to 40% (average 6%)	2.2 m	30 to 1000 tons (average 150 tons)
Southern belt of slight peat accumulation	0.01% to 1% (average 0.2%)	1.5 m	0.25 to 25 tons (average 5 tons)
Southern belt of negligible peat accumulation	Up to 0.01%	1.0 m	Less than 0.25 tons

countries within the taiga zone (Finland, Canada, etc.). In Finland, for example, about 52,000 km^2 of boggy forested areas alone were drained in 1979 ("Finnish-Soviet Symposium on Forest Drainage," 1980).

The surface area of peatland in the USSR was estimated on the basis of calculations for the Bakchar Bog, one of the largest bogs of western Siberia, with an area of 2268 km^2 (Table 19-3) (Neustadt et al., 1977).

The development of the Chistik peatland of Kalinin Province, with an area of 8000 ha (Neustadt, 1976), is traced by a series of four maps showing the extent of the bog during the Paleoholocene (12,000-9800 yr B.P.), Eoholocene (9800-7700 yr B.P.), Mesoholocene (7700-2700 yr B.P.), and Neoholocene (2700-0 yr B.P.). During the Paleoholocene (Figure 19-2A), there were 50 separate lakes and a number of swamps. In the Eoholocene (Figure 19-2B), the picture changed substantially. Forty-eight lakes developed into bogs. Individual centers of bog formation merged into 17 approximately large paludal bodies. During the Mesoholocene (Figure 19-2C) all except 3 paludal areas merged, and during the Neoholocene (Figure 19-2D) the remaining areas fused together, creating one continuous bog with an area of about 8000 ha. The dynamics of this development are represented in Table 19-4).

The thickness of peats varies widely. In the European USSR, the greatest peat thickness, 12.5 m, was recorded in a small area of the Panfilovskiy raised *Sphagnum* bog. A similar maximum figure of 12.4 m was noted for the Imnatskiy bog of the Georgian SSR (Provorkin, 1957). Other bogs have maximum depths of 11 m and 10 m (Dubakh, 1938). In western Siberia, depths of 10.4 m near Nizhnevartovsk, 9.2 m in the Urals, and 10 m on Kamchatka were recorded (Neustadt et al., 1936). A depth of 7.9 m is not uncommon, although peat thicknesses do not generally exceed 5 to 6 m in large bogs. The average peat depth usually is not greater than 3 to 4 m.

For comparison, I will cite data on the thickness of other Holocene deposits. Lacustrine deposits (sapropel) reach a thickness of 20 to 25 m in some freshwater lakes; and at Lake Somino of Yaroslavl' Province 40 m has accumulated in a funnellike depression, the largest known thickness of Holocene sediments in the world (Neustadt et al., 1965). Holocene delta deposits of the Severnaya Dvina attain a thickness of 26 m (Jousé, 1939). Marine Holocene deposits on the Caspian Shelf are 25 m thick (Shcherbakov et al., 1977), and those in the Sea of Japan are 25.2 m thick (Troitskaya, 1975).

Peatlands represent a special landscape with a unique hydrologic regime and an unusual flora and fauna. Plants indigenous to bog environments give rise to bog formation. Large peatlands can influence the character of the climate of neighboring regions. The accumulation of moisture in the bog soils causes infertility, just as drought affects desert soils. Because bogs are capable of independent development and utilization, we should consider the reclamation of boggy lands and control of bog formation just as we plan for desertification. For several centuries, people have been trying to convert bogs into cultivated areas to be used as agricultural fields and pastures and areas for extracting peat for fuel and for fertilizer. During

Table 19-3. Increase in Boggy Areas during the Holocene

Date (yr B.P.)	Total Boggy Area (%)	Increase in Boggy Area (km^2)	Total Area of Boggy Lands (km^2)
2000-0	20.3	497,350	2,450,000
4000-2000	35.3	864,850	1,952,650
6000-4000	28.3	693,350	1,087,800
8000-6000	14.7	360,150	394,450
9000-8000	1.4	34,300	—
Total	100	2,450,000	2,450,000

Table 19-4. Holocene Peat Formation in the Chistik Bog

Period of Holocene	Area Occupied by Bog-Forming Processes (ha)	Remark
Neoholocene	About 8000	—
Mesoholocene	4700	—
Eoholocene	2500	Average yearly increase: 0.6 to 0.8 ha
Paleoholocene	1200	—

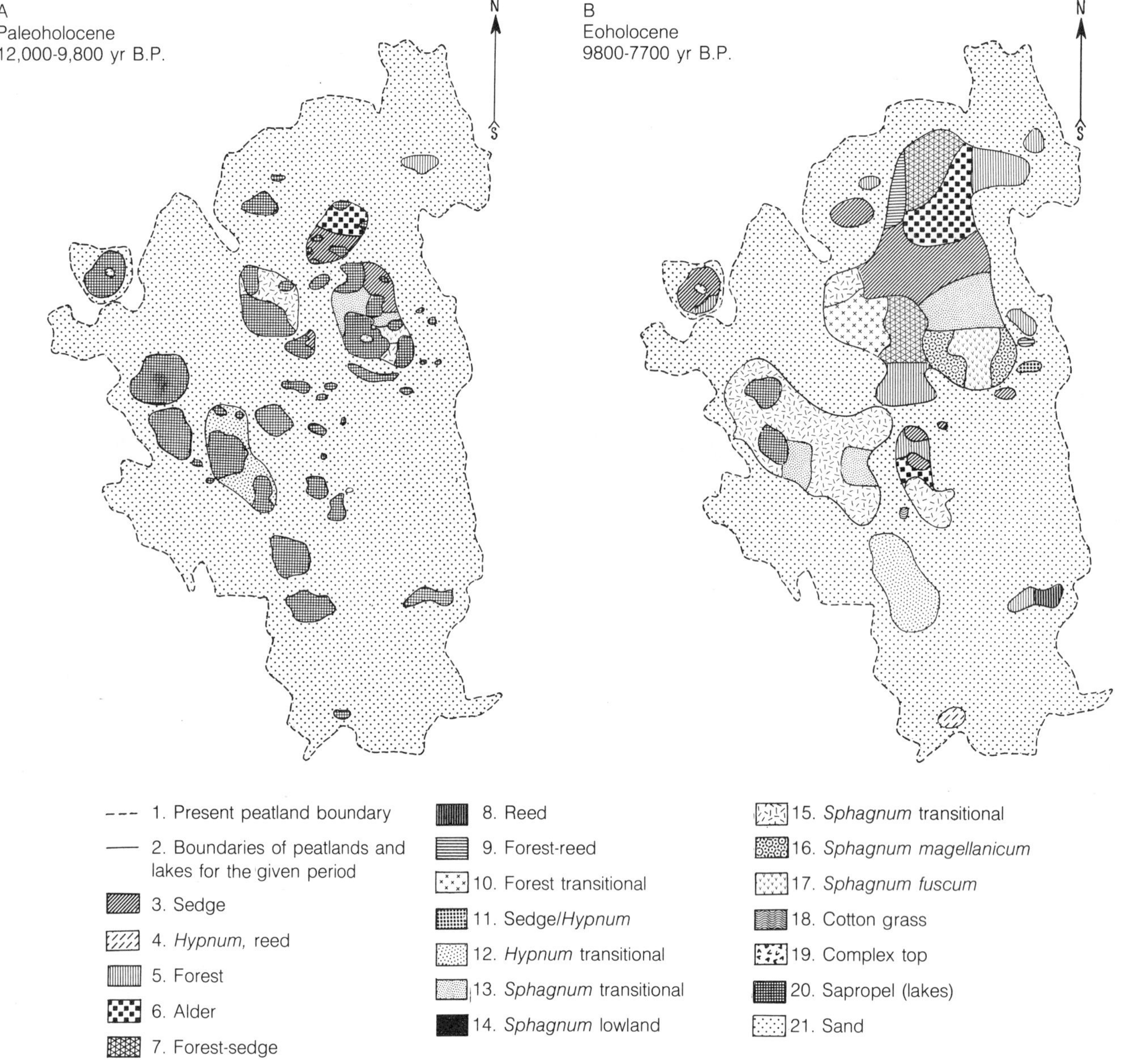

Figure 19-2. The development of the Chistik peatland during the Holocene.

the 20th century, peatlands have been actively drained in many regions, and this practice undoubtedly will continue in the future. However, part of the bogs must be left alone as an example of natural geosystems and as a special type of natural landscape.

It should be noted that, in addition to negative impacts, bog formation also has a positive side—the accumulation of organic matter in the form of peat. In the USSR, 200 billion tons of peat (based on a 40% moisture content), amounting to 66% of the world reserves, accumulated during the Holocene (Markov and Koroshev, 1975).

Bog formation occurs mainly in the forest zone of the temperate belt, where the climate favors peat accumulation. However, it can develop in a number of other vegetation zones, including areas of Cuba, Indonesia, and tropical Africa, providing there exists favorable climatic, geomorphic, and hydrologic conditions.

In most cases, bog formation is irreversible under present conditions. The frequent occurrence of well-decomposed peat in bogs, often associated with large stumps and trunks, indicates extreme climatic and hydrologic conditions in which either the rate of bog formation decreased, thus allowing well-decomposed peat to accumulate, or previously deposited peat decomposed. In either case, the bog is maintained by internal water reserves during a period of arid climate. Following some disturbances, such as forest clearance or fire, bog formation has ceased and a new forest cover has developed as a result of increased water loss through transpiration.

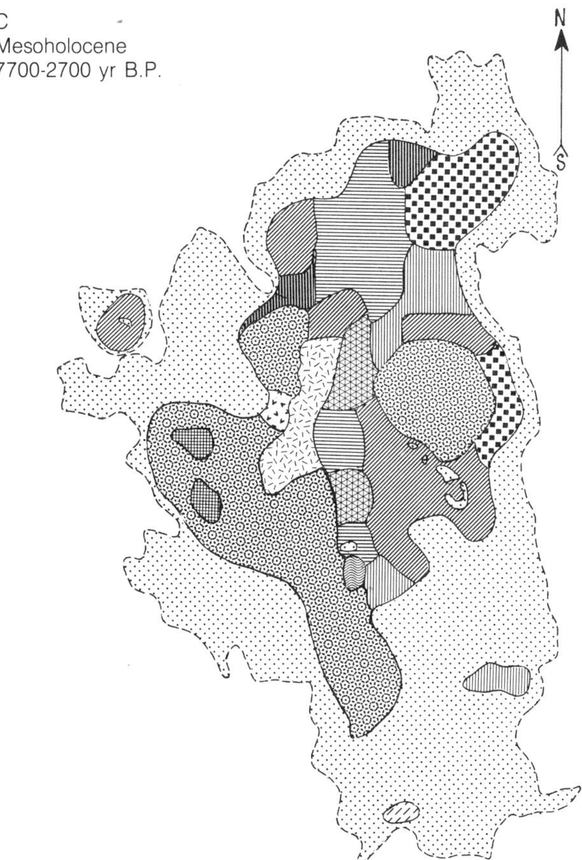

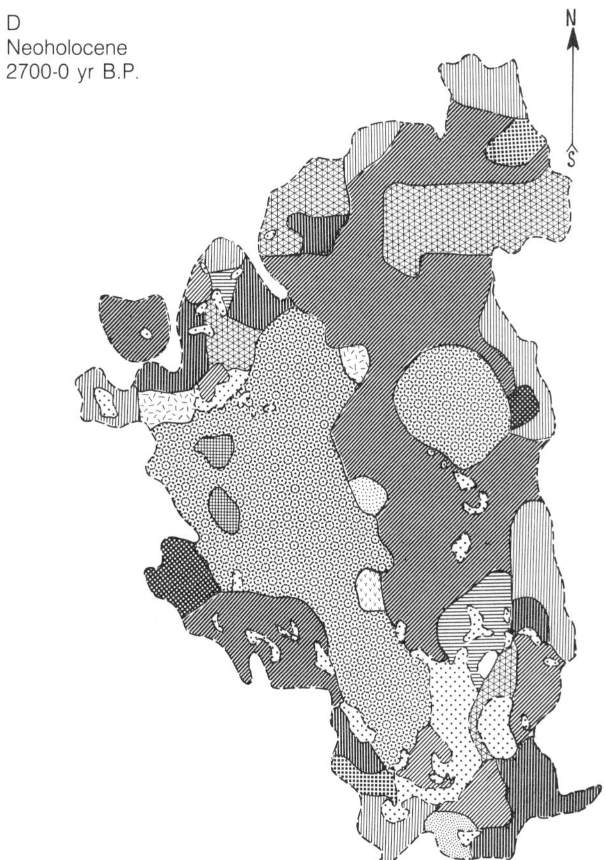

References

Dubakh, A. (1938). Largest depths of peat deposits. *Gosudarstvennogo Geograficheskogo Obshchichestva, Izvestiya* 1, 121-23.

Finnish-Soviet Symposium on Forest Drainage, Hygtiälä, Finland, September 17, 1979. (1980). *Bulletin of the International Peat Society* 11, 14-15.

Jousé, A. P. (1939). Paleogeography of bodies of water based on diatomaceous analysis. *Transactions of the Upper Volga Expedition, Trudy* 4, 86-90. Leningrad University Press, Leningrad.

Kats, N. Ya. (1948). "Types of Bogs in the USSR and Western Europe and Their Geographical Distribution." Nauka Press, Moscow.

Kats, N. Ya. (1971). "Bogs of the Earth." Nauka Press, Moscow.

Markov, V. D., and Khoroshev, P. I. (1975). An evaluation of probable peat reserves in the USSR. *Torfyanaya Promyshlennost* 6, 20-24.

Matveyev, A. K. (1960). "Geology of Coal Deposits." Gosgortekhizdat Press, Moscow.

Neustadt, M. I. (1936). Peat bogs of western Kamchatka. *In* "Peat Bogs of the Far North and Asian USSR" (M. I. Neustadt, ed.). Transactions of the Central Experimental Peat Station (TsTOS), Vol. 1, 41-45. ONTI Press, Moscow.

Neustadt, M. I. (1976). Regional patterns in the history of phytocenoses of the USSR in the Holocene from palynological data. *In* "History of Biogeocenoses of the USSR in the Holocene" (L. G. Dinesman, ed.), pp. 79-91. Nauka Press, Moscow.

Neustadt, M. I., Khotinskiy, N. A., Devirts, A. L., and Markova, N. G. (1965). Somino Lake. *In* "Paleogeography and Chronology of the Upper Pleistocene and Holocene Based on Data of the Radiocarbon Method" (M. I. Neustadt, ed.), pp. 91-97. Nauka Press, Moscow.

Neustadt, M. I., and Korotkina, M. Ya. (1936). Peat bogs of southeastern Kamchatka. *In* "Peat Bogs of the Far North of Asian USSR" (M. I.

Neustadt, ed.), 7-30. Transactions of the Central Experimental Peat Station (TsTOS), Vol. 1. ONTI Press, Moscow.

Neustadt, M. I., et al. (1977). "Scientific Premises of the Development of Bogs of Western Siberia." Nauka Press, Moscow.

Nikonov, M. N. (1946). Certain distribution characteristics of peat deposits. *In* "Proceedings of the Anniversary Session Devoted to the One Hundredth Birthday Anniversary of V. V. Dokuchayev" (L. I. Prasolov, ed.), pp. 602-7. USSR Academy of Sciences Press, Moscow and Leningrad.

Nikonov, M. N. (1948). Role of exploration of recent peat deposits for the purpose of determining the genesis of fossil coals. *Bulletin of the Moscow Society of Naturalists, Geology Section* 23, 93-102.

Nikonov, M. N. (1955). Regionalization of peat bogs in connection with their utilization in the national economy. *Forest Institute, Trudy* 31, 49-63. Moscow.

Provorkin, A. S. (1957). Utilization of peat in Georgian SSR. *In* "Collected Papers on the Study of Peat Reserves" (M. I. Neustadt, ed.), Issue 2, pp. 221-34. Main Administration of Peat Reserves Press, Moscow.

Pyavchenko, N. I. (1980). The bog-forming process in the forest zone. *In* "The Importance of Bogs in the Biosphere" (N. I. Pyavchenko, ed.), pp. 7-16. Nauka Press, Moscow.

Sabo, Ye. D. (1980). Marshland and forest reclamation reserves of the USSR and prospects for draining them. *In* "Importance of Bogs in the Biosphere" (N. I. Pyavchenko, ed.), pp. 16-24. Nauka Press, Moscow.

Shcherbakov, F. A., Kuprin, P. N., and Shatov, A. S. (1977). Stratigraphy of Late Quaternary deposits of the continental terrace and some aspects of paleogeography of the Black and Caspian Seas. *In* "Geology of the Quaternary" (A. T. Aslanyan, ed.), pp. 208-16. Armenian Academy of Sciences Press, Yerevan.

Troitskaya, T. S. (1975). Development and distribution of foraminifer complexes from Holocene sediments of the Sea of Japan. *In* "Way of Life and Principles of Dispersal of Modern and Fossil Microfauna" (A. F. Fursenko, ed.), pp. 89-94. Nauka Press, Novosibirsk.

Development of Animal Populations

CHAPTER 20

Late Pleistocene Mammal Fauna of the Russian Plain

A. K. Markova

A systematic investigation of fossil mammals of eastern Europe began in the past century and included the entire Russian Plain. The works of V. I. Gromov, I. M. Gromov, N. K. Vereshchagin, K. K. Flerov, Ye. I. Belyayeva, V. I. Gromova, I. G. Pidoplichko, V. A. Topachevskiy, V. I. Bibikova, K. A. Tatarinov, I. Ye. Kuz'mina, V. E. Garutt, A. I. David, L. P. Aleksandrova, L. I. Alekseyeva, A. K. Agadjanian, and A. I. Shevchenko discuss the composition, phylogeny, morphology, distribution, and type localities of Pleistocene species. The most detailed faunal records from the Russian Plain are of Late Pleistocene age, although data are unevenly distributed. For the Mikulino Interglaciation only a very few mammal sites are known; most sites date to the Valdai Glaciation.

Mammals of the Mikulino Interglaciation and Krutitsa Interstade

The vertebrate fossils from the Mikulino Interglaciation are poorly preserved and often decomposed, undoubtedly as a result of their burial under conditions of a warm and moist climate; active decomposition proceeded in particular in fossil soils. The majority of Late Pleistocene vertebrate sites are associated with Paleolithic campsites. Late Acheulian campsites are tentatively assigned to the Mikulino Interglaciation in this chapter, although some archaeologists and geologists believe that they date to earlier stages of the Pleistocene (Praslov, 1968; Gromov, 1948). Thus, artifacts attributed to Acheulian culture probably span a very broad period of time from the beginning of the Pleistocene to the beginning of the Valdai Glaciation.

The vertebrate remains associated with Late Acheulian campsites cannot provide precise correlations and should be used in conjunction with palynologic, paleopedologic, or stratigraphic data. On the Russian Plain there are no multidisciplinary studies of archaeological sites of that age.

At the present time, all vertebrate sites of the Mikulino Interglaciation are found on the Russian Plain and in the Caucasus. Most sites on the Russian Plain are dated by geology and palynology rather than by archaeology. The northernmost sites contain only small mammals (Table 20-1 and Figure 20-1, sites 1, 2, 3) such as squirrel, bank vole, European pine vole, and widely distributed species such as common vole, narrow-headed vole, northern vole and Eurasian water vole (Aleksandrova and Tseytlin, 1965; Kalinovskiy, 1979; Agadjanian, 1977). Steppe and arctic species are absent. This fauna indicates a forested landscape, particularly the presence of the European pine vole, which at present is common only in broad-leaved forests. The present-day northern boundary of the broad-leaved forest on the Russian Plain lies at about latitude 54°N to 56°N, which is south of the northernmost interglacial sites. Thus, the broad-leaved forests must have expanded northward during Mikulino time.

Small mammals have been found in the molehill horizon of the Mezin soil complex (Figure 20-1, site 4), located in the present forest-steppe of the Middle Dnepr River basin near the town of Gaydach (Markova, 1975a). The first phase of the Mezin soil complex formed in Mikulino time (Velichko, 1975); the second phase formed during the Krutitsa Interstade in Early Valdai time. The fossil vertebrate remains are thought to be associated with the second phase. The steppe species, which dominate the fauna, include steppe lemming, steppe pika, gray hamster, and ground squirrel. Only one forest species was noted, the field vole. Apparently, during the Krutitsa Interstade the Middle Dnepr region was an open landscape, with minor areas of forest vegetation probably confined to river valleys, very similar to that of today.

To the east in the Don Basin south of Voronezh Province, remains of both large and small mammals have been found at the Shkurlat site (Figure 20-1, site 5, and Table 20-1) (Raskatov et al., 1977; Alekseyeva, 1980). Large mammals found by L. I. Alekseyeva include an advanced form of straight-tusked elephant, an early form of

Table 20-1. Occurrence of Late Pleistocene Mammalian Fossils at Sites on the Russian Plain.

Taxon	Common Name	Mikulino Interglaciation						Early and Middle Valdai Glaciation									Bryansk Interstade												Late Valdai Glaciation								
		1	2	3	4	5	6	7	8	9	10	11	12	13	14	15	16	17	18	19	20	21	22	23	24	25	26	27	28	29	30	31	32	33	34	35	36
Insectivora																																					
Desmana moschata Pall.	Russian desman	•	•	•	•	•	•	•	•	•	•	•	•	•	•	•	•	•	•	•	•	•	•	•	•	•	•	•	•	•	X	•	•	•	•	•	•
Desmana sp.	Desman	•	•	•	•	•	•	•	•	•	•	•	•	•	•	•	•	•	X	•	•	•	•	•	•	•	•	•	•	•	•	•	•	•	•	•	•
Sorex araneus L.	Eurasian common shrew	•	•	•	•	•	•	•	X	•	•	•	•	•	•	•	•	•	•	•	•	•	•	•	•	•	•	•	•	•	X	•	•	•	•	•	•
Sorex sp.	Red-toothed shrew	X	X	X	•	•	•	•	X	•	•	•	•	•	•	•	•	•	•	•	•	•	•	•	•	•	•	•	•	•	•	•	•	•	•	•	•
Soricidae gen.	Shrews	•	•	•	•	•	•	X	•	•	•	•	•	•	•	•	•	•	•	•	•	•	•	•	•	•	•	•	•	•	•	•	•	•	•	•	•
Lagomorpha																																					
Lepus europaeus Pall.	Brown hare	•	•	•	•	•	•	•	•	•	•	•	•	•	•	X	•	•	•	•	•	•	•	•	•	•	•	•	•	•	•	•	•	•	•	•	X
Lepus timidus L	Arctic hare	•	•	•	•	•	•	•	•	•	•	•	•	•	•	•	•	X	•	•	•	•	•	•	X	•	•	•	•	X	X	•	X	•	•	•	X
Lepus sp.	Hare	•	•	•	•	•	•	•	X	X	•	•	•	•	•	•	•	•	X	•	•	•	•	X	X	X	X	•	•	•	X	•	•	X	X	X	•
Ochotona pussilla Pall.	Steppe pika	•	•	•	X	•	•	•	X	X	•	•	•	•	•	X	•	•	•	•	•	•	•	X	X	•	•	•	•	•	•	•	•	•	•	•	•
Ochotona sp.	Pika	•	•	•	•	X	•	X	•	•	•	•	•	•	•	•	•	•	X	•	•	•	•	X	X	•	•	•	•	•	X	•	•	•	X	•	X
Rodentia																																					
Alactagulus acontion Pall.	Little jerboa	•	•	•	•	X	•	•	•	•	•	•	X	•	•	X	•	•	•	•	•	•	•	•	•	•	•	•	•	•	•	•	•	•	•	•	X
Allactaga jaculus Pall.	Great jerboa	•	•	•	•	•	•	•	•	•	•	•	X	•	•	X	•	•	•	•	•	•	X	•	•	•	•	•	•	•	X	•	•	•	•	•	X
Allactaga sp.	Jerboa	•	•	•	•	X	•	•	•	•	•	•	•	•	•	•	•	•	•	•	•	•	•	•	•	•	•	•	•	•	•	•	•	•	•	•	•
Apodemus sylvaticus L.	Wood mouse	•	•	•	•	•	•	•	•	•	•	•	•	•	•	X	•	•	•	•	•	•	•	•	•	•	•	•	•	•	•	•	•	•	•	•	•
Arvicola terrestris L.	European water vole	•	•	•	•	•	•	X	•	•	•	•	•	•	•	X	•	•	X	•	X	•	•	X	X	X	•	•	•	•	X	•	•	X	•	•	X
Arvicola cf. *terrestris* L.	European water vole	X	X	X	•	X	•	•	•	•	•	•	•	•	•	•	•	•	•	•	•	•	•	•	•	•	•	•	•	•	•	•	•	•	•	•	•
Arvicola sp.	Water vole	•	X	•	•	•	•	•	•	•	•	•	•	•	•	•	•	•	•	•	•	•	•	•	X	•	•	•	•	•	•	•	•	X	•	•	•
Castor fiber L.	Eurasian beaver	•	•	•	•	•	•	•	•	•	•	•	•	•	•	•	•	•	•	•	•	•	•	X	•	•	•	•	•	•	•	•	•	•	•	•	•
Citellus birulai J. Grom.		•	•	•	•	•	•	•	•	X	•	•	•	•	•	X	•	•	•	•	•	•	•	•	•	•	•	•	•	•	•	•	•	•	•	•	X
Citellus major Pall.	Great suslik	•	•	•	•	•	•	•	•	•	•	•	X	•	•	•	•	•	•	•	•	•	•	•	•	•	•	•	•	•	•	•	•	X	•	•	•
Citellus pygmaeus Pall.		•	•	•	•	•	•	•	•	•	•	•	•	•	•	•	•	•	•	•	•	•	•	X	•	•	•	•	•	•	•	•	•	•	•	•	•
Citellus rufescens Keys. et Blas.	Red-cheeked ground squirrel	•	•	•	•	•	•	•	•	•	•	•	•	•	•	•	•	X	•	•	•	•	•	•	•	•	•	•	•	•	•	•	•	X	•	•	•
Citellus cf. *severescensis*		•	•	•	•	•	•	•	•	X	•	•	•	•	•	•	•	•	•	•	•	•	•	•	•	•	•	•	•	•	•	•	•	•	•	•	•
Citellus suslicus Güld.	Spotted suslik	•	•	•	•	•	•	•	•	•	•	•	•	•	•	•	•	•	•	•	•	•	•	•	•	•	•	•	•	X	X	•	•	X	•	•	•
Citellus sp.	Ground squirrel (suslik)	•	•	•	X	X	•	X	•	•	•	•	•	•	•	•	•	X	•	•	•	•	X	•	X	•	•	•	X	•	X	X	•	X	X	•	•
Clethrionomys glareolus Schreb.	Bank vole	•	•	X	•	•	•	X	X	•	•	•	•	•	•	•	•	•	X	•	•	•	•	•	•	•	•	•	•	•	•	•	•	•	•	•	X
Clethrionomys sp.		X	X	•	•	•	•	•	•	•	•	•	•	•	•	•	•	•	•	•	•	•	•	•	•	•	•	•	•	•	•	•	•	•	•	•	•
Cricetulus migratorius Pall.	Gray hamster	•	•	•	X	•	•	•	•	•	•	•	•	•	•	•	•	•	•	•	•	•	•	•	•	•	•	•	•	•	X	•	•	•	•	•	X
Cricetus cricetus L.	Common hamster	•	•	•	•	X	•	•	•	X	•	•	•	•	•	X	•	•	•	•	•	•	•	X	X	X	•	•	•	•	X	•	•	X	•	•	X
Dicrostonyx ex gr. *gulielmi-henseli*	*Pied lemming*	•	•	•	•	•	•	X	X	•	•	•	•	•	•	•	•	X	X	•	•	•	X	•	•	•	•	•	X	•	X	•	•	•	•	•	•
Dicrostonyx torquatus Pall.	Pied lemming	•	•	•	•	•	•	•	X	X	•	•	•	•	•	•	•	•	•	•	•	•	•	•	•	•	•	•	•	•	•	•	X	•	•	•	•
Ellobius talpinus Pall.	Northern mole-vole	•	•	•	•	•	•	•	•	•	•	•	•	•	•	•	•	•	•	•	•	•	•	•	•	•	•	•	•	•	X	•	•	•	•	•	X
Eolagurus luteus Eversm.	Yellow vole	•	•	•	•	X	•	•	•	•	•	•	X	•	•	•	•	•	•	•	•	•	•	•	X	•	•	•	•	•	X	•	•	X	X	•	X
Lagurus lagurus Pall.	Steppe lemming	•	•	•	X	X	•	•	X	•	•	•	•	•	•	X	•	X	X	•	•	•	X	X	X	•	•	•	•	•	X	•	•	•	•	•	X
Lemmus obensis Brants.	Ob' lemming	•	•	•	•	•	•	X	X	X	•	•	•	•	•	•	•	•	X	•	•	•	•	•	•	•	•	•	X	X	X	•	•	•	•	•	•
Marmota bobak Müll.	Bobak marmot	•	•	•	•	•	•	•	X	X	X	•	•	•	•	X	•	•	•	•	•	•	•	X	•	•	•	•	•	•	X	•	•	X	•	•	X
Marmota bobak palerossica J. Gromovi		•	•	•	•	X	•	•	•	•	•	•	•	•	•	•	•	•	•	•	•	•	•	•	•	•	•	•	•	•	•	•	•	•	•	•	•
Microtus agrestis Pall.	Field vole	•	•	X	X	•	•	X	X	•	•	•	•	•	•	X	•	•	•	•	•	•	•	•	•	•	•	•	X	•	X	•	•	•	•	•	•
Microtus arvalis Pall.	Common vole	•	•	•	•	•	•	X	•	X	•	•	•	•	•	•	•	•	•	•	•	•	•	X	•	•	•	•	•	•	•	•	•	X	X	•	X
Microtus arvalis/agrestis Pall.		•	X	•	•	•	•	•	•	•	•	•	•	•	•	•	•	•	•	•	•	•	•	•	•	•	•	•	•	•	•	•	•	•	•	•	•
Microtus hyperboreus Vinigr.		•	•	•	•	•	•	•	X	•	•	•	•	•	•	•	•	•	•	•	•	•	•	•	•	•	•	•	•	•	•	•	•	•	•	•	•
Microtus (Stenocranius) gregalis Pall.	Narrow-headed vole	X	•	•	X	X	•	X	X	X	•	•	•	•	•	X	•	•	X	•	•	•	•	•	•	•	•	•	X	•	X	•	X	•	•	•	X
Microtus oeconomys Pall.	Tundra vole (root vole)	•	X	•	•	X	•	•	•	•	•	•	•	•	•	X	•	•	X	•	•	•	•	•	•	•	•	•	•	•	•	•	•	•	•	•	X
Microtus (Pitymys) cf. *subterraneus* DeSelys-Longchamps	European pine vole	•	X	•	•	•	•	•	•	•	•	•	•	•	•	•	•	•	•	•	•	•	•	•	•	•	•	•	•	•	•	•	•	•	•	•	•
Microtus sp.	Vole	•	•	•	•	•	•	•	X	•	•	•	•	•	•	•	•	•	•	•	•	•	•	X	X	•	•	•	•	•	•	X	•	•	•	•	•
Sciurus sp.	Tree squirrel	•	X	•	•	•	•	•	•	•	•	•	•	•	•	•	•	•	•	•	•	•	•	•	•	•	•	•	•	•	•	•	•	•	•	•	•

Spalax microphtalmus Güld.	Greater mole-rat	•	•	•	•	•	•	•	•	•	•	•	•	•	•	•	•	•	•	•	•	•	•	X	X	X	•	•	•	•	•	•	•	•	•	•	•
Spalex sp.	Mole-rat	•	•	•	•	•	•	•	X	•	•	•	•	•	•	•	•	•	•	•	•	•	•	•	•	•	•	•	X	•	•	•	X	X	X	•	•
Carnivora																																					
Alopex lagopus L.	Polar fox (arctic fox)	•	•	•	•	•	•	•	•	X	•	•	•	•	•	X	•	X	•	X	X	•	•	X	X	X	•	•	•	X	X	X	•	X	•	•	X
Canis lupus L.	Wolf	•	•	•	•	•	•	•	X	•	X	•	X	•	X	X	•	X	•	•	X	•	•	X	X	X	•	•	•	X	X	•	X	X	•	•	X
Crocuta spelaea Goldf.	Cave hyena	•	•	•	•	•	•	•	•	•	•	•	•	•	•	•	•	•	•	•	•	•	•	•	•	•	•	•	•	X	X	•	•	•	•	•	X
Felis (Leo) spelaea Goldf.	Cave lion	•	•	•	•	•	•	•	X	•	X	X	•	•	•	X	•	•	•	•	•	•	•	X	X	•	•	•	•	•	•	•	X	X	•	•	•
Gulo gulo L.	Wolverine	•	•	•	•	•	•	•	•	•	•	•	•	•	•	X	•	X	•	•	•	•	•	•	•	X	•	•	•	X	X	•	•	X	•	•	•
Hyaena spelaea Goldf.	Striped hyena	•	•	•	•	•	•	•	X	•	X	X	•	•	•	X	•	•	•	•	•	•	•	•	•	•	•	•	•	•	•	•	•	•	•	•	•
Lynx lynx L.	Lynx	•	•	•	•	•	•	•	•	•	•	•	•	•	•	•	•	•	•	•	•	•	•	•	•	•	•	•	•	•	X	•	•	X	•	•	•
Martes martes L.	Pine marten	•	•	•	•	•	•	•	•	•	•	•	•	•	•	•	•	X	•	•	•	•	•	•	•	•	•	•	•	•	•	•	•	•	•	•	•
Mustela nivalis L.	Weasel	•	•	•	•	•	•	•	X	X	•	•	•	•	•	•	•	•	•	•	•	•	•	•	•	•	•	•	•	•	X	•	•	•	•	•	X
Mustela (Putorius) sp.	Polecat	•	•	•	•	•	•	•	•	•	•	•	•	•	•	•	•	•	•	•	•	•	•	•	•	•	•	•	•	•	X	•	•	•	•	•	•
Panthera (Leo) spelaea Goldf.	Cave lion	•	•	•	•	X	•	•	•	•	•	•	•	•	•	•	•	X	•	•	•	•	•	X	•	•	•	•	•	•	X	•	X	•	•	•	•
Putorius sp.	Polecat	•	•	•	•	•	•	•	•	X	•	•	•	•	•	•	•	•	•	•	•	•	•	•	•	•	•	•	•	•	•	•	•	•	•	•	•
Ursus arctos L.	Brown bear	•	•	•	•	•	•	•	X	•	X	•	•	•	•	X	•	X	•	•	•	•	•	•	X	•	•	•	•	X	X	•	X	X	•	•	•
Ursus spelaeus Ros. et Hein.	Cave bear	•	•	•	•	•	•	X	X	•	•	•	•	X	•	X	•	•	•	•	•	•	•	•	•	•	•	•	•	X	X	•	•	•	•	•	X
Vulpes corsac L.	Corsac fox	•	•	•	•	•	•	•	•	•	•	•	•	•	•	X	•	•	•	•	•	•	•	•	•	•	•	•	•	•	•	•	•	•	•	•	X
Vulpes cf. *corsac* L.	Corsac fox	•	•	•	•	•	•	•	•	•	•	•	•	•	•	•	•	•	•	•	•	•	•	X	•	•	•	•	•	•	•	•	•	•	•	•	•
Vulpes vulpes L.	Red fox	•	•	•	•	•	•	•	X	•	X	•	•	•	•	X	•	•	•	•	•	•	•	•	•	•	•	•	•	X	•	•	•	X	•	•	X
Vulpes sp.	Fox	•	•	•	•	•	•	•	•	•	•	•	•	•	•	•	•	•	•	•	•	•	•	X	•	•	•	•	•	•	•	•	•	•	•	•	•
Artiodactyla																																					
Alces alces L.	Moose	•	•	•	•	•	•	•	X	•	X	•	X	•	•	•	•	•	•	•	•	•	•	•	•	•	•	•	•	•	•	•	X	X	X	•	X
Bison priscus Boj.	European large-horned bison (steppe wisent)	•	•	•	•	X	•	•	X	•	X	•	X	X	X	X	•	X	•	•	•	•	•	•	•	•	•	•	•	•	X	X	X	X	X	X	X
Bison priscus longicornis W. Grom.		•	•	•	•	•	X	•	•	•	•	•	•	•	•	•	•	•	•	•	•	•	•	•	•	•	•	•	•	•	•	•	•	•	•	•	•
Bison aut *Bos*		•	•	•	•	•	•	•	•	•	•	•	•	•	•	•	•	•	•	•	•	•	•	X	X	X	•	•	•	•	•	•	•	•	X	•	•
Bos primigenius Boj.	Aurochs	•	•	•	•	•	•	•	•	•	•	•	•	•	•	•	•	•	•	•	•	•	•	•	•	•	•	•	•	X	X	X	X	X	X	•	•
Bos trochoceros Meyer		•	•	•	•	•	X	•	•	•	•	•	•	•	•	•	•	•	•	•	•	•	•	•	•	•	•	•	•	•	•	•	•	•	•	•	•
Bos sp.		•	•	•	•	•	•	X	•	•	•	•	•	•	X	•	•	•	•	•	•	•	•	•	X	X	•	•	•	•	•	•	X	•	•	•	•
Capreolus sp.	Roe deer	•	•	•	•	•	•	•	•	•	•	•	•	•	•	•	•	•	•	•	•	•	•	•	•	•	•	•	•	•	•	•	X	•	•	•	X
Cervus elephas L.	Red deer (wapiti)	•	•	•	•	•	X	X	X	X	X	•	X	•	•	X	•	•	•	•	•	•	•	X	X	•	•	•	•	•	X	•	X	X	X	•	X
Cervidae gen.	Red deer	•	•	•	•	X	•	•	•	•	•	•	•	•	•	•	•	•	•	•	•	•	•	•	•	•	•	•	•	•	•	•	•	•	•	•	•
Megaloceros giganteus Blüm.	Giant deer	•	•	•	•	•	•	•	X	X	X	•	•	•	X	X	•	•	•	•	•	•	•	X	•	•	•	•	•	•	•	•	X	•	X	•	•
Ovibos moschatus Zimmer	Barren-ground musk ox	•	•	•	•	•	•	•	•	•	•	•	•	•	•	•	•	•	•	•	•	•	•	•	•	•	•	•	•	X	X	•	X	X	•	•	•
Ovis sp.	Sheep	•	•	•	•	•	•	•	•	•	•	•	•	•	•	X	•	•	•	•	•	•	•	•	•	•	•	•	•	•	•	•	•	•	•	X	•
Rangifer tarandus L.	Reindeer	•	•	•	•	•	•	•	X	X	•	•	X	•	•	X	X	X	•	•	X	•	•	X	X	X	X	X	•	X	X	X	X	X	X	X	X
Saiga tatarica L.	Saiga antelope	•	•	•	•	•	•	•	•	•	•	•	•	•	•	•	•	X	•	•	•	•	•	X	X	X	•	•	•	•	•	•	X	X	X	•	X
Sus scrofa L.	Eurasian wild hog	•	•	•	•	•	•	•	•	•	•	•	•	•	•	X	•	•	•	•	•	•	•	•	•	•	•	•	•	•	•	•	•	•	•	•	X
Perissodactyla																																					
Coelodonta antiquitatis Blüm.	Woolly rhinoceros	•	•	•	•	X	•	X	X	X	X	X	•	•	•	X	•	•	•	•	X	•	•	X	X	•	•	•	•	X	X	X	X	X	X	•	•
Equus caballus L.	Caballine horse	•	•	•	•	X	•	X	X	X	X	X	X	X	X	X	•	X	•	X	X	•	•	X	X	X	X	X	•	X	X	X	X	X	X	X	X
Equus hidruntinus Reg.		•	•	•	•	•	•	•	•	•	•	•	•	•	X	X	•	•	•	•	•	•	•	•	•	•	•	•	•	•	•	•	•	X	X	•	•
Equus (Asinus) sp.	Donkey	•	•	•	•	•	•	•	•	•	•	•	•	•	•	•	•	•	•	•	•	•	•	•	•	•	•	•	•	•	•	•	•	•	X	•	•
Proboscidae																																					
Mammuthus primigenius Blüm.	Woolly mammoth	•	•	•	•	•	•	X	X	X	X	X	X	•	X	X	X	X	•	X	X	X	•	X	X	X	X	X	•	X	X	X	X	X	X	•	•
Mammuthus primigenius Blüm. (early type)	Woolly mammoth	•	•	•	•	X	•	•	•	•	•	•	•	•	•	•	•	•	•	•	•	•	•	•	•	•	•	•	•	•	•	•	•	•	•	•	•
Mammuthus primigenius chasaricus Dub.	Khazar mammoth	•	•	•	•	•	X	•	•	•	•	•	•	•	•	•	•	•	•	•	•	•	•	•	•	•	•	•	•	•	•	•	•	•	•	•	•
Palaeoloxodon antiquus (Falc. et Gaut.)	Straight-tusked elephant	•	•	•	•	X	X	•	•	•	•	•	•	•	•	•	•	•	•	•	•	•	•	•	•	•	•	•	•	•	•	•	•	•	•	•	•

Note: Site numbers correspond to numbers in the text and are not necessarily the same as those on Figures 20-1 to 20-4.

Key: Mikulino Interglaciation (Russian Plain):
(1) Cheremoshnik, Yaroslav' Province (Agadjanian, 1977).
(2) Ulovka, Klyaz'ma River basin (Aleksandrova and Tseytlin, 1965).
(3) Timoshkovichi, Neman River basin (Kalinovskiy, 1979).
(4) Gadyach, Psel River basin (Markova, 1975a).
(5) Shkurlat, Don River basin (Alekseyeva, 1980).
(6) Karagash, Dnestr River basin (David and Lungu, 1972).
Early and Middle Valdai Glaciation (Russian Plain and Crimea):
(7) Belorussia (Kalinovskiy, 1979). Key sites: Rumlovka, Borisoya Mountain. (According to the mammologist A. N. Moruzko, who also worked on this section, the fauna of Borisova Mountain is older and dates back to the beginning of the Mikulino Interglaciation [verbal communication].)
(8) Prut and Dnestr River basins (Chernysh, 1973; Ketraru, 1973; Ivanova, 1977a, 1977b). Key sites: Vykhvatitsy, Buteshty, Korman' IV (layers 9-12), Molodova (layers 9-11). Radiocarbon dates: 40,300 yr B.P. (LU-4017), 45,000 yr B.P. (LU-3659), 44,000 yr B.P. (LU-3659).
(9) Desna River basin (Tarasov, 1977; Zavernyayev, 1978). Key sites: Betovo, Khotylevo I. Radiocarbon date: >36,100±550 yr B.P.
(10) Dnepr River basin (Pidoplichko, 1936; Boriskovskiy, 1953). Key sites: Kodak, Grigor'yevka.
(11) Don and Severskiy Donets River basins (Gromov, 1935; Yefimenko, 1953). Key sites: Shubnoye, Krasnyy Yar.
(12) Volga River basin (Vereshchagin and Kolbutov, 1957). Key sites: Sukhaya Mechetka.
(13) Northern region along the Black Sea (Beregovaya, 1960). Key sites: Il'inka.
(14) Region along the Sea of Azov (Boriskovskiy and Praslov, 1964). Key sites: Rozhok 1, P.
(15) Crimea (Vekilova, 1957; Vereshchagin and Baryshnikov, 198[illegible]). Key sites: Kiik-Koba, Kosh-Koba, Volchiy Grotto, Chokur[illegible]a, Bakhchisarayaskaya, etc.
Bryansk Interstade (Russian Plain):
(16) Byzovaya, Pechora River basin (Guslitser and Kanivets, 1965). Radiocarbon date: 25,400±380 yr B.P. (TA)
(17) Sungir', Klyaz'ma River basin (Bader, 1977). Radiocarbon dates: 25,500±200 yr B.P. (GRN 54/25), 24,300± 400 yr B.P. (GRN 54/46).
(18) Troitsa II, Oka River basin. Radiocarbon dates: 27,700±360 yr B.P. (IGAN-203), 32,500±70 yr B.P. (IGAN-20).
(19) Yurovichi, Pripyat' River basin (Polikarpovich, 1968). Radiocarbon date: 26,470±420 yr B.P. (LU-125).
(20) Pushkari, Desna River basin (Velichko, 1961).
(21) Pogon, Desna River basin (Velichko, 1961).
(22) Arapovichi, Desna River basin (Markova, 1972). Radiocarbon date: 24,000±300 yr B.P. (IGAN-46).
(23) Kostenki VII (layers 2-4), Don River basin (Rogachev, Anikovich, and Artemova, 1979; Rogachev, Anikovich, and Belyayeva, 1979; Rogachev and Sinitsyn, 1979; Vereshchagin and Kuz'mina, 1977).
(24) Kostenki XIV (layers 2-4), Don River basin (references same as #23). Radiocarbon dates: 26,400±600 yr B.P. (LU-59A), 28,200±700 yr B.P. (LU-59B).
(25) Kostenki I (layers 4-5), Don River basin (references same as #23).
(26) Molodova V (layers 8-10), Dnestr River basin (Chernysh, 1973; Ivanova, 1977a, 1977b). Radiocarbon dates: 28,100±1000 yr B.P. (LG-17g), 29,650±1320 yr B.P. (LG-17a).
(27) Korman' IV (layer 7), Dnestr River basin (references same as #26). Radiocarbon dates: 24,500 yr B.P. (GIN-1099), 25,140±350 yr B.P. (LU-586).
Late Valdai Glaciation (Russian Plain and Crimea):
(28) Belorussia (Kalinovskiy, 1979; Motuzko and Kalinovskiy, 1976; Arslanov et al., 1971a, 1971b). Key sites: Pashino, Drychaluki, Diskeninovo. Radiocarbon dates: 17,750±170 yr B.P. (LU-95a), 17,900±160 yr B.P. (LU-95b), 18,370±150 yr B.P. (LU-96a), 23,630±150 yr B.P. (LU-97a).
(29) Upper Dnestr River basin, Volyno-Podoliya (Chernysh, 1973; Savich, 1975; Bogucki et al., 1974). Key sites: Lipa IX, Gorodok, Vyshnivets, Kulichivka. Radiocarbon date: 17,900±2250 yr B.P.
(30) Upper Dnepr and Desna River basins (Gromov, 1948; Velichko et al., 1977; Voznyachuk and Kalechets, 1969). Key sites: Khotylevo II, Novgorod-Severskaya, Mezin, Yeliseyevichi, Berdyk. Radiocarbon dates: 23,430±180 yr B.P. (LU-104), 17,340±170 yr B.P. (LU-360), 23,660± 270 yr B.P. (LU-359).
(31) Oka River basin (Milanov, 1958). Key sites: Karacharovo, Yaltunovo-Poltnoye.
(32) Middle Dnestr, Prut, Yuzhnyy Bug River basins (Ketraru, 1973; Chernysh, 1968, 1973). Key sites: Korman' IV (layers 2-6), Molodova V (layers 2-6), Ataki I, Staryye Druitory, Vladimirovka. Radiocarbon dates: 18,000±400 yr B.P. (GIN-719), 18,560± 2000 yr B.P. (SOAN-145), 17,100±180 yr B.P. (MO-II), 13,370±540 yr B.P. (GIN-9), 15,375±850 yr B.P. (SOAN-143), 16,600±750 yr B.P. (SOAN-144).
(33) Middle Dnepr, Don, and Severskiy Donets River basins (Pidoplichko, 1969; Tarasov, 1967; Rogachev, Anikovich, and Artemova, 1979; Rogachev, Anikovich, and Belyayev, 1979; Rogachev and Sinitsyn, 1979). Key sites: Dobranichivka, Mezherichi, Avdeyevo, Kostenki, Gagarino, Osokorivka I, Gontsy. Radiocarbon dates: 13,900±200 yr B.P. (IGAN-78), 22,700±700 yr B.P. (GIN-1571), 21,800±300 yr B.P. (GIN-1872), 14,300±460 yr B.P. (GIN-79), 19,100±150 yr B.P. (LE-1436).
(34) Middle Volga River basin (Panichkina, 1953; Abramova, 1958; Bryusov, 1940). Key sites: Mysy, Yunga-Kusherga, Pervoloki.
(35) Northern Black Sea region, Sea of Azov region, lower Don River basin (Boriskovskiy and Praslov, 1964; Gvozdover, 1964). Key sites: Akkarzha, Amvrosiyevka, Kamennaya Balka II.
(36) Crimea (Vekilova, 1957; Vereshchagin and Baryshnikov, 1980b). Key sites: Syuren' I (Upper Paleolithic layers), Adzhikoba (layer 2).

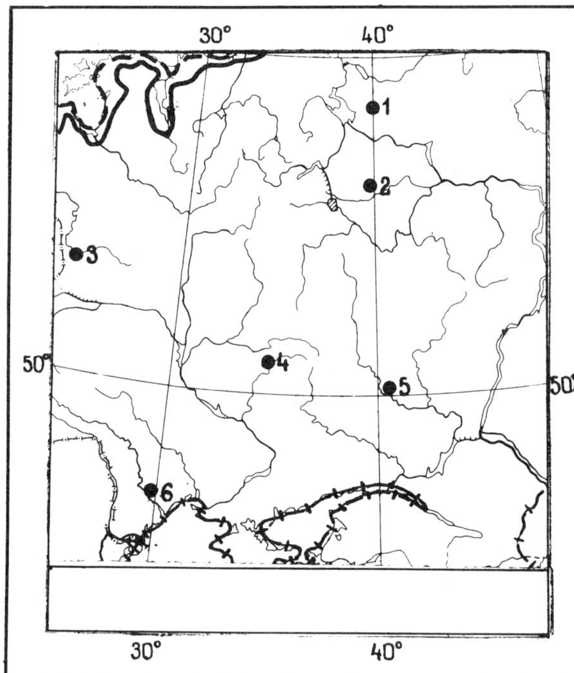

30° 40°
50° 50°
30° 40°

Sites:
1. Cheremoshnik
2. Ulovka
3. Timoshkovichi
4. Gadyach
5. Shkurlat
6. Karagash

Coastline of Mikulino Interglaciation:
According to the International Map of Quaternary Deposits of Europe, 1969, with additions by V. P. Grichuk
According to N. S. Blagovolin, O. K. Leont'yev, V. M. Muratov, and L. R. Serebryanny
Coast line of the Karahgatskiy and Pozdnekhazarskiy Basins, according to A. L. Chepalyga

Figure 20-1. Sites of mammal remains from the Mikulino Interglaciation of the Russian Plain.

woolly mammoth, steppe wisent, woolly rhinoceros, and other species. The small mammals include ground squirrel, little jerboa, Eurasian water vole, steppe lemming, yellow lemming, narrow-headed vole, northern vole, and pika (as determined by A. K. Markova).

With the exception of the straight-tusked elephant, the composition of this fauna suggests a steppe landscape. Ground squirrel, jerboa, pika, steppe lemming, and yellow lemming currently inhabit the steppe and semidesert zones. The straight-tusked elephant apparently entered these regions along floodplain forests.

Finally, the Karagash site is located on the southern Russian Plain, in the alluvium of the second terrace of the Dnestr River (David and Lungu, 1972). This terrace has been dated as Mikulino by several methods. The fauna includes straight-tusked elephant, Khazar mammoth, and *Bos trochoceros.*

The presence of the straight-tusked elephant (a forest species) and the Khazar mammoth (a steppe species) suggests a forest-steppe landscape along the lower Dnestr River. The location of this site on the western Russian Plain supports this interpretation, because at present this region is located on the boundary between forest-steppe and steppe.

Thus, even the scarce vertebrate data from Mikulino time give a definite idea of the faunal composition and ecology. Among the large mammals are ancient species such as straight-tusked elephant, an early woolly mammoth, steppe wisent, woolly rhinoceros, and cave-dwelling predators, and yet on the whole the fauna is very similar to the modern one. The interglacial rodents do not differ from modern ones above the subspecies rank. Obviously, the basic ecologic requirements of the mammals were also the same as those of present mammals; therefore, changes in the Mikulino mammal communities reflect changes in the landscape from north to south. The fossils suggest that broad-leaved forests occurred north to latitude 56°N to 57°N. From there the forest-steppe fauna had a range extending from southwest to northeast just as it does today. The steppe mammals were widely distributed over the Russian Plain, and the northern boundary of the steppe was nearly the same as the present one, as the findings in the south Voronezh region indicate.

Mammals of the Early and Middle Valdai

Rich material is available for the first half of the Valdai. Fossil mammals of Early and Middle Valdai age are found in Mousterian campsites, but some nonarchaeological sites also occur. The latest Mousterian sites are dated between 45,000 and 30,000 yr B.P. A Valdai age for some Mousterian campsites is suggested by the composition of the mammal fauna, which is dominated by forms tolerant of the cold. More than 40 Early Valdai mammal sites are available.

The Early and Middle Valdai mammals record a sharp rearrangement in the faunas of eastern Europe, particularly a very wide distribution of periglacial animals (Table 20-1). Mammoth remains are found at sites from the Neman and Desna River basins (latitude 53°N) to the Crimea (Figure 20-2). Woolly rhinoceros remains are found in the Middle Valdai alluvium of the Neman River (Kalinovskiy, 1979), at Mousterian campsites of the Middle Dnestr and Dnepr regions, on the Don, and in the Crimea. Reindeer, which extended to the Crimea in the South, were also very widely distributed. The areas of pied and common lemmings expanded considerably southward, and their remains are known at Belorussian sites, in Mousterian campsites of the Middle Dnestr (Chernysh, 1973), and on the Desna River (Tarasov, 1977), 1300 km from their present habitats.

Thus, in the Early and Middle Valdai, the so-called

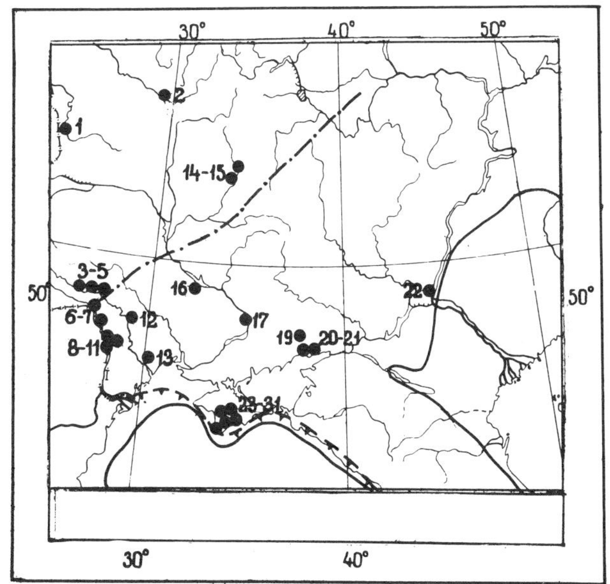

Sites:
1. Rumlovka
2. Borisova Mountain
3. Kasperovtsy
4. Molodova I and V (Mousterian layers)
5. Korman (Mousterian layers)
6. Trinka
7. Buzduzhany
8. Kokuneshty
9. Skulyany
10. German-Dumeny
11. Vasilika
12. Vykhvatintsy
13. Il'inka
14. Betovo
15. Khoylevo I
16. Grigor'yevka
17. Kodak I
18. Shubnoye
19. Antonovka
20. Rozhok I and II
21. Gerasimovka
22. Sukhaya Mechetka

23-31. Crimean sites (Kiik-Koba, Kosh-Koba, Chagorak-Koba, Bolchiy Grot, Chokurcha, Shaytan-Koba, Kabazi, Bakhchisarayskaya

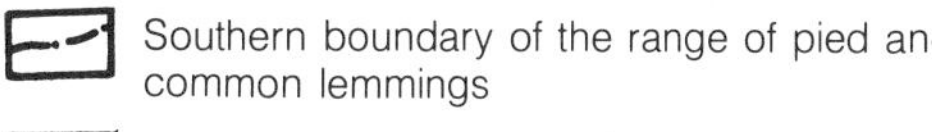

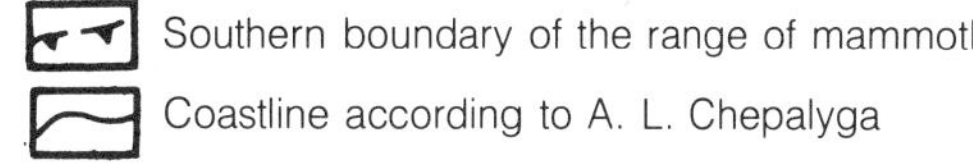

Figure 20-2. Sites of mammal remains from the Early and Middle Valdai of the Russian Plain and Crimea.

mammoth faunal complex (Vereshchagin, 1979; Vereshchagin and Baryshnikov, 1980c), which included mammoth, woolly rhinoceros, reindeer, polar fox, and pied lemming, was widespread over the Russian Plain. The presence of pied and common lemmings suggests that particularly severe conditions prevailed up to latitude 49°N on the western Russian Plain, to latitude 51°N in the Desna Basin, and apparently still farther north in the East. In Belorussia and the Prut, Dnestr, and Desna basins, however, woodland species such as red deer, moose, brown bear, weasel, and wood and bank vole were found along with arctic species. A major role was played in these regions by steppe species including the bobac, marmot, ground squirrels of several species, narrow-headed vole, steppe wisent, and horse. To the southeast in the Middle Dnepr and Don Basins, the Severskiy Donets Basin, and the Middle Volga Basin, extinct periglacial species including mammoth, reindeer, woolly rhinoceros, cave lion, and bear, as well as reindeer, were also widely distributed. Arctic animalș such as lemmings and polar fox have not been observed in these regions. The role of the steppe species increases from west to east as a result of increased continentality. In the Volga Basin, and particularly at the Mousterian campsite Sukhaya Mechetka, bones of great suslik, little jerboa, and saiga have been found (Vereshchagin and Kolbutov, 1957).

During the Early and Middle Valdai, the southernmost Russian Plain was inhabited mainly by steppe species, including horse, bison, and saiga antelope, but mammoth and reindeer apparently also existed there, as indicated by their remains in numerous Crimean Mousterian campsites (Vereshchagin and Baryshnikov, 1980b). The link between the Crimean Highland and the Russian Plain during the Early and Middle Valdai was stronger than it is at present, for the level of the Black Sea during that period was lower than it is today. (See Figure 20-2.)

Thus, the mammals indicate a marked environmental change in eastern Europe during the Early and Middle Valdai. A considerable expansion of arctic species (lemmings, polar fox) had already taken place at that time. The periglacial mammoth faunal complex (mammoth, woolly rhinoceros, cave-dwelling predators) spread south to the Crimea and Caucasus; this finding indicates an appreciable cooling and aridity. Steppe mammals penetrated far to the northwest during more arid conditions, and remains of ground squirrels and pikas have been noted in the modern forest zone of Belorussia. Woodland mammals found in most sites on the Russian Plain (except in the southern present-day steppe) suggest the existence of localized forest vegetation. The role of these woodland species increases in the western Russian Plain and mountainous Crimea and decreases eastward.

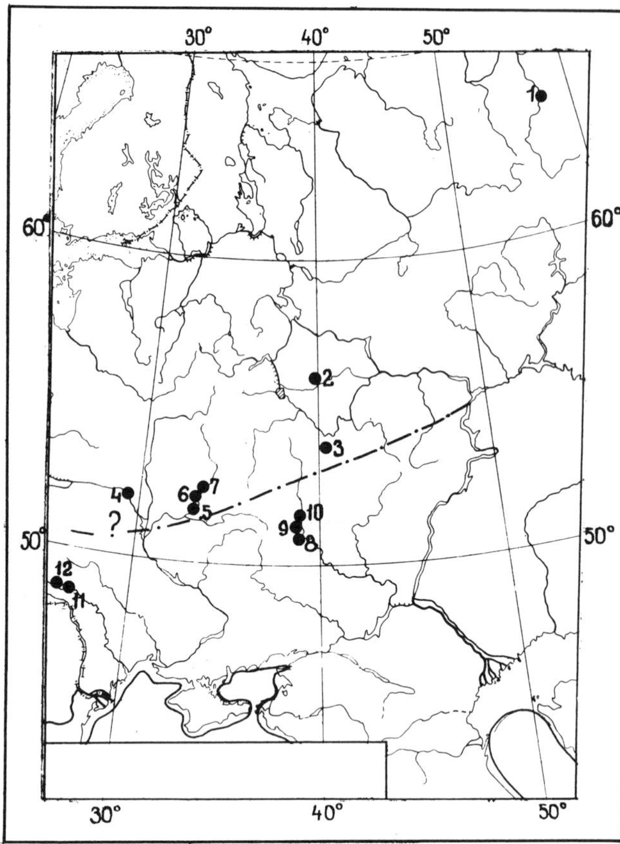

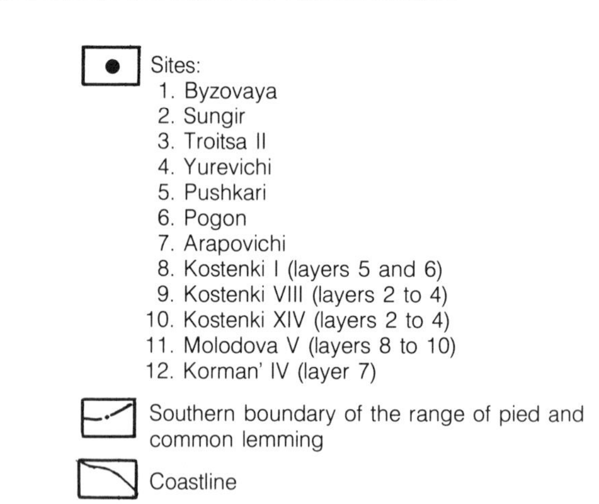

Figure 20-3. Sites of mammal remains from the Bryansk Interstade of the Russian Plain.

Mammal Fauna of the Bryansk Interstade

The Bryansk (Paudorf) Interstade is recorded throughout the Russian Plain as the Bryansk fossil soil (Velichko and Morozova, 1972), dated at 30,000 to 25,000 yr B.P. The interstadial fauna is known from a series of radiocarbon-dated sites (Figure 20-3 and Table 20-1). The Arapovichi site lies in the soil itself (Figure 20-3, site 7), but correlative strata are found at the Pushkari (site 5), Pogon (site 6), Kostenki VIII (layers 2-4) (site 9), Kostenki I (layers 4 and 5) (site 8), and Kostenki XIV (layers 2-4) (site 10) (Velichko, 1961), as well as Molodova V (layers 8-10) (site 11), Korman' IV (layer 7) (site 12) (Ivanova, 1977a, 1977b; Chernysh, 1973), Yurevichi (site 4) on the Pripyat' River, and Sungir' (site 2) on the Klyaz'ma River. Mammal bones were found in alluvium of the Oka River near the village of Troitsa (site 3) and at the Byzovaya campsite on the Pechora River (site 1) near the Arctic Circle (Guslitser and Kanivets, 1965).

The mammal sites of Bryansk time reveal the following basic features. (1) Mammoth and reindeer remains were found everywhere, from the Pechora Basin in the north to the middle course of the Dnestr (Figure 20-3). Bones of woolly rhinoceros were found at only two sites. Remains of other tundra-steppe periglacial species, such as cave-dwelling predators and giant deer, were noted at a number of sites. (2) Remains of arctic species (pied lemming and Ob' lemming) were found in the Desna, Oka, and Klyaz'ma River basins. The polar fox was found in the Don Basin at Kostenki. (3) Steppe mammals, including pika, ground squirrels, mole-rat, steppe lemming, yellow lemming, narrow-headed vole, horse, saiga, bison, and aurochs, were found at all sites. (4) Forest animals were less common, but brown bear, marten, and wolverine were found at Sungira; bank vole at Troitsa II; beaver, wolverine, brown bear, and red deer in Kostenki VIII; and red deer at the Dnestr campsite. In all cases, forest animals were no less common than arctic, periglacial, and steppe species. At sites farther north in the Oka and Klyaz'ma Basins, red deer bones were absent, and the group of forest animals included taiga species such as wolverine, marten, and brown bear. Judging from the fossil record, the Bryansk warming was slight and did not greatly affect the faunal composition. If the data are complete, only lemmings show a restricted distribution. During the Early Valdai they spread all the way up to the middle course of the Dnestr, whereas during the Bryansk Interstade they were not observed south of the Desna Basin. On the central Russian Plain, open landscapes of periglacial forest-steppe were inhabited by a large group of arctic, periglacial, steppe, and, to a lesser degree, woodland species. Judging from the latter, conifers must have been part of the vegetation in northern regions such as the Klyaz'ma and Oka Basins, while broad-leaved species grew farther south in the Don and Dnestr Basins. Forest islands were probably confined to moist areas near rivers and other bodies of water.

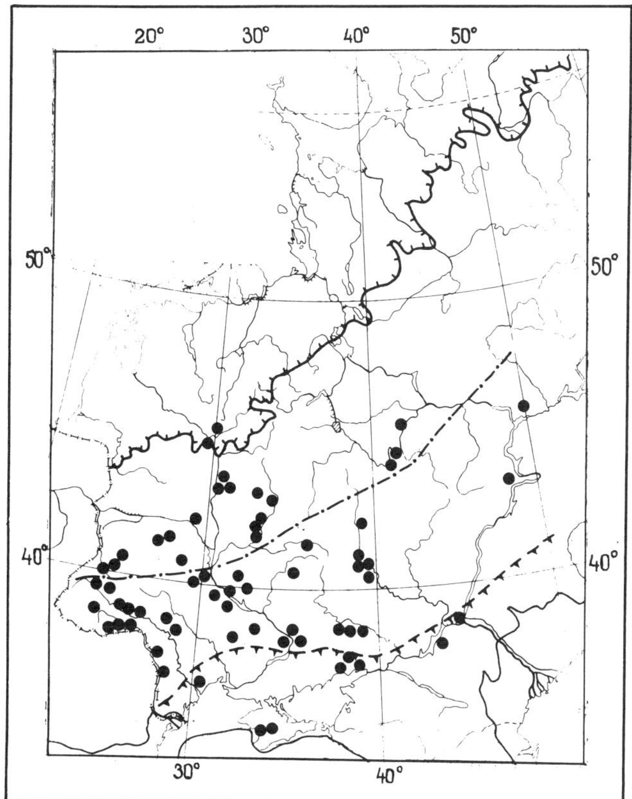

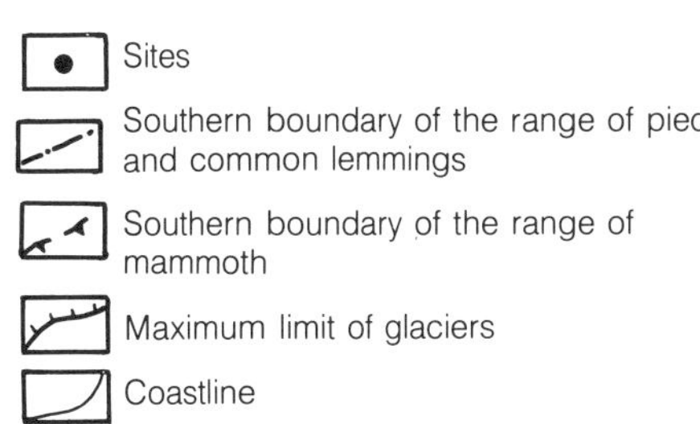

Figure 20-4. Sites of mammal remains from the Late Valdai of the Russian Plain and Crimea.

Mammal Faunas of the Late Valdai

The Late Valdai (24,000 to 11,000 yr B.P.) is represented by more than 60 sites, most of which are found in association with Upper Paleolithic campsites (Table 20-1). Sites extend from latitude 55°N in the North to the southern Russian Plain (Figure 20-4) and to Crimea, which in Late Valdai time was connected to the Russian Plain as a result of lowered sea level (Figure 20-4).

The northernmost sites located in the Zapadnaya Dvina Basin date to the period of maximum glaciation at 18,000 yr B.P. Thus, the animals lived near the glacier margin prior to its maximum advance. The arctic and steppe species indicate severe natural conditions, and only one forest animal, the field vole, was noted.

Somewhat more temperate conditions are indicated by fossils from Volyno-Podoliya and along the upper reaches of the Dnepr, Don, and Oka Rivers. Remains of mammoth, woolly rhinoceros, cave lion, hyena, and cave bear were found at these sites, as well as reindeer, whose remains dominate Late Valdai sites (in contrast to Early Valdai sites, in which mammoth is dominant). All the sites contained pied lemming, Ob' lemming, and polar fox; and remains of the musk ox were found at the Volyno-Podoliya and Desna sites. Steppe species include pika, bobac marmot, ground squirrels, little jerboa, northern mole-vole, common hamster, steppe lemming, yellow lemming, narrow-headed vole, horse, and bison.

Woodland mammals are less common, although remains of field vole, brown bear, weasel, wolverine, and lynx were noted at these sites. Even remains of red deer have been found at the Novgorod-Severskaya campsite (Desna Basin). On the whole, taiga mammals dominate among the forest species, but they are not abundant, although they indicate the vegetation types. Tundra-steppe animals prevailed from the ice border south to latitude 55°N (in the Oka Basin), to 52°N in the Desna Basin, and to 50°N on the western part of the Russian Plain in an open landscape 300 to 400 km wide, with localized taiga vegetation in the river valleys.

Farther south along the Middle Dnestr and Prut, the Yuzhnyy Bug, the Middle Dnepr, the Severskiy Donets, and the Middle Volga River basins, as well as in regions farther north, there were mammoth, woolly rhinoceros, steppe wisent, and aurochs. Giant deer, which was not found in the periglacial strip, has been noted there. During the Late Pleistocene, this animal lived in the periglacial steppe that extended from northern Eurasia and the British Isles to southern Siberia (Vereshchagin and Baryshnikov, 1980a). Musk ox was uncommon. Red lemming remains are reported within this region in the middle course of the Prut River. Polar fox was more common there than it was farther north. Steppe species included bobac marmot, ground squirrel, mole-rat, yellow lemming, narrow-headed vole, and horse. Donkey remains were found in the Volga River basin.

The woodland species include brown bear, wolverine, lynx, and moose. Species of the present-day broad-leaved forests, including red and roe deer, were more widespread than they are farther north. Thus, extinct periglacial, steppe, and woodland mammals jointly inhabited a considerable portion of the Russian Plain north of latitude 47°N to 48°N.

Along the north coast of the Black Sea and near the Sea of Azov, the faunas are dominated by such steppe species as bison, horse, wild ass, and sheep. Reindeer remains also are characteristic for this area during the Late Valdai, but remains of mammoth or woolly rhinoceros have not been found. Either these large mammals could not find an adequate food source in the impoverished steppe landscapes, or they were exterminated by ancient hunters.

Finally, the mountain faunas of the Crimea are diverse as a result of the variety of landscapes at different altitudes. There is also a considerable component from the Russian Plain, for example, arctic hare, arctic fox, and reindeer (Vereshchagin and Baryshnikov, 1980b). By Late Valdai time, mammoth and woolly rhinoceros had already disappeared from the Crimea, apparently because of Paleolithic hunters. Cave-dwelling predators were widespread, and steppe species attained a great diversity. The following mammals have been reported in Crimean faunas: pika, several species of ground squirrel, little jerboa, common hamster, gray hamster, steppe lemming, yellow lemming, narrow-headed vole, small corsac fox, Caballine horse, wild ass, and saiga. Woodland species included brown bear, wild boar, red deer, beaver, wood vole, and wood mouse. Moose was absent.

Thus, the Late Valdai fauna of the Russian Plain and Crimea reflects considerable environmental change during Late Valdai time. Cold-tolerant species such as mammoth, woolly rhinoceros, giant deer, and reindeer inhabited the periglacial tundra-steppe and forest-steppe. Steppe species ranged far north and west of their present habitats. On the other hand, northern tundra species, such as lemming, polar fox, and musk ox, penetrated far south. The reindeer, one of the animals most extensively hunted by ancient humans, ranged over the entire Russian Plain to the Black Sea and Crimea.

Woodland species were few and inhabited riparian areas. Practically none has been found adjacent to the ice margin or on the southern Russian Plain. For many woodland species, the forested low mountains of Crimea provided refugia.

Discussion

On the whole, the Late Pleistocene fauna reflects the general cooling, decreased precipitation, hyperzonal cryoxerotic conditions, and associated expansion of permafrost and cryoxerophytic vegetation (Velichko, 1973).

A number of absolute climatic parameters were obtained by means of climatograms applied to data on mammals (mainly small mammals). This method, proposed by Iversen (1944), is discussed in Chapter 25 of this monograph. The climatic reconstructions depend on the quality of the original data. Faunal remains furnish good results only when species with a clearly manifest ecology and a relatively narrow distribution are used from nonredeposited sediments. Even so, only approximate values can be derived, because the data are incomplete.

For the Mikulino Interglaciation, we obtained climatic reconstructions for the Shkurlat site (about latitude 51°N) (Alekseyeva, 1980) by combining the climatic areas of the yellow and steppe lemmings, great and little jerboa, narrow-headed vole, and common hamster. Average July temperature was practically the same as the present, about +20°C, and the average January temperature was −14°C, about 5°C lower than at present. Total annual precipitation was 230 mm (now 450 mm), and the total snowfall was 130 mm (now 150 mm). Thus, the climate at the Shkurlat site apparently was cooler and drier (more continental), probably corresponding to the beginning of the interglaciation.

The indexes reconstructed for the Gadyach site at latitude 51°N presumably indicate the climate during the Krutitsa Interstade. Average July temperature was +20°C; average January temperature, −14°C; total annual precipitation, 420 mm (now 500 mm); and snowfall, about 100 mm (now 130 mm). Hence, climatic conditions were somewhat more continental then than they are now, confirming a Krutitsa age for this site.

Climatic indexes for the Early Valdai were obtained from the fauna of the Betovo Mousterian campsite (Desna Basin; see Figure 20-2) (Tarasov, 1977). The climatograms for pied and Ob' lemmings, steppe pika, bobac marmot, narrow-headed vole, common vole, and other species were used to derive an average July temperature of +15°C, an average January temperature of −17°C, a total annual precipitation of 350 mm, and a snowfall of 120 mm. Winter temperatures were more than 10°C lower than today, and summer temperatures were 5°C to 15°C lower, at +15°C. The annual precipitation was 200 mm less than present values.

For Bryansk time, climatic parameters were derived for two sites, Arapovichi and Troitsa II (Figure 20-3). They showed winter temperatures of +8°C to +10°C. July temperatures were also lower than today. The precipitation was 200 to 250 mm less, and snowfall was about 100 mm less.

The climatic characteristics for Late Valdai time are based on fossils from the Upper Paleolithic campsites of the Desna Basin. The severe continental climate (Markova, 1975b) is reflected in the decreased precipitation (200 to

400 mm lower than today) and snowfall (reduced by almost one-half). The January temperatures dropped to −22°C, more than 10°C below the present, but summer temperatures were similar to present ones.

Thus, climatograms based on local faunal data (for small mammals) suggest fairly stable but severe continental conditions during the Valdai. These results have been confirmed by the results of other paleoecologic methods.

References

Abramova, Z. A. (1958). Explorations of the Paleolithic in the middle course of the Volga in 1954. *Krayevedcheskiye Zapiski Ul'yanovskogo Oblastnogo Krayevedcheskogo Muzeya,* Issue II. Ulyanovsk.

Agadjanian, A. K. (1977). Quartäre Kleinsäuger aus der Russischen Ebene. *In* "Quartär," Vol. 27/28, pp. 111-45. L. Röhrscheid Verlag, Bonn.

Aleksandrova, L. P. and Tseytlin, S. M. (1965). Occurrence of fossil remains of small mammals in Quaternary deposits of the Nerl' River basin (Vladimir Province). *In* "Stratigraphic Importance of the Anthropogene Fauna of Small Mammals" (K. V. Nikiforova, ed.), pp. 158-61. Nauka Press, Moscow.

Alekseyeva, . I. (1980). Characteristics of the mammal complex of the last interglacial of the Russian Plain. *In* "Mammals of Eastern Europe in the Anthropogene" (O. A. Skarlato, ed.). USSR Academy of Sciences Zoologicheskogo Instituta, Trudy 93, pp. 68-74.

Arslanov, Kh. A., Voznyachuk, L. N., Velichkovich, F. Yu., et al. (1971a). Age of the maximum stage of the last glaciation at the interfluve of the Zapadnaya Dvina and Dnepr. *USSR Academy of Sciences Doklady* 196, 161-64.

Arslanov, Kh. A., Voznyachuk, L. N., Velichkovich, F. Yu., et al. (1971b). Paleogeography and geochronology of the Middle Valdai Interstadial on the territory of Belorusskoye Poozer'ye. *USSR Academy of Sciences, Doklady* 201, 661-64.

Bader, O. N. (1977). The paleoecology and people of the Sungir' campsite. *In* "Paleoecology of Ancient Man" (I. V. Ivanova and N. D. Praslov, eds.), pp. 31-40. Nauka Press, Moscow.

Beregovaya, N. A. (1960). "Paleolithic sites of the USSR." USSR Academy of Sciences, Institute of Archaeology, Moscow and Leningrad.

Bogucki, A. B., Savich, V. P., and Tatarinov, K. A. (1974). Nature and development of a primitive community on the territory of southwestern Volynya. *In* "Primitive Man and the Environment" (I. P. Gerasimov and A. A. Velichko, eds.), pp. 143-48. USSR Academy of Sciences, Institute of Geography, Moscow.

Boriskovskiy, P. I. (1953). "The Paleolithic of the Ukraine." USSR Academy of Sciences, Institute of Archaeology, Moscow and Leningrad.

Boriskovskiy, P. I., and Praslov, N. D. (1964). "The Paleolithic of the Dnepr Basin and the Sea of Azov Region. Nauka Press, Moscow and Leningrad.

Bryusov, A. Ya. (1940). Traces of a Paleolithic campsite near the village of Ulyank (Chuvash ASSR). *Bulletin of the Commission on the Study of the Quaternary* 6-7, 76-77.

Chernysh, A. P. (1968). The Ataki I Paleolithic campsite. *Bulletin of the Commission on the Study of the Quaternary* 35, 102-112. Moscow.

Chernysh, A. P. (1973). "The Paleolithic and Mesolithic of the Dnestr Region." Nauka Press, Moscow.

David, A. I., and Lungu, A. N. (1972). Remains of mammals from the Karagash quarry. *In* "Vertebrates of the Neogene and Pleistocene of Moldavai," pp. 19-24. Shtiintsa Press, Kishinev.

Gromov, V. I. (1935). New data on the fauna and geology of the Paleolithic of Eastern Europe and Siberia: The Paleolithic of the USSR. *State Academy of History and Material Culture, Izvestiya* 118, 246-70.

Gromov, V. I. (1948). Paleontological and archaeological substantiation of the stratigraphy of continental deposits of the Quaternary on the territory of the USSR. *USSR Academy of Sciences, Institute of Geology, Trudy, seria geologicheskaya* 17.

Guslitser, V. I., and Kanivets, V. I. (1965). Paleolithic monuments on the Pechora. *In* "Stratigraphy and Periodization of the Paleolithic of Eastern and Central Europe." Nauka Press, Moscow.

Gvozdover, N. D. (1964). Late Paleolithic monuments of the Lower Don. *In* "The Paleolithic of the Dnepr and Sea of Azov Region" (B. A. Rybakov, ed.), pp. 37-41. Nauka Press, Moscow and Leningrad.

Ivanova, I. K. (1977a). Geology and paleogeography of the Korman' IV site against the general background of the Stone Age geological history of the middle Dnestr Region. *In* "The Multilayer Paleolithic Campsite Korman' IV" (G. I. Goretsky and S. M. Tseitlin, eds.), pp. 126-52. Nauka Press, Moscow.

Ivanova, I. K. (1977b). Natural habitation conditions of Stone Age people in the Dnestr River Basin. *In* "Paleoecology of Ancient Man" (I. V. Ivanova and N. D. Praslov, eds.), pp. 7-17. Nauka Press, Moscow.

Iversen, J. (1944). *Vicium, Hedera,* and *Ilex* as climate indicators. *Geologiska Föreningens i Stockholm Förhandlingar* 66 (3).

Kalinovskiy, P. F. (1979). Paleoteriofaunal sites in Belorussia. *In* "Advances in Anthropogene Geology of Belorussia" (E. A. Levkov and T. L. Yacubskaya, eds.), pp. 54-63. Nauka i Tekhnika Press, Minsk.

Ketraru, N. A. (1973). "Monuments of the Paleolithic and Mesolithic." Shtiintsa Press, Kishinev.

Markova, A. K. (1972). Fossil rodents from the soil of the Bryansk Interstadial. *USSR Academy of Sciences, Izvestia, seria geograficheskaya* 2, 109-16.

Markova, A. K. (1975a). Fossil rodents from soils of the Mezin Complex. *USSR Academy of Sciences, Izvestiya, seria geograficheskaya* 5, 82-89.

Markova, A. K. (1975b): Paleography of the Upper Pleistocene based on data from an analysis of fossil small mammals of the Upper and Middle Dnepr Region. *In* "Problems of Paleogeography of Loessial and Periglacial Regions" (A. A. Velichko, ed.), pp. 59-68. Moscow.

Milanov, N. P. (1958). The Historico-archaeological work of the Department of History of the USSR in the last fifteen years (1942-1957). *Uch. Zap. Ryazan. Pedag. Inst.* Issue VI, Vol. XIX.

Motuzko, A. N. and Kalinovskiy, P. F. (1976). New finds of Anthropogene rodents in Belorussia. *Belorussia Academy of Sciences, Doklady 20* (10), 928-30.

Panichkina, M. Z. (1953). Exploration of the Paleolithic on the middle Volga. *Sovet. Arkheol.,* XVIII, 233-64.

Pidoplichko, I. G. (1936). Fauna of the Kodak Paleolithic campsite. *Priroda,* 6, 118-20.

Pidoplichko, I. G. (1969). "Late Paleolithic Mammoth-Bone Dwellings in the Ukraine." Naukova Dumka Press, Kiev.

Polikarpovich, K. U. (1968). "Paleolithic of the Upper Dneipr Basin." Nauka i Technika Press, Minsk.

Praslov, N. D. (1968). "The Early Paleolithic of the Northeastern Region of the Sea of Azov and Lower Don." USSR Academy of Sciences, Institute of Archaeology, Moscow and Leningrad.

Raskatov, G. I., Shevyrev, L. T., Antsiferova, G. A., and Alekseyeva, L. A. (1977). New location of large mammal fauna in the basin of the Upper Don. *In* "Lithology and Stratigraphy of the Sedimentary Cover of the Voronezh Anticline." Issue 4, pp. 83-90. Voronezh State University, Voronezh.

Rogachev, A. N., Anikovich, M. V., and Artemova, V. D. (1979). Kostenki VIII (Tel'manovskaya campsite). *In* "The Upper Pleistocene and Development of Paleolithic Culture at the Center of the Russian Plain" (G. I. Goretsky, ed.), pp. 88-91. Voronezh State University, Voronezh.

Rogachev, A. N., Anikovich, M. V., and Belyayeva, V. I. (1979). Kostenki I (Polyakov campsite). *In* "The Upper Pleistocene and Development of Paleolithic Culture at the Center of the Russian Plain" (G. I. Goretsky, ed.), pp. 68-74. Voronezh State University, Voronezh.

Rogachev, A. N., and Sinitsyn, A. A. (1979). Kostenki XIV (Markina

Mountain). *In* "The Upper Pleistocene and Development of Paleolithic Culture at the Center of the Russian Plain," (G. I. Goretsky, ed.), pp. 88-91. Voronezh State Univesity, Voronezh.

Savich, V. P. (1975). Late Paleolithic settlements on the Kulichivka Mountain in the town of Kremenets. *Bulletin of the Commission on the Study of the Quaternary* 44, 41-51.

Tarasov, L. M. (1967). The Gagarino settlement and its place in the Upper Paleolithic of Europe. Candidate Dissertation in Historical Sciences, University of Leningrad.

Tarasov, L. M. (1977). The Betovo Mousterian campsite and its natural surroundings. *In* "Paleoecology of Ancient Man" (I. K. Ivanova and N. D. Praslov, eds.), pp. 40-50. Nauka Press, Moscow.

Vekilova, Ye. A. (1957). "The Syuren' I Campsite and Its Place among the Paleolithic Sites of the Crimea and Neighboring Areas." USSR Academy of Sciences, Institute of Archaeology, Moscow and Leningrad.

Velichko, A. A. (1961). "The Geological Age of the Upper Paleolithic of the Russian Plain's Central Region." Nauka Press, Moscow.

Velichko, A. A. (1973). "The Natural Process in the Pleistocene." Nauka Press, Moscow.

Velichko, A. A. (1975). Problems of correlation of Pleistocene events in the glacial, periglacial-loessial, and maritime regions of the eastern European Plain. *In* "Problems of Regional and General Paleogeography of Loessial and Periglacial Regions" (A. A. Velichko, ed.), pp. 7-25. USSR Academy of Sciences, Institute of Geography, Moscow.

Velichko, A. A., Gribchenko, Yu. N., Markova, A. A., and Udartsev, V. P. (1977). Age and conditions of habitation of the Khotylevo II campsite on the Desna. *In* "Paleoecology of Ancient Man" (I. K. Ivanova and N. D. Praslov, eds.) pp. 40-50. Nauka Press, Moscow.

Velichko, A. A., and Morozova, T. D. (1972). The Bryansk fossil soil and its stratigraphic importance and natural conditions of formation. *In* "Loesses, Buried Soils and Cryogenic Phenomena on the Russian Plain" (A. A. Velichko, ed.), pp. 71-114. Nauka Press, Moscow.

Vereshchagin, N. K. (1979). "Why the Mammoths Died." Nauka Press, Leningrad.

Vereshchagin, N. K., and Baryshnikov, G. F. (1980a). Areas of hoofed fauna of the USSR in the Anthropogene. *In* "Mammals of Eastern Europe in the Anthropogene" *USSR Academy of Sciences, Institute of Zoology, Trudy* 93, Leningrad.

Vereshchagin, N. K., and Baryshnikov, G. F. (1980b). Mammals of the piedmont of North Crimea during the Paleolithic. *In* "Mammals of Eastern Europe in the Anthropogene" (E. O. Skarleto, ed.). *USSR Academy of Sciences, Institute of Zoology, Trudy* 93, pp. 26-49. Leningrad.

Vereshchagin, N. K., and Baryshnikov, G. F. (1980c). Paleoecology of the late mammoth fauna in the Arctic zone of Eurasia. *Byull. Mosk. obshch. isp. prirody. Otd. biolog.*, 85 (2), 3-25.

Vereshchagin, N. K., and Kolbutov, A. D. (1957). Remains of mammals at a Mousterian campsite near Stalingrad, and stratigraphic location of the Paleolithic layer. *USSR Academy of Sciences, Institute of Zoology, Trudy* 22, pp. 75-89. Moscow and Leningrad.

Vereshchagin, N. K., and Kuz'mina, I. Ye. (1977). Remains of mammals from Paleolithic campsites on the Don and Upper Desna. *In* "The Mammoth Fauna of the Russian Plain and Eastern Siberia." *USSR Academy of Sciences, Institute of Zoology, Trudy* 72, pp. 77-110. Leningrad.

Voznyachuk, L. N., and Kalechets, Ye. G. (1969). Some results of zoological studies at the Berdyzhskiy Upper Paleolithic campsite in 1959-1968. "Transactions of the Third Scientific Conference of Young Geologists of Belorussia," pp. 36-42. Nauka i Tekhnika Press, Minsk.

Yefimenko, P. P. (1953). "The Primitive Society." Naukova Dumka Press, Kiev.

Zavernyayev, F. M. (1978). "The Khotylevka Paleolithic Site." Nauka Press, Leningrad.

CHAPTER 21

Late Pleistocene Mammal Fauna of Siberia

N. K. Vereshchagin and I. Ye. Kuz'mina

The Late Pleistocene mammal fauna of Arctic and southern Siberia became known at the turn of the century from the fragmentary data of Academicians P. S. Pallas, I. F. Brandt, A. F. Middendorf (1869), I. D. Cherskiy (1891), M. V. Pavlova (1910), and others. By the 1950s, new light had been shed on the fauna of central and southern Siberia by the remarkable excavations of Paleolithic campsites by Soviet archaeological work and through the paleozoological studies of Gromov (1948).

A series of new studies on paleontologic materials now makes it possible to characterize the Late Pleistocene mammal fauna of individual regions of northern Asia on the basis of the remains of 65 species recovered from geologic sites (Alekseyeva, 1980; Vangengeym, 1977; Vereshchagin, 1959a, 1959b; Vereshchagin and Barishnykov, 1980; Galkina, 1975; Yermolova, 1978; Kuz'mina, 1971, 1977; Sher, 1971). The indicator and widely distributed species in this region (Table 21-1) include wolf, red fox, arctic fox, brown bear, sable, wolverine, cave lion, northern pika, Eurasian beaver, bank vole, woolly mammoth, woolly rhinoceros, Siberian roe deer, red deer, moose, reindeer, and steppe wisent. Some of these species, inhabitants of forested and rocky biotopes, lived in intrazonal elements of the landscape zone of cold Pleistocene steppe.

Western Siberia

Sites with skeletal remains of the following Late Pleistocene animals are confined in western Siberia to the eastern slopes of the southern Urals, the Ishim-Irtysh interfluve, the Kuznetsk Basin, and the Western Altay: corsac fox, small cave bear, steppe ferret, gray marmot, Siberian mole-rat, and wide-hoofed horse.

In Zyryanka and Sartan time, as the northern half of western Siberia became free of the sea, it was settled from the west, south, and east. On the Yamal Peninsula, in the lower reaches of the Ob', and on the Gydan Peninsula were found wolf, brown bear, polar bear, woolly mammoth, Lena horse, reindeer, and musk ox. Remains of saiga antelope were found in the lower reaches of the Ob'. Remains of woolly rhinoceros and bison were not found on the Gydanskiy Peninsula (Kuz'mina, 1977). No remains of arctic fox have been found thus far in northwestern Siberia in the Late Pleistocene.

Farther south, cave lion, mammoth, Lena horse, woolly rhinoceros, red deer, and moose inhabited the basin of the lower course of the Irtysh during Zyryanka time (Vangengeym, 1977). Giant deer occurred 100 km north of Tomsk near Krasnyy Yar on the Ob' (Alekseyeva, 1980). The Kuznetsk Basin, according to the data of Galkina (1975), was inhabited by bank vole, collared lemming, Siberian brown lemming, steppe lemming, water vole, narrow-headed vole, tundra vole, and Siberian mole-rat. Among large mammals the following also inhabited the Kuznetsk Basin and the western slopes of the Altay: cave hyena, cave lion, Asiatic wild ass, argali sheep, Siberian ibex, yak, and aurochs.

Thus, three regions with a greatly similar mammal fauna can be distinguished in western Siberia in the Late Pleistocene as follows: a northern part, with a predominance of cold-tolerant species (musk ox, Lena horse, polar bear); a southern region, with a predominance of warmth-loving species (red deer, giant deer, corsac fox, wide-hoofed horse, Asiatic wild ass) and with species associated with forest biotopes (sable, wolverine, roe deer, beaver, water and bank voles, etc.); and a southeastern part, with a fauna very similar to that of eastern Siberia.

Eastern Siberia

During the Late Pleistocene, the mammals of eastern Siberia constituted a single complex that in the present state of knowledge is difficult to divide according to landscape-geographic zones. Deposits dated as Zyranka and Sartan

Table 21-1. Late Pleistocene Mammals of Siberia (Including the Arctic Zone)

Taxon	Common Name	Western Siberia	Eastern Siberia	Far East, South
LAGOMORPHA				
Caprolagus brachyurus Temm.	Manchurian hare			x
Lepus tanaiticus Gureev.	Don hare		x	
Lepus timidus L.	Arctic hare	x		x
Ochotona alpina Pall.	Alpine pika	x	x	x
RODENTIA				
Arvicola terrestris L.	Water vole	x	x	
Castor fiber L.	Eurasian beaver	x	x	x
Citellus glacialis Vinog.	Indigirka long-tailed ground squirrel		x	
Citellus undulatus Pall.	Long-tailed ground squirrel		x	
Clethrionomys sp.	Bank vole	x	x	x
Dicrostonyx torquatus Pall.	Collared lemming	x	x	
Lagurus lagurus Pall.	Steppe lemming	x	x	
Lemmus sibiricus Kerr	Siberian brown lemming	x	x	
Marmota baibacina Kastsch.	Gray marmot	x		
Marmota camtschatica Pall.	Black-capped marmot		x	
Marmota sibirica Radde	Tarbagan marmot		x	
Microtus gregalis Pall.	Narrow-headed vole	x	x	
Microtus oeconomys Pall.	Tundra vole	x		
Myospalax myospalax Laxm.	Siberian mole-rat	x		
CARNIVORA				
Alopex lagopus L.	Arctic fox	x	x	
Canis lupus L.	Gray wolf	x	x	x
Crocuta spelaea Gold.	Cave hyena	x		
Cuon alpinus Pall.	Dhole		x	x
Felis bengalensis Kerr.	Leopard cat			x
Felis lynx L.	Eurasian lynx		x	x
Gulo gulo L.	Wolverine	x	x	x
Lutra lutra L.	Eurasian otter			x
Martes flavigula Bodd.	Yellow-throated marten			x
Martes zibellina L.	Sable	x	x	x
Meles meles L.	Eurasian badger			x
Mustela eversmanni Less.	Steppe ferret	x		
Mustela sibiricus Pall.	Siberian weasel			x
Nyctereutes procyonoides Gray	Raccoon dog			x
Panthera pardus L.	Snow leopard			x
Panthera spelaea Gold.	Cave lion	x	x	x
Panthera tigris L.	Tiger			x
Spelaearctos rossicus Boris.	Small cave bear	x		
Ursus arctos L.	Brown bear	x	x	x
Ursus maritimus Phipps	Polar bear	x	x	
Vulpes corsac L.	Corsac fox	x		
Vulpes vulpes L.	Red fox	x	x	x
ARTIODACTYLA				
Alces alces L.	Moose (Eurasian elk)	x	x	x
Bison priscus Boj.	Steppe wisent	x	x	x
Bos baikalensis N. Ver.	Baikal yak	x	x	
Bos primigenius Boj.	Aurochs	x	x	
Capra sibirica Pall.	Siberian ibex	x		
Capreolus pygargus Pall.	Siberian roe deer	x	x	x
Cervus elaphus L.	Red deer (wapiti)	x	x	x
Gasella gutturosa Gmel.	Mongolian gazelle		x	
Megaloceros giganteus Blum.	Giant deer	x	x	
Moschus moschiferus L.	Musk deer		x	x
Nemorrhaedus caudatus Milne-Edw.	Common goral			x
Ovibos moschatus Zimm.	Tundra musk ox	x	x	
Ovis ammon L.	Argali sheep	x	x	
Ovis nivicola Eschsch.	Snow sheep		x	
Parabubalis capricornis V. Gromova	Goat-horned antelope		x	
Rangifer tarandus L.	Reindeer	x	x	
Saiga tatarica L.	Saiga antelope	x	x	
Spirocerus kiakhtensis M. Pavl.	Twisted-horned antelope		x	
Sus scrofa L.	Wild boar			x
PERISSODACTYLA				
Coelodonta antiquitatis Blum.	Woolly rhinoceros	x	x	x
Equus sp.	Wild horse			x
Equus hemionus Pall.	Asiatic wild ass	x	x	
Equus latipes Gromova	Wide-hoofed horse	x		
Equus lenensis Russ.	Lena horse	x	x	
PROBOSCIDEA				
Mammuthus primigenius Blum.	Woolly mammoth	x	x	x

(Vangengeym, 1977; Sher, 1971) in the basins of the Nizhnyaya Tungska, Angara, Aldan, Lena, Vilyuy, and Markha Rivers were found to contain skeletal remains of wolf, red fox, arctic fox, brown bear, wolverine, cave lion, Don hare, alpine pika, long-tailed ground squirrel, Indigirka ground squirrel, black-capped marmot, beaver, collared lemming, Siberian brown lemming, steppe lemming, water vole, and narrow-headed vole.

In addition, complete or partial carcasses of mammoth, horse, woolly rhinoceros, and steppe wisent have been preserved in the frozen state in deposits of eastern Siberia (Rusanov, 1968; Vereshchagin, 1981; Lazarev, 1980; Vereshchagin and Lazarev, 1977).

Mountain animals were abundantly represented. The region was inhabited by the dhole, musk deer, twisted-horned antelope, Siberian ibex, argali sheep, snow sheep, and Baikal yak.

Somewhat apart faunistically stood the Transbaikal region, in southeastern Siberia, inhabited by animals of xerophytic landscapes and a sharply continental climate: tarbagan marmot, Mongolian gazelle, Asiatic wild ass, and saiga antelope. These species have survived in the steppes and semideserts of Kazakhstan and Mongolia to this day, and during the Late Pleistocene the saiga antelope occurred widely in western and eastern Siberia up to the shores of the Polar Basin. Its fossil remains are also known on the Novosibirsk Islands. Such a wide distribution of saiga antelope remains can be explained not only by the considerable spread of open steppe landscapes in the Late Pleistocene but also by the distant migrations very typical of this animal.

Thus, 43 species of mammals known from fossil remains (Table 21-1) inhabited the territory of eastern Siberia during the Late Pleistocene. Skeletal remains of 8 species were found only in eastern Siberia: black-capped and tarbagan marmots, long-tailed and Indigirka ground squirrels, Mongolian gazelle, twisted-horned antelope, goat-horned antelope, and snow sheep.

An extensive exchange of species between Eurasia and North America occurred during the Pliocene and Pleistocene (Hibbard et al., 1965; Vereshchagin, 1971). Eastern Siberia was last connected to North America during the Late Pleistocene via the Bering region. Apparently, cave lion, northern pika, moose, musk ox, snow sheep, saiga antelope, and yak entered North America at that time. Mammoth and short-horned bison moved back into America. In addition, collared lemming, reindeer, and, apparently, horse and marmot reentered America, their place of origin (Kuz'mina, 1977).

Bering Region and Asian-American Faunistic Links

Skeletal remains of Late Pleistocene mammals are particularly abundant in deposits of Zyryanka-Sartan (Wisconsin) age over all of northeastern Siberia, including the Novosibirsk Islands, Wrangel Island, Maritime Lowland, Chukotka Peninsula, and Kamchatka. There, the remains of soft tissues, horn sheaths, and fur from many species of Pleistocene wild animals are found in frozen ground more often than anywhere else. The shelf of the Laptev and East Siberian Seas in some places (Oyagosskiy and Khaptashinskiy Yar, Ayon Island, etc.) is covered with bones of horse, reindeer, bison, musk ox, and mammoth washed up by tidal waters. Occasionally, bones of wolf, brown bear, cave lion, and (rarely) woolly rhinoceros, moose, and saiga antelope are also found.

Sites in Alaska and in the valley of the Yukon River and its tributaries, similar in age and taxonomy, contain essentially similar fauna (Guthrie, 1968; Irving et al., 1977) but not woolly rhinoceros, which apparently did not live east of the Kolyma Basin. This similarity of Pleistocene mammal complexes from the plains of the two adjacent continents, now separated once again by the sea, confirms the existence, during the epoch of maximum cooling, of wide continental links at the location of the shelf zone and Bering Straits, that is, the Bering Land Bridge.

Similar features are also preserved in the modern mammal fauna of extreme northeastern Siberia and Alaska. This fauna includes Bering brown bear, arctic fox, red fox, wolverine, stag moose (*Cervalces*), reindeer, and long-tailed and Indigirka ground squirrels among other species.

At the same time, secondary and marked differences in the mammal fauna of these regions are apparent as a result of ecologic barriers and older migration routes. Such are the subspecies differences among the modern snow sheep and Dall sheep of Chukotka and Alaska, the absence of the mountain goat of Alaska on the Chukotka Peninsula and the Koryak Highland, species differences among otters, and so on (Chernyavskiy, 1976).

South of the Far East

The south of the Far East was faunistically the most unusual region of the Late Pleistocene. Deposits in caves of the Maritime Territory have been found to contain skeletal remains of species not found farther northeast: raccoon dog, Siberian weasel, yellow-throated marten, badger, otter, leopard cat, tiger, leopard, Manchurian hare, horse, wild boar, and mountain antelope. In addition, species common in southeastern Siberia also inhabited the Far East, namely, dhole, lynx, and musk deer in addition to 15 species widely distributed in the Paleoarctic: wolf, red fox, brown bear, sable, wolverine, cave lion, northern pika, beaver, red-backed vole, mammoth, woolly rhinoceros, roe deer, red deer, moose, and bison (Table 21-1). Remains of cave hyena are known from cave deposits of western Siberia and the Far East.

The mammal fauna of the Far East indicates that the climate there during the Late Pleistocene was milder than that in Siberia. The species diversity was due to the penetration from the south of certain species more characteristic of the Indo-Malayan zoogeographic region.

Conclusions

Comparison of species compositions of the mammal fauna of western and eastern Siberia and the south of the Far East showed that differences can be traced even for the Late Pleistocene. Remains of corsac fox, small cave bear, steppe ferret, gray marmot, Siberian mole-rat, and wide-hoofed horse were found only in western Siberia. Remains of tarbagan marmot, black-capped marmot, long-tailed ground squirrel, Indigirka arctic ground squirrel, Mongolian gazelle, and snow sheep were found only in eastern Siberia. There were many more species in the south of the Far East, namely, raccoon dog, Siberian weasel, yellow-throated marten, badger, otter, leopard cat, tiger, leopard, Manchurian hare, horse, wild boar, and mountain antelope.

Paleontologic data indicate a later Late Pleistocene settlement by mammals of northern regions of western Siberia, where cold-tolerant species predominated widely. The southern part was inhabited by warmth-loving species, associated either with forest biotopes or with steppe and semidesert biotopes. The Late Pleistocene mammal fauna of eastern Siberia was more uniform. In this complex one can distinguish only the faunas of the Transbaikal region and southeastern Siberia, including species of mountain and plains landscapes adapted to a dry, sharply continental climate. The composition of Far Eastern mammals in the Maritime Territory indicates a milder and more humid climate than in Siberia, and the penetration from the south of representatives of the Indo-Malayan zoogeographic region. All the latest studies confirm the hypothesis offered by Sushkin (1925) that there were different conditions of formation and development of western Paleoarctic, eastern Paleoarctic, and Far Eastern faunas.

References

Alekseyeva, E. V. (1980). "Mammals of the Pleistocene of Western Siberia's Southeast." Nauka Press, Moscow.

Chernyavskiy, F. B. (1976). Systematic relationships of certain land mammals of the Old and New World in connection with the problem of Beringia. *In* "Beringia in the Cenozoic" (V. L. Kontrimavichus, ed.), pp. 147-50. USSR Academy of Sciences, Far East Scientific Center, Vladivostock.

Cherskiy, I. D. (1891). "Description of Collections of Postquaternary Mammals Obtained by the Novo-Siberian Expedition of 1885-1886." Russian Geographical Society, St. Petersburg.

Galkina, L. I. (1975). Fauna of Anthropogene rodents and leporids of the plateau along the Ob' River and Kuznetskaya Basin. *USSR Academy of Sciences, Siberian Branch, Institute of Biology, Trudy* 23, 155-64.

Gromov, V. I. (1948). "Paleontological and Archeological Substantiation of the Stratigraphy of Quaternary Continental Deposits in the USSR (Mammals, Paleolithic)." *USSR Academy of Sciences, Institute of Geology, seria geologicheskaya, Trudy* 17.

Guthrie, R. D. (1968). Paleoecology of the large-mammal community in interior Alaska during the Late Pleistocene. *American Midland Naturalist* 79, 346-63.

Hibbard, C. W., Ray, D. E., Savage, D. E., Taylor, D. W., and Guilday, J. E. (1965). Quaternary mammals of North America. *In* "The Quaternary of the United States" (H. E. Wright, Jr., and D. G. Frey, eds.), pp. 509-26. Princeton University Press, Princeton, N.J.

Irving, W. N., Mayhall, I. T., Melbye, F. I., and Beebe, B. F. (1977). A human mandible in probable association with a Pleistocene faunal assemblage in eastern Beringia: A preliminary report. *Canadian Journal of Archeology* 1, 81-93.

Kuz'mina, I. Ye. (1971). Formation of the theriofauna of the Northern Urals in the Late Anthropogene. *USSR Academy of Sciences, Institute of Zoology, Trudy* 49, 44-122.

Kuz'mina, I. Ye. (1977). On the origin and history of the theriofauna of the Siberian Arctic. *USSR Academy of Sciences, Institute of Zoology, Trudy* 63, 18-55.

Lazarev, P. A. (1980). "Anthropogene Horses of Yakutiya." Nauka Press, Moscow.

Middendorf, A. F. (1869). "A Voyage to the North and East of Siberia," Vol. 2, Sect. 5, "Siberian Fauna." Russian Geographical Society, St. Petersburg.

Pavlova, M. V. (1910). Description of fossil remains of mammals of the Troitsko-Kyakhtinskiy Museum. *Troitsko-Kyakhtinskiy Section of the Russian Geographical Society, Trudy* 13 (1), 21-64.

Rusanov, B. S. (1968). "Biostratigraphy of Cenozoic Deposits of Southern Yakutiya." Nauka Press, Moscow.

Sher, A. V. (1971). "Mammals and Stratigraphy of the Pleistocene of the Extreme Northeast of the USSR and North America." Nauka Press, Moscow.

Sushkin, P. P. (1925). Zoological regions of Central Siberia and neighboring parts of mountainous Asia. *Bulletin of the Moscow Society of Naturalists* New Series 34, 50-71.

Vangengeym, E. A. (1977). "Paleontological Substantiation of the Stratigraphy of Central Asia's Anthropogene." Nauka Press, Moscow.

Vereshchagin, N. K. (1959a). "The Mammals of the Caucasus." USSR Academy of Sciences, Moscow and Leningrad.

Vereshchagin, N. K. (1959b). Remains of mammals of the mammoth epoch of the Taimyr Peninsula. *Bulletin of the Moscow Society of Naturalists* Biological Series 64 (5), 5-16.

Vereshchagin, N. K. (1971). The cave lion and its history in the Holarctic region and within the confines of the USSR. *USSR Academy of Sciences, Institute of Zoology, Trudy* 49, 123-99.

Vereshchagin, N. K. (1981). "The Magadan Mammoth Cub." Nauka Press, Moscow.

Vereshchagin, N. K., and Baryshnikov, G. F. (1980). Areas of the hoofed fauna of the USSR in the Anthropogene. *USSR Academy of Sciences, Institute of Zoology, Trudy* 93, 3-20.

Vereshchagin, N. K. and Lazarev, P. A. (1977). Description of parts of a carcass and skeletal remains of the Selerikanskiy fossil horse. *USSR Academy of Sciences, Institute of Zoology, Trudy* 63, 85-185.

Yermolova, N. M. (1978). "Theriofauna of the Angara Valley in the Late Anthropogene." Nauka Press, Novosibirsk.

CHAPTER 22

Late Pleistocene Insects

S. V. Kiselev and V. I. Nazarov

Systematic studies of fossil insects in the USSR did not begin until the 1960s. Since that time, data have been published on the composition of Pleistocene insect faunas of the European USSR (Panfilov, 1965; Medvedev, 1968a, 1968b, 1976; Nazarov, 1979; Voznyachuk, Makhnach, et al., 1979; Voznyachuk, San'ko, et al., 1979), Siberia (Kiselev, 1973), Yakutiya (Grushevskiy and Medvedev, 1962, 1970; Grunin, 1973; Medvedev and Voronova, 1977; Sher et al., 1979; Kiselev, 1981; Kaplina et al., 1980), Chukotka (Kiselev, 1980a, 1980b; Boyarskaya and Kiselev, 1981), and some other regions of the Soviet Union (Kiselev et al., 1981). The locations of the main sites of fossil insect remains in this country are shown in Figure 22-1).

To date, only two regions of the USSR—Belorussia and northeastern Siberia, primarily the Kolyma Lowland—have been studied extensively in the paleoentomological sense. This makes it possible to compare the development of the insect fauna in the West and East of the country, that is, in regions of markedly different environmental history.

Northeastern Siberia

In the Kolyma Lowland, the Late Pleistocene entomofaunas can be grouped into tundra, tundra-taiga, and steppe species. Among the most typical members of the steppe group are the weevil genera *Coniocleonus* and *Stephanocleonus*. Today, species of the subfamily Cleoninae, which includes these weevils, are confined to the steppe and mountain-steppe regions of southern Siberia and Mongolia, although some penetrate into central Yakutiya through steppe-type extrazonal areas (*Stephanocleonus eruditus* Faust, *S. fossulatus* Disch.) and as relicts known even from areas of Taimyr, western Chukotka, and Wrangel Island (*Coniocleonus ferrugineus* Fahr., *C. astragali* Ter-Min. et Korot., *C. cinerascens* Hochh.) (Korotyayev, 1977; Korotyayev and Ter-Minasyan, 1977). Almost all fossil sites of these weevils in the Northeast are quite distant from their modern ranges, indicating a previous greater extent of steppe and meadow-steppe than at present. Analysis of weevil fossils indicates cyclic climatic changes in the occurrence of steppe and tundra-taiga species.

Steppe assemblages dominate the fossil fauna; tundra assemblages consist almost exclusively of inhabitants of comparatively dry biotopes, such as grass and herb meadows or shrub-herb gentle slopes (e.g., the carabids *Pterostichus sublaevis* Sahlb., *Curtonotus alpinus* Payk., and *Amara glacialis* Munch. and the leaf beetles *Chrysolina septentrionalis* Dej., *C. subsulcata* Munh., and *C. cavigera* Sahlb.). Taiga and tundra (*sensu stricto*) beetle species play a secondary role in this type of fauna. Thus, a steppe-tundra must have predominated in a severe continental and arid or semiarid climate similar to, or more severe than, the present ultracontinental climate of the intermontane basins of Yakutiya (Verkhoyansk, Oymyakon).

The opposite type of Late Pleistocene entomofauna includes taiga insects (e.g., *Trachypachus zetterstedti* Gyll., *Pissodes gyllenchali* Gyl., and *Camponotus herculeanus* L.) and tundra mesophiles (e.g., *Pterostichus costatus* Men., *P. agonus* Horn., and *Cryobius* spp.) typical of forest-tundra or sparse larch forests. The presence of a few steppe insects in such faunas indicates a somewhat more severe climate than the present one in the Kolyma Lowland even during the "warm" interstades of the Late Pleistocene.

Belorussia

The situation in Belorussia was entirely different. During the Mikulino Interglaciation, the entomofauna differed little from the present one, although it did include insects that now live farther south and west, such as the carabids *Epaphius rivularis* Gyll., *Oodes gracillis* Villa, and *Chisenius tristis* Schall.; the click beetles *Ampedus nigrinus* Hbst.

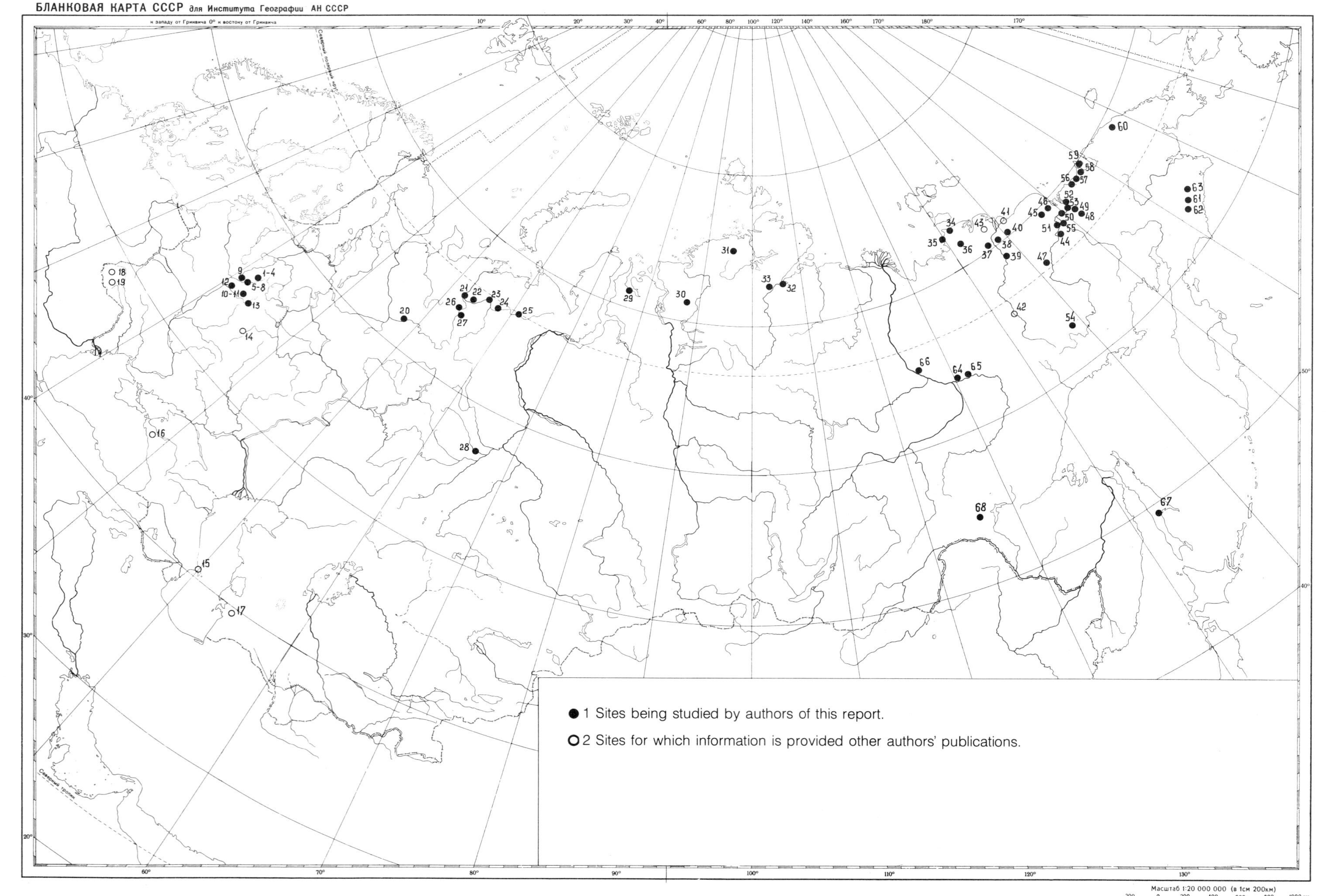
БЛАНКОВАЯ КАРТА СССР для Института Географии АН СССР
к западу от Гринвича 0° к востоку от Гринвича
Северный полярный круг
Северный тропик
1–4
5–8
9
10–11
12
13
14
15
16
17
18
19
20
21
22
23
24
25
26
27
28
29
30
31
32
33
34
35
36
37
38
39
40
41
42
43
44
45
46
47
48
49
50
51
52
53
54
55
56
57
58
59
60
61
62
63
64
65
66
67
68
1 Sites being studied by authors of this report.
2 Sites for which information is provided other authors' publications.
Масштаб 1:20 000 000 (в 1см 200км)
200 0 200 400 600 800 1000 км

Figure 22-1. Principal sites of Late-Cenozoic insects in the USSR.

1. Smolenskiy Ford
2. Panfilovo
3. Bol'shoye Verkhov'ye
4. Konevich
5. Rogatka
6. Kasplyane
7. Drichaluki
8. Sloboda Dvinskaya
9. Gralevo
10. Mikhalinovo
11. Rubezhnitsa
12. Chizhovka
13. Verkhniye Nemykari
14. Chekalinskiy Section
15. Binagady
16. Il'skaya
17. Monzhukly
18. Borislav
19. Starun'
20. Kur'yador
21. Garevo
22. Kushshor
23. Kipiyevo
24. Os'van'
25. Rogovaya
26. Rodionovo
27. Bol'shoy Aranets
28. Mal'kovo
29. Yuribey
30. Agapa
31. Shrenk
32. Khatanga
33. Romanikha
34. Oyogosskiy Ravine
35. Entrykayskiy Ravine
36. Khroma
37. Achchagyy-Allaikha
38. Shamanovo
39. Sypnoy Ravine
40. Keremsit
41. Shandrin
42. Tirekhtyakh
43. Berelekh
44. Krestovka
45. Chukoch'ya
46. Kon'kovaya
47. Sededema
48. Utkinskiy Kamen'
49. Molotkovskiy Kamen'
50. Duvannyy Ravine
51. Aleshkina Zaimka
52. Kray Lesa
53. Tretiy Ruchey
54. Kirgilyakh
55. Omolon
56. Mil'kera
57. Primorskiy
58. Ust'-Rauchua
59. Ayon
60. Leningrad
61. Mamontov Obryv
62. Ledovyy Obryv
63. Ust'-Alganskiy
64. Chuyskoye exposure
65. Dygdal
66. Lepiske
67. Leonidovka
68. Tynda

and *Orithales serraticornis* Payk.; and the weevils *Hylobius abietis* L., *Curculio nucum* L., and *Acales* sp. This fauna indicates mixed coniferous/broad-leaved forests, with average July temperatures 1°C to 3°C higher than at present. Only in the last stages of the Mikulino Interglaciation was the climate similar to the present one or even somewhat cooler.

Insect faunas of the Valdai Glaciation in Belorussia include numerous tundra beetles, such as the carabids *Pelophila borealis* Payk., *Blethisa catenaria* Brown, *Diacheila polita* Fald., *Pterostichus vermiculosus* Men., *P. haematopus* Dej., *P. tundrae* Tschitsch., *P. sublaevis* Sahlb., *Curtonotus alpinus* Payk., and *Amara glacialis* Mann.; the rove bettle *Tachinus arcticus* Motsch.; and the leaf beetle *Chrysolina septentrionalis* Dej. The very few steppe forms include the weevils *Coniocleonus ferrugineus* Fahr., *Stephanocleonus eruditus* Faust, and *Aelia frigida* Kir. The forest-tundra apparently changed gradually to tundra as the ice advanced and as the continentality and aridity of the climate increased.

Conclusions

The fossil entomofaunas of Belorussia and Kolyma Lowland thus changed in different ways, although in both areas the changes reflect the alternation of cold and warm epochs. During interglaciations, there were regional differences of approximately the same order as those at the present time. The cold and sharply continental climate of the Valdai (Zyryanka-Sartan, Wisconsin) caused the appearance of functionally similar insect communities reflecting the development of open landscapes, with a patchy and heterogeneous grass cover and an almost total absence of tree vegetation. However, the composition of these communities was fundamentally different. Whereas on the Kolyma Lowland the entomofaunas of cold epochs were characterized by an abundance and variety of steppe insects, the latter were extremely scarce in Belorussia. Their place was mostly taken by inhabitants of interzonal meadow areas of the forest belt, such as *Meligethes aeneus* Fabr., *M. viridescens* Fabr., *Trichalophus Korotyaevi* Zher. et Nar., *Ceuthorhynchus rapae* Gill., and *C. chalybeus* Desbr. Thus, provincial differences were considerably smoothed out during the glacial epochs, but they did not completely disappear.

References

Boyarskaya, T. D., and Kiselev, S. V. (1981). Certain characteristics of the history of Holocene biocenoses of northeastern Asia. *In* "Anthropogenic Factors in the History of Development of Present Ecosystems" (L. G. Dinesman, ed.), pp. 195-202. Nauka Press, Moscow.

Grunin, K. Ya. (1973). First find of larvae of the mammoth botfly—*Cobboldia* (Mamontia, subgen. n.) *rusanovi* sp. n. (Diptera, Gasterophidae). *Entomologicheskoe Obozrenie* 52, 228-30.

Grushevskiy, I. I., and Medvedev, L. N. (1962). "Preliminary Data on the Application of Coleopterological Analysis to the study of Quaternary Deposits of Northern Yakutiya." *In* Collected Articles on Paleontology and Biostratigraphy of the Scientific Research Institute of Arctic

Geology (N. H. Shvedov, ed.), *Vypad* 28, 38-72. Leningrad.

Grushevskiy, I. I., and Medvedev, L. N. (1970). Importance of coleopterological analysis in the study of continental deposits of the Arctic. *In* "Biostratigraphic and Paleobiofacial Studies and Their Practical Importance" (L. G. Dinesman, ed.), pp. 252-56. Nedra Press, Moscow.

Kaplina, T. N., Sher, A. V., Giterman, R. E., Zazhigin, V. S., Kiselev, S. V., Lozhkin, A. V., and Nikitin, V. P. (1980). A key section of Pleistocene deposits on the Allaikha River (lower course of the Indigirka River). *Bulletin of the Commission on the Study of the Quaternary* 50, 73-95.

Kiselev, S. V. (1973). Late Pleistocene Coleoptera of Transuralia. *Paleontological Journal* 7, 507-10.

Kiselev, S. V. (1980a). A section of the Ayon Island: Analysis of the entomofauna. *In* "Latest Deposits and Paleogeography of Chukotka's Pleistocene" (P. A. Kaplin, ed.), pp. 194-96. Nauka Press, Moscow.

Kiselev, S. V. (1980b). A section of the Main River: Analysis of the entomofauna. *In* "Latest Deposits and Paleogeography of Chukotka's Pleistocene" (P. A. Kaplin, ed.), pp. 149-50. Nauka Press, Moscow.

Kiselev, S. V. (1981). "Late Cenozoic Coleoptera of the Northeast of Siberia." Nauka Press, Moscow.

Kiselev, S. V., Korotyayev, B. A., Golosova, L. P., Laskova, L. M., and Druk, A. Ya. (1981). Change in the fauna of insects and armor mites of the South Sakhalin in the Holocene. *In* "Anthropogenic Factors in the History of the Development of Present Ecosystems" (L. G. Dinesman, ed.), pp. 186-94. Nauka Press, Moscow.

Korotyayev, B. A. (1977). An ecological-faunistic survey of snout beetles (Coleoptera, Curculionidae) of the Northeast of the USSR. *Entomologicheskoe Obozrenie* 56 (1), 60-70.

Korotyayev, B. A., and Ter-Minasyan, M. Ye. (1977). A survey of snout beetles of genus *Coniocleonus* Motsch. (Coleoptera, Curculionidae) of the fauna of eastern Siberia and the Far East. *Entomologicheskoe Obozrenie* 56 (4).

Medvedev, L. N. (1968a). The finding of a new fossil beetle (Coleoptera, Curculionidae) from the Likhvinskiy Interglaciation of the Moscow region. *In* "History of the Development of the Vegetative Cover of Central Regions of the European USSR in the Anthropogene." (N. I. Pyavchenko, ed.), pp. 124-28. Nauka Press, Moscow.

Medvedev, L. N. (1968b). Method and prospects for application of coleopterological analysis to the study of Quaternary deposits and history of formation of faunas. *In* "History of the Development of the Vegetative Cover of Central Regions of the European USSR in the Anthropogene" (H. I. Pyavchenko, ed.), pp. 115-23. Nauka Press, Moscow.

Medvedev, L. N. (1976). Composition of entomocomplexes from Holocene badger coprolites in the Moscow region. *In* "History of Biogeocenoses of the USSR in the Holocene" (L. G. Dinesman, ed.), pp. 183-93. Nauka Press, Moscow.

Medvedev, L. N., and Voronova, N. N. (1977). Coleoenterological analysis of geological sections of mammoth cemeteries in northern Yakutiya. *In* "The Mammoth Fauna of the Russian Plain and Eastern Siberia" (O. A. Scarlato, ed.), pp. 72-76. Nauka Press, Leningrad.

Nazarov, V. N. (1979). Coleoptera from the Rubezhnitsa locality and their environment. *Paleontological Journal* 13, 462-72.

Panfilov, D. V. (1965). Subfossil remains of insects from Serebryanny pine forests. *Bulletin of the Moscow Society of Naturalists, Biological Section* 70, 115-16.

Sher, A. V., Kaplina, T. N., Giterman, R. E., Lozhkin, A. V., Arkhandelov, A. A., Virina, Ye. I., Zazhigin, V. S., Kiselev, S. V., and Kuznetsov, Yu. V. (1979). "Guide for the Scientific Excursion on the Problem 'Late Cenozoic Deposits of the Kolyma Lowland,' 14th Pacific Scientific Congress (USSR, Khabarovsk, August 1979)." All-Union Institute of Scientific and Technical Information, Moscow.

Voznyachuk, L. N., Makhnach, N. A., San'ko, A. F., Motuzko, A. N., Nazarov, V. I., and Shalaboda, V. L. (1979). Interglacial deposits of the Smolenskiy Ford Tract on Zapadnaya Dvina in the Velizh region of Smolensk Province. *In* "Advances in the Geology of Belorussia's Anthropogene" (T. V. Yakubovskaya and E. A. Leskov, eds.), pp. 64-80. Nauka i Tekhnika Press, Minsk.

Voznyachuk, L. N., San'ko, A. F., Velichkevich, F. O., Litvinyuk, G. I., and Nazarov, V. I. (1979). New data on Mikulino deposits of Smolenskoye Podvin'ye. *In* "Advances in the Geology of Belorussia's Anthropogene" (T. V. Yakubovskaya and E. A. Leskov, eds.), pp. 87-101. Nauka i Tekhnika Press, Minsk.

Inland Sea Basins

CHAPTER 23

Inland Sea Basins

A. L. Chepalyga

The USSR is surrounded by thirteen seas characterized by different degrees of isolation from oceans. The following types of basin can be distinguished: (1) marginal seas with fairly wide connections to the ocean (for example, the Sea of Japan, the Sea of Okhotsk, the Bering Sea, and the Arctic seas); one distinguishes shelf seas (for example, the Barents Sea) and marginal basin seas (for example, the Sea of Japan); (2) semiisolated seas with a restricted connection through one or several narrow and shallow straits; the most typical of these is the Black Sea; periglacial variants of such a basin are the Baltic and White Seas; (3) completely isolated intracontinental basins (for example, the Caspian Sea), and in the past certain basins in the Black Sea depression.

The degree of isolation has a considerable effect on the ecologic state of the basin, in particular its salinity, temperature, circulation of waters, oxygen regime, composition of fauna and flora, and distribution of biocenoses. On the basis of these data, the following ecologic types of basin are distinguished (Table 23-1): marine, semimarine, brackish-water, semifreshwater, freshwater, and hyperhaline. Each basin taken as a whole is regarded as a single major ecosystem consisting of smaller biocenoses. Frequent changes of basin types have occurred over the course of geologic time, and, therefore, the history of each sea is regarded as a succession of various paleoecosystems in time, under the influence of external conditions, mainly climatic changes and eustatic fluctuations in sea level. Each such change in conditions was associated with a rearrangement and loss of some ecosystems and the formation of new ones, accompanied by regional ecologic changes (Chepalyga, 1980).

Reconstructions of inland sea basins are based on the detailed study of the geomorphology (terraces, coastal features, etc.), sediment composition, fauna and flora, absolute datings, and so on. Three basin zones are recognized (Shcherbakov et al., 1978): (1) a coastal zone (study of marine terraces, overdeepening phases, etc.), (2) a shelf zone (study of bottom sediments by dredging, vibratory piston coring, and drilling in the shoal zone), and (3) a deep-sea zone (study through deep-sea drilling and vibratory piston coring for the upper part of the section).

Reconstructions of the following basin parameters are given on the basis of all these data. First, the distribution of sediments, terraces, and offshore bars is used to determine the boundaries of the basin, its level, coastline, and connections with other basins; absolute dates or the geologic ages are determined. Then, on the basis of the lithology of the sediments and the faunal and floral remains, the ecologic type of the basin is reconstructed, together with the distribution of the biocenoses and ecologic conditions of the basin. These studies involve soils (according to the composition of rocks), salinity and its spatial distribution, water temperature at the surface and in the bottom layer (on the basis of the faunal composition and oxygen isotopes), and gaseous regime (hydrogen sulfide contamination according to the presence of authigenic pyrite and sapropel). In addition, one can establish the boundary of ice-rafted stones as well as the direction of warm and cold currents on the basis of isolines of paleotemperatures and analyses of sediments.

An important method for studying basins, especially isolated ones, consists in reconstructing their water balance. The starting data that can be used for this method are the boundaries of the water area and drainage basin, approximate precipitation (based on the composition of the vegetation), runoff volume, and evaporation from the water area (based on paleotemperatures). Such calculations have been made for the Caspian and Black Seas (Kalinin et al., 1976; Kvasov, 1975).

The principal factor determining the history of the seas is climate. Different types of basins reacted differently to the same climatic conditions. Given below are reconstructions of marine basins for five major climatic phases of the Late Quaternary: the Mikulino Interglaciation, Early Valdai Glaciation, Middle Valdai Interstade (or Interglacia-

Table 23-1. Ecological Types of Closed and Semiclosed Basins (Chepalyga, 1980; Nevesskaya, 1970)

Types of Basins	Examples of Basins	Type of Fauna	Trophic Zones	Salinity ‰	Prevailing Ions	Degree of Endemism of Fauna and Biocenoses	Degree of Isolation of Basins
Marine	Mediterranean Sea, Karangatian	Marine stenohaline and euryhaline	Filter-feeding organisms, detritophages	30-40	K^+ Cl^-	Very few endemic subspecies; normally marine biocenoses	Free connection
Semimarine	Black Sea, Akchagylian, Surozh, Paleouzunlarian, Uzunlarian	Marine euryhaline of Euxinian type	Filter-feeding organisms	5-30		Endemic species, subgenera; endemic biocenoses	Restricted connection
Brackish-water	Caspian Sea, Apsheronian, Bakuan Chaudian, Gurian, Old Euxinian, Khazar, Khvalyn	Caspian-type brackish-water	Filter-feeding organisms	5-12	Ca^{++}	Endemic genera, subfamilies, families, all biocenoses endemic	Complete or nearly complete isolation
Semi-freshwater	Aral Sea, Early Apsheronian, New Euxinian; Dnepr-Bug, Dnester, Miuss limans	Brackish-water + freshwater of liman type	Filter-feeding organisms	0.5-5		Endemic subspecies, species; endemic biocenoses	Isolation, unidirectional runoff
Freshwater	Rivers, lakes, freshened portions of major basins	Freshwater	—	0-0.5			Running water
Hyperhaline	Kara-bogazgol, Sivash, Dead Sea	—	—	40	Ca^{++},Mg^{++}, CO^{--},SO_4^{--}	—	Complete isolation

tion), Late Valdai Glaciation, and Holocene Interglaciation.

The Mikulino Interglaciation

BLACK SEA: KARANGAT MARINE BASIN

The Karangatian transgressive basin was formed as a result of a glacioeustatic rise of the ocean level and a penetration of the salt waters of the Mediterranean through the Bosphorus into the basin of the Black Sea. The absolute age of mollusk shells from terrace deposits of the Kerch' and Caucasus coasts, determined by the uranium-ionium method, was found to range from 74,000±3000 (LU-404V) to 88,000±3000 (LU-403) years (Zubakov, 1974).

The level of the Karangat Sea at the transgression maximum (Figure 23-1) exceeded the present level of the Black Sea by 8 to 12 m (Fedorov, 1978; Ostrovskiy, Izmaylov, Shcheglov et al., 1977). In the south, a wide connection through the Bosphorus with the Tyrrhenian Basin of the Mediterranean existed. In the northeast, in the Manych Trough, where a relationship was noted between Karangat and late-Khazar (Girkanian) sediments, a connection with the Caspian may have existed. This could have been fostered by a deep pre-Karangatian regression and an overdeepening of the straits. The boundaries of this sea were close to the present ones, but considerable ingressions of the basin are indicated by sediments in the Kolkhida Lowland, the Manych Trough, and the delta of the Danube. Coastal terrace deposits up to several meters thick with a mollusk fauna are known over the entire perimeter of the Black Sea basin. Most typical are sections of a 10 to 12 m terrace of the Kerch' Peninsula (El'tigen, Chokrak, Karangat) and the Taman Peninsula (Cape Tuzla). On the Caucasus coast, the Karangat terrace was raised by neotectonic movements to a height of 20 to 25 m (Fedorov, 1978) or even 30 to 37 m (Ostrovskiy, Izmaylov, Shcheglov, et al., 1977). The Karangat deposits are widely developed in the Rioni Lowland, in Guriya, and in Bulgaria and Turkey.

In the deep depression of the Black Sea, deposits of the Karangat Basin were uncovered by holes drilled by the *Glomar Challenger*. Karangat terrigenous sapropel up to 15 m thick was uncovered in holes 379A (depth 83 to 97 m) and 380 (60 to 75.5 m). It was dated by means of diatom flora and spore-pollen complexes (Usher and Sepko, 1978).

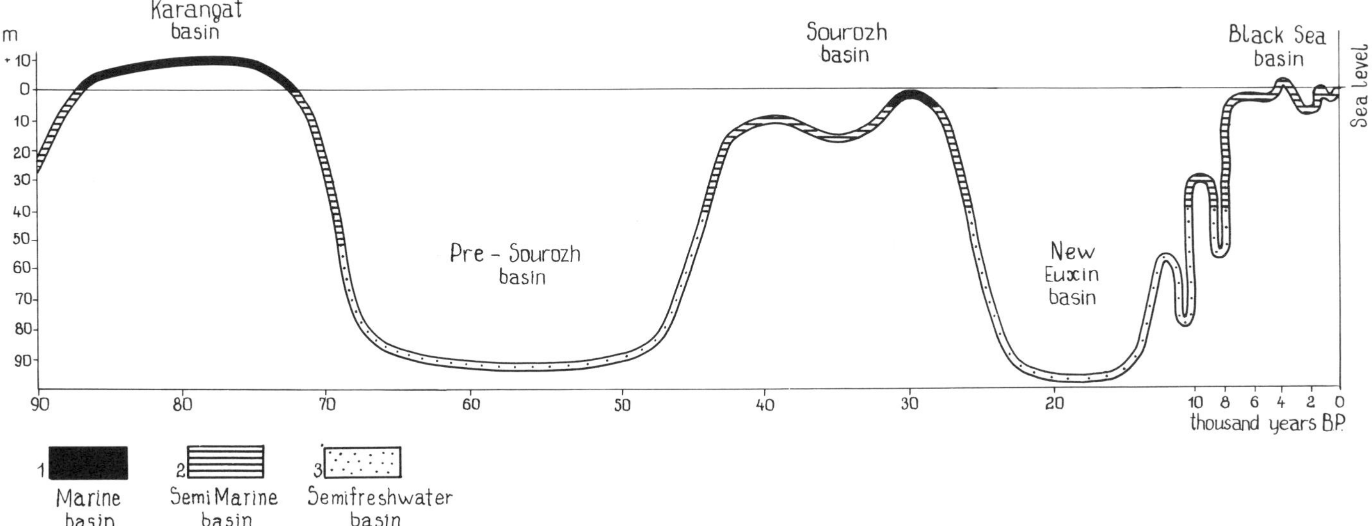

Figure 23-1. Fluctuations of level and alternation of basin types of the Black Sea during the Late Pleistocene and Holocene.

Coastal waters of the littoral and sublittoral of the Karangat Sea were populated by a rich fauna of Mediterranean-type mollusks (120 species, including 41 bivalves and 62 gastropods) (Nevesskaya, 1965; Il'ina, 1965; Trashchuk, 1974). However, in composition this fauna is closer to that of the North Atlantic; it contains only 40% of the species inhabiting the Mediterranean and 10% of the species inhabiting the Tyrrhenian Basin. The composition of the Karangat fauna includes characteristic mollusks with 25% of species that do not inhabit the Black Sea at the present time (e.g., *Cardium tuberculatum* Poirr., *Paphia senescens* Coc., *Ensis ensis* (L.), *Chlamys varia* (L.), and *Cerithium vulgatum* Brug.). It is postulated that on the shelf of the Karangat Basin and in the Mediterranean Sea there existed ecologic zones of filter-feeding organisms living at shallow depths and feeding on detritus from aqueous suspensions and collecting detritophages (Nevesskaya, 1965).

In the open sea, particularly in the region of the deep-sea depression, planktonic organisms lived only in the upper layer of the water, which was not subject to hydrogen sulfide contamination. A rich complex of diatoms of marine and brackish-water species is known there, including *Thalassiosira oestrupii* (Ostf.) Pr.-Lavr., *T. subsalina* Pr.-Lavr., *Cyclotella caspia* Grun., *Coscinodiscus jahischi, C. perforatus* Ehr., and *Thalassionema nitzschoides* Grun. (Jousé and Mukhina, 1980). In sediments of the Karangat Basin in the deep-sea trench, microfloras of nannofossils were abundant, with a predominance of one species, *Gephyrocapsa caribbeanica* Boudr. et Hay, along with *G.* cf. *oceanica* Kampt. (Shumenko and Ushakova, 1980). These species are characteristic of the zone of *Emiliania huxley* (Lohm.) of Eemian age, although *E. huxley* itself was not observed there.

On the basis of the above data on the fauna and flora, one can reconstruct the ecologic conditions of the Karangat Basin. This marine basin was similar in salinity and temperature to the modern Mediterranean. Its salinity, on the basis of the mollusk composition, reached 30 ‰ (Nevesskaya, 1965). The dominantly marine diatom flora confirms the high salinity of the waters of the Karangat Basin.

The water temperature in the shallow part of the basin was somewhat higher than the present one, for the content of warm-water Lusitanian and Canarian species of bivalve mollusks reached 25% (15 species out of 59) whereas in the Black Sea today it does not exceed 20% (7 species out of 39) (Nevesskaya, 1965). The surface waters in the open part of the basin were also warmer than the modern ones, judging from the presence of a recent tropical diatom species, *Thalassionema oestrupii*, in its sediments (Jousé and Mukhina, 1980). These data are in agreement with the conclusions from spore-pollen complexes both in deep-sea boreholes (Koreneva, 1980) and on the surrounding dry land. At the same time, periglacial steppes disappeared, and the coast was mainly covered by broad-leaved forests (white beech, beech, oak, elm, linden, and alder) as well as coniferous forests (spruce, fir, and pine) in the mountains.

The presence of an appreciable content of sapropel and pyrite in the deep-sea sediments of the Karangat Sea (Neprochnov, 1980) indicates the existence of a zone of hydrogen sulfide contamination at great depths.

THE CASPIAN SEA: LATE KHAZAR BRACKISH-WATER BASIN

An isolated late Khazar brackish-water basin was formed in the closed Caspian depression. Its coastal deposits in tectonically relatively stable regions (lower Volga, Mangyshlak) occur at elevations of 10 to 20 m below sea level, whereas in uplift zones (Dagestan) they form terraces with elevations of 40 to 50 m above sea level (Fedorov, 1978).

The level of the late Khazar basin was probably at 10 to 15 m below sea level, that is, above the present level of the Caspian, but it still cannot be regarded as transgressive (Figure 23-2). This was more probably a regression of small

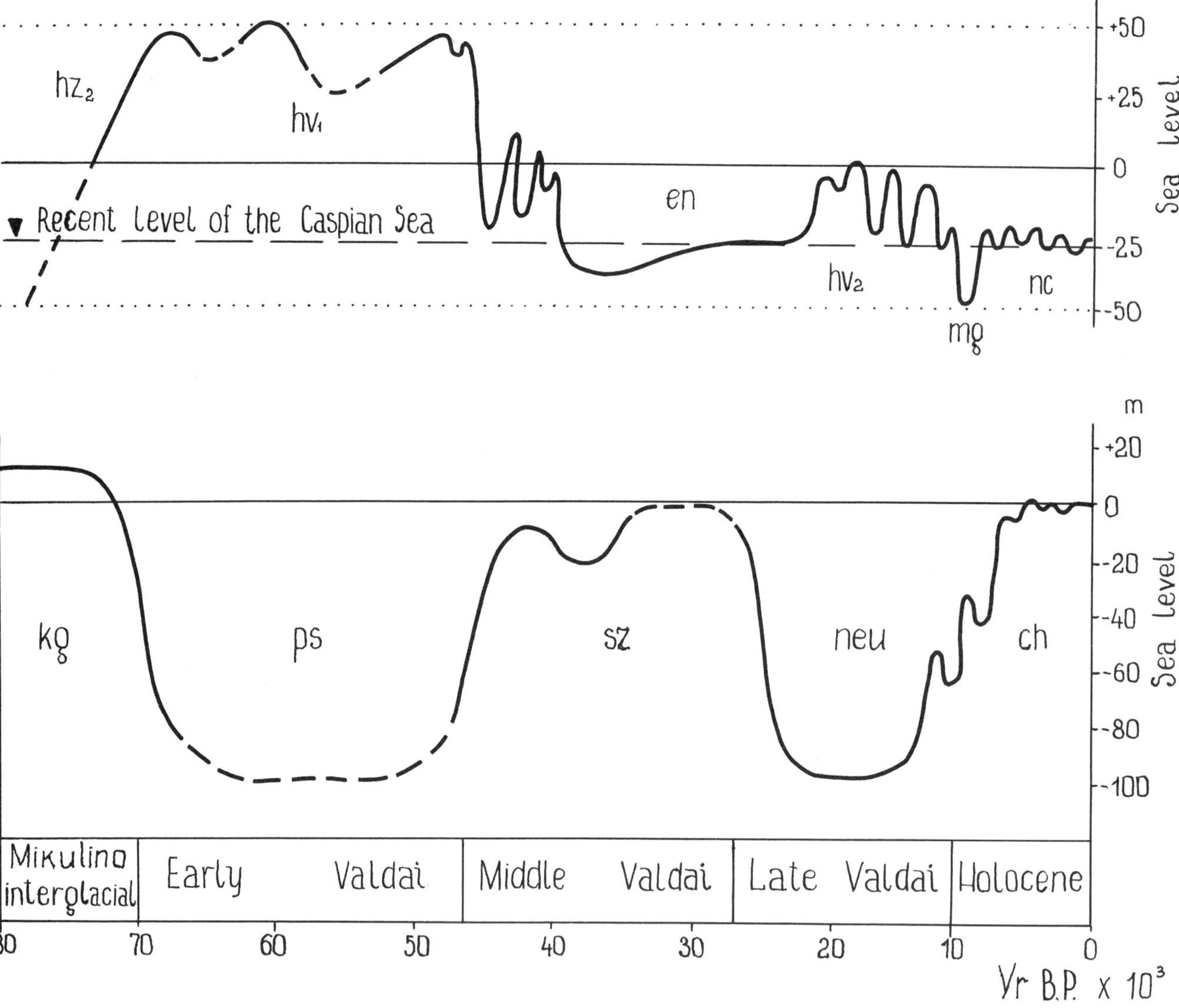

Figure 23-2. Relationship of transgressions and regressions of the Black and Caspian Seas to glaciations on the Russian Plain. (Based on data from Rychagov, 1977, and Ostrovskiy, Izmaylov, Shcheglov, et al., 1977.) (Abbreviations: hz_2, late Khazar; hv_1, early Khvalyn; en, Yenotayevka; hv_2, late Khvalyn; mg, Mangyshlak; nc, New Caspian; kg, Karangatian; ps, pre-Surozh; sz, Surozh; neu, New Euxinian; ch, Black Sea.

dimensions, associated with a more favorable water balance than during other interglacial epochs. The Mikulino Interglaciation was distinguished not only by a warmer climate but also by more humid conditions in the arid zone. Thus, in the basin of the Volga, which is the main tributary of the Caspian, forest spread eastward at the expense of steppe, and mixed grass steppe occupied large areas of the northern Caspian region. The expansion of areas with more humid conditions considerably increased the inflow from the basins of the Volga and Ural and offset the increased evaporation under warm climatic conditions. Therefore, the regression, usually very deep for interglaciations, did not reach any considerable magnitude.

The water balance of late Khazar time can be reconstructed as follows. The expansion of forest areas into the Volga River drainage basin and of mixed grass steppes into the Caspian region makes it possible to postulate a 33% increase in runoff in comparison with the present runoff, that is, an increase of up to 430 km^3. The water area of the late Khazar Basin was approximately 500,000 km^2. The sediments in this water area were probably also 33% greater than present ones; that is, a layer of 230 mm over this area makes a volume of 115 km^3. The total gain of the water basin thus amounts to 545 km^3. The loss consisted of evaporation from the water area, which could have been 20% higher than at present, that is, a layer of 1100 mm or 550 km^3, a value close to the gain.

The waters of the late Khazar basin were inhabited by a rich molluscan fauna dominated by the endemic genus *Didacna* (*D. surachanica* Andr., *D. subovalis* Prav., *D.*

pallasi Prav.), as well as by *Adacna plicata* Eichw., *Monodacna caspia* Eichw., *Dreisena polymorpha* Pall., and *Corbicula fluminalis* Müll. The abundance of fauna with warmth-loving elements reflects the variety of ecologic niches and the warm-water conditions of this basin, as confirmed by the spore-pollen characteristic of its sediments (Rychagov, 1977). The mean annual temperature of the water based on data of the calcium/strontium method reached 18°C in Transcaucasia.

ARCTIC BASIN: BOREAL TRANSGRESSION

During the Mikulino (Kazantsevo) Interglaciation an appreciable transgression of the marine basin to the adjoining Arctic dry land occurred, a transgression most extensive in the regions of continental glaciation. In northern Europe this basin flooded considerable areas around the White Sea, in the valleys of Severnaya Dvina, Mezen', Onega, Pechora Rivers, and in the Karelian lakes, where the basins of Onega Lake and Ladoga Lake probably provided a connection with the Mya basin of the Baltic and farther west with the Eemian basin of the North Sea. In northern Siberia, the boreal transgression also flooded the Yamal, Gydan, and Tazovskiy Peninsulas; the southern part of Taimyr; and the lower parts of the Ob', Yenisey, and Khatanga River valleys, as well as lower parts of the Arctic islands. Within the confines of Chukotka its deposits are known along the shores of Provideniya Bay and Krest Bay and stretch in a wide band around Anadyrskiy Bay and Kolyuchinskiy Inlet. According to Troitskiy (1979), the north-Siberian field of the marine Pleistocene extended over 1400 km from north to south and 1900 km from west to east, occupying an area of about 1 million km^2.

The marine basin of the boreal transgression was settled by a rich and varied fauna of mollusks (111 species), crustaceans (6 species), brachipods (3 species), and foraminifers (66 species). The zoogeographic composition of the molluscan fauna—including 14% boreal, 6% primarily boreal, 37% boreal-arctic, and 43% arctic species—reflects a significant warming and deep penetration of boreal species from both west and east into the center of the Arctic (Troitskiy, 1979). Thus, the region of northern Siberia was penetrated by the following western boreal immigrants presently living in the zone influenced by the Gulf Stream: *Balanus balanoides, Cingula aculenus, Pagurus pubescens, Lacuna divaricata, L. pallidula, Buccinum undatum, Neptunea despecta, Nucula tumidula, Modiola modiolus, Cerastoderma edule, Astarte sulcata, Arctica islandica, Mya arenaria,* and *Zyrphaea crispata.* Among eastern immigrants, only two species, *Liomesus nassula* and *Pseudopyrita compressa,* made their way from the Bering Sea, that is, 4000 km farther east.

The fauna of benthic foraminifers is composed quantitatively and qualitatively of a rich complex containing 66 species in the European USSR, 42 in western Siberia, 35 in the Taimyr Lowland, and 19 on Chukotka (Gudina, 1976). The number of warm-water elements decreases from west to east: 1 boreal-Lusitanian, 22 boreal, 19 arcto-boreal, 10 boreal-arctic, and 9 arctic species; in northern Siberia, 1, 11, 20, 7, and 7, respectively; on Chukotka, 1, 7, 3, 3, and 2, respectively. In the boreal sea, there first appears a new faunal element, *Elphidium excavatum,* and in Siberia and Europe, *E. propinquum, Triloculina trihedra*, and *Pseudopolymorphina novangliae*. In addition, *Lagena hispida, L. laevis, L. semilineata*, and *Ammonia beccarii* appear only in Europe; *Elphidium boreale* and species that previously inhabited the Pechora Basin (*Quinqueloculina agglutinata, Q. deplanata, Pyrulina cylindroides*, and *Elphidella arctica*) appear in Siberia and *Bulimina marginata* on Chukotka (Val'katlen complex) (Gudina, 1976).

The climatic conditions of the boreal sea were much warmer than the present ones. Thus, in the early phase of Kazantsevo time, the water temperature in the southeast of the Kara Sea had already risen 0.5°C to 1°C above the present value. At the interglacial optimum, the air temperature rose by 2.5°C to 4.5°C on the Yamal Peninsula, by 5°C to 6°C in the mouth of the Yenisey, and by 4°C to 6°C in the Ob' River basin (Troitskiy, 1979). The water temperature within the boundaries of dispersal of boreal species did not descend below 0°C, and the mean annual temperature was 2°C to 4°C above the present one. The major portion of this sea did not freeze, but ice-rafted detritus indicates that floating ice could have been present there (Troitskiy, 1979). These data indicate that the influence of the Gulf Stream extended far east and covered a considerable portion of the Arctic basin.

Early Valdai

BLACK SEA: PRE-SUROZH (PITSUNDA) SEMIFRESHWATER BASIN

According to data from a study of overdeepenings in valleys of the Caucasian coast, the post-Karangatian regression of the Early Valdai (Figure 23-1) reached strandlines of −100 to −110 m (Ostrovskiy, Izmaylov, Shcheglov, et al., 1977). Deposits of this basin were successfully uncovered by borings on Cape Pitsunda at depths of 90 to 180 m. The uranium-thorium absolute age of mollusk shells is 47,000 ± 1700 yr B.P. (LU-413). These deposits were also observed in deep-sea boreholes 379A (64-m to 78-m interval) and 380 (47-m to 58.5-m interval), and also in core sample 1857-2 (230- to 274-cm interval) (Neprochnov, 1980).

As for the ecologic conditions of this basin, the presence of the mollusks *Dreissena rostriformis* Desh., D. *polymorpha* Pall., *Micromelania caspia* Eichw., *M. elegantula* Grimm determines that this basin was of semifreshwater type (Ostrovskiy, Izmaylov, Shcheglov, et al., 1977). However, the presence of infrequent *Didacna* suggests that at the peak of the regression the basin could have been of brackish-water type. The composition of the diatom flora from deep-sea boreholes (*Stephanodiscus astraea, S. hantzschii, Cyclotella kützingiana*) is characteristic of moderately cold slightly mineralized bodies of oligotrophic water (Jousé and Mukhina, 1980).

It is possible that the regression of this basin was associated with Early Valdai glaciation and the glacioeustatic regression of the ocean at that time. A sea-level depression could have resulted in the drying of the Bosphorus, the complete isolation of this basin, and the formatin of a closed Caspian-type basin. This was probably a normally aerated basin with a salinity of up to 5‰ to 10‰. The climatic conditions were severe, and, judging from pollen data from deep-sea boreholes, marked changes took place from periglacial steppes to coniferous forests (Koreneva, 1980).

CASPIAN SEA: EARLY VALDAI BRACKISH-WATER BASIN

A brackish-water early Khvalyn Basin formed in the Caspian depression of that time (Figure 23-2). This was one of the major transgressions of the Caspian (up to 48 m above sea level), its level was 76 m above the present one, and the water area was 950,000 km^2 (2.5 times the area of the present Caspian). This transgression reached these enormous dimensions in its early phase. It was probably short-lived, for a discharge is postulated through the Manych Strait, where the present discharge threshold is +26 m. At a later stage, the level stabilized at lower strandlines, +20 to +25 m (Buynak stage), and at the end of the Early Valdai it first dropped to +14 to +15 m (Turkmen stage), then to +4 to +6 m. These stages were separated by appreciable regressions.

Radiocarbon datings of mollusk shells are contradictory and obviously too young (some dates of early Khvalyn are younger than those of late Khvalyn). Dates obtained by the thermoluminescence method give the age of the early Khvalyn transgression as 40,000 to 70,000 years, the age of the maximum stage being assumed to be about 60,000 years, and they give the age of the Buynak transgression as 42,000 years (Rychagov, 1977).

The water balance of the early Khvalyn Basin can be most satisfactorily explained, in particular, by assuming conditions of an ice-free regime on the Russian Plain (Chebotareva and Makarycheva, 1974). The absence of an ice sheet did not prevent moist air masses from moving from the Atlantic into the basins of the Volga and Ural, and, therefore, the precipitation was not much less than it is at the present time. If one considers that, because of the low temperatures, the evaporation was insignificant and that almost the entire Volga River basin was occupied by permafrost (Velichko, 1973), the river discharge could have exceeded the present one by a factor of at least 1.5 and reached 500 km^3. The precipitation on the water area could have been close to the present amount of 170 mm, giving a water volume of 162 km^3. The total gain was 662 km^3. Because of low temperatures and increased sea-ice cover, the chief loss component of the water balance (evaporation from the water area) could have dropped by a factor of 1.5 to 710 mm and thus a loss of 650 km^3. This excess of about 12 km^3 could have discharged through the Manych Strait into the Black Sea. However, the maximum level of the Caspian, +48 m, was probably short-lived, for the large difference in the levels of these seas (+48 m and −100 m) must have resulted in rapid overdeepening of the strait between them, and the level could have stabilized at +20 to +25 m. Moreover, the water area decreased to 800,000 km^2, and under constant climatic conditions the discharge through the Manych Strait should have increased and reached a volume of 100 km^3. However, this basin could not have been a through-flow basin for a long time, for balance calculations show that it should have been completely freshened after 2000 to 3000 years. This figure is not confirmed by the molluscan fauna, however, which continued to reflect fairly high salinity. Thus, on the basis of the presence of the mollusks *Didacna cristata* (Bog.), *D. ebersini* Fed., *D. protracta* Eichw., *D. parallela* Bog., and others, the salinity of the Early Khvalyn Basin appears to have been around 12‰ to 13‰ (Svitoch, 1976). Therefore, the depression of the level of the Early Khvalyn Basin was caused by a change in climatic conditions, possibly during the phases of climatic warming.

The water temperature of this basin was comparatively low, although at times it may have risen almost to the present value. This is indicated by a low $\delta\ ^{18}O$ value (up to −12‰ to −14.5‰ in comparison with the present value of about 0), possibly as a result of the influence of cold glacial waters of light isotopic composition (Nikolayev and Popov, 1973). However, the presence of *Corbicula fluminalis* Müll. shells attests to short warming phases.

Middle Valdai

BLACK SEA: SUROZH SEMIMARINE BASIN

In recent years, increasing amounts of data have appeared on the post-Karangatian transgression of the Late Pleistocene, which was earlier referred to as the Surozh transgression (Popov and Zubakov, 1975). The Surozh transgression probably differed little in size from the Karangatian transgression and the present Black Sea. Its deposits are located mostly below sea level (about −10 m) or in overdeepenings of valleys (on the shores of the Kerch' Strait, in the valley of Zapadnyy Manych, and in the lower reaches of Salgir and Mius [Livensadovka]). In the water area of the Black Sea, these deposits probably include lithified sands with a molluscan fauna, recently uncovered by marine drilling west of the Karkinitskiy Peninsula (Golitsyn Rise) at the Shagana liman and Kyzaul'skaya bank (Pazyuk et al., 1974; Shnyukov and Trashchuk, 1976; Trashchuk and Boltivets, 1978). In tectonically active areas of the Caucasian coast, the height of the Surozh Terrace is 18 to 20 m (Ostrovskiy, Izmaylov, Shcheglov, et al., 1977). Two phases of the Surozh transgression are observed (Figure 23-1): an early phase with a height of −10 to −15 m and a late phase almost up to the present level or even higher (Ostrovskiy, Izmaylov, and Shcheglov, et al., 1977).

In the deep-sea depression, Surozhian deposits were uncovered in boreholes 379A (64 - to 78-m interval) and 380 (47- to 58.6-m interval) (Neprochnov, 1980).

Absolute radiocarbon dates of molluscan shells are available for sections of the Surozh Terrace on Cape Tuzla (Zubakov, 1974)—30,450±2200 yr B.P. (LG-90)—and for the Surozh stratotype on Zapadnyy Manych—32,350±5200 yr B.P. (LU-534-V). The uranium-thorium method for these two sections gave an age of 34,030±900 yr B.P. (LU-450). In addition, analysis by the uranium-thorium method of Surozh mollusks on the Chushka Spit (Kerch' Strait) gave 40,700±1200 yr B.P. (LU-449) for samples from a depth of 11.8 to 12.5 m and 41,250±1340 yr B.P. (LU-448) from a depth of 14.6 to 17.3 m (Ostrovskiy, Izmaylov, Shcheglov, et al., 1977).

In the shallow zone of the shelf, deposits of the final phase of development of the Surozh basin include Tarkhankut layers with marine fauna (sands with *Cerastoderma lamarcki* Reeve, *Abra ovata* Phil., and *Dreissena polymorpha* Pall.), as well as Karkinit layers with the latest marine elements (silty sands with *Cerastoderma larmarcki* and *Dreissena polymorpha*).

In the deep-sea depression, these deposits are uncovered by the deepest borings and contain infrequent shells of *Cerasoderma lamarcki* in a stratum of ooze with hydrotroilite interlayers (Shcherbakov et al., 1978).

The molluscan fauna of the sublittoral is similar in composition to the present Black Sea fauna, but in the final maximum stage of the Surozh transgression (after Ostrovskiy, Izmaylov, Shcheglov, et al., 1977) it differed little from the Karangatian fauna, and its composition includes not only Black Sea but also Mediterranean species: *Paphia senescena* (Coc.), *Chlamys glabra* (L.), and *Cerithium vulgatum* Brug. A special flora of planktonic diatoms inhabited the surface layer of the open sea. In addition to the brackish-water *Cyclotella caspia* Grun., *Stephanodiscus astraea* Kütz., and *S. hantzschia* Grun., the following marine species appear: *Thalassiosira oestrupii* (Ost.) Pr.-Lavr. and *T. subsalina* Pr.-Lavl. (Jousé and Mukhina, 1980).

On the basis of the composition of the fauna and flora, the salinity of the Surozh basin was similar to the present one or slightly lower. At the maximum phase of the transgression, the salinity could have risen almost to the level of that of the Karangatian Basin (Ostrovskiy, Izmaylov, Balabanov, et al., 1977).

The high content of pyrite and sapropel in the sediments (Neprochnov, 1980) suggests hydrogen sulfide contamination of deep waters of the Surozh Basin.

On the basis of the mollusks and diatoms from that time, the climate appears similar to the present one or somewhat colder. Pollen spectra in the deep-sea depression were characterized by frequent changes of steppe and forest associations, with a predominance of the latter (Koreneva, 1980). The Surozh transgression is correlated with the warming of the Middle Valdai (Bryansk, Mologa-Sheksna Interstade).

CASPIAN SEA: YENOTAYEVKA REGRESSIVE BASIN

A considerable regression, marked by the formation of Yenotayevka deposits of the lower Volga, is notable in the isolated Caspian Basin of that time (Figure 23-2). The level of the Caspian dropped by 100 m in comparison with the maximum of the preceding Early Valdai Transgression to strandlines of −45 to −60 m (Rychagov, 1977; Varushchenko et al., 1980). However, the ecologic conditions and composition of the fauna underwent little change. The regression of the Yenotayevka Basin should be attributed to a warming of the climate in the Middle Valdai and a markedly increased evaporation from the water area of the basin.

THE ARCTIC BASIN: KARGINSKIY TRANSGRESSION

The transgression of the Arctic basin encompassed vast areas of northern Asia, particularly in regions of continental glaciation. This transgression has been studied most closely on the North Siberian Lowland, where it penetrated into the residual uncompensated glacioeustatic trough that remained after the Zyryanka Glaciation (Andreyeva, 1980). The maximum extent of the marine basin occurred during the first phase of the interglaciation: the coast of Khatanga Bay, the upper reaches of the Kheta and Dudypta Rivers, the basin of the Pyasina and Verkhnyaya Taimyra Rivers, the shore of Lake Taimyr and Yeniseyskiy Bay, and the lower course of the Yenisey River. Marine deposits occur mainly on Zyryanka glacial sediments. Radiocarbon dates give their ages as 24,000 to 50,000 years (Andreyeva, 1980).

The shallow marine basin was settled by boreal and arctic species of mollusks: *Mytilus edulus, Macoma baltica, Portlandia arctica, P. arctica siliqua, Mya truncata*, and *Astarte borealis*. The zone of deeper waters was inhabited by *Bathiarca glacilis* and *Yoldiella intermedia*. The foraminiferal fauna of the middle portion of the marine sections is represented by an arctoboreal complex (35 species) with a predominance of Ephidiidae and an occurrence of Islandiellidae and Cassidulinidae (Gudina, 1976). Faunistic and palynologic data indicate a warmer climate than the present one (Andreyeva, 1980).

Late Valdai

BLACK SEA: NEW EUXINIAN SEMIFRESHWATER BASIN

The eustatic regression of the world ocean caused by the Late Valdai Glaciation led to a drop in the level of the Black Sea to strandlines of −90 to −110 m and to a loss of the two-way connection with the Mediterranean (Figure 23-1). As a result, a considerable portion of the shallow-water shelf was drained, particularly the northwestern water area and the entire Sea of Azov. The influx of waters through the Bosphorus from the Mediterranean ceased, and the basin was freshened.

Regressive sediments of offshore bars (pebble gravel and sands with coquinas) marking the coastline of the New Euxinian Basin were observed at depths to 90 m (Shcherbakov et al., 1978). These sediments record the lowest level of the New Euxinian Basin. According to data on strandlines of overdeepenings of river valleys (Ostrovskiy, Izmaylov, Shcheglov, et al., 1977), the regression could have reached strandlines of −110 m.

The oldest radiocarbon dates for mollusk shells of New Euxinian offshore bars (17,780±200 yr B.P.) were obtained at a depth of 80 to 90 m at the shelf edge south of the Crimea (Shcherbakov et al., 1978). In a deep-sea depression, deposits of terrigenous clayey ooze of the "lacustrine stage" are dated at 16,900±270 and 13,850±200 yr B.P. (Degens and Ross, 1972). The youngest dates of the New Euxinian deposits in the shelf zone are 8550±130 yr B.P. and in the deep-sea zone 8600±200 yr B.P. (Shcherbakov et al., 1978).

The shallow-shelf zone of the New Euxinian Basin was occupied by a fauna of mainly brackish-water mollusks, consisting of species that tolerate maximum freshening: *Dreissena polymorpha* Pall., *D. rostriformis rostriformis* Desh., *D. r. distincta* Andr., *Monodacna caspia* Eichw., *Adacna vitrea euxinica* Nev., *Hypanis plicatus relictus* (Mil.), and *Micromelania caspai* Eichw., as well as purely freshwater species of gastropods, probably brought into the sea by river waters: *Viviparus viviparus* (L.), *Lithoglypus naticoides* C. Pf., and *Valvata pulchella* Stud. (Nevesskaya, 1965; Fedorov, 1978).

In the deep part of the sea, paleontologic remains are represented by a microflora of cold-water diatoms, with a predominance of a depleted freshwater complex of *Stephanodiscus astraea*, and others (Jousé and Mukhina, 1980).

In either case, whether a semifreshwater or a brackish-water basin, salt differentiation of the water mass should not have taken place, so that one can postulate the absence of the zone of hydrogen sulfide contamination and an adequate aeration of the entire water mass of the New Euxinian Basin. This hypothesis is confirmed by the low content of organic carbon and pyrite in the deep-sea sediments.

The climatic conditions and temperature regime at that time were the most severe of the Pleistocene. Pollen spectra of New Euxinian deposits from deep-sea boreholes and core samples reflect the development of periglacial vegetation and pine forests in the mountains. *Artemisia* pollen is dominant, with much chenopod and some grass pollen; pine and birch pollen predominate among tree species (Neprochnov, 1980).

The New Euxinian Basin was a special water ecosystem that differed markedly in ecology from the present Black Sea. It was a flow-through lacustrine type of basin. Its size was almost 30% smaller, but the depths were only 100 m less and exceeded 2000 m in its central part. Its salinity approached that of freshwater lakes but still amounted to 3‰ to 7‰. The presence of infrequent *Didacna moribunda* Andr. (Nevesskaya, 1965) may indicate phases of complete isolation of the New Euxinian Basin. Indeed, the discharge threshold in the Bosphorus has a strandline of −36 m; therefore, it can be postulated that, if the level had dropped below this mark, the New Euxinian Basin could have become completely isolated. However, the majority of investigators believe that it was always a flow-through basin (Fedorov, 1978; Shcherbakov et al., 1978; Nevesskaya, 1965). For this purpose, it is necessary to postulate a much greater depth of the Bosphorus than the present one. The latest studies at the gauging site of the Istanbul Bridge overpass (Degens and Ross, 1974) reveal that the bottom of the Bosphorus consists of loose, presumably very young deposits to a depth of 90 m. This confirms the possibility of a flow-through character of the New Euxinian Basin even at its lowest level. This conclusion is supported by data from a reconstruction of the water balance of this basin (Kvasov, 1975).

The water balance of the basin could have been as follows. It is assumed that, because of a drop in temperature, the evaporation and precipitation for the water area (320,000 km^2) were lower than the present ones by a factor of 1.5 and that the river discharge was 40% lower than the present one (Kalinin et al., 1976). In addition, glacier waters passed through the Dnepr River basin from a sector of the ice sheet with an area of 500,000 km^2 and an annual precipitation of 150 mm (Kvasov, 1975).

Gain:	
River discharge	230 km^3
Precipitation on the water area	65 km^3
Glacier waters	75 km^3
	370 km^3
Loss:	
Evaporation	180 km^3

Excess water with a volume of 190 km^3 (approximately equal to the present one) was discharged through the Bosphorus into the Mediterranean. Thus, the balance calculations confirm the paleontologic data to the effect that during the existence of the New Euxinian Basin the latter was a flow-through basin irrespective of its level and that the water was almost fresh.

CASPIAN SEA: LATE KHVALYN BRACKISH-WATER BASIN

In the Caspian depression, there existed at that time a late Khvalyn isolated brackish-water basin with maximum strandlines at about present sea level (28 m above the present level of the Caspian) and an area of about 600,000 km^2 (1.5 times that of the present Caspian). When compared with the preceding basin of Yenotayevka time (−43 to −45 m) and the present level (−28 m), this basin can be considered transgressive, but, when compared with other transgressions of the Caspian, the late Khvalyn transgression appears minor (Figure 23-2).

The history of the late Khvalyn Basin (Figure 23-3) shows up to four major transgressive-regressive stages marked by coastlines and terraces (Rychagov, 1977): a premaximum stage with strandlines slightly below 0 m (18,500 years ago), a maximum (Makhachkala) stage at a height of 0 m (16,000 years ago), the Kuma stage at −5 to −6 m (14,500 years ago), and the Sartas stage at −5 to −6 m (12,000 years ago).

The late Khvalyn Basin was settled by a rich fauna, particularly of brackish-water *Didacna* mollusks. The shallow parts of the bottom were inhabited by *Didacna subcatillus* Andr. and *D. praetrigonoides* Nal., and deeper parts in the range of 70 to 110 m were occupied by *D. protracta* Eichw., *Adacna (Hipanis) plicata* Eichw., and *Dreissena rostriformis distincta* Andr. (Lebedev and Glazunova, 1972).

The salinity of this basin was similar to that of the present one, about 12‰ to 13‰ (Svitoch, 1976). Judging from faunistic data, the water temperature was below the present one. This conclusion is confirmed by isotopic data on the content of ^{18}O. In shells of the late Khvalyn mollusk *Didacna*, the values of $\delta\ ^{18}O$ are 33% to 50% of the present values and amount to -4‰ to -4.8‰ (Nikolayev and Popov, 1973), which may indicate an influence of glacier waters. The size and level of the late Khvalyn Basin were determined by its water balance and climatic conditions. During the maximum cooling of the Late Valdai, under low-temperature and permafrost conditions over most of the drainage basin, the precipitation and volume of evaporation from the water area were 1.5 times lower than at present (Kalinin et al., 1976). In addition, a vast section of the Scandinavian ice sheet with an area of 1.2 million km^2 discharged into the Volga River basin, the atmospheric precipitation being 100 mm per year (Kvasov, 1975).

Under these conditions, the water balance of the late Khvalyn Basin was as follows.

Gain:	
River discharge	240 km^3
Precipitation on the water area (120 mm)	72 km^3
Glacial runoff	120 km^3
	432 km^3
Loss:	
Evaporation from the water area (710 mm)	426 km^3

THE SEA OF JAPAN—SEMICLOSED MARINE BASIN

During the epoch of maximum cooling 20,000 years ago, the level of the sea dropped to 130 m below present sea level (Pletnev, 1979). Correspondingly, the depths of the sea decreased considerably, causing major changes in the outlines and external connections of the basin. The shallow Tatarskiy, Laperouse, Tsugaru, and East Korea Strait completely emerged. The only connection with the ocean was through the deeper West Korea Strait. The marginal marine basin was converted into a semiclosed one. Considerable areas of the shallow shelf were converted into dry land.

The marginal sea basin, with a fairly wide connection with the ocean, was converted into a semiisolated marine-type basin with only one connection with the ocean, as in the case of other semiisolated basins of the type of the Black and Baltic Seas. This change was immediately reflected in the ecologic state of the basin. The water temperature dropped by 8°C to 10°C in comparison with the present one. The field of surface temperatures shows a distinct directional drop from south to north; the mean annual temperatures in the area of the Korea Strait dropped from 12°C to 4°C in the central part of the basin (Pletnev, 1979).

A distinct decrease in salinity from south to north is also apparent. In the southern and central parts of the basin, the salinity was close to the normal marine value of 30‰ to 35‰, but in the northern part, in the region of Tatarskiy Bay, it decreased considerably, probably because of the freshening influence of the Amur and other rivers of this region.

The content of floating ice in the sea increased appreciably. On the basis of the extent of berg-rafted stones, one can conclude that a major portion of the water area was covered by floating ice, the southern boundary of which ran much farther south of the present freezing line—to the region of the Yamato Seamount at a latitude of about 39°-40°N (Pletnev, 1979).

Despite the increased isolation, the connection with the ocean through the West Korea Strait remained fairly wide. Isotherms plotted on the basis of the foraminiferal fauna and oxygen-isotope analysis make it possible to reconstruct the branch of the warm Kuroshio Current that entered the Sea of Japan. However, this warm current did not reach as far north as it does now; it ended at about 37°N, where it encountered a cold current coming from the northwest (Pletnev, 1979).

The Holocene

BLACK SEA: BLACK SEA SEMIMARINE BASIN

At the start of the Holocene, as a result of a glacioeustatic rise of the ocean level, the waters of the Mediterranean Sea entered the Black Sea depression through the Bosphorus, and there they formed a semimarine sea that still exists today (Figure 23-4). The oldest elements of marine fauna are dated at 9000 to 7000 yr B.P. (Degens and Ross, 1972). In its initial stage, the transgression developed very rapidly; in 3500 years, the level rose by 18 m at an average rate of 5 mm per year (Nevesskiy, 1967). At that time, relicts of a New Euxinian semifreshwater basin with a Caspian fauna were still preserved in the shallow-water zone, whereas a major portion of the water area of the Black Sea was occupied by a semimarine basin. The maximum rise in level occurred 4000 to 3500 years ago, in the Subboreal, when the level rose above the present one. At that time, a terrace 3 to 5 m high, known by the name Old Black Sea Terrace, was formed; it had the richest Holocene fauna of mollusks. The absolute radiocarbon age of this terrace was determined in the area of the city of Sochi: 5500 ± 380 yr B.P. (LU-195) from a depth of 18 to 19 m, 4500 ± 160 yr B.P. (LU-199) from a depth of 13 to 14 m, and 4170 ± 90 yr B.P. (LU-507) from a depth of 3 to 4 m in the area of Pitsunda (Ostrovskiy, Izmaylov, Balabanov, et al., 1977).

In the Late Holocene, a rapid lowering of the sea level to strandlines of 6 to 8 m below sea level occurred (Fanagorian regression of Fedorov, 1978). At that time, ancient Greek settlements appeared on the Old Black Sea Terrace,

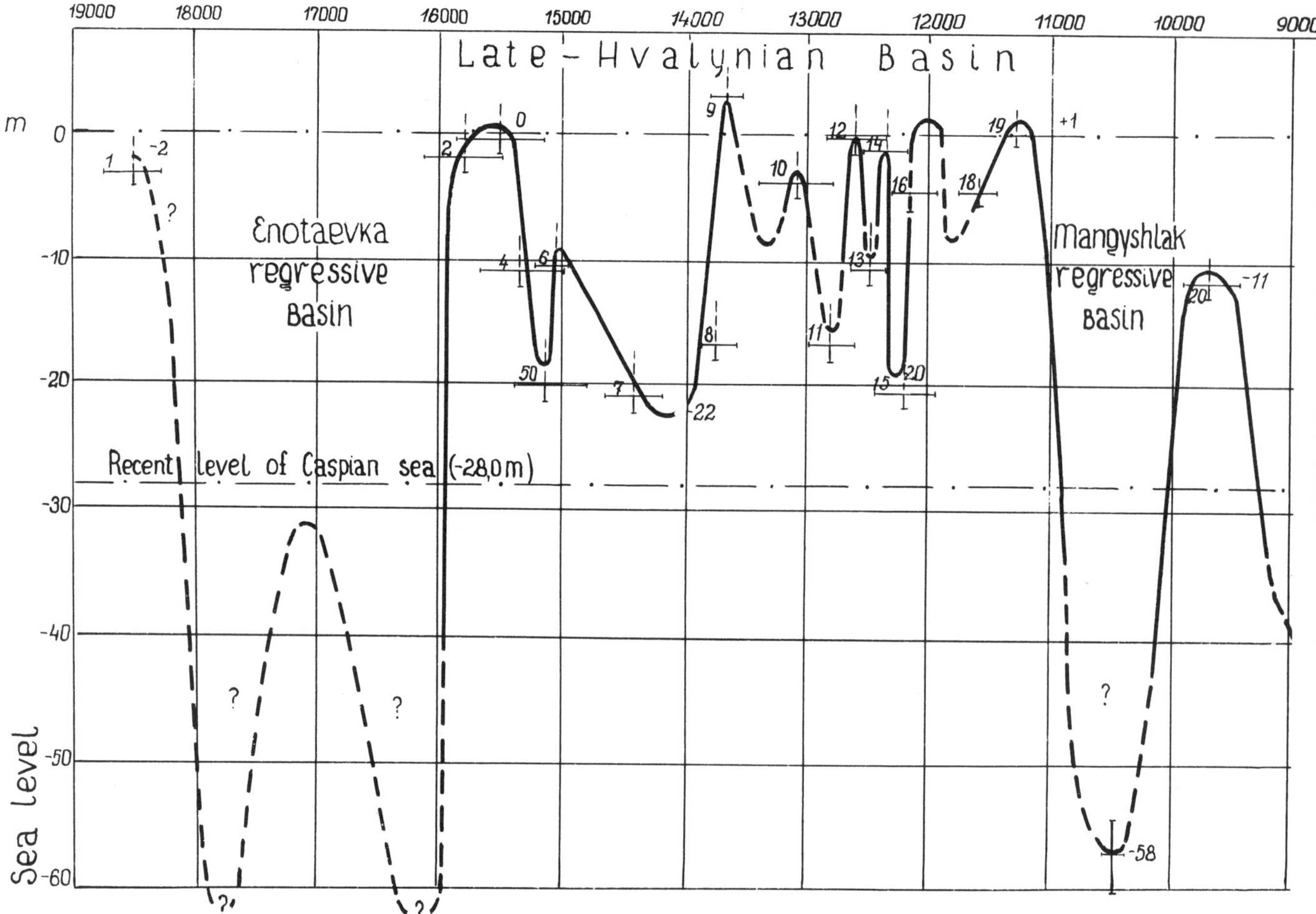

Figure 23-3. Fluctuations in the level of the Caspian Sea over the last 18,000 years. (Based on data from Varushchenko et al., 1980.) (Points dated according to the radiocarbon age of mollusks; horizontal and vertical lines indicate, respectively, the range of error in the radiocarbon determination and in the geomorphic data for the level.)

for example Olsia at the mouth of the Southern Bug River, which existed in the seventh through first centuries B.C. (Shilik, 1977). According to archaeological data, the beginning of the Fanagorian regression dates back to the middle of the second millennium B.C. (about 3500 years ago).

According to P. V. Fedorov, the following Nymphaean transgression (Istrian according to M. Blyakh and Gemetinian according to Nevesskaya, 1965) rose to 2 to 3 m above the present level. According to the data of Shilik (1977), obtained from excavations in Olsia, this transgression was short-lived and did not rise above 0.7 m below sea level. After the Nymphaean transgression, a new lowering of the level to 3 m below sea level took place (Shilik, 1977), starting approximately in the 10th century A.D. (Tsereteli's [1966] Lazian transgression).

The Black Sea Holocene basin is a semimarine-type basin. The molluscan fauna in the littoral zone on the shelf is very abundant but considerably poorer than the Mediterranean fauna; shells of *Chione gallina* L., *Cerastoderma lamarcki*, and *Divaricella divaricata* predominate there.

A biocenosis of *Chione gallina* and *Cerastoderma glaucum* is characteristic of the sublittoral zone with a sandy substrate. The shallow zone of the shelf is inhabited to depths of 30 m by a "mytiloid ooze" biocenosis with a predominance of *Mytilus galloprovincialis*. The deeper part of the shelf up to its very edge (at depths of 60 to 90 m) is inhabited by a "phaseoline ooze" biocenosis with shells of only one species, *Modiolus phaseolinus*, which lives in comparatively cold and deep waters.

In the Sea of Azov, where the salinity was considerably lower than in the Black Sea, a specific biocenosis, *Cerastoderma glaucum* Poirr, was formed on the sandy substrates.

In the deep-sea zone, the bottom fauna was absent, but the upper water mass was inhabited by a rich warm-water microflora of oceanic-type diatoms with a predominance of euryhaline marine species: *Rhizosolenia setigera, R. calcaravis* Schul., *R. alata* Bright., *Thalassionema nitschioidies* Grun., *T. oestrupii* (Ost.) Lav.-Pr., and others.

The latest detailed studies of the littoral zone of the Caucasian coast reveal a more complex pattern of fluctuations in the level of the Black Sea (Ostrovskiy, Izmaylov, Balabanov, et al., 1977). The transgressive rise in sea level

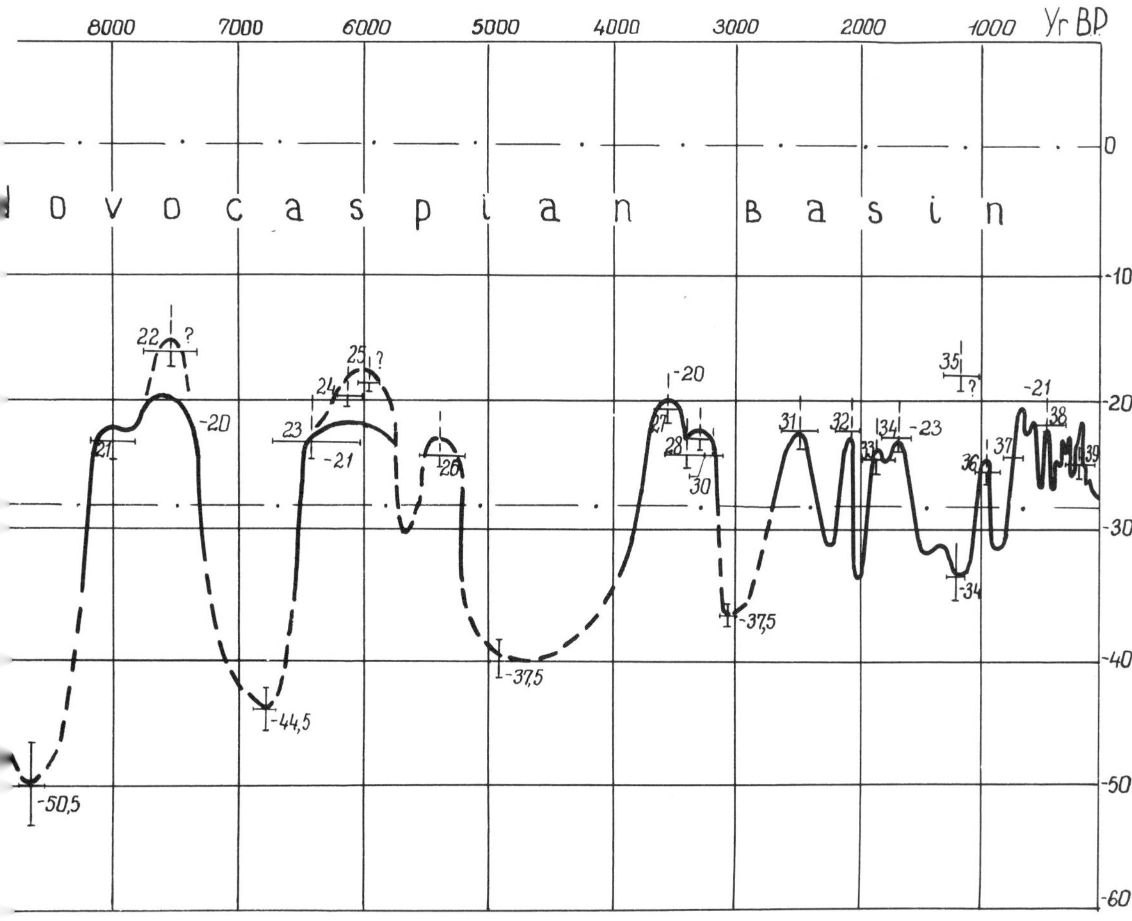

was complicated by at least six major regressive-transgressive phases (Figure 23-1). These phases are reflected in the migration of sediments of beach facies and along beach ridges, both in plan and in height within the confines of the upper part of the shelf, as well as in the phases of overdeepenings of river valleys. The phases have been adequately dated by numerous radiocarbon analyses and archaeological data (remains of the Neolithic and of Greek and Roman colonization).

The first transgressive phase, with a maximum between 12,000 and 11,500 years ago, reached an isobath of 60 m below sea level, but it was not associated with a salinization of the waters, for it contained only a brackish-water New Euxinian fauna of the mollusks *Dreissena polymorpha, Glessiniola variabilis, Micronelania caspia,* and *Monodacna caspia*. This was probably still a New Euxinian semifreshwater basin that was backed up by waters of an incipient late-glacial transgression. The regressive phase that followed later was associated with a drop in sea level to isobaths of 80 to 85 m.

The first transgressive phase with the fauna of a semimarine basin reached a level of 45 m below sea level 10,700 to 9700 years ago. At that time, the waters of the Mediterranean penetrated through the Bosphorus. This is indicated by the presence of larvae and young shells of the marine mollusks *Cerastoderma lamarcki, Chione galline,* and *Spisula subtuncata* in the region of Adler-Pitsunda, where remains of this phase have several radiocarbon dates ranging from 10,530 ± 190 yr B.P. (LU-367) to 10,130 ± 180 yr B.P. (Vsegingeo-11/124). The regression that replaced this phase led to a lowering of the sea level to depths of 65 to 70 m below sea level; it is possible that a semifreshwater basin regime was reestablished at that time.

The next transgressive phase occurred around 9200 to 8400 years ago and reached an isobath of 30 m below sea level. The molluscan fauna in the Kerch' Strait was represented by a Bugaz complex with *Cerastoderma lamarcki, Abra ovata,* and similar species and at depths of 35 m or more below sea level by a fauna of "mytiloid ooze" with *Mytilus galloprovincialis.* The regressive phase that replaced this phase was associated with a drop to 55 to 60 m below sea level.

The new transgressive phase (about 7900 to 6800 years

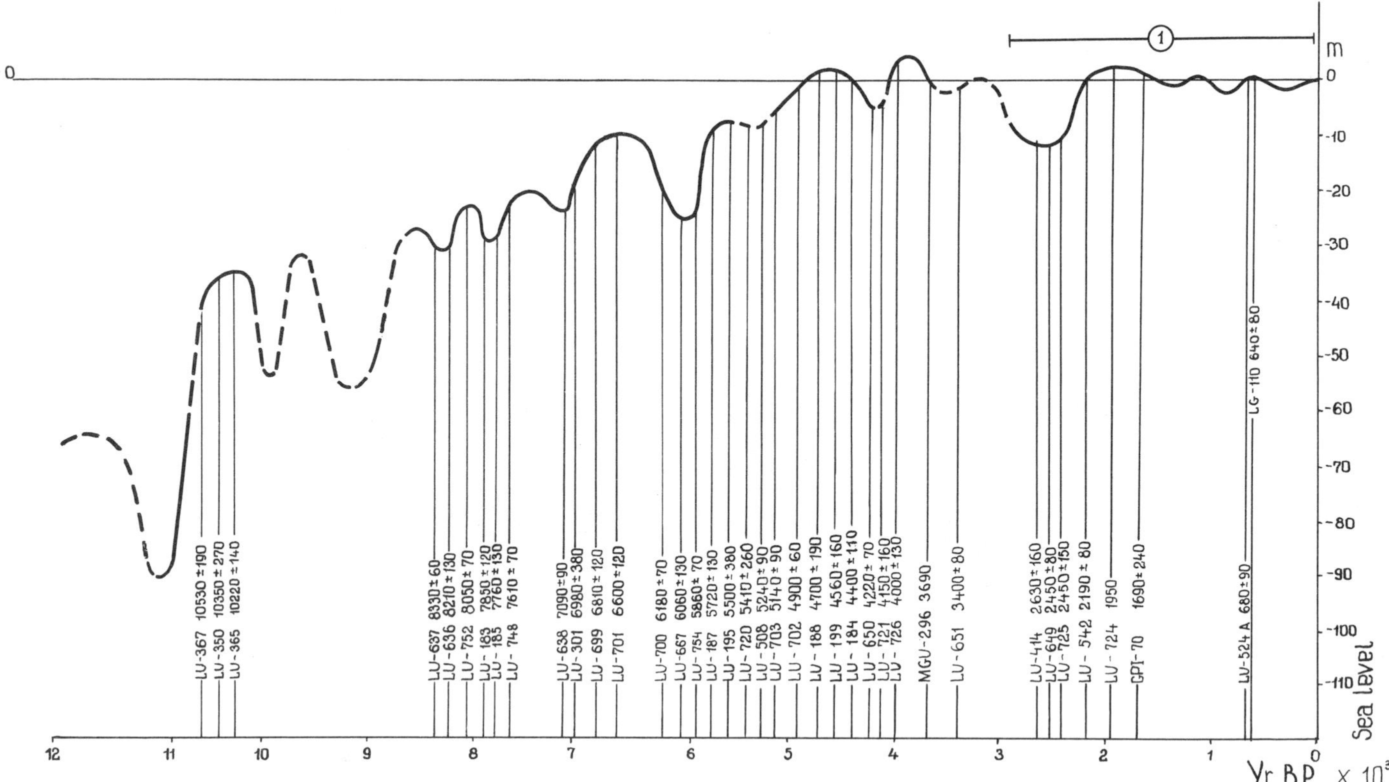

Figure 23-4. Fluctuations in the level of the Black Sea in the Holocene. (Based on data from Ostrovskiy, Izmaylov, Balabanov et al., 1977.) (The interval back to 3000 yr B.P. is also supported by archaeological and historical dates.)

ago) came close to the present level (10 m below sea level) and was characterized by a Vityaz molluscan fauna in the Kerch' Strait. On the Caucasian coast, the semimarine basin was populated by biocenoses that were practically indistinguishable from the present ones. Ecologic conditions similar to the present ones were established at that time, including the salinity and zone of hydrogen sulfide contamination (Degens and Ross, 1972). Littoral deposits of the Old Black Sea Terrace were being formed on the coast at that time (Fedorov, 1978). The regressive phase in the interval of 6200 to 5800 years ago led to a lowering of the level to isobaths of 25 to 27 m below sea level.

The maximum transgressive phase in the Holocene (5700 to 4000 years ago) reached levels of 3 to 4 m above sea level. It was accompanied by halts and even slight drops in level (of a few meters), when peat bogs were being formed. This was a period of maximum-width connection with the Mediterranean, maximum salinity, and richest fauna with Mediterranean elements (Gemetinian complex). Sediments of Fedorov's (1978) New Black Sea Terrace were being formed in the littoral zone.

The Fanagorian regression that followed led to a decrease in level—of about 5 to 7 m (Fedorov, 1978) or 10 to 15 m (Ostrovskiy, Izmaylov, Shcheglov, 1977). At that time, the New Black Sea Terrace was settled by ancient Greeks in the seventh through fifth centuries B.C.

The last (Nymphaean) transgression occurred about 2000 to 1000 years ago and scarcely exceeded the present level (1 to 2 m above sea level). The faunal composition and salinity were no different from those of the present Black Sea. In post-Nymphaean time, slight oscillations at about the present level are noted. A small (2- to 5-m) Medieval regression in the 14th and 15th centuries A.D. was followed by the start of a rise in the level of the Black Sea that still continues at the present time.

CASPIAN SEA

Mangyshlak Regressive Basin

In the Caspian region, at the very beginning of the Holocene, about 10,000 years ago, a regression occurred wherein the level decreased to 50 to 60 m below sea level and a major portion of the shallows in the northern part of the basin was drained (Figure 23-3). At these depths, the relict offshore bar "Derbent bank" was exposed, composed of coarse-grained sediments with a fauna of the mollusks *Didacna barbotdemarnyi* (Grimm.), *D. parallela* (Bog.), *D. protracta novocaspia* Glaz., *D. baeri* (Grimm.), and *Dreissena alata* Andr. (Mayev et al., 1977). The Mangyshlak threshold was inhabited by mollusks of a shallow-water complex, freshened by Volga waters, with *Didacna trigonoides* (Poll.), *D. barbotdemarnyi* (Grimm), *Hypanis plicata* Eichw., *H. vitrea* Eichw., and *Dreissena polymorpha* Pall. (Artamonov, 1976). Judging from the fauna, the salinity of this basin as a whole has remained almost unchanged. As in the course of other regressions, the contrast in the salinity of different parts of the Caspian markedly increased. The mineralization of a major portion of the basin (central and southern Caspian) increased somewhat, and the salinity of the northern Caspian dropped as a result of a decrease in water area and dilution by waters of the Volga (Rychagov, 1977).

The Mangyshlak regression was caused by a sharp aridization of the climate at the beginning of the Holocene and by the formation in the Caspian region of desert and semidesert landscapes with eolian landforms, the so-called Baer knolls. The climatic conditions, nevertheless, remained severe, as evidenced by pebbles carried by floating ice found not only in the northern but also in the central Caspian up to the Apsheron threshold (Mayev et al., 1977). Bottom and suspension currents associated with high rates of sedimentation were intensified.

New Caspian Brackish-water Basin

In the Middle Holocene, the level of the Caspian was restored, and a New Caspian Basin that exists to this day was formed. Its maximum level reached 20 m below sea level; that is, 8 to 9 m above the present level, and the fluctuation amplitude of the level exceeded 20 m (Figure 23-3). Several transgressive stages separated by regressions have been noted, as follows: first stage −25 m (9000 years ago); maximum stage −19 to −20 m (8000 years ago); third stage, −21 m (6500 to 5500 years ago); fourth stage, −22 to −23 m (3500 to 3000 years ago); and fifth stage, −24 to −25 m (from the 14th century A.D. to the beginning of the 19th century) (Rychagov, 1977). The complete picture of fluctuations of the Caspian level is considerably more complex (Figure 23-2).

The waters of the New Caspian Basin were inhabited by a rich fauna of mollusks, mainly of the genus *Didacna*; in the Middle Holocene *Cerastoderma lamarcki* Reeve also appears, but the routes and causes of dispersal of this marine species in the Caspian are thus far unknown. The mollusk complexes had the following distribution with depth: the shallow-water complex of *Didacna barbotdemarnyi* Grimm/*Cerastoderma lamarcki* Reeve (5 to 20 m); the complex of *Didacna protracta protracta* Eichw. (20 to 50 m); and the complex of *D. protracta submedia* Andr. (50 to 80 m) crossing the complex of *D. barbotdemarhyi*/*D. protracta protracta* on sandy soils (20 to 30 m) (Lebedev and Glazunova, 1972; Artamonov, 1976). Judging from the fauna, the salinity and depths of the New Caspian Basin were close to the present ones (salinity, 12‰ to 13‰).

BALTIC SEA

Baltic Glacial Lake: A Freshwater Basin

A single basin in the Baltic depression first appeared during the late-glacial about 12,000 years ago (Figure 23-5 and Table 23-2). Up to that time, a system of disconnected ice-dammed proglacial lakes had existed there. The Baltic glacial lake developed at the edge of the Scandinavian ice sheet, south of the Stockholm-Turku line. This was an isolated lacustrine-type basin with a one-way connection with the North Sea through central Sweden. Near the town of Billingen, the ice sheet's margin formed a dam that raised the level of the Baltic glacial lake to 20 to 30 m above sea level (Sauramo, 1958). Because the sea level at that time could have had a strandline of about 50 m below present sea level, the actual level of the Baltic glacial lake was much lower than the present (possibly about −20 m). Shore-terrace deposits of this basin now occur at strandlines of +100 to +20 m in Finland, +70 m in northern Estonia, +35 m in southwestern Estonia, and at a depth of −60 m on the sea floor in the South Baltic (Kolp, 1974).

The waters of the Baltic glacial lake were inhabited only by cold-resistant arctic and boreoalpine species of a *Melosira* flora of diatoms. However, the flora was rich, consisting of about 90 taxa (Kessel and Pork, 1971; Sauramo, 1958). Among planktonic species, *Melosira islandica helvetica* predominated (up to 45% of the specimens), along with *Stephanodiscus astraea* (3.1%). Benthic and epiphytic diatoms were represented by the freshwater *Opephora marthi, Fragilaria inflata,* and others. At the end of the basin's existence, the marine species *Thalassiosira balthica, Diploneis interrupta, Nitzschia navicularis,* and others appeared. There are no reports of mollusk shells.

The depths in the northern part of the Baltic glacial lake were 100 m greater than present ones and amounted to 550 m or more in the Landshort depression. In the southern part of the lake, however, they were much shallower (by 50 to 60 m) than present ones.

The stratigraphic distribution of diatoms indicates that initially the Baltic glacial lake was completely isolated from the ocean as a fresh and through-flowing lake dammed by ice. Later, because of fluctuations of the glacier margin in central Sweden, the level of the lake twice dropped to the contemporaneous sea level; these changes were associated with a brief inflow of saltwater (Sauramo, 1958). However, on subsequent isolation, the increased salinity was probably preserved only in the deep layers (Järvekülg, 1979).

A tundra vegetation of the Younger Dryas phase existed in the surrounding dry land at that time. The July temperature in Sweden did not rise above 10 °C. The cold climatic conditions and influx of large masses of glacier waters caused very low temperatures in the water masses of the Baltic glacial lake, so that many groups of organisms (e.g., mollusks) were unable to live in them. The Baltic glacial lake basin was highly oligotrophic.

Yoldia Sea: Semimarine Basin

As the climate warmed, the glacier receded and the ocean level rose; a free exchange opened through the Central Sweden Strait, and a semimarine basin of the Yoldia Sea was formed. Its northern coast was bounded at the beginning of the stage by the margin of the Scandinavian ice sheet along the Uppsala-Pori-Oslofjord line. The waters of this basin flooded southern Finland, the coast of the Gulf of Finland, and western Estonia, as well as considerable areas of central and southern Sweden.

Table 23-2. Basins and Coastlines of the Baltic Sea in Estonia (Kessel, 1975)

Horizon	Subhorizon	Basin	Coastline	Correlation with pollen zones of continental organic deposits and ^{14}C age of zones
Holocene (Hl)	Upper (Hl_3)	Limnaea (Lim)		Pine and birch SA 2
			Limnaea V (Lim V)	--1100
				Upper spruce SA 1^a
			Limnaea IV (Lim IV)	--1700
			Limnaea III (Lim III)	Birch and alder SA 1^a
			Limnaea II (Lim II)	--2800
	Middle (Hl_2)			Lower spruce SB 2
			Limnaea I (Lim I)	--3700
				Oak SB I
		Littorina (L)	Littorina IV (L IV)	--4800
			Littorina III (L III)	
			Littorina II^b (L II^b)	Linden AT 2
			Littorina II^a (L II^a)	--6600
			Littorina I	Elm AT 1
			Mastogloian (M)	
		Ancylus (A)	Ancylus III (A III)	--7800
	Lower (Hl_1)			Hazel BO 2
			Ancylus II (A II)	--8100
			Ancylus I (A I)	Pine BO 1
			Echeneisian (E)	--9100
		Yoldia (Y)	Yoldia I (Y I)	Birch PB
				10,200
Pleistocene	Upper	Baltic (B)	Baltic III (B III)	Younger Dryas DR 3
				10,800

The level of the Yoldia Sea was the same as that of the ocean at that time, that is 40 to 50 m below present sea level. Coastal landforms (terraces, scarps, coastal dunes) are located at a height of 40 m above sea level in Tallinn and at a depth of 60 m in the southern Baltic. Deformation of the coastline in central Sweden reached 200 m. Sediments of deeper parts of the sea (Gdansk depression) are represented by varved clays with interlayers of carbonate clays (Kessel et al., 1973).

The fauna and flora also included arctic and boreal-arctic species. Molluscan fauna with *Yoldia islandica, Cerastoderma lamarcki, Mytilus edulus, Hydrobia ulvae, H. ventrosa,* and others is known in central Sweden and the Bornholm depression.

The diatom flora is poorer than in the Baltic glacial lake (72 taxa), with a predominance of marine benthic species: *Diploneis smithii* (35% of the specimens) and the brackish-water *Nitzschia navicularis*, as well as freshwater species. This complex is characteristic of the central portion of the basin, which is affected by saltwater. Near the Gulf of Finland, the Gulf of Riga, and other gulfs, a different complex diatoms is notable, with a predominance of *Melosira baltica, Stephanodiscus astraea,* and other species.

The depth of the sea in the northern part exceeded the present depth by 100 m or more and could have reached 600 m in the Landshort depression. In the southern part, the depths were less than the present ones, and the coastline ran along a 60-m isobath (Kolp, 1974). The salinity in the central part of the Yoldia Sea was 5‰ to 15‰, the maximum salinzation being observed in the Central Sweden Strait (up to 15‰). In the remaining water area, the salinity was much lower and did not exceed 5‰.

The existence of the Yoldia Sea coincides with a marked postglacial warming of the Preboreal (10,500 to 9000 years ago), when birch forests appeared on its shores and the

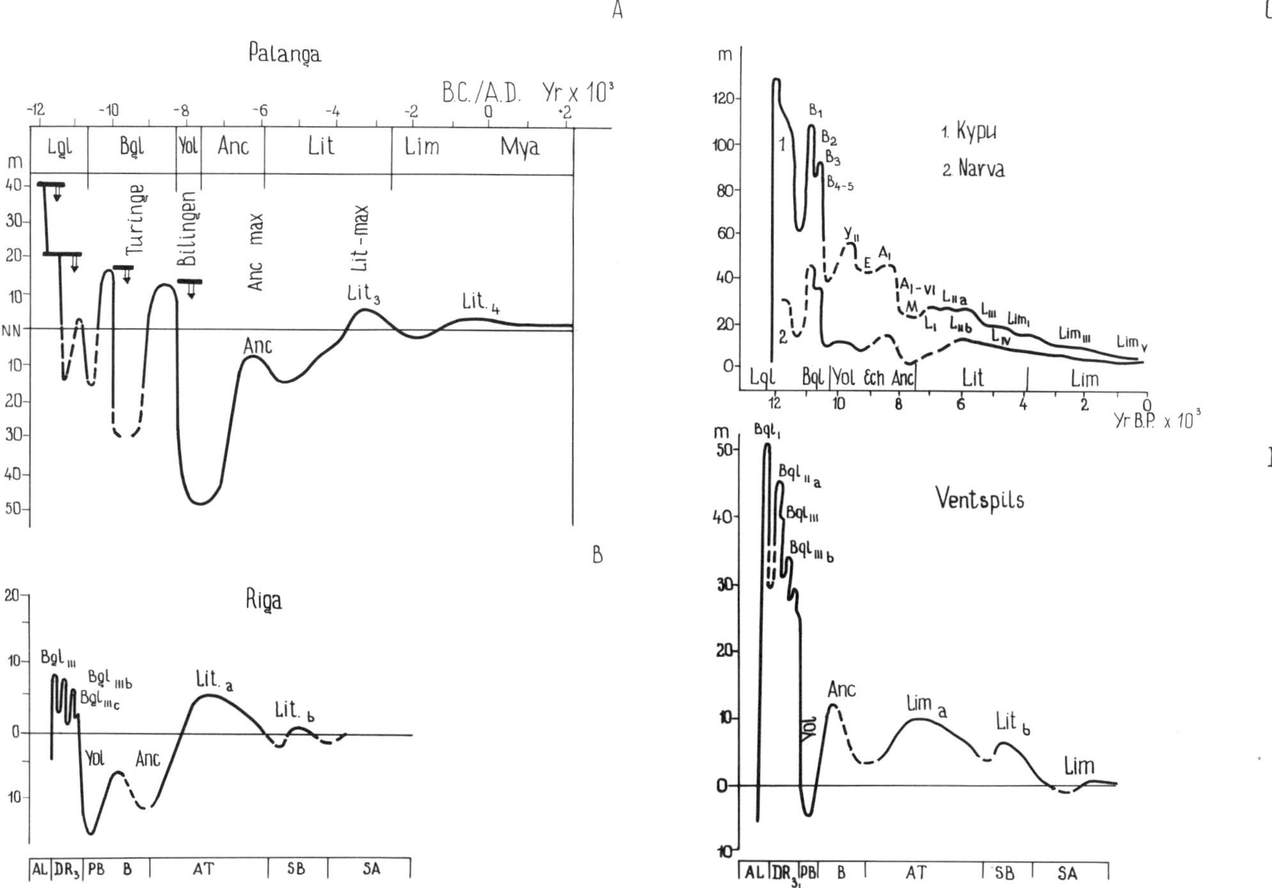

Figure 23-5. Curves of vertical displacement of the coastline within the confines of Estonia, Latvia, and Lithuania. (Based on data by V. K. Gudelis, Kh. Ya. Kessel, and M. A. Veynbergs [Gudelis, 1975].) (Abbreviations: Lgl, local glacial lakes, Bgl, Baltic glacial lake; Yol, Yoldia Sea; Ech, Echeneisian Basin; Anc, Ancylus Sea; Lit, Littorina Sea; Lim, Limnaea Sea; Mya, Mya Sea; Al, Alleröd; DR_3, Younger Dryas; PB, Preboreal; B, Boreal; AT, Atlantic; SB, Subboreal; SA, Sub-Atlantic.

average July temperature in Sweden rose to 16°C. However, the temperature of the water masses remained very low because of the influence of the melting Scandinavian ice sheet. The basin of the Yoldia Sea remained oligotrophic.

Lake Ancylus: A Semifreshwater Basin

About 9300 years ago, as a result of an isostatic uplift of the Earth's crust in central Sweden and a simultaneous lowering of the world ocean level, the two-way connection with the North Sea through the Central Sweden Strait gradually ended. A freshening of the sea began, as reflected in the appearance of the diatom *Campilodiscus echeneis*, typical of the early (Echeneisian) stage of Lake Ancylus. By that time, the ice sheet had almost completely disappeared from Scandinavia, and the littoral deposits of Lake Ancylus extended far from the present shores of the Baltic, particularly in the Gulfs of Bothnia and Finland and in eastern Sweden. Almost all the islands, including the Moosund Archipelago, were under water. In central Sweden, a one-way connection with the North Sea existed through the basin of Lakes Vättern and Vänern. The narrowest part of this connection, the so-called Svea River, about 3 km wide, ran past the foot of Mount Billingen.

The level of Lake Ancylus was 20 m higher than the ocean level, and, since the latter was about 65 to 70 m below present sea level at that time, the strandlines of Lake Ancylus amounted to about 45 to 50 m below sea level.

The coastline of Lake Ancylus is presently located at strandlines that are 90 m above sea level (Degerfors), +50 m (Helsinki), +40 m to +45 m (northern Estonia), and −20 m to −30 m (southern Baltic region).

The molluscan fauna is represented by freshwater and slightly brackish-water species: *Ancylus fluviatilis, Bithynia tentaculata, Lymnaea ovata, Valvata piscinalis, V.*

cristata, Bathiompalus contortus, Hydrobia, and others. A complex with *Unio tumidus* has been observed at the mouths of rivers and freshened portions of the Gulfs of Finland and Riga (Kessel et al., 1973).

The diatom flora is represented by freshwater species with a predominance of *Melosira arenaria* (dominant), *Stephanodiscus astraea*, and others. The low temperatures and the poor fauna and flora kept the lake oligotrophic.

The depth of Lake Ancylus in the northern and central parts of the basin was 50 to 200 m greater than at the present time and reached 530 m in the Landshort depression and more than 400 m (central part) and 300 m (northern part) in the Gulf of Bothnia.

The salinity of the central part of Lake Ancylus was 1‰ to 3‰; it reached 3‰ only in central Sweden; the peripheral parts of the area were freshwater (up to 1‰). It should be noted that, during the short period of the lake's existence (about 500 years), its deep layers were not freshened as much during the existence of this basin; this could have been responsible for the lake's mineralization.

Littorina Sea: Semimarine Basin

The transgression of the Littorina Sea took place under conditions of the postglacial Atlantic interval of maximum warmth, the rise in sea level, the isostatic uplift of central Sweden, and the compensation submergence in the region of the Denmark Straits. This set of factors led to the final closing of the Central Sweden Strait and to the opening of a connection with the North Sea in the south through the Eresunn and Great Belt Straits.

The level of this sea approached the present one, and a maximum rise in level by 2 to 3 m above the present level is noted for the interval of 5000 to 4000 years ago. The littoral deposits of the Littorina Sea are considerably deformed. In the northern part of the Gulf of Bothnia the uplift was 100 m; in northern Estonia, 25 m; in southwestern Estonia, 12 m. The deposits were almost undeformed in Latvia, and in the southern Baltic region the submergence was 13 m below sea level.

The molluscan fauna in the central and southern parts of the Littorina Sea is represented by euryhaline marine species, which penetrated from the North Sea through the Denmark Straits: *Littorina littorea, L. rudis, L. saxatilis, Macoma balthica* L., *Cerastoderma lamarcki* Reeve, *Mytilus edulus* L., *Rissoa inconspicua* (Alder), and others. Brackish-water species inhabited the freshened portions: *Theodoxus fluviatilis* f. *baltica, Hydrobia ventrosa*, and so on.

The diatom flora contains the most saline and oceanic elements, particularly abundant in the southern part of the basin: *Grammatophora oceanica, Terpsinoe americana, Campelodiscus clipaeus,* and *Mastogloia schmidti.* For the first time in the postglacial history of the Baltic region, its basin became eutrophic.

The depths of the Littorina Sea differed insignificantly from the present ones, with the exception of the Gulf of Bothnia, where they exceeded the present ones by 100 m, amounting to 350 m. The salinity of the central part of the basin from the Denmark Straits to Åland Islands and the Gulf of Finland was 6‰ to 10‰ higher than the present one. For the first time, a sharp gradient of decreasing salinity from south to north is noted: from 25‰ in the Denmark Straits to 15‰ in the region of Åland Islands and down to 8‰ in the northern part of the Gulf of Bothnia (Järvekülg, 1979). In the large gulfs, the salinity reached 10‰, and the salinity of deep waters was much higher than at the surface. One can, therefore, postulate a wider development of the zone of hydrogen sulfide contamination in the deep-sea zone of the basin, particularly in the Gotland and other depressions of the central Baltic region and possibly also in the Gulf of Finland.

The climate of the Atlantic optimum was warmer than the present one in central Sweden. The average July temperature rose as high as 18°C. However, the maximum of the Littorina transgression does not coincide with the Atlantic climatic optimum but was shifted to the Subboreal.

Limnaea Sea: Semimarine Basin

This last stage in the history of the Baltic region (the last 4200 years) is associated with a regression, a shoaling of the straits, and a freshening of the basin. The coastline of the basin was already close to the present outline; the level was close to the present one, with slight variations. Thus, a regression is noted during phase III of the Limnaea Sea. A stable two-way connection with the North Sea existed through the Denmark Straits.

Littoral terrace deposits (up to five steps of coastlines in Estonia) occur in Sweden at a height of 0 m to 30 m, in Finland from 2 m to 16 m, in Estonia from 2 m to 13 m, and in the southern Baltic region at the present sea level. On the level of step III (2500 to 2000 years ago), a slight regression of the sea, 2 to 3 m, is observed.

A weakening of the connection with the ocean and a freshening caused significant changes in faunal composition. The most polyhaline marine species were forced out, and of marine mollusks only *Cerastoderma lamarcki* (Reeve), *Mytilus edulus* (L.), and *Macoma baltica* L. remained. Freshwater mollusks of the genus *Lymnaea* (*L. ovata, L. peregra,* and *L. stagnalis*), as well as brackish-water *Theodoxus fluviatilis* (L.), *Hydrobia*, and others, came to replace the marine mollusks. The marine species *Mya arenaria* appeared about 500 years ago.

The complex of diatoms is similar to that of the Littorina complex, but it contains more freshwater species and has no oceanic elements. Thus, in bottom deposits, the content of freshwater species reached 65%, that of brackish-water species was as high as 25%, and that of marine species was 30%. The ecologic conditions of the Limnaea Sea were similar to th epresent ones. The depths remained practically unchanged. The salinity field reflected a distinct gradient of decreasing salinity from south to north, from 15‰ in the Denmark Straits to 2‰ to 3‰ in the upper parts of bays. In the central part, a salinity of less than 12‰ was preserved. In the deep-sea zone, it was possible for the conditions of contamination with hydrogen sulfide to be preserved.

The climatic conditions worsened somewhat in comparison with the optimum, but they were favorable, and the basin remained eutrophic.

The last phase of development of the Limnaea Sea occurred under the influence of anthropogenic action. Initially, it consisted in the introduction of immigrants due to the increase in navigation, for example, the mollusk *Mya arenaria* L. (Hessland, 1946). In recent years, contamination and anthropogenic eutrophication of the waters have been observed.

SEA OF JAPAN: MARGINAL SEA BASIN

In the Middle Holocene, the level of the Sea of Japan approached the present one, and a wide connection with the ocean was restored through the Tatarskiy, Laperouse, Tsugaru, and Vostochno-Tsusimskiy Straits. A branch of the warm Kuroshio Current penetrated far north, up to latitude 43°-44°N (Pletnev, 1979). The temperature of the surface waters was higher than the present one by 2°C to 3°C in the south, and by as much as 3°C to 4°C in the central portion of the sea; it amounted to 13°C to 14°C. The arrival of large masses of warm ocean waters caused an equalization of salinity over almost the entire water area at 30‰ to 35‰, and an appreciable dilution is observed only in the northern part of the Tarazy Strait. For the same reason, the ice content of the sea decreased, and the boundary of floating ice moved to the region of the Tatarskiy Strait up to latitude 49°N, that is, 2° north of the present boundary and 9° to 10° north of the boundary of floating ice 20,000 years ago (Pletnev, 1979).

Conclusions

The above discussion suggests that different types of basins and their modifications reacted differently during the interglacial and glacial epochs.

First, in marginal seas during the glacial epochs, the level dropped by 100 to 130 m; they lost their wide connection with the ocean and were converted into intracontinental semiisolated basins. For example, in the Sea of Japan, only one strait (the West Korea Strait) connects it with the ocean, and the ecologic characteristics were typical of intracontinental basins: a sharp salinity and temperature gradient, starting from the strait northward, and a marked freshening of the peripheral parts of the sea. Marginal shelf seas (Laptev, East Siberian, Chukchi, and Azov) were partially or completely drained during regressive phases.

Second, intracontinental semiisolated basins reacted to climatic fluctuations with sharper changes in level, salinity, faunal composition, and other ecologic indexes. Particularly pronounced changes in conditions were observed in semiisolated basins of the Black Sea type, with shallow connecting straits whose depth was less than the amplitude of glacioeustatic regressions. Over the course of the interglacial-glacial climatic cycle, an alternation of several types of basins took place, namely, marine, semimarine, semifreshwater, and brackish-water basins (Figure 23-1). During the interglacial epochs, transgressions of marine and semimarine basins occurred to a level 8 to 10 m above sea level, and a rise in salinity to 20‰ to 30‰, which was accompanied by the stratification of the water mass and the appearance of a zone of hydrogen sulfide contamination of deep waters.

Third, during the glacial epochs, semiisolated seas regressed, lost their connection with the ocean, were transformed into isolated basins of brackish-water and semifreshwater types, and developed according to climate and water balance, that is, in the manner of closed drainage basins. In regions with a passive water balance, as the level dropped below the threshold in the connecting straits, the basins became completely isolated (Old Euxinian Basin). In the case of an active water balance, the level of such a basin decreased after the drop in ocean level, and a permanent discharge freshened the sea, converting it into a through-flowing basin with a semifreshwater lake (New Euxinian Basin).

Fourth, basins of glacial and periglacial regions are a special type of semiisolated sea, for example, the Baltic and White Seas in Europe and Hudson Bay in North America. Their regime was additionally complicated by isostatic uplifts of deglaciated areas and by compensation downwarping of the periglacial regions. Immediately after the glacier left, the bottom of the sea was 200 to 300 m below the present level, and its depths and area were correspondingly greater than the present depths. Connecting straits disappeared in some places and appeared in others, depending on the isostatic movements of the Earth's crust. Thus, the Central Swedish Strait of the Yoldia Sea closed at the end of the Ancylus phase, and the Denmark Straits opened when the Littorina Sea began. The fluctuations in the level of these seas were superimposed on the isostatic movements and differed markedly in different parts of the basin (Figure 23-5).

Fifth and finally, the regime of isolated inland basins of the Caspian Sea type was independent of fluctuations in ocean level and completely determined by the climate and water balance in the drainage basin and water area. Fluctuations in their level (with an amplitude up to 100 m or more) and their salinity occurred out of phase with the fluctuations of the ocean and associated seas (Figure 23-3). The transgressive phases of the Caspian are confined to the glacial epochs, particularly to their early cryohygrotic phases, when as a result of a marked drop in temperature evaporation from the water area decreased. At the same time, the atmospheric precipitation decreased somewhat, but the discharge volume from the drainage basin increased as a result of a decrease in evaporation and a blocking of groundwaters by permafrost. An example of such a situation is the early Khvalyn transgression in the Early Valdai.

Regressions of isolated basins of the Caspian occurred in the interglacial epochs, particularly in thermoxerotic phases, when a substantial rise in temperature led to an increase in evaporation so great that it was not offset by the increase in atmospheric precipitation. As a result, the level of the basins dropped by tens of meters until a new equilibrium of the water balance was reached.

References

Andreyeva, S. M. (1980). The North Siberian Lowland in Karginskiy time: Paleogeography, radiocarbon chronology. *In* "Chronology of the Quaternary" (J. K. Ivanova and N. V. Kind, eds.), pp. 183-90. Nauka Press, Moscow.

Artamonov, V. I. (1976). Late Quaternary regressions of the Caspian Sea based on biostratigraphic and geomorphic studies of the Dagestan Shelf of the central Caspian. Author's abstract of dissertation, Moscow State University, Moscow.

Chebotareva, N. A., and Makarycheva, I. A. (1974). "The Last Ice Sheet." Nauka Press, Moscow.

Chepalyga, A. L. (1980). Paleogeography and paleoecology of the Black and Caspian Seas (Ponto-Caspian) in the Pliopleistocene. Author's abstract of dissertation, USSR Academy of Sciences, Institute of Geography, Moscow.

Degens, E. T., and Ross, D. A. (1972). Chronology of the Black Sea over the last 25,000 years. *Chemical Geology* 10, 1-16.

Degens, E. T., and Ross, D. A. (1974). "The Black Sea: Geology, Chemistry, and Biology." American Association of Petroleum Geologists, Tulsa, Oklahoma.

Fedorov, P. V. (1978). "The Pleistocene of the Ponto-Caspian." Nauka Press, Moscow.

Gudelis, V. K. (1975). Some pressing problems of the chronostratigraphy and paleogeography of the Baltic Sea. *In* "Information Bulletin 3" (V. K. Gudelis, ed.), pp. 25-35. Coordination Center of Member Countries of the Council for Economic and Mutual Assistance, Moscow.

Gudina, V. I. (1976). "Foraminifers, Stratigraphy, and Paleozoogeography of the Marine Pleistocene of the North of the USSR." Nauka Press, Novosibirsk.

Hessland, J. (1946). On the Quaternary Mya period in Europe. *Arkiv für Zoologi* 37A (8).

Il'ina, L. B. (1965). "History of Gastropods in the Black Sea." Nauka Press, Moscow.

Järvekülg, F. A. (1979). "The Bottom Fauna of the Eastern Part of the Baltic Sea." Valgus Press, Tallinn.

Jousé, A. P., and Mukhina, V. V. (1980). Stratigraphy of Late Cenozoic deposits based on diatoms. *In* "Geological History of the Black Sea Based on Results of Deep-Sea Drilling" (Yu. P. Neprochnov, ed.), pp. 52-64. Nauka Press, Moscow.

Kalinin, G. P., Klige, R. K., Leont'yev, O. K., and Shleynikov, V. A. (1976). Analysis of the change in the level of the Caspian Sea as one of the indices of global water exchange. *In* "Problems of Paleohydrology." Nauka Press, Moscow.

Kessel, H. J. (1975). A brief survey of the stratigraphy of Baltic Sea deposits in Estonia. "Information Bulletin 3" (V. K. Gudelis, ed.). Coordination Center of Member Countries of the Council for Economic and Mutual Assistance, Moscow.

Kessel, H. J., Davydova, N. N., and Blashchizhin, A. I. (1973). Pollen and diatoms from deep-sea depressions of the Baltic. *Estonian Academy of Sciences, Izvestiya, Chemistry and Geology* 4.

Kessel, H. J., and Pork, M. I. (1971). Stratigraphy of bottom sediments of the Baltic within the confines of Estonia. *In* "Palynological Studies in the Baltic Region" (T. D. Bartosh, ed.), pp. 93-110. Zinatne Press, Riga.

Kolp, O. (1974). Submarine Uferterrassen in der südlichen Ost und Nordsee als Marken eines stufenweise erforlgten Holozänen Meeres—ansfiggers. *Baltika* 5. Vil'nyus.

Koreneva, E. V. (1980). Palynological studies of Late Cenozoic deposits. *In* "Geological History of the Black Sea from Results of Deep-Sea Drilling" (Yu. P. Nepruchnov, ed.), pp. 65-70. Nauka Press, Moscow.

Kvasov, D. D. (1975). "Late Quaternary History of Major Lakes and Inland Seas of Eastern Europe." Nauka Press, Leningrad.

Lebedev, L. I., and Glazunova, K. N. (1972). Stratigraphy and faunistic complexes of Late Quaternary deposits of the eastern shelf of the central Caspian Sea. *Bulletin of the Commission on the Study of the Quaternary* 39, 65-75.

Mayev, Ye. G., Lebedev, L. I., and Artamonov, V. I. (1977). Certain features of the paleogeography of the Caspian Sea in the Upper Quaternary based on data from a geologic-stratigraphic study of sediments. *In* "Paleogeography and Deposits of the Pleistocene of Southern Seas of the USSR" (P. A. Kaplin and F. A. Shcherbakov, eds.), pp. 78-84. Nauka Press, Moscow.

Neprochnov, Yu. P. (ed.) (1980). "Geological History of the Black Sea from Results of Deep-Sea Drilling," pp. 52-77. Nauka Press, Moscow.

Nevesskaya, L. A. (1965). "Late Quaternary Bivalve Mollusks of the Black Sea: Their Systematics and Ecology." Nauka Press, Moscow.

Nevesskaya, L. A. (1970). Contribution to the classification of ancient closed and semiclosed bodies of water on the basis of the character of their fauna. *In* "Modern Problems in Paleontology" (D. V. Obruchev and V. N. Shimansky, eds.), pp. 258-78. Nauka Press, Moscow.

Nevesskiy, Y. N. (1967). "Sedimentation Processes in the Littoral Zone of the Sea." Nauka Press, Moscow.

Nikolayev, S. D., and Popov, S. V. (1973). Use of the oxygen isotope method in the study of closed and semiclosed basins (of the type of the Black and Caspian Seas). *In* "Paleogeography and Deposits of the Pleistocene of the Southern Seas of the USSR" (P. A. Kaplin and F. A. Shcherbakov, eds.). Nauka Press, Moscow.

Ostrovskiy, A. B., Izmaylov, Ya. A., Balabanov, I. P., Skiba, S. I., Skryabina, N. G., Arslanov, S. A., Gey, N. A., and Suprunova, N. I. (1977). New data on the paleohydrological regime of the Black Sea in the Upper Pleistocene and Holocene. *In* "Paleogeography and Deposits of the Pleistocene of the Southern Seas of the USSR" (P. A. Kaplin and I. A. Shcherbakov, eds.), pp. 131-41. Nauka Press, Moscow.

Ostrovskiy, A. B., Izmaylov, Ya. A., Shcheglov, A. P., Arslanov, Kh. A., Tretichnyy, N. I., Gey, N. A., Piotrovskaya, T. Yu., Muratov, V. M., Shchelinskiy, V. Ye., Balabanov, I. P., and Skiba, S. I. (1977). New data on the stratigraphy and geochronology of Pleistocene marine terraces of the Black Sea coast, Caucasus, and Kerch-Taman region. *In* "Paleogeography and Deposits of the Pleistocene of the Southern Seas of the USSR" (P. A. Kaplin and I. A. Shcherbakov, eds.), pp. 61-69. Nauka Press, Moscow.

Pazyuk, L. I., Rychkovskaya, N. I., Samsonov, A. I., Tkachenko, G. G., and Yatsko, I. Ya. (1974). History of the northwestern margin of the Black Sea in light of new data on the stratigraphy and lithology of Plio-Pleistocene bottom rocks in the area of Karkinitskiy Bay. *Baltika* 5, 86-92. Vil'nyus.

Pletnev, S. P. (1979). Dissection of bottom sediments and paleogeographic stages of development of the Sea of Japan in the Late Pleistocene-Holocene (based on planktonic foraminifers). Author's abstract of dissertation, Moscow State University, Moscow.

Popov, G. I., and Zubakov, V. A. (1975). Age of the Surozh transgression of the Black Sea region. *In* "Fluctuations in the level of the World Ocean in the Pleistocene,' pp. 115-28. Proceedings of the Twenty-Third Session of the International Geographical Congress, Leningrad.

Rychagov, G. I. (1977). The Pleistocene history of the Caspian Sea. Author's abstract of dissertation, Moscow State University Press, Moscow.

Sauramo, M. (1958). Die Geschichte der Ostsee. *Annales Academiae Scientiarum Fennicae A.3: Geologica-geographica* 51.

Shcherbakov, F. A., Kuprin, P. N., Potapova, L. I., Polyakov, A. S., Zabelina, E. K., and Sorokin, V. M. (1978). "Sedimentation on the Continental Shelf of the Black Sea." Nauka Press, Moscow.

Shilik, K. K. (1977). Changes in the level of the Black Sea in the Late Holocene, and paleotopography of archeological remains of the northern Black Sea region of antiquity. *In* "Paleogeography and Deposits of the Pleistocene of Southern Seas of the USSR" (P. A. Kaplin and I. A. Shcherbakov, eds.), pp. 158-63. Nauka Press, Moscow.

Shnyukov, Ye. F., Orlovskiy, G. N., Usenko, V. P., et al. (1974). "Geology of the Sea of Azov." Naukova Dumka Press, Kiev.

Shnyukov, Ye. F., and Trashchuk, N. N. (1976). A new region of occurrence of Karangatian deposits on the southeastern slope of the Kerch Peninsula. *Ukrainian Academy of Sciences, Doklady B* 12, 1078-80.

Shumenko, S. I., and Ushakova, M. G. (1980). Limestone nannofossils in deep-sea drill cores. *In* "Geological History of the Black Sea Based on Results of Deep-Sea Drilling" (Yu. P. Neprochnov, ed.), pp. 71-72. Nauka Press, Moscow.

Svitoch, A. A. (1976). Salinity of the Caspian Sea in the Pleistocene (based on data from a study of the mollusk *Didacna* Eichwald). *In* "Problems in Paleohydrology," pp. 223-28. Nauka Press, Moscow.

Trashchuk, N. N. (1974). "Marine Pleistocene Sediments of the Black Sea Region in the Ukrainian SSR," pp. 699-701. Naukova Dumka Press, Kiev.

Trashchuk, N. N., and Boltivets, V. A. (1978). A new area of occurrence of Karangatian deposits on the NW coast of the Black Sea. *Ukrainian Academy of Sciences, Doklady B* 8.

Troistkiy, S. L. (1979). "The Marine Pleistocene of Siberian Plains." Nauka Press, Novosibirisk.

Tsereteli, D. V. (1966). "Pleistocene Deposits of Georgia." Mitsniyereba Press, Tbilisi.

Usher, J. L., and Sepko, P. (eds.) (1978). "Initial Reports of the Deep-Sea Drilling Project," Vol. 42 (2). National Science Foundation, Washington, D.C.

Varushchenko, A. I., Varushchenko, S. I., and Klige, R. K. (1980). Change in the level of the Caspian Sea in the Late Pleistocene-Holocene. *In* "Variations in the Moisture Supply of the Aral-Caspian Region in the Holocene" (B. A. Andrianov, L. V. Zorin, and R. V. Nikolayeva, eds.), pp. 74-90. Nauka Press, Moscow.

Velichko, A. A. (1973). "The Natural Process in the Pleistocene." Nauka Press, Moscow.

Zubakov, V. A. (ed.) (1974). "Geochronology of the USSR," Vol. 30, pp. 111-24, 134-45. Nedra Press, Leningrad.

Paleoclimatic Reconstructions

CHAPTER 24

Methods and Results of Late Pleistocene Paleoclimatic Reconstructions

V. P. Grichuk, Ye. Ye. Gurtovaya, E. M. Zelikson, and O. K. Borisova

As was shown by V. P. Grichuk in the chapter on vegetation (Chapter 17), three types of flora occur in the USSR: migration, orthoselection, and relict. The formation and evolution of these floras took place differently, and their reactions to environmental change were also dissimilar. Therefore, it is necessary to use a different approach to the paleoclimatic interpretation of fossil-plant data for each type of flora in question.

The method of reconstructing climate from fossil-plant data was developed by Grichuk (1969, 1973), who used concepts from Szafer (1946) and Iversen (1944). This method consists of determining the limiting climatic values that allow for the existence of all the species of plants determined in the composition of a given fossil flora. These values are attained either by identifying a region where all the species grow at the present time or by determining the climatic ranges of each species and combining them to establish common climatic fields. It is important to consider only those plants that really existed simultaneously in an area, so only those pollen spectra from a single section or from rigorously correlated sections are used.

Migration floras are richer in number of species than orthoselection floras and are ecologically more diverse. Ecologic differences among species can usually be determined through pollen analysis. The climatic indexes that define an entire group, therefore, are usually narrow, and the region where they all grow at the present time is comparatively small.

As an example of paleoclimatic reconstructions based on the analysis of fossil migration floras, we cite data from a section near the village of Posudichi (Figure 24-1, point 20) (Gurtovaya and Faustova, 1977). The following species were identified in two samples from the thermoxerotic to the thermohygrotic phase of the Mikulino Interglaciation: *Alnus glutinosa* (L.), Gaertn., *A. incana* (L.) Moench., *Atriplex patula* L., *Betula pubescens* Ehrh., *B. pendula* Roth., *Botrychium lunaria* (L.) Sw., *Carpinus betulus* L., *Chenopodium album* L., *Corylus avellana* L., *Lycopodium selago* L., *Nuphar luteum* (L.) Smith, *N. pumilum* (Hoffm.) D.C., *Nymphaea alba* L., *Picea abies* Karst., *Pinus silvestris* L., *Polygonum amphibium* L., *Quercus petraea* (Matt.) Liebl., *Q. robur* L., *Salvinia natans* (L.) All., *Tilia cordata* Mill., *T. platyphyllos* Scop., *Typha angustifolia* L., *T. latifolia* L., *T. minima* Funk., and *Ulmus campestris* L.

It is apparent that this list reflects the species of zonal plant formations. The region currently inhabited by the listed species, determined cartographically by superimposing their ranges (Grichuk, 1978), lies in the Sudetes, between the headwaters of the Oder and Elbe Rivers (Figure 24-2). Because climate is a key factor restricting the range of plants, the climatic indexes at the time when these species grew together in the Desna Basin are identical to the present indexes of the region.

Another method of paleoclimatic reconstruction consists of adding together the present climatic ranges (climagrams of taxa found in the fossil assemblage (Iversen, 1944). Superimposed climagrams of seven selected species from the Posudichi section are used to define the minimum limits of temperature for the complex as a whole (Figure 24-3). The climagrams show that simultaneous growth of the complex is possible at mean July temperatures of 19°C or 20°C and mean January temperatures between 0°C and −4°C. The maximum January temperature is determined by spruce, but all other values are defined by two or three species with similar climatic ranges. Total annual precipitation and duration of the frost-free period are determined in the same manner.

Climatic reconstructions obtained from other areas in the European USSR by the two methods are very similar (Table 24-1).

Paleoclimatic reconstructions can be determined by orthoselection floras that extend from Siberia to the plains of Central Asia. This range lacks major refugia, where plants could have survived glacial epochs, so species that

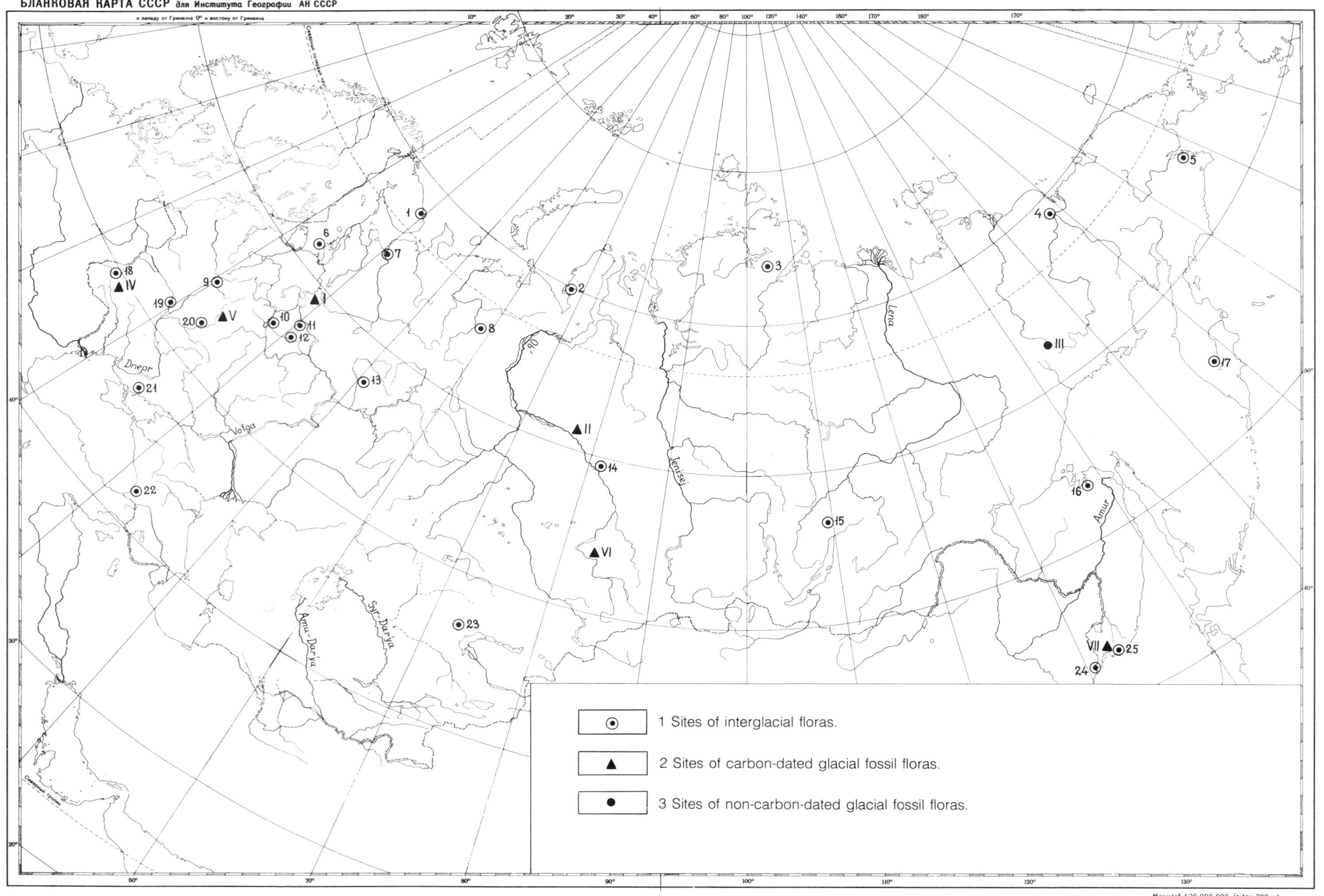
БЛАНКОВАЯ КАРТА СССР для Института Географии АН СССР
Lena
Jenisej
Ob'
Volga
Dnepr
Amur
Syr-Dar'ya
Amu-Dar'ya
1 Sites of interglacial floras.
2 Sites of carbon-dated glacial fossil floras.
3 Sites of non-carbon-dated glacial fossil floras.
Масштаб 1:20 000 000 (в 1см 200км)

disappeared during extreme climatic conditions never reappeared.

Migration undoubtedly occurred in areas dominated by orthoselection floras, especially in transitional regions such as western Siberia, but it did not play an important role in the development of the flora. Orthoselection floras were generally more depleted in the Pleistocene than migration floras, and as a result they consisted of species with fairly broad ecologic amplitudes. These species were able to survive despite different climatic conditions. Orthoselection floras reacted to climatic change primarily by changing species abundance and, to a lesser degree, species composition. In addition, the Pleistocene climatic changes themselves were much less severe in the range of orthoselection floras than in the range of migration floras. Paleofloristic information from regions of orthoselection floras is in no way comparable to the volume of information available on the history of European migration floras. It is necessary to use indicator species that have a narrow ecological amplitude and a distinct relationship to climate (Iversen, 1944).

A section from an Ob' River terrace near the village of Kozyulino and the mouth of the Tomi River is used as an example (Figure 24-1, point 14) (Arkhipov, Votakh, and Levina, 1973). The interglacial Kazantsevo deposits are overlain by Late Pleistocene lacustrine sediments dated at 44,990 ± 2100 yr B.P. (SOAN-335). The climatic optimum of the Kazantsevo Interglaciation contained eight pollen taxa: *Abies sibirica, Alnus, Osmunda, Picea, Pinus sibirica, P. silvestris, Quercus*, and *Ulmus*. Obviously, this limited complex does not represent the entire interglacial flora, and it includes taxa of only generic rank. However, this complex can be used for paleoclimatic reconstructions, because the broad-leaved species, such as *Quercus robur* and *Ulmus scabra*, were growing at the limits of their ranges and hence had the greatest climatic tolerances. These species now grow along with *Abies* and *Pinus sibirica* in western Siberia.

Climagrams for temperature, precipitation, and frost-free period (Figure 24-4) show that for the climatic optimum of the Kozyulino section simultaneous growth of tree species can occur only in a fairly narrow range of values. Similar values are now observed where these species occur together in the mountains of the southern Urals.

Thus, superposition of the climatic ranges of indicator species permits characterization of the climate at a time when these species grew together. The use of a small number of indicator species results in less precise reconstructions, because a certain amount of randomness may be possible in the assemblage and because only fairly wide climatic parameters can be determined. In fact, although conditions most unfavorable for the complex can be delimited, the climatic limits may be narrower.

Certain complications also arise in reconstructing past climate from relict floras. Of all the types of Quaternary floras, relict floras changed the least as a result of their confinement to regions characterized by little climatic change. Hultén (1937) defined as "rigid" those plants that have a depleted set of ecotypes despite their occurrence in small populations, favorable habitats, and small areas. Because

Figure 24-1. Map showing locations of fossil-plant sites for which the paleoclimatic reconstructions were made.

Sites with interglacial floras:
1. Cape Svyatoy (Nikonov and Vostrukhina, 1964)
2. Marre-Sale (Gurtovaya and Troitskiy, 1968)
3. Bol shaya Rassokha (Nikol'skaya, 1980)
4. Berelekh River (Giterman, 1963)
5. Andyrskiy liman (Muratova, 1973)
6. Petrozavodsk (Devyatova and Starova, 1970)
7. Arkhangel'sk (Pleshivtseva, 1972)
8. Bol'shoy Aranets (Ariskina and Zakirova, 1963)
9. Boyarshchina (Chebotareva, 1954); additional data by V. P. Grichuk)
10. Il'inskoye (data by E. M. Zelikson)
11. Levina Mountain (Gorlova, 1968)
12. Yakimanka (Metel'tseva and Sukachev, 1961)
13. Suvod' (Ivanova, 1973)
14. Kozyulino (Arkhipov, Votakh, and Levina, 1973)
15. Chaya River (Rindzyunskaya and Pakhomov, 1977)
16. Nikolay Bay (Karevskaya et al., 1981)
17. Krutoy Ravine (Skiba, 1975)
18. Kolodiyev (data by Ye. Ye. Gurtovaya)
19. Semikhody (Zerov, 1946)
20. Posudichi (Gurtovaya and Faustova, 1977)
21. Mironovka (Artyushenko, 1970)
22. Lia (Mamatsashvili, 1975)
23. Kaktas (Chupina, 1978)
24. Tumangan (Alekseyev and Golubeva, 1980)
25. Vostok Bay (Korotkiy et al., 1980)

Sites with glacial floras (radiocarbon-dated except for III):
I. Puchka River (data of V. P. Grichuk, E. M. Zelikson, and M. Kh. Monoszon)
II. Mega (Levina, 1979)
III. Khudzhakh River (Grichuk et al., 1975)
IV. Boyanichi (Gurtovaya, 1981)
V. Khotylevo (Zelikson and Monoszon, 1974)
VI. Kargopolovo (Panychev, 1979)
VII. Chernaya River (Korotkiy et al., 1980)

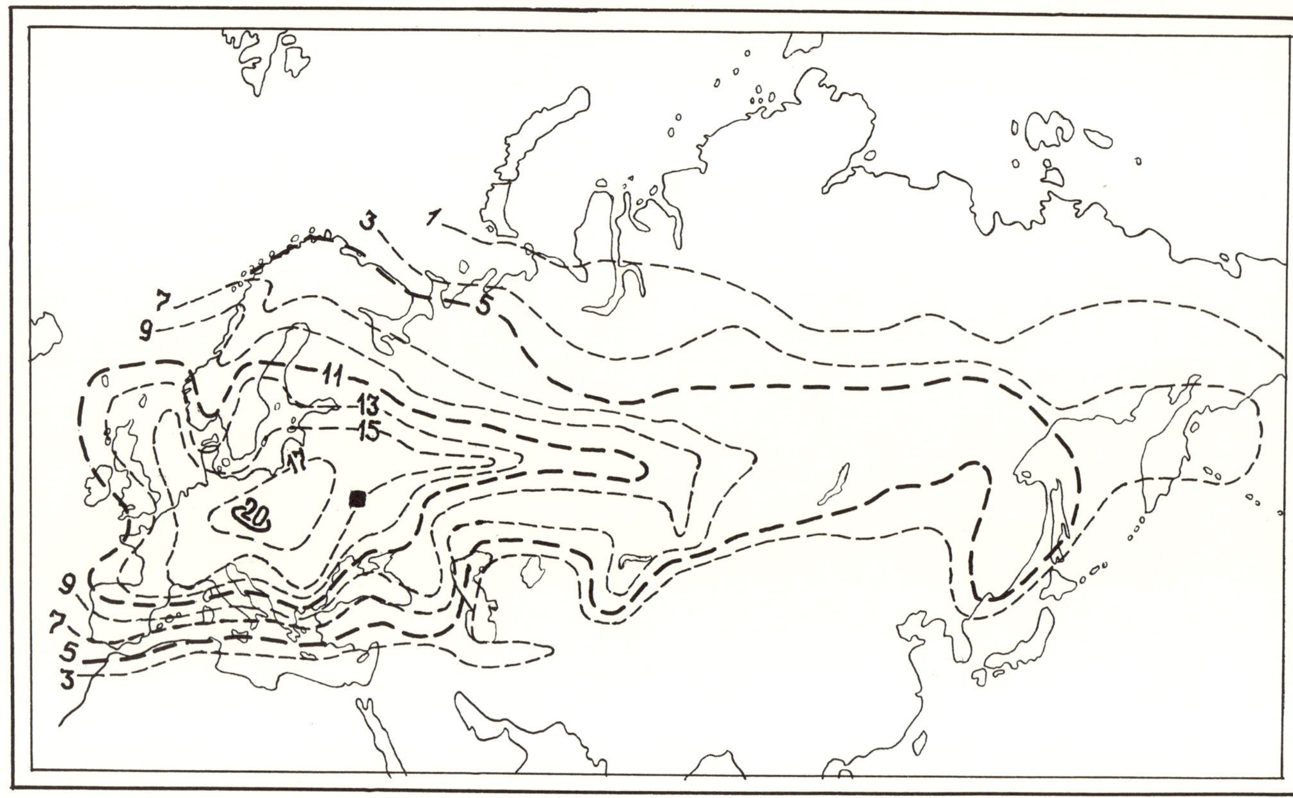

Figure 24-2. Location of the center of present-day concentration of species of the Posudichi fossil flora. (The square denotes the location of the section; numbers indicate the number of species.)

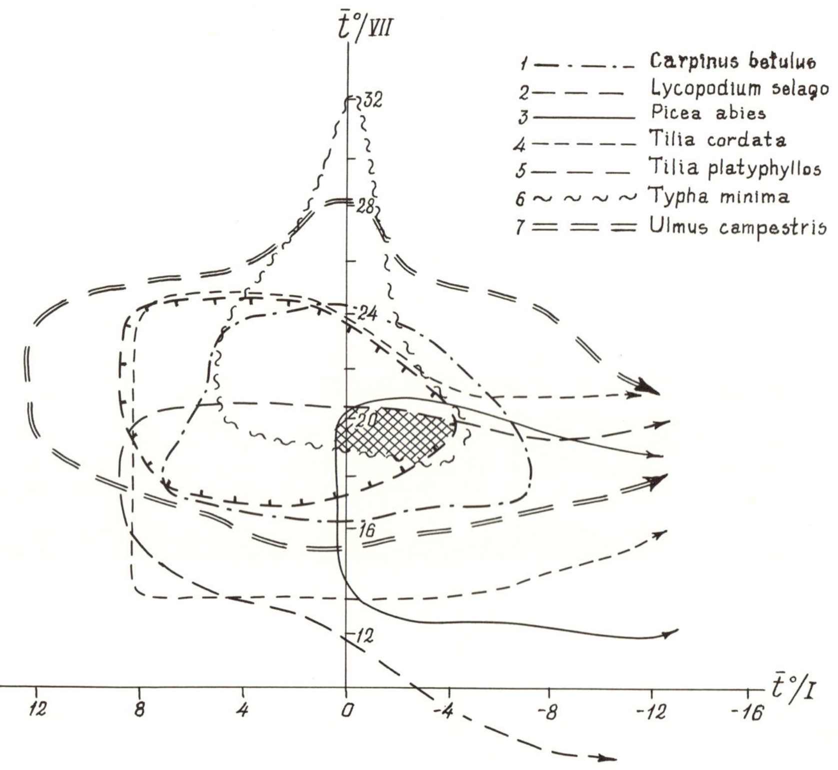

Figure 24-3. Climagram of species whose pollen was identified in the Posudichi section. Region of overlap is shown by shaded pattern. Axes are for January (I) and July (VII) mean temperatures (°C).

Table 24-1.
Reconstructed Values of Principal Climatic Indexes of the Optimum Interval of the Mikulino (Kazantsevo) Interglaciation

Sites of Fossil Floras	January Temperature (°C)		July Temperature (°C)		Annual Temperature Amplitude (C°)		Total Annual Precipitation (mm)		Duration of Frost-Free Period (days)	
	Reconstructed Values	Deviations from Present Values	Reconstructed Values	Deviations from Present Values	Reconstructed Values	Deviations from Present Values	Reconstructed Values	Deviations from Present Values	Reconstructed Values	Deviations from Present Values
1. Svyatoy nos	− 7	+ 2	+ 9	+1	16	− 1	750	+100	110	0
2. Marre-Sale	−21	+ 3	+14	+8	35	+ 5	450	+200	75	+30
3. Bol'shaya Rassokha	−21	+13	+14	+8	35	− 5	450	+200	75	+25
4. Berelekh River	−26	+12	+14	+2	40	−10	220	+ 40	85	+35
5. Anadyrskiy liman	−25	− 1	+15	+4	40	+ 5	400	0	105	+45
6. Petrozavodsk	− 3	+ 7	+17	0	20	− 7	700	+130	180	+60
7. Arkhangel'sk	−10	+ 3	+16	+1	26	− 2	570	+100	130	+40
8. Bol'shoy Aranets	−18	+ 2	+17	+3	35	+ 2	—	—	—	—
9. Boyarshchina	− 1	+ 6	+18	0	19	− 6	650	0	190	+40
10. Il'inskoye	− 3	+ 8	+19	0	22	− 8	600	+ 30	180	+40
11. Levina Mountain	− 2	+10	+18	0	20	−10	600	0	150	+30
12. Yakimanka	− 1	+11	+19	0	20	−11	570	− 20	150	+30
13. Suvod'	− 4	+11	+19	+1	23	−10	650	+ 50	160	+40
14. Kozyulino	−16	+ 7	+18	0	33	− 7	550	+150	110	0
15. Chaya River	−20	+ 8	+20	+4	40	− 4	400	+ 50	110	+40
16. Nikolay Bay	−18	+ 4	+17	+3	35	− 1	700	+200	110	+20
17. Krutoy Ravine	−15	0	+14	0	29	0	700	0	110	+20
18. Kolodiyev	− 1	+ 3	+20	+1	22	− 2	1000	+300	—	—
19. Semikhody	− 1	+ 5	+18	−1	19	− 6	600	+ 30	180	+20
20. Posudichi	− 3	+ 4	+19	0	22	− 4	650	+ 50	180	+30
21. Mironovka	− 3	+ 1	+22	−2	25	− 3	600	+300	200	0
22. Lia	+ 4	0	+22	0	18	0	2000	0	275	0
23. Kaktas	−16	− 1	+20	−3	36	− 2	300	+100	130	− 5
24. Tumangan	−14	0	+17	0	31	0	—	—	—	—
25. Vostok Bay	−14	0	+17	0	31	0	700	0	195	+15

the ecologic amplitude of such species was probably broader in the past, the reconstructed climatic indexes may be narrower than the actual values. In all cases, the errors and inaccuracies are minimized when a group of species is used in the analysis. For this reason, we tried to obtain as much paleobotanic data as possible to derive the reconstructed climatic indexes presented in Tables 24-1 and 24-2.

The Mikulino (Kazantsevo) Interglaciation

We selected 25 fossil floras with a restricted stratigraphic range and typical pollen spectra (Figure 24-1). Climatic indexes were derived for the warmest, wettest interval of the Mikulino (Kazantsevo) Interglaciation—the transition from the thermoxerotic to the thermohygrotic stage.

In the northern European USSR, Mikulino-age marine sediments that correlate with the Eemian transgression of western Europe (Lavrova, 1961; Yevzerov, 1970) are dated precisely by a boreal and boreal-Lusitanian molluscan fauna containing *Cyprina islandica* L. and *Cardium edule* L. Pollen assemblages from marine units deposited farther west at Svyatoy nos (Figure 24-1, point 1) in the present zone of shrub-tundra show a period of open forest-tundra (Nikonov and Vostrukhina, 1964).

In northwestern Siberia, on the western coast of the Yamal Peninsula, strata 50 to 70 m thick exposed on the third marine terrace above sea level contain *Cyprina islandica* L., *Macoma calcarea* (Chemn), and *Neptunea borealis* Phil., which are diagnostic Kazantsevo mollusks. Near the settlement of Marre-Sale (Figure 24-1, point 2), in the zone of shrub-tundra, interglacial pollen spectra record the development of a narrow northern taiga during the climatic optimum of the interglaciation (Gurtovaya and Troitskiy, 1968).

Kazantsevo deposits are widespread in the Northern Siberian Lowland (Taimyr Peninsula). They reach a thickness of 30 m and contain a relatively shallow-water thermophilous fauna, which includes *Cyprina islandica* L., *Zirphaea crispata* L., and *Neptunea despecta* L. (Troitskiy,

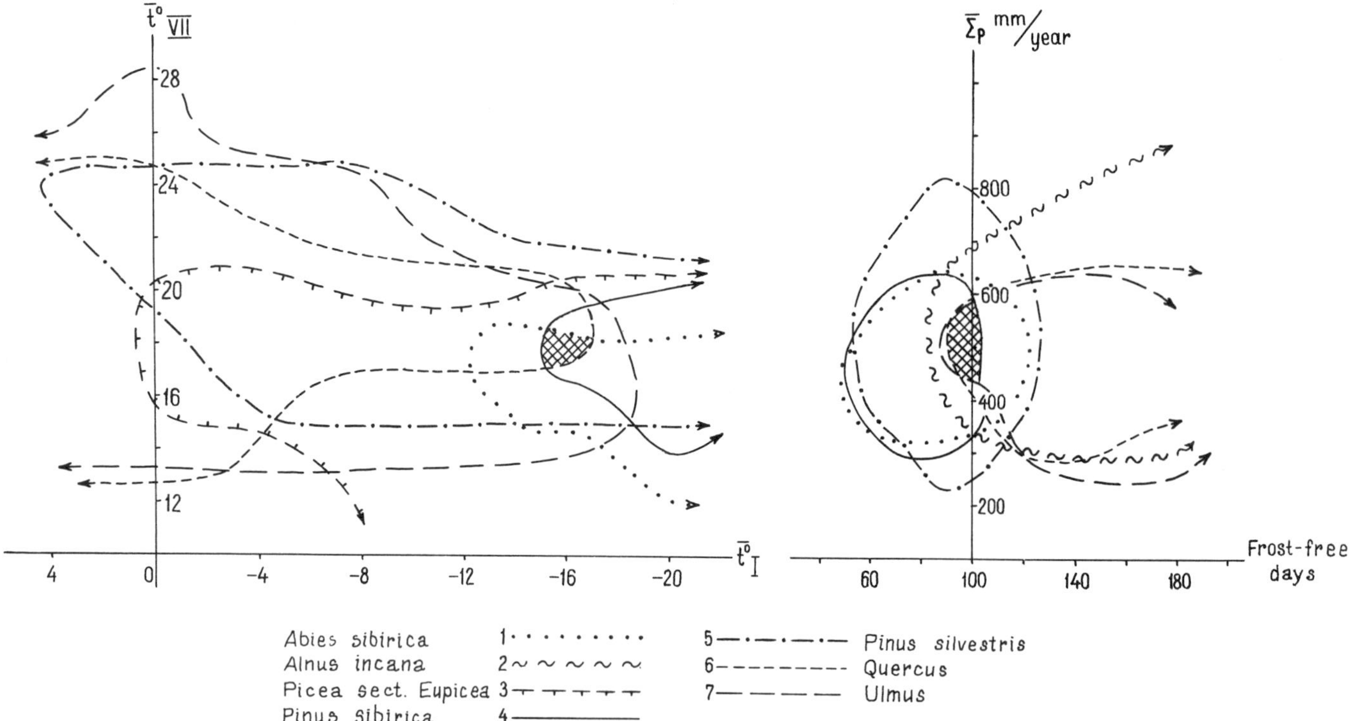

Figure 24-4. Climagrams of species whose pollen was identified in the Kozyulino section. Region of overlap is shown by shaded pattern. Axes on left diagram are for January (I) and July (VII) mean temperatures (°C). Right diagram shows total mean annual precipitation (mm) and number of frost-free days.

1966, 1979). Pollen contained in the marine deposits of the Bol'shaya Rassokha section definitely indicate the presence of northern open forests in an area presently occupied by shrub-tundra (Nikol'skaya, 1980).

Despite compositional differences in the fossil floras from the Marre-Sale and Bol'shaya Rassokha sections, they had similar ecologic-cenotypic components and similar principal climatic indexes (Table 24-1). This similarity contrasts with the significant differences in average January temperatures that now occur on the Yamal and Taimyr Peninsulas (−24°C and −34°C, respectively).

Farther east, in the lower Indigirka River, the alass-lake deposits of the Omukseenskiy section (Lavrushin and Giterman, 1961) date back to the Kazantsevo Interglaciation on the basis of stratigraphic position between deposits of the Vorontsovskiy suite (Taz or Middle Pleistocene glaciation) and deposits containing the remains of a later type of mammoth (Zyryanka Glaciation). Climatic reconstructions were derived from pollen data of an interglacial section from the Berelekh River (Giterman, 1963) (Figure 24-1, point 4), which shows that during the climatic optimum a shrub-forest-tundra of larch and tree birch grew in the region of present-day moss-tundra.

Petrov (1966) assigned to the last interglaciation the marine Val'katlenskiy and alluvial and lacustrine Konnerginskiy deposits on the northern coast of the Chukotskiy Peninsula, which now has a moss and cotton-grass tundra. They are associated with Middle Pleistocene glacial outwash and are overlain by Late Pleistocene glacial and glaciofluvial sediments. Val'katlenskiy deposits are found on the third marine terrace (30 m) on the Chukotskiy Peninsula and correlate with the Sangamon-age Pelukskiy beds in Alaska on the basis of the molluscan fauna (Merklin et al., 1964). Pollen data (Muratova, 1973) indicate a forest-tundra vegetation and a climate similar to those of today.

Mikulino deposits in the non-Arctic European USSR are exposed in river terraces and ancient lake basins. Pollen spectra in the forest zone can be distinguished from other interglacial assemblages (Grichuk, 1961). Climatic reconstructions from the broad-leaved and coniferous/broad-leaved forests of eastern Europe are based on floristic materials from a large number of sections (Figure 24-1, points 9-13, 18-20). In the southern European USSR, the Mironovka section is a buried soil near the Sea of Azov, in the *Artemisia*-fescue-feathergrass formation. Pollen data show that broad-leaved forests of oak, hornbeam, and hazel existed along with steppe in that area (Artyushenko, 1970).

The Lia section, located in the Kolkhida Lowland (Figure 24-1, point 22) characterizes the vegetation of the western Caucasus, which at present is represented by a mixed broad-leaved forest. The age of the sediments is determined by their location on the second terrace of the Inguri River and by palynologic data that correlate with Karangatian deposits, dated by mollusks. Similar to other interglacial sections from the Kolkhida Lowland, the Lia section contains pollen and spores of species that also grow in the region today (Mamatsashvili, 1975), and one cannot postulate any significant differences betwee the interglacial climate and that of today.

In the middle Ob' River basin of western Siberia, allu-

Table 24-2.
Reconstructed Values of Principal Climatic Indexes of the Epoch of the Glacial Maximum

Sites of Fossil Floras	January Temperature (°C)		July Temperature (°C)		Annual Temperature Amplitude (°C)		Total Annual Precipitation (mm)		Duration of Frost-Free Period (days)	
	Reconstructed Values	Deviations from Present Values	Reconstructed Values	Deviations from Present Values	Reconstructed Values	Deviations from Present Values	Reconstructed Values	Deviations from Present Values	Reconstructed Values	Deviations from Present Values
I. Puchka River	−21	− 9	+11	−7	32	+ 2	550	− 30	70	−60
II. Mega	−29	− 5	+17	0	46	+ 5	285	−165	90	−30
III. Khudzhakh River	−35	+ 1	+10	0	45	− 1	350	+125	60	0
IV. Boyanichi	−21	−15	+14	−5	35	+10	350	−250	60	−105
V. Khotylevo	−18	−10	+17	−1	35	+10	—	—	—	—
VI. Kargopolovo	−25	− 6	+18	−1	43	+ 5	350	− 30	95	−25
VIII. Chernaya River	−24	−10	+19	−2	43	+ 8	600	+ 30	135	−50

vial and lacustrine plains developed during Kazantsevo time (Arkhipov, Votakh, and Levina, 1973). The best section is at Kozyulino (Figure 24-1, point 14) in the middle taiga, where flat divides and river valleys are covered by dark fir-spruce-cedar forests dominated by Siberian cedar. Pollen spectra of this section show a period of spruce-fir forests and mixed coniferous forests of spruce, oak, and fir.

Late Pleistocene interglacial deposits in Kazakhstan are confined to the alluvium of the first floodplain terraces. At present, Kazakhstan is an area of *Artemisia*-grass steppe and semidesert, but the Kaktas section from the Ishim River valley (Figure 24-1, point 23) shows a penetration of broad-leaved trees into this region during the last interglaciation. Judging from the present vegetation, steppe was widespread there as well (Chupina, 1978).

In eastern Siberia, in the central North Baikal Upland, Kazantsevo interglacial sediments exposed in the upper part of the third terrace are mainly alluvial-lacustrine in origin. Pollen spectra from a section along the Chaya River (Figure 24-1, point 15) show the development of dark fir-spruce-cedar forests in an area presently occupied by larch taiga (Rindzyunskaya and Pakhomov, 1977). The interglacial climate of southwestern Okhotsk was determined from paleofloristic data from a section near Nikolay Bay (Figure 24-1, point 16) (Karevskaya et al., 1981). On the Kamchatka Peninsula near the Central Kamchatka Depression, Late Pleistocene Kazantsevo interglacial deposits at Krutoy Ravine are underlain by Middle Pleistocene tills and fluvioglacial sediments and are overlain by a thick (up to 20-m) unit containing an Upper Paleolithic mammalian fauna (Kuprina, 1970). The Krutoy Ravine Interglaciation of Kamchatka correlates with the Val'katlenskiy and Konerginskiy Interglaciation of Chukotka (Petrov and Khoreva, 1968; Khoreva, 1974) on the basis of mollusks, foraminifers, and diatoms. Interglacial pollen spectra from the Central Kamchatka Depression, including the type section of Krutoy Ravine (Figure 24-1, point 17), show a wide distribution of spruce, fir, and birch (Skiba, 1975). At present, both spruce and fir survive on the peninsula only in isolated areas.

The Nakhodkinskiy section, correlated with the Kazantsevo Interglaciation, is a deposit with marine, lagoonal, alluvial, and continental facies in the littoral zone of the southern Maritime Territory. Paleobotanic data from the best-studied sections, in particular the sections at Tumangan (Figure 24-1, point 24) (Alekseyev and Golubeva, 1980) and Vostok Bay (Figure 24-1, point 25) (Korotkiy et al., 1980), show a mixed Manchurian-type forest, identical in composition to the present one during the interglacial optimum. Consequently, the climatic indexes for this region are identical to the present ones.

Maximum of Late Pleistocene Glaciation

Few well-dated pollen spectra in the USSR span the Valdai (Sartan, Late Varyanka) glacial maximum of 20,000 to 18,000 yr B.P. Paleoclimatic reconstructions are further complicated in that only fossil floras that are representative of major geographic regions are used in this discussion. The principal climatic elements for the glacial maximum are based on seven fossil floras and encompass the transition from the cryohygrotic to the cryoxerotic stage of glaciation (Table 24-2).

The Puchka section (Figure 24-1, point 1) is located near Lake Kubenskiy, 20 to 30 km north of the Valdai glacial maximum. A layer of plant detritus provided radiocarbon dates of 21,410 ± 150 yr B.P. (LU-18V) and 21,880 ± 110 yr B.P. (LU-18A) (Arslanov et al., 1971). A palynologic study by V. P. Grichuk, M. Kh. Monoszon, and E. M. Zelikson revealed a predominance of herbaceous pollen, as well as pollen of *Betula nana, B. humilis*, pine, spruce, and larch. The pollen composition reflects the end of the cryohygrotic stage and the beginning of the cryoxerotic stage. Spores, pollen, and plant macrofossils (Kolesnikova and Khomutova, 1971) from the layer of vegetative detritus contained 28 species, which at present are concentrated in the polar Urals and the Khamar-Daban Range of Transbaikalia. In both regions, they grow mainly in the montane forest-tundra belt. The average January and July temperatures, the duration of the frost-free period, and the total annual precipitation within this vegetation belt are very similar in the two regions, although they lie at different altitudes.

The Mega section (Figure 24-1, point II) is a set of peat lenses dated at about 21,900 yr B.P. (SOAN-324) and located on a 20-m terrace of the Ob' River. This section is lined with cryoturbated loams dated at 33,100±2300 yr B.P. (MGU SOAN-132) and covered by loams with tree stumps dated at about 10,650 yr B.P. (SOAN-323) (Levina, 1979). Pollen spectra from the peat show a predominance of herbaceous pollen taxa, in particular *Ephedra*, a xerophilic pioneer plant that now grows from south of the steppe zone to the steppe of central of Yakutiya. The arboreal pollen includes tree birch, *Betula* sect. *Nanae*, and an appreciable amount of spruce pollen. Nine taxa have been identified in the spectra. Thus, periglacial forest-steppe developed in this region approximately 20,000 years ago, similar to that growing in northeastern Transbaikalia today. Total annual precipitation was 165 mm less than present values, and the frost-free period was shorter.

The Khudzhakh section (Figure 24-1, point III), located in the Kolyma River basin, has a fossil flora contained in deposits along the 10- to 15-m terrace of the Khudzhakh River (Grichuk et al., 1975). The age has been clearly established from the stratigraphy and the paleoflora, but it is the only flora without radiocarbon dating. Nevertheless, it is included because of the scarcity of detailed paleofloristic data from this region. The 10 species of plants identified all grow in the southern Anyuyskiy Range, a considerable distance northeast of the site. Thus, the January and July temperatures and the duration of the frost-free period were close to the present ones, but the total annual precipitation was much higher.

The Boyanichi section, located in the northern Volynskaya Upland (Figure 24-1, point IV), is a gyttjalike deposit, with a date of 22,500±400 yr B.P. (IGAN-113) (Gurtovaya, 1981). Pollen of the microthermic shrubs *Betula nana, B. humilis,* and *Alnaster fruticosus* dominate, and a total of 16 species was identified. Cold-tolerant species, in particular *Selaginella selaginoides* and *Botrychium boreale*, are especially conspicuous. Most of these species grow in the northern Ural Mountains in the golets and subgolets altitudinal belts at present, and the climatic parameters are nearly identical to those from the Boyanchi section, which were determined from climagrams. Characteristically, July temperatures differed little from present ones, but January temperatures were lower by 15°C.

The Khotylevo Late Paleolithic campsite (Figure 24-1, point V) is located in the middle Desna River basin. It is confined to the marginal part of a plateau and underlain by Middle Pleistocene fluvioglacial sands. The age of the campsite was determined to be 23,660±270 yr B.P. (LU-359) (Arslanov and Kurenkova, 1975). Pollen spectra from the cultural layer have an overwhelming abundance of pollen of grasses and composites (Zelikson and Monoszon, 1974), although 28 plant taxa were identified. The present-day distribution is the Altay. The principal climatic indexes suggest an appreciable lowering of winter temperature compared with the present, but summer temperatures were similar to the present values. The Khotylevo II campsite does not date back to the maximum stage of glaciation but rather to the preceding interval, that is, the transition from the Bryansk Interval to the Late Valdai proper.

The Kargopolovo section (Figure 24-1, point VI) is located in southwestern Siberia, on a 25- to 27-m alluvial terrace of the Ob' River dated at 33,450±550 yr B.P. Pollen analysis by M. R. Votakh (Panychev, 1979), paleofloristic data, and climatic reconstructions suggest that the sandy loam encompasses the transition from the cryohygrotic to the cryoxerotic climatic stage of Sartan Glaciation, that is, the maximum phase. Six species are now found near Lake Baikal. As in regions of western Siberia farther north, the greatest differences between glacial-age and present-day climate are in the total annual precipitation and the duration of the frost-free period.

The section on the Chernaya River of the southeastern Maritime Territory (Figure 24-1, point VII) contains a layer of peat dated at 22,500±240 yr B.P. (SOAN-551) from a 3- to 4-m terrace (Korotkiy et al., 1980). Pollen spectra show mainly boreal types (*Picea* sect. *Eupicea, Larix*) and microtherms (*Betuloa* sect. *Nanae, Alnaster fruticosus*). Nine taxa were identified. Pollen of Tertiary relicts and broad-leaved taxa characteristic of temperate and interglacial intervals were not noted. The spectra suggest less favorable conditions than at present, for today these plants grow much farther north, in the lower Amur River region. The reconstructed climate in the southern Far East differs from present conditions mainly by having lower winter temperatures and a frost-free period a month and a half shorter than today's.

References

Alekseyev, M. N., and Golubeva, L. V. (1980). Stratigraphy and paleogeography of the Upper Pleistocene of the southern Maritime Territory. *Bulletin of the Commission on the Study of the Quaternary* 50, 96-107.

Ariskina, N. P., and Zakirova, N. F. (1963). Micropaleobotanical studies of buried peat bogs from the basin of the B. Aranets River (middle Pechora). *In* "Uchenye Zapiski Kazanskogo Gosudarstvennogo Universiteta 123," Book II, pp. 128-38. Kazan' University, Kazan'.

Arkhipov, S. A., Votakh, M. R., and Levina, T. P. (1973). Palynological characterization of Riss-Würm (Kazantsevo) and lower-middle-Würm deposits of the middle Ob' Valley. *In* "The Pleistocene of Siberia and Adjacent Regions" (V. N. Saks, ed.), pp. 143-50. Nauka Press, Moscow.

Arslanov, Kh. A., and Kurenkova, Ye. I. (1975). Radiocarbon datings of certain Late Paleolithic campsites of the Desna Basin. *Bulletin of the Commission on the Study of the Quaternary* 44, 165-66.

Arslanov, Kh. A., Auslender, V. G., Gromova, V. I., et al. (1971). Paleogeographical characteristics and absolute age of the maximum stage of Valdai Glaciation in the area of Lake Kubenskoye. *USSR Academy of Sciences, Doklady* 195, 1395-98.

Artyushenko, A. T. (1970). "Vegetation of the Forest-Steppe and Steppe of the Ukraine in the Quaternary (Based on Data of Spore-Pollen Analysis)." Naukova Dumka Press, Kiev.

Chebotareva, N. S. (1954). A new section of Dnepr Valdai interglacial deposits on the Kasplya River near the village of Boyarshchina. *In* "Data on Paleogeography" (V. P. Grichuk and K. K. Markov, eds.), Vypad 1, pp. 69-81.

Chupina, L. N. (1978). Palynological characterization of Late Pleistocene deposits of central and southern Kazakhstan. *Kazakhstan Academy of Sciences, Izvestiya, seriya geologicheskaya* 3, 58-63.

Devyatova, E. I., and Starova, N. N. (1970). The Late Quaternary history of the Onega and Ladoga Basins based on data of spore-pollen and diatom analyses. *In* "History of Lakes" (M. V. Kabailene, ed.), pp. 82-100. Lithuanian Geographical Society. Vil'nyus.

Giterman, R. Ye. (1963). Stages in the development of Yakutiya's Quaternary vegetation and their importance in stratigraphy. *USSR Academy of Sciences, Geological Institute, Trudy* 78, 1-193.

Gorlova, R. N. (1968). "Change of Vegetation as a Component of Biocenoses in the Penultimate Interglaciation." Nauka Press, Moscow.

Grichuk, V. P. (1961). Fossil floras as a paleontological basis of the stratigraphy of Quaternary deposits. *In* "Relief and Stratigraphy of Quaternary Deposits of the Northwest of the Russian Plain" (K. K. Markov, ed.), pp. 25-71. USSR Academy of Sciences Press, Moscow.

Grichuk, V. P. (1969). Experience in the reconstruction of certain climatic elements of the Northern Hemisphere. *In* "The Holocene" (M. I. Neustadt, ed.), pp. 41-57. Nauka Press, Moscow.

Grichuk, V. P. (1973). Climatic conditions of the Northern Hemisphere during the Atlantic period of the Holocene. *In* "Thermal Reclamation of the Northern Latitudes" (G. A. Avsyuk, ed.), pp. 107-27. Nauka Press, Moscow.

Grichuk, V. P. (1978). Method of interpretation of paleobotanical data in solving problems of stratigraphy and correlation of the Late Cenozoic. *In* "Palynological Studies in the Northeast of the USSR" (V. P. Grichuk, ed.), pp. 29-42. Far East Research Center, Vladivostok.

Grichuk, M. P., Karevskaya, I. A., Polosukhina, Z. V., Ter-Grigoryan, Ye. P. (1975). "Paleobotanical Substantiation of the Age Correlation of Late Cenozoic Deposits in the Indigirka-Kolyma Region." VINITI Press, Moscow.

Gurtovaya, Ye. Ye. (1981). Reconstruction of the natural conditions of the Bryansk interval of the last glacial epoch for the southwest of the Russian Plain. *USSR Academy of Sciences, Doklady* 257, 1225-28.

Gurtovaya, Ye. Ye., and Faustova, M. A. (1977). On the Mikulino stage of formation of alluvium in the basin of the middle course of the Desna (as exemplified by a section near the village of Posudachi). *USSR Academy of Sciences, Izvestiya, seriya geograficheskaya* 2, 69-75.

Gurtovaya, Ye. Ye., and Troitskiy, S. L. (1968). Palynological characterization of Sangompan deposits of western Yamal. *In* "Neogene and Quaternary Deposits of Western Siberia" (V. N. Saks, ed.), pp. 131-40. Nauka Press, Moscow.

Hultén, E. (1937). "Outline of the History of Arctic and Boreal Biota during the Quaternary Period." Bokförlags Aktiebolaget Thule, Stockholm.

Ivanova, N. G. (1973). Experience with the dating of alluvial deposits of the Vyatka River and reconstructions of vegetation from palynofloristic data. *In* "Palynology of the Pleistocene and Pliocene" (V. P. Grichuk, ed.), pp. 64-69. Nauka Press, Moscow.

Iversen, J. (1944). *Viscum, Hedera,* and *Ilex* as climate indicators. *Geologiska Föreningens i Stockholm Förhandlingar* 66, 463-83.

Karevskaya, I. A., Boyarskaya, T. D., Grichuk, M. P., and Makhova, Yu. V. (1981). Provincial characteristics of Late Pleistocene vegetation in the Amur Basin. *In* "Transactions of the Third Interdepartmental Conference on Palynological Studies in the Far East" (M. P. Grichuk and A. M. Korotkiy, eds.). Far East Geological Institute, Vladivostok.

Khoreva, I. M. (1974). Stratigraphy and foraminifers of marine Quaternary deposits of the west coast of the Bering Sea. *USSR Academy of Sciences, Trudy* 225, 1-152.

Kolesnikova, T. D., and Khomutova, V. I. (1971). New data on the history of the development of vegetation of the Valdai ice age on the territory of Vologda Province. *USSR Academy of Sciences, Doklady* 196, 413-16.

Korotkiy, A. M., Karaulova, L. P., and Troitskaya, T. S. (1980). Quaternary deposits of the Maritime Territory. *USSR Academy of Sciences, Siberian Branch, Institute of Geology and Geophysics, Trudy* 429, 1-234.

Kuprina, N. P. (1970). Stratigraphy and history of sedimentation of Pleistocene deposits of central Kamchatka. *USSR Academy of Sciences, Geological Institute, Trudy* 216, 1-148.

Lavrova, M. A. (1961). Relationship of the interglacial boreal transgression of the northern USSR to the Eemian transgression in western Europe. *Estonian Academy of Sciences, Geological Institute, Trudy* 8, 65-88.

Lavrushin, Yu. A., and Giterman, R. Ye. (1961). Principal stages of the development of vegetation in the lower course of the Indigirka River in the Quaternary. *USSR Academy of Sciences, Doklady* 130, 681-87.

Levina, T. P. (1979). Palynological characterization of ice age deposits in the valley of the middle Ob'. *In* "Stratigraphy and Palynology of Siberia's Mesozoic and Cenozoic." *USSR Academy of Sciences, Siberian Branch, Institute of Geology and Geophysics, Trudy* 396, 74-98.

Mamatsashvili, N. S. (1975). "Palynological Characterization of Quaternary Continental Deposits of Kolkhida." Metsnisreba Press, Tbilisi.

Merklin, R. L., Petrov, O. M., Hopkins, D. M., and McNeil, F. S. (1964). An attempt at correlating Late Cenozoic marine sediments of Chukotka, northeastern Siberia and western Alaska. *USSR Academy of Sciences, Izvestiya, seria geologicheskaya* 10, 45-47.

Metel'tseva, Ye. P., and Sukachev, V. N. (1961). New data on the Pleistocene flora of the central part of the Russian Plain (interglacial peat bog in Vladimir Province). *Bulletin of the Commission on the Study of the Quaternary* 26, 50-60.

Muratova, M. V. (1973). "History of the Development of the Vegetation and Climate of Southeastern Chukotka in the Neogene-Pleistocene." Nauka Press, Moscow.

Nikol'skaya, M. V. (1980). Paleobotanical characterization of the Upper Pleistocene and Holocene deposits of Taimyr. *In* "Paleopalynology of Siberia" (V. N. Saks, ed.), pp. 97-111. Nauka Press, Moscow.

Nikonov, A. A., and Vostrukhina, T. M. (1964). Stratigraphy of the Anthropogene of the northeastern part of the Kola Peninsula. *USSR Academy of Sciences, Doklady* 158, 104-7.

Panychev, V. A. (1979). Radiocarbon chronology of alluvial deposits of the cis-Altay Plain. *USSR Academy of Sciences, Siberian Branch, Institute of Geology and Geophysics, Trudy* 451.

Petrov, O. M. (1966). Stratigraphy and fauna of marine mollusks of Quaternary deposits of the Chukotka Peninsula. *USSR Academy of Sciences, Geological Institute, Trudy* 155.

Petrov, O. M., and Khoreva, I. M. (1968). Correlation of Late Neogene and Quaternary deposits of the extreme northeast of the USSR and Alaska. *In* "Boundary of the Tertiary and Quaternary" (R. L. Merklin, ed.), pp. 1-290. Nauka Press, Moscow.

Pleshivtseva, E. S. (1972). Palynological characterization of a key section of sediments of the boreal transgression in the northwest of Arkhangel'sk Province (region of Severo-Dvinskaya depression). *In* "Palynology of the Pleistocene" (V. P. Grichuk, ed.), pp. 93-104. USSR Academy of Sciences, Institute of Geography, Moscow.

Rindzyunskaya, N. M., and Pakhomov, M. M. (1977). Stratigraphy of Quaternary deposits of the North Baikal Highland. *USSR Academy of Sciences, Izvestiya, seria geologicheskaya* 4, 146-49.

Skiba, L. A. (1975). History of the development of Kamchatka's vegetation in the Late Cenozoic. *USSR Academy of Sciences, Geological Institute, Trudy* 276, 1-72.

Szafer, W. (1946). Pliocene flora from Kroszenek on the Dunajec. *Polska Akademia umiejetnosci, Wydzial matematyczno-przyredniczy* 72 (B).

Troitskiy, S. L. (1966). "Quaternary Deposits and Topography of the Flat Shores of Yenisey Bay and Adjacent Area of Byrranga Mountains." Nauka Press, Moscow.

Troitskiy, S. L. (1979). The marine Pleistocene of the Siberian plains: Stratigraphy. *USSR Academy of Sciences, Siberian Branch, Institute of Geology and Geophysics, Trudy* 430.

Yevzerov, V. Ya. (1970). The question of the age of Kola Peninsula's interglacial deposits. *In* "Proceedings of a Scientific Session of the Kola Branch of the USSR Academy of Sciences, Institute of Geology" (V. G. Zagorodny, ed.), pp. 88-93. USSR Academy of Sciences, Institute of Geology, Moscow.

Zelikson, E. M., and Monoszon, M. Kh. (1974). Conditions of human habitation at the Khotylevo II campsite based on palynological data. *In* "Primitive Man: His Material Culture and Environment in the Pleistocene and Holocene" (I. P. Gerasimov, ed.), pp. 137-43. USSR Academy of Sciences, Institute of Geography, Moscow.

Zerov, D. K. (1946). Fossil peat moss in the vicinity of the village of Semikhody in the lower course of the Pripyat' River. *Ukrainian Academy of Sciences, Botanichesky Zhurnal* 3 (1-2).

CHAPTER 25

Late Pleistocene Spatial Paleoclimatic Reconstructions

A. A. Velichko

Optimum of the Mikulino Interglaciation

As is evident from the preceding chapter, through the use of indicator paleofloristics it is possible to determine past climatic conditions for 25 localities. Despite the relatively small number of "weather stations of the past," the 25 points are representative of their geographic locations and are scattered more or less uniformly across the main regions, except for Central Asia, which has almost no points. Fourteen of the sites are concentrated in the European part of the USSR. The points encompass almost all the present-day climatic zones, including arctic tundra, temperate forest, steppe, subtropics (Lia), and desert (Kaktas). In addition, the points cover regions influenced by diverse air masses. The West is affected primarily by maritime Atlantic air masses, most of Siberia by continental air masses, and the East by alternating maritime and continental air masses. Thus, the climatic characteristics of the past can be compared with present conditions in these regions. Finally, the accuracy of reconstructed temperature and precipitation is established by the relationship between vegetation and climate (Budyko, 1971) and by the rough similarity between interglacial vegetation and that of the present day.

JANUARY TEMPERATURE

In most regions, the temperature of the coldest month differed markedly from that of the present. North of latitude 50°N, temperatures were higher, particularly in the Arctic (Figure 25-1). In the western Arctic along the Barents Sea coast of the Kola Peninsula, the warming was only 2°C above today, and the average monthly temperature was close to −7°C. A temperature rise of 3°C is reconstructed for the Yamal Peninsula. In contrast, a marked warming of the Arctic coast took place in eastern Siberia. On the Taimyr, the temperatures rose by 13°C to a value of −21°C. Farther east, in the lower course of the Indigirka River, a temperature of −26°C was established, whereas at the present time the local January temperature is −38°C. Temperature departures were minimal and possibly negative (toward cooling) in the far-eastern Arctic near the Pacific Ocean.

The warming south to latitude 50°N in the intracontinental regions was less than along the eastern Siberian Arctic coast. In the western European USSR, the January temperatures were 4°C to 6°C higher, close to 0°C. The eastern European part was substantially warmer. On the left bank of the Volga, the temperatures rose by 11°C, to −4°C. Temperatures in central western Siberia were 7°C higher than today, and in southern central Siberia 8°C higher. A marked warming of 10°C occurred at these latitudes even on the Okhotsk coast. Only on Kamchatka were no departures from present noted.

South of latitude 50°N, temperature deviations were slight. In the southern Russian Plain, a slight warming of 1°C is estimated. Still farther south, temperatures were similar to present values in the Black Sea subtropics. In the northern Kazakhstan-Central Asia region, temperatures might have dropped by 1°C, but no deviations from present temperatures have been established for the coastal regions of the Far East.

These are the relations reconstructed in a latitudinal direction. In a meridional direction (Figure 25-1) it is apparent that in the European USSR positive deviations gradually increase from the Arctic to the central part of the plain and then decrease again. A similar trend is noted in western Siberia. However, if one follows the deviation in a submeridional direction (from Taimyr through the Middle Ob' in the Balkhash area), the greatest positive deviations are noted in the North and decline southward with the exception of the coast along the Sea of Okhotsk (Figure 25-2).

On the whole, the pattern of January isotherms for the Mikulino Interglaciation resembles the present one (Figure 25-3), except for the more latitudinal position of the isotherms in the Northwest and the higher isotherm values.

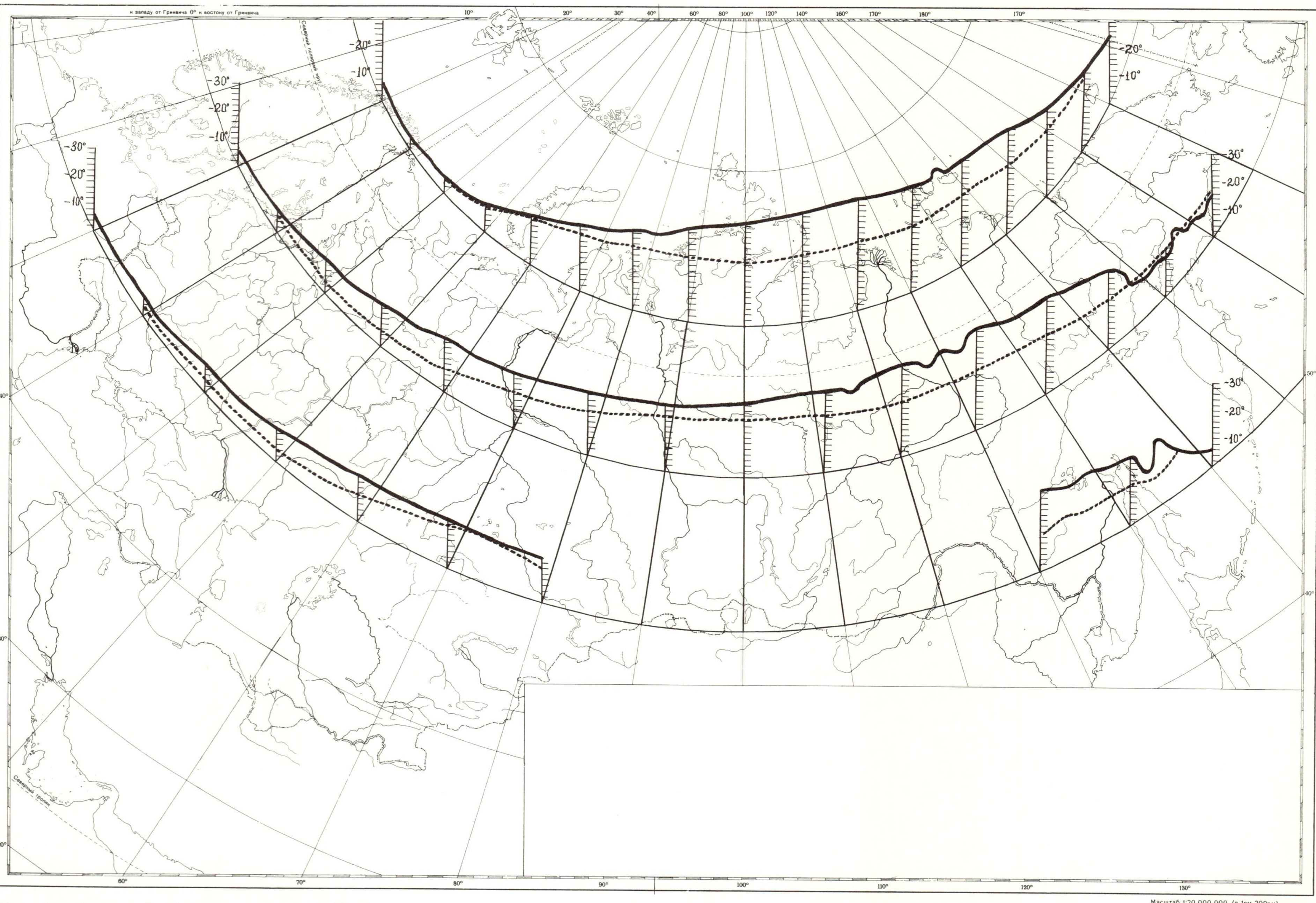

Figure 25-1. January latitudinal temperature (°C) profiles for the present (solid line) and for the Mikulino Interglaciation optimum (dotted line). Note reversed vertical temperature scales.

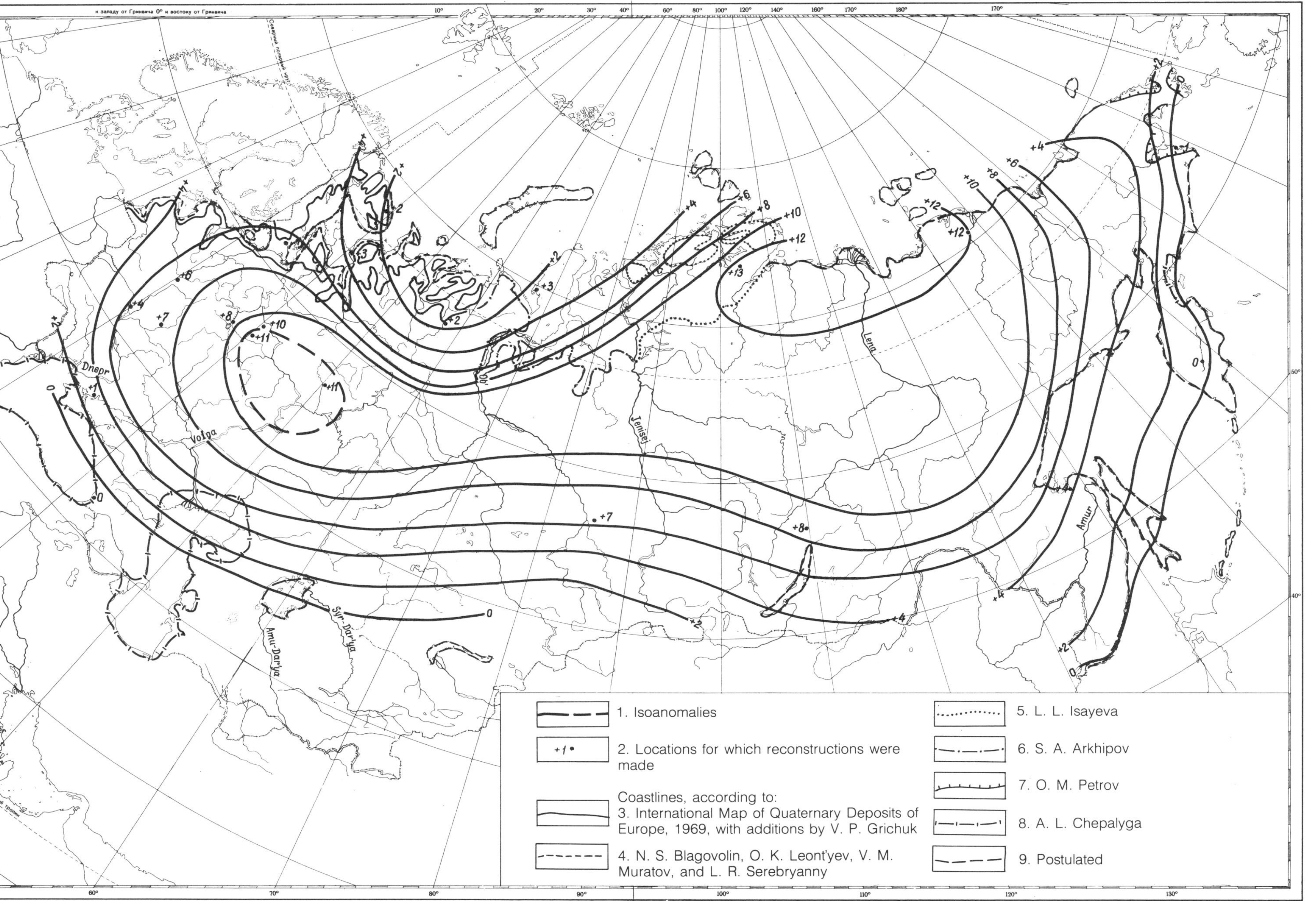

Figure 25-2. January deviations of temperatures (°C) for the Mikulino Interglaciation optimum from their present values.

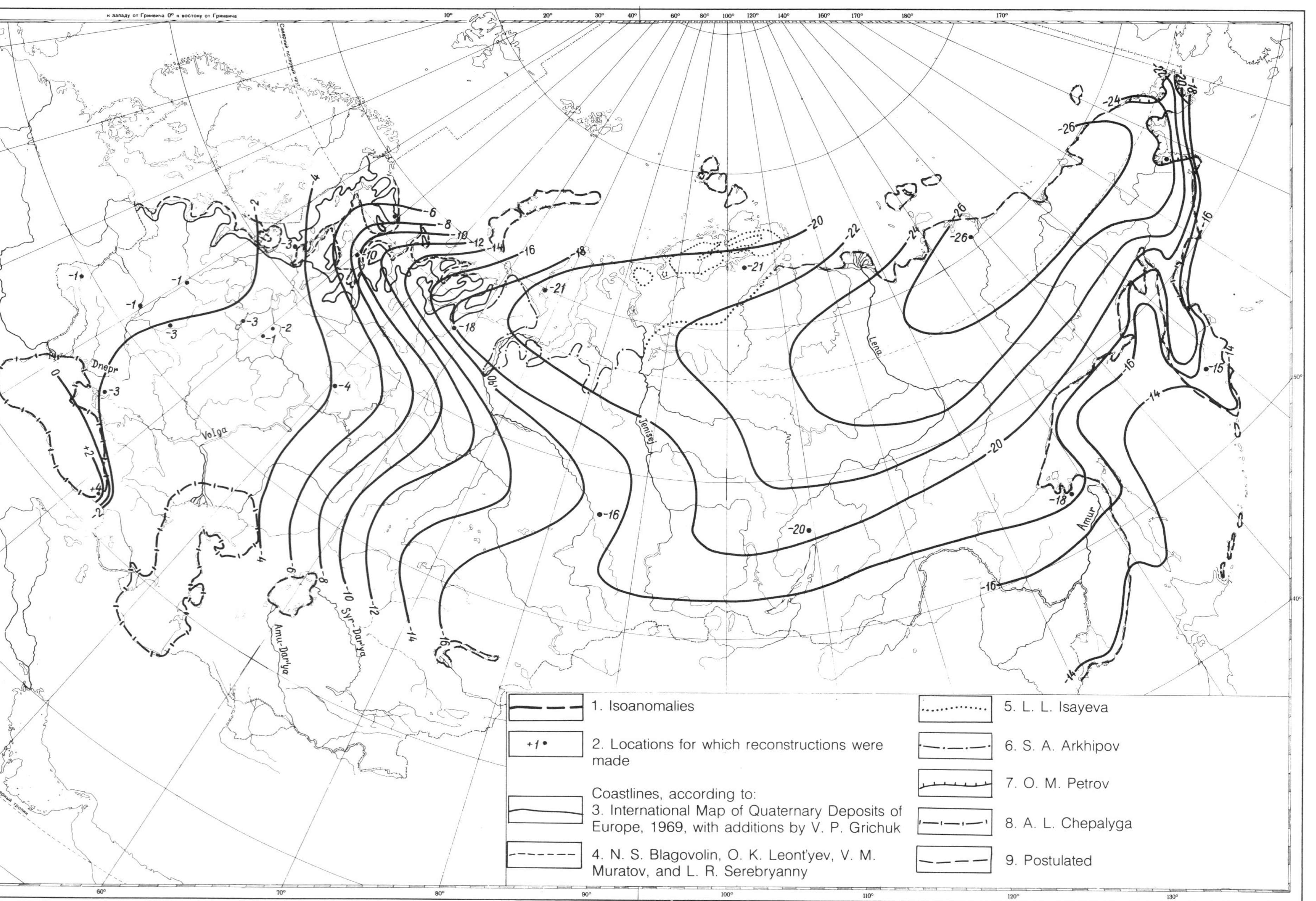

Figure 25-3. January temperatures (°C) for the Mikulino Interglaciation optimum.

JULY TEMPERATURES

July temperatures were more similar to present values than January temperatures are, and their distribution over the country was fairly simple. The greatest positive anomalies are recorded for the northernmost Arctic latitudinal belt (Figures 25-4 and 25-5), and the smallest in the western Barents Sea sector of the Arctic. Increased temperatures mainly affected a part of the Arctic coast east from Yamal to Taimyr, where the temperatures were 8°C above the present ones. Toward the Yana-Indigirka Lowland and Chukotka, the deviations decreased to +2°C and +1°C.

Farther south, July temperatures differed very little from present one. Within a broad band between the Arctic and latitude 50°N, no deviation is recorded in most regions. Only in the eastern European sector was the July temperature 1°C higher. Temperatures were 3°C to 4°C higher in the Baikal region and along the Sea of Okhotsk.

South of latitude 50°N, decreased temperatures occurred in the steppe regions of the European USSR and the steppe and semidesert regions of northern Kazakhstan. Temperatures were −1°C less in the West and −3°C less in Kazakhstan. No changes in temperature have been recorded in the extreme Southwest (Black Sea subtropics) or the extreme Southeast (shore of Petra Velikogo Bay).

The July isotherm pattern was similar to the present one across a large part of the USSR (Figure 25-6). However, in the Arctic belt, summer was appreciably warmer than at present, and temperatures were more similar in the western and eastern Arctic than they are today. A +14°C isotherm paralleled most of the Arctic coast during the optimum of the Mikulino Interglaciation, whereas at the present time summer temperatures increase from west to east along the same band (from +6°C to +12°C). Another possible center of warming, although one not as significant, may be the area from Lake Baikal to the Sea of Okhotsk. However, the climate cannot be reconstructed precisely because of the complex topography and the scarcity of data points for that region. For example, the Chaya section north of Lake Baikal reflects only the local climatic parameters of the intermontane basin.

Isotherms in the steppe and semidesert from the Black Sea region to the Balkhash region were possibly somewhat lower. On the Pacific coast, a distinct warming occurred only in the north, adjacent to the Bering Sea.

ANNUAL TEMPERATURE RANGE

It is well known that a rise in winter temperatures for temperate and polar regions signifies a decrease in the annual temperature range, which is a measure of climatic continentality. Despite the climatic warming during both warm and cold seasons, the annual temperature range differed throughout the country during the Mikulino Interglaciation. In the Arctic, the range decreased by 1°C in the Barents Sea sector, but farther east on the Yamal Peninsula it increased by 5°C. This reflects the greater local temperature increase in summer than in winter. On Taimyr and in the Yana-Indigirka lowland the range decreased, whereas on the Pacific coast (Anadyrskiy Bay) it increased as a result of warmer summers and probably slightly colder winters.

Elsewhere in the country, the continentality of the climate was mostly less. In the band between the Arctic Circle and latitude 50°N, the ranges decreased everywhere, particularly on the Russian Plain, along the Volga (−10°C), and in western Siberia (−7°C). Only on Kamchatka were the ranges similar to the present ones.

The decreased temperature ranges in the past are also noted in regions that now have a sharply continental climate (south of latitude 50°N). During the optimum of the Mikulino Interglaciation, ranges were reduced by 3°C in the present steppe region near the Black Sea, and by 2°C in the Balkhash region. The temperature ranges in the Black Sea subtropics and in the southern Far East did not change.

FROST-FREE PERIOD

The duration of the frost-free period increased nearly everywhere except along the Barents Sea, in the Black Sea subtropics, on the coast of the southern Far East, and in central western Siberia. The duration of the frost-free period changed most dramatically (up to 50% longer) in the Siberian Arctic. Elsewhere it was 20% to 30% longer than the present duration. A 60% increase in the Baikal region may reflect local conditions. An insignificant value was obtained for the Balkhash region.

TOTAL ANNUAL PRECIPITATION

The distribution of precipitation during the Mikulino Interglaciation was more uniform than at the present time (Figures 25-7, 25-8, and 25-9). A general trend toward increased precipitation occurred both in northern regions, which are now characterized by excess precipitation, and southern regions (south of latitude 50°N), where the precipitation is scant at present. Naturally, the absolute values of increased precipitation cannot characterize their influence on the water balance of an area. For example, a 50-mm increase in present total annual precipitation will represent an 8% increase in precipitation for the central regions and a 70% increase for the Arctic coast of Siberia.

In the middle belt of the European USSR, the increase in precipitation was insignificant, about 7% to 8% from 600 mm per year to 650 mm per year. In the Smolensk region, the total precipitation was similar to the present amount.

Somewhat greater was the 15% to 23% increase in precipitation from 570 to 700 mm in Karelia and from 650 to 750 mm on the Kola Peninsula. There was a sharp 70% increase in total precipitation from 250 to 450 mm on the Arctic coast of western and central Siberia. There was also an increase in precipitation in areas that at the present time are marked by the greatest aridity.

The Yana-Indigirka Lowland in the Northeast, the flatland of Central Asia, and southern Kazakhstan in the Southwest have the lowest annual precipitation today.

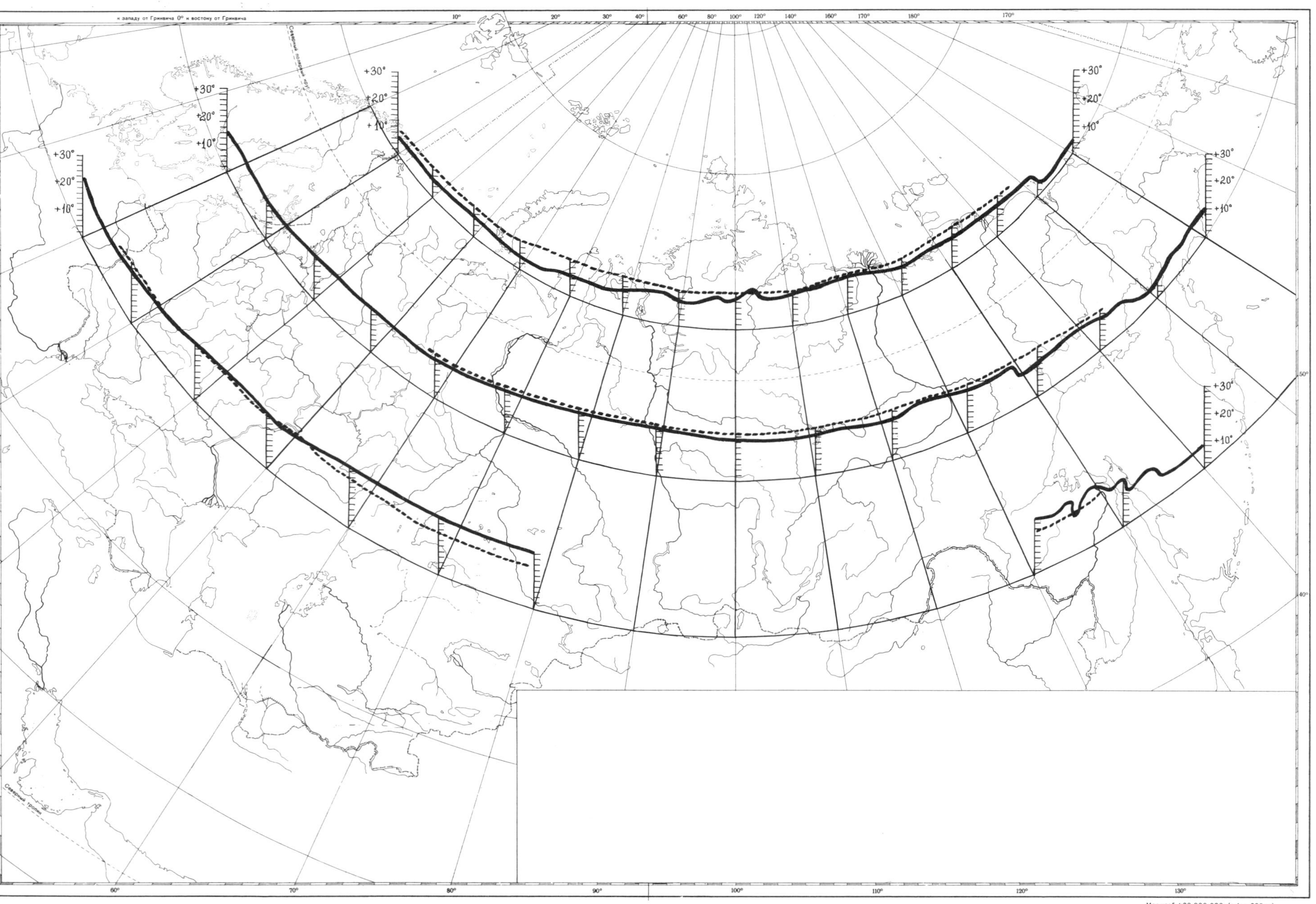

Figure 25-4. July latitudinal temperature (°C) profiles for the present (solid line) and for the Mikulino Interglaciation optimum (dotted line). Note reversed vertical temperature scale.

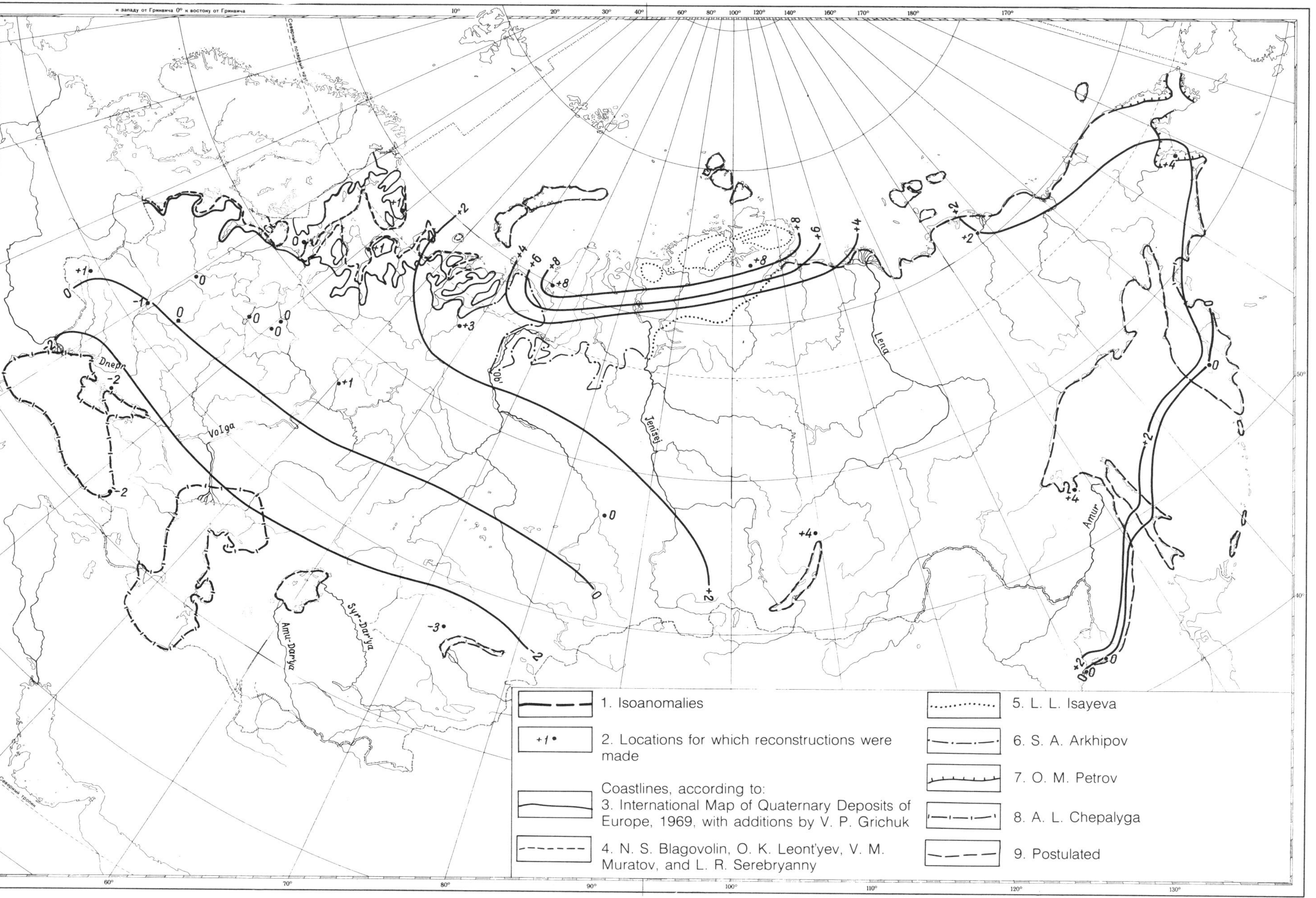

Figure 25-5. July deviations of temperatures (°C) for the Mikulino Interglaciation optimum from their present values.

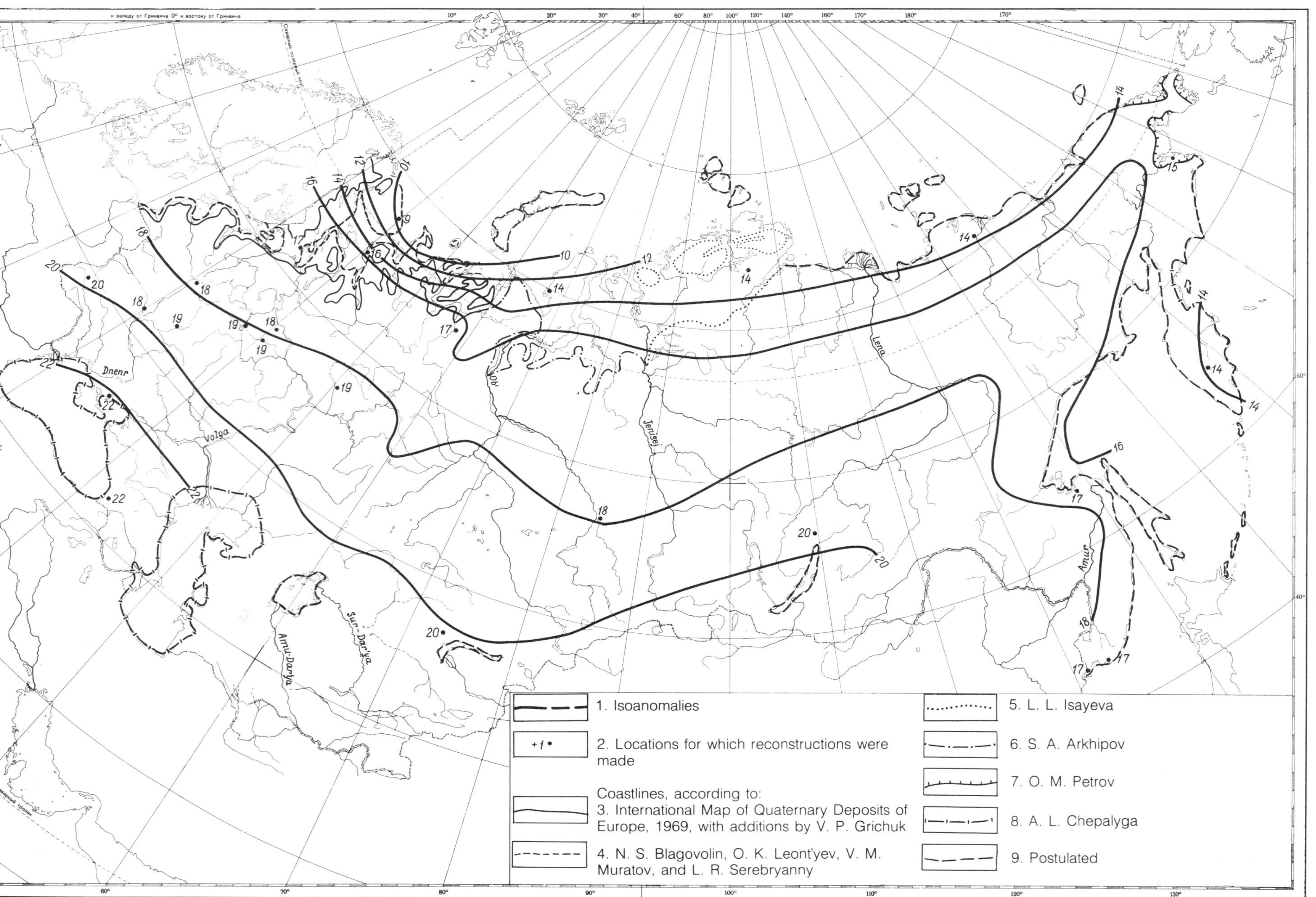

Figure 25-6. July temperatures (°C) for the Mikulino Interglaciation optimum.

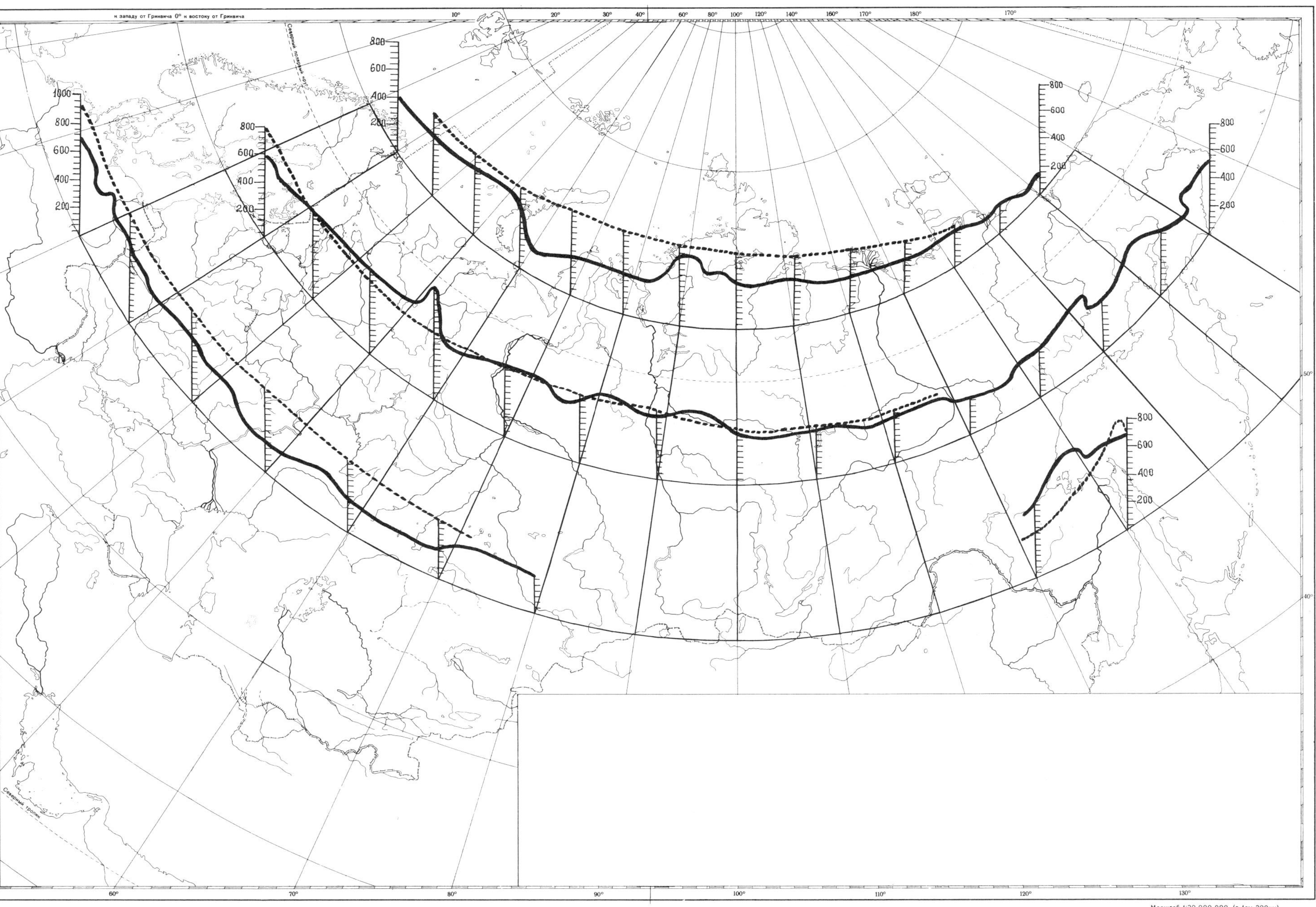

Figure 25-7. Mean annual precipitation (mm), latitudinal profiles for the present (solid line) and for the Mikulino Interglaciation (dotted line).

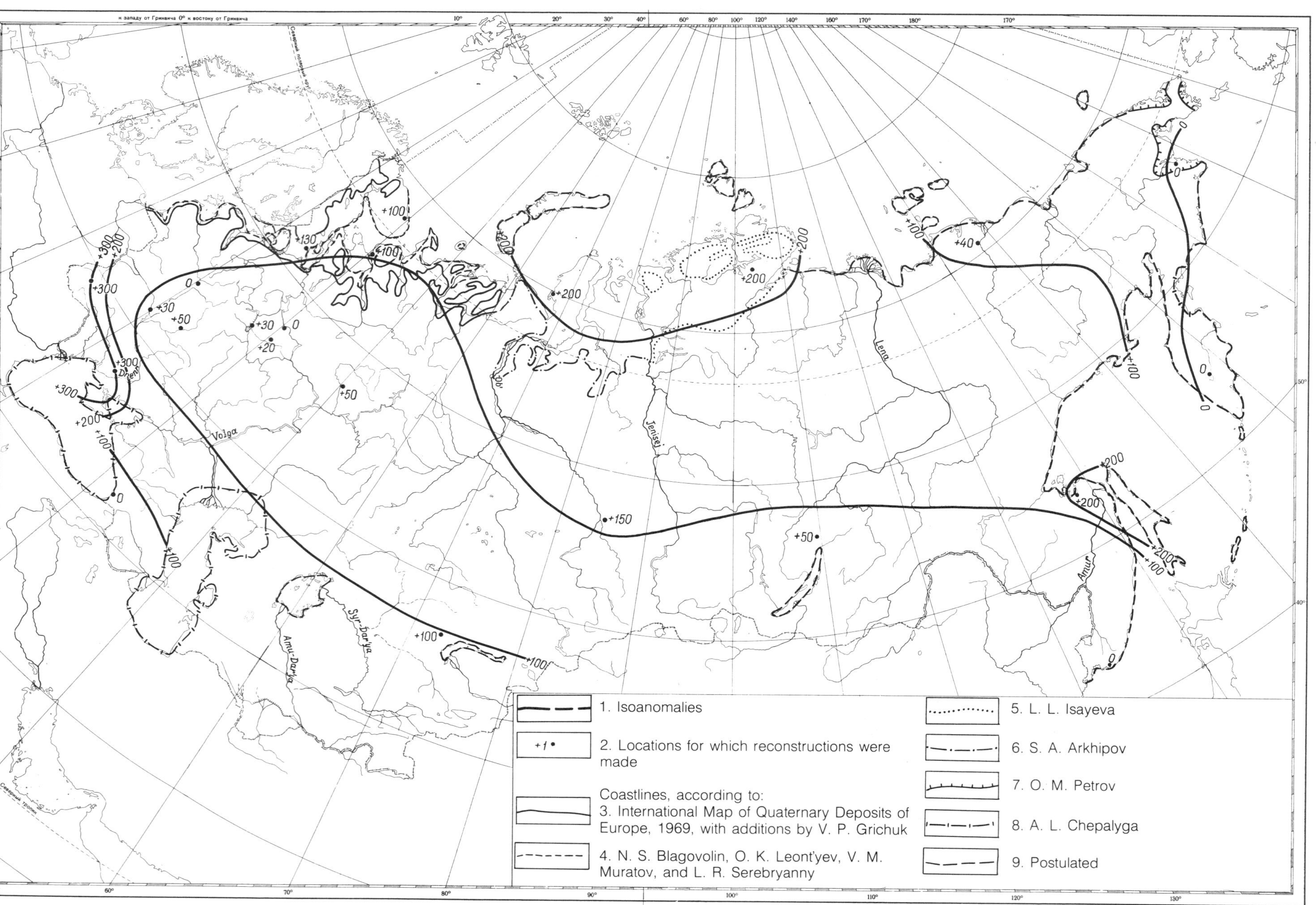

Figure 25-8. Mean annual precipitation (mm) for the Mikulino Interglaciation optimum, deviations from present values.

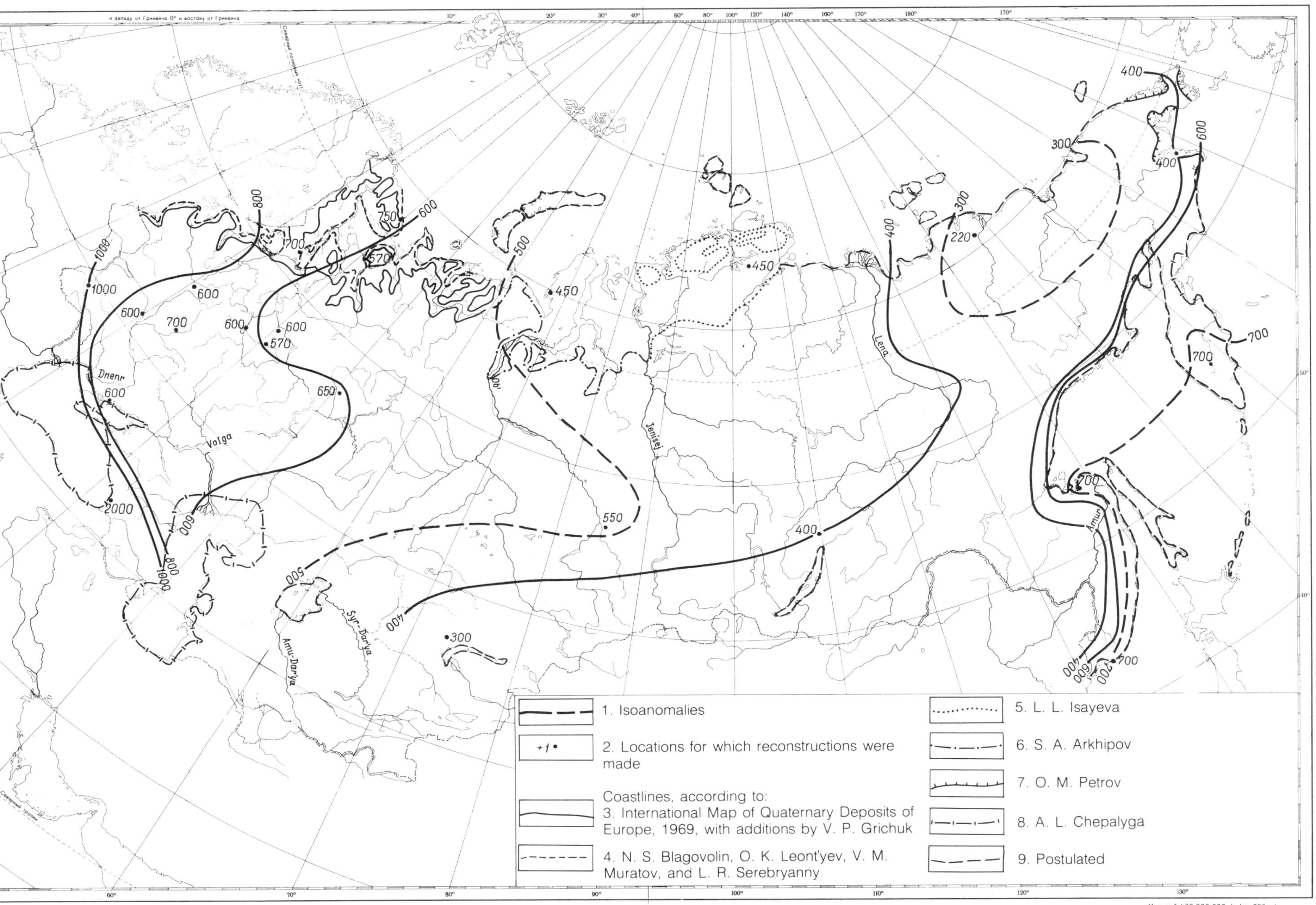

Figure 25-9. Distribution of mean annual precipitation (mm) in the Mikulino Interglaciation optimum.

During the Mikulino Interglaciation optimum on the Yana-Indigirka Lowland, the total annual precipitation was 20% higher than today's values, reaching 220 mm. Data from the Balkhash region suggest that northern Central Asia and Kazakhstan received 50% more precipitation than at the present time, an increase from 200 to 300 mm.

Arid regions located south of latitude 50°N from Kazakhstan to the Black Sea experienced a general increase in precipitation, particularly in the west, where the greatest positive precipitation anomaly of the entire country occurred. There in the Black Sea region the total precipitation increased by 100%, from 300 to 600 mm. In addition, there were areas such as the Black Sea subtropics and the eastern Pacific coast (with the exception of the coast of the Sea of Okhotsk) where the total precipitation was practically the same as it is today.

General Characteristics of the Climate of the Mikulino Interglaciation

In comparison with the present time, the climate was warmer over most of the USSR during the Mikulino Interglaciation. Winter temperatures increased, but summer temperatures remained either close to or slightly higher than the present ones. A marked rise in both summer temperature and winter temperature occurred throughout the Arctic and eastern Siberia. South of latitude 50°N, the temperature changes were dissimilar. In the southern Russian Plain, the winters were warmer and the summers cooler. In northern Central Asia, both summer and winter temperatures decreased slightly. Along Pacific margins, temperatures changed only slightly. Most of the country during the Mikulino Interglaciation had more precipitation that it does at present.

These temperature and precipitation fields and deviations from present conditions permit an analysis of the past climate. The relative importance of the different factors in determining local climate varies widely. Total solar radiation and radiation balance determine the regional characteristics of climate in the Arctic regions and account for the latitudinal increase in temperature from north to south, but it is the ratio of land to sea that controls the regional characteristics of the climate over a given area, along with the general atmospheric circulation and pressure patterns (Khromov, 1968). Several main climatic regions are distinguished (Borisov, 1975; Khromov, 1968).

In many respects, the climate of the European USSR is controlled by such pressure patterns as the Icelandic Low and the Azores High. This region is characterized by a consistent influx of moist, warm Atlantic air brought by cyclones primarily from the Northwest and West but also from the Southwest. With meridional circulation in winter, cold Arctic air comes from the North and by extension of the Siberian anticyclone.

A major factor responsible for warming northern Europe is the Gulf Stream, which raises the air temperature above the Barents Sea by 15°C to 20°C. In summer, the southern European USSR is subjected to a subtropical high-pressure region, and an extension of the Azores High brings dry, clear weather.

Most of Siberia is affected by a different climatic regime. In winter, cold masses of continental air are formed as a result of a vast Central Asian and Siberian anticyclone. The temperature is somewhat higher on the Arctic coast because of air masses brought by cyclones that penetrate eastward from the west coast and by a diffuse low anticyclone from the Arctic. The southwesterly cyclones ameliorate the climate of southwestern Siberia as well. Cyclones from the northwest and southwest increase precipitation in these areas.

In summer, reduced pressure prevails over a large portion of Siberia. Temperate, dry continental air masses are formed there. Cyclonic precipitation is particularly low in interior eastern Siberia.

The Siberian Pacific coast, on the other hand, is affected by the Aleutian Low, which causes a temperature rise and fairly high amounts of precipitation on Chukchi and Kamchatka Peninsulas. In winter, the Siberian anticyclone is important, creating a monsoonal climate similar to that of the Soviet Far East. Finally, the winter climate of Central Asia is determined by the Siberian anticyclone, with its cold, dry air. In summer, however, a low-pressure area with masses of dry, intracontinental tropical air is formed.

Thus, several major climatic regions of differing character are recognized in the USSR. The climate of the European USSR is controlled by air masses coming from other regions, mainly the Atlantic and the Mediterranean but in winter from Siberia. In the European USSR, advection thus plays a major role in climate formation. In Siberia, air masses are formed locally, and air arriving from other regions is transformed into temperate continental air. Thus, the climate of the European part could be defined as allochthonous (advective), and the climate over a major portion of Siberia is autochthonous (autonomous). The climate of Asia is also autochthonous; only within the meridional band along the Pacific coast does it become advective. Finally, a monsoonal climate occurs in the Far East.

A specific relationship exists between the climatic regions and the various floristic provinces identified by Grichuk. (See Chapter 17). The advective type of climate in the European USSR is associated with a transitional flora in the Mikulino Interglaciation. The autochthonous type of climate in Siberia is associated with the orthoselection (i.e., locally developing) vegetation, and the monsoonal type of climate in the Far East is associated with the relict flora. This distinctive correlation suggests stability of climatic types (as well as their spatial combinations) throughout interglacial periods (see also Boyarskaya, 1983). In addition, the similarity between the main meterological parameters of the Mikulino Interglaciation and those of the present day suggests the same position of atmospheric centers during the last interglaciation as well. This similarity explains the regional characteristics of past climate. In the European USSR, the climate was controlled by Atlantic air masses, not only in summer but (unlike the present) also in winter, resulting in increased winter temperatures. The influence of the cold air masses from the Siberian anticyclone was markedly reduced, particularly in the eastern Russian Plain. The area east of the Volga was markedly warmer and moister than it is today. Apparently, stable Atlantic air masses brought by western winter cyclones shifted to the Urals and pushed Siberian anticyclonic systems eastward.

The most distinctive feature of the Mikulino Interglaciation climate was a warming of the north. The Barents Sea sector of the Arctic, surprisingly, underwent the least warming. The climate of this region has always been controlled by the same circulatory system, with a strong influence from the Gulf Stream. Increased winter temperatures in the Siberian sector of the Arctic indicate increased activity in the Gulf Stream. The warming on the Yamal Peninsula was much less than on Taimyr or the Yana-Indigirka Lowland to the east, for Novaya Zemlya blocked the warm current, which should have penetrated farther east, enveloping Novaya Zemlya from the north (Figure 25-10). Thus, a marked warming of the coast from Taimyr to Kolyma could occur. Confirmation of this reconstruction comes from the discovery of a thermophilous molluscan fauna of Mikulino age on the Taimyr coast, similar to that inhabiting the Gulf Stream waters in the Barents Sea today (Troitskiy, 1979).

In summer, however, when the solar radiation increased at high latitudes, the warming uniformly covered the entire coast. Positive summer temperature anomalies were reduced only in the far west, where cloud cover was significant. Thus, the entire band of the Arctic Ocean adjacent to Eurasia was ice free.

Certain data indicate increased winter temperatures in interior Siberia, where the Siberian anticyclones prevailed, as well as in the Okhotsk region. Therefore, statistically, anticyclonic conditions may not have been as stable as today. Frequent invasions of cyclonic systems caused both increased temperature and precipitation. The increased winter temperatures in the Okhotsk region (which was affected in winter by the Siberian anticyclone) indicate a warming of intracontinental air masses. The Siberian High became weaker and hardly influenced the eastern Russian Plain. As is the case today, Chukotka and Kamchatka, along the Pacific coast, were affected by the Aleutian Low, and the southern Far East experienced a monsoonal climate.

Slight but very notable deviations from present climatic conditions occurred south of latitude 50°N. Increased winter temperatures, decreased summer temperatures, and increased precipitation in the western regions indicated an increased influence of Atlantic air masses and western or southwestern summer cyclones, which penetrated farther east into Central Asia than they do at present.

Thus, the large-scale warming that occurred north of latitude 50°N., particularly in the Arctic, was accompanied by increased precipitation everywhere. This finding does not support the view that a warming in the North and a consequent decrease in contrasts from north to south necessarily leads to reduced westerly transport and hence to decreased precipitation in arid regions. Instead, intensified westerly transport caused an increase in total precipitation, especially in arid regions.

Climatic Characteristics of the Valdai Pleniglacial

The greatest cooling within the Late Pleistocene started in post-Bryansk (post-Dunayevo) time, that is, after 27,000 to 24,000 years ago, and lasted until 16,000 to 15,000 years ago. No substantial changes in environmental conditions occurred during this time interval in the extraglacial zone, although one obviously cannot speak of complete stability, for the glacier had reached its maximum (Bologoe stage in eastern Europe 20,000 to 18,000 years ago) and began to retreat. Nevertheless, no clear-cut chronologic series of paleoclimatic fluctuations can be established in this interval. In some cases, indexes of a more rigorous climate do not correspond to the glacial maximum at 20,000 to 18,000 years ago but either precede or follow it. There is little likelihood that such a discrepancy reflects actual changes in nature, and instead it is related to statistical errors in radiocarbon measurements. However, it would be inappropriate to criticize the radiocarbon method in this case, for it is precisely this method that makes it possible to correlate in time points located in different regions for the maximum of the last glaciation.

The chief difficulties involved in paleoclimatic reconstructions for glacial epochs result from other causes. The paleofloristic indication method has limited applicability to a glacial epoch. Only six sites with radioisotopic datings and species determinations in palynologic spectra are available. Paleoclimatic information from other natural components must be used to supplement the paleobotanic data. Paleoclimatic determinations based on the ecology of fossil insects and small mammals are beginning to be widely recognized. The migratory capacity of animals with different ecologies must be considered, however. Ice-wedge casts permit the determination of a certain limit of mean annual subzero temperatures. The mutual correlation and calibration of such methods make a substantial contribution to reconstructions of the climate of glacial epochs.

Reconstructions of past climates must not be based solely on data for individual points. An important framework for such reconstructions is the zonal-provincial structure of natural components. For the Mikulino Interglaciation, a major factor was the fundamental similarity of the zonal climatic structure to that of the present time. This makes it possible to improve the reliability of reconstructions of temperature and precipitation, and so on. For glacial epochs, however, the absence of a modern quasi-analogous natural structures constitutes a serious difficulty in paleoclimatic reconstructions based on empirical data rather than on models. The completely unique character of soil and plant formations and the transformation of a zonal system into a hyperzonal one create a situation that is

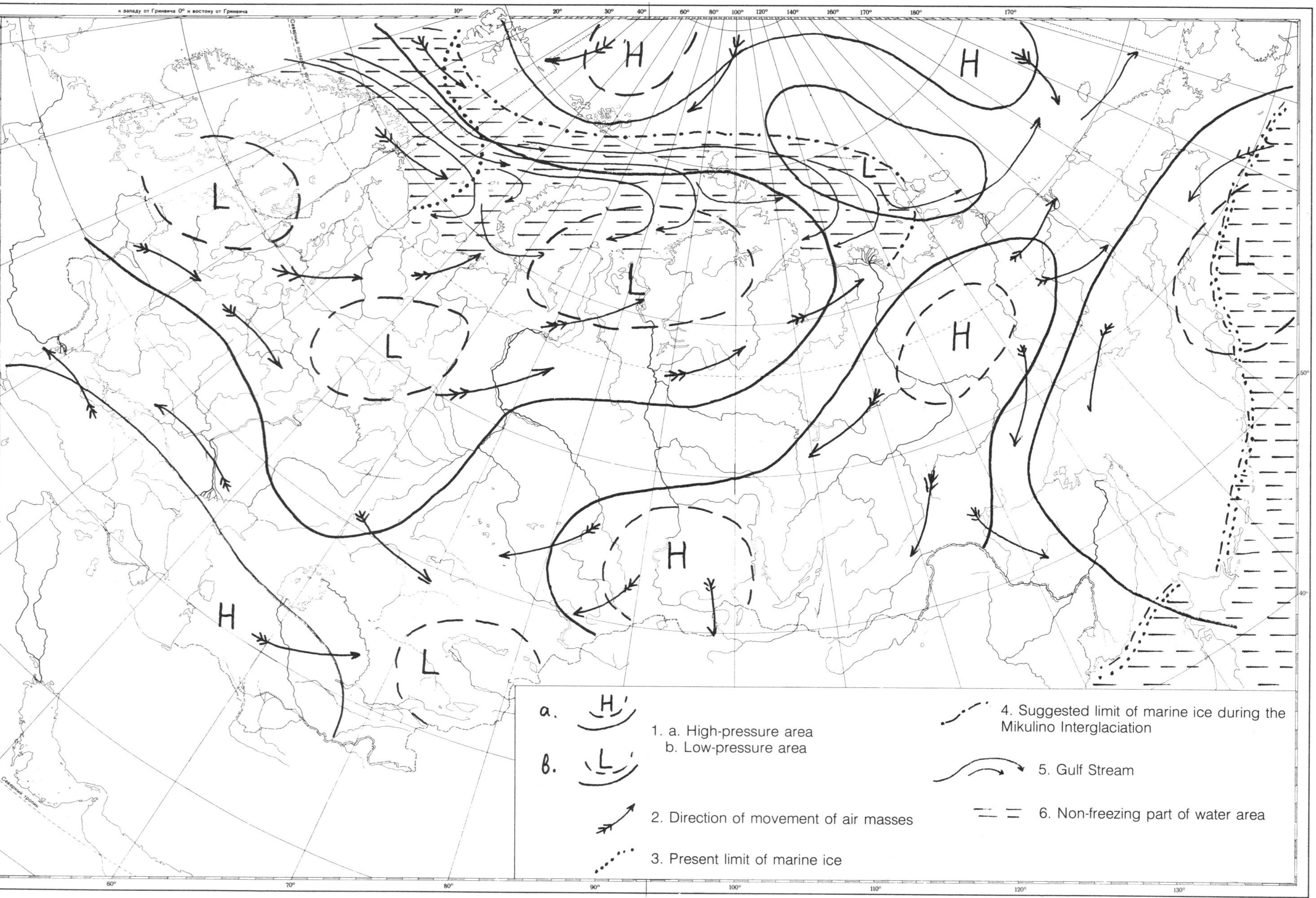

Figure 25-10. Circulation and predominant location of pressure fields in winter for the Mikulino Interglaciation optimum.

almost terra incognita for a paleoclimatologist. Such a comparison obviously must be conditional. Paleogeographic material contains very important information on the condition of the underlying surface and on the formation of vast areas of ice sheets and permafrost. In addition, such material itself makes it possible to extract some direct information on climate. In particular, it is well known that the boundary of permanent ice corresponds to mean annual isotherms not above −10°C (Budyko, 1974), and the mean annual isotherm at the stable southern boundary of the permafrost region cannot be higher than 0°C.

There exist some calculations on air temperature at ground level in glacial and cryogenic regions during the maximum of Late Valdai Glaciation for eastern Europe. I will touch on these data a little later. First, however, on the basis of the arguments given above, I will examine the results of estimates of paleoclimatic parameters obtained by the floristic indicator method.

Although the floristic indicator method has been used to treat the data for only six points, the geographic and paleogeographic location of these points is such that they are of unquestionable interest in a comparative evaluation of climate in the country's major regions.

The two westernmost points are located in the European USSR, one of them—Puchka—near the margin of the Scandinavian ice sheet and the other near the southern boundary of the permafrost region during its Late Valdai maximum. Two other points are located in the western Siberian region: Mega is in the center of the West Siberian Plain, and Kargopolovo is in the south of the region in the basin of the upper Ob' River, in the Altay territory. Finally, the last pair of points is located in the country's eastern edge: the northern point—Khudzhakh—is located in the Okhotsk region northwest of Magadan (Upper Kolyma Highlands), and the southern point is located in the extreme Southeast near the Pacific coast. The location of these points also makes it possible to estimate, albeit in the simplest form, the latitudinal and meridional variations in temperature and precipitation.

Near the margin of the Scandinavian glacier, winter temperatures were 9°C below the present ones (Figure 25-11). Near the southern edge of the Late Pleistocene region of permafrost, the depression was 15°C. Farther east, the negative deviations do not exceed −5°C to −6°C for central and southern western Siberia. Essentially, no deviations of winter temperatures are observed for the Northeast of the country. Only in the south of the Far East does the negative anomaly again reach the impressive value of −10°C. For summer (July) temperatures, as for winter ones, the largest negative deviations are seen in the West (Figure 25-12). Near the glacier margin the temperatures were 7°C lower than present ones (18°C), and at the southern edge of the permafrost region they were 5°C lower than present ones (19°C). For the central and southern parts of western Siberia, there are practically no deviations of summer temperatures from present ones. Nor are any deviations from present temperatures observed for the Northeast. A very slight cooling by 2°C has been determined for the southern Far East.

The duration of the frost-free period generally is reduced, in accordance with the indicated temperature changes. In the northern European USSR, the cold season was two months longer than at present, and in the South, three months longer. In western Siberia, it was one month longer, and in the southern Far East, one and a half months longer. The annual temperature range increased 7°C to 10°C in the western sector and the extreme Southeast, only 5°C in western Siberia, and not at all in the Northeast. With increased continentality, precipitation decreased, except in the Northeast, where it increased by 125 mm per year, that is by 51%.

On the whole, the estimates obtained by the floristic indicator method exhibit trends that definitely agree with the general concepts of the climatic state of glacial epochs. At the same time, one should not ignore the different temperature relations in the periglacial regions of eastern Europe and Siberia. Anomalies of summer temperatures (−7°C and less) recorded for eastern Europe correlate well with data from paleoglaciologic models (Khodakov, 1979). These anomalies are consistent with the development of the Scandinavian glacier at its maximum. In Siberia, however, the summer temperature anomalies necessary for a broad glacial expanse did not exist.

The data from both methods point to a general trend toward a fan-shaped increase in summer temperature anomalies from east to west. The paleoglaciologic approach indicates that in Verkhoyan'ye and the Urals the location of the ends of mountain glaciers during the Late Pleistocene may be explained by a 2°C to 8°C depression of summer temperatures. When comparing these data, however, one must take into account the complexities involved in the determination of the absolute age of moraines in mountainous regions.

Obviously, further study of the question of summer temperatures will be very important in solving the problem of the magnitude of glaciation in western and central Siberia, which is believed by some to have been the site of very extensive glaciation in the Late Valdai (Sartan) and by others the site of a very slight development of such glaciation. (These problems are discussed in the chapters on glaciation in this area, Chapters 2 and 3.)

Even with the most advanced method, that is, the floristic indicator method, the estimation of the paleoclimatic parameters of a glacial epoch on the scale of such a vast territory is a major and important problem. Further methodologic studies will probably contribute certain corrections. However, the possibilities of this method in the determination of summer temperature from glacial floras show up clearly in the European USSR. There the July temperatures for the transition from the Bryansk interval to the Late Valdai epoch at the Late Paleolithic campsite Khotylevo II in the Desna Basin, dated at 23,660 ± 270 yr B.P. (LU-359), were found to be appreciably higher (only 1°C lower than today) than the estimates for the glacial maximum (Zelikson and Monoszon, 1974). On the other hand, according to studies of the small-mammal fauna by the method of areagrams and climagrams, a drop in July temperature of 4°C, from 18°C to 14°C, is noted for this

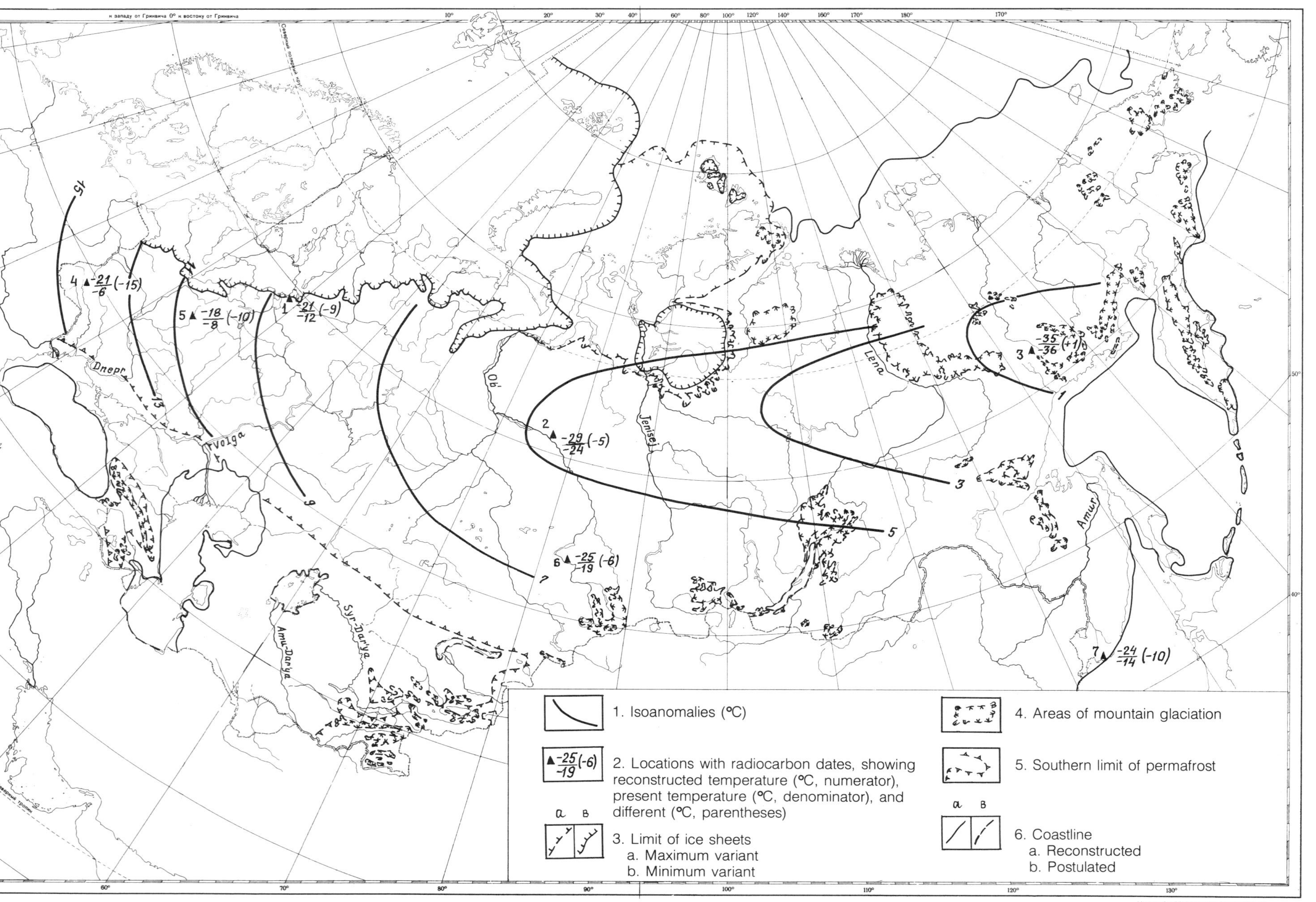

Figure 25-11. Late Valdai cooling maximum for January. Deviations of temperatures (°C) from present values (based on data of the floristic indicator method).

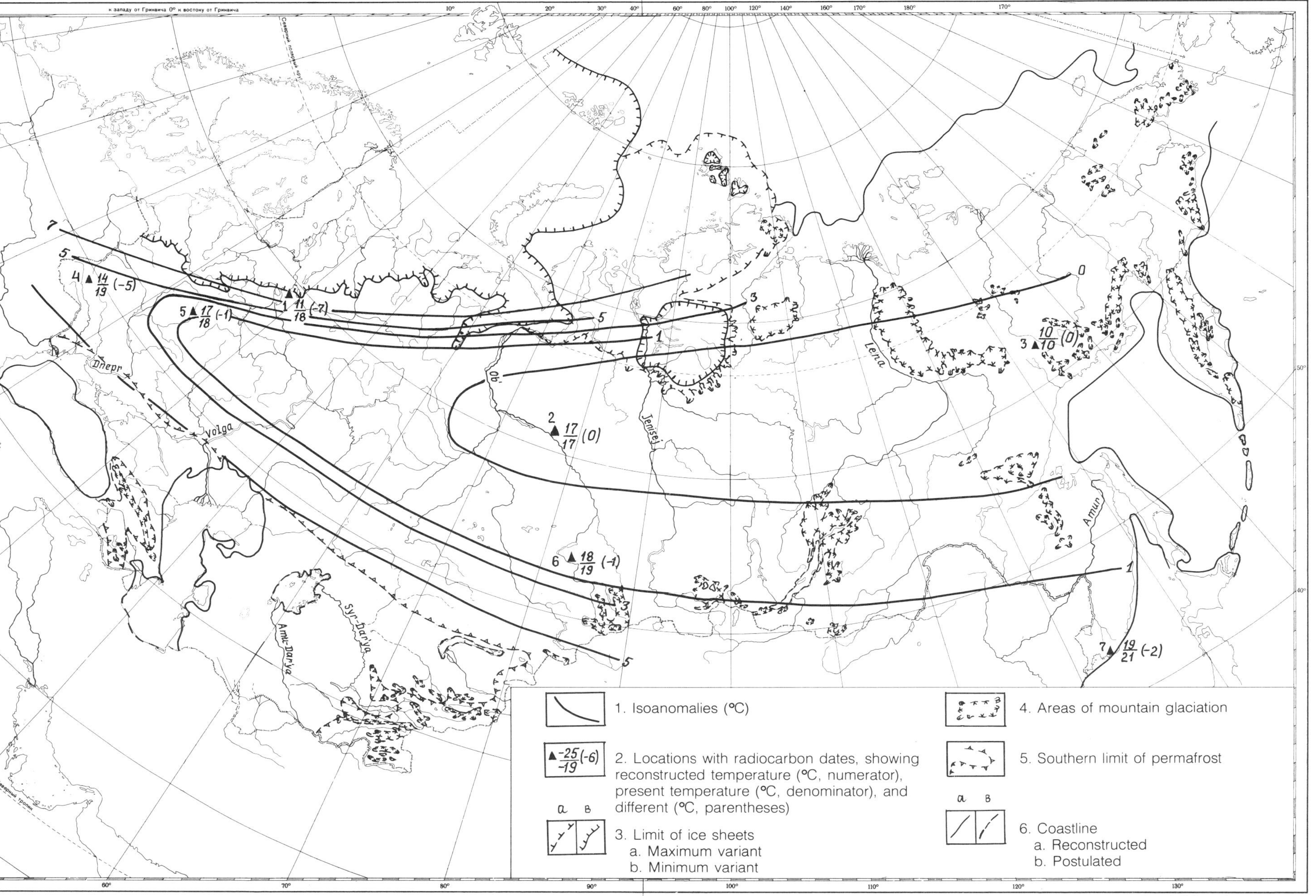

Figure 25-12. Late Valdai cooling maximum for July. Deviations of temperatures (°C) from present values (based on data of the floristic indicator method).

campsite (Markova, 1975). A similar drop of July temperatures (by 3°C) was also determined from the fauna for the later (Magdalenian) campsite of Mezin, also located in the Desna Basin, whereas for the cooling maximum, even for regions farther south (Boyanichi), the floristic indicator method gives a greater drop in summer temperatures (−5°C). All these data generally confirm the results obtained by the floristic indicator method, at least for eastern Europe.

More questions arise in the estimates of winter (January) temperatures. These questions are related to the ecologic tolerance of plants living in areas with low winter temperatures. Can one rule out the assumption that certain perennial plants (for example, tree species from the northern taiga), which are dormant in winter, can withstand not only the present temperatures (e.g., −20°C) but also lower ones (e.g., −30°C or −40°C)? Such a possibility is supported by a table of extreme values of average temperatures at the northern boundaries of ranges, which was prepared by Z. P. Gubonina (Velichko et al., 1977). (See Table 25-1.)

Table 25-1.
Mean Temperatures (°C)

Species	Regions of Oceanic Climate		Regions of Continental Climate	
	January	July	January	July
Picea abies	−12	+14	−38	+12
Pinus silvestris	− 8	+12	−40	+14
Betula pubescens	−12	+14	−38	+12
Betula verrucosa	−12	+14	−40	+12

The conclusion that January temperatures are not always estimated at the lowest limit is in accord with the results of species abundance graphs, used by Z. P. Gubonina for the Late Pleistocene campsite of Timonovka (also in the Desna River basin). They show that the January temperatures there could have dropped to between −35°C and −40°C. It is true that the radiocarbon age of the campsite was determined at 15,110±530 (LU-358), when deglaciation had begun. Temperatures must have been lower at the glacial maximum, according to paleoglaciologic models (Khodakov, 1973). Low January temperatures close to the indicated ones, about −30°C (22°C lower than at present), were obtained by means of the small-mammal method for the aforementioned Late Paleolithic campsite Khotylevo II.

With the paleocryogenic method, ice-wedge pseudomorphs in any region indicate not only low average yearly temperatures but also low winter temperatures. Well-developed epigenetic ice veins can form only under very severe conditions. Polygonal ice wedges are formed in those regions of Alaska where the ground temperature below the level of annual fluctuations is −5°C or lower, and where the average yearly air temperatures drops to −6°C, −8°C, or even −12°C (Péwé, 1966). In western Siberia, the ground temperature at the level below the level of annual fluctuations is estimated as −7°C or lower, and the average yearly air temperatures are estimated at −8°C, −9°C, or lower (Baulin et al., 1967). A temperature regime in which frost cracking develops practically everywhere on all relief forms and in all types of deposits exists in northern Yakutiya, where the ground temperatures range from −7°C to −12°C and the average yearly air temperatures from −10°C or −12°C to −14°C or −15°C (Kaplina and Kuznetsova, 1975). Similar conditions during the last glacial maximum should be assumed for the southwestern part of the permafrost region (western Ukraine). It is obvious from Chapter 9 on permafrost that in Volyno-Podol'ya well-developed pseudomorphs up to 3 to 4 m thick occur on water-divide surfaces and in loess of low ice content. Accordingly, the estimate that the average yearly air temperature was close to −8°C in this region is probably accurate. Because Boyanichi falls within this area for which calculations of temperatures by the floristic indicator method are available, one can try to estimate the winter temperature for this region by using a summer temperature of 14°C for Boyanichi and the average yearly temperature of −8°C indicated above. This calculation results in an estimate of −30°C for the January temperature, a temperature 9°C below the value given by the floristic indicator method and 24°C below the present January temperature.

This paleocryogenic estimate of winter temperatures is confirmed by fossil insects, which indicate that in Belorussia (Kasplyane section, Vitebsk Province) during the glacial maximum the conditions resembled those that presently characterize Taimyr, with January temperatures of −25°C or −30°C (Voznyachuk et al., 1978).

Thus, for the western sector (Russian Plain) winter temperatures were very low during the last glacial maximum. Even in the southern part at latitude 50°N, these temperatures were close to −30°C near the southern boundary of the permafrost region (Figure 25-13). The depression of not only winter but also summer (July) temperatures obtained for the Russian Plain with the aid of fossil insects is largely in agreement with data obtained through the floristic indicator method (Figure 25-14), although the deviation according to the first method is greater (−10°C) than according to the second (−5°C to −7°C).

The paleocryologic data suggest an appreciable drop in winter temperatures over a major portion of Siberia, including its southern regions, as indicated by ice-vein pseudomorphs in the upper course of the Yenisey River and in the Irtysh Basin. (See Chapter 8.) If one assumes (according to the floristic indicator method) that summer temperatures there remained close to the present ones (17°C to 18°C) and that the average yearly temperature dropped to −8°C, winter temperatures in these regions must have ranged between −30°C and −40°C. The spread of permafrost polygonal systems that far south also makes it possible to postulate a drop in summer temperatures of several degrees. Winters at the end of the Late Pleistocene were even more severe at high latitudes of Siberia, where the January temperatures in the Northeast are the lowest in the Northern Hemisphere today as well. Calculations based on the structure of Late Pleistocene ice wedges in the Kolyma Lowland (Duvannyy Ravine section) give very low values of average yearly temperatures (from −20°C to −32°C) and extremely low values for the coldest months of the winter (down to −70°C or −80°C).

Calculated temperatures above the Scandinavian ice sheet, based on paleoglaciologic models (Khodakov, 1973), show that the average summer temperatures at the glacier margin at its maximum were close to 6°C to 8°C. The average yearly temperatures above the central part of the dome were very low (estimated at −32°C).

If the July isotherm reduced to sea level in the central region of the glacier was not higher than 8°C (the present July temperature in southern Taimyr), one can presume that, when the temperature jump above the glacier (−3°C) and the vertical temperature gradient (−0.7°C per 100 m) are taken into account, the July temperatures above the center of the glacier were close to −16°C and the January ones close to −48°C or −50°C. The annual temperature ranges were larger (Figure 25-15).

It is evident from a survey of the available data that one can arrive at a very general representation of the temperature fields during the cold maximum in the Late Pleistocene on the basis of experimental data.

The cryogenic features, which are directly dependent on winter temperatures, suggest much lower temperatures than the present ones, especially in Arctic regions. Above the central areas of the Scandinavian ice sheet, the January temperatures were as low as −50°C.

Similar temperatures in the eastern part of the Eurasian Arctic imply a very intense cooling at high Arctic latitudes. In the South, the temperatures were somewhat higher. The zone of active formation of large polygonal patterns south to latitude 50°N indicates that the temperature came close to −25°C or −30°C. It was somewhat higher in the south; the position of the permafrost limit south of latitude 50°N suggests, however, very low January temperatures over the whole USSR at that time. On the whole, all this confirms the previous conclusion (Velichko, 1977) that a major portion of extratropical Eurasia at the climatic minimum of the Late Pleistocene had subzero values for average yearly air temperatures.

This conclusion is in good agreement not only with data from the continents on the extent of ice sheets, regions of permafrost, and degradation of forest zones. A key factor in cooling at high latitudes in Eurasia was the fact that the Gulf Stream ceased to penetrate into northern Europe and the Arctic Basin (CLIMAP Project members, 1976). This alone resulted in a profound and progressive cooling of all extratropical Eurasia and caused a change in the atmospheric pressure patterns. The spread of marine ice far into the south during the cooling maximum (Barash, 1974), the elimination of the Gulf Stream at high latitudes, and the distribution of air-pressure fields show that major features such as the Icelandic Low weakened very markedly during the long, cold seasons or ceased to act altogether. At the same time, there was a decrease in cyclonic activity, which in the winter brings precipitation and higher temperatures not only to the present zone of advective climate but also to the autochthonous climatic region of the eastern part of the continent.

A cessation or very sharp reduction in the eastward routing of cyclones led to the formation of powerful and stable anticyclonic air masses: a low polar anticyclone in the North and the Siberian anticyclone in the East. The Arctic continental air masses that entered Siberia from ice-covered polar regions, as well as the continental temperate air itself, which formed locally, were progressively cooled and remained practically unaffected by cyclones from the west, which today promote a temperature increase in this region. Today, in the western sector of the USSR, the intense cyclonic activity and the warm current of the Gulf Stream prevent a stable expansion of the Siberian anticyclone westward, but at the glacial maximum the cold anticyclonic masses from Siberia and Central Asia spread freely far to the west. In addition, this process was greatly enhanced by the powerful anticyclonic masses forming above the ice sheets. Models of pressure systems (Gates, 1976) show that even in July the pressure reached 1045 mb above the center of the Scandinavian ice sheet, and that it was 1035 mb (i.e., close to the present January value) in the zone of the Siberian High. Considering that in Siberia now the difference in pressure between summer and winter amounts to 35 to 40 mb, one can postulate that in the wintertime of the glacial maximum in the anticyclonic centers the pressure reached high values that today are completely nonexistent on earth. The merging of the polar, Siberian, Central Asian, and Scandinavian anticyclones produced a single powerful anticyclonic system that predominated in the USSR during the cold season and gave rise to low temperatures, little precipitation, and reduced cloud cover. The general character of the winter pressure field above the USSR was sharply different from the present one (Figure 25-16). Whereas at the present time the winter pressure gradient is directed from the Southeast to the Northwest of the country, at the last glacial maximum the high pressure also predominated in the Northwest. One can also assume a reversal of the gradient and its direction from north to south. On the whole, the pressure field was more even than it is at the present time. There is no doubt, however, that it also had its own mesorelief at that time.

Let us turn to the characteristics of the summer climate during the last glacial maximum in the USSR. Various climate-forming factors as well as the data on July temperatures suggest a very specific climate for summer. Indeed, a marked contrast in pressure and temperatures appeared at that time between the Arctic region and the extraglacial, periglacial region located farther south.

Above the Scandinavian ice sheet, the temperature profile from the center to the periphery indicates temperatures as low as −16°C in the central part of the dome, around 11°C in the proglacial zone, and 14°C and possibly somewhat higher in the central part of the Russian Plain. Thus, two major air masses of contrasting temperatures existed side by side—a cold air mass of high pressure (according to Gates's model, up to 1045 mb) above a vast ice sheet and an air mass of low pressure in the extraglacial zone that

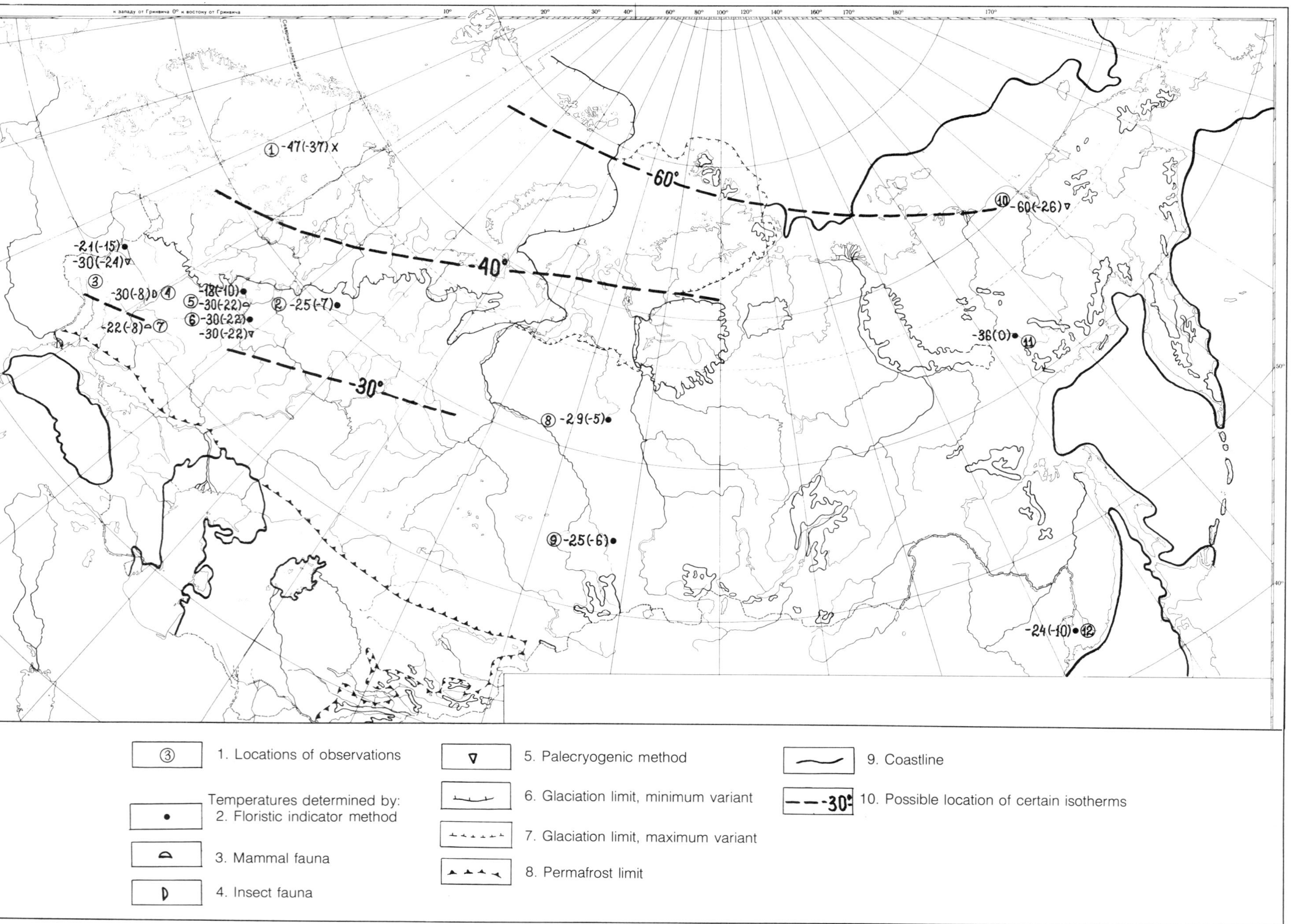

Figure 25-13. January temperatures (°C) and their deviations from present values (given in parentheses) for the Late Valdai cooling maximum (based on various methods) for the following locations (encircled numbers): [1] center of Scandinavian ice sheet, [2] Puchka, [3]

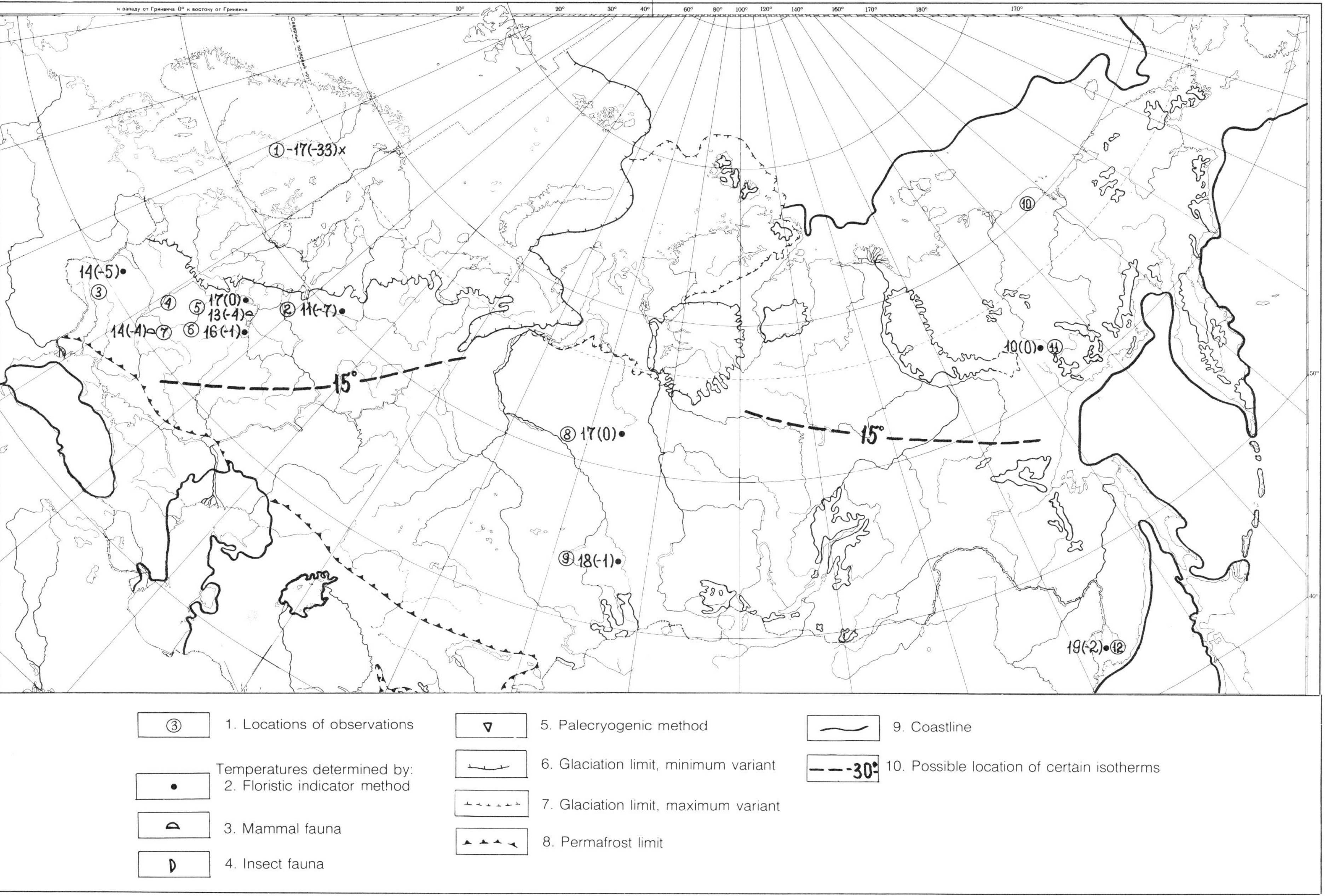

Figure 25-14. July temperatures (°C) and their deviations from present values (given in parentheses) for the Late Valdai cooling maximum.

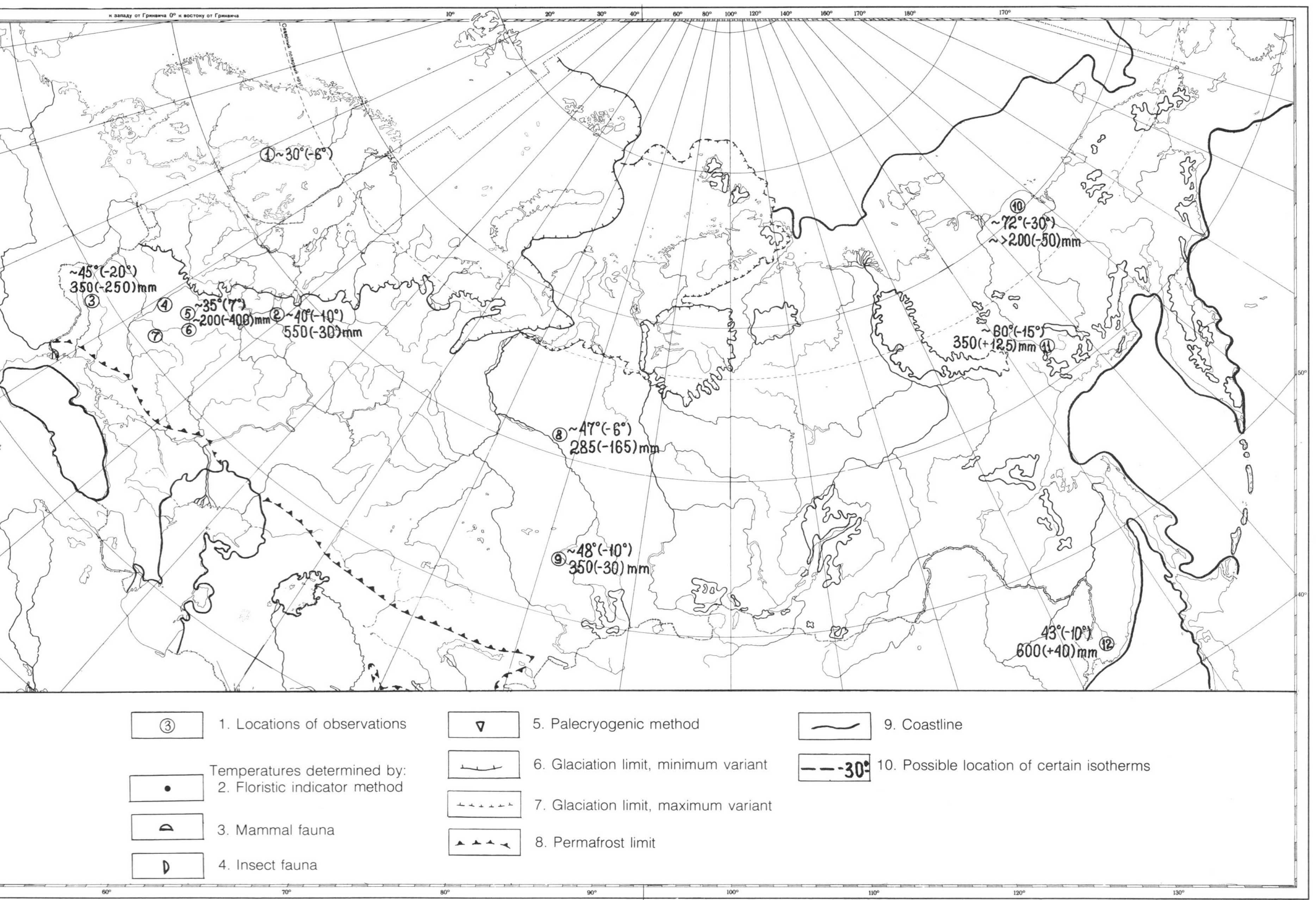

Figure 25-15. Annual temperature ranges (°C) and total annual precipitation (mm) of the Late Valdai cooling maximum, and their deviation from present values (in parentheses).

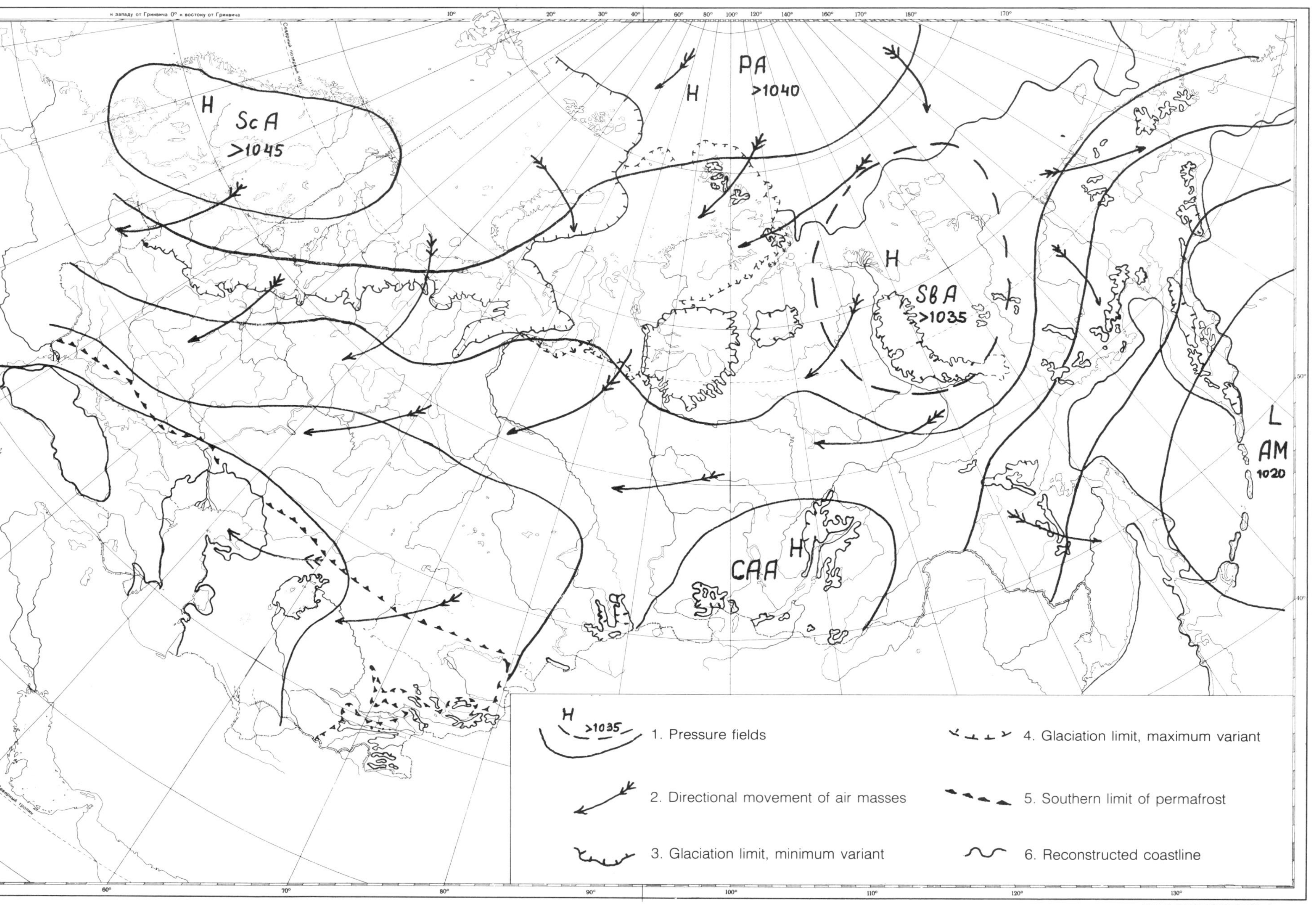

Figure 25-16. Circulation and predominant location of pressure fields in winter for the Late Valdai cooling maximum.

was well heated. In its type and scale, the resulting circulatory system cannot be compared to outflow of air masses from modern glaciers, for warm air masses in the southern regions also played an important role. This unique system, which no longer exists today, could be labeled a summer glacial monsoon (Figure 25-17). Vast plains to the south were invaded from the north by cold air masses that were heated adiabatically during their descent from the glacier; as they were further heated above the periglacial plain, they became increasingly dry. In the loess regions of the Russian Plain, the precipitation was 40% to 60% lower than at present; according to floristic indicator data, it was close to 300 mm per year, and, according to pedolithologic data, it was 200 mm per year or less. (See Figure 25-15.) This type of circulation, precipitation, and temperature could have led to another unique phenomenon of the cold epochs of the Pleistocene—the loess regions. In particular, air masses descending from the ice sheet produced vast latitudinally oriented loessial ramparts on the Russian Plain, although other factors were also involved. (See Chapter 11 on loess.) In particular, fine dust could have settled out during the quiet anticyclonic conditions of cold seasons. For climates of cold loessial epochs, it is important to consider the dust content of the atmosphere.

The glacial summer monsoon system was probably most clearly manifested in the West. In its central part (western and central Siberia), the system was reduced because of the decreasing size of the ice sheet. In the Northeast, continental air masses similar to present-day ones were most probably formed, reinforced in the coastal regions by the vast ice fields of the Arctic Ocean. The so-called loess-ice complex was formed precisely in these regions, on the Yana-Indigirka and Kolyma Lowlands, within the confines of the then-exposed coastal shelf.

The climate of Central Asia was also similar to the present one, although with even less precipitation. Cyclones from the West must have been greatly reduced, especially in winter. In summer, the penetration of cyclones from the West could have been blocked not only by a region of subtropical high pressure but also by the meridional exchange of air masses between the glacial and periglacial regions. The northward retreat of the marine ice in the Atlantic during the summer, as well as the warming of the Atlantic waters, could have given rise to a feature like the Icelandic Low and the generation of westerly cyclonic storms. However, one can scarcely postulate an enhanced influence of summer westerly cyclones in the Siberian Arctic. The Scandinavian ice sheet, reaching up to 3000 m high, and its associated high-pressure system substantially blocked this process. On the whole, the pressure gradient in summer was of the same character as it is now, from Northwest to Southeast.

At the cooling maximum, the distribution of climate zones, as classified according to air-mass genesis, was different from today's in the West, where the advective (allochthonous) system was replaced by an anticyclonic (autochthonous) region, with the features of a summer glacial monsoon. In the East, it was autochthonous in the Northeast and monsoonal in the Southeast, as today, but the role of cold air above dry land in winter was enhanced.

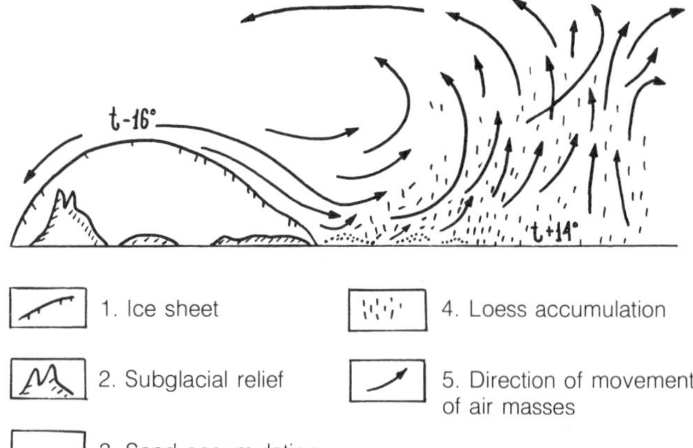

Figure 25-17. Summer "glacial monsoon" and its influence on sedimentation in the periglacial zone of eastern Europe. Suggested temperatures (°C) over the ice sheet and in the periglacial area are indicated.

Thus, the climate of the last glacial epoch in the USSR differed markedly from the interglacial climate. Whereas the interglacial was characterized by a latitudinal type of circulation and a clear and stable manifestation of the role of transport from the west, the glacial epoch was dominated by meridional circulation accompanied by a summer glacial monsoon. Anticyclonic regions predominated for a substantial part of the year above a major portion of the country. As a result, the predominant direction of movement of air masses in the surface layer was from the east, which was a characteristic feature of the climate of the Northern Hemisphere's glacial epochs (Velichko, 1980). During such time intervals, steadily acting transport from the west was found to be displaced upward from the Earth's surface.

References

Barash, M. S. (1974). Boundary of floating ice in the North Atlantic in the Upper Pleistocene. *Okeanologiya* 14, 846-51.

Baulin, V. V., Belopukhova, Ye. B., Dubikov, G. I., and Shmelev, L. M. (1967). "Geocryological Conditions of the West Siberian Lowland." Nauka Press, Moscow.

Borisov, A. A. (1975). "Climates of the USSR in the Past, Present and Future." Leningrad State University, Leningrad.

Boyarskaya, T. D. (1983). Comparison of changeability of the Pleistocene vegetation in the regions of the USSR with different types of climate. International Union for Quaternary Research, XI Congress (Moscow, 1982). Abstracts 3, 56.

Budyko, M. I. (1971). "Climate and Life." Hydrometeorological Press, Leningrad.

Budyko, M. I. (1974). "Climatic Change." Hydrometeorological Press, Leningrad.

CLIMAP Project Members (1976). The surface of the ice-age Earth. *Science* 191, 1131-37.

Gates, W. L. (1976). Modeling the ice-age climate. *Science* 191, 1138-44.

Kaplina, T. N., and Kuznetsova, I. L. (1975). Geotemperature and climatic model of the epoch of accumulation of the edoma suite of Yakutiya's Maritime Lowland. *In* "Problems of Regional and General Paleogeography of Loessial and Periglacial Regions" (A. A. Velichko, ed.), pp. 170-74. USSR Academy of Sciences, Institute of Geography, Moscow.

Khodakov, V. G. (1973). Construction of a model of the European cover glacier, based on the actualistic approach. *In* "Paleogeography of Europe in the Late Pleistocene: Reconstructions and Models" (I. P. Gerasimov and A. A. Velichko, eds.), pp. 27-32. USSR Academy of Sciences, Institute of Geography, Moscow.

Khodakov, V. G. (1979). Paleoglaciological reconstructions for the epoch of the maximum of Late Würmian Glaciation of the territory of the USSR and some paleoclimatic consequences. *USSR Academy of Sciences, seria geograficheskaya, Izvestiya* 6, 27-32.

Khromov, S. P. (1968). "Meteorology and Climatology for Geography Departments." Hydrometeorological Press, Leningrad.

Markova, A. K. (1975). Paleogeography of the Late Pleistocene based on data from an analysis of small mammals of the upper and central Dnepr region. *In* "Problems of Regional and General Paleogeography of Loessial and Periglacial Regions" (A. A. Velichko, ed.), pp. 59-68. USSR Academy of Sciences, Geographical Institute, Moscow.

Péwé, T. L. (1966). Paleoclimatic significance of fossil ice-wedges. *Biuletyn peryglacjalny* 15, 65-73.

Troitskiy, S. L. (1979). The marine Pleistocene of the Siberian plains: Stratigraphy. *Institute of Geology and Geophysics, Vyrad* 430, 1-293.

Velichko, A. A. (1977). Experience in paleogeographical reconstruction of the nature of the Late Pleistocene for the territory of eastern Europe and the USSR. *USSR Academy of Sciences, seria geograficheskaya* 4, 28-47.

Velichko, A. A. (1980). Latitudinal asymmetry in the state of natural components of the Northern Hemisphere's glacial epochs. *USSR Academy of Sciences, seria geograficheskaya* 5, 32-39.

Velichko, A. A., Gubonina, Z. P., and Grekhova, L. V. (1977). "The Habitat of Primitive Man of Timonovka Campsites." Nauka Press, Moscow.

Voznyachuk, L. N., Nazarov, V. I., and San'ko, A. F. (1978). First finds of fossil entomofauna in submorainic deposits of the maximum stage of the last glaciation and their paleogeographical significance. *In* "Studies of Belorussia's Anthropogene" (E. A. Zevkov, ed.), pp. 75-84. Nauka i Teknika Press, Minsk.

Zelikson, E. M., and Monoszon, M. Kh. (1974). Conditions of human habitation at the Khotylevo II campsite, based on palynological data. *In* "Primitive Man: His Material Culture and Environment in the Pleistocene and Holocene" (I. P. Gerasimov, ed.), pp. 137-43. USSR Academy of Sciences, Institute of Geography, Moscow.

CHAPTER 26

Holocene Paleoclimatic Reconstructions Based on the Zonal Method

S. S. Savina and N. A. Khotinskiy

The Quaternary is characterized by alternating glacial and interglacial epochs and a general increase in cooling, which reached an extreme at the end of the Pleistocene (Markov, 1960; Velichko, 1973). Analysis of the Holocene as an interglaciation may yield considerable new information for retrospective-prognostic constructions and answer vitally important questions pertaining to present and future relationships between nature and modern civilizations. The growing importance of the anthropogenic factor makes the Holocene very special compared to preceding phases of the Quaternary.

Reconstruction of Holocene climates is based mainly on the vegetational history of northern Eurasia. Sernander (1908) first reported evidence for a transition from cold arctic and subarctic conditions at the end of the last glaciation to a postglacial warming and subsequent cooling. All later regional schemes (von Post, Firbas, Neustadt) and global schemes (Hafsten, Mörner) of the Holocene reflect an essentially similar trend.

The frequently used term "climatic optimum" refers to a period that was originally defined in Sweden as the most favorable stage in the development of the plant and animal world (Rudolph and Firbas, 1924). The timing of the thermal maximum itself was not clearly established, and some authors (Sernander, Faegri) referred it to the Subboreal, while others (Andersson, Gams) dated it back to the Atlantic.

In the USSR, the climatic optimum represents the warmest stages of the Holocene, either the Atlantic and Subboreal (Neustadt, 1957) or the Boreal and Atlantic (Kind, 1974). Recent palynologic, radiocarbon, and other data permit a refinement of Holocene climatic history based on past vegetation. The zonal distribution of such natural components as vegetation and soil, at the present time and during most of the Holocene, is determined by fairly definite climatic characteristics (Budyko, 1977; Gerasimov, 1979).

Modern Climate and Vegetation

Each modern zone, subzone, or group of plant formations of northern Eurasia is characterized by definite climatic indexes corresponding to the optimum and extreme conditions of their existence. Reconstruction of past vegetation during various stages of the Holocene and identification of zonal analogues in the modern vegetation permit an extension of present-day indexes into the past. Past vegetation does not always have modern analogues, but in the USSR modern equivalents exist partly for past assemblages since the Boreal and completely since the Atlantic.

We calculated the present climatic characteristics on the basis of climatic data from an extensive network of meteorologic stations. Conditions of plant growth were characterized as the sums of air temperatures above 5 °C ($\Sigma_t > 5$ °C) and 10 °C ($\Sigma_t > 10$ °C), the evaporativities (E_o), and the mean temperatures for January and July. The moisture conditions were characterized by a humidity factor, (H/E_o), the ratio of total annual precipitation (H) to the annual evaporativity (E_o). Evaporativity is the maximum evaporation that is possible given sufficient soil moisture and specified meteorologic conditions. The evaporation values (E) were also used.

Data on precipitation, evaporativity, and evaporation were taken from Korotkevich et al. (1974), and the January and July temperatures were taken from Davitaya (1960). All the climatic characteristics were correlated with the USSR vegetation map (scale 1:15,000,000) from Gerasimov (1964). For this purpose, the USSR was divided into a grid measuring 2°30′. In each square, the dominant vegetational and climatic characteristics were determined. Climatic characteristics corresponding to a specific vegetational formation were then plotted. This made it possible to establish the optimum (highest occurrence) and extreme (lowest occurrence) conditions for the vegetational formation (Tables 26-1, 26-2, and 26-3). Some of the values are

Table 26-1.
Optimum Characterization for Vegetation Formations in Eastern European USSR

Zones, Subzones, and Groups of Plant Formations	H (mm/year)	January T (°C)	July T (°C)	H/E_0	$\Sigma_t > 5$°C	$\Sigma_t > 10$°C
Typical tundra	650	−8/−17	7	2.0	450	100
Thin birch forest and forest-tundra	700	−8/−16	9	2.0	650	300
Northern taiga dark coniferous forest	700	−11	12	1.8	1150	700
Northern taiga pine forest	700	−14	14			900
Middle taiga dark coniferous forest	750*/650	−11/−15	16/16	1.6	1750	1350
Southern taiga dark coniferous forest	750/650	−12/−15	17/17	1.4	2100	1700
Subtaiga broad-leaved, coniferous forest	750/650	−8/−15	18/18	1.3	2300	1900
Broad-leaved forest	650	−6	19	1.1	2700	2300
Forest-steppe	625	−7	20	0.82	3000	2400
Steppe	500	−5	22	0.70	3100	2800
Semidesert	350	−14	24	0.40	3400	3100
Desert	250	−9	26	0.25	3600	3200

*Fraction indicates climatic characteristics corresponding to western (numerator) and eastern (denominator) distribution regions of the corresponding formation.

Table 26-2.
Optimum Characterization for Vegetation Formations in Western European USSR

Zones, Subzones, and Groups of Plant Formations	H (mm/year)	January T (°C)	July T (°C)	H/E_0	$\Sigma_t > 5$°C	$\Sigma_t > 10$°C
Arctic tundra	350	−26	5	2.5	440	100
Typical tundra	450	−28	8	2.2	500	150
Shrubby tundra	550	−27	12	1.8	540	200
Thin larch forest and forest-tundra	550	−28	13	1.8	1060	600
Northern taiga larch and spruce-larch forest	550	−26	15	1.8	1340	1000
Northern taiga pine forest	550	−23	15	1.54	1450	1100
Northern taiga larch-spruce-cedar forest	750	−23	15	1.8	1240	900
Middle taiga spruce-fir-cedar forest	650	−22	17	1.6	1860	1500
Southern taiga spruce-cedar-fir forest	550	−20	17	1.3	2040	1700
Middle and southern taiga, western Siberian pine forest	550	−21	17	1.2	1940	1600
Subtaiga aspen-birch forest	550	−18	17	0.9	2200	1900
Meadow-steppes and steppe-covered meadows combined with birch and aspen groves (forest-steppe)	450	−18	18	0.7	2420	2000
Steppe	350	−17	21	0.5	2620	2300
Northern semidesert	350	−16	22	0.3	2700	2400
Southern semidesert	300	−15	21	0.3	3000	2700
Artemisia desert	150	−14	24	0.14	3500	3200
Halophytic desert	150	−14	25	0.1	3900	3600
Saxaul and shrubby desert on sands	150	−2	29	0.2	4250	4000

Table 26-3.
Optimum Characterization for Vegetation Formations in Eastern Siberia

Zones, Subzones, and Groups of Plant Formations	H (mm/year)	January T (°C)	July T (°C)	H/E_0	$\Sigma_t > 5$°C	$\Sigma_t > 10$°C
Arctic tundra	250	−36	4	1.3	540	180
Typical tundra	250	−36	6	1.1	580	230
Shrubby tundra	350	−37	12	1.2	660	300
Ural and southern Siberian mountain-taiga, dark coniferous forest	750	−22	14	1.5	1800	1400
Middle and southern taiga, central Siberian larch forest	500	−28	17	1.1	1600	1200
Middle and southern taiga, central Siberian pine and larch forest	450	−22	18	0.9	1800	1400
Larch thin forest and forest-tundra		−31	10	1.2	740	390
Northern taiga light larch forest	400	−40	15	1.04	1150	800
Middle-taiga larch forest:						
(a) Mossy-grassy-shrubby	350	−31	16	0.84	1560	1300
(b) Same, with moist meadows	350	−40	17	0.74	1560	1100
(c) Same, with steppe areas on alasses	250	−39	18	0.66	1640	1300
Middle taiga mountain larch forest of Transbaikalia	450	−30	16	1.2	1400	1100
Forest-steppe of Transbaikalia	400	−19	16	0.65	1600	1250
Southern Siberian tansy meadow-steppe	300	−19	16	0.64	2220	1900
Mountain tundra	350	−38	12	1.2	640	280
Subtaiga broad-leaved, coniferous forest		−25	17	1.0	2120	1800
Dark coniferous mountain taiga forest	750	−15	17	1.6	1860	1500
Oak forest	750	−18	19	1.1	2420	2100
Broad-leaved mountain forest	750	−19	19	1.2	2520	2200

approximate, especially those for the mountainous regions of Siberia, where meteorologic data are poor.

The relationship between vegetation and climate as characterized by temperature sums above 5°C and by humidity values is clearly shown in the eastern European plains (Figure 26-1). Points on the graph correspond to optimum conditions of growth for each formation, and vertical and horizontal lines correspond to their possible ranges. Similar constructions were made for western and eastern Siberia.

The plains of the USSR are characterized by a decrease in the humidity factor and an increase in evaporitivity from north to south. Against this background, the evaporation values are the most representative for determining the conditions of vegetational development. Evaporation is a complex characteristic dependent on the average soil moisture (humidity factor) and the thermal resources (evaporativity). In the northern USSR the evaporation process is restricted by heat deficiency and in the southern regions by moisture deficiency. In the "middle belt" the humidity factor values are close to 1.0 and indicate an uninterrupted supply of soil moisture to the plants. Under such conditions, the evaporation reaches its highest values.

A plant-cover productivity map (Budyko, 1971) shows that the belt of maximum evaporation values coincides with the belt of maximum productivity and corresponds to the optimum conditions of plant-cover development. Experimental data (Lavrenko et al., 1955; Khodasheva, 1966) from eastern Europe confirm that the maximum biologic productivity as determined by above-ground biomass is found in the broad-leaved forests and forest-steppe (evaporation of 500 to 600 mm per year). In western Siberia the optimum belt includes subtaiga aspen-birch forests (evaporation, 430 mm per year), whereas in southwestern Siberia it is the broad-leaved and broad-leaved/coniferous subtaiga forests of the Far East (evaporation 500 mm per year).

Climate of the Boreal Period

On the basis of these relationships, the paleoclimate of the Boreal period (9000 to 8000 yr B.P.) was reconstructed from the map of Boreal vegetation presented in Chapter

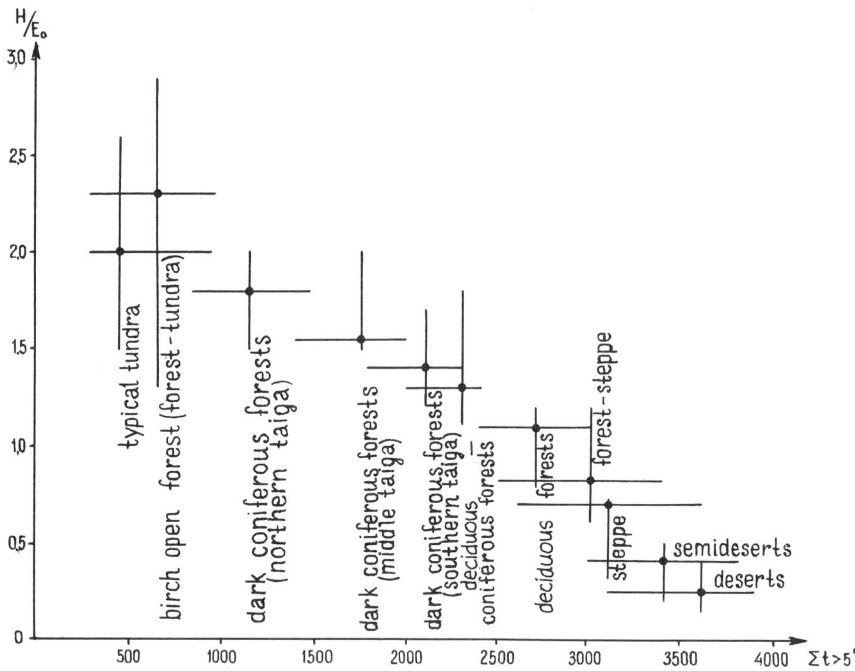

Figure 26-1. Relation of humidity factor values (H/E_o) and sums of air temperatures above 5°C ($\Sigma_t > 5°$) to the vegetational formations of eastern European USSR.

18. This map shows an early stage in the formation of zonal landscapes in the USSR. Decomposition of the hyperzonal (after Velichko, 1973) Late Pleistocene vegetation was essentially completed, and the main vegetation zones—tundra, forest, and steppe—had been formed. In addition, significant differences between plant cover during the Boreal and that of today have been identified. A map of air temperature for January (Figure 26-2) and July (Figure 26-3) and a map of precipitation (Figure 26-4) are presented for the Boreal period. Comparison of the Boreal January and July isotherms with present isotherms shows (1) an appreciable January cooling in the European USSR and increased temperatures in the Asian part and (2) similar altitudinal positioning of the July isotherms, with a slight general northward shift of isotherms in both the West and East.

The temperature differences are reflected more clearly by the fields of anomalies, which show the location of positive and negative deviations from the present during the Boreal period. January temperature anomalies (Figure 26-2) divide the USSR into two regions along a line from the Kola Peninsula toward the Urals to latitude 60°N, eastward along the parallel and then southward at the longitude of Lake Baikal. North of this boundary, the winter was warmer during the Boreal than it is today, especially in the Northeast. South of this boundary, on the other hand, a cooling was noted, especially in the western European USSR. In summer (July), the temperatures in northwestern Siberia (up to latitude 60°N) and in central Siberia were lower than they are now (Figure 26-3). Elsewhere, a warming occurred, particularly in the northeastern maritime regions. Total annual precipitation was greater in the northern and northeastern USSR and less south of latitude 60°N (Figure 26-4).

These climatic conditions were determined by the atmospheric circulation characteristics, which can be reconstructed if one assumes that the basic multiyear circulation fluctuations are governed by the physical mechanisms that control seasonal changes (Willett, 1953; Dzerdzeyevskiy, 1975). Thus, the generalized winter and summer circulation processes may be regarded as limiting variants of multiyear changes in atmospheric circulation. Hence, an analogy can be drawn between present-day seasons and the long-term circulation conditions of the past. Such an analogy is supported by the hydrodynamic theory of similarity (Sergin, 1974).

Dzerdzeyevskiy (1975) identifies for the Northern Hemisphere 13 macroscale circulation patterns, or "elementary circulation mechanisms" (ECMs), with 41 varieties. This finite number of patterns is determined by the constant influx of solar heat and by the location of the oceans and continents. Varieties are caused by seasonal changes in the properties of the underlying surface.

Continuously developing atmospheric circulation is divided into individual periods of 1 or 2 to 10 days, when a specific ECM is operating. The alternation of ECMs and the sequence, recurrence, and duration of action determine circulation structure and associated weather conditions not only for individual months, seasons, and years but also for several decades. Thus, it becomes possible to discuss the climatic regime.

Analysis of the recurrence of ECMs in the annual development of atmospheric circulation shows that ECMs are clearly confined to specific seasons. On this basis, Dzerdzeyevskiy distinguished six circulation seasons: winter, late winter, spring, summer, autumn, and late autumn. For each circulation season, schemes of general circulation in the extratropical latitudes of the Northern Hemisphere are

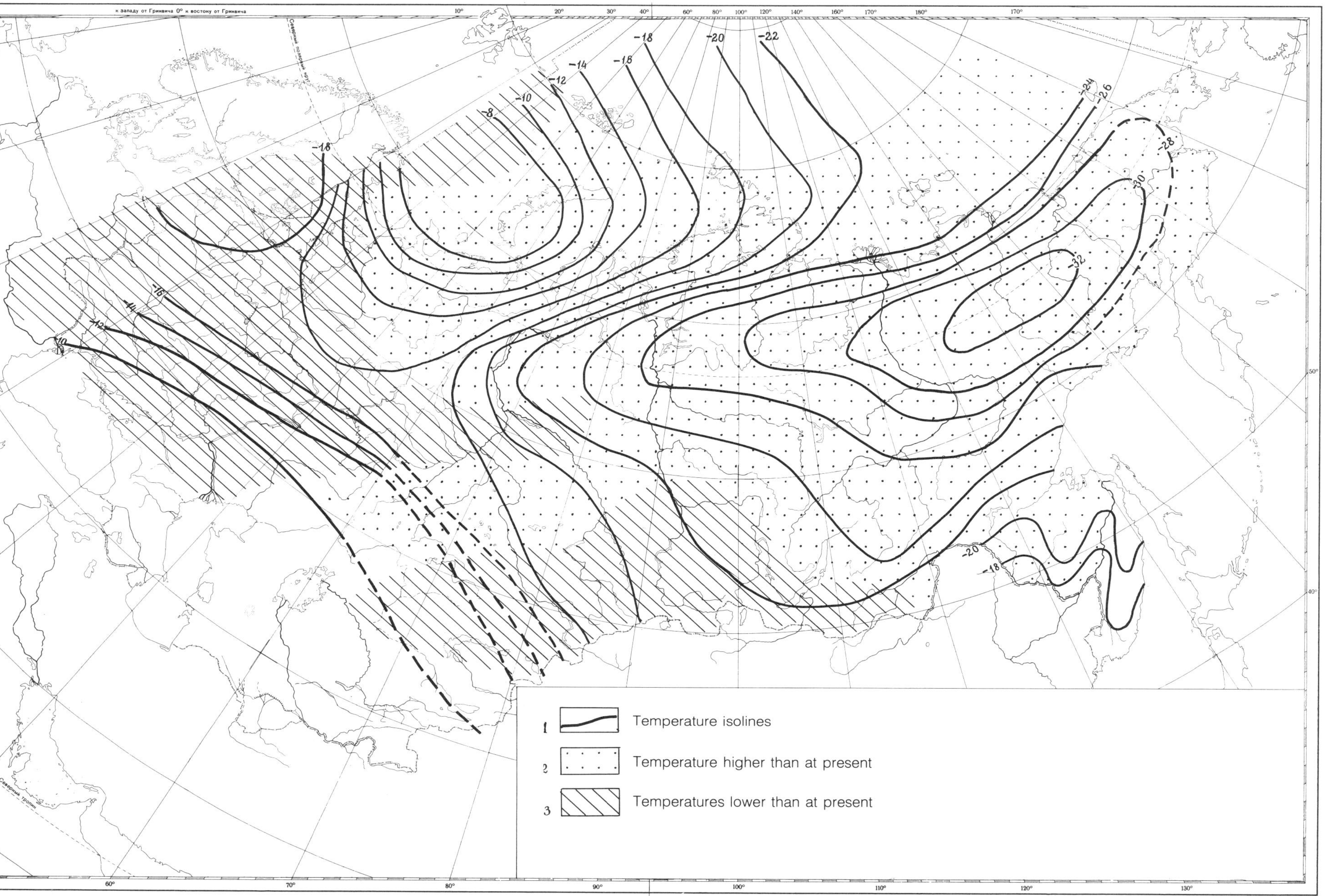

Figure 26-2. January temperatures (°C) during the Boreal period.

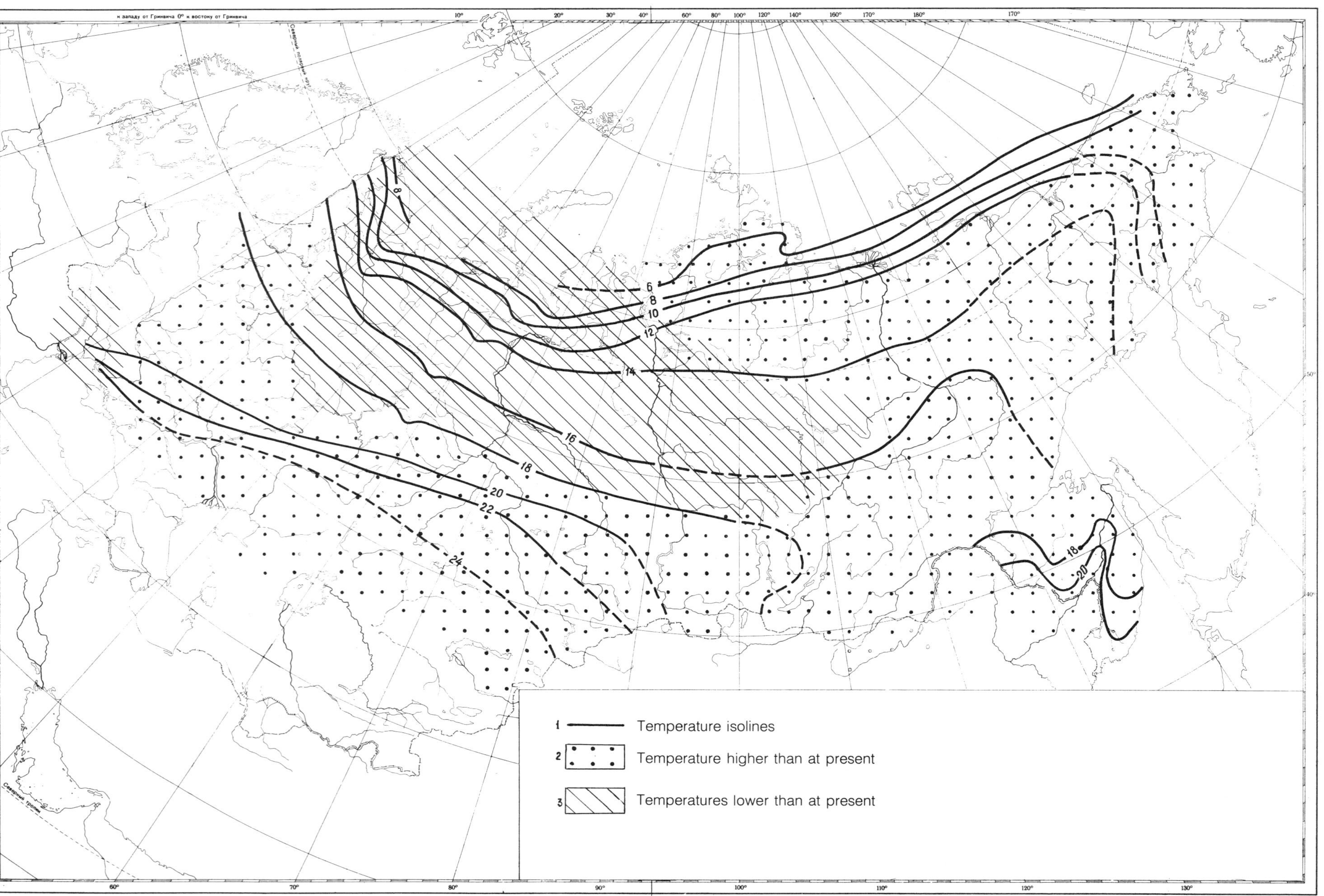

Figure 26-3. July temperatures (°C) during the Boreal period.

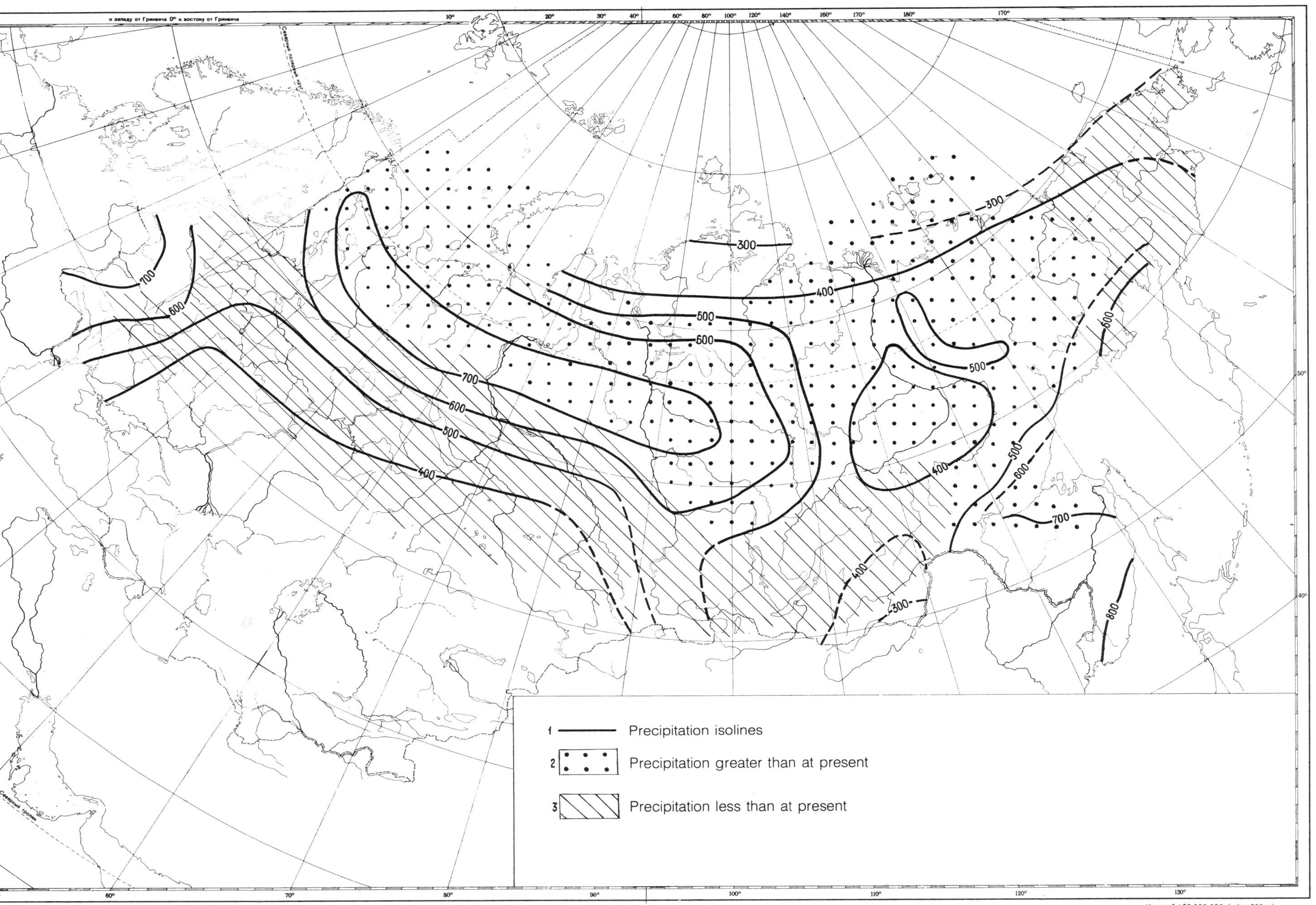

Figure 26-4. Annual precipitation (mm) during the Boreal period.

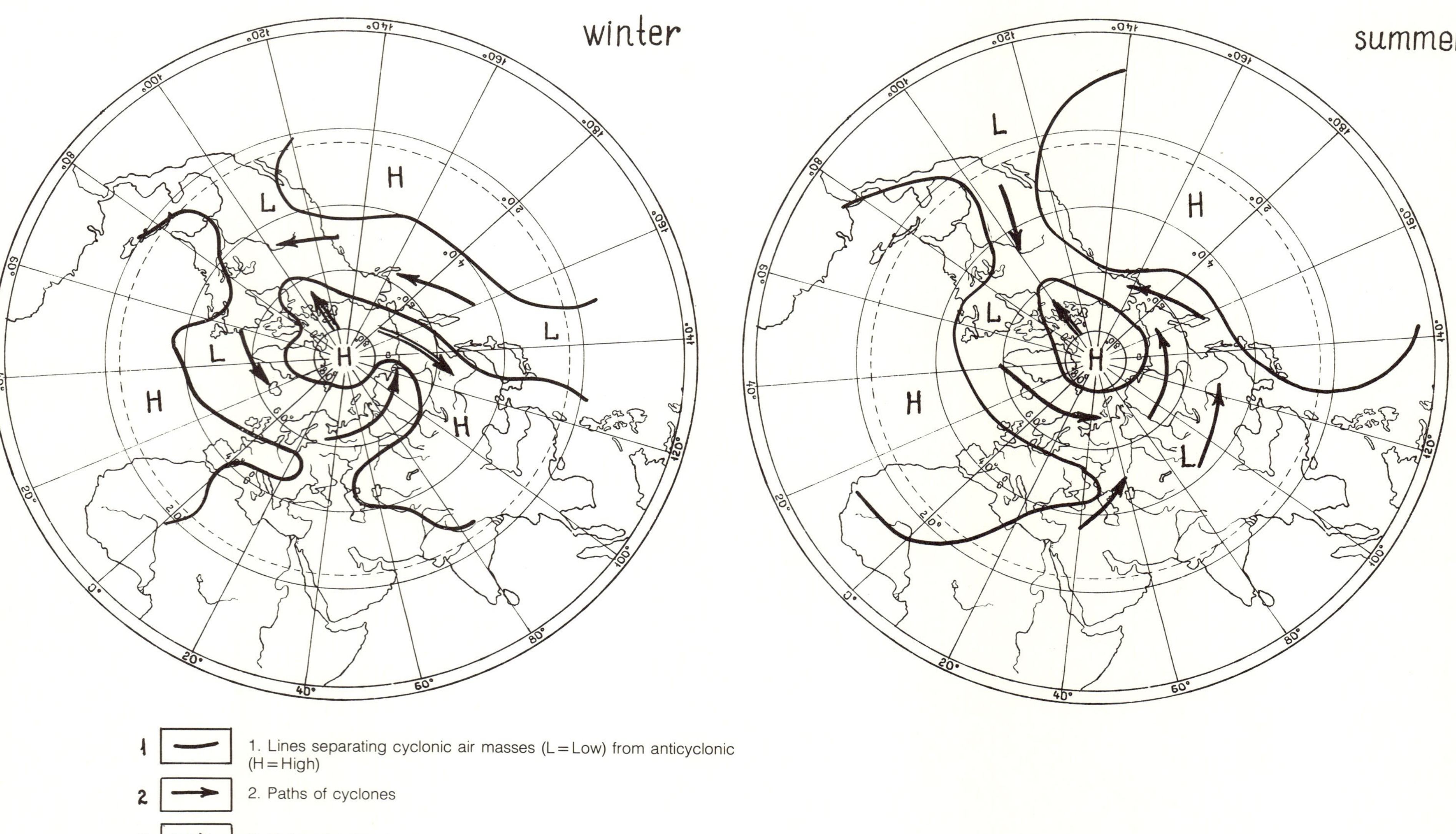

Figure 26-5. General atmospheric circulation of the Northern Hemisphere for winter and summer, common to the present and the Boreal period (after B. L. Dzerdzeyevskiy, with additions by S. S. Savin).

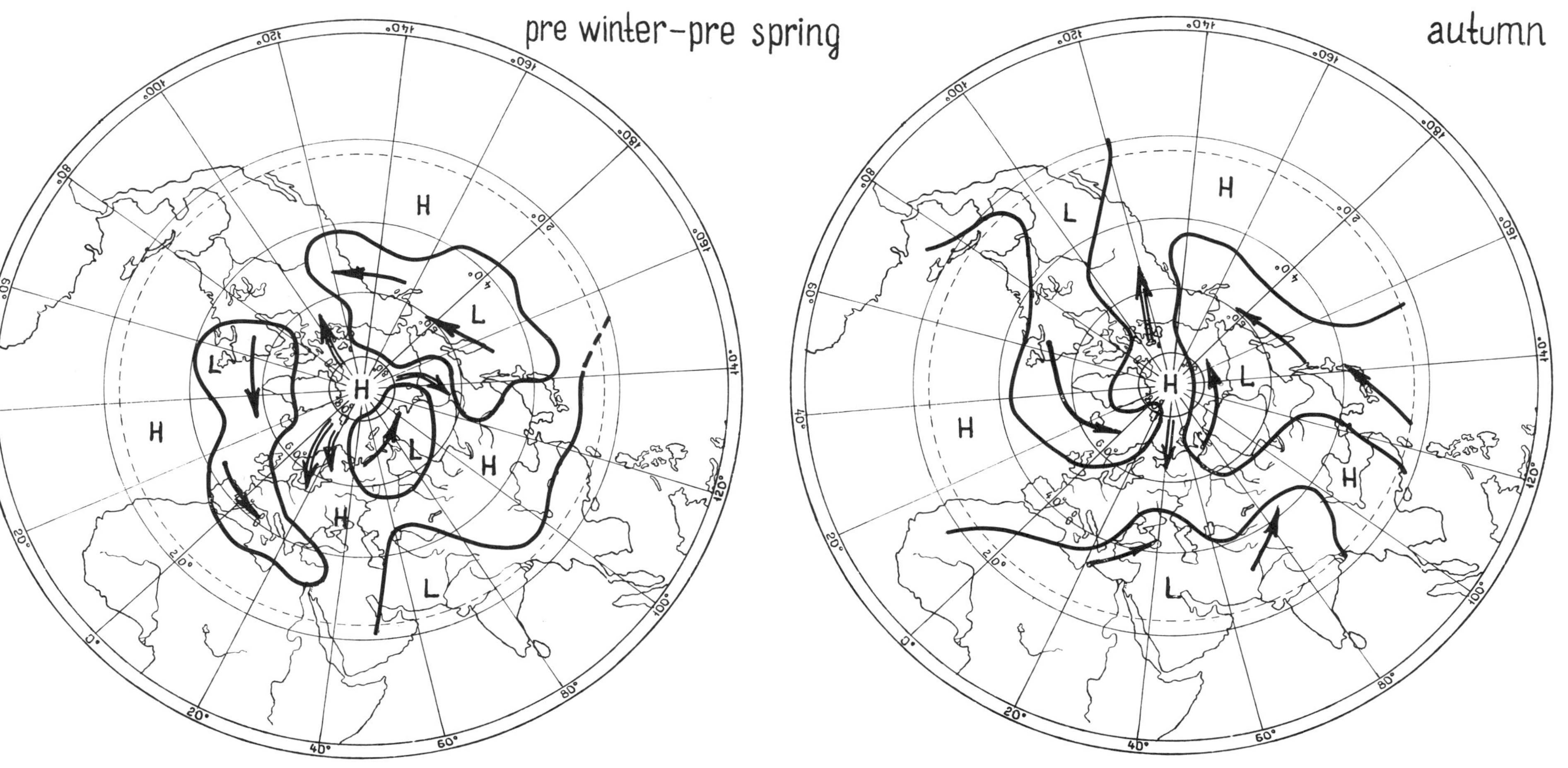

Figure 26-6. "Prewinter-prespring" and autumn schemes of general atmospheric circulation of the Northern Hemisphere characteristic, respectively, of the winter and summer of the Boreal period (after B. L. Dzerdzeyevskiy). For notations, see caption for Figure 26-5.

presented on the basis of a correlation of ECM characteristics during these seasons. A total of 10 schemes was derived, 1 each for winter and autumn circulation seasons and 2 each for summer, late winter, spring, and late autumn (Gerasimov, 1964).

The climatic characteristics reconstructed for the USSR suggest that the pressure patterns for winter and summer during the Boreal period were generally similar to those of the present. Hence, one can also assume similarity in the development of fundamental circulation processes during the Boreal period. The observed differences in climatic indexes of the Boreal as compared to the present were apparently a result of changes in the activity of individual links of this pressure system. In this connection, it is desirable to distinguish two types of generalized circulation schemes for both winter and summer.

The first type describes the main processes that predetermine the similarity between the circulation pattern of the present day and that of the Boreal (and thus the temperature distribution controlled by that pattern) (Figure 26-5). For example, in Siberia the decreased continentality in the Boreal corresponds to the present-day circulation processes that cause warm winters and cool summers. Circulation schemes for cold or warm winters can be studied with data on extreme decades in the 20th century.

The second type of circulation scheme, characteristic of the transitional seasons today, complements the main patterns. For example, the "late autumn-late winter" (termed prewinter/prespring in Figure 26-5) complements the winters in the Boreal, with intensification of cyclonic activity in northern Siberia and a weakening of the Polar Anticyclone and a northeastern extension of the Siberian High (Figure 26-6). This was associated with a warming in the Arctic regions, in the northern European USSR, and over a large part of western and eastern Siberia. Cooling resulted from intensified effects on the western periphery of the Siberian High in southwestern Siberia and Central Asia as well as in much of the European USSR, with intensive discharges of cold air masses from the Polar Basin at the rear of the cyclonic series.

The other transitional pattern is the "autumn" scheme, which gives information in addition to the main scheme for the Boreal summer season. The intensification of cyclonic activity in northeastern Europe and western Siberia caused decreased summer temperatures and increased precipitation in these regions.

Thus, the circulation processes that developed according to the "late autumn/late winter" and "autumn" schemes lessened the stability of the winter and summer circulation processes. This ultimately caused a reduction in the annual temperature range in Siberia. At the same time, in the European sector the contrast between winter and summer increased somewhat, indicating a diminished oceanic climate in these territories in comparison with the present time.

The paleoclimatic reconstruction for the Atlantic and Subboreal periods by this method will make it possible to form a general concept of the climatic and circulation regimes of the Holocene.

References

Budyko, M. I. (1971). "Climate and Life." Gidrometeoizdat Press, Leningrad.

Budyko, M. I. (1977). "Global Ecology." Mysl' Press, Moscow.

Davitaya, F. F., ed. (1960). "Climatic Atlas of the USSR." Vol. 1. Gidrometeoizdat Press, Moscow.

Dzerdzeyevskiy, B. L. (1975). "General Circulation of the Atmosphere and Climate." Nauka Press, Moscow.

Gerasimov, I. P. (1964). "A Physicogeographical Atlas of the World." Central Geodetic and Cartographic Service (GUGK), Moscow.

Gerasimov, I. P. (1979). Climates of past geological epochs. *Meteorologiva i gidrologiya* 7, 37-53.

Khodasheva, K. S. (1966). Geographical characteristics of the structure of the population of terrestrial vertebrates. *In* "Zonal Characteristics of the Terrestrial Mammal Population" (Yu. A. Isakov, ed.), pp. 7-38. Nauka Press, Moscow.

Kind, N. V. (1974). "Geochronology of the Late Anthropogene from Isotope Data." Nauka Press, Moscow.

Korotkevich, E. S., Budyko, M. I., Sokolov, N. N., and Drozdov, O. A., eds. (1974). "Atlas of the World Water Balance." Gidrometeoizdat Press, Moscow and Leningrad.

Lavrenko, Ye. M., Andreyev, V. N., and Leont'yev, V. L. (1955). Productivity profile of the above-ground part of the natural plant cover of the USSR from tundras to deserts. *Botanicheskiy zhurnal* 40, 415-19.

Markov, K. K. (1960). "Paleogeography." Moscow State University Press, Moscow.

Neustadt, M. I. (1957). "History of Forests and Paleogeography of the USSR in the Holocene." USSR Academy of Sciences Press, Moscow.

Rudolph, K., and Firbas, F. (1924). Paläofloristische und stratigraphische Untersuchungen böhmischer Moore. *Botanisches Zentralblatt Beiheft* 41, Abt. II, Heft I/2, 1-162.

Sergin, V. Ya. (1974). Analysis of the state of general atmospheric circulation during annual and climatic variations on the basis of the hydrodynamic theory of similarity. *In* "Studies of the Glacier-Ocean-Atmosphere System" (S. Ya. Sergin, ed.), pp. 43-53. USSR Academy of Sciences, Far Eastern Scientific Center, Vladivostok.

Sernander, K. (1908). On the evidences of postglacial changes of climate furnished by the peat mosses of northern Europe. *Geologiska Föreningens i Stockholm* 30, 465-73.

Velichko, A. A. (1973). "The Natural Process in the Pleistocene." Nauka Press, Moscow.

Willett, H. S. (1953). Atmospheric and oceanic circulation as factors in glacial-interglacial changes of climate. *In* "Climate Change: Evidence, Causes, and Effects" (H. Shapley, ed.), pp. 51-72. Harvard University Press, Cambridge, Mass.

CHAPTER 27

Paleoclimatic Reconstructions Based on the Information Statistical Method

V. A. Klimanov

Paleoclimatic reconstructions from pollen data are based on the correlation of subrecent pollen spectra with present vegetation and climatic conditions. Various mathematical methods have been used to reconstruct paleoclimates from these correlations and fossil pollen data (Muratova et al., 1972; Klimanov, 1976; Geleta and Spiridonova, 1979; Klimanov et al., 1980; Webb and Bryson, 1972). The statistical information method used here (Klimanov, 1976) is adapted from that proposed in physical geography by Puzachenko and Moshkin (1969).

Every natural system has a certain degree of uncertainty or unpredictability, for example, the variability of mean annual temperature at a particular location. In statistical information theory, the uncertainty of a system is calculated from the Shannon-Weiner formula,

$$H_n(A) = -\sum_{i=1}^{n} P(a_i) = \log_2 P(a_i), \quad (1)$$

where $P(a_i)$ is the probability of event a_i in system A. The larger the value of H_n, the greater the uncertainty of the system.

In many systems, the uncertainty can be reduced if another event in a different system has occurred. For example, climatic variables of a region can be predicted with less uncertainty if the percentages of particular pollen taxa are known. In mathematical terms, if the conditional uncertainty of phenomenon A (in the example, a particular climate variable) is combined with known state of factor b_k (the percentage of the kth pollen taxon), then

$$H_n(A/b_k) = -\sum_{i=1}^{n} P(a_i/b_k) \cdot \log_2 P(a_i/b_k), \quad (2)$$

where $P(a_i/b_k)$ is the conditional probability of event a_i, provided that event b_k has occurred. In the calculations, the values of $P \cdot \log_2 P$ are taken from special tables. If the conditional uncertainty $[H_n(A/b_k)$ is less than the unconditional value $(H_n(A)]$, then event b_k decreases the uncertainty of phenomenon A, as in the example. The difference of these quantities is a measure of the extent to which the estimation of A becomes more certain when b_k is known, and it is referred to as the information (Y) on A contained in b_k:

$$Y(A/b_k) = H_n(A) - H_n(A/b_k). \quad (3)$$

Each state b_k carries only part of the information transmitted in the two-component system $A-B$ as a whole, with all of the information transmitted determined from the formula

$$T(A,B) = \sum_{k=1}^{m} P(b_k) \cdot Y(A/b_k). \quad (4)$$

In this context, information transmitted means the measure of uncertainty that can be eliminated. The magnitude of $T(A,B)$ depends not only on the strength of the correlation between the phenomenon A and the factor B but also on the magnitude of their uncertainties.

The dimensionality of information determined in a given system is eliminated by means of the normalized coefficient of contingency

$$K(A,B) = \frac{2T(A,B) - 1}{2H_{min} - 1}, \quad (5)$$

where H_{min} is the minimum uncertainty of phenomenon A or factor B.

Still another characteristic of correlation between two systems (e.g., climatic data and pollen spectra) is the partial measure of correlation, or the so-called classification criterion (C), which is defined as the ratio of the conditional probability of the phenomenon to its unconditional probability $C = [P(a_i/b_k)]/[P(a_i)]$. When C is maximized, the probability of phenomenon a_i having occurred is maximized at state b_k. In the example, C, and hence the probability of a particular mean July temperature (a_i), is maximized at the location where the particular pollen percentage of a taxon most highly correlated with that particular temperature occurs.

The reliability of the differences between the conditional and unconditional distributions at each state of factor B can be tested by using Pearson's chi-square test.

Table 27-1.
Classification Criterion (*C*) for Tree Pollen Types and Mean July Temperature

	Pollen Percentage Classes	Mean July Temperature (°C)										
		6	6-10	10-14	14-16	16-18	18-20	20-22	22-24	24-28	28	$Y(A/b_k)$
Arboreal pollen (%)	0.1-10	1	0	0	0	—	0	1	1	1	1″	0.468
	10.1-20	1″	1	0	0	—	0	1	1	1	1″	0.365
	20.1-30	1	1	1	0	0	0	1	0	0	—	0.671
	30.1-40	1	1″	1	0	0	1	1″	0	0	—	0.202
	40.1-50	0	1	1″	0	0	1″	1	0	0	—	0.677
	50.1-60	0	0	1″	1	1	1″	0	0	0	—	0.545
	60.1-70	0	0	0	1″	1	1″	0	—	—	—	0.511
	70.1-85	—	—	0	1	1″	1	0	—	—	—	0.767
	85.1-100	—	—	0	1	1″	0	—	—	—	—	1.352
Nonarboreal pollen (%)	0.1-10	—	—	0	1	1″	0	0	0	—	—	0.871
	10.1-20	0	0	1	1″	1	0	0	0	—	—	0.569
	20.1-30	0	0	1″	1	1	1″	0	0	—	—	0.587
	30.1-40	1″	1	1	0	0	1″	0	0	0	—	0.640
	40.1-50	1	1″	1	0	0	1″	1	0	0	—	0.472
	50.1-60	1	1″	1	0	0	1	1″	0	0	—	0.227
	60.1-70	1	1	0	0	0	1	1″	0	0	—	0.842
	60.1-80	—	1″	0	0	0	0	1	1″	0	0	0.846
	80.1-90	—	1	—	—	—	0	1	1	1″	1	0.572
	90.1-100	—	—	—	—	—	0	0	1	1″	1	0.991
Spores (%)	0-5	—	—	0	0	0	1	1	1	1	1″	0.021
	5.1-10	—	—	0	1	1″	1	0	0	0	0	0.319
	10.1-20	1	0	0	1″	1	1	0	0	0	—	0.431
	20.1-30	1	1″	1	1	0	1	0	—	—	—	0.463
	30.1-40	1″	1	1	0	0	1	0	—	—	—	0.788
	40.1-50	1″	1	1	0	0	1	0	—	—	—	0.082
	>50	—	—	1″	0	0	0	0	—	—	—	0.403
Quercus	0	1	1	1	1″	1	0	0	0	1	1	0.074
	0.1-5	—	—	0	0	1	1	1″	1	0	0	0.678
	5.1-10	—	—	0	—	0	1″	1	1	0	—	0.860
	10.1-20	—	—	0	—	—	1″	1	0	0	—	0.884
	20.1-35	—	—	0	—	—	1	1″	0	0	—	0.992
	35.1-55	—	—	—	—	—	1	1″	0	—	—	1.041
	55.1-75	—	—	—	—	—	1	1″	0	—	—	1.416
	75.1-100	—	—	—	—	—	0	1″	1	—	—	0.953
Tilia	0	1	1	1	1	0	0	0	1	1	1	0.060
	0.1-5	—	—	0	0	1	1	1″	0	0	0	0.664
	5.1-10	—	—	0	—	0	1″	1	—	—	—	1.117
	10.1-20	—	—	0	—	—	1″	1	—	—	—	1.370
	20.1-35	—	—	0	—	—	1″	—	—	0	—	1.443
	35.1-50	—	—	0	—	—	1″	—	—	—	—	2.150
	>50	—	—	—	—	—	1″	—	—	—	—	2.742
Carpinus	0	1	1	1	1	1	1	0	0	1	1	0.062
	0.1-10	—	—	—	0	0	1	1	1″	—	—	0.576
	10.1-25	—	—	—	—	0	1	1	1″	—	0	0.899
	25.1-45	—	—	—	—	—	0	0	1″	—	0	1.282
	>50	—	—	—	—	—	—	—	1″	—	0	1.930
Fagus	0	1	1	1	1	1	1	1	0	1	1	0.037
	>0	—	—	—	—	0	0	0	1″	—	0	0.799
Ulmus	0	1	1	1	1	1	0	0	1	1	1	0.019
	0.1-5	—	—	0	0	0	1	1″	0	0	0	0.517
	>5	—	—	—	—	—	1	1″	—	—	0	1.220
Betula S. *nana*	0	0	1	0	0	1	1	1	1	1	1	0.027
	0.1-5	1″	1	1	1	0	0	—	—	—	—	1.005
	5.1-10	1″	1	1	1	0	0	—	—	—	—	1.032

Table 27-1.
continued

	Pollen Percentage Classes	Mean July Temperature (°C)										
		6	6-10	10-14	14-16	16-18	18-20	20-22	22-24	24-28	28	$Y(A/b_k)$
	10.1-20	—	—	1″	1	0	0	—	—	—	—	1.123
	20.1-30	—	—	1″	0	0	0	—	—	—	—	1.490
	30.1	—	—	1″	0	—	—	—	—	—	—	1.820
Larix	0	1	1	0	0	1	1	1	1	1	1	0.012
	0.1-5	—	—	1	1″	1	1	0	—	—	—	0.586
	5.1-10	—	—	1″	1	0	—	—	—	—	—	0.930
	10.1-20	—	—	1″	1	0	—	—	—	—	—	1.370
	20.1	—	—	1″	1	—	—	—	—	—	—	1.930
Abies	0	0	1	1	0	1	0	0	0	1	1	0.009
	0.1-5	0	0	0	1″	1	1	1	1	0	—	0.251
	5.1-10	—	—	—	0	0	1	1″	—	—	—	0.985
	10.1-20	—	—	—	—	—	1	1″	—	—	—	2.742
	20.1	—	—	—	—	—	1″	1	—	—	—	2.742
Pices	0	1	1	1	—	0	0	1	1	1	1	0.055
	0.1-5	0	1	1	1	1	1″	1	1	0	—	0.364
	5.1-10	0	0	1	1	1″	1	0	0	0	—	0.572
	10.1-15	0	0	1	1″	1	1	0	0	0	—	0.476
	15.1-20	—	—	0	1	1″	0	0	0	—	—	1.028
	20.1-30	—	—	0	1	1″	0	0	—	—	—	0.965
	30.1-40	—	—	0	1″	1	0	—	—	—	—	1.284
	40.1-50	—	—	0	1″	1	0	—	—	—	—	1.443
	50.1	—	—	0	1″	1	—	—	—	—	—	1.930
Pinus	0	—	1	1″	—	0	0	—	—	0	0	1.174
	0.1-10	1	1	1	1	0	0	0	0	1	1″	0.048
	10.1-20	1	1	1″	1	1	1	0	0	1	1″	0.254
	20.1-30	1	1″	1	1	1	1	0	0	1	1	0.236
	30.1-40	1	1	1″	1	1	1	0	0	0	—	0.580
	40.1-60	1	1	0	0	1″	1	0	0	0	—	0.377
	60.1-80	1	1	0	1	1	1	1	1	1″	—	0.116
	80.1-100	—	1	0	0	1	1	1	1	1″	—	0.148
Betula	0	—	1	1	0	0	0	0	1	1	1″	−0.163
	0.1-10	0	0	0	0	0	0	1	1″	1	1	0.054
	10.1-20	1	0	0	1	1	1″	1	1	0	0	0.135
	20.1-30	1	1	1″	1	1	1	1	0	0	—	0.503
	30.1-40	1	1	1	1″	1	1	1	0	0	—	0.311
	40.1-60	1	1″	1	1	1	1	1	0	—	—	0.364
	60.1-80	1″	1	1	0	0	1	1	0	—	—	0.161
	80.1-100	—	—	1″	0	0	1	—	—	—	1	1.250
Salix	0	0	0	0	0	1	1	1″	1	1	1	0.003
	0.1-5	0	1	1″	1	1	1	1	1	0	0	0.220
	5.1-10	1	1	1″	1	0	0	0	0	0	—	0.572
	10.1-20	1	1″	1	1	0	0	0	0	0	—	0.066
	20.1-30	1	1	1″	0	0	0	0	0	—	—	2.742
	30.1	1″	1	1	0	0	0	0	0	—	—	1.242
Alnus	0	0	1″	1	0	0	0	0	1	1	0	0.058
	0.1-5	1	0	0	1	1	1″	1	0	0	1	0.419
	5.1-10	1	0	0	1	1	1	1	1″	0	1	0.112
	10.1-15	1	1	1″	1	1	1	1	1	1	1	−0.067
	15.1-20	1″	1	1	1	1	1	0	0	0	1	0.050
	20.1-30	—	—	1	1″	1	0	0	0	0	1	0.340
	30.1-50	—	—	1″	1	0	0	0	0	0	1	1.824
	50.1	—	—	0	—	—	—	—	—	0	1	1.139

Table 27-2.
Correlation (K) of Pollen Types with Climatic Parameters

Composition of Pollen and Spores	Normalized Contingency Coefficient K			
	Mean Annual Temperature	Mean July Temperature	Mean January Temperature	Total Annual Precipitation
Total Pollen Composition				
Arboreal pollen	0.047	0.093	0.032	0.050
Nonarboreal pollen	0.052	0.104	0.040	0.043
Spores	0.059	0.052	0.038	0.024
Arboreal Pollen Types				
Beech	0.384	0.333	0.343	0.314
Larch	0.325	0.124	0.351	0.141
Dwarf birch	0.273	0.164	0.267	0.103
Hornbeam	0.251	0.257	0.267	0.207
Fir	0.236	0.067	0.203	0.120
Oak	0.142	0.132	0.117	0.049
Linden	0.142	0.117	0.131	0.088
Elm	0.112	0.125	0.087	0.073
Willow	0.105	0.051	0.099	0.046
Spruce	0.075	0.079	0.029	0.050
Alder	0.046	0.050	0.039	0.055
Birch	0.046	0.027	0.039	0.028
Pine	0.040	0.036	0.042	0.029

Thus, information analysis enables one to obtain several numerical indices reflecting the correlation of subrecent spectra with modern climatic conditions: (1) the partial measures of the correlation (C) (classification criterion) between pollen percentages and the magnitude of a climatic variable, (2) the amount of information (Y) contained in pollen percentages (b_k) on particular climatic parameters, and (3) the general correlation between the pollen of any species and climatic variables (K).

Climatic variables used in the information analysis are the mean annual, January, and July temperatures and the total annual precipitation. Numerical values for these variables were taken from a climatic atlas of the USSR with a scale of 1:12,500,000. Data on subrecent pollen spectra were taken from the literature; about 800 spectra from 220 sites in plains areas of the USSR were used in the analysis. The pollen spectra were divided into 2 to 10 percentage classes according to the number of pollen grains of a given taxon, the variability of the percentages, and the reliability of data as established by the chi-square criterion. The climatic indices were divided into 9 to 11 classes.

Treatment of these data by information analysis resulted in the preparation of tables that revealed the correlation of subrecent pollen data and the four climatic variables. For example, Table 27-1 summarizes the classification criterion (C) obtained for subrecent tree pollen percentages and present-day mean July temperatures. In the table, unity indicates pollen percentages typical of a given mean July temperature, unity with a double prime indicates the most typical pollen percentages for a particular mean July temperature, and zero denotes pollen percentages not typical for that mean July temperature. The right side of the table shows the magnitude of climatic information $Y(A/b_k)$, contained in the pollen spectra. The greater $Y(A/b_k)$, the more reliable is the climatic estimate from a given pollen spectrum. Similar tables prepared for the other climate variables are not presented here.

Table 27-2 shows the values of the normalized coefficient of contingency (K) for the data, that is, the value of the correlation between the pollen types and the four climatic variables. Pollen of different tree species are arranged in order of decreasing correlation with mean annual temperature. On the whole, according to the value of K, the pollen of different tree species may be divided into three groups: (1) pollen strongly correlated (high value of K) with climatic variables (high-significance pollen): beech, hornbeam, larch, dwarf birch, oak, linden; (2) medium-significance pollen: elm, fir, spruce; (3) low-significance pollen: willow, alder, birch, pine. These groups indicate which arboreal pollen provide the best estimate of each climatic characteristic.

Thus, a statistical relationship exists between subrecent spectra and present climatic conditions, and the pollen of each tree species may contain much or little information about climate. This relationship can be used to reconstruct past climates. The most probable value of a paleoclimatic variable for a particular pollen spectrum corresponds to the largest sum of the classification criterion (C) obtained from all the pollen in the spectrum. If two climatic classes give equal summations, the more probable value is the one corresponding to the larger sum of maximum classification criterion (double prime unity). In the case of a dual solu-

Table 27-3.
Example of a Reconstruction of Mean July Temperatures

Pollen Types	Pollen Percentage	Mean July Temperatures (°C)									
		6	6-10	10-14	14-16	16-18	18-20	20-22	22-24	24-28	28
Arboreal pollen	31	1	1″	1	0	0	1	1″	0	0	—
Nonarboreal pollen	13	0	0	1	1″	1	0	0	0	—	—
Spores	56	—	—	1	0	0	0	0	—	—	—
Oak	16	—	—	0	—	—	1″	1	0	0	—
Elm	2	—	—	0	0	0	1	1″	—	—	—
Linden	37	—	—	0	—	—	1″	—	—	—	—
Spruce	3	0	1	1	1	1	1″	1	1	0	—
Pine	35	1	1	1″	1	1	1	0	0	0	—
Birch	7	0	0	0	0	0	0	1	1	1	1
Sum of classification criteria		2	3	5	3	3	6	5	2	1	1

tion in which the reconstructed climatic parameter values are separated by one class, the more probable value of the climatic parameter lies between them. A negative value (empty boxes in Table 27-1) is changed to a positive absolute value in the case of high-significance pollen. For low-significance pollen, an absolute value is given when two or more kinds of pollen are involved.

An example of a climatic reconstruction is presented in Table 27-3, in which mean July temperature is determined from a pollen spectrum. First, the highest-significance pollen in the spectrum is considered, and the July temperature classes for which it is characteristic, uncharacteristic, and negative determined from the tables. In the example, the most significant pollen is that of oak, and the probability characteristics in Table 27-1 for 16% oak pollen indicate that this percentage is characteristic of temperatures between 18°C to 20°C and 20°C to 22°C, uncharacteristic of temperatures between 10°C to 14°C and 22°C to 28°C, and negative at temperatures below 10°C, at 14°C to 18°C, and above 28°C. These estimates are made for each component of the pollen spectrum (Table 27-3). The maximum sum of classification criteria is calculated to be at temperatures of 18°C to 20°C. These temperatures are also not negated by any of the pollen taxa. Hence, this temperature range is most probable for the given spectrum. In this example, the most probable temperature was selected only with respect to the sum of classification criteria, but there are cases in which the sum would apply to two or more temperature values. In such a case, the value corresponding to the largest sum of maximum classification criteria is selected.

The reliability and accuracy of the prepared tables for reconstructing climatic variables were determined by reconstructing present-day mean July temperatures from modern spectra. In the reconstruction, 190 spectra were analyzed. An incorrect result was obtained from 39 spectra, and from 27 spectra out of the 39 the reconstructed and actual climatic values were in adjacent classes. Because the values of all the climatic variables are discrete and modern spectra could be taken at the boundary of neighboring percentage classes, such discrepancies are not significant. Appreciable discrepancies of more than one class were predicted by 12 (6.3%) of the assemblages. In the reconstruction of mean annual temperatures, appreciable discrepancies were 6.7%, for mean January temperatures 8.2%, and for total annual precipitation 7.0%. The mean statistical errors for the reconstructions were ±0.6°C for mean annual temperature, ±1°C for mean January temperature, ±0.6°C for July temperature, and ±25 mm for total annual precipitation. From these estimates it appears that the reliability and accuracy of our reconstruction system is satisfactory.

The calculated paleoclimatic data are useful in constructing paleoclimatic maps for specific time intervals and areas. Three possible methods of constructing such maps are used by various workers: theoretical, theoretical-empirical, and empirical. Examples of the theoretical method include a paleotemperature model of dry-land areas between 20,000 and 18,000 yr B.P., used by American scientists (Gates, 1976), and paleoclimatic maps of the USSR for the climatic optimum of the Holocene, based on recent climatic fluctuations caused by changes in the general circulation of the atmosphere (Avenarius et al., 1978). The theoretical-empirical methods include those based on a small amount of factual data, which are subsequently extrapolated over vast areas by means of modeling. In particular, there are paleoclimatic maps (Muratova et al., 1980) for the climatic optimum of the Holocene in the USSR, which are based on trend analysis of pollen data from 26 sites in the USSR. Empirical methods are based on a large body of factual data. Paleoclimatic maps are constructed by drawing isolines of paleoclimatic variables through numerous points at which the quantitative paleoclimatic indices were determined.

In this chapter the empirical method was used to construct paleoclimatic maps of the European USSR during the climatic optimum of the Holocene. Calculations of quantitative indices of paleoclimate were based on 90

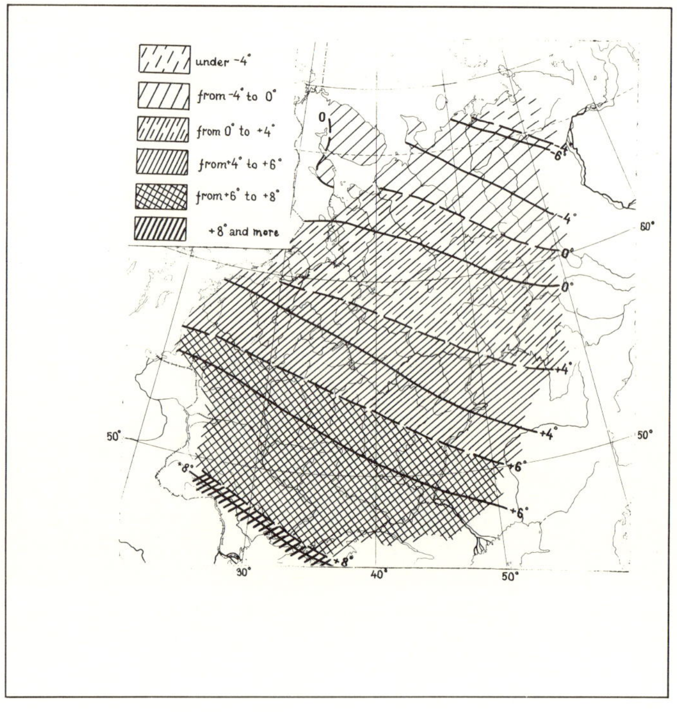

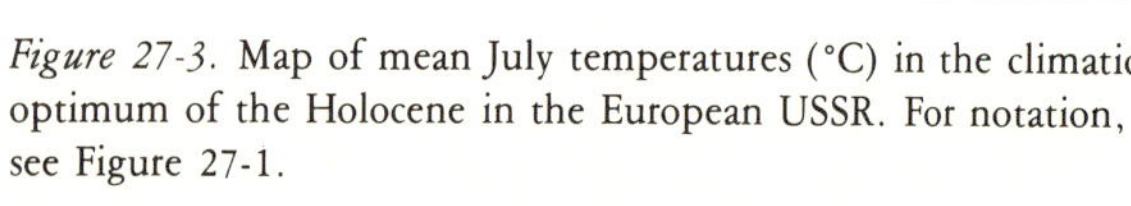

Figure 27-1. Map of mean annual temperatures (°C) in the climatic optimum of the Holocene in the European USSR.

Figure 27-2. Map of mean January temperatures (°C) in the climatic optimum of the Holocene in the European USSR. For notation, see Figure 27-1.

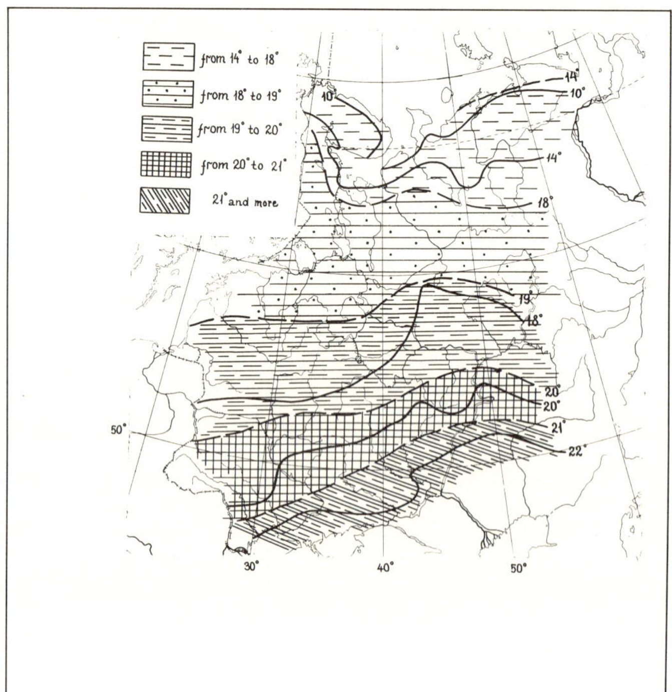

Figure 27-3. Map of mean July temperatures (°C) in the climatic optimum of the Holocene in the European USSR. For notation, see Figure 27-1.

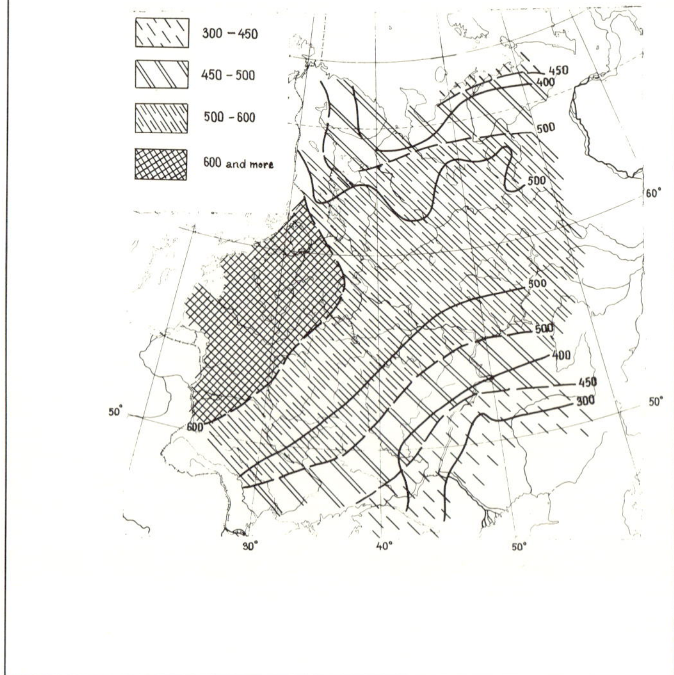

Figure 27-4. Map of mean total annual precipitation (mm) in the climatic optimum of the Holocene in the European USSR. For notation, see Figure 27-1.

Holocene spectra (see Figure 27-2 for location of spectra) taken from the literature using the statistical-information method outlined above. The calculations used spectra that were radiometrically dated, of which, unfortunately, not enough were available, as well as those spectra inferred to be from the Atlantic period. The time interval of the reconstructed paleoclimatic conditions is thus estimated at 5500 ± 500 yr B.P.

Maps of mean annual paleotemperatures (Figure 27-1) show that a rise in temperature took place over the entire European USSR. The maximum rise in temperature was observed north of latitude 60°N and amounted to between 2°C and 2.5°C. South of latitude 60°N the increase was less, and in the region of latitude 50°N the deviations from present-day values were less than 1°C. Thus, increased temperatures in the north apparently resulted from activation of the North Atlantic Current (Gulf Stream).

Changes in mean January temperatures (Figure 27-2) were qualitatively similar to the changes in annual temperatures. However, it is difficult to detect a south-to-north trend because of the almost meridional direction of the January isotherms. One does see a clear trend in temperature variation from west to east. In the Baltic region the January temperatures at the climatic optimum were 1°C to 1.5°C higher than present-day ones, whereas in the region of the middle course of the Ural River they were 3°C to 3.5°C higher. Thus, mean January temperatures increased from west to east, perhaps as a result of a diminished influence of the Siberian High during the winter.

The mean July temperature during the climatic optimum increased (Figure 27-3) over most of the territory. The maximum increase in July temperatures was north of latitude 65°N in the northeast amounting to between 3°C and 4°C, whereas south of latitude 45°N almost no increase in temperatures was observed.

The map of mean total annual precipitation (Figure 27-4) during the climatic optimum shows localized increases in precipitation. Almost no change in precipitation is noted for the Baltic and some central regions. The most appreciable increases in precipitation occurred in the northern European USSR and in southeastern European USSR, where precipitation was 50 to 100 mm greater than today.

Thus, the proposed method of paleoclimatic reconstruction shows that during the climatic optimum of the Holocene a universal rise in temperatures occurred in the European USSR. The greatest rise in temperatures was observed in the northern regions, with lower increases to the south. The total mean annual precipitation increased mainly in the northern and southeastern regions of the European USSR. On the whole, the climate of the European USSR was less continental and warmer than it is at the present time.

References

Avenarius, I. G., Muratova, M. V., and Spasskaya, I. I. (1978). "Paleogeography of Northern Eurasia in the Late Pleistocene and Holocene and Geographical Forecasting." Nauka Press, Moscow.

Gates, W. L. (1976). Modeling the ice-age climate. *Science* 191, 1138-44.

Geleta, I. F., and Spiridonova, Ye. A. (1979). New possibilities for application of palynological studies of lacustrine deposits using multivariate statistical analysis to reconstruct the Holocene climate. *In* "History of Lakes," Part I (N. A. Florensov, ed.), pp. 13-17. Irkutsk University Press, Irkutsk.

Klimanov, V. A. (1976). A method of reconstruction of quantitative characteristics of past climate. *Moscow State University, seria geograficheskaya, Vestnik* 2, 92-98.

Klimanov, V. A., Liberman, A. A., and Muratova, M. V. (1980). Reconstruction of paleoclimatic conditions of the Pleistocene and Holocene from data of palynological analysis using mathematical methods. *In* "Most Recent Tectonics: Most Recent Deposits and Man, No. 7" (N. I. Nikolayev, ed.), pp. 178-82. Moscow State University Press, Moscow.

Muratova, M. V., Boyarskaya, T. D., and Liberman, A. A. (1972). Application of probability theory to the reconstruction of paleoclimatic conditions from data of palynological analysis. *In* "Most Recent Tectonics: Most Recent Deposits and Man, No. 3," pp. 239-46. Moscow State University Press, Moscow.

Muratova, M. V., Suyetova, I. A., Burashnikova, T. A., and Krolichenko, Ye. I. (1980). Climate and vegetation of a zone on the territory of the USSR 5000-6000 years ago. *Priroda* 7, 42-45.

Puzachenko, Yu. G., and Moshkin, L. V. (1969). Information logic analysis in medicogeographical research. *In* "Meditsinskaya Geografiya," No. 3, pp. 5-74. Moscow.

Webb, T., III, and Bryson, R. A. (1972). Late- and postglacial climate change in northern Midwest, USA: Quantitative estimates derived from fossil pollen spectra by multivariate statistical analysis. *Quaternary Research* 2, 70-115.

CHAPTER 28

Holocene Climatic Change

N. A. Khotinskiy

Data obtained on the history of the Holocene vegetation and climate show that past climatic fluctuations were intermittent and nonuniform across northern Eurasia. The three main types of Holocene climatic fluctuations are the Atlantic-continental type in the European USSR, the continental type in Siberia, and the oceanic or Pacific Ocean type in the Far East (Figure 28-1). There is also the Atlantic type in northwestern Europe, which was very similar to the Atlantic-continental type except for having more precipitation. An interglacial climate developed everywhere during the Holocene, and synchronous climatic changes caused abrupt and simultaneous responses over vast areas. Such sudden changes, alternating with periods of relative stability, occurred at the late-glacial/postglacial boundary, at the Boreal-Atlantic boundary, at the Atlantic-Subboreal boundary, and at other boundaries of the Blytt-Sernander scheme.

Holocene Climate

The main Holocene thermal maxima of northern Eurasia are the Boreal (8900 to 8300 yr B.P.), the Late Atlantic (6000 to 4600 yr B.P.), and the Middle Subboreal (4100 to 3200 yr B.P.). The Boreal thermal maximum affected Siberia and the Far East, the Middle Subboreal maximum affected the northeastern European USSR, and the Late Atlantic maximum affected the forest zones of northern Eurasia. Precipitation increased in Siberia during the Boreal, whereas northern Europe and the Far East were dry. However, the high precipitation in Europe and the Far East during the last 3200 years has been accompanied across Siberia by an increase in aridity and continentality.

These phenomena are related to significant rearrangements of atmospheric circulation above northern Eurasia and to appreciable changes in the distribution of Arctic ice. The unexpected increase in precipitation in Siberia during the Boreal (Figure 28-2A) could have resulted from reduced ice cover in the Arctic, penetration of the Gulf Stream from the Atlantic into the Barents Sea, and northwestward movement of moist air masses around Scandinavia and to Siberia.

Direct transport of moist air masses from the Atlantic across Europe to Siberia is not supported by paleobotanic data, which indicate that in northern Europe the climate during the Boreal was relatively dry and cool. A more or less stable anticyclone associated with glacial remnants in Scandinavia probably predominated, and it partially blocked air masses from the west.

Following the disapperance of the glaciers and the disintegration of the North European High during the Atlantic period, moist Atlantic air masses penetrated farther east than they do today (Figure 28-2B). The winter Siberian High probably decreased during the Boreal and Atlantic periods.

The cooling and increased continentality of the Siberian climate during the Early Subboreal is explained by the increased Arctic ice cover, the intensified Siberian High, and the decreased transport of moist air masses from the west (Figure 28-2C). This cooling also occurred in northern Europe and in the Far East, although it was somewhat attenuated. Moist air masses penetrated into these regions, and together with a general cooling they led to increased precipitation, which has persisted to the present day.

The optimum combination of temperature and precipitation for most of northern Eurasia coincided with the Late Atlantic (AT-2) thermal maximum (6000 to 4600 yr B.P.), which is the Holocene climatic optimum in the USSR. At that time, the interaction of temperature and precipitation was balanced (i.e., the humidity factor approached 1.0) over vast areas. The broad-leaved forest underwent its maximum expansion, and evaporation reached its highest values. As discussed earlier, maximum biologic productivity can occur under such conditions. Optimum climatic conditions also appeared during the Boreal thermal maximum in parts of Siberia and during the Middle Subboreal thermal maximum in the northeastern European USSR.

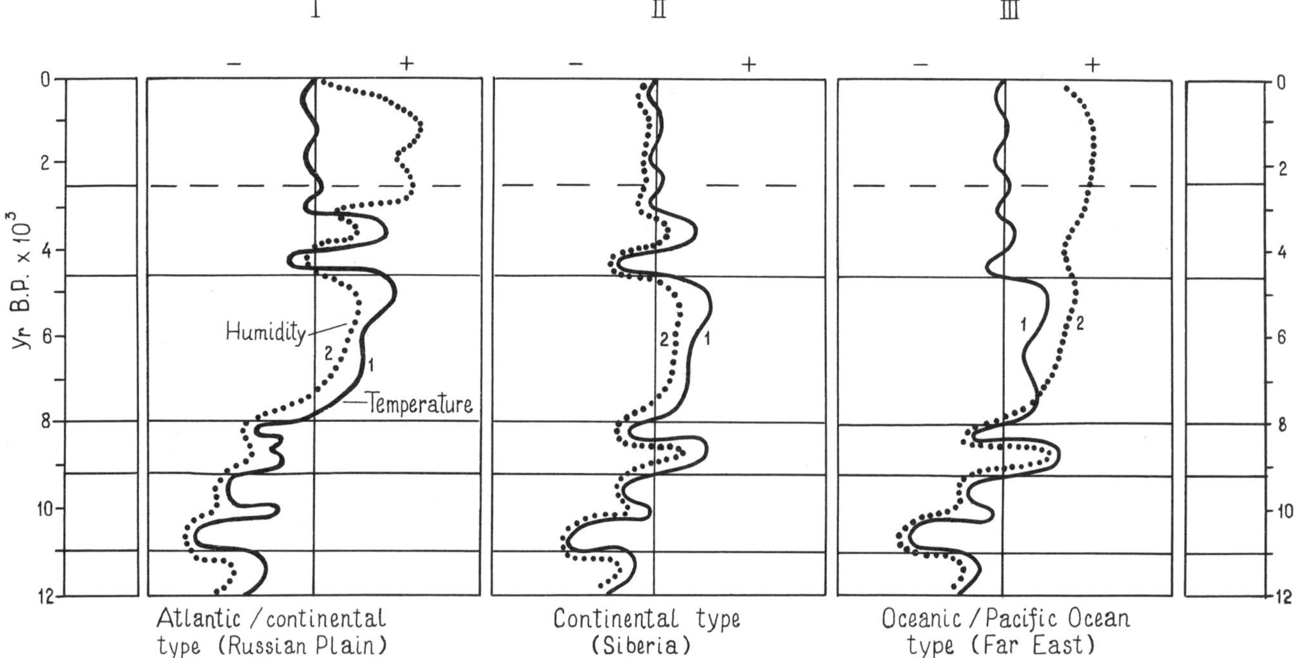

Figure 28-1. Three types of climatic oscillation in the Holocene.

Thus, the climatic optimum may have occurred in different Holocene stages in individual areas, but on a regional level it was limited to the Late Atlantic phase.

Today, we are living at the end of a warming that has been similar to those of past interglacial epochs. Each glacial epoch can be divided into an initial cryohygrotic stage of cold and moist climate and a succeeding cryoxerotic stage of cold and dry climate (Grichuk and Grichuk, 1960). Interglaciations can be subdivided into an intitial thermoxerotic (warm and dry) stage and a thermohygrotic (warm and moist) stage (Figure 28-3).

This general scheme can be compared with the three types of Holocene climatic fluctuations identified for northern Eurasia. The late-glacial climate was generally very continental, cold, and dry, as part of the cryoxerotic stage of glaciation. The postglacial interval is divided into two stages. On the Russian Plain (European USSR) and in the Far East, the thermoxerotic stage occurred in the first half of the postglacial and the thermohygrotic stage in the second half. In Siberia, on the other hand, the thermohygrotic stage occurred first, followed by the thermoxerotic stage.

The general scheme of Quaternary climatic stages shows that modern natural conditions on the Russian Plain (Figure 28-3) and in the Far East (Figure 28-1) coincide with the end of an interglacial thermohygrotic stage, which is usually followed by a glacial cryohygrotic stage. In support of this conclusion, I note that the phytogeographic and climatic stages of the Mikulino (Eemian) Interglaciation and the Holocene are generally similar on the Russian Plain. The thermoxerotic stage of the Mikulino Interglaciation (phases M_2, M_3, M_4), according to V. P. Grichuk's scheme, generally corresponds to the Preboreal (PB), Boreal (BO), and in part the Atlantic (AT) periods of the Holocene. In both cases, a rise in temperature caused the degradation of continental ice sheets and a unidirectional change in the vegetation by which birch-pine forests of boreal type gave way to thermophilous broad-leaved forests. These trends continued until the end of the thermoxerotic stage, at approximately the climatic optimum of the Mikulino Interglaciation and Holocene.

The thermohygrotic stage of the Mikulino Interglaciation (phases M_5, M_6, M_7) closely resembles the Subboreal and Sub-Atlantic periods of the Holocene. Both intervals are characterized by decreased temperatures, a unidirectional decline in thermophilous vegetation, the development of moisture-loving forests, and in particular the expansion of dark spruce forests.

Finally, the M_8 phase between the Mikulino Interglaciation and the subsequent Valdai Glaciation may be compared with the latest Sub-Atlantic phase (SA-3) of the last 700 to 800 years. In these phases, one notes a uniform degradation of spruce forests, an almost complete disappearance of broad-leaved species, and a development of birch and pine forests.

Such a comparison may lead to the conclusion that we have already entered the cold and humid cryohygrotic stage of a new ice age, a conclusion that requires well-grounded verification. Changes in forest composition in the central Russian Plain at the end of the Sub-Atlantic period could have been caused not only by environmental changes but also by human activity. However, the anthropogenic factor, frequently cited in the literature, is probably unimportant. Spruce forests began to degrade actively

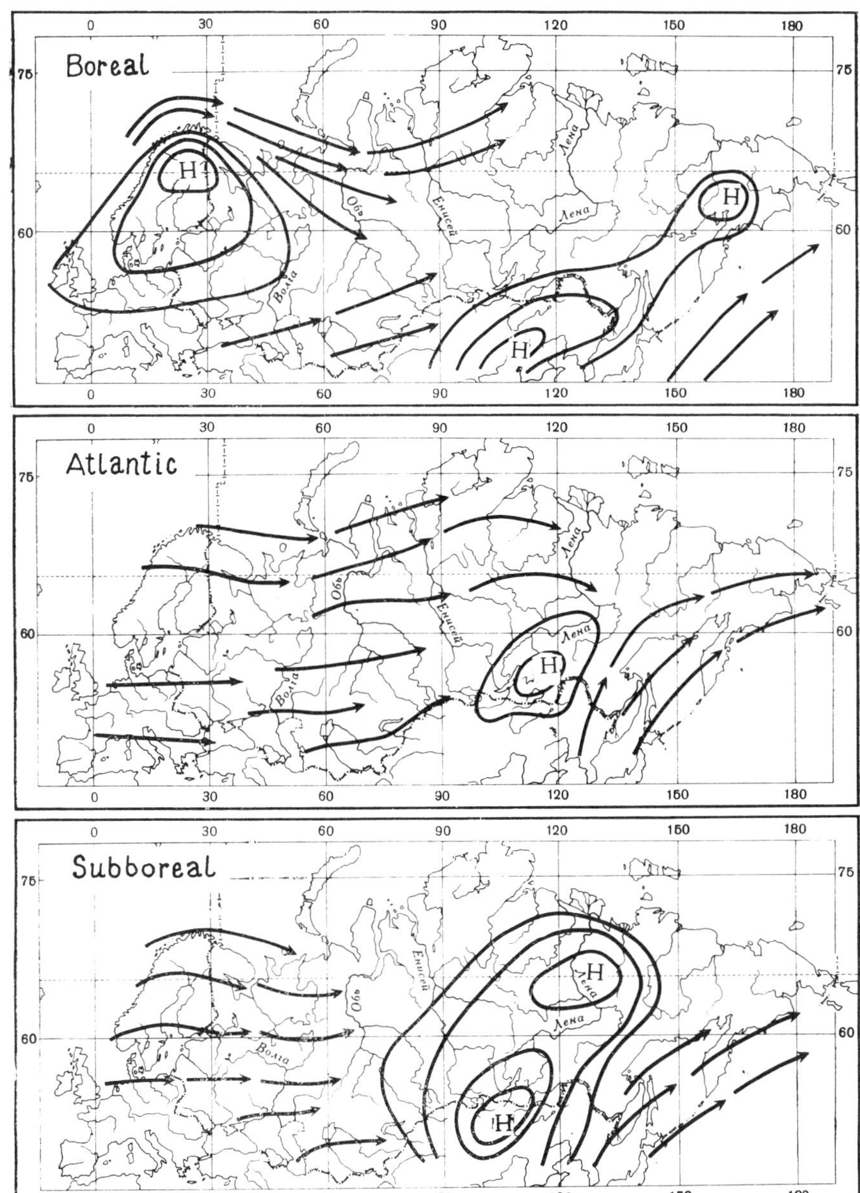

Figure 28-2. Anticyclones (H) and paths of cyclones in northern Eurasia in the Holocene (Khotinskiy, 1977).

in these regions around 700 to 800 yr B.P., that is, in the 13th century A.D., when the human inhabitants were incapable of causing such major, rapid, and widespread changes in the vegetation. Intensive tree felling and plowing of the land on a large scale did not occur in the central Russian Plain until the 17th and 18th centuries A.D. In addition, the spruce taiga began to shrink in forests throughout the northern European USSR in areas inhabited by small tribes of hunters and fishermen. Intensive cutting of virgin forests in these northern regions did not begin until the 20th century. Thus, the replacement of spruce forests by birch groves in the 13th century resulted from environmental changes on the Russian Plain rather than from anthropogenic causes, just as earlier changes in forest vegetation were the result of climatic cooling. This cooling resulted from the onset of the so-called Little Ice Age, which held many extratropical regions of the Northern Hemisphere in its cold grip from the end of the 15th until the middle of the 19th century A.D.

The climate then warmed, reaching a maximum in the 1930s and 1940s. During this warm period, Arctic ice cover shrank, and the transport of moist, warm westerly air masses increased. In the northern European USSR, the climatic warming caused a northward advance of spruce forests some 25 to 30 km into the tundra zone.

Today, a new intracentury cooling probably has not been completed. Since the 1940s, the global mean annual temperature has dropped by 0.6°C, resulting in increased Arctic ice cover, reduced growing season, increased meridional transport of air masses, and increased temperature difference between the poles and the equator (Borisenkov, 1976). These trends, related to a natural climatic cooling,

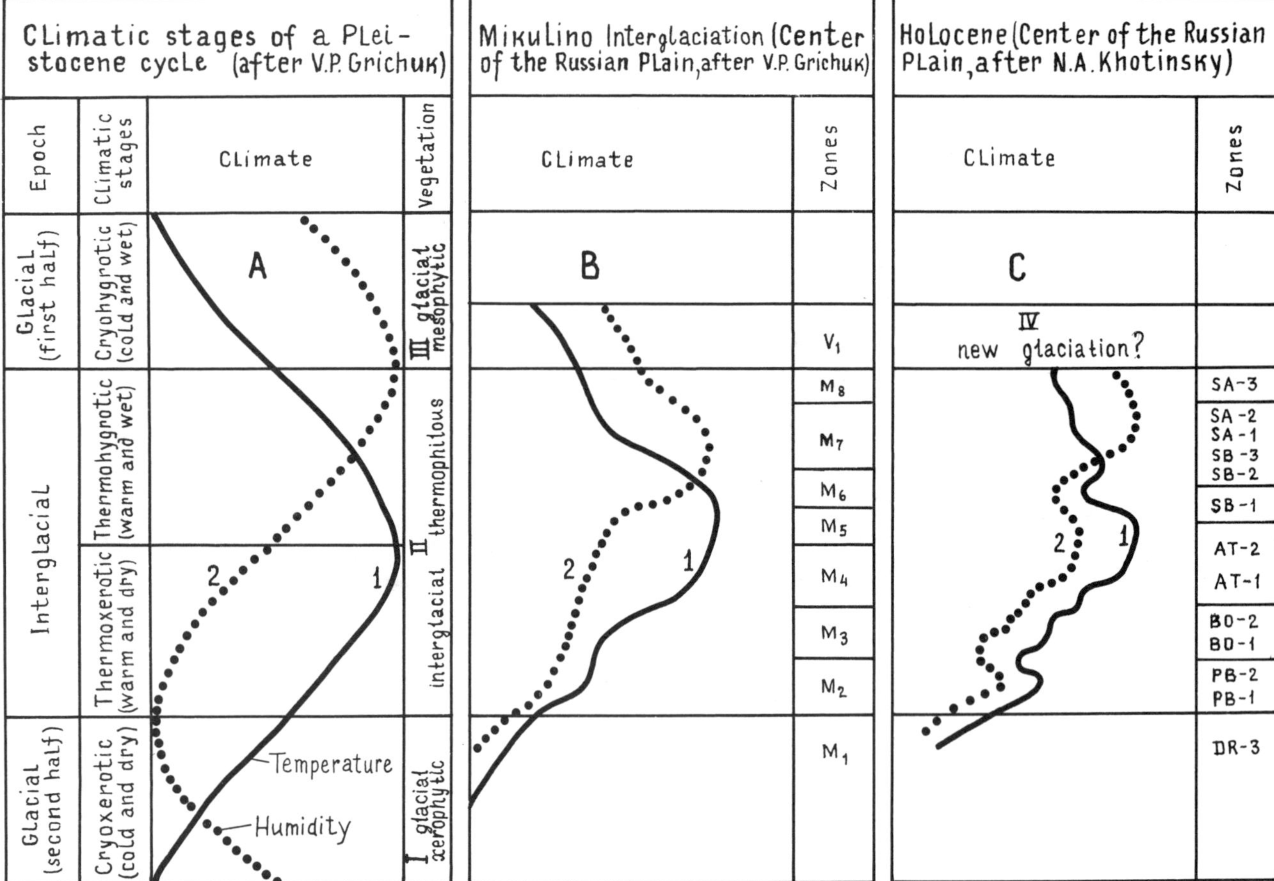

Figure 28-3. Correlation of (A) climatic and phytogeographic stages of Pleistocene cycles, (B) Mikulino Interglaciation, and (C) Holocene. (Abbreviations: M_1, periglacial complex; M_2, birch-pine forests; M_3, birch-pine forests with an admixture of broad-leaved species; M_4, broad-leaved forests of *Quercus* and *Tilia*; M_5, broad-leaved forests with a maximum of *Tilia*; M_6, broad-leaved forests with a maximum of *Carpinus*; M_7, spruce forests; M_8, pine forests with spruce and birch; V_1, birch forests with elements of tundra vegetation; DR-3, Younger Dryas periglacial complex; PB, Preboreal birch-pine forests; BO, Boreal birch-pine forests with an admixture of broad-leaved species; AT, Atlantic broad-leaved forests of *Quercus, Ulmus,* and *Tilia*; SB-1, Early Subboreal birch-pine forests; SB-2, Middle Subboreal broad-leaved forests; SB-3, SA-1, SA-2, Late Subboreal, Early Subatlantic, Middle Subatlantic spruce forests; SA-3, Late Subatlantic birch-pine forests.)

may continue in the future, although they may not signify a definitive transition to a new ice age. Situations similar to "glacial" ones repeatedly arose in earlier stages of the Holocene, for example, around 8300 to 8000 yr B.P. At 4900 and 4600 yr B.P., a general cooling occurred in northern Eurasia, and "glacial conditions" prevailed in North America during the Late Holocene. Although these coolings caused a reactivation of glaciers, they did not exceed the limits set for an interglaciation. A true epoch of continental glaciation can begin only with the onset of a prolonged cold and moist (cryohygrotic) stage.

Future Climate

We have already entered a period of highly unstable climate, with a general cooling trend in the Sub-Atlantic period. In other words, the Holocene as an interglacial epoch has "exhausted itself," and at the present time it is close to a new glaciation (Khotinskiy, 1977). As a further indication of an approaching ice age, in northern Eurasia the rate of peat accumulation, which started about 10,000 yr B.P., has decreased in the Late Holocene as a result of climatic cooling. For a similar reason, the accumulation of organic matter in ocean sediments has slowed down appreciably in the last 4000 to 5000 years (I. V. Grakova, V. M. Kuptsov, and A. S. Moskalev, personal communication). When will the transition to the new ice age occur? How will the environment change? The answer to these complex questions can be arrived at only after further study of the natural cyclicity in the Pleistocene and Holocene. Of particular interest is the study of transitions from interglacial to glacial conditions. Variations in solar activity, which probably determine climatic fluctuations, are cyclic but not periodic (Eygenson, 1963). The cycles do not resemble each other in either amplitude or duration. Solar activity is multicyclic, and high-order cycles affect low-order ones. Therefore, one should not expect a rigorous periodicity in paleoclimatic oscillations.

In the Holocene, the durations of warm and cold or moist and arid phases vary widely and do not fall into equal chronologic intervals. The nonuniformity makes it difficult to obtain precise predictive reconstructions and in particular to determine the onset time of a future glaciation.

In addition, climatic development shows a gradual change of the system in a specific direction. During the Quaternary, there has been a general cooling, with each succeeding glacial or interglacial epoch being slightly colder than its preceding paleogeographic analogue (Velichko, 1973). The Holocene is the "coldest" interglaciation yet, and, if this trend is maintained, there is no doubt that future coolings will be even more pronounced than during the last ice age.

At the same time, paleogeographic analysis of situations favorable to humans may be the basis for carrying out major environmental transformation projects in the future. Such projects must be approached with great care, for the improvement of conditions in some regions may result in undesirable consequences in other areas. The paleogeographic paradox of the Holocene of northern Eurasia indicates that situations may arise in which, as the precipitation increases in some regions, other areas will become more arid. The second half of the Atlantic period—the Holocene climatic optimum—was marked by the greatest flourishing of the plant and animal world and the most favorable conditions for human habitation in all of northern Eurasia. This period deserves particular attention, for it can serve as a paleogeographic standard for the development of models and transformational projects of the natural environment.

References

Borisenkov, Ye. P. (1976). "Climate and Its Variation." Znaniye Press, Moscow.

Eygenson, M. S. (1963). "The Sun, Weather, and Climate." Gidrometeoizdat Press, Leningrad.

Grichuk, M. P., and Grichuk, V. P. (1960). On the periglacial vegetation of the territory of the USSR. *In* "Periglacial Phenomena in the Territory of the USSR" (K. K. Markov and A. Popov, eds.), pp. 66-100. Moscow State University Press, Moscow.

Khotinskiy, N. A. (1977). "The Holocene of Northern Eurasia." Nauka Press, Moscow.

Velichko, A. A. (1973). "The Natural Process in the Pleistocene." Nauka Press, Moscow.

Dispersal of Primitive Cultures

CHAPTER 29

Paleolithic Cultures in the Late Pleistocene

N. D. Praslov

Determining the first appearance of Paleolithic man in the USSR is hampered by the scarcity of sites. Paleolithic settlements are frequently buried and are discovered by accident. Except for cave sites, they are not well suited to systematic investigation. Discoveries in northern Eurasia during the last two decades have made it necessary to reexamine the time of man's arrival in the USSR, as well as the nature of Paleolithic culture in the European USSR, Central Asia, Kazakhstan, Siberia, and the Far East.

The first people in the USSR came from various centers of earlier habitation and arrived at different times. The natural environment and the level of cultural development played an enormous role in the settlement process. The temperate zone, with its seasonal climatic oscillations, abrupt temperature changes, and lasting snow cover, did not become settled until man had learned to use fire. During the last glaciation, the periglacial zone could not be settled without warm clothing and well-heated shelters. Fire, clothing, and dwellings were the main achievements of Paleolithic peoples, which along with the development of hunting weapons permitted expansion of human populations.

Early and Middle Paleolithic

The southernmost regions of the USSR (the Caucasus, the southern Russian Plain, Central Asia, and Kazakhstan) were occupied during the Early and Middle Pleistocene (Praslov, 1968; Lyubin, 1970). Acheulian sites have been found in several caves in the Caucasus (Azykh, Tsona, and Kudaro I and III), and Early Paleolithic open-air sites in the Caucasus (Satani-Dar and Ignatenkov Kutok), the southern part of the Russian Plain (Gerasimovka, Khryashchi, and Mikhaylovskoye), Transcarpathia (Korolevo), and Central Asia and Kazakhstan (Karatau, Lakhuti I, etc.). Stone implements are reported in Early and Middle Pleistocene deposits in the Altay and in the Amur region (Okladnikov, 1968), but their age has not been confirmed. Apparently, in Early and Middle Pleistocene Paleolithic humans did not go north of latitude 48°N, possibly because of the presence of large continental ice sheets.

Favorable climatic conditions during the Mikulino Interglaciation enabled primitive people to migrate as far as latitude 54°N. Sites from that period include Khotylevo I on the Desna River near Bryansk and Sukhaya Mechetka on Volga River outside Volgograd. The Urta-Tyube campsite in the southern Urals also dates to that time (Bader and Matyushin, 1973).

At the Khotylevo site, numerous stone implements and faunal remains were collected from the basal pebble horizon along the 20-m terrace of the Desna River (Velichko, 1961; Zavernyayev, 1978). This horizon lies about 5 m above the present level of the Desna River and is overlain by 15 m of sand and loesslike loam containing traces of soil formation. The associated molluscan fauna suggests that climatic conditions during the time of settlement were much better than present ones (Motuz, 1967). The stone inventory, including spearheads sharpened on both sides and numerous scrapers, is typical of the Mousterian culture.

The Sukhaya Mechetka site dates back to the Mousterian culture. The cultural layer occurs in a fossil soil, about 20 m below the present surface. It is overlain by Atelian-age sandy loam and stratified sediment deposited during an Early Khvalyn transgression of the Caspian Sea 54,000 to 35,000 years ago. The cultural horizon lies above the Upper Khazar sands, which are more than 100,000 years old (Kaplin et al., 1977). Hence, the thick fossil soil and cultural layer probably date to the Mikulino Interglaciation. Faunal (Zamyatnin, 1961) and pollen analyses (Ye. S. Malyasova, unpublished data) suggest a steppe environment with some trees. The stone inventory of over 6000 items contains bilaterally worked weapons including numerous knives, pointed instruments, and scrapers.

The onset of the Valdai cold epoch apparently did not restrict the area of habitation, since Mousterian sites have

Figure 29-1. Kostenki I: Paleolithic dugouts of the upper cultural layer (1972 excavation). (Photograph by N. D. Praslov.)

been found throughout the region. Most interesting are those at Betovo near the Khotylevo site (Tarasov, 1977). Paleolithic hunters adapted to colder conditions by improving their clothing and dwellings. Siliceous piercing tools used to sew hides together were found in some Mousterian sites from that time (Praslov and Semenov, 1967), including Rozhok I on the Sea of Azov (Praslov, 1968). Remains of Mousterian-age dwellings were discovered at Molodova I and V on the Dnestr River (Chernysh, 1965).

The range of Paleolithic people expanded into northern Asia during Mousterian times, and sites have been discovered in the Altay, the Kuznetskiy Ala-Tau and Khakasiya, the Sayan, and the Angara region. Campsites include the Ust'-Kanskaya and Dvuglavka grottos and the Strashnaya and Denisov caves. In the future, Mousterian sites may also be found in other regions of Siberia and the Far East.

Late Paleolithic

The treatment of hides improved as a result of the advent of the prismatic technique of splitting stone and the creation of new types of scrapers, as did the construction of hearths and dwelling structures. These improvements enabled primitive man to move much farther north, despite progressive cooling in the Late Paleolithic (40,000 to 10,000 yr B.P.) (Velichko, 1973).

In eastern Europe, the Late Paleolithic campsites of Sungir', located slightly north of latitude 56°N and Talitskaya on the Chusovaya River, record the next stage in the settlement of the Russian Plain. These sites probably date to the Bryansk Interstade. The northernmost Late Paleolithic campsite near the village of Byzovaya on the middle Pechora River, approximately 175 km from the Arctic Circle, also dates to that interstade (Kanivets, 1976). Mammoth bones from this site were dated at 25,450 ± 380 yr B.P. (Guslitser and Liyva, 1972). The date for the Sungir' site is similar: 25,500 ± 200 (GrN-5425). The environmental conditions in the Central Russian Plain during the Bryansk Interstade were similar to those of the present day near central Yakutiya (Velichko and Morozova, 1975). The clothing of Paleolithic hunters preserved in burials at Sungir' also indicates that the climate was very cold (Bader, 1977).

The climatic minimum after the Bryansk Interstade apparently caused Late Paleolithic populations to move south to latitude 56°N. However, the extent of the migration cannot yet be determined.

Figure 29-2. Kostenki I: remains of mammoth tusks overlying a dugout. (Photograph by N. D. Praslov.)

In loess regions of the Russian Plain, a rich culture developed, with dwellings built from the large bones of animals, mainly mammoth (Figures 29-1 through 29-4). Dwellings made of bones and earth, dated between about 20,000 and 17,000 yr B.P., were found in loesslike loams on the second terrace of the Don River at Kostenki II, Borshchevo I, and other sites. In the Dnepr Basin, similar dwellings have been studied at Dobranichevka, Mezin, and Mezhirichi. Paleobotanic analyses show that periglacial steppes and in some areas semideserts were widespread on the Russian Plain at that time (Grichuk, 1969). Small islands of forest grew only in river valleys. What would seem to be a very unfavorable time for large grazing animals is characterized, in fact, by the largest quantity of mammoth, horse, and bison bones. Judging from the fact that many of the remains are of adult and elderly mammoth specimens, living conditions must have been favorable for them, and the total amount of biomass required to feed enormous herds of large animals was apparently adequate.

Mammoth hunting was important in the central Russian Plain, and bison hunting predominated in the southern regions. Settlements of bison hunters include the Amvrosiyevka campsite in the Azov region and Zolotovka I in the basin of the lower Don near the mouth of Severskiy Donets. At both campsites, the skeletal remains are almost exclusively those of short-horned bison. A date of 17,400 ± 700 yr B.P. (GIN-1998) was obtained from charred bone at Zolotovka I. Primitive people transferred their settlements from river banks to higher areas at that time.

During the Late Paleolithic, the habitation area expanded into northern Asia, that is, the southeastern margin of western Siberia (Achinsk and Tomsk campsites), the Krasnoyarsk Territory, the Baikal and Transbaikal regions, the Maritime Territory, and Yakutia. Particularly well studied are Late Paleolithic settlements in the basins of the Angara and Yenisey Rivers.

The Mal'ta and Buret' campsites contain remnants of dwellings and superb carved-bone art, including sculptural representations of women. The first engraved image of a mammoth from the USSR was found in Mal'ta. These sites are confined to a fossil soil that apparently correlates with the Bryansk Soil in eastern Europe, although a date (probably erroneous) of 14,750 ± 120 (GIN-97) was obtained from Mal'ta. Mal'ta is considerably older than the Kokorevo culture, which has dates of 15,460 ± 320 (LE-540), 14,450 ± 150 yr B.P. (LE-628), 13,300 ± 50 yr B.P. (GIN-91), and 12,840 ± 270 yr B.P. (LE-526). Unlike Mal'ta, the

Figure 29-3. Kostenki II: remains of a Paleolithic dwelling made of mammoth bones. (Photograph by A. N. Rogachev.)

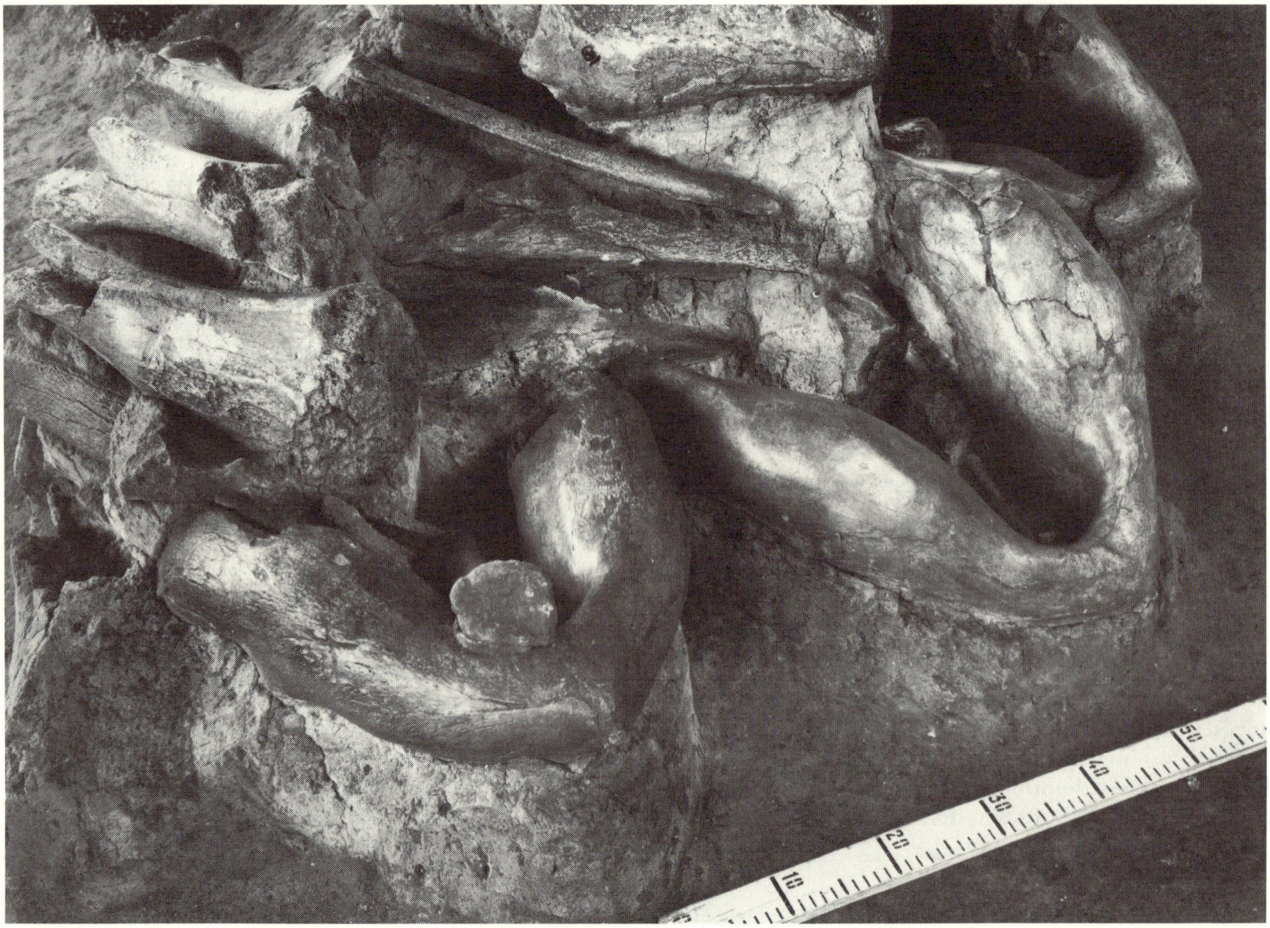

Figure 29-4. Kostenki II: edge of a dwelling made from the lower jaws of a mammoth. (Photograph by A. N. Rogachev.)

Kokorevo settlements (Abramova, 1979a, 1979b) have no evidence of permanent dwellings, and aurochs rather than mammoth was the prime target of hunting.

In the last decade, new Late Paleolithic sites were found in the basin of the Don, the Aldan, and even the Indigirka Rivers up to latitude 71°N (Mochanov, 1977). Of these, the sites Berelekh and Dyuktay Cave are of greatest interest. The Berelekh site was discovered by the zoologist N. K. Vereshchagin in 1970. After studying an enormous accumulation of mammoth bones on the Ugamysh site of the Berelekh River in the Indigirka Basin, he found several stone and bone implements. Excavation by Yu. A. Mochanov in 1971 uncovered an excellently preserved cultural layer of a Late Paleolithic campsite, with bone fragments of hare, wolf, and deer and with stone flakes and drilled pendants of nephrite (Vereshchagin and Mochanov, 1972: 336). Dates of 12,930 ± 80 yr B.P. (GIN-1021) and 13,420 ± 200 yr B.P. (IM-152) were obtained at the site. An engraving of a long-legged mammoth carved on the surface of a tusk apparently comes from the same site (Figure 29-5). The mammoth is represented in profile with disproportionately long legs and trunk (Bader and Flint, 1977). The short tail is raised high, and the belly shows long fur. There is no doubt that the primitive artist had excellent knowledge of this mighty beast.

Dyuktay Cave (Mochanov, 1977) is located on the right bank of the Dyuktay River 112 km from the point where it empties into the Aldan River. In the cave and on the bench in front of it, 317 m² have been excavated. Three Pleistocene horizons with Paleolithic finds were identified below the Holocene deposits. Similar faunal remains were found in all the Pleistocene layers, including bones of mammoth, reindeer, moose, bison, horse, snow leopard, cave lion, wolf, fox, hare, and various rodents, among them lemming. The stone objects consist mainly of production waste. There are approximately 100 implements in all three layers. The most characteristic are bifacial spearheads, dart tips, and knives. Radiocarbon dates from wood charcoal are 12,100 ± 120 yr B.P. (LE-907) for the upper layer, 13,070 ± 90 yr B.P. (LE-784) and 12,690 ± 120 yr B.P. (LE-860) for the middle layer, and 13,110 ± 90 yr B.P. (LE-908) for the bottom layer.

Paleolithic finds in the northeastern USSR have particular bearing on the migration of humans in North America across Beringia, and for this reason the data must be considered carefully. For example, the lower layers of the Ushka campsite in Kamchatka have been attributed to the Late Paleolithic (Dikov, 1977). The stratigraphy of the campsite, however, is unclear; the stone inventory from the Paleolithic layer is no different from that of the Mesolithic layer. It has the same shapes of arrowheads, and the dates are out of sequence: 13,600 ± 250 yr B.P. (GIN-167), 21,000 ± 900 yr B.P. (GIN-186), and 10,360 ± 350 yr B.P. (MO-345). The artifacts do not provide a firm sequence, and hence this site cannot yet be definitely identified as Paleolithic.

The peopling of Siberia during the Paleolithic probably was not gradual and continuous as some workers have suggested. Undoubtedly, such a hypothesis has resulted from inconsistencies in the investigation of different regions, that is, the poor dating control, the complexity of paleogeographic reconstructions, and the sometimes hasty conclusions made regarding the antiquity of a given site. In

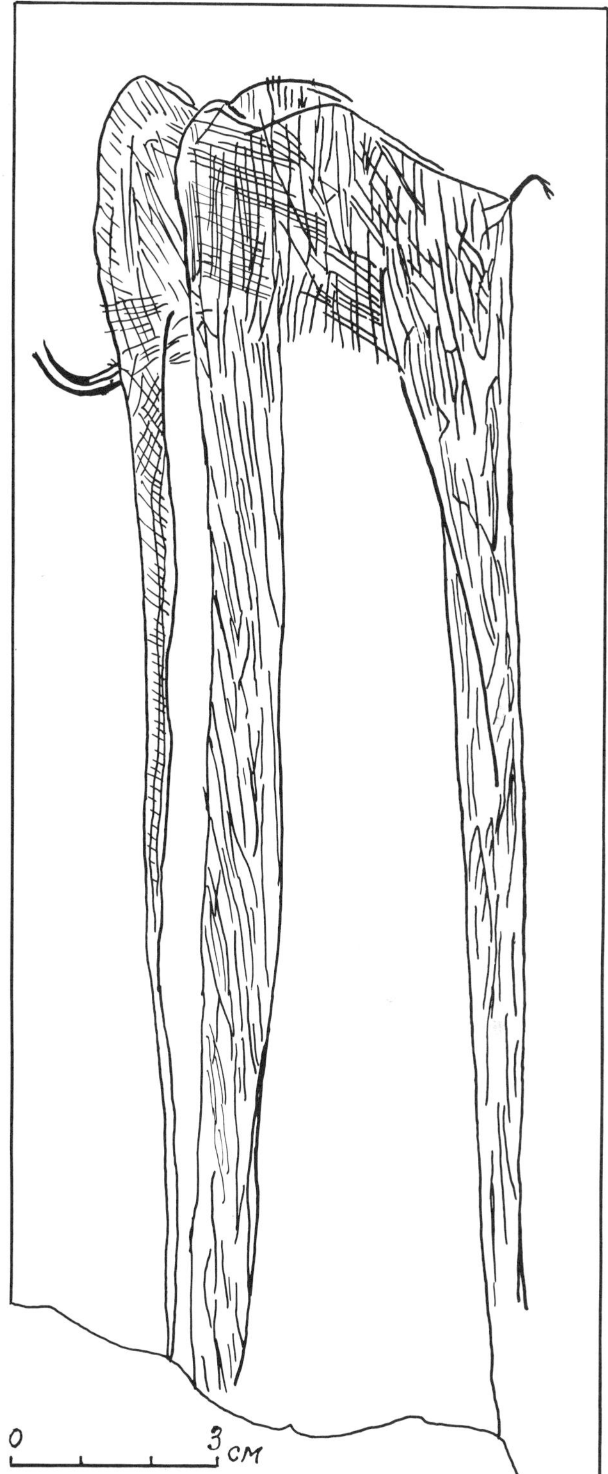

Figure 29-5. Engraved image of a mammoth on a tusk from the Berelekh site. (After Bader and Flint, 1977.)

the Early Holocene, Mesolithic hunters appeared north of latitude 72°N in Taimyr; this finding provides a minimum age for the settling of northern Asia. A similar pattern occurs at the end of the Pleistocene in eastern Europe. An expansion of human occupation onto the Russian Plain occurred during the late glacial (the Altynovo and Zolotoruch'ye campsites of the Upper Volga). In the Baltic region, primitive hunters first appeared at beginning of the Holocene. In northeastern Belorussia, Lithuania, and Latvia, there are no confirmed Paleolithic sites, and Late Paleolithic sites in Lithuania (Rimantiene, 1971) require additional study. Dates based on a typological analysis and comparison with sites in other regions are not reliable. Apparently, during the Mesolithic the peopling of eastern Europe was complete, since many sites of that period are known in the Baltic region and in the northeastern Russian Plain.

References

Abramova, Z. A. (1979a). "The Paleolithic of the Yenisey: The Afontovskiy Culture." Nauka Press, Novosibirsk.

Abramova, Z. A. (1979b). "The Paleolithic of the Yenisey: The Kokorevo Culture." Nauka Press, Novosibirsk.

Bader, O. N. (1977). Paleoecology and people of the Sungir' campsite. *In* "The Paleoecology of Ancient Man" (I. K. Ivanova and N. D. Praslov, eds.), pp. 31-40. Nauka Press, Moscow.

Bader, O. N., and Flint, V. Ye. (1977). Engraving on a mammoth tusk from Berelekh. *In* "The Mammoth Fauna of the Russian Plain and Eastern Siberia" (N. K. Vereshchagin, ed.), pp. 68-72. Institute of Zoology, Trudy 72, Leningrad.

Bader, O. N., and Matyushin, G. N. (1973). New Middle Paleolithic in the Southern Urals. *Sovetskaya Arkheologiya* 4, 135-42.

Chernysh, A. P. (1965). "The Early and Middle Paleolithic of the Dnestr Region." Commission on the Study of the Quaternary, Trudy 25. Nauka Press, Moscow.

Dikov, N. N. (1977). "Archeological Monuments of Kamchatka, Chukotka and Upper Kolyma: Asia at the Junction with America in Antiquity." Nauka Press, Moscow.

Grichuk, V. P. (1969). The vegetative cover in the Late Pleistocene. *In* "The Forest-Periglacial Paleolithic in the Territory of Central and Eastern Europe" (I. P. Gerasimov and A. A. Velichko, eds.), pp. 448-58. USSR Academy of Sciences, Institute of Geography. Moscow.

Guslitser, B. I., and Liyva, A. A. (1972). Age of the site of remains of Pleistocene mammals and Paleolithic campsite Byzovaya on the Middle Pechora. *Estonian Academy of Sciences, Izvestiya* 121 (Biology 3), 250-53.

Kanivets, V. I. (1976). "Paleolithic of the Extreme Northeast of Europe." Nauka Press, Moscow.

Kaplin, P. A., Leontyev, O. K., Rychagov, G. I., Parunin, O. N., Svitoch, A. A., and Shlyukov, A. I. (1977). Chronology and paleogeography of the Pleistocene Pont-Caspian region (according to the absolute dating results). *In* "Paleogeography and Pleistocene Deposits of the Southern Seas of the USSR," pp. 33-42. Nauka Press, Moscow.

Lyubin, V. P. (1970). The Lower Paleolithic. *In* "The Stone Age in the Territory of the USSR" (A. A. Formozov, ed.), pp. 19-42. Nauka Press, Moscow.

Mochanov, Yu. A. (1977). "The Oldest Stages of Man's Settlement of Northeastern Asia." Nauka Press, Novosibirsk.

Motuz, V. M. (1967). Quaternary mollusks of the Lower Paleolithic site of the area of Bryansk. *Bulletin of the Commission on the Study of the Quaternary* 33, 150-54.

Okladnikov, A. P. (1968). Siberia in the early Stone Age: The Paleolithic epoch. *In* "History of Siberia," Vol. 1 (A. P. Okladnikov, ed.). Nauka Press, Leningrad.

Praslov, N. D. (1968). "The Early Paleolithic of the Northeastern Azov Region and Lower Don." Nauka Press, Leningrad.

Praslov, N. D., and Semenov, S. A. (1967). Functions of Mousterian siliceous implements from campsites of the Azov region. *Short Communications of the USSR Academy of Sciences, Institute of Archeology, Kratkiye soobshcheniya* 117, 13-21.

Rimantiene, R. K. (1971). "The Paleolithic and Mesolithic of Lithuania." Mintis Press, Vilnius.

Tarasov, L. M. (1977). The Betovo Mousterian campsite and its natural environs. *In* "Paleoecology of Ancient Man" (I. K. Ivanova and N. D. Praslov, eds.), pp. 18-31. Nauka Press, Moscow.

Velichko, A. A. (1961). "Geological Age of the Upper Paleolithic of Central Regions of the Russian Plain." USSR Academy of Sciences Press, Moscow.

Velichko, A. A. (1973). "The Natural Process in the Pleistocene." Nauka Press, Moscow.

Velichko, A. A., and Morozova, T. D. (1975). Change in the development and paleogeographic inheritedness of indicators of recent soils of the center of the Russian Plain. *In* "Problems of Paleogeography of Loessial and Periglacial Regions" (A. A. Velichko, ed.), pp. 102-21. Institute of Geography. Moscow.

Vereshchagin, N. K., and Mochanov, Yu. A. (1972). The world's northernmost traces of the Paleolithic (Berelekh site in the lower course of the Indigirka River). *Sovetskaya Arkheologiya* 3, 332-36.

Zamyatnin, S. N. (1961). The Stalingrad Paleolithic campsite. *USSR Academy of Sciences, Institute of Archeology, Kratkiye soobshcheniya* 82, 5-36.

Zavernyayev, F. M. (1978). "The Khotylevo Paleolithic Site." Nauka Press, Leningrad.

CHAPTER 30

Human Cultures and the Natural Environment in the USSR during the Mesolithic and Neolithic

P. M. Dolukhanov and N. A. Khotinskiy

Introduction

The development of primitive society was a complex process determined by socioeconomic, cultural, biologic, and ecologic factors, which were of varying importance at different stages of prehistoric human development. Human society as a whole may be regarded as an adaptive system (Markaryan, 1976). In the case of primeval tribes, adaptation was expressed by the choice of economic strategies, tool kits, demographic processes, settlement patterns, and so on. We attempt here to evaluate the ecologic factors that determined human settlement and cultural evolution in the USSR.

The Mesolithic and Neolithic epochs from approximately 10,300 to 4000 yr B.P. correspond to the Preboreal (PB), Boreal (BO), Atlantic (AT), and Early Subboreal (SB-1) periods of the Blytt-Sernander scheme. Maps for the Mesolithic (9000 to 8000 yr B.P., Figure 30-1) and the Neolithic (6000 to 5000 yr B.P., Figure 30-2) show the location of reference campsites, principal cultures, and types of vegetation.

The Mesolithic

The Mesolithic period spanned the years between 10,000 and 6000 yr B.P. in the forest zone and between 10,000 and 8000 (or 7000) yr B.P. in the southern regions of the USSR. The beginning of the Holocene (10,300 yr B.P.) was a period of general warming, decreased continentality, and reduced hyperzonality (Velichko, 1973). At the same time, steppe, forest, and tundra vegetation zones became rapidly consolidated.

This transition to zonal conditions was accompanied by significant faunal changes. By the end of the Pleistocene (Vereshchagin, 1971), at least 10 species were extinct on the Russian Plain, including the mammoth (*Mammuthus primigenius*), woolly rhinoceros (*Rhinoceros tichorhinus*), and cave bear (*Ursus spelaeus*). Some species, including reindeer (*Rangifer tarandus*), died out in the South but multiplied in the North. Steppe and semidesert species of ungulates retreated to the semideserts and foothills of Central Asia. In the forest zone, populations of elk, duck (Anseriformes), grouse (*Tetrao tetrix*), fishes, and sea mammals (Pinnipeda, Cetacea) increased.

These far-reaching floral and faunal changes were accompanied by substantial changes in primitive economies. The Late Paleolithic tribes hunted primarily the large herd animals (mammoth, reindeer, aurochs, saiga, etc.) of the periglacial landscapes. However, at the beginning of the Holocene, Mesolithic tribes hunted animals that inhabited the steppe, forest, and tundra zones, and fishing and gathering became more important. The Mesolithic in the USSR was a transitional period that incorporated adaptations to the differentiated Holocene landscapes. Small groups of nomadic hunters traveled great distances, building temporary campsites on the shores of lakes and rivers. The thin, inorganic cultural layers that frequently are the only remains of these campsites confirm the indistinct character of Mesolithic culture.

The vegetational reconstruction for the period between 9000 and 8000 yr B.P. and the locations of Mesolithic campsites are shown in Figure 30-1. The discontinuous distribution of Mesolithic sites across the USSR probably reflects the limited number of studies. Over 90% of the Mesolithic sites have not been radiocarbon-dated, and many age estimates are based solely on typology. Nevertheless, it appears that Mesolithic people occupied the entire country.

MOUNTAIN AND FOOTHILL REGIONS OF THE SOUTHERN USSR

In western Georgia and the Trialeti Plateau, Mesolithic campsites, usually located on the lower river terraces (Gabuniya and Tsereteli, 1977), have been dated to the Boreal period on the basis of typological and geomorphologic data. Mesolithic inhabitants of the Caucasus hunted such forest animals as brown bear, wild boar, and red deer.

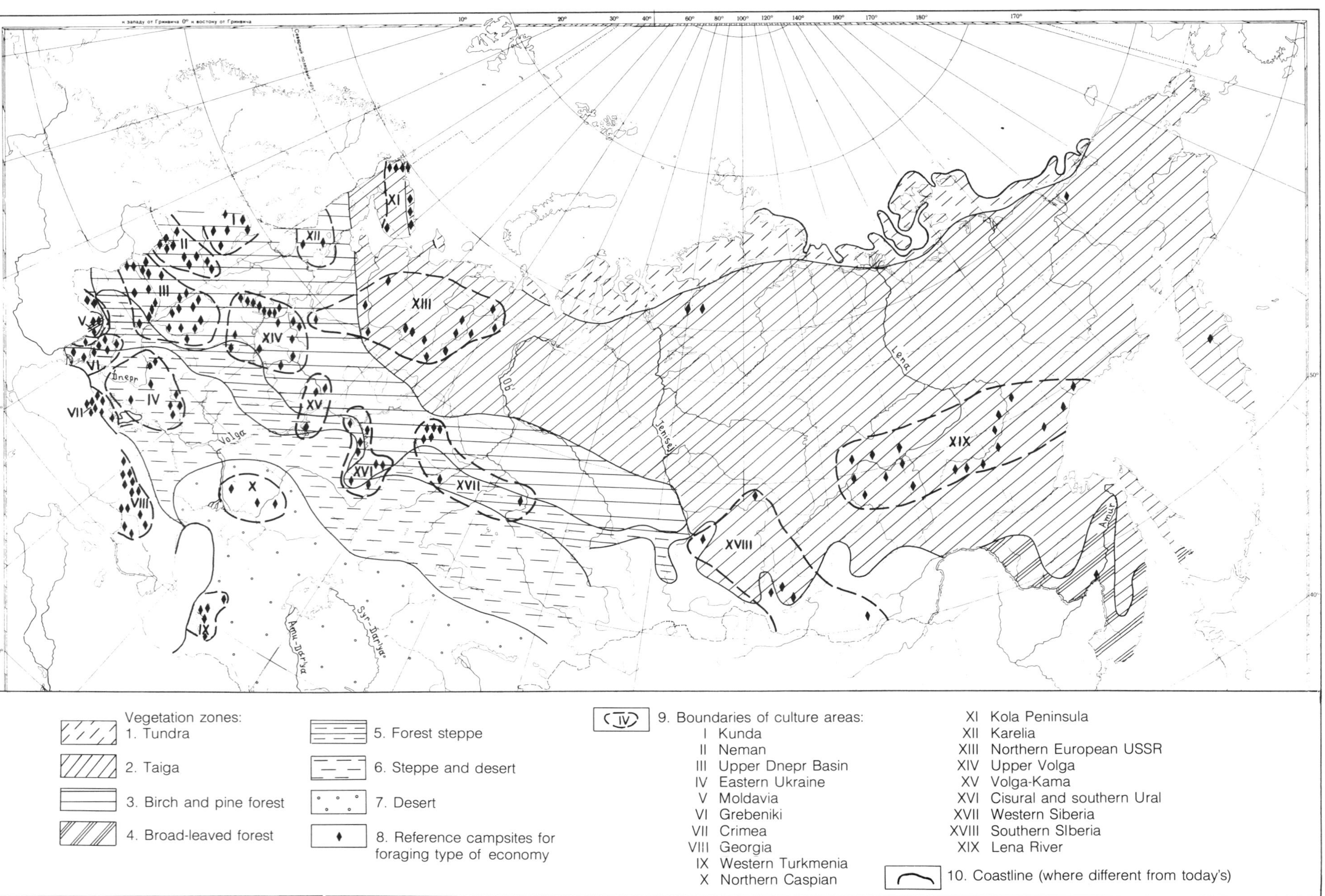

Figure 30-1. Reference campsites and areas of principal Mesolithic cultures against the background of vegetation zones in the seventh millennium B. C. (9000 to 8000 yr B.P.).

In the mountain and foothill regions of western Turkmenia, Mesolithic campsites are known from the shores of the Caspian Sea (Kaylyu Cave) and in the Malyy Balkhan Range (Djebel, Dam-Dam-Cheshme). The Mesolithic layers of Dzhebal Cave contain the remains of gazelle, goat, and sheep (Okladnikov, 1956). In southern Uzbekistan, Mesolithic people hunted Siberian goat (*Capra falconeri*), moufflon, marten, hare, red deer, fox, gazelle, and brown bear. Domestic cattle have also been reported (Islamov, 1975).

In western Tadzhikistan, Mesolithic campsites are located in the foothills of Pamir and Tien-Shan. The Mesolithic layer at the Tutkaul campsite has dates of 8020 ± 140 and 7100 ± 140 yr B.P. The Mesolithic layer of Ak-Tan'ga cave, dated at 8785 ± 130 yr B.P., contains bones of fox, red deer, ox, goat, and sheep (Ranov and Korobkova, 1971).

STEPPE ZONE OF THE EUROPEAN USSR

A substantial number of Mesolithic sites are known in the steppe zone of the European USSR. Sites are usually confined to limans or river valleys that cross the Black Sea coastal lowland. In some cases, these sites represent distinctive archaeological cultures; an example is the Grebeniki culture, which spread over the western Odessa Province (Stanko, 1972). Pollen analysis from the Mirnoye campsite records steppe vegetation (Dolukhanov and Pashkevich, 1977) and indicates that mountain goat and wild horse were the primary game animals.

A number of Mesolithic open settlements are known from the Kerch Peninsula in the eastern Crimea (Matskevoy, 1977). Pollen analyses have established that xerotic-type steppe vegetation was widespread there during the Mesolithic (Matskevoy and Pashkevich, 1973). The wild horse and wild donkey dominate the faunal remains, but remains of the gazelle, mountain goat, aurochs, red deer, and dolphin were also found. Numerous Mesolithic cave sites are known from the mountainous Crimea, in which saiga, cave lion, and great deer dominate the local fauna (Kolosov, 1971).

FOREST ZONE OF THE USSR

Within the forest zone, the remains of temporary campsites are usually confined to floodplains or low river terraces. Many campsites are confined to riparian dunes. Mesolithic sites have been discovered in some parts of the Upper Volga, the Neman Basin, and the Upper Dnepr River valleys. Within glaciated regions, numerous Mesolithic sites are found along periglacial lakes and lagoons along the Baltic Sea.

Within the forest zone of the European USSR, several Mesolithic cultures are distinguished. The Kunda culture existed in Estonia and Latvia. The earliest site, Pulli, has been dated between 9700 and 9200 yr B.P.; the latest site, Osa, between 7200 and 7100 yr B.P. (Dolukhanov and Timofeyev, 1972). The Neman Mesolithic culture occurred in Lithuania and northwestern Belorussia (Rimantiene, 1971). An Upper Volga Mesolithic culture was identified in the Upper Volga and Oka Basins (Kol'tsov, 1965), and a Volga-Kama culture was found in the Middle Volga Basin (Khalikov, 1969). Other Mesolithic sites have no distinctive typology and are distinguished mainly by their location, that is, the Mesolithic of Karelia, Kola Peninsula, and so on. Radiocarbon dates of 8080 ± 90 yr B.P. (LE-776) and 7820 ± 80 yr B.P. (LE-616) were obtained for the Mesolithic layer of the Vis peat bog in the basin of the Middle Dvina (Burov et al., 1972).

The primary game animal of the northern and central forest zone was the reindeer, an inhabitant of tundra and forest-tundra (Kol'tsov, 1974). In the Baltic region, Mesolithic tribes hunted elk, wild boar, brown bear, and red deer, whose appearance was associated with the spread of broad-leaved forests (Paaver, 1965). In the maritime regions, sea mammals, particularly seals, were hunted. Fishing and gathering were of great economic importance everywhere.

Numerous Mesolithic sites are known in forested regions of the Urals and Siberia. As in the European USSR, Mesolithic campsites in these regions were confined to low river terraces and floodplains. In southern Siberia at sites with dates of 8960 ± 60 (GIN-96) and 7890 ± 235 yr B.P. (SOAN-580), the principal game animal was the roe deer, but red deer, elk, aurochs, and wild horse were also hunted (Yermolova, 1978).

An extensive Mesolithic culture was also distributed throughout the Lena Basin (Mochanov, 1969). There the main game animal was elk, but reindeer, water fowl, and fishes were also present. Numerous radiocarbon dates from the Bel'kachi I campsite indicate that Mesolithic tribes were in the area from 9200 to 6000 yr B.P.

In the Far East, Mesolithic sites are reported from the Lower Amur near the city of Khabarovsk, in the Maritime Territory near the mouth of the Zerkal'naya River, and on Sakhalin Island (Okladnikov, 1977). On Kamchatka, layers IV and III of the site near Ushkovskoye Lake [above the layer radiocarbon-dated at 10,360 ± 350 yr B.P. (Mo-345)] are Mesolithic (Dikov, 1977).

The Mesolithic was characterized by a foraging economy. Reports of domesticated animals require verification. Hunting was more important than fishing or gathering. The Mesolithic populations adjusted to natural zones by hunting whatever game animals lived in a particular area.

The Neolithic

The Neolithic in the USSR occurred generally between 7000 and 4000 yr B.P., during the Atlantic and Early Subboreal periods. The transition to the Atlantic period around 8000 yr B.P. was characterized by substantial warming, increased humidity, and initial landscape differentiation. The northern boundary of the forest zone moved northward into the tundra, and its southern boundary shifted into the steppe. In the European USSR, broad-leaved forests appeared, and the fauna became more diverse as populations of thermophilous species

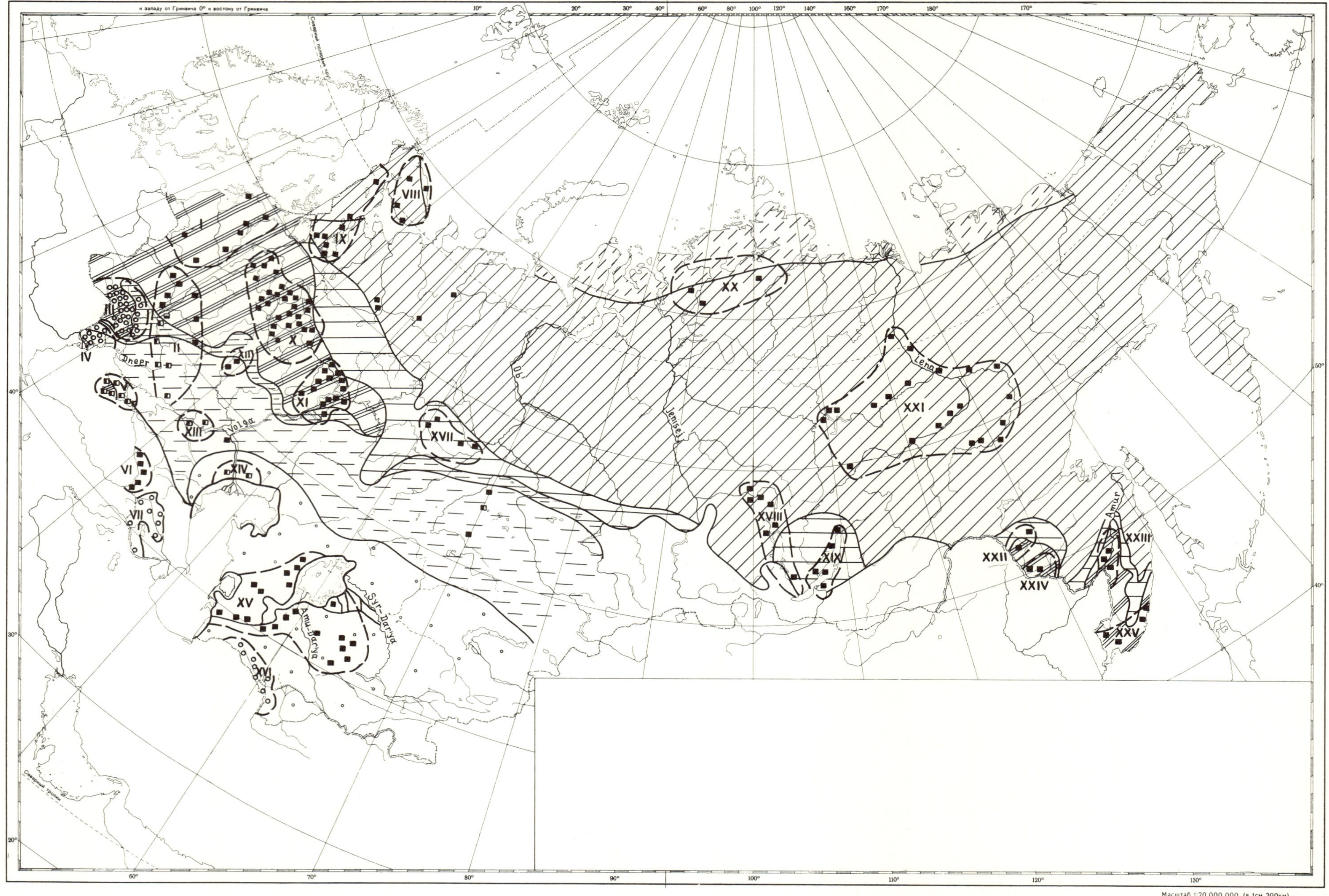
к западу от Гринвича 0° к востоку от Гринвича
Северный полярный круг
Северный тропик
Dnepr
Volga
Ob
Jenisej
Lena
Amur
Syr-Dar'ya
Amu-Dar'ya
I
II
III
IV
V
VI
VII
VIII
IX
X
XI
XII
XIII
XIV
XV
XVI
XVII
XVIII
XIX
XX
XXI
XXII
XXIII
XXIV
XXV
Масштаб 1:20 000 000 (в 1см 200км)
200 0 200 400 600 800 1000 км

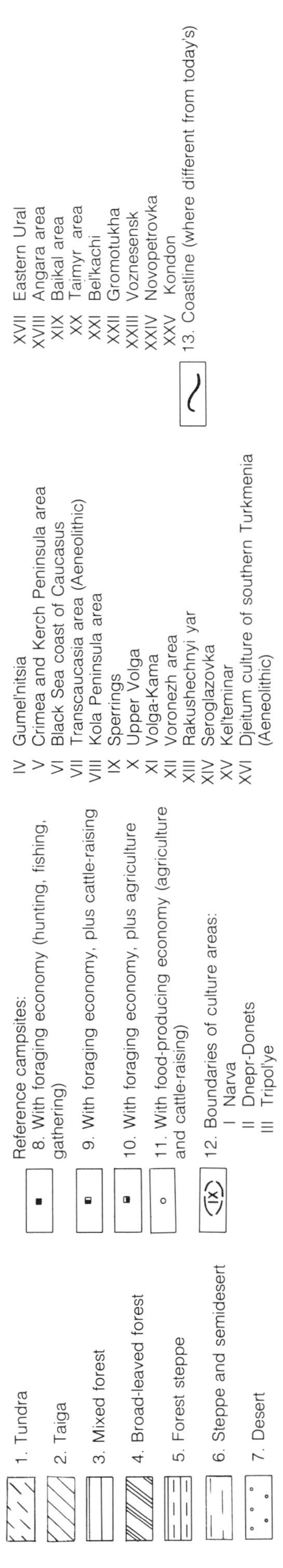

Figure 30-2. Reference campsites and areas of principal Neolithic cultures against a background of vegetation zones in the fourth millennium B. C. (6000 to 5000 yr B.P.).

increased. During the climatic optimum between 6000 and 4600 yr B.P., the vegetation of the steppe acquired a more mesophytic character as compared to that of the Mesolithic. In Central Asia, humidity increased, salinity of groundwater decreased, soils developed, and halophytic and mesophytic vegetation spread, especially tugai-type forests. This stage is referred to as the "Lyavlyakan pluvial" (Vinogradov and Mamedov, 1975).

At the start of the Subboreal (4700 to 4100 yr B.P.), a climatic cooling led to a decline in the role of thermophilous tree species in the forest zone. In the arid regions of Central Asia, dry conditions set in, rivers disappeared, and the vegetation developed a definite desert character.

During the Neolithic, the zonal structure of natural landscapes resulted in increased economic differences among regions. The most significant event was the advent of domestication of plants and animals in the foothills of Central Asia, Transcaucasia, and the Carpathian Mountains. A food-producing economy spread northward into the present forest-steppe zone during the Aeneolithic Age. At the same time, a foraging economy was retained over much of the forest zone. Hence, there are no universal criteria for identifying the Neolithic. In the southern regions, the period is marked by the appearance of a producing economy, whereas elsewhere the start of the Neolithic is determined by the first appearance of ceramic products, improved working tools, new types of hunting, and more reliance on fishing.

Neolithic cultures from 6000 to 5000 yr B.P. are shown in Figure 30-2. In contrast to the Mesolithic, a large amount of archaeological and radiocarbon data is available. We will examine the development of Neolithic cultures in different natural zones.

MOUNTAIN AND PIEDMONT REGIONS OF THE SOUTHERN USSR

The earliest Neolithic sites with evidence of agriculture and stock breeding are in the central Kopet Dagh piedmont plain of Central Asia. These sites of the Djeitun culture have been dated at 8000 to 6500 yr B.P. (Masson, 1971) and are found on alluvial fans. The existence of primitive irrigation systems is postulated. The economy was based on the cultivation of wheat (*Triticum vulgare, T. compactum*) and barley (*Hordeum disticum*). Domesticated animals included goats and sheep, although cattle appear in younger sites. In addition to agriculture, hunting was important, and such animals as the bezoar goat (*Capra hirsus aegagrus*), wild goat (*Capra lbex*), and Central Asian gazelle (*Ovis orientalis*) were taken (Masson, 1971).

The Aeneolithic in southern Turkmenia (6500 to 4500 yr B.P.) was characterized by the further development of an agrarian culture. In the submontane belt of Kopet Dagh, which was probably occupied by savanna vegetation, settlements reached considerable dimensions (Namazga-depe: 50 ha). Settlements appeared in the Tedjen River delta around 5700 yr B.P. and in the Murghab River delta around 5000 yr B.P. A fairly complex irrigation network was established (Lisitsyna, 1978). At the start of the third millennium B.C., a gradual decline of agrarian culture occurred in southern Central Asia as a result of a more arid climate.

At the start of the sixth millennium B.C., Aeneolithic agricultural settlements appeared in the areas of thin broad-leaved forest in the intermontane depressions of Transcaucasia (Kushnareva and Chubinashvili, 1970). The economy of these settlements was based on the cultivation of wheat, barley, oats, millet, and vegetables and on stock breeding (cattle, sheep, goat, and pig) (Lisitsyna and Prishchepenko, 1971). At the end of the third millennium B.C., farming gradually declined, cattle raising increased and acquired a nomadic or seminomadic character, while the number of settlements in the intermontane depressions decreased. These changes are attributed to a drier climate.

DESERT ZONE OF THE USSR

Contemporaneous with the agrarian culture in southern Turkmenia, tribes with diversified foraging economies existed in large areas of deserts and semideserts. Numerous campsites of the Kel'teminar culture are found in the Balkhan Range, on the Krasnovodsk Peninsula, and along the Amu-Darya River and its extension the Uzboy River, which empties directly into the Caspian Sea (Vinogradov, 1968). This entire territory was well supplied with water, and forests grew along the rivers and streams. Gazelle, saiga, sheep, and water fowl were hunted. The increased precipitation during the Lyavlyakan pluvial enabled hunters to settle the previously lifeless regions of Kyzylkum and Karakum in search of game, aurochs being one of the most important animals (Vinogradov and Mamedov, 1975).

STEPPE AND FOREST-STEPPE ZONES OF THE USSR

Several Neolithic cultures of different age, range, and economy are recognized in the steppe and forest-steppe zones. The Bug-Dnestr culture (7000 to 6000 yr B.P.) was centered in the floodplains of the Dnestr and Western Bug Rivers (Danilenko, 1969; Markevich, 1974). The economy of these settlements was based on the hunting of red deer, wild boar, and roe deer and on fishing and gathering. Bones of pig and imprints of wheat on ceramics have been observed at the campsites (Yanosevich, 1976). Around 6500 to 5800 yr B.P. in the Moldavian SSR and in the foothills of the Carpathian Mountains, settlements of the Linear Pottery culture appeared. These settlements were usually located on the margins of loessial plateaus and on low river terraces in the Dnestr Basin. The inhabitants of these settlements cultivated wheat, millet, oats, and barley; raised cattle, sheep, goat, and pig; and hunted aurochs, wild boar, red deer, and roe deer (Tsalkin, 1970).

From 5800 to 5000 yr B.P., the Tripol'ye-Cucuteni culture appeared in the forest-steppe and steppe zone. The economy of early Tripol'ye settlements (5800 to 5600 yr B.P.) was of a mixed nature largely based on hunting, fishing, and gathering. During the middle and late stages (5600 to 5000 yr B.P.), an agrarian economy developed; it was based on the cultivation of wheat, barley, millet, oats, and legumes and on cattle raising (Figure 30-2). The often large settlements were usually located in the margins of loessial plateaus (Bibikov, 1965).

Within the steppe zone of the western Odessa Province and southern Moldavia, settlements of the Gumel'nitsa culture existed between 5800 and 5300 yr B.P. (Figure 30-2). The economy was based on the cultivation of wheat, barley, oats, and millet and on cattle raising (Tsalkin, 1970; Yanushevich, 1976).

At the end of the fifth millennium B.C. and during the fourth millennium B.C., settlements of the Neolithic Dnepr-Donets culture were found in the Dnepr and Seversky Donets River basins (Telegin, 1968), thus this culture extended across the southern part of the broad-leaved forest and the steppe zone. Settlements were covered with broad-leaved forest and extended along the fluvial plains south to the Black and Azov Seas. The economy of the Dnepr-Donets settlements was of a foraging nature. Some cattle raising was established in the steppe settlements. At the Vita Litovskaya settlement near Kiev, an imprint of barley was found on a ceramic object (Telegin, 1968).

Between 5600 and 4500 yr B.P., in the steppe zone of the central Ukraine, the economy of the Sredniy Stog culture was based on horse breeding (Telegin, 1973).

In the middle and second half of the third millennium B.C., agrarian cultures disappeared in the forest-steppe and steppe zone as a result of a more arid climate. Around 4500 yr B.P., the number of late Tripol'ye settlements decreased, and the economy assumed a nomadic or seminomadic character. This transition to nomadic cattle raising was one of the principal factors that gave rise to the Pit-Grave culture (4500 to 3900 yr B.P.) in the steppe zone (Merpert, 1974; Telegin, 1977).

FOREST ZONE OF THE USSR

Several cultures, including the Narva in Estonia, Latvia, and Lithuania, the Upper Volga on the central Russian Plain, the Volga-Kama on the Middle Volga, and the Sperrings in Karelia, existed in the broad-leaved-forest regions of the USSR rich in biomass (Figure 30-2) between approximately 6300 and 5000 yr B.P. Settlements of these early Neolithic cultures were usually located on fluvial plains and along lakes when the water level was higher than it is now. The location of the settlements enabled Neolithic settlers to utilize the natural resources of lake basins, floodplains, and stratified broad-leaved forests on watershed uplands. Settlements subsisted on hunting of wild boar, red deer, aurochs, elk, brown bear, and waterfowl; on fishing; and on gathering edible fruits and berries. This foraging economy promoted a substantial population increase as compared to the Mesolithic and led to the establishment of permanent settlements in some regions. The similarity of Neolithic and Mesolithic stone implements indicates that early Neolithic cultures in the forest zone had an autochthonous development.

A foraging type of economy survived in the forest zone of the European USSR throughout the Neolithic. The onset of the Bronze Age around 4000 yr B.P. coincided with a climatic warming during the Middle Subboreal period (Khotinskiy, 1978). At that time, Corded Ware cultures spread throughout the forest zone in the northwestern and central Russian Plain. The appearance of the domesticated pig and cattle also occurred, but in general the economy retained a foraging character.

In the Urals, no reliably dated sites from the Early Neolithic are yet known. Large foraging settlements on the Shigirskiy and Gorbunovo peat bogs are middle (or late) Neolithic in age. The most-hunted animals were elk, brown bear, red deer, waterfowl, and forest birds. Fishing tools attest to the great importance of fishing (Starkov, 1980).

The Neolithic of Siberia is not well studied. Sites in the taiga zone of the Angara Valley (Yermolova, 1978) contain the remains of brown bear, roe deer, and elk. In the forest-steppe of southern Siberia, roe deer was the most commonly hunted animal, but other game animals included red deer, aurochs, and wild boar. In steppe intermontane basins, for example at the Onon campsite southeast of Chita, were found remains of aurochs, culan (*Equus hemionus*), and Mongolian gazelle (*Procapra gutturosa*) (Yermolova, 1978). Sites of the Bel'kachi (Early Neolithic) culture dated at between 6000 and 4000 yr B.P. occurred along the floodplain and terraces of the Lena River in Siberia (Mochanov, 1969). Everywhere the economy was based on the hunting of the elk, reindeer, roe deer, and brown bear, whereas fishing and gathering were of secondary importance (Mochanov, 1969; Fedoseyeva, 1980).

TUNDRA AND FOREST-TUNDRA ZONES OF THE USSR

Neolithic populations on the Kola Peninsula (5580 ± 80 yr B.P.) and northern Norway subsisted on wild boar, bear, lynx, marten, reindeer, seal, and waterfowl as well as on edible plants and mollusks and on fishing (Gurina et al., 1974).

Neolithic inhabitants of Taimyr built their camps in elevated areas of river valleys, along the migration routes of deer herds. The economy was based on the hunting of reindeer and marine animals (mainly pinnipeds), although whales were also hunted in Late Neolithic time (Khlobystin and Levskovskaya, 1974).

In general, the Neolithic is characterized by increased population size and density and a more complex economy than the Mesolithic, as the landscape itself became differentiated. Stable food-producing economies appeared in the well-watered foothills and on loessial plateaus covered with forest-steppe or savanna vegetation. The centers of ancient plant cultivation and animal domestication were the Carpathian-Black Sea, Transcaucasia, and the Kopet Dagh. Neolithic cultures of the steppe and forest-steppe zones of flatland areas show an early development of stock breeding. In contrast, a foraging type of economy persisted throughout the entire forest zone as a result of the low population density and high biomass content there (Figure 30-3).

The transition from the Neolithic to the Bronze Age coincided with the boundary between the Early Subboreal cool phase (SB-1) and the Middle Subboreal warm phase (SB-2). This interval was marked by a decline of agriculture, a switch to nomadic pastoralism in a number of regions, and increased migration. Some investigators attribute these processes to the northward expansion of steppe vegetation during the Subboreal, but pollen and radiocarbon data do not record a shift in the position of the forest/steppe boundary.

The migration of the population during the Early Bronze Age may have been caused by increased aridity, recurrent droughts, and lowered water tables in the steppe and desert zones, all of which caused reduced agricultural production and a switch to nomadic pastoralism.

No less important a problem is the degree to which the primitive populations affected the natural landscapes. Studies by western European palynologists (Iversen, 1941; Berglund, 1969; van Zeist, 1955) suggest an appreciable change in the forest vegetation as a result of Neolithic agriculture. However, similar studies from the forest zone of the USSR do not show any significant disturbance of natural landscapes attributable to prehistoric economic activity, even in the immediate vicinity of campsites. The foraging nature of the economy (hunting, fishing, gathering) apparently preserved the natural landscapes almost in their original form.

Major disturbance of the landscape occurred surprisingly late in the European USSR. According to historians, Russia contained isolated population centers confined to river valleys even as late as the 16th and 17th centuries A.D. The endless forests of Muscovy staggered the imaginations of foreigners who had visited there. Cereal and weed pollen suggest that active forest clearance did not occur in the center of the Russian Plain until the 17th and 18th centuries A.D. and much later in regions farther north (Khotinskiy et al., 1979). In western Europe, which since the Neolithic had a high population density and a food-producing economy, major anthropogenic changes in the landscape began earlier.

In general, ecologic crises were not anthropogenically induced. Rather, they were the result of marked landscape and climatic changes, which either caused the population to migrate to more favorable regions or stimulated the selection of new economic strategies better adapted to new conditions. This selection depended on both the magnitude of the natural change and the level of the social and cultural development within the population groups.

Among the natural factors affecting the development of prehistoric society, climate played a major role, for climatic fluctuations caused a rearrangement in the zonal structure of natural landscapes and determined the resources available to the society, thus affecting its economic development. However, despite their importance, these environmental changes did not determine the development of

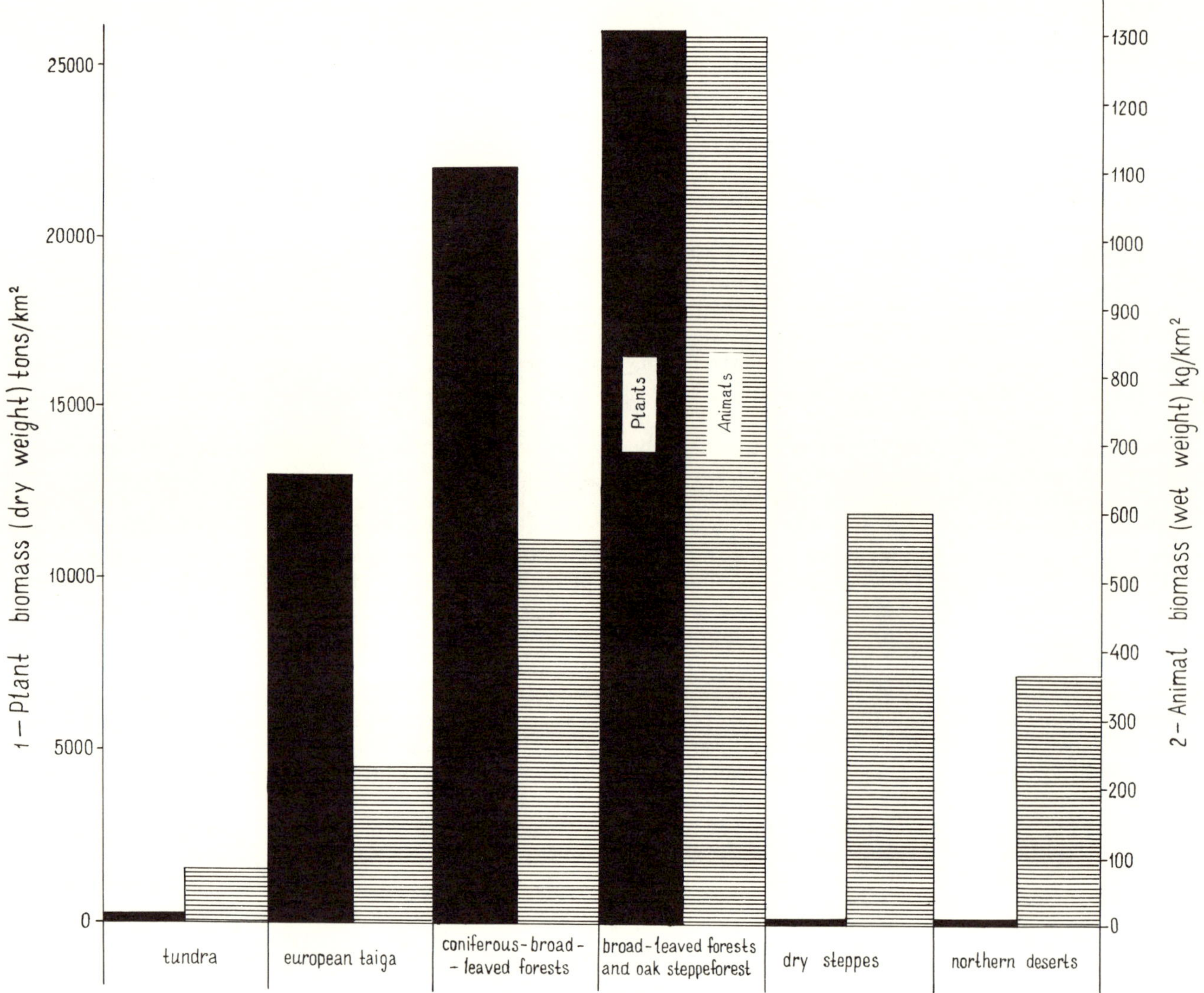

Figure 30-3. Plant and animal biomass in different zones and subzones. (Based on data by K. S. Khodashova, Ye. M. Lavrenko, V. N. Andreyev, and V. L. Leont'yev.)

primitive society but merely determined the options that were available in the presence of specified socio-economic and cultural conditions.

References

Berglund, B. E. (1969). Vegetation and human influence in south Scandinavia during prehistoric time. *Oikos Supplement* 12, 9-28.

Bibikov, S. N. (1965). The economic complex of developed Tripol'ye. *Sovetskaya Arkheologiya* 1, 48-62.

Burov, G. M., Romanova, Ye. N., and Sementsov, A. A. (1972). Chronology of wooden implements and objects found in the Severnaya Dvina Basin. *In* "Problems of Absolute Dating in Archeology" (B.A. Kolchin, ed.), pp. 76-79. Nauka Press, Moscow.

Danilenko, V. N. (1969). "The Neolithic of the Ukraine." Naukova Dumka Press, Kiev.

Dikov, N. N. (1977). On the problem of the "Mesolithic" of Kamchatka. *USSR Academy of Sciences, Institute of Archeology, Kratkiye Soobshcheniya* 149, 120-123.

Dolukhanov, P. M., and Pashkevich, G. A. (1977). Paleogeographic dividing lines of the Upper Pleistocene-Holocene and development of economic types in the southeast of Europe. *In* "Paleoecology of Ancient Man" (I. K. Ivanoya and N. D. Praslov, eds.), pp. 134-45. Nauka Press, Moscow.

Dolukhanov, P. M., and Timofeyev, V. I. (1972). Absolute chronology of the Neolithic of Eurasia (based on radiocarbon data). *In* "Problems of Absolute Dating in Archeology" (B. A. Kolchin, ed.), pp. 28-75. Nauka Press, Moscow.

Fedoseyeva, S. A. (1980). "Ymyyakhtakhskiy Culture of Northeastern Asia." Nauka Press, Novosibirsk.

Gabuniya, M. K., and Tsereteli, L. D. (1977). The Mesolithic of Georgia. *USSR Academy of Sciences, Institute of Archeology, Kratkiye Soobshcheniya* 149, 34-40.

Gurina, N. N., Koshechkin, B. I., and Strelkov, S. A. (1974). Primitive cultures and evolution of the ecological situation in the Upper Pleistocene and Holocene on the shores of the European Arctic. *In* "Primitive Man and the Natural Environment" (I. P. Gerasimov and A. A. Velichko, eds.), pp. 231-35. USSR Academy of Sciences, Institute of Geography, Moscow.

Islamov, U. I. (1975). "The Machay Cave." Fan Press, Tashkent.

Iversen, J. (1941). Landnam i Danmarks Stenalder. *Danmarks Geologiske Undersögelse* 2 (66), 1-68.

Khalikov, A. Kh. (1969). "Ancient History of the Middle Volga Region." Nauka Press, Moscow.

Khlobystin, L. P., and Levskovskaya, G. M. (1974). The role of social and economic factors in the development of archeological cultures of Eurasia. *In* "Primitive Man and the Natural Environment" (I. P. Gerasimov and A. A. Velichko, eds.), pp. 235-42. USSR Academy of Sciences, Institute of Geography, Moscow.

Khotinskiy, N. A. (1977). "The Holocene of Northern Eurasia." Nauka Press, Moscow.

Khotinskiy, N. A. (1978). Paleogeographic principles of the dating and periodization of the Neolithic of the European USSR's forest zone. *USSR Academy of Sciences, Institute of Archeology, Kratkiya Soobshcheniya* 153, 7-14.

Khotinskiy, N. A., Falomeyev, B. A., and Guman, M. A. (1979). Archeological and paleogeographic studies on the Middle Oka. *Sovetskaya Arkheologiya* 3, 63-81.

Kolosov, Yu. G. (1971). The Mesolithic. *In* "Archeology of the Ukrainian SSR" (D. Ya. Telegin, ed.), pp. 64-77. Naukova Dumka Press, Kiev.

Kol'tsov, L. V. (1965). Some results of a study of the Mesolithic of the Volga-Oka interfluve. *Sovetskaya Arkheologiya* 4, 17-26.

Kol'tsov, L. V. (1974). Environment and material culture of the Late Paleolithic and Mesolithic in the Volga-Oka interfluve. *In* "Primitive Man and the Natural Environment" (I. P. Gerasimov and A. A. Velichko, eds.), pp. 275-79. USSR Academy of Sciences, Institute of Geography, Moscow.

Kushnareva, K. Kh., and Chubinashvili, T. N. (1970). "Ancient Cultures of the Caucasus." Nauka Press, Leningrad.

Lisitsyna, G. N. (1978). "Onset and Development of Irrigated Cultivation in Southern Turkmenia." Nauka Press, Moscow.

Lisitsyna, G. N., and Prishchepenko, L. V. (1971). "Paleoethnobotanical Finds of the Caucasus and Near East." Nauka Press, Moscow.

Markaryan, E. S. (1976). Toward an understanding of the specific nature of human society as an adaptive system. *In* "Geographical Aspects of the Ecology of Man" (A. D. Lebedev, ed.), pp. 139-49. Nauka Press, Moscow.

Markevich, V. I. (1974). "The Bug-Dnepr Culture on the Territory of Moldavia." Shtiintsa Press, Kishinev.

Masson, V. M. (1971). The Settlement of Djeitun. *Materials and Studies on the Archeology of the USSR*, 180.

Matskevoy, L. G. (1977). Some results of a study of Crimea's Mesolithic. *USSR Academy of Sciences, Institute of Archeology, Kratkiya Soobshcheniya* 149, 41-45.

Matskevoy, L. G., and Pashkevich, G. A. (1973). Contribution to the paleogeography of the Kerch Peninsula of the Mesolithic and Neolithic. *Sovetskaya Arkheologiya*, 2, 123-38.

Merpert, N. Ya. (1974). "The Oldest Cattle Breeders of the Volga-Ural Interfluve." Nauka Press, Moscow.

Mochanov, Yu. A. (1969). "The Bal'kachi I Multilayered Campsite and Periodization of the Stone Age of Yakutiya." Nauka Press, Moscow.

Okladnikov, A. P. (1956). The Djebel Cave: A site of the ancient culture of Caspian tribes. *Transactions of the South Turkestan Combined Archeological Expedition* 7, 11-129.

Okladnikov, A. P. (1977). The Mesolithic of the Far East (preceramic sites). *USSR Academy of Sciences, Institute of Archeology, Kratkiya Soobshcheniya* 149, 115-19.

Paaver, K. L. (1965). Formation of the teriofauna and variability of mammals. *In* "The Baltic Region in the Holocene" (K. Klensky, ed.), pp. 1-497. Estonian Academy of Sciences, Tallinn.

Ranov, V. A., and Korobkova, G. F. (1971). Tutkaul: A multilayered settlement of Ghissar Culture in Southern Tadzhikistan. *Sovetskay Arkheologiya* 2, 133-47.

Rimantiene, R. K. (1971). "The Paleolithic and Mesolithic of Lithuania." Mintis Press, Vilnius.

Stanko, V. N. (1972). Types of sites and local cultures in the Mesolithic of the northern Black Sea Region. *Materials and Studies in Archeology of the USSR* 115, 46-52.

Starkov, V. F. (1980). "The Mesolithic and Neolithic of the Forested Transural Region." Nauka Press, Moscow.

Telegin, D. Ya. (1968). "The Dnepr-Donets Culture." Naukova Dumka Press, Kiev.

Telegin, D. Ya. (1973). "The Sredniy Stog Culture of the Bronze Age." Naukova Dumka Press, Kiev.

Telegin, D. Ya. (1977). Absolute age of the Pit-Grave culture and certain aspects of Aeneolithic chronology of the south of the Ukraine. *Sovetskaya Arkheologiya* 2, 5-19.

Tsalkin, V. I. (1970). "The Oldest Domestic Animals of Eastern Europe." Nauka Press, Moscow.

Velichko, A. A. (1973). "The Natural Process in the Pleistocene." Nauka Press, Moscow.

Vereshchagin, N. K. (1971). Primitive man's hunts and the extinction of Pleistocene mammals in the USSR. *In* "Material on Faunas of the USSR's Anthropogene." *USSR Academy of Sciences, Institute of Zoology, Trudy* 49, 200-232.

Vinogradov, A. V. (1968). "Neolithic Sites of Khorezm." Nauka Press, Moscow.

Vinogradov, A. V., and Mamedov, E. A. (1975). "Primeval Lyavlyakan: Stages of Oldest Settlement and Development of Internal Kyzylkum." Nauka Press, Moscow.

Yanosevich, Z. V. (1976). "Cultivated Plants of the Southwest of the USSR, Based on Paleobotanical Data." Shtiintsa Press, Kishinev.

Yermolova, N. M. (1978). "Teriofauna of the Angara Valley in the Late Anthropogene." Nauka Press, Novosibirsk.

van Zeist, W. (1955). "Pollen Analytical Investigations in the Northern Netherlands with Special Reference to Archeology." North-Holland, Amsterdam.

A. A. Velichko is on the staff of the Institute of Geography of the USSR Academy of Sciences, Moscow. **H. E. Wright, Jr.**, is Regents' Professor of Geology, Ecology, and Botany at the University of Minnesota. He served as editor, with D. G. Frey, of *Quaternary of the United States*, published in 1965, and as general editor of *Late-Quaternary Environments of the United States* (Minnesota, 1983). **Cathy W. Barnosky** is a research associate in the Division of Earth Sciences, Carnegie Museum of Natural History, Pittsburgh.